Maximizing AutoCAD
Release 13

Maximizing AutoCAD
Release 13

Rusty Gesner

Mark Middlebrook

Tony Tanzillo

Press

I(T)P® An International Thomson Publishing Company

Albany • Bonn • Boston • Cincinnati • Detroit • London • Madrid
Melbourne • Mexico City • New York • Pacific Grove • Paris • San Francisco
Singapore • Tokyo • Toronto • Washington

NOTICE TO THE READER

Publisher does not warrant or guarantee any of the products described herein or perform any independent analysis in connection with any of the product information contained herein. Publisher does not assume and expressly disclaims, any obligation to obtain and include information other than that provided to it by the manufacturer. The reader is expressly warned to consider and adopt all safety precautions that might be indicated by the activities described herein and to avoid all potential hazards. By following the instructions contained herein, the reader willingly assumes all risks in connection with such instructions. The publisher makes no representations or warranties of any kind, including but not limited to, the warranties of fitness for particular purpose or merchantability, nor are any such representations implied with respect to the material set forth herein, and the publisher takes no responsibility with respect to such material. The publisher shall not be liable for any special, consequential, or exemplary damages resulting, in whole or in part, from the readers' use of, or reliance upon, this material.

Trademarks

AutoCAD,® AutoCAD LT® and the AutoCAD® logo are registered trademarks of Autodesk, Inc.
Windows is a trademark of the Microsoft Corporation.
All other product names are acknowledged as trademarks of their respective owners.

COPYRIGHT © 1997
By Delmar Publishers Inc.
An International Thomson Publishing Company
The ITP logo is a trademark under license

Printed in the United States of America

For more information, contact:

Delmar Publishers Inc.
3 Columbia Circle, Box 15015
Albany, New York 12212-5015

International Thomson Publishing Europe
Berkshire House
168-173 High Holborn
London, WC1V7AA
England

Thomas Nelson Australia
102 Dodds Street
South Melbourne 3205
Victoria, Australia

Nelson Canada
1120 Birchmont Road
Scarborough, Ontario
M1K SG4, Canada

International Thomson Editores
Campos Eliseos 385, Piso 7
Col Polanco
11560 Mexico D F Mexico

International Thomson Publishing GmbH
Konigswinterer Str. 418
53227 Bonn
Germany

International Thomson Publishing Asia
221 Henderson Road
#05-10 Henderson Bldg.
Singapore 0315

International Thomson Publishing Japan
Kyowa Building, 3F
2-2-1 Hirakawa-cho
Chiyoda-ku, Tokyo 102
Japan

All rights reserved. No part of this work covered by the copyright hereon may be reproduced or used in any form or by any means—graphic, electronic, or mechanical, including photocopying, recording, taping, or information storage and retrieval systems—without the written permission of the publisher.

1 2 3 4 5 6 7 8 9 XXX 02 01 00 99 98 98 97 96

Library of Congress Cataloging-in-Publication Data
Gesner, Rusty.
    Maximizing AutoCAD R13 / Rusty Gesner, Mark Middlebrook, Tony Tanzillo.
      p. cm.
    Includes index.
    ISBN 0-8273-7993-5
    1. Computer graphics. 2. AutoCAD (Computer file)
I. Middlebrook, Mark. II. Tanzillo, Tony. III. Title.
T385.G468 1996
620'.0042'02855369—dc20   96-31103
CIP

Acknowledgments

Although the authors of this book are listed in alphabetical order, in no way does that reflect on individual contributions or efforts, and his co-authors thank Tony for going beyond the call of duty in developing LispPad, the Maximizing AutoCAD Windows Runtime Extension (MaWIN.ARX), and several other valuable development tools.

Rusty thanks his co-authors, Tony and Mark, for their selfless cooperation, perseverance, insight, and dedication to quality and comprehensiveness that make this the best ever book on customizing AutoCAD. A very special thanks is once again due Margaret Berson for her invaluable editing and for establishing and maintaining the formatting. Rusty also thanks numerous friends at Autodesk for the help and support which make this effort possible, and William Blackmon and David Pitzer for their technical editing and checking the exercises and code. And Rusty thanks his family, Kathy, Alicia, and Ricky, the dogs, Damon and Dubs, the cats, Callie, Kelsie, and Bandie, the horses, C.J. and Shatze, and even the raccoon family who didn't follow through on their threat to ambush him at 2 AM, for their patience and understanding.

Mark thanks fellow authors Rusty and Tony for their camaraderie and enthusiasm during the writing of this book. And he's not sure whether to bless or curse his father, Ron Middlebrook, for giving him his start in CAD consulting back in 1988 by agreeing to be his first client.

Tony thanks fellow authors Mark and Rusty for their energetic participation and patience. Tony also thanks his parents for their support. Finally, Tony extends a special thanks to Duff Kurland, senior programmer and Autodesk co-founder, for his generous help and encouragement in the early days.

Quick Table of Contents

Chapter 1: AutoCAD System Setup and Optimization 1
Chapter 2: Creating a Customization Environment 39
Chapter 3: Creating Keyboard Macros and Toolbars 75
Chapter 4: Support Files 117
Chapter 5: Introduction to Menu Macros 145
Chapter 6: Text Fonts and Styles 195
Chapter 7: Linetypes and Hatch Patterns 247
Chapter 8: Slides, Scripts, and PCP Files 295
Chapter 9: Introduction to AutoLISP Programming 335
Chapter 10: Running on DIESEL 451
Chapter 11: Introduction to Dialog Boxes 517
Chapter 12: Designing Dialog Boxes 569
Chapter 13: Understanding Menu Structure 625
Chapter 14: Creating an Integrated Professional Menu System 713
Appendix A: AutoLISP Function Quick Reference 775
Appendix B: Diesel Function Quick Reference 803
Appendix C: DCL Quick Reference 807
Appendix D: Errors, Problems, and Solutions 823
Appendix E: System Variables 833
Appendix F: ASCII and Unicode Tables 859
Appendix G: The Maximizing AutoCAD CD-ROM 867

Table of Contents

Chapter 1 AutoCAD System Setup and Optimization

About This Chapter 1
Understanding Intel-Based AutoCAD Platform Choices 2
   DOS vs. the Windows Family 3
   Windows 3.1 vs. NT and 95 4
      Windows NT and 95 Advantages over 3.1 6
      Windows NT vs. 95 6
   Deciding between Windows NT and 95 6
      Windows NT Advantages 8
      Windows 95 Advantages 8
   Customization Platform vs. Production Platform 9
   Maintaining Multiple Operating Systems 10
   Setting up a Multiple Boot System 10
   Running Release 12 in Windows 95 or NT 12
      Running R12 DOS in Windows 95 12
Networking Issues 13
   Storing Drawing Files on the Network 13
   Keeping the AutoCAD Program on the Network 14
   Keeping Support Files on the Network 15
Third-Party Application Issues 15
   System Requirements 15
   Installation 15
   Directories and Support File Conflicts 16
Installing and Configuring AutoCAD 16
   Streamlined Installation Procedure Outline 17
   Installing Release 13 17
      New Installation, Reinstallation, or Update 18
      Upgrade Strategies 20
      Typical, Compact, or Custom Installation 21
   Configuring Release 13 21
      Basic Windows Configuration 21
      Basic DOS Configuration 22

Uninstalling AutoCAD in Windows 95 22
Optimizing System Performance 22
    Platform-Independent Optimization 23
        Configuration Menu Settings 23
        AutoCAD Pager Settings 24
        Other System Variable Settings 25
    Network Optimization 27
    Windows Optimization: Common Issues 28
        Display Drivers 28
        Digitizers 28
        Plotters and Printers 29
        Swap File 29
    Windows 95 and NT Optimization 30
        ADI Plotter and Printer Drivers 30
        Plotting to a Port Name 31
        Plotting to AUTOSPOOL 31
        Long File Names 33
        Swap File 33
    Windows 3.1 Optimization 33
        Swap File 33
        32-Bit File and Disk Access 34
    MS-DOS Optimization 35
        Display Drivers 35
        Digitizers 35
        ADI Plotter and Printer Drivers 36
        Phar Lap Swap File 36
Hardware Issues 36
Conclusion 37

Chapter 2 Creating a Customization Environment

About This Chapter 39
Organizing Your AutoCAD Setup: The Benefits 40
Directories (or Folders) 40
    The Maximizing AutoCAD Directory Structure 41
    The AutoCAD R13 Directory Structure 43
Installing the MaxAC CD-ROM 45
AutoCAD Configuration Files 48
    AutoCAD Support Path 49
    Multiple AutoCAD Configurations 51
Using Different AutoCAD Configurations in Windows 52
    Creating a New Program Item or Shortcut in Windows 52
    Directing AutoCAD to Different Support Files with /C 54
    Modifying the Support Path 56
Using Different AutoCAD Configurations in DOS 58
Selecting and Using Text Editors with AutoCAD 61
    Selecting an Editor for Windows 61
    Associating Registered File Types with Your Editor 64

Table of Contents

    Selecting and Configuring an Editor for DOS 67
    Testing Your Editor 68
Avoiding ACAD.LSP Conflict 69
Setting the Default Menu 70
Finishing Up 72
Conclusion 73

Chapter 3 Creating Keyboard Macros and Toolbars

About This Chapter 75
Understanding Keyboard Macros 76
Defining Keyboard Aliases in the ACAD.PGP File 77
    Reloading and Testing ACAD.PGP 80
Defining External Commands in the ACAD.PGP File 82
    Using ACAD.PGP External Commands 82
    ACAD.PGP External Command Definition Syntax 84
    Sample External Command Definitions 86
Creating Simple AutoLISP Command Definitions 86
Defining Accelerator Keys in the Menu File 91
    Creating Accelerator Key Definitions 93
    Reloading and Testing a Menu File 95
    Accelerator Key Tips 97
Deciding among Aliases, AutoLISP, and Accelerators 98
Understanding Windows Toolbars 100
Showing and Hiding Toolbars 101
Creating Your Own Toolbars 105
    Creating Toolbars with Existing Icons 105
    Creating Toolbars with New Icons 109
Finishing Up 116
Conclusion 116

Chapter 4 Support Files

About This Chapter 117
AutoCAD's Stock Support Files 118
    Release 13 Support Directories 119
    Release 13 Support Files 119
        Configuration Files: ACADNT.CFG, ACAD.CFG, ACAD.INI 119
        ACAD.PGP 122
        ACADR13.LSP and ACAD.LSP 123
        Menu Files: ACAD.MNC, MNR, MNS, MNU, MNX, MNL 123
        Programs: ADS, ARX, LSP, DCL 127
        ACAD.ADS and ACAD.RX 128
        Scripts: SCR Files 129
        Definition Files: ACAD.LIN, MLN, PAT 129
        Fonts: PFB, PFM, SHP, SHX, TTF 130
        Font Mapping Files: FMP 131
        PostScript Support Files: ACAD.PSF and FONTMAP.PS 131
        ACAD.UNT 131

Plot Configuration Parameters: *.PCP 132
  Help Files: ACAD.AHP, HDX, and HLP 132
  ACAD.LOG 133
 Understanding AutoCAD's Initialization Sequence 133
  AutoCAD Load Initialization 134
  Drawing Load Initialization 135
 Loading, Listing, and Unloading AutoLISP, ADS, and ARX Programs 136
 The Support Path and Library Search Path 139
 Third-Party Application Issues 140
 Customizing for Portability 142
 Conclusion 143

Chapter 5   Introduction to Menu Macros

 About This Chapter 145
 Defining Macros and Menus 147
  Macros and Menus in This Chapter 147
 Menu Structure Overview 148
 Writing a Simple Menu Macro 149
 Pausing Macros for User Input 154
  Ensuring Compatibility with Foreign Versions and Redefined Commands 157
 Using Special Characters in Macros 157
  Semicolons: A Special Case 159
  Semicolons in Menu Text Strings 161
  Calling and Avoiding Dialog Boxes in Macros 161
 Setting Modes with Control Characters, and Using Blocks in Macros 167
  Using Blocks with Path Names 170
 Making Long Macros 172
 Polyline Macro Tricks 173
 Object Snaps in Macros 177
  Object Sort Method Control 179
  Controlling Aperture in Macros 180
 Controlling Layers with Macros 180
 Repeating Macros and Multiple Commands 183
  Indefinite Repetitions 185
 Using Selection Sets in Macros 186
  Object Selection Settings and Menus 188
  Selection Set Modes in Menus 190
 Finishing Up 192
 Conclusion 193

Chapter 6   Text Fonts and Styles

 About This Chapter 196
 New Text Features in Release 13 197
 Understanding Text Fonts and Styles 200
 Using AutoCAD Text Styles and Fonts 201
  Native AutoCAD SHX Fonts and Unicode 205
   Special Characters 207

Big Fonts	207
PostScript and TrueType Fonts	208
Custom Fonts	209
Mapping Fonts	210
PostScript Font Mapping	212
Stacked Text	213
Customizing Text Font Files	**214**
The ROMANS.SHP Font	214
A Line-Feed Character Definition	219
Making an Angle Character	221
Font Definition Instruction Codes	223
Font Vector Codes	226
Adding Big Font Files	**227**
Big Font Escape Characters	228
Creating a Big Font File	228
Using Subshapes to Combine Characters	229
A Center Line Character	230
Custom Fraction Characters	233
Extended Big Fonts	241
Extended Big Font Subshape Code Format	241
Understanding Shape Files	**242**
Finishing Up	**244**
Conclusion	**245**

Chapter 7 Linetypes and Hatch Patterns

About This Chapter	**248**
Linetypes, Hatch Patterns, and Programs in This Chapter	249
New Linetype and Hatching Features in Release 13	**249**
Simple Dash-Dot Linetypes	**251**
Understanding Simple Linetypes	251
Creating Simple Linetypes	252
Understanding the Linetype File Format	256
Setting Linetype Endpoint Alignment and Scale	258
Scaling Linetype Patterns	260
Scaling ISO Linetypes	262
Using 2D Polyline Linetype Generation	263
R13 Complex Dash-Dot Linetypes	**263**
Understanding Complex Linetypes	264
Embedding Text in Linetypes	265
Embedding Shapes in Linetypes	268
Hatch Patterns	**271**
Using Hatch Patterns	271
Understanding Hatch Patterns	273
Creating a Simple Hatch Pattern	274
Creating a Complex Hatch Pattern	277
PostScript Fills	**289**
Finishing Up	**293**

Conclusion 293

Chapter 8 Slides, Scripts, and PCP Files

About This Chapter 296
Slides 297
    Making and Viewing Slides with MSLIDE and VSLIDE 298
        Using Slides in Model Space and Paper Space 302
        Making Slides for Image Tile Menus and Dialog Boxes 303
    Using Slide Library Files 305
        Adding Hatching Slides to ACAD.SLB 307
Script Files 308
    Understanding Scripts 308
    Planning and Writing Scripts 309
    Creating Scripted Slide Shows 311
    Pausing Scripts for Input 314
    Plotting with Scripts 314
    Batch Plotting Drawings 325
    Other Batch Processing Methods 328
    Other Batch Processing Ideas 329
Plot Configuration Parameters (PCP) Files 329
    Plotting with Scripts and PCP Files 331
    Batch Plotting Summary 332
Finishing Up 333
Conclusion 333

Chapter 9 Introduction to AutoLISP Programming

About This Chapter 336
    Programs and Data in this Chapter 336
Programming with AutoLISP 337
    Interpreted and Compiled Languages 337
    Platform and Operating System Independence 338
    AutoLISP versus Mainstream Programming Languages 338
    AutoCAD Version Independence 338
    Limitations of AutoLISP and Alternatives 339
Fundamentals of AutoLISP 340
    Variables and Expressions 340
        Prefix versus Infix Notation 341
        Variables 342
        Variable Assignment 342
    Expressions and Evaluation 344
    AutoLISP Syntax and Semantics 347
        Regional Arguments 347
        Optional Arguments 348
        Repeating Arguments 349
        Formatting AutoLISP Code 350
    AutoLISP Data Types 350
        Symbols 351

Table of Contents

      Integers 352
      Strings 352
      Reals 353
      Lists 354
      Entity Names 355
      Selection Sets 355
      File Descriptors 355
Using AutoLISP at the AutoCAD Command Prompt 355
    Working with Coordinates 359
    The COMMAND Function 360
    AutoCAD System Variables 361
AutoLISP Program Files 363
    Automatically Loading AutoLISP Programs 364
    Using MNL Files to Load AutoLISP Code 365
    Loading ADS and ARX Programs 366
    Using the Maximizing AutoCAD Code Editor 366
User-Defined Functions 369
    Anatomy of a User-Defined Function 371
    Defining New AutoCAD Commands 378
    Calling Command Functions Transparently 379
Arguments, Bound Variables, and Rules of Scope 381
Basic Programming Operations 384
    Math Operations 384
    String Operations 387
    List Operations 392
      Creating Lists with the Quote Function 395
      Putting the List Functions to Work 396
      Other List Functions 398
    Geometric Functions 399
Executing AutoCAD Commands 403
    The AutoLISP COMMAND Function 403
Conditional Branching and Program Logic 407
    Branching Tests 409
      Predicates and Relational Functions 409
      Logical Functions 413
      The PROGN Function 414
      Conditional Program Branching with COND 418
    Program Looping and Control Structures 422
      The REPEAT Function 423
      The WHILE Function 424
Acquiring and Validating User Input 428
    INITGET Input Control 429
    INITGET Bit Modes and Keywords 430
    Selecting Objects 435
    Formatting Input Prompts 437
Error Handling 440
    The *ERROR* Function 442
    Debugging AutoLISP Programs 445

Coding Practices and Standards 448
   Symbol Naming Conventions 448
   Comments and Annotations 449
Finishing Up 450
Conclusion 450

Chapter 10 Running on DIESEL

About This Chapter 452
DIESEL Mechanics 453
   Differences between DIESEL and AutoLISP 454
      DIESEL Syntax versus AutoLISP Syntax 454
   Testing DIESEL Expressions 455
   Debugging DIESEL Macros 458
Creating Custom Status Line Displays 459
   Simple Status Line Displays 460
   Time and Date Format 461
   Complex Status Line Displays 462
   Conditional Branching in DIESEL 466
   The Color Status 468
   Complex Data Manipulation 469
   DIESEL Data Storage and Retrieval 470
      MODEMACRO String Length Limit 471
   Custom Status Displays in AutoCAD for Windows 472
   Dynamic Context-Sensitive Status Lines 477
Controlling Pull-Down and Cursor Menu Labels with Diesel 479
   Check Marks in Menu Labels 480
      Setting Check Marks with DIESEL 482
Using DIESEL in Menu Macros 487
   DIESEL Expression Evaluation 488
Calling DIESEL Expressions from AutoLISP 491
Disabling Menu Items with DIESEL 494
   Bitwise Logical Operators 498
Creating Dynamic Menus with DIESEL 505
   Creating Dynamic Menu Captions with DIESEL 505
   Creating Dynamic Menus with DIESEL and AutoLISP 508
   Creating Toolbar Drop-Down List Menus with DIESEL and AutoLISP 512
Extending Custom Applications with DIESEL Power 513
Debugging and DIESEL Error Messages 514
Finishing Up 515
Conclusion 515

Chapter 11 Introduction to Dialog Boxes

About This Chapter 518
Introduction 518
Dialog Box Primer 519
Common Dialog Boxes 522
   The Message Dialog Box 523

An Enhanced Message Dialog Box	524
File Dialog Boxes	526
Using File Dialog Boxes in AutoLISP Programs	528
Selecting Drawing Files with a Dialog Box	532
The Library Search Path	532
Selecting Directories and Folders	533
The Color Dialog Box	535
Custom MTEXT Edit Dialog Box	536
Dialog Control Language	536
Dialog Control Language Basics	537
Subassemblies	538
Prototype Tiles	538
DCL Syntax and Semantics	539
Tile Attributes	540
Tile References	540
DCL Learning and Development Tools	543
DCL Code Formatting	545
Tile Parametrics	546
Using Spacers	549
Clusters	550
Prototypes	558
Prototypes and Inheritance	559
Structured Design Techniques	563
Finishing Up	567
Conclusion	567

Chapter 12 Designing Dialog Boxes

About This Chapter	569
Programs and Functions in this Chapter	570
Introduction	570
Dialog Box Tile Access and Control	574
Accessing Dialog Box Tile Data	574
Controlling Tile States	580
Accessing Tile Attributes from AutoLISP	585
Populating List Boxes and Popup Lists	588
Handling Dialog Evnets with Action Expressions	591
Event-Driven Programming	591
Dialog Box Events	592
Action Expression Variables	593
Validating Edit Box Input and Handling Errors	601
Validating Numerical Input	601
Handling Edit Box Callbacks	605
Validating Distance and Angle Input	612
Handling List Boxes	613
Putting Dialog Boxes to Work Finding Attribute Blocks	617
Tools for Dialog Development	622
Finishing Up	623

Conclusion 623

Chapter 13 Understanding Menu Structure

About This Chapter 626
   Menus and Support Files in This Chapter 628
Overview of New Menu Features in R13 Windows 629
Menu Structure 631
   Structure of the Windows MNS File 632
   Structure of the DOS MNU File 633
   Menu Sections 634
      Using Submenus 642
   Menu Item Labels 643
   Menu Item Name Tags 645
   Menu Macros 648
Controlling Command-Line Echoing 648
Menu Section Details 651
   Button and Auxiliary Menus 652
   Pull-Down and Cursor Menus 659
      Pull-Down Menu Titles and Item Labels 661
      Displaying and Swapping Pull-Down Menus Automatically 662
      The Cursor Menu 663
      Special Codes in Pull-Down and Cursor Menus 664
      Disabling and Checking Menu Items 669
   Toolbar Menus 669
      Toolbar Syntax 670
      Toolbar Titles 671
      Buttons 672
      Flyouts 672
      Controls 674
      Icons and DLLs 675
      Developing a Custom Toolbar 676
   Image Tile Menus 679
      Making Image Tile Menu Slides 680
      Designing Image Tile Menus 683
      Using Image Tile Menus versus Dialog Boxes 687
   Accelerator Menus 688
   Help Strings 690
   Side-Screen Menus 690
      Overlaying Side-Screen Menus 694
      Context-Sensitive Side-Screen Menu Paging 698
   Tablet Menus 699
Combining Menus by Loading Partial Menus 703
   The MENULOAD Command 704
   Partial Menu Loading with AutoLISP 707
   Partial Menus and MNL Files 709
Deciding Which Menu Interfaces to Use 710
Finishing Up 711

Conclusion 711

Chapter 14 Creating an Integrated Professional Menu System

About This Chapter 715
Application Initialization 716
    The ACAD.LSP File 716
    The `s::startup` Function 719
    Supporting Menus with MNL Files 720
        MNL "Helper" Functions 721
        ACAD.MNL 723
        Using MNL Files and ACAD.LSP with Partial Menus 723
        MNL Files vs. ACAD.LSP 723
    Partial Menu Loading Techniques 725
    Automatic Program Loading 732
        Automatic Loading from Menu Macros 735
    Redefining Commands 736
        Redefining PLOT 737
        Redefining File-Handling Commands 740
Using DIESEL and AutoLISP in Menus 740
    Controlling Scale Factors in Macros 740
    Controlling Cursor Menus with DIESEL 744
        Creating a Context-Sensitive Cursor Menu 744
    Extending the Object Snap Menus with AutoLISP 750
    Comparison of AutoLISP and DIESEL in Macros 753
Supporting Menus and AutoLISP Programs with an ARX Application . . . 753
    MAWIN.ARX Functions 754
Creating Help Files 764
    Other Means of Providing Application Help 766
    Developing Windows Help Files 766
    Understanding the Platform-Independent Help File Format 767
    Developing Platform-Independent Help Files 770
Finishing Up 773
Conclusion 773

Appendix A AutoLISP Function Quick Reference

(Each AutoLISP function is listed alphabetically, with a brief description of the function's action, the results returned, and its syntax, showing number and data type of arguments.)

Appendix B Diesel Function Quick Reference

(Each DIESEL function is listed alphabetically, with a brief description of the function's action, the results returned, and its syntax, showing number and data type of arguments.)

Appendix C DCL Quick Reference

    DCL Tile Attributes 807
    DCL Color Reserved Words 810
    DCL Tile Reference 811
    Predefined Buttons and Button Assemblies 820

Appendix D

Errors, Problems, and Solutions

    (Descriptions of common errors and problems encountered in customizing and programming AutoCAD, and ways to avoid or solve them.)

Appendix E System Variables

Appendix F ASCII and Unicode Tables

    ASCII Table 859
    Unicode Table 865

Appendix G The Maximizing AutoCAD CD-ROM

    (Describes the programs and utilities contained on the bundled CD-ROM. These include unique new tools and program libraries to enhance your customization efforts and make them quicker and easier.)

Introduction

Welcome to *Maximizing AutoCAD R13*

Welcome to the most creative part of using AutoCAD—customizing it. The goal of this book is to bring you the information and tools to make AutoCAD do exactly what you want it to do. Whether you simply want to add a few toolbar macros, design a sophisticated user interface controlled by menus and dialog boxes, or write a complete "system" that executes your application with AutoLISP assistance, *Maximizing AutoCAD R13* gives you the knowledge you need to make AutoCAD work for you.

Maximizing AutoCAD R13 has a companion book titled *Maximizing AutoLISP for AutoCAD R13*. Together, these two books cover every aspect of AutoCAD customization in detail, except ADS and ARX.

Maximizing AutoCAD R13 provides you with the knowledge to take complete control of AutoCAD and its user interface. Topics include menus, macros, toolbars, all support libraries, DIESEL programming, and introductory DCL dialog box development and AutoLISP programming. It also features ways to organize and optimize your system's performance, and presents the accepted standards for proper application and interface presentation and behavior. The book explains, in detail, every aspect of AutoCAD customization except ADS, ARX, and advanced AutoLISP and dialog box development. As an added bonus, the MaxAC CD-ROM includes several valuable unique new development tools, to make AutoCAD customization quicker and easier.

Maximizing AutoLISP for AutoCAD R13 offers in-depth coverage of AutoCAD's own programming language, AutoLISP. *Maximizing AutoLISP for AutoCAD R13* picks up where *Maximizing AutoCAD R13* leaves off by explaining the full scope and use of AutoLISP and DCL dialog box development and control. It provides the external programming tools and techniques that you need to develop a well-integrated and controlled application system.

You may think that learning to customize AutoCAD would be a formidable task, but it is not. You may also harbor the suspicion that you have to be a programmer to use AutoLISP, but you do not. The *Maximizing AutoCAD R13* and *Maximizing AutoLISP for AutoCAD R13* books are written for typical AutoCAD users like you, as well as for advanced users, CAD administrators, and application developers.

 Note: This book was developed in Windows, which is far superior to DOS as a customization environment. If you use DOS, you can follow most of the book, but not the advanced customization topics and tools.

The History of Maximizing AutoCAD R13

Maximizing AutoCAD R13 dates back to 1988. The first edition, *Customizing AutoCAD* (by Rusty Gesner, Pat Haessly, and Joe Smith, from New Riders Publishing), was one big book! It fully explained AutoLISP, as well as the intricacies of customizing menus, macros, and support files. When AutoCAD R10 was introduced, the material became so extensive that *Customizing AutoCAD* was expanded and split into two volumes, originally published by New Riders Publishing under the titles *Customizing AutoCAD* and *Inside AutoLISP*, and in later editions under various Maximizing titles. Now, Autodesk Press is pleased to bring you the current editions, *Maximizing AutoCAD R13* and *Maximizing AutoLISP for AutoCAD R13*.

Customizing AutoCAD

Most users adapt AutoCAD to their own needs at some level. They often start by writing macros that they add to toolbars or menus. Macros are a natural extension of AutoCAD. Many users move on to a second stage of customizing AutoCAD by making their macros more efficient with AutoLISP or DIESEL expressions or by adding new drafting and calculation functions, commands, and subroutines to their systems. They integrate these commands and functions into their menus. Finally, some users and groups move on to develop "full-blown" integrated application systems. These systems have everything from complete custom user interfaces for AutoCAD to the capability to integrate and run external programs.

Maximizing AutoCAD R13 addresses the key issues you encounter in planning, developing, programming, and testing simple custom tools or professional AutoCAD application systems. The book assumes that you are fairly experienced with using the AutoCAD program. If not, you might want to begin with one of the many available tutorial and reference books on AutoCAD.

Who Should Read This Book

Whether you are a user ready to try the first stages of customization, an administrator needing to control and standardize your office CAD systems, or a developer putting a complete custom application together, this book helps you assess your application needs and choose the best avenues for customizing. The book provides menu macros and AutoLISP-enhanced macros to help you gain greater control over AutoCAD and better access to your programs and data. The book helps you develop a user interface, enabling you to create your own prompts, input criteria, and default options. The book covers issues such as integrating drawing standards and drawing scales to take advantage of AutoLISP and creating the robust, controlled programs and applications you need to develop an application.

You will find that this book covers AutoCAD customization more completely than any other single source.

Intermediate or Advanced AutoCAD Users

Maximizing AutoCAD R13 starts out easy. It introduces intermediate or advanced AutoCAD users who are ready to increase their productivity to the power of toolbar and menu macros. It provides concise and well-planned exercises and examples showing how to build time-saving macros quickly. Many of the macros can be immediately integrated into your existing system or applications without modification.

No prior AutoCAD customizing experience is necessary for this book. If you have already started writing macros of your own, you will learn many new tricks, including how to organize your AutoCAD interface for easy and efficient drafting. All the user interface and menu devices are covered—toolbars, dialog boxes, keyboard shortcuts and aliases, and all of the menus: side-screen, pull-down, popup cursor, toolbar, tablet, image tile (icon), and auxiliary or button. In addition to user interface customization, the book also shows you how to create custom hatch patterns, line types, and fonts to make your drawings more clear and more professional in appearance.

AutoCAD System Administrators

In addition to the information that is beneficial to intermediate and advanced AutoCAD users, AutoCAD system administrators will learn techniques for establishing and enforcing drawing and office standards, creating interactive macros, automating drawing setup, and providing a consistent, organized menu system to your AutoCAD users. The book also includes techniques for improving AutoCAD and system performance.

The Benefits of This Book to AutoCAD Developers

Maximizing AutoCAD R13 shows how to develop all aspects of an integrated application system from start to finish. This book also shows how to make customized help files, complete with graphics and color for users of your applications. The authors have experience as professional AutoCAD developers, consultants, and dealers, and they share their secrets, tips, and techniques for creating robust AutoCAD applications. The book also includes tools and techniques for rapid development and testing of your menu systems.

How This Book Is Organized

Maximizing AutoCAD R13 takes you from basic system setup all the way through creating custom applications.

Part One: Getting Started

Chapter 1 explains the pros and cons and installation and configuration procedures for a properly configured AutoCAD installation in Windows 95, Windows NT, Windows 3.1, or MS-DOS. It also covers the additional demands of networks, and techniques for optimizing AutoCAD performance on your system. These concepts and techniques prepare you to establish a customization environment in Chapter 2.

Chapter 2 shows how to organize directories for customizing AutoCAD and for this book, set up and control multiple AutoCAD configurations, and avoid ACAD.LSP and prototype drawing conflicts. It also covers installing the MaxAC CD-ROM and selecting and configuring a suitable text editor. The book's separate customization environment eliminates conflicts with your existing system and ensures that the book's exercises match your Maximizing AutoCAD setup.

Part Two: The Essentials of Customization

Chapter 3 is all about keyboard macros and toolbars, two easy ways to improve productivity. It covers three different keyboard shortcuts: command aliases, simple AutoLISP command definitions, and menu accelerator keys, as well as how to easily customize the new R13 for Windows toolbars.

Chapter 4's purpose is to ensure that you understand the big picture before you create and modify AutoCAD support files. It will help you control AutoCAD better and understand the consequences of your customization changes. It covers the AutoCAD initialization sequence, what kinds of support files AutoCAD uses, what they are called, where they are located, how AutoCAD loads them, and what happens behind the scenes when you launch AutoCAD or open a drawing.

Chapter 5 covers menu macros in detail. It explains enough about menu structure to enable you to create and edit pull-down menu sections. You will learn the details of menu macro syntax, how to write macros, and how to add a custom pull-down menu page to a menu.

Chapter 6 teaches you how to control and customize text with AutoCAD's font and style features. You will learn to create and modify new text styles, how and when to use PostScript and TrueType fonts, how to map fonts, how to create custom characters such as a centerline character and fractions, and all about shapes, which can be used as symbols in R13's new complex linetypes.

Chapter 7 is all about linetypes, hatch patterns, and PostScript fills. You will learn to create simple and complex (symbolic) custom linetype definitions; to control linetype scaling with global, object-specific, and paper space options; to create simple and elaborate custom hatch pattern definitions; and to apply PostScript fills to polyline boundaries.

Chapter 8 explores slides, script files, and plot configuration parameter (PCP) files. You will learn to create and view slides, assemble slide libraries, create a scripted slide show, develop scripts for plotting and device configuration, use batch processing to plot multiple drawings with a single command, and use PCP files and integrate them into plotting scripts.

Part Three: Introduction to Programming AutoCAD

Chapter 9 is an introduction to AutoLISP's full power. You will learn to use AutoLISP to create and call user-defined AutoLISP functions; add new commands to AutoCAD; obtain user input and assign it to program variables; handle errors in user input or program execution; perform basic math, string, and list operations; and query and control the state of the AutoCAD drawing editor and AutoCAD commands.

Chapter 10 covers programming with DIESEL. With DIESEL, you'll learn to create custom dynamic status displays; create menu item labels whose contents change dynamically; and to conditionally disable, enable, check, and uncheck menu items in response to current settings. You'll also create intelligent menu macros that can automate responses to input solicited by any command, including those written in AutoLISP and ADS. You'll learn to use DIESEL expressions in AutoLISP programs, add dynamic popup cursor menus to AutoLISP programs, and use DIESEL to dynamically manage and control popup cursor menus.

Part Four: Improving the User Interface with Dialog Boxes

Chapter 11 first teaches you about predefined dialog boxes and the functions that access them, and then introduces Dialog Control Language (DCL) and explains how to use it to create and modify custom dialog boxes.

Chapter 12 teaches the basics of controlling dialog boxes with AutoLISP, including loading and initialization, writing callback functions, techniques for handling callback events and input errors, how to write reusable callback functions, and loading and managing list boxes.

Chapters 11 and 12 together provide a gentle introduction to DCL and how to control dialog boxes with AutoLISP, but for complete coverage of DCL and driving dialog boxes with AutoLISP control, you need the book *Maximizing AutoLISP for AutoCAD R13*.

Part Five: Designing an Integrated User Interface

Chapter 13 covers menu structure, including all of the AutoCAD menu interfaces and their numerous new features in R13 for Windows. You'll learn to create menu files and sections; use menu item labels, menu groups, and name tags; control command-line echoing of menu macros and AutoLISP code; and create and control menu items in each type of menu section. You'll develop pointing device menus that interact with cursor menus, cascading pull-down menus, flyout toolbars, and image tile, side-screen, and tablet menus. You'll learn to use two kinds of accelerator key syntax; define help strings to accompany pull-down, cursor, and toolbar items; attach partial menus and use their menu sections; and decide which menu interfaces to use.

Chapter 14 brings it all together by showing you how to integrate your customization efforts, especially menus and AutoLISP, into more sophisticated application components. It explores some additional partial menu loading techniques and problems, and covers the new R13 help files. You'll learn to integrate DIESEL, AutoLISP, and menus into unified application functions, develop a menu LISP (MNL) file that supports your custom menu, and to automate AutoLISP program loading on demand. This chapter also presents an extremely convenient and powerful context-sensitive popup cursor menu.

Appendixes

Maximizing AutoCAD R13 has seven appendixes. Appendix A is an AutoLISP function quick reference. Appendix B is a DIESEL function quick reference. Appendix C is a DCL (dialog box) language quick reference. Appendix D covers common problems and solutions encountered in customizing AutoCAD. Appendix E provides a handy reference table of AutoCAD system variables. Appendix F provides a handy reference table of ASCII character and Unicode font codes. And Appendix G covers all of the valuable tools and programs on the MaxAC CD-ROM.

Using This Book

If you just read the book, you learn a great deal about customizing AutoCAD. But to make your knowledge concrete, you need to sit down at a computer, work through the exercises, and try the programs.

The best way to work through the book is in sequence, but most of the material is independent so that you can pick and choose topics. The MaxAC CD-ROM makes it possible to "jump in" nearly anywhere. Wherever you start, you should first do the setup and MaxAC CD-ROM installation exercises in Chapter 2, so that your system setup corresponds with the directions in the exercises.

Fulfilling Your Customization Needs

This book covers a wide variety of AutoCAD customization topics, most of them in-depth. You don't need to read every chapter, or read them in order. Table I.1 tells you which part of the book covers what you want to customize, and suggests some prerequisite chapters.

Welcome to Maximizing AutoCAD R13

Table I.1
Where to Go from Here

If You Want to Customize	Read Chapters	Prerequisite Chapters
Menus	5 and 14	1 through 4
Text	6	1, 2, and 4
Linetypes	7	1, 2, and 4
Hatch patterns	7	1, 2, and 4
PostScript support	7	1, 2, and 4
Slides	8	1, 2, and 4
Scripts	8	1, 2, and 4
Plot Configuration Parameters	8	1, 2, and 4
Help files	8	1, 2, and 4
AutoLISP	9	1 through 5
DIESEL	10	1 through 5 and 9
Dialog boxes	11 and 12	1 through 5 and 9
Menus integrated with AutoLISP	13	1 through 10
Professional application menus	14	1 through 13

Using The Exercises

The process of customizing AutoCAD may seem complex at times, but the authors have tried to make this rather comprehensive book as easy to use and to follow as possible. To avoid errors and misunderstandings, please read the following instructions before you start working with the book. Also, be sure to do the setup exercises in Chapter 2 before doing the exercises in the rest of the chapters.

Chapters 1 and 2 explain how to set up your system and AutoCAD for use with *Maximizing AutoCAD R13*. The book's setup and exercises are designed so that they do not interfere with any AutoCAD settings or with other work that you may be doing with AutoCAD.

The Exercise Format

The following exercise is a sample of the format followed throughout the book. AutoCAD's prompts or screen display text and your input are in computer-style type on the left of the exercise. Explanatory comments are in the right column. Prompts, including operating system and AutoLISP prompts, are shown like the `Command:` prompt in the following exercise. Prompts are shown as they appear on your screen, although the exercises abbreviate or omit portions of the command dialog for simple commands, unimportant command dialog, or code that scrolls by, and obvious or repetitive sequences. Necessary commands, prompts, and input are always shown. Boldface type indicates what you need to enter. The symbol [Enter] or the instruction *press ENTER* tells you to press the RETURN or ENTER key on your keyboard. AutoCAD prompts refer to this key as the RETURN key, but most keyboards label it as Enter, so the text refers to it as ENTER.

 Exercise

Sample Exercise

Continue in the previous drawing, or begin a NEW drawing named CA-INTRO.

Command: *Choose* **O**ptions, **P**references	Issues the **P**references command from the **O**ptions pull-down menu, and opens the Preferences dialog box
Press CANCEL	Closes the Preferences dialog box
Command: **(+ 2 3)** Enter	Uses the AutoLISP **plus** function to add two numbers

LISP returns: 5

In your text editor, open the \MAXAC\DEVELOP\ACADMA.MNS file. Add the bubble sequence to ACADMA.MNS below the Circle and Text lines:

```
MA_MaxAC        [MaxAC]
[Circle]_CIRCLE
[Text]_TEXT
[Bubble]_.CIRCLE \24 _.TEXT _M @ 18
0 \ Enter
```

Exercise Notes and Tips:

- Type input exactly as shown, watching carefully for the difference between the numbers 0 or 1 and the letters O or l. The @ is the @ character, located above the 2 on your keyboard. Keyboard keys are shown similarly to their labels, such as F9 for function key 9 and CTRL+C for the Control-C key combination.

- When you see an instruction such as Choose **O**ptions, **P**references, you should choose the **O**ptions pull-down menu, then the **P**references item from it, using any means you prefer. Menu hotkeys are shown in a bold, underline font. If a pull-down menu or dialog box is currently displayed, Choose refers to an item on it.

▶ In Windows, rather than retyping something if you make a mistake and have to cancel, you can use the Edit menu on the AutoCAD Text Window to copy/paste from/to the command line.

▶ When you are to add code to an existing menu or AutoLISP file, the new code is shown in bold.

▶ Input shown in bold and in parentheses, like **(+ 2 3)**, is AutoLISP code that you enter with the parentheses. You will not see the words "*LISP returns:*" on your AutoCAD *text console* (the command lines and text window), but when you see the words *LISP returns:* in the exercise, what immediately follows is what AutoLISP should "echo" back to the text console. This will usually be the result of evaluating the expression you entered. If what you see echoed differs from the exercise text, you've probably made an error or mistyped the input. When this happens, it is best to cancel input by pressing CANCEL and then re-enter the input.

▶ When you see an instruction to *Press CANCEL*, you should press CTRL+C or ESC, depending on your operating system and settings. Use CTRL+C in DOS or if you have *AutoCAD Classic* keystroke mapping enabled. Use ESC in Windows, unless you have *AutoCAD Classic* keystroke mapping enabled. The *AutoCAD Classic* setting is on the System page of the Preferences Dialog.

▶ Typographical conventions and the physical dimensions of this book often prevent an entire line of input from being displayed on a single line in the exercise text. Where this occurs, the input will be broken into two or more lines in the exercise text, even though it is a single line of input which should not be interrupted by pressing ENTER. In order to make each line of input more explicit, all exercises will have a [Enter] symbol at the point where you should press ENTER. Do not press ENTER when you reach the end of a line if there is no [Enter] symbol; just continue entering the input shown on the next line(s) and press ENTER when you reach a [Enter] symbol.

▶ Some exercises contain bubbles, like ②. These bubbles refer to corresponding bubbles in accompanying illustrations or code listings.

▶ We use the terms directory and folder interchangeably.

Exercises and Your Display

This book's AutoCAD illustrations were created in a 800x600-pixel window. If you are using a different size AutoCAD window, your AutoCAD images may vary, and you may need to zoom to get an appropriate view for an exercise. You may also need to zoom or adjust your aperture size when using object snap if your screen resolution is extremely high or extremely low.

Your Digitizer or Mouse Buttons

Button assignments vary with the type of digitizer puck (that is, cursor), mouse, or other pointing device you use. The pick button usually is the left-most, upper-most, or lowest-numbered button on your pointing device. On a digitizer puck, it might be marked with a 1, a zero, or an asterisk, or might be not be labeled at all. You can't customize the pick button's operation—it always acts as the pick button. We call this the pick button throughout this book, and an instruction to just click or pick refers to this button.

Note: If you've used the Windows or AutoCAD driver configuration program to alter the pointing device button mapping (for example, swapping the left and right mouse buttons for a left-hander), then the button assignments will be different than what's described here.

The first button *after* the pick button is the first button that you can customize. On a mouse, this usually is the right mouse button. On a digitizer puck, it might be labeled with a 2 or a 1 (depending on whether the pick button is labeled 1 or zero). Regardless of what it is marked with, we call this first configurable button "Button1". The next button (for example, third digitizer puck button or middle button of a three-button mouse) is the second one that you can customize ("Button2"), and so on ("Button3", etc.). In the diamond-shaped four-button layout that's common on many digitizer pucks, the pick button is at 12 o'clock, and they proceed counter-clockwise: Button1 at 9 o'clock, Button2 at 6 o'clock, and Button3 at 3 o'clock.

When you are supposed to press and hold down the CTRL, ALT, and/or SHIFT keys while clicking a button, you will see an instruction like *Click* SHIFT+Button1.

Code Listings

All menus, AutoLISP programs, DCL code, and other text files used in the book are listed in monospace type, as shown in the following example. **Bold, monospace** type may be used to call your attention to a modification or addition you are instructed to make to a file, like the `[Control(_Layer)]` and `[-]` lines in the following listing. Some listings contain lines that are too long to fit on a single printed line in the book. These long lines are shown on two lines, broken with a ▷ symbol, as the example in the following listing shows. Although the book shows such long lines on two lines, they are single lines in the files, and if you are entering such a line, you should enter it unbroken on a single line.

```
**TB_OBJECT_PROPERTIES
ID_TbObjpro     [_Toolbar("Object Properties", _Top, _Show, 0, 100, 1)]
ID_Layers       [_Button("Layers", ICON_16_LAYERS, ▶]
ICON_32_LAYERS)]'_ddlmodes
                [-]
                [_Control(_Layer)]  [Enter]
                [-] [Enter]
                [_Control(_Color)]
                [-]
ID_Linety       [_Button("Linetype", ICON_16_LINETY, ▶]
ICON_32_LINETY)]'_ddltype
                [-]
                [_Control(_Linetype)]
                [-]
```

Sometimes files may include code not yet discussed or needed in the corresponding exercises. Just ignore such code until it is mentioned or discussed.

AutoLISP, DIESEL, and DCL Syntax

When this book presents an AutoLISP or DIESEL function it is shown in **`bold and monospace`** type. Required literal information is also shown in **`bold and monospace`** type. Required arguments are shown in ***`bold, italic, monospace`*** type and optional arguments are shown in ***`bold, italic, monospace`*** type in square brackets *[ ]*. Whenever the arguments are referred to in text, they are shown in the appropriate format. References in text to DCL elements such as labels, dialog names, identifiers, and attributes, shown in `monospace` type, but not bold. The following is a sample AutoLISP definition:

`if`

`(if testexpr thenexpr [elseexpr])`

Evaluates **`testexpr`**. *If the result is non-nil(true),* **`thenexpr`** *is evaluated and the result is returned by* **`if`**, *and* **`elseexpr`** *(if present) is ignored. If* **`testexpr`** *is* **`nil`** *(false) and* **`elseexpr`** *is present,* **`elseexpr`** *is evaluated and its result is returned by* **`if`**. *If* **`testexpr`** *is* **`nil`** *and* **`elseexpr`** *is not present,* **`if`** *returns* **`nil`**.

In the previous example, the opening parenthesis, the **`if`**, and the closing parenthesis are required literal information. The ***`testexpr`*** and ***`thenexpr`*** *arguments are required, and the* ***`elseexpr`*** argument is optional. If any number of ***`elseexpr`*** arguments were permitted, it would have been shown with ellipses, like *[****`elseexpr...`****]*.

Using the MaxAC CD-ROM

The MaxAC CD-ROM comes packaged with this book. The CD-ROM includes all the files used in *Maximizing AutoCAD R13* exercises, as well as other example files and programs.

The MaxAC CD-ROM saves you time and energy. It ensures that your customization menus, macros, and AutoLISP routines are accurate. The MaxAC CD-ROM also releases you from tedious typing and lets you focus your attention on the real point of the material when you work through the exercises. Instructions on installing and using the files from the CD-ROM are given in Chapter 2.

The MaxAC CD-ROM also includes several programs and tools to enhance your customization efforts and make them quicker and easier. These are covered in Appendix G.

What You Need for This Book

You should have enough experience to feel comfortable using AutoCAD. You should be familiar with most of AutoCAD's commands, but you do not need to be an expert. You should know how to configure your AutoCAD system, and you should understand your operating system fairly well. Beyond that, you just need a desire to make your AutoCAD system more effective and productive.

The book makes the following assumptions:

- You can use the basic features and utility commands of your operating system.
- You have a CD-ROM drive on your system or available over a network.
- You have a copy of AutoCAD Release 13 c4 or later installed and configured.
- You have a system with at least 20M of free disk space after AutoCAD is installed and configured.

AutoCAD Versions, and Windows 95 versus Other Operating Systems

This book was developed with the North American unlocked single-user version of AutoCAD Release 13 c4 under Windows 95. We highly recommend you obtain the c4 version if you do not already use it. It is a free update to all registered users of R13 and fixes many problems with earlier versions.

Windows 95 is recommended for complete compatibility with this book. Windows NT is a good alternative, particularly NT 4.x. Windows 95 and Windows NT offer valuable features for a customizing environment and can easily deal with memory and environment space limitations.

If you are using Windows 3.xx, Windows for Workgroups, or Windows NT 3.x, you should be able to perform all exercises and follow everything in the book, but you will see minor differences between your system and some illustrations and instructions.

If you are using DOS, you can perform many of the exercises and follow most of the book, but the more advanced customization features and tools are not available to you, and customization will take greater effort than in the Windows environments.

We strongly encourage Windows 3.xx, Windows for Workgroups, and DOS users to upgrade to Window 95 or NT.

Solving Problems

- Make sure you are using AutoCAD R13 c4 or later.

- Try again. Use the UNDO command to clean up the drawing and restore system variables. Go back to the previous exercises to see if you made an error that did not show up immediately.

- Check the settings for defaults such as snap, object snaps, aperture, entity selection system variables, current layer, and layer visibility.

- See Appendix D.

- Check the *AutoCAD Reference Manual*, the *AutoCAD Customization Guide*, and the *AutoCAD Installation Guide*.

- Call your AutoCAD dealer. If you have no current dealer, Autodesk can find you one.

- Try the ACAD forum on CompuServe, the world's largest CAD user group and the most knowledgeable source of free support.

- See the README.TXT file for any possible updated information.

- If you are using AutoCAD R13 c4 or later in a Windows environment, and only if there is a *specific* problem with this book, particularly a program bug or problem with an exercise, you can contact Rusty Gesner by e-mail (preferred) at rusty@teleport.com via the Internet or at 76310,10 via CompuServe, by fax at (503) 663-3043, or by mail c/o Autodesk Press. You must have the page number where you are having a problem ready—we cannot give free general AutoCAD customization support. Remember, we cannot provide help unless you are using AutoCAD R13 c4 or later in a Windows environment.

▶ If you need customization services or help with an AutoCAD customization problem that is not a *specific* problem or bug in this book, all three of the authors of this book are available for consultation for a reasonable fee.

Rusty Gesner specializes in problem-solving and troubleshooting, consulting to new third-party developers, and writing tutorial documentation. You can contact Rusty by e-mail (preferred) at rusty@teleport.com via the Internet or at 76310,10 via CompuServe or by fax or telephone at (503) 663-3043. Look for Rusty's home page on the World Wide Web at http://www.group-a.com/~rusty.

Tony Tanzillo provides full-service consulting and application development of all types, and is probably the leading expert on AutoLISP. You can contact Tony by e-mail at tonyt@compuserve.com or 71241,2067 via CompuServe or telephone at (201) 523-7256. You can also visit Tony's home page on the World Wide Web at http://ourworld.compuserve.com/homepages/tonyt.

Mark Middlebrook provides a range of consulting, training, and development services. You can contact Mark by e-mail at 73030.1604@compuserve.com or telephone at (510) 547-0602 (Oakland, CA).

Moving On

Turn to Chapter 1 to explore system set-up and optimization for AutoCAD R13.

chapter 1

AutoCAD System Setup and Optimization

AutoCAD customization begins with proper installation, configuration, and optimization of AutoCAD on your computer. An improperly configured AutoCAD installation not only slows down AutoCAD performance, but also makes your customization work slower and more error-prone. AutoCAD is one of the most demanding applications you can run on a desktop computer, so it pays to spend some time ensuring that your system is properly configured for it.

The job of installing, configuring, and optimizing AutoCAD has many new facets in Release 13, thanks to the family of Windows operating systems (Windows 3.1, Windows NT, and Windows 95) that AutoCAD now supports. Some aspects of configuration (for example, display drivers) are easier in R13, but others are potentially more complicated (plotter and digitizer drivers), or different (swap files). Thus, even if you've mastered the mysteries of configuring previous versions of AutoCAD under DOS or perhaps Windows 3.1, there is much to learn when you make the move to one of the newer Windows operating systems.

About This Chapter

In this chapter, you'll learn how to do the following:

- Establish a properly configured AutoCAD installation in Windows 95, Windows NT, Windows 3.1, or MS-DOS

- Learn some of the merits and shortcomings of each operating system

- Consider the additional demands of networks and third-party applications

- Walk through the installation and configuration procedure for each operating system

- Learn techniques for optimizing AutoCAD performance on your system

The concepts and techniques you will learn in this chapter pave the way for establishing a customization environment in Chapter 2.

Note: If you've already installed, configured, and optimized AutoCAD on your system and you're satisfied with its performance, you can skim this chapter and move on to Chapter 2, where we establish a customization environment that matches the one used in the book's exercises.

You don't need any tools from the *Maximizing AutoCAD* CD-ROM for this chapter, but you will need your R13 CD-ROM or disks. You might also find the other items listed here helpful:

- AutoCAD R13c4 (or later) installation CD-ROM or disks

- Autodesk fax-back system documents: call (415) 507-5595 or check CompuServe's ACAD forum libraries

- *Microsoft Windows NT Workstation Installation Guide*

- *Microsoft Windows 95 Resource Kit*

Understanding Intel-Based AutoCAD Platform Choices

If you're customizing AutoCAD, there's a good chance that you also have some responsibility for platform (that is, hardware and operating system) decisions. Whether you're deciding for yourself, making recommendations to your company, or advising your clients, you need to understand the AutoCAD R13 platform choices. Also, if you customize AutoCAD for other users, they might use your menus and programs on different platforms, in which case you should have at least some familiarity with those platforms.

In this book, we concentrate on the operating systems that run on Intel (or Intel-compatible) microprocessors: MS-DOS, Windows 3.1, Windows NT, and Windows 95. For most people, these platforms offer the best combination of mainstream acceptance, widespread support by third-party applications, utilities, dealers, and consultants, and a promising future (at least for the Windows operating systems).

Until recently, MS-DOS was the overwhelmingly favorite AutoCAD platform because of its excellent performance and familiarity to long-time AutoCAD users. With Release 12 for Windows, the tide began to turn away from MS-DOS, especially among new users. Nonetheless, DOS 386 AutoCAD remained the popular choice for those upgrading to R12 from earlier versions.

AutoCAD System Setup and Optimization

 With R13, Autodesk hopes that the choice of Windows operating systems and enhancements to AutoCAD for Windows will persuade a large percentage of MS-DOS users to make the switch. The eventual migration of most AutoCAD users from MS-DOS to some flavor of Windows is all but inevitable, but when you should make the switch depends on many factors.

 Because AutoCAD LT Release 3 is based on R13, the discussion of R13 in this chapter generally applies to AutoCAD LT Release 3.

DOS vs. the Windows Family

The first decision for many companies is whether to use MS-DOS or one of the Windows operating systems (regardless of which one). R13 for DOS continues to offer the best raw performance and the least amount of change to long-time AutoCAD users. On the other hand, DOS AutoCAD is becoming increasingly archaic. In R13, MS-DOS hasn't received the same level of attention to interface and feature improvements as have the Windows platforms. Some of the new customization features in R13 are available only on the Windows platforms.

Figure 1.1 shows the DOS AutoCAD Release 13 interface.

The advantages of DOS AutoCAD are:

- Lower hardware requirements (processor, memory, and disk space)
- Better raw performance on the same hardware
- Few interface changes for long-time DOS AutoCAD users

The advantages of Windows AutoCAD are:

- Multitasking
- Interoperability with other Windows applications
- Up-to-date interface consistent with other Windows applications
- Additional customization features: partial menu loading, toolbars, accelerator keys, and Windows-style help
- More flexible text window (cut-and-paste capability; resizable, and relocatable)

In sum, R13 DOS continues to be attractive to long-time users, owners of older computers (for example, a 486 with 8 MB of RAM), and those for whom raw performance (for example, loading very large drawings) is paramount. R13 Windows will appeal to users

Figure 1.1
AutoCAD R13
for DOS.

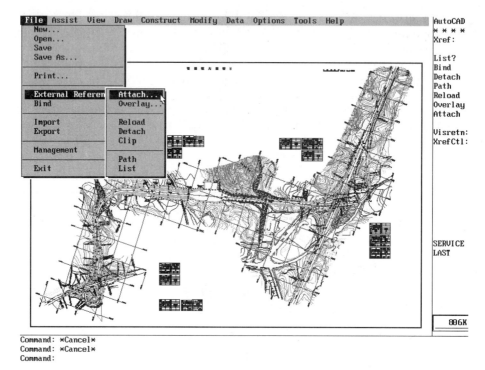

who are conversant with Windows, those who want to multitask other applications with AutoCAD, owners of modern hardware (Pentium with at least 16 MB of RAM), and those for whom the additional productivity features of Windows outweigh the lower raw performance compared with MS-DOS. These Windows AutoCAD advantages make Windows, particularly Windows 95 or NT, a much easier and more productive environment for customizing and programming AutoCAD.

Windows 3.1 vs. NT and 95

If you've decided in favor of Windows over DOS AutoCAD, the next question is "which Windows?" With R13c4, AutoCAD now supports all of Microsoft's Windows operating systems: Windows 3.1, Windows NT Workstation, and Windows 95.

Windows for Workgroups 3.11 is fundamentally the same as Windows 3.1, with the addition of peer-to-peer networking services and 32-bit file access (an improved caching system for reading and writing files). For the purposes of this book, Windows for Workgroups 3.11 and Windows 3.1 are essentially the same, so when we say "Windows 3.1", we're talking about both.

Windows NT comes in two flavors: Windows NT Workstation and Windows NT Server. NT Workstation is the version that is run on users' computers. It includes peer-to-peer

AutoCAD System Setup and Optimization

networking services similar to those in Windows for Workgroups 3.11, but its primary role is as an operating system for running a user's applications. NT Server is a network operating system for file, print, and application servers (computers whose sole function is providing network resources to other computers). NT Server is similar in concept to Novell NetWare, with which it competes. When we say "Windows NT" (or simply "NT") in this book, we mean Windows NT Workstation.

Autodesk developed R13 for Windows as a 32-bit Windows application, which means that it's best-suited to 32-bit versions of Windows (NT and Windows 95). When you run R13 for Windows in Windows 3.1, it relies on Microsoft's Win32S, a piece of software that translates the 32-bit AutoCAD program functions to 16-bit instructions that Windows can understand. As you would expect, the translation process slows down performance. Win32S also seems to cause stability problems, at least on some computers.

As a result, Windows 3.1 is not the ideal operating system for R13 Windows. If you're wedded to Windows, performance and stability will be higher—and the "fuss factor" lower—if you choose Windows 95 or NT. Of course, for many people, that decision entails a change of operating system, which is not to be undertaken lightly. Nonetheless, the Microsoft juggernaut is propelling everyone to Windows 95 or NT, so you're likely to be headed in that direction soon anyway.

Figure 1.2 shows the interface for AutoCAD R13 for Windows 3.1.

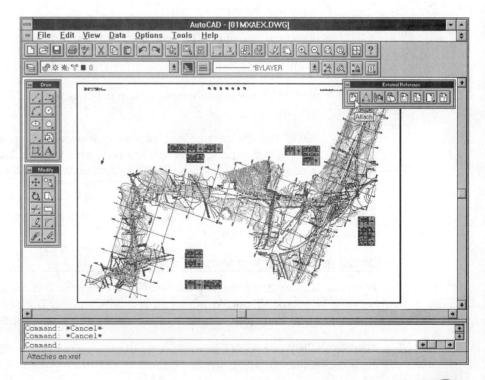

Figure 1.2
AutoCAD R13 for Windows 3.1.

Windows NT and 95 Advantages over 3.1

The advantages of Windows NT and Windows 95 compared to Windows 3.1 are:

- Better performance
- Better stability
- Multiple R13 sessions
- Better memory management, including elimination of most low Windows resources problems
- Better multitasking
- Better support for other new applications
- Better network connectivity

Windows NT vs. 95

The advantages of Windows NT over Windows 95 are:

- Excellent stability
- Excellent security
- Excellent scalability (for example, additional memory or processors)
- With NT 3.51, familiar interface for Windows 3.1 users
- Full 32-bit operating system (which is better for Intel's Pentium Pro processor)

Note: This chapter's observations about Windows NT are based on version 3.51, which is the shipping version as this book goes to press. Microsoft plans to release a new version of Windows NT with a Windows 95 style interface in mid-1996.

As you can see, there's little reason to use R13 with Windows 3.1. That's especially true for customization work, where greater operating system stability makes development and debugging easier.

Deciding between Windows NT and 95

For those who've made the commitment to Windows and AutoCAD R13, the decision between Windows NT and Windows 95 is a difficult one. Both operating systems offer advantages, and there's no simple formula for deciding between them.

Microsoft bills Windows NT as its most robust and secure operating system for high-end engineering and so-called "mission-critical" applications. Windows 95 is the mainstream operating system for business applications, and it is designed to perform at least as well as Windows 3.1 on average hardware. From these characterizations, you might expect Windows NT to be the more appropriate operating system for AutoCAD, but unfortunately it isn't that simple. Many users fall in between the two classifications, and some of the newer features of Windows 95 will appeal to AutoCAD users.

"Stability" is the first advantage that most Windows NT users mention. NT's crash protection is excellent. Windows NT goes to great lengths to prevent dangerous operations by application programs or unauthorized actions by users. That doesn't mean that programs can't crash, but NT prevents ill-behaved applications from crashing other applications or the operating system. The additional stability and security does have a price, though: NT demands more hardware resources for equivalent performance with Windows 95 or 3.1. Autodesk recommends a minimum of 32 MB of RAM for R13 in Windows NT, compared with 16 MB for Windows 95 or Windows 3.1. In addition, NT is not compatible with some Windows 3.1 and MS-DOS applications. For example, AutoCAD R12 for DOS won't run at all in an NT DOS window, and R12 for Windows is very slow when opening drawing files in NT.

"Compatibility" is the watchword for Windows 95. The design requirements for Windows 95 included a high degree of compatibility with existing hardware, Windows 3.1 applications, and MS-DOS applications. As a result, Windows 95 runs well on memory-constrained 486 computers, and it runs more MS-DOS and 16-bit Windows 3.1 applications than does NT. These advantages are important to companies with a mix of newer and older hardware (or lower-powered laptops), and to users who rely on older applications. Windows 95's greater compatibility results in somewhat lower stability than NT (although still much improved over Windows 3.1). Windows 95 is a mixed 32-bit and 16-bit operating system, which causes some performance degradation with the new 32-bit optimized Pentium Pro (P6) microprocessor.

Both operating systems have excellent disk performance, although NT's native NTFS file system is more robust and secure (assuming that you choose to use it). Both Windows 95 and NT use vastly improved memory management schemes compared to Windows 3.1. In particular, the low system resources problems that plagued Windows 3.1 (especially with complex applications like R13) go away in both Windows 95 and NT. You can open many more applications at the same time without getting out-of-memory errors or strange program behavior due to low system resources. Both operating systems include client support for multiple network protocols, which can make network installation and support simpler. For those hard-to-please DOS applications, you can install Windows NT or 95 in a "dual boot" configuration that allows booting to plain MS-DOS (this feature is more likely to be necessary with NT). For the time being, Windows 95 has a more modern user interface than Windows NT, but Microsoft will release a similar interface for NT soon.

Windows NT Advantages

The advantages of Windows NT are:

- Excellent stability
- Excellent security
- Excellent scalability (for example, additional memory or processors)
- Familiar interface for Windows 3.1 users
- Full 32-bit operating system (which is better for Intel's Pentium Pro processor)

Figure 1.3 shows the interface for AutoCAD R13 in Windows NT.

Windows 95 Advantages

The advantages of Windows 95 are:

- Excellent DOS and Win 3.1 compatibility
- Lower hardware requirements (especially memory)

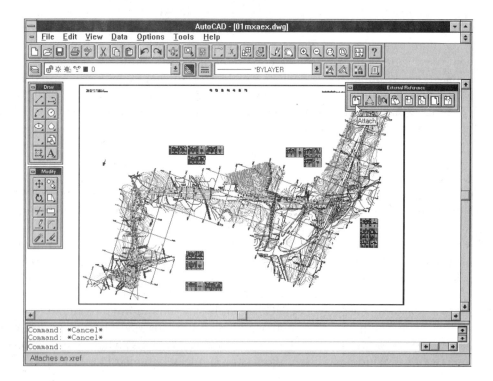

Figure 1.3
AutoCAD R13 for Windows NT.

AutoCAD System Setup and Optimization

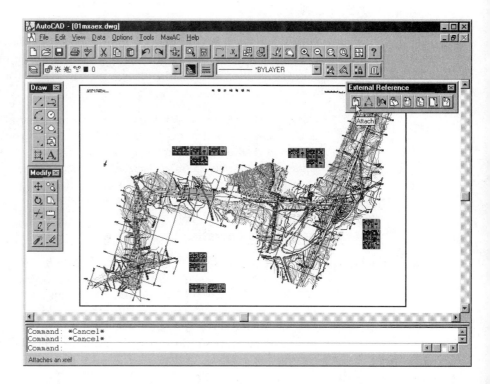

Figure 1.4
AutoCAD R13 for Windows 95.

- Good stability

- Works well with AutoCAD R12 for Windows and DOS

- Improved (but also different) user interface

- Plug-and-Play support for easier peripheral installation and management

Figure 1.4 shows the interface for AutoCAD R13 in Windows 95.

If you're still confused about whether Windows NT or 95 is better for you, Autodesk has published several documents that discuss the differences and the decision process, including *AutoCAD Release 13 Windows Platforms: Windows NT Workstation and Windows 95* and *Windows NT & Windows 95 at a Glance: A Step-by-Step Guide for AutoCAD Users Considering the Move*. Check with the Autodesk fax-back system, your dealer, or the ACAD forum on CompuServe for these and other documents.

Customization Platform vs. Production Platform

Bear in mind that the best platform for your work will depend on how you use AutoCAD. And if, like many people who customize AutoCAD, you wear several hats, that might mean that you'll choose to use different platforms for your different duties. One of the Windows platforms is a natural choice for customization work, since it allows you to have AutoCAD and a text editor open at the same time. Documenting your customization efforts is also much more convenient in Windows. On the other hand, if you're also responsible for producing large, complex drawings, you might choose to use DOS AutoCAD for production work.

Note: In this book, we use Windows 95 for our screen shots. This choice doesn't imply that Windows 95 is better than Windows NT for customization work. In fact, NT's crash protection is advantageous for more advanced AutoCAD development, especially by C programmers who use ADS (AutoCAD Development System) and ARX (AutoCAD Runtime eXtension). But Windows 95 is becoming the standard mainstream operating system, and a popular choice for AutoCAD R13 users.

Maintaining Multiple Operating Systems

Fortunately for those of us who can't decide among operating systems, AutoCAD R13 ships with a multiplatform license for MS-DOS, Windows 3.1, Windows NT, and Windows 95. You can install R13 for any (or all) of these platforms from one CD-ROM. Although installing four different operating systems and flavors of AutoCAD probably isn't the best way to increase drafting productivity, it does help you evaluate and perhaps customize AutoCAD on the four platforms.

There are several ways to run multiple operating systems and AutoCAD platforms. The obvious one is to obtain one computer for each operating system, but few of us are that lucky. The more practical schemes rely on multiple boot capabilities built into Windows 95 and NT.

Setting up a Multiple Boot System

The following procedure outlines a method of running MS-DOS, Windows 3.1, Windows NT, and Windows 95 on one computer. You might want to perform this procedure if you need to test your AutoCAD customization on multiple platforms. Obviously, you can omit the procedures for any operating systems that you don't need.

Before you embark on this project, check whether your hardware is on the Windows NT and 95 Hardware Compatibility Lists (HCLs), which are available from Microsoft. Windows NT and to a lesser degree Windows 95 are more finicky than MS-DOS about hardware compatibility, in part because many of the drivers are now included in the operating system. If you try to install on oddball hardware or use older 16-bit drivers with Windows 95, you may run into problems. More detailed AutoCAD installation and configuration procedures appear later in this chapter.

1. Make sure that you have a hard disk large enough for all of the operating system files, AutoCAD installations, and your other applications and

AutoCAD System Setup and Optimization

data. You might want to buy a new hard disk for this purpose. If so, use the MS-DOS FDISK utility to create one or more DOS partitions.

2. Install MS-DOS if you haven't done so already. Install and configure R13 for DOS.

3. Install Windows 3.1 if you haven't done so already. Install and configure R13 for Windows 3.1, including the required Win32S and WinG components.

4. Install Windows NT into its own directory, so that your Windows 3.1 installation remains in place. If you want to use NTFS with NT, tell the installation program to put NT on a separate partition or disk. Reinstall and then configure R13 for Windows NT in the same directory as you used in step 3.

5. Now when you boot your computer, NT presents a menu that includes the choice of Windows NT or MS-DOS. Of course, if you choose MS-DOS, you'll be able to run Windows 3.1.

6. Install Windows 95 in its own directory so that your Windows 3.1 and NT installations remain in place. Reinstall R13 for Windows 95 in the same directory as you used in step 3.

7. Now when you boot, you'll still see the NT boot menu. If you select MS-DOS instead of Windows NT, you'll have the choice of booting to Windows 95 or MS-DOS. By default, you'll boot to Windows 95, but if you press the F8 key, you'll see Windows 95's boot menu, which includes a choice for "Previous version of DOS," which really will take you to your old version of MS-DOS.

Tip: By installing R13 in the same directory three times, you cut down drastically on disk space requirements. You do need to go through the motions of reinstalling, though, in order to add AutoCAD information to the Windows 95 and NT Registries. Of course, if you want to install AutoCAD in separate directories (for example, on an NTFS partition for NT only), you can do so.

Warning: When you install R13c4 for more than one Windows operating system, you can encounter conflicts because of different versions of the same Microsoft DLL files (for example, MSVCRT20.DLL). Autodesk's fax-back document #508 (FAX508.DOC or FAX508.TXT in Library 19 of CompuServe's ACAD forum) describes the potential problems and how to work around them.

For more information about multiple booting, see the *Microsoft Windows NT Workstation Installation Guide* that comes with NT and the *Microsoft Windows 95 Resource Kit* published by Microsoft Press.

Running Release 12 in Windows 95 or NT

Depending on who uses your customization, you might need to maintain compatibility with R12 and R13. In order to test your customization with R12, you'll need to load R12, which Autodesk doesn't support in Windows 95 or NT.

Fortunately, most people have had success running R12 Windows in both Windows 95 and Windows NT (although loading drawings is somewhat slow in NT). You should be able to install and run it in the usual way. The one common problem is with plotting, because neither Windows 95 nor NT allow R12's ADI drivers to access the printer ports directly. The workaround is to use R12's System Printer driver or an ADI driver with the AUTOSPOOL mechanism described in the R12 and R13 documentation.

Many people also have run R12 and R13 DOS successfully as full-screen applications from Windows 95. DOS AutoCAD will not run from Windows NT; the Phar Lap DOS extender is incompatible with NT. However, you can dual boot to MS-DOS and run DOS AutoCAD from there.

Running R12 DOS in Windows 95

Here's the installation procedure if you're going to run AutoCAD R12 DOS in Windows 95:

1. Boot to MS-DOS.

2. Install R12c3 DOS, configuring it for Autodesk's plain VGA display driver.

3. Copy PHARLAP.386 from the \ACAD to the \WIN95\SYSTEM directory.

4. Add the line `device=pharlap.386` to the [386Enh] section of the \WIN95\SYSTEM.INI file.

5. Boot to Windows 95.

6. Set up a shortcut to the ACADR12.BAT file. (Open Explorer, right-click and drag ACADR12.BAT onto the desktop, or anywhere else you want to park it. When you let up on the right button, pick Create Shortcut(s) Here.

7. Configure the shortcut to run in MS-DOS Mode: Right-click on the shortcut and choose Properties. Click on the Program tab, then click on Advanced, and choose MS-DOS Mode.) Note that MS-DOS Mode causes a mini-reboot of Windows 95 when you launch AutoCAD, so it takes longer and prevents multitasking while you're running R12 DOS.

8. Launch R12 DOS by double-clicking on the shortcut that you created. Test the display, pointing device, and printing.

9. If everything works fine, you can try other display drivers and higher resolutions. Your luck will depend on your hardware, drivers, and system configuration.

In order to run DOS AutoCAD reliably under Windows 95, you must run it in MS-DOS Mode. If you don't use MS-DOS Mode, the Phar Lap DOS extender can't create its swap file, and you'll risk crashing AutoCAD, especially in large drawings. If your computer has plenty of memory and you limit yourself to working with smaller drawings (for example, when testing customization), you might be able to get away with not using MS-DOS Mode. As long as AutoCAD can keep everything in memory and the Phar Lap DOS extender doesn't need to create a swap file, you won't have a problem. But AutoCAD will crash without warning if it does use up physical memory.

Note: This book does not cover cross-version (for example, R13 and R12) customization issues in depth. We'll point out important customization differences between R13 and R12, but we won't go into all of the details of developing customization to support both R13 and R12.

Networking Issues

Networks are an increasingly common part of AutoCAD installations. Your AutoCAD customization should take account of the possibility that drawing files, support files, and the AutoCAD program files might be spread across multiple disks, some of which are located on network servers. Plotters and printers also may be connected to servers rather than directly to AutoCAD workstations. In addition, you might be called upon to give advice on or perform AutoCAD network installations.

Networking AutoCAD really breaks down into two questions: Where are the drawing files, and where are the AutoCAD program files?

Storing Drawing Files on the Network

Centralized storage of drawing files on a server's hard disk has many benefits, especially when two or more people need to work on the same project. The benefits include:

- Making drawings easily available without time-consuming file transfers
- Avoiding duplicate versions of drawings and the attendant revision control problems
- Centralizing backup

If your AutoCAD workstations are connected to a reliable file server, there's rarely any reason not to keep drawings there. You do not need to use the networked version of AutoCAD to realize these benefits, but the networked version can provide additional benefits.

Keeping the AutoCAD Program on the Network

The trend towards centralizing common files leads naturally to the desire to keep the AutoCAD program files on the file server as well. The obvious advantages are:

- Sharing of AutoCAD licenses among users who don't need full-time access to AutoCAD

- Smaller disk storage requirements for all workstations (the R13 program files can consume 50 MB or more)

- Centralized program updates

There are disadvantages as well, though, including:

- Complications and potential problems caused by R13's network hardware lock

- Additional network traffic when users launch AutoCAD

- Slower AutoCAD launch

R13's network hardware lock replaces the R12 ACAD.PWD file as AutoCAD's method for metering licenses on network installations of AutoCAD. The hardware lock eliminates numerous problems that plagued the ACAD.PWD scheme, but it introduces its own problems. The time required for the lock to free a license after a user exits AutoCAD varies, which can lead to licenses that are unavailable when users need them. The lock requires a dedicated parallel port, preferably on the server, and an NLM (NetWare Loadable Module) that might conflict with other server processes. Network administrators are understandably nervous about complicating the server any more than they have to.

Because of performance and hardware lock concerns, many companies continue to load AutoCAD on local hard disks rather than on the server. This approach is simpler and probably makes sense in companies that don't need to share licenses among part-time users.

Note: If your company has many people who use AutoCAD part-time, then it might be worthwhile to endure the additional complexities of the network version in order to save money on AutoCAD licenses. We don't cover network installation of AutoCAD in this book. Refer to Autodesk's documentation, including fax-back documents, for information. If you aren't a networking expert (or your network administrator isn't an AutoCAD expert), you might want to enlist the help of a competent dealer or consultant to perform and support the network installation.

AutoCAD System Setup and Optimization

Keeping Support Files on the Network

A subset of the "where are the AutoCAD program files?" question is "where are the support files?" Some companies load the AutoCAD program files on each workstation's local hard disk, but they keep support files (menus, fonts, AutoLISP programs, and so on) on the file server. This approach avoids the hardware lock complications, but it centralizes the files that are most likely to change as you customize AutoCAD. By centralizing these files, you ensure that everyone is using the same up-to-date versions.

Another advantage of centralizing support files is that it can help coordinate customization by more than one person in the office. When each customizer has copies of the same support files on a local hard disk, it's easy to end up with different changes to the same files. A single repository for the approved office support files gives you some hope of controlling this problem.

In order to use support files in network directories, each AutoCAD workstation must be configured to look for the files in those directories. You make this configuration change by adding the network directories to the AutoCAD support file search path. See Chapters 2 and 4 for details.

Third-Party Application Issues

AutoCAD third-party applications (also called *companion applications*) bring along their own installation, configuration, and optimization issues. It's impossible to generalize about all third-party applications or all issues, but you should at least consider platform support, additional system requirements, installation, and support file conflicts.

System Requirements

As third-party applications have become more ambitious, their hardware and other system requirements have increased. Some applications won't run well with the minimum or standard AutoCAD system recommendations from Autodesk. Running out of system resources in Windows 3.1 is a common problem with applications that include complex toolbars or menus (which is another reason not to run R13 in Windows 3.1). Ask the application developer or dealer for system recommendations, and make sure that you don't skimp on hardware.

Platform support is another concern. Many third-party applications now support AutoCAD in DOS and Windows 3.1, but some developers haven't yet ported their applications to Windows 95 or NT. Find out about operating system compatibility before you buy or upgrade a third-party application.

Installation

Installing third-party applications raises the same questions as installing AutoCAD. Some applications can reside on the network server, while others must be loaded on the local hard disk. Even if the application supports network installation, you'll need to decide whether it makes sense for your company.

Many applications create their own data files (for example, point databases in civil engineering applications) in addition to the AutoCAD drawing files. If multiple users are working on one project, you'll probably want those data files to be commonly accessible on the network. Look into how the application manages multiple user access to its data files and whether these files can be shared reliably.

Some applications require a hardware lock on every workstation, even for local installations. The lock may increase the "hassle factor" of installing and using the application on your system, but for highly discipline-specific applications (like engineering programs), you may not have a choice.

Directories and Support File Conflicts

A well-designed AutoCAD application should install itself into its own set of directories, separate from the AutoCAD program files. The application shouldn't overwrite AutoCAD support files, and ideally, it shouldn't require its own versions of the files that users commonly customize (especially ACAD.PGP and ACAD.LSP: see Chapter 4). Custom menus should be available in their source form (MNU or MNS: see Chapter 4) so that you can add your own custom menu macros. In general, you want to avoid applications that take over the AutoCAD system and make it difficult for you to further modify them.

Autodesk has codified these and other rules of application behavior in the Application Interoperability Guidelines (AIG). Compliance with AIG helps minimize conflicts between two or more applications, and between applications and your own customization. AIG isn't the final word on whether an application is well-behaved, but a developer's compliance with it at least represents some attempt to deliver an application that doesn't stomp all over your AutoCAD system and other applications. In practice, few applications are completely well-behaved, but look for ones with a high degree of AIG compliance and an overall sense of "good citizenship."

Installing and Configuring AutoCAD

We won't rehash the installation instructions that come with AutoCAD. Instead, we'll suggest a streamlined installation procedure that validates system operation and establishes a reliable baseline for customization.

Autodesk has gone a long way toward making basic installation of AutoCAD easier, especially in R13 for Windows. The standard Windows SETUP.EXE installation program will be familiar to anyone who's installed other Windows applications. The default Windows display and pointing device drivers will work well for many people. Setting up plotter and printer drivers can still be complex, but the Windows System Printer option at least gets most users started quickly.

AutoCAD System Setup and Optimization

Tip: If at all possible, use the CD-ROM version of R13. It's much faster than shuffling a huge number of disks into the floppy drive. If your workstation doesn't have a CD-ROM drive, now is the time to get one. Be sure to get a drive that's supported by Windows 95 and NT if you plan to use those operating systems. If your company has a large number of AutoCAD workstations, it might be more economical to use a networked CD-ROM drive.

Streamlined Installation Procedure Outline

The essential steps in our suggested installation procedure are:

1. Verify system requirements and operating system functionality.
2. Install AutoCAD R13 from CD-ROM or disks.
3. Launch AutoCAD using its menu choice, program icon, or batch file.
4. Configure AutoCAD for lowest common denominator drivers.
5. Modify configuration parameters for sharing drawings on a network, if required.
6. Test AutoCAD display, pointing device, and printer operation.
7. Optimize configuration with advanced or third-party drivers, and test any changes.
8. Install, configure, and test any third-party applications.
9. Add and test custom support directories and files.

Installing Release 13

Take a moment to skim the R13 *Installation Guide* for the platform(s) you'll be installing, as well as the README files and INSTALL.TXT located on your R13 CD-ROM or disks. Then launch the installation by running SETUP.EXE for any of the Window platforms or INSTALL.EXE for DOS.

Warning: Many of the directory names and some of the installation directions listed in the R13 *Installation Guides* are wrong, so be sure to glance at the README files and INSTALL.TXT before you install.

Tip: In Windows 95, you can use the Add/Remove Programs Properties applet (under Start, **S**ettings, **C**ontrol Panel) instead of launching the SETUP.EXE program. See Figure 1.5. Add/Remove Programs Properties automatically searches your floppy and CD-ROM drives, in that order, for the application's installation program. If you use Add/Remove Programs Properties, dont put the Personalization Disk in the drive until after Windows 95 has searched for the installation program. Otherwise, Windows 95

Figure 1.5
The Windows 95 Add/Remove Programs Properties applet.

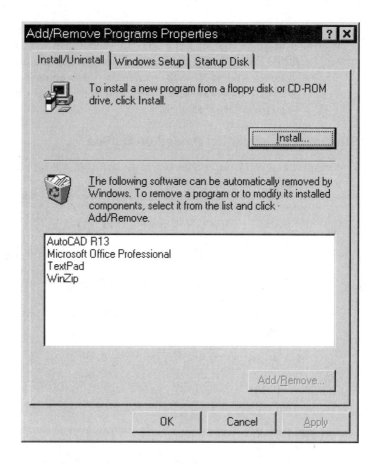

will find the SETUP.EXE program on the Personalization Disk, rather than the one on the CD-ROM. If you have a Personalization Disk from an older release (for example, R13c0), the wrong SETUP.EXE program will run.

Note: For versions of R13 prior to R13c4, you could install Windows and DOS AutoCAD in one fell swoop from the Windows-based SETUP.EXE installation program. R13c4 does away with this convenience. Use SETUP.EXE to install for any of the Windows operating systems (see Figure 1.6) and INSTALL.EXE from DOS or a DOS window to install for MS-DOS.

New Installation, Reinstallation, or Update

At the beginning of the installation process, you're asked to specify the kind of installation, as shown in Figure 1.7. The two buttons cover three distinct installation procedures, as the R13c4 INSTALL.TXT file discusses:

AutoCAD System Setup and Optimization

Figure 1.6
Running AutoCAD R13 for Windows SETUP.EXE.

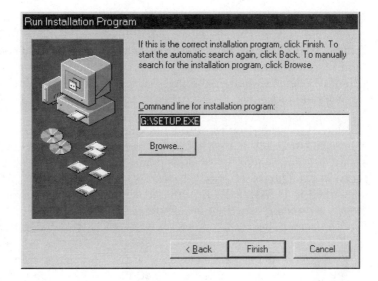

- New installation

- Reinstallation

- Update

New installation happens during a completely new installation in which you haven't yet "branded" the Personalization Disk. A new installation requires that you go through

Figure 1.7
Choosing the type of installation.

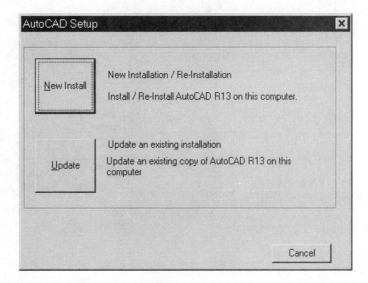

phone registration with Autodesk. Reinstallation is for when you want to do a complete installation, but you've already branded the Personalization Disk and gone through the phone registration process. With reinstallation, AutoCAD keeps your old serial number and registration information, so you don't have to call again for a registration number. AutoCAD detects whether the Personalization Disk has been branded and chooses the appropriate procedure.

Update is for when you want to leave your existing R13 pre-c4 AutoCAD installation in place and upgrade it to the current version. Upgrade is equivalent to patching the AutoCAD files. When you choose Update, the installation program offers to back up your current configurable support files (ACAD.ADS, ACAD.MNU, ACAD.PGP, and so on), as described in INSTALL.TXT. AutoCAD copies the files that you choose to back up to a separate directory and then installs new versions. It leaves alone any files that you choose not to back up. Figure 1.8 shows the AutoCAD Setup dialog box.

Tip: You might want to leave ACAD.CFG in place (by not backing it up), so that you don't have to reconfigure AutoCAD after the update.

Upgrade Strategies

Chapter 2 recommends that you segregate customized support files from AutoCAD's standard support files and shows you how to do so. One of the benefits of this approach is that it makes upgrading AutoCAD less painful. You don't have to worry about the

Figure 1.8
Backing up configurable support files.

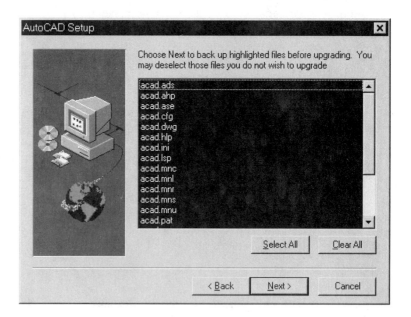

AutoCAD System Setup and Optimization

installation program overwriting custom support files, and you can see at a glance which files you've modified.

If you've already modified support files in the AutoCAD program directories, then you'll need to sort out the changed files. As the previous section mentions, the new R13c4 installation program helps by offering to back up existing support files. If you have any doubts, let the installation program back up everything and then compare the old and new versions afterwards. Another technique is to sort the files by date. Usually the files that you've modified will be at the beginning of the directory listing (assuming a sort order of newest to oldest).

Once you've identified the support files with customization changes, you need to decide how to reconcile them with the new standard versions in the update. If there are no changes (or modest changes) to the new version of a file, you can get away with using your custom version as is. If the new version is substantially different, or if you're not sure whether it's different (for example, in a long menu file), you might need to do a careful comparison of the files and then copy appropriate regions from one file to another. Software tools such as the Visual Diff shareware program can be helpful.

Typical, Compact, or Custom Installation

Another early decision you must make is how many files to install. The Windows installation includes Typical, Compact, and Custom installation options. The typical installation consumes about 60 MB in Windows, 50 MB in DOS, and 95 MB for both. Compact installation reduces these numbers to 30 MB in Windows, 25 MB in DOS, and 50 MB for both. Typical installation does allow you to select the installation directories, so you don't need to do a custom installation unless you want to specify precisely what gets installed and what doesn't.

Configuring Release 13

The first time you launch AutoCAD after installing, you're taken directly into the configuration program, which has changed little in R13. You can save yourself a lot of hassles, especially on a new system, by configuring for the simplest driver choices at first. After you confirm that AutoCAD works with the basic drivers, then you can try more optimized drivers and configuration settings.

After you've configured, always test the three peripherals—display, pointing device, and printer—in order to confirm operation. Choose an AutoCAD sample drawing from R13\COM\SAMPLE as your test bed and open it, zoom in and out, select a few entities, and then plot the drawing.

Basic Windows Configuration

Autodesk's WHIP Accelerated Display Driver offers excellent performance in most conditions and is the one you should select when configuring R13 in any Windows operating system. If you encounter display problems, try configuring for the Accelerated

Display Driver by Rasterex or Windows Display Driver by Autodesk temporarily. If the problem goes away, then there is a compatibility problem between the WHIP driver and your hardware or system.

When configuring the pointing device in Windows, choose `Current System Pointing Device` for an ordinary mouse or `Wintab Compatible Digitizer` if you have a digitizer and a Windows Wintab driver installed. The other digitizer drivers are of limited usefulness because they don't allow access to all parts of the screen.

During initial configuration, choose `System Printer ADI 4.2` as your first printer. You might want to add printer-specific ADI drivers later, but System Printer should work as a starting point for testing all of the printers that are configured under Windows.

Tip: When you add the System Printer driver, AutoCAD asks `Do you want dithered output?` and defaults to yes. Dithering attempts to reproduce colors on a monochrome device with patterns of shaded dots. This effect is usually not very appropriate for line work in technical drawings. If you always map AutoCAD colors to solid lines of different weight, then answer no to the dithering question.

Basic DOS Configuration

Because there is no such thing as a "system pointing device" or "system printer" in MS-DOS, the initial configuration of R13 DOS must be more refined. When you configure R13 DOS for the first time, choose the standard VGA display driver, an appropriate digitizer driver for your pointing device, and an Autodesk printer or plotter driver. If any of these don't work with your peripherals, then you'll need to get DOS ADI drivers from the hardware manufacturer.

Uninstalling AutoCAD in Windows 95

R13c4 supports the uninstall feature of Windows 95. When you install R13c4 in Windows 95, the operating system records information about the installed directories and files. You then can use Windows 95's Add/Remove Programs Properties applet (Start, **S**ettings, **C**ontrol Panel) to remove R13c4, or most of it anyway. See Figure 1.9.

Tip: Windows 95's uninstall utility isn't foolproof—it sometimes can't determine that certain files should be deleted with the application, especially if you've added those files after installing the application. In such cases, the uninstall utility will tell you that you need to remove files manually.

Optimizing System Performance

After you've installed and configured AutoCAD and confirmed that it works, you should optimize the peripheral drivers and other settings. Doing so will improve performance and stability and ensure a solid foundation for your customization.

AutoCAD System Setup and Optimization

Figure 1.9
Uninstalling R13.

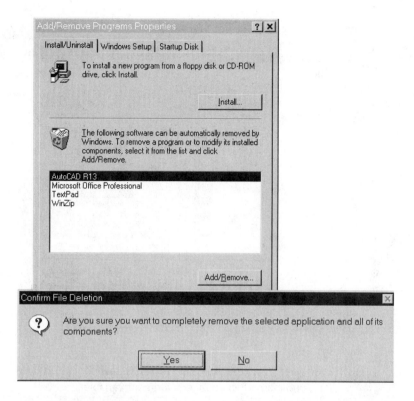

Platform-Independent Optimization

Whether you're installing in DOS or one of the Windows operating systems, there are several settings controlled by the Configuration menu and system variables that you should consider changing. Use the CONFIG command or the **O**ptions, **C**onfigure menu choice to enter the Configuration menu.

Configuration Menu Settings

The last choice on AutoCAD's Configuration menu (Configure operating parameters) displays a submenu containing many of the AutoCAD optimization settings (Figure 1.10). The platform-independent settings that are worth changing are:

- Placement of temporary files
- Automatic-save feature

By default, AutoCAD puts its temporary files in \R13\WIN. Normally these files are deleted when you exit AutoCAD, but if AutoCAD crashes, the files remain on your hard disk. You might prefer to direct the swap files to a separate directory, so that they stay

Figure 1.10
The Operating parameters sub-menu.

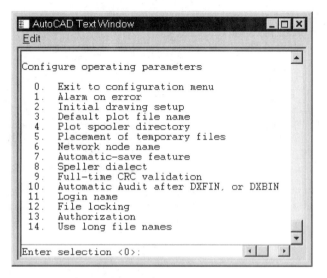

out of the way of other files and so that you can find and delete them easily after a crash. Specify C:\TEMP or another temporary file directory.

Tip: The `Placement of temporary files` configuration menu setting is overridden by any ACADPAGEDIR DOS environment variable setting. See Chapter 4 for more information about environment variables.

AutoCAD's automatic save time defaults to 120 minutes, during which time you can lose a lot of work. Set the automatic save time to a more reasonable 10 or 15 minutes.

Tip: If you forget to change the automatic save time in the configuration menu, you can do it in the drawing editor. Just type `SAVETIME` and enter a new time. Or equivalently, change the Automatic Save field on the System tab in the Preferences dialog box. The SAVETIME system variable value is stored in ACAD.CFG or ACADNT.CFG, so any changes you make to it, whether from the configuration menu or the drawing editor, are global.

Warning: Don't rely on AutoCAD's automatic save as your primary means of saving drawings. Automatic save only kicks in when there is command-line activity, so if you aren't actually doing something in AutoCAD, the automatic save doesn't occur. Save frequently with the AutoCAD QSAVE or SAVE command and use automatic save as a secondary backup.

AutoCAD Pager Settings

AutoCAD uses its own memory paging systems, or *pagers*, to help manage the memory allocated to the current drawing. The normal pager is controlled by the Memory section of the Preferences dialog's <u>E</u>nvironment tab, or by DOS environment variables. The

AutoCAD System Setup and Optimization

Maximum memory setting (or ACADMAXMEM environment variable) specifies the amount of memory in bytes that AutoCAD's pager can request from Windows before the pager must begin writing parts of the drawing to a page file. The Maximum Bytes in a Page setting (or ACADMAXPAGE environment variable) specifies the maximum size in bytes of the first page file. Normally you leave this setting at zero, which indicates no limit on the size of the first page file.

You can turn on an alternative AutoCAD paging system called the *object pager*. This pager is controlled by the MAXOBJMEM AutoCAD system variable, or the ACADMAXOBJMEM DOS environment variable. By default, the object pager is turned off (MAXOBJMEM = 0). If you turn it on by setting it to a number between zero and 2,147,483,647, AutoCAD will use the setting to determine how much virtual memory the drawing can use before writing some of the drawing data to a page file. Note that if you make a change to MAXOBJMEM that turns the object pager on or off (for example, change MAXOBJMEM from 0 to 32,000,000), the change doesn't take effect until you leave the current drawing.

Both pagers use the ACADPAGEDIR setting to determine where to put their page files. The normal pager creates PG* page files, and the object pager creates MGPR* page files.

There isn't much data on how various pager settings affect performance. The question is complicated by the fact that AutoCAD's pagers are always interacting with the Windows or DOS Phar Lap virtual memory managers, which are doing their own paging. Autodesk recommends leaving ACADMAXPAGE set to zero and not reducing (and perhaps increasing) ACADMAXMEM. If you want to try the object pager, Autodesk recommends setting MAXOBJMEM to the amount of physical memory in your computer. See Chapter 4 of the R13 *Installation Guide* for more information about AutoCAD's pagers.

Other System Variable Settings

Because AutoCAD save times increased so dramatically in R13, Autodesk did additional work, especially in R13c3, to whittle away at them. The user-controllable options are:

- Incremental saving (ISAVEPERCENT)

- Backup file creation (ISAVEBAK)

- Raster preview creation (RASTERPREVIEW)

- Undo file location (UNDOONDISK)

Note: To view or change a system variable setting, enter the system variable name at AutoCAD's command prompt. The Options, System Variable, List menu choice sends a list of most system variables and settings to the text window.

Before R13c3, AutoCAD always rewrote the entire drawing to the disk when you saved. Incremental saving allows AutoCAD to write only changed data to the DWG file during each save. This technique cuts down dramatically on the time it takes to save a large drawing with modest changes. The only disadvantage is that the DWG file temporarily contains some wasted space after an incremental save. You can control how much wasted space AutoCAD tolerates with the ISAVEPERCENT system variable. The default value of 50 means that when the estimated wasted space would exceed 50% of the total DWG file size, AutoCAD will do a full save instead, thus eliminating all of the wasted space.

If your system is cramped for disk space, set ISAVEPERCENT to a smaller value, or turn off incremental saves entirely by setting it to 0. If your drawing saves take a long time, try larger values of ISAVEPERCENT.

Warning: Incremental saving was unstable in R13c3, especially with the VISRETAIN system variable set to 1. Make sure you're using R13c4 or later.

The ISAVEBAK system variable gives you control over whether AutoCAD creates a BAK file of the old version each time you save a drawing. The default value of 1 causes AutoCAD to create BAK files, as it always has in the past. ISAVEBAK=0 turns off BAK file creation, which speeds up saves, especially in Windows. Obviously you should change this option with care because the lack of a BAK file can complicate recovery attempts in the event of DWG file corruption or an AutoCAD crash.

The new drawing preview feature in R13 relies on a preview raster image of the drawing that's stored in the DWG file along with the normal drawing data. Writing the preview image with each drawing save takes time, so the RASTERPREVIEW system variable gives you control over what kind of preview, if any, gets written. Table 1.1 shows the valid values.

The default value of 0 causes AutoCAD to create its normal Windows bitmap preview image (which is used by DOS and Windows AutoCAD). RASTERPREVIEW=3 can speed up saves for visually complex drawings, but it eliminates the drawing preview. The other values, 1 and 2, aren't very useful. They create a WMF preview that AutoCAD doesn't use. Both values result in larger DWG files than RASTERPREVIEW=0, and RASTERPREVIEW=2 causes the preview image to be distorted.

AutoCAD normally keeps its undo information (a log of each action since you opened the drawing) in memory. The UNDOONDISK system variable allows you to redirect the undo log to a file. This option is a trade-off between the speed gained by creating each

Table 1.1 RASTERPREVIEW Values	Value	Meaning
	0	Create a Windows bitmap (BMP) only
	1	Create a Windows bitmap (BMP) and metafile (WMF)
	2	Create a Windows metafile (WMF) only
	3	Don't create a preview image

AutoCAD System Setup and Optimization

undo entry in memory (rather than on disk) and the potential speed gained by leaving more memory available for drawing data and program code. The UNDOONDISK default of 1 puts the temporary file on the disk. UNDOONDISK=0 leaves the undo log in memory. If you have plenty of memory, you might want to try setting UNDOONDISK to 0 and see whether it speeds up performance on your system and drawings.

Network Optimization

If your computer is connected to a network and you might be saving drawings to network drives or plotting over the network, there are four other configuration options you should change or verify. All of these settings are in the Configuration menu's `Configure operating parameters` submenu:

- Network node name
- Automatic-save feature
- Login name
- File locking

The network node name, AC$ by default, controls the extension for spooled plot files. If your spool files get written to a common network directory, set the network node name to a unique three-character extension on each workstation. You might use first and last initials or sequential numbers followed by the dollar sign ($), like MM$ or 01$. You don't have to use a dollar sign, but it's a convenient convention for identifying temporary files.

AutoCAD's default file name for automatic saves is AUTO.SV$. If more than one network user works on drawings in the same network directory, the users will overwrite each other's automatic save files. To avoid this problem, specify a unique automatic save name or path on each workstation. The extension must be SV$. Specifying a local path also cuts down on network traffic.

 Tip: In Windows 95 and NT, you can use a long automatic save file name, but AutoCAD still appends the SV$ extension to it.

Your default login name is a concatenation of the individual and company name that you specified for the AutoCAD Personalization Disk (truncated to 30 characters). This name is what other network users will see when they try to open a drawing that you're working on. You might want to shorten the login name to something shorter and more identifiable, such as your e-mail name.

If you're saving drawings to a network disk, make sure that file locking is turned on (which it is by default). AutoCAD's file locking prevents two users from working on the same drawing at the same time.

27

Tip: If you keep all of your drawings on your local disk, you can turn file locking off. This step makes recovery from an AutoCAD crash somewhat simpler.

Windows Optimization: Common Issues

Several optimization issues are common to all three Windows operating systems.

Display Drivers

The default display driver for R13 Windows (Autodesk's WHIP Accelerated Display Driver) works well and provides excellent performance on most systems, so in most cases there's no additional tweaking necessary. There are a few operations, such as removing an object from a large selection set, that bog down with the WHIP driver, but usually the overall performance gain you see will outweigh these aberrations.

Tip: The WHIP driver includes new RTZOOM and RTPAN commands for real-time panning and zooming.

Digitizers

If you use a digitizer instead of a mouse, you must have a current Wintab driver for your digitizer and operating system (Windows 3.1, Windows NT, or Windows 95). The Wintab driver is supplied by your digitizer manufacturer, not by Autodesk. Some of the early versions of Wintab drivers for R13 didn't work very well in Windows 95 or NT, so make sure you have the most recent version and talk to your digitizer manufacturer if you run into problems.

Warning: Using a digitizer effectively in R13 Windows requires an up-to-date Wintab driver from your digitizer manufacturer. Wintab is an independent driver specification for Windows pointing devices. A Wintab driver allows Windows applications to define enhanced pointing device features, such as AutoCAD's tablet and button menus. Wintab also enables the digitizer to act as both the general Windows system pointing device and the digitizer in AutoCAD. The Wintab driver must match your operating system, AutoCAD version, digitizer model. Most digitizer manufacturers have reliable Wintab drivers for Windows and AutoCAD R12, but some manufacturers have been slow to upgrade their drivers for Windows NT, Windows 95, and AutoCAD R13.

If you're forced to wait for a Wintab driver update, you should be able to use AutoCAD's Current System Pointing Device driver for the time being, although your digitizer will act like nothing more than a big mouse.

Once you have the appropriate Wintab driver, install it according to the instructions and test it as a system pointing device in Windows. You might need to use the driver's Windows configuration program to configure button mappings and a comfortable pointing area.

AutoCAD System Setup and Optimization

In AutoCAD, use CONFIG to configure AutoCAD for the Wintab Compatible Digitizer driver. When you return to the drawing editor, use the TABLET command to configure the tablet menu and pointing areas.

Tip: It usually is easier to use your digitizer, with a good Wintab driver, as the only pointing device in Windows and AutoCAD. Although it's sometimes possible to use both a mouse and a digitizer in Windows, this approach tends to complicate configuration. It also uses up an extra serial port and more desk space.

Plotters and Printers

If the general-purpose Windows drivers for your plotters and printers work well, you don't have to do anything other than configure AutoCAD for the System Printer driver. This driver communicates with the Windows printing system and allows AutoCAD to communicate with all of the printer drivers (including any fax drivers).

Tip: To change devices with the System Printer driver, use the PLOT command. In the Plot dialog box, choose <u>D</u>evice and Default Selection, then <u>C</u>hange Device Requirements. In the Print Setup subdialog (Figure 1.11), use the <u>N</u>ame drop-down list to select among your available Windows printers.

For some output devices, especially plotters, AutoCAD-specific ADI drivers provide faster printing and more features than do the general-purpose Windows drivers. ADI drivers can be more trouble to configure in Windows 95 or NT, but if you do a lot of plotting or need special features, ADI drivers are worth the small extra hassle. Later in this chapter, we describe some configuration techniques.

Tip: Autodesk's fax-back documents #561 and 562 describe plotting options for R13c4 in Windows 95. Document #561 covers plotting to a local device and document #562 covers plotting to a network device.

R13c4 includes ADI drivers for popular plotters from manufacturers like CalComp and Hewlett-Packard. Check the *AutoCAD Installation Guide* for a list of ADI drivers and configuration instructions. Many plotter manufacturers also distribute their own ADI drivers, some of which are more current and include more features than the ones that come with R13. Check with the plotter manufacturer for information.

Swap File

All of the Windows operating systems include virtual memory managers, which means in part that they create and manage swap files on your local hard disk. When Windows and the applications running in it fill up physical memory (RAM), Windows uses a swap file as additional, "virtual" memory. Windows takes care of all the details of creating and managing its swap file. AutoCAD depends on this swap file for storing program code and drawing data when there isn't enough physical memory to hold all of both. As a result, it's important that your Windows swap file be configured correctly.

Figure 1.11
Selecting a printer with the System Printer driver.

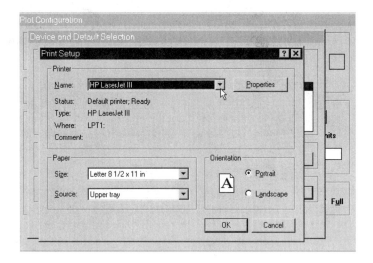

The virtual memory managers and swap file requirements are a bit different in each of the Windows operating systems, so we'll cover swap file configuration in the operating system-specific sections that follow. In all cases, though, you need to allow plenty of free disk space for swap file activity. Don't let your hard disk get down to only a few free megabytes.

Windows 95 and NT Optimization

The primary AutoCAD optimization issues in Windows 95 are ADI plotter and printer drivers and long file names.

Tip: Windows system drivers are the Achilles heel of Windows 95. Windows 95 runs applications at a lower privilege level than itself, where it can control them if they misbehave. Drivers, though, run at the same privilege level as the operating system, so a flaky driver can bring the whole system down.

Although system drivers aren't specific to AutoCAD, you should make sure that you're using reliable and up-to-date drivers. Always try to use 32-bit drivers written specifically for Windows 95. Avoid older 16-bit drivers.

ADI Plotter and Printer Drivers

AutoCAD ADI plotter and printer drivers attempt to control printer ports directly in order to improve performance. Windows 95 and NT prevent AutoCAD's ADI drivers from accessing the ports directly in order to improve operating system robustness. As a result, if you configure AutoCAD for an ADI driver and attempt to print to a device on a port that the operating system is managing, you'll see an error message like `Unable to set port parameters` or `Cannot open port`. AutoCAD's System Printer driver doesn't have this problem because it sends the print job to the operating system, which in turn sends it to the port.

AutoCAD System Setup and Optimization

In order to avoid the port conflict problem with ADI plotter drivers, you have two options:

- Plot to a "file" whose name is the same as the port name.
- Plot to a "file" called AUTOSPOOL and define a plot spool command or batch file for AutoCAD to run with each plot file.

The port name method is easier to configure, but the AUTOSPOOL method can be faster and more reliable on some systems, especially when plotting to a network plotter.

Plotting to a Port Name

To plot to a port name, turn on Plot to File in the Plot dialog box, click File Name, and type the name of the port (for example, LPT1) in the Create Plot File dialog (Figure 1.12). Don't add a colon to the port name; if you do, AutoCAD will try to plot to a file named LPT1:.PLT and generate an error.

Plotting to AUTOSPOOL

AutoCAD's AUTOSPOOL feature causes AutoCAD to plot to a spool file and then automatically run a user-defined command that sends the plot file to the plotter (or to the network queue that feeds the plotter).

Note: AUTOSPOOL is covered in the ACADPLCMD section of the R13 *AutoCAD Installation Guides.* You should use the Plot Spooling setting (under PREFERENCES, Misc, Plot) instead of ACADPLCMD to set the plot spool command in AutoCAD for Windows.

Tip: AUTOSPOOL is a good way to plot to a network plotter.

There are a variety of ways to configure the AUTOSPOOL spool command, one of which we outline for you:

1. Run the CONFIG command and choose `Configure operating`

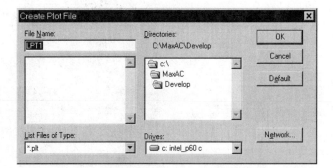

Figure 1.12
Plotting to a port name "file."

Maximizing AutoCAD R13

parameters. Choose `Plot spooler directory` and change the name to C:\SPFILES (or any other directory name on a local hard disk. If you haven't done so already, use `Configure plotter` to add the appropriate ADI plotter driver definition. Return to the drawing editor, saving the configuration changes.

2. Create the C:\SPFILES directory if it doesn't already exist.

3. Run the PREFERENCES command and choose the **M**isc tab (Figure 1.13). In the **P**lot spooling edit box, type `COPY /B %S LPT1:`, or an equivalent command for sending the plot file to a network queue. %S designates the plot file, and AutoCAD will substitute the appropriate file name at plot time. Click OK to exit the Preferences dialog, saving your changes.

4. Turn on Plot to **F**ile in the Plot dialog box, click File **N**ame, and type the magic name **AUTOSPOOL** in the Create Plot File dialog.

If the plot doesn't come out, then you probably didn't type a valid command sequence for copying the plot file. In that case, plot to an ordinary file (for example, TEST.PLT), then shell out to DOS and copy the plot file from DOS. Once you get the command sequence right, transfer it to the Preferences dialog.

Tip: You might want to use AUTOSPOOL with more than one spool command and ADI driver (for example, to send output to different ports or network queues). In that case, you should create a batch file that employs some of the additional % options shown in the *AutoCAD Installation Guides*. Autodesk's fax-back document #759, *AutoCAD R13 Plotting under Windows NT* (which also applies to Windows 95), contains useful information and an example batch file.

Figure 1.13
Specifying the AUTOSPOOL spool command.

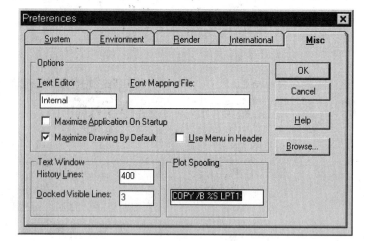

AutoCAD System Setup and Optimization

Tip: You can make a port name or AUTOSPOOL the default plot file name by specifying it in the Configuration menu, `Configure operating parameters`, `Default plot file name`. Be aware, though, that the name you specify becomes the default for all AutoCAD plotter and printer drivers.

Long File Names

If you've installed R13c4 in Windows 95 or NT, you have the option of using long file names. By default this option is turned on, but you can disable it from the Configuration menu's `Configure operating parameters` submenu. For compatibility reasons, you'll probably want to disable long file names if anyone in your offices is using R13 on a platform that doesn't support them (for example, DOS or Windows 3.1), especially if you use xrefs.

Tip: The R13c4 README file contains additional information on long file name compatibility issues.

Swap File

Windows 95's and NT's virtual memory managers do a good job of managing the swap file by themselves, so you don't need to make any changes for AutoCAD. If you do want to experiment with swap file changes, refer to the *Microsoft Windows 95 Resource Kit* or *Windows NT Resource Kit* for details.

Windows 3.1 Optimization

As we said earlier, R13's performance and reliability suffers in Windows 3.1 compared with Windows 95 and NT, but if you're forced to run this combination, you can minimize problems by configuring the Windows swap file and 32-bit access properly.

Swap File

R13 needs plenty of room in the Windows 3.1 swap file in order to run at all reliably. Autodesk recommends a permanent (as opposed to temporary) swap file of at least 64 MB. AutoCAD benefits from a huge swap file. If you work on very large drawings, you may need to make the swap file even larger, up to 128MB.

In order to view the swap file size, open Windows Control Panel, choose 386 Enhanced, and click on <u>V</u>irtual Memory. Use the <u>C</u>hange button to change the size.

Note: Common symptoms of a too-small swap file include errors such as heap, segmentation violation, insufficient memory, general protection exception, and general protection fault. A large swap file won't cure all of R13's problems in Windows 3.1, but it's a good start. If you encounter any of these errors, the first thing to do is to increase the swap file size.

Tip: Windows normally limits the swap file size to four times your installed RAM memory, and although it may let you set it larger, it will not use the extra space unless you set the PageOverCommit setting in the [386Enh] section of the Windows SYSTEM.INI file to a larger value, such as 8. Autodesk's fax-back document #775, *How to Configure and Optimize the Windows Swapfile*, contains complete instructions.

32-Bit File and Disk Access

Both Windows 3.1 and Windows for Workgroups 3.11 offer 32-bit file access. It improves data transfer performance between memory and disk by allowing Windows to talk directly to the hard drive, bypassing the slower and more limited INT 13h BIOS routines. It also lets Windows swap non-Windows applications to disk.

32-bit disk access is available only in Windows for Workgroups 3.11. It implements a 32-bit cache for reading and writing hard disk files. 32-bit disk access does most of what DOS's 16-bit SMARTDRV does, only faster.

Both 32-bit file access and 32-bit disk access improve performance, but 32-bit file access has a much greater effect than 32-bit disk access. Also, 32-bit file access is compatible with almost all systems, while 32-bit disk access is incompatible with some hard drive controllers.

Autodesk recommends turning off 32-bit file access and 32-bit disk access when using R13 in Windows 3.1 or Windows for Workgroups 3.11 because of potential conflicts between these performance-enhancing options and Microsoft's Win32S extender. However, 32-bit file access should not cause problems with most systems. Unfortunately, turning off both options slows down Windows and all of its applications.

Some users have run R13 successfully with both 32-bit file and disk access turned on, so you might want to turn them on and do some testing. If you encounter any problems, immediately turn off both options and see whether the problems go away. Then try again with just 32-bit file access on.

Tip: The normal Windows installation leaves 32-bit file and disk access turned off by default. You must turn them on if you want to use them.

The 32-bit file and disk access options live in the same Virtual Memory dialog as the swap file setting (Control Panel, 386 Enhanced, **V**irtual Memory). Use the **C**hange button and click the appropriate check box to change either setting.

Warning: Some disk controllers don't work with 32-bit file or disk access, and some require special drivers.

AutoCAD System Setup and Optimization

MS-DOS Optimization

Display driver optimization is the most important performance improvement you can make for R13 DOS. You should also configure the Phar Lap swap file location.

Display Drivers

After you've confirmed that R13 DOS works with the VGA display driver on your system, you'll want to move up to a display list driver with higher resolution and faster display performance.

R13 includes a DOS Accelerated Display Driver (ADD) by Vibrant Graphics. This display list driver is much faster than the standard AutoCAD VGA, SVADI, or VESA drivers, so you'll want to use it if possible. The Accelerated Display Driver supports a large number of video cards and chipsets, but it's possible that yours isn't on the list, especially if you have a very new card. See the Video Driver Support appendix of the *AutoCAD Installation Guide for DOS* for a list of supported cards, configuration instructions, and display driver commands.

Tip: The _AV command starts the Accelerated Display Driver's Aerial View.

Tip: Despite the DOS Accelerated Display Driver included in R13, there still are third-party display list drivers for R13 (including from Vibrant Graphics). Many video cards come with their own R13 DOS display list drivers, and several companies sell third-party drivers. These drivers offer support for video cards that aren't supported by the Accelerated Display Driver, more speed, and additional resolutions and features.

Digitizers

Digitizer drivers really haven't changed in R13, so you shouldn't be in for any surprises here. Try the Autodesk-supplied drivers, and if you encounter any problems, ask your digitizer manufacturer for an updated driver. If you also use your digitizer as a mouse in other MS-DOS programs, make sure that the digitizer's mouse emulation driver loads in your AUTOEXEC.BAT file. Test digitizer and mouse operation with the following sequence:

1. Launch AutoCAD and then shell out with the SH command.

2. Launch a DOS mouse-aware program (for example, EDIT) and test mouse operation.

3. Exit the DOS application, type **QUIT** to return to AutoCAD, and test digitizer operation.

Most digitizer drivers can handle context switching, but some older drivers have problems when you shell out of or return to AutoCAD.

ADI Plotter and Printer Drivers

You shouldn't have any special problems configuring ADI plotter and printer drivers in R13 DOS. As with Windows, the manufacturer of your peripheral may be able to provide a more up-to-date driver than the ones that come with R13. If you're plotting over a network, you might need to use either the "Plotting to a Port Name" or "Plotting to AUTOSPOOL" technique we described. Even if you don't need to use these techniques, you might find that AUTOSPOOL is faster or more reliable.

Phar Lap Swap File

Although MS-DOS doesn't have a virtual memory manager or swap file, the Phar Lap DOS extender built into R13 DOS does. The Phar Lap swap file is always temporary, so you usually don't need to control its size. You should make sure that it's directed to a local hard drive, in order to avoid unnecessary network traffic as the DOS extender reads and writes the swap file. The default swap file location is in the root directory of the drive that's current when you launch AutoCAD.

AutoCAD deletes the Phar Lap swap file when you exit AutoCAD. As with AutoCAD's temporary files, though, the swap file gets left around if AutoCAD crashes. It makes sense to direct the Phar Lap swap file to the same directory you specified for AutoCAD's temporary files (for example, C:\TEMP—see the "Configuration Menu Settings" section).

You use the CFIG386.EXE program, located in \R13\DOS, to change the Phar Lap swap file location. In DOS, change to the \R13\DOS directory and type the following:

```
CFIG386 ACAD -SWAPDIR C:\TEMP
```

Substitute your temporary directory for C:\TEMP.

Tip: You can change many other Phar Lap settings with the CFIG386 program, as described in the "DOS Extender and VMM" appendix of the *AutoCAD Installation Guide for DOS*, but the default values work fine on most systems.

Hardware Issues

Throughout this chapter, we've spoken mostly from a hardware-independent perspective, except to assume that you're using a computer with an Intel or Intel-compatible processor. Of course your hardware has a large effect on AutoCAD performance, and therefore on your productivity.

Adequate memory is the most important hardware issue for any computer. For R13c4, Autodesk specifies a minimum of 12 MB for DOS, 16 MB for Windows 95 and Windows

3.1, and 32 MB for Windows NT. These minimums are likely to be adequate if you work on smaller drawings (under 1 MB) and don't run more than a couple of applications at a time in Windows. If you routinely work on larger drawings and/or multitask more applications (especially multiple AutoCAD sessions), you should increase these numbers by at least 4 to 8 MB. If you notice AutoCAD becoming sluggish and a rapid increase in hard disk activity, that's a good sign that more memory would reduce disk swapping and therefore increase performance. You can use the memory report provided by the STATUS command to keep an eye on disk swapping activity.

Intel's Pentium Pro is the processor of choice for Windows NT. Its optimization for 32-bit code matches well with NT's all 32-bit architecture. A 512 KB processor cache with the 200 MHz Pentium Pro boost performance considerably over a 256 KB processor cache. The Pentium Pro may not be the best choice for Windows 95, though. Compared with the ordinary Pentium, the Pentium Pro suffers some performance degradation when running 16-bit code, of which Windows 95 still has some.

A fast hard disk ensures that disk performance doesn't bog down file I/O operations such as opening and saving drawings. A fast disk also improves overall operating system performance and AutoCAD swapping performance. Enhanced IDE drives are adequate for many applications, but SCSI disks may have an advantage for multitasking in NT. Also, don't neglect hard disk size. AutoCAD, third party applications, and other Windows applications consume an ever-larger amount of disk space. R13 drawing files are, on average, about a third larger than R12 files. Don't settle for anything less than a 1 GB disk drive on a new computer, and consider a larger disk if you'll be storing a large number of DWG files on your local disk rather than a network disk.

Conclusion

R13 will mark the shift from DOS to Windows for many AutoCAD users, but the decision is now three times as complicated. Autodesk has streamlined some aspects of AutoCAD installation and configuration in R13, but complexities still abound, and it pays to spend some time optimizing your AutoCAD installation.

Once you have R13 properly installed, configured, and optimized on your system, you're ready to set up a customization environment in Chapter 2.

chapter 2

Creating a Customization Environment

This chapter helps you set up your system to create better drawings faster and to create an environment for customization. As you customize AutoCAD, you'll be modifying and adding support files. This process is iterative and involves editing, testing, and finally distributing many files. In order to maintain control over the process, you should establish a customization environment that's separate from your normal production AutoCAD environment. A separate customization environment ensures that you don't accidentally cripple your AutoCAD production environment right before some important drawings are due!

The customization environment that you create in this chapter also ensures that the exercise instructions in subsequent chapters will match your system setup.

The details of setting up a customization environment in Windows are different than in DOS, because Windows AutoCAD launches from a program shortcut rather than a batch file. This difference isn't specific to R13—it also exists in R12 Windows—but many people will encounter the new procedures when they switch to R13 and Windows simultaneously. Even if you're familiar with starting DOS AutoCAD from multiple batch files, which was the technique used in previous editions of *Maximizing AutoCAD*, be sure to read up on the new Windows procedures.

The information in this chapter about directory structure and multiple configuration files applies to AutoCAD LT as well. The text editor information also applies, except where AutoLISP is discussed. You should configure a text editor for editing menu files and scripts. You don't need to install and configure LispPad, since you can't use AutoLISP in LT. (However, you can use LispPad's Tools, Test script option to develop and debug scripts, so there's no harm in installing LispPad if you want to experiment with it.)

> **About This Chapter**
>
> In this chapter, you will learn how to do the following:
>
> - Organize directories for customizing AutoCAD and for this book
> - Set up and control multiple AutoCAD configurations
> - Install the MaxAC CD-ROM
> - Select and configure suitable text editors for customizing AutoCAD
> - Avoid ACAD.LSP file conflicts and configure for a custom prototype drawing

If you haven't done so already, please read the book's introduction before you start working through this chapter. The information there helps you to avoid mistakes and misunderstandings and to gain the maximum benefits from this book. Also make sure that you have AutoCAD properly installed, configured, and optimized as described in Chapter 1.

The resources you'll need in this chapter are:

- The MaxAC CD-ROM
- The TextPad editor from the MaxAC CD-ROM
- The LispPad editor from the MaxAC CD-ROM
- ACADMA.DWG, ACADMA.MNS, and ACADMA.MNL from the MaxAC CD-ROM
- Windows WordPad, Windows Write, or DOS EDIT (optional)
- Windows Explorer or File Manager

Organizing Your AutoCAD Setup: The Benefits

You'll find that development and management are easier in a controlled AutoCAD operating-system environment. A central tenet of good file management is organizing your files into meaningful groups, which you can do with directories. A well-managed system helps you, AutoCAD, and other programs find your application files. You can also use subdirectories to run multiple AutoCAD configurations for different working applications or to do applications development.

Your text editor is your key developer's tool. You use it to write and edit menus, macros, AutoLISP programs, and other support files. This chapter discusses text

Creating a Customization Environment

editors and guides you through setting them up so that you can use them efficiently from AutoCAD.

Warning: You need to perform the exercises in this chapter to conform to the book's assumptions, even if you already are familiar with the topics in the introductory chapters of the book. You need to create, copy, or verify operating and support files and directories. You also need to create an AutoCAD shortcut or program item especially for working on the book's exercises.

Directories (or Folders)

A well-organized directory structure provides flexibility in organizing files. Efficiency increases if you limit the files in each directory to a reasonable number (say, 100 or 200). AutoCAD supports a well-organized directory structure by enabling you to set search paths for its support and configuration files. The directory structure you will use with this book ensures that the exercises do not interfere with your current AutoCAD setup.

Note: In Windows 95 terminology, directories are called "folders," as in the Macintosh operating system. Because most AutoCAD users and the R13 documentation still use the terms "directory" and "subdirectory,", we'll stick with those except where we're referring to a program such as Explorer that relies on the folder metaphor.

The Maximizing AutoCAD Directory Structure

The instructions in this book assume that your hard disk is C: and that you have a directory structure similar to that shown in the following listing. The following listing is for a Windows 95 system; listings on Windows 3.xx or NT systems will be similar. The directories whose names are shown in italics are those created for the book, and the others are the standard Release 13 and Windows directories. You can use a different directory structure, but you'll need to remember to translate the instructions throughout the book for your configuration.

```
C:\
    MAXAC\
        BOOK\
            CH02\
            CH03\
            CH04\
            ...and so on...
        DEVELOP\
        TOOLS\
            LISPPAD\
            BUTTONS\
            BIGDCL
            ...and others...

    PROGRAM FILES\
        TEXTPAD\
        ...and others...

    R13\
        COM\
            FONTS\
            SUPPORT\
        WIN\
            DRV\
            SUPPORT\

    WINDOWS\
```

The AutoCAD installation process creates the C:\R13 directory structure (or C:\ACADR13 in early versions of Release 13), which we will examine more closely later in this chapter. Your C:\WINDOWS directory might be named something different (for example, \WIN95 or \WINNT), depending on how Windows was installed on your hard disk. The Windows subdirectories are not shown in the preceding directory listing.

Note: Depending on your operating system, system configuration, and version of Release 13, your drive letter or subdirectory names may vary from those shown. For example, early versions of R13 used \ACADR13 and \ACADR13\COMMON instead of \R13 and \R13\COM. If your system is set up differently, substitute your prompt, drive letter, and names wherever you encounter the C:\path prompts (such as C:\R13), C: drive, or various directory names in the book.

You should be using R13c4 or later in order to correspond with this book's descriptions.

Creating a Customization Environment

C:\MAXAC is the directory that will hold "finished" custom AutoCAD support files for this book. The subdirectories of C:\MAXAC\BOOK will contain files from the MaxAC CD-ROM. You will copy files from each of the C:\MAXAC\BOOK\CHxx directories into C:\MAXAC\DEVELOP as you work on the exercises in each chapter. C:\MAXAC\DEVELOP is your working customization directory, containing works-in-progress that you're in the process of modifying and testing. Keeping finished and unfinished custom support files separate helps ensure that incomplete or untested customization doesn't foul up your production AutoCAD environment.

C:\PROGRAM FILES\TEXTPAD is the default Windows 95 installation directory for the shareware TextPad editor that we use in the exercises to edit menus and other support files. In Windows NT and 3.11, the default directory is C:\TEXTPAD.

C:\PROGRAMS\TOOLS contains subdirectories with useful utility programs that you can use in some of the exercises and for your own customization work. One of these programs is LispPad, a text editor by Tony Tanzillo that's optimized for AutoLISP and DCL (Dialog Control Language) programming. It provides the ability to easily test and preview AutoLISP code and DCL dialog box designs. LispPad is the default editor for the AutoLISP and DCL exercises in Chapters 9 through 12.

Tip: Some users keep custom support files with the standard AutoCAD support files in the \R13 directory tree. We recommend against this practice because it prevents you from seeing at a glance which support files you've modified or added. In addition, it makes AutoCAD updates more difficult, because you have to worry about which modified support files might be overwritten during the update process.

The AutoCAD R13 Directory Structure

The R13 directory structure shown in the following listing is only the tip of the iceberg. A standard R13 for Windows installation adds 24 directories below C:\R13, and R13 for DOS adds 10 more! You don't need to understand the purpose and contents of every directory, but you should have a general understanding of the AutoCAD directory structure.

Warning: Documentation of the AutoCAD directory structure in some versions of the R13 manuals is incomplete and incorrect.

```
C:\
    R13\                                AutoCAD R13
        COM\                            Common directory for Windows and DOS
            ADS\                        Common ADS header files
                SAMPLE\                 Common ADS programming samples
            ADSRX\
                INC\                    ARX header files
                LIB\                    ARX library files
                SAMPLE\                 ARX programming samples
            FONTS\                      AutoCAD, PostScript, and TrueType fonts
            SAMPLE\                     Sample drawings and support files
                DBF\                    Sample databases for tutorial
            SUPPORT\                    Common support files
        DOS\                            DOS program files
            ADS\                        DOS ADS library files
                SAMPLE\                 DOS ADS compile batch files
            ASE\                        DOS ASI programming support files
                LANG\               DOS ADI language files
                SAMPLE\                 DOS ASI programming samples
            DRV\                        DOS drivers
            SAMPLE\                     DOS sample utilities
            SUPPORT\                    DOS support files
            TUTORIAL\                   DOS tutorial drawings
        WIN\                            Windows program files
            ASE\                        Windows ASI programming support files
                LANG\               Windows ADI language files
                SAMPLE\                 Windows ASI programming samples
            DRV\                        Windows drivers
            EDOC\                       "Dynatext" Electronic documentation
                BIN\                    Dynatext viewer program
                TMP\                    Dynatext temporary directories
                    PRIVATE\
                    PUBLIC\
            SAMPLE\                     Windows sample application
            SUPPORT\                    Windows support files
            TUTORIAL\                   Windows on-line tutorials and drawings
```

The R13 directory structure is organized in three main branches: COM (for common), WIN, and DOS. Of course, if you don't install R13 for DOS or Windows, you won't have the DOS or WIN branches. Only a few of these directories are important for the customization you perform in this book. \R13\COM\FONTS and \R13\COM\SUPPORT contain the fonts and other support files that are common to Windows and DOS AutoCAD. \R13\WIN\SUPPORT contains Windows-specific support files, and \R13\DOS\SUPPORT contains DOS-specific support files. The \R13\WIN (or \R13\DOS) directory contains, in addition to the AutoCAD executable program, default configuration files.

 Tip: Support files that were in C:\ACAD\SUPPORT in R12 are divided between C:\R13\COM\SUPPORT and C:\R13\WIN\SUPPORT in R13.

Creating a Customization Environment

If you're familiar with the R12 directory structure and support file locations, spend a few minutes in Explorer or File Manager browsing through the directory tree to see where things have moved in R13.

Installing the MaxAC CD-ROM

Now you are ready to install the CD-ROM disk that came packaged with this book. As mentioned in the introduction, this disk can save you a lot of typing and debugging time in the exercises. It also includes several valuable tools that will make AutoCAD development easier and reduce errors. You can choose to install everything contained on the CD-ROM, or install selected components. The choices that the MaxAC CD-ROM setup offers are listed here:

- **Typical.** This is the recommended choice (Figure 2.1). It installs everything you need for all of the exercises in the book and all required or recommended tools. You can later use the following two choices if you want to examine or use other tools or demos. If you don't choose this option, you will need to use the following options to install the files for specific chapters or tools as you come to them in the book. (DOS users should XCOPY the entire MAXAC\BOOK directory tree from the CD-ROM contents to install this option. There are no required tools for DOS users.)

- **Compact.** This is the minimum installation. It only installs the minimum required tools (such as LispPad) and the exercise files for Chapter 2. You will have to later use the Custom setup choice to install exercise files

Figure 2.1
The Maximizing AutoCAD R13 main setup options.

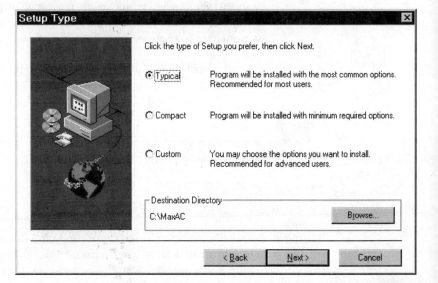

Maximizing AutoCAD® R13

for other chapters and any other tools you need or want. (DOS users should XCOPY only the MAXAC\BOOK\CH02 directory from the CD-ROM to install this option. There are no required tools for DOS users.)

- **Custom.** This setup choice displays gives you the following setup options:

    - **Everything.** Installs everything contained on the CD-ROM. This will copy many files you may not use and requires a lot of disk space. (DOS users should XCOPY the entire \MAXAC CD-ROM directory and its subdirectories to install this option.)

    - **Selected Chapters.** Installs the files you need for the exercises in one or more specific chapters. (DOS users should XCOPY the MAXAC\BOOK\CH*nn* directory from the CD-ROM contents to install a specific chapter's files, where *nn* is the chapter number.)

    - **Selected Tools.** Installs the files for one or more selected tools. These tools include development tools used in the exercises, various other shareware and freeware utilities, and demos of commercial programs. See Appendix G for more information on all of these tools. (DOS users should XCOPY the MAXAC\TOOLS*name* directory from the CD-ROM to install a specific tool. See Appendix G for the directory names of specific tools.)

The CD-ROM setup will list the approximate disk space required for the option you choose.

Note: You should install the MaxAC CD-ROM in the default \MAXAC directory so that the path names match those in the book. If you decide instead to install it in another directory, you'll need to substitute the name you use for the MAXAC directory throughout the book. The book assumes you are installing on drive C. If not, substitute your drive letter wherever drive C or C: is shown throughout the book.

If you are using any version of Windows, use the following exercise. If you are using DOS, skip to the next exercise.

 Exercise

Installing the MaxAC CD-ROM in Windows

Put the MaxAC CD-ROM in your CD-ROM drive.

If using Windows 95, from the Taskbar choose Start, Run and enter **d:SETUP** (where *d:* is your CD-ROM drive letter), then choose OK. Otherwise, in Program Manager, choose File, Run, and enter **d:SETUP**, then choose OK.

Creating a Customization Environment

Figure 2.2
The Maximizing AutoCAD R13 setup window.

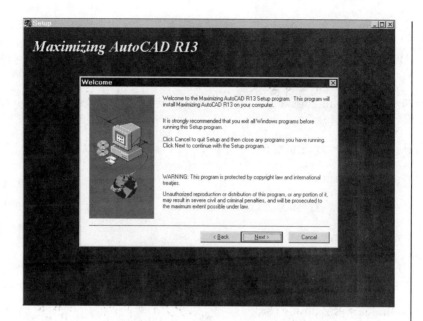

The Maximizing AutoCAD R13 setup window should appear (Figure 2.2). Follow its instructions and choose from its options to complete the installation.

When the installation is complete, you will find that it created a program group or Start menu folder named Maximizing AutoCAD R13. When you open this program group or folder, it will display a window or submenu with one or more program items or shortcuts. At the very least, you will see an Uninstall program item or shortcut. You may also see other program items or shortcuts, depending on the tools you installed. Later in this chapter, you will add an AutoCAD program item to this group or submenu.

See the README.TXT file in the root directory of the MaxAC CD-ROM or in your \MAXAC directory if you have any problems.

 Note: When you type a directory or file name in all uppercase and without any spaces, Windows 95 changes the name to initial capital only. Thus the MAXAC folder will appear as Maxac. In this book, we always show directory names in all uppercase in order to make them easily identifiable.

Windows is a much better development environment, but if you are using DOS, use the following instructions and exercise to install the MaxAC CD-ROM. Otherwise, skip to the next section.

You use the DOS XCOPY command to copy the components you want to install. The syntax is as follows:

```
C:\> XCOPY d:\MAXAC\component\*.* C:\MAXAC\component /s  [Enter]
```

where **d:** is your CD-ROM drive letter, and **component** is the path of the component you want to copy.

For example, to copy all of the chapter files used for exercises in the book, you would use the following:

```
C:\> XCOPY d:\MAXAC\BOOK\*.* C:\MAXAC\BOOK /s  [Enter]
```

Or, to copy only the Chapter 2 files, use:

```
C:\> XCOPY d:\MAXAC\BOOK\CH02\*.* C:\MAXAC\BOOK\CH02 /s  [Enter]
```

Or, to copy only the DOSLIB utilities, use:

```
C:\> XCOPY d:\MAXAC\TOOLS\DOSLIB\*.* C:\MAXAC\TOOLS\DOSLIB /s
[Enter]
```

Or, to copy everything, use:

```
C:\> XCOPY d:\MAXAC\*.* C:\MAXAC /s  [Enter]
```

Exercise

Installing the MaxAC CD-ROM in DOS

Put the MaxAC CD-ROM in your CD-ROM drive, and change to the root directory of the hard disk you want to install it on.

At the C:\> DOS prompt, enter the XCOPY instructions to copy the components you want installed.

See the README.TXT file in the root directory of the MaxAC CD-ROM or in your \MAXAC directory if you have any problems.

AutoCAD Configuration Files

When you go through the initial configuration process after installing AutoCAD, the program creates a CFG (configuration) file in the executable program directory (\R13\WIN or \R13\DOS). Subsequent configuration changes, including some system variable settings, will be written to this file.

Creating a Customization Environment

 There are three major changes to CFG files in R13:

- INI-style CFG files
- ACADNT.CFG for Windows 95 and NT
- Additional configuration settings in the ACAD.INI file

The CFG file is now an ASCII file arranged in the format of a Windows-style INI (initialization) file. Before R13, ACAD.CFG was a binary file that you couldn't inspect or edit easily. Now you can view, print out, and even edit the CFG (although you should do the latter with caution, and make a backup copy first).

The configuration file's name still is ACAD.CFG when R13 runs in Windows 3.1 or DOS, but it's now ACADNT.CFG when R13 runs in Windows 95 or NT. In addition, on all platforms, some configuration settings are now stored in a separate file called ACAD.INI. The ACAD.INI file also existed in R12 Windows, but there's more in the R13 version.

 Tip: ACAD.INI contains directory, font mapping, and toolbar location information. ACADNT.CFG (or ACAD.CFG) contains hardware configuration and global system variable settings, such as MAXSORT, MTEXTED, and PICKBOX. AutoCAD adds to ACAD.INI and ACADNT.CFG as you make changes to global system variables and use additional toolbars.

Although R13 for Windows normally keeps its configuration information in the ACADNT.CFG and ACAD.INI files, there are alternative ways of controlling some of the settings contained in ACAD.INI. See Chapter 4 for details.

AutoCAD Support Path

One of the most important settings stored in ACAD.INI is the *support path*, which is a series of directories in which AutoCAD looks for its support files (menus, fonts, AutoLISP, ADS, ARX programs, and so on). This setting is the key to maintaining multiple AutoCAD system configurations. By controlling which directories AutoCAD searches for support files, you control how the program works and what customized features are available.

 Tip: In fact, AutoCAD looks for support files in other places besides the support path that you specify. It looks in the directory that was current when AutoCAD was launched first, the DWG file directory second, the support-path directories you specify third, and the AutoCAD executable (ACAD.EXE) directory fourth. The complete list of directories in which AutoCAD looks for support files is called the *library search path*.

To find out what your current AutoCAD library search path is, enter a name that does not exist as a block or file at the block name prompt of the INSERT command. An AutoCAD Message dialog box will appear and display the library search path, as shown in Figure 2.3.

See Chapter 4 for more information on library search paths.

Figure 2.3
The AutoCAD library search path.

In Windows, you set the support path in the S̲upport edit box in the E̲nvironment page of the Preferences dialog box, as shown in Figure 2.4 ①. To access this dialog box, you can enter **PREFERENCES** or choose O̲ptions, P̲references. The list of directories is formatted like a DOS path setting: directories are separated by semicolons, with no spaces in between. The directory list is usually much longer than the edit box, so you need to click in the box and scroll to see the entire setting. A more convenient way to view or change the support path is to move the cursor into the S̲upport edit box and choose the B̲rowse button. AutoCAD then displays an Edit Path subdialog with the currently configured support directories in a list box. When you modify the support path, or most of the other settings in the Preferences dialog, AutoCAD writes the new values to ACAD.INI.

Warning: When you use the B̲rowse button, AutoCAD re-sorts the directory names, and thus might change the order in which directories are searched for support files. The change in search order can have undesirable effects if you have different versions of support files in different directories. See Chapter 4 for details.

Note: The equivalent in DOS AutoCAD to the support path setting accessed through O̲ptions, P̲references, S̲upport in Windows is the environment variable ACAD. See the R13 *AutoCAD Installation Guide for DOS* for details about AutoCAD environment variables. You can use environment variables with AutoCAD for Windows, but you should avoid doing so. Resetting environment variables requires modifying system files and rebooting Windows.

Figure 2.4
Setting the support path.

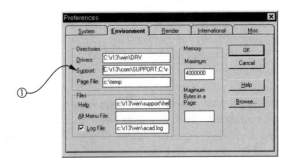

Creating a Customization Environment

The default R13 support path comprises four directories:

- C:\R13\COM\SUPPORT
- C:\R13\WIN\SUPPORT
- C:\R13\WIN\TUTORIAL
- C:\R13\COM\FONTS

Tip: There's no reason to keep \R13\WIN\TUTORIAL on the support path. You can delete it and its trailing semicolon. When you remove or add directories to the search path by typing in the Support edit box, make sure that you use one semicolon (and no spaces) between each adjacent pair of directories. If you use the Browse button instead, you can select and delete this directory without having to worry about the semicolons—AutoCAD takes care of the details.

Multiple AutoCAD Configurations

AutoCAD supports the use of multiple configuration files. Each set of configuration files can have different settings for all of the information stored in ACADNT.CFG and ACAD.INI: peripherals, global system variable settings, font mappings, and so on. Most importantly for our purposes, each configuration can have a different support path (because it's stored in ACAD.INI).

Use the following exercise to set up a separate set of configuration files by copying your regular configuration files to \MAXAC. The copied files will establish the AutoCAD configuration that you use for this book's exercises.

Note: If your configuration files are not in the C:\R13\WIN directory, find them and substitute their directory name where you see C:\R13\WIN in the following exercise. If you have trouble finding your ACADNT.CFG file, check the properties of the program shortcut you used to launch AutoCAD. If the program's launch command includes a /C (configuration directory) switch, then look in that directory. See the section "Directing AutoCAD to Different Support Files with /C" later in this chapter for information.

Exercise

Copying Configuration Files to the MAXAC Directory

Use Explorer or File Manager to copy ACADNT.CFG and ACAD.INI from \R13\WIN to \MAXAC. (Substitute ACAD.CFG for ACADNT.CFG in Windows 3.11 or DOS.)

Open the \MAXAC folder and confirm that both files are there.

In subsequent exercises, you'll create an AutoCAD program shortcut that uses the new configuration files.

Note: Some drivers, especially third-party DOS video drivers, use their own additional support or configuration files. For example, the R13 HP-GL/2 driver creates an HPDRIVER.CFG file. The VGA and SVADI drivers included with DOS AutoCAD use an SVADI.CFG file. If you are using such a driver, you may need to copy its configuration file or set a DOS environment variable in the ACADR13.BAT file.

Using Different AutoCAD Configurations in Windows

Now that you have two sets of AutoCAD configuration files, you have to tell AutoCAD to use the new set, which requires three steps:

1. Create a new program item or shortcut in Windows.

2. Change the shortcut's properties and add the /C switch to the program launch command.

3. Start AutoCAD with the new shortcut, and modify the settings in AutoCAD's Properties dialog.

Note: Windows 3.1 used the term "program item icon" to describe what's now called a "shortcut" in Windows 95. We'll use "shortcut" as the generic term throughout this book.

Creating a New Program Item or Shortcut in Windows

AutoCAD installation creates a program item or shortcut that you can copy for alternate configurations. If you are using Windows 95 or NT 4.0, use the following exercise to make a new shortcut that you'll use for this book's exercises. If you are using Windows 3.1x or NT 3.51, skip to the next exercise. You'll create the new program item or shortcut in the Maximizing AutoCAD R13 program group or Start menu folder that was created earlier by the MaxAC CD-ROM installation. Skip ahead to the next exercise if you're using Windows 3.1 or NT 3.51. The following instructions are for Windows 95 and NT 4.0.

 Exercise

Creating a New Program Shortcut in Windows 95 or NT 4.0

First, you need to open the R13 program folder, as follows.

Creating a Customization Environment

Right-click Start	The alternate Start menu appears
Choose **O**pen	The Windows 95 Start Menu group appears
Double-click on Programs, *then double-click on the* AutoCAD R13 *and the* Maximizing AutoCAD R13 *folders*	The AutoCAD R13 and Maximizing AutoCAD R13 groups appear

Next, you'll copy the program shortcut.

While holding down the CTRL key, use the right mouse button to drag the AutoCAD R13 *program shortcut to a blank spot in the* Maximizing AutoCAD R13 *program group*

If you prefer, you can drag the copy onto the desktop, for faster access.

Release the mouse button	Windows makes a copy of the shortcut

Close the AutoCAD R13, Programs, and Start Menu groups.

You'll configure the new shortcut's properties in the Modifying Program Shortcut Properties exercise.

If you are using Windows 3.1x or NT, use the following exercise to make a new program item and icon that you'll use for this book's exercises. The following instructions are for Windows 3.1 or NT. Skip them if you're using Windows 95 or DOS.

 Exercise

Creating a New Program Item in Windows 3.1 or NT

Open Program Manager.

Locate the AutoCAD R13 *program group, and if it's not open, double-click on it, then locate and double-click on the* Maximizing AutoCAD R13 *program group*	The AutoCAD R13 and Maximizing AutoCAD R13 groups appear

Maximizing AutoCAD® R13

Next, you'll copy the program item.

While holding down the Ctrl key, click on the AutoCAD R13 *program item and drag it to a blank spot in the* Maximizing AutoCAD R13 *program group*

Release the mouse button — Windows makes a copy of the program item

Close the AutoCAD R13 program group.

You'll configure the new program item's properties in the next exercise, Modifying Program Shortcut Properties.

The next section shows you how to configure the new program item or shortcut.

Directing AutoCAD to Different Support Files with /C

So far all you have are two different program items or shortcuts that do the same thing. In order to direct the new one to use your new set of configuration files, you need to add the /C switch and configuration directory after the AutoCAD launch command (which is called **T**arget in Windows 95). Skip to the next exercise for Windows 3.1 and NT 3.51 instructions.

 Exercise

Modifying Program Shortcut Properties in Windows 95 and NT 4.0

First, you'll change the properties.

Right-click on the new shortcut — A context-sensitive menu appears

Choose Prope**r**ties — The Windows 95 Properties dialog appears

Click on the Shortcut *tab*

In the **T**arget *edit box (Figure 2.5, ①), type* **/C C:\MAXAC** *after* C:\R13\WIN\ACAD.EXE — /C is AutoCAD's configuration directory switch

Creating a Customization Environment

Figure 2.5
Modifying program shortcut properties

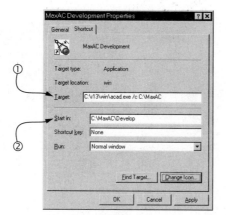

Change the directory in the **S***tart in edit box to*
`C:\MAXAC\DEVELOP` ②

If you want to give the MaxAC shortcut a distinctive look, use the Change Icon button to choose a different icon. You can browse to the ACAD.EXE file to choose from various ACAD icons.

The final step is to rename the new shortcut.

Right-click on the new shortcut

Choose Rename   The shortcut's label is highlighted
                                                  for editing

Type `MaxAC Development` [Enter]   The shortcut's label changes

The next exercise shows the procedure for adding the /C switch in a Windows 3.1 or NT 3.51 program item. Skip this exercise if you're not using Windows 3.1 or NT 3.51.

 Exercise

Modifying Program Item Properties in Windows 3.1 or NT

First, you'll change the properties.

Single-click on the new program item   Windows selects it, but the
                                                 program doesn't start

If AutoCAD started, you double-clicked by mistake. Exit AutoCAD.

Choose **F**ile, **P**roperties   The Program Item Properties dialog appears

In the **D**escription *edit box, type*
`MaxAC Development`

In the **C**ommand Line *edit box, type*
`/C C:\MAXAC` *after* `C:\R13\WIN\ACAD.EXE`   /C is AutoCAD's configuration directory switch

Change the directory in the **W**orking Directory *edit box to* `C:\MAXAC\DEVELOP`

If you want to give the MaxAC program item a distinctive look, use the Change **I**con button to choose a different icon. Use the Browse button to choose C:\R13\WIN\ACAD.EXE and select from the standard AutoCAD icons defined therein.

Choose OK

Additional AutoCAD program program items or shortcuts aren't just for multiple AutoCAD configurations. You can create a program item or shortcut for each project and set the **S**tart in directory (that is, working directory) for each program item or shortcut to that project's main DWG directory. By doing so, you ensure that AutoCAD's drawing file dialogs (open, save, and so on) default to the project directory. This procedure is the Windows equivalent of launching DOS AutoCAD from different project directories. In order to distinguish program item or shortcuts for different projects, you should give each one a descriptive name.

In Windows 95 and NT 4.0, another use for AutoCAD shortcuts is to launch AutoCAD with different prototype drawings (instead of having to enter **NEW** and specify the prototype drawing in the Create New Drawing dialog). In order to specify a prototype drawing, you use the `dwg_name=prototype` syntax in the shortcut's AutoCAD launch command. For example, you could create one shortcut with C:\R13\WIN\ACAD.EXE UNNAMED=PROTO1 /c C:\MAXAC and another with C:\R13\WIN\ACAD.EXE UNNAMED=PROTO2 /c C:\MAXAC.

Modifying the Support Path

The final step in setting up your alternate AutoCAD configuration is adding C:\MAXAC to AutoCAD's support path. As you saw earlier in the chapter, you make this change in the Preferences dialog. You'll also make sure that the AutoCAD log file is turned on and direct it to the \MAXAC directory.

Creating a Customization Environment

 Exercise

Modifying the Support Path

Start AutoCAD with the MaxAC Development shortcut.

Choose **O***ptions,* **P***references*	The AutoCAD Preferences dialog appears
Choose the **E***nvironment tab*	The Environment page appears
In the **S***upport edit box, add* `C:\MAXAC;` *to the beginning of the support path*	You can also remove C:\R13\WIN\TUTORIAL;, as described earlier

If you have AutoVision installed along with AutoCAD, you must leave its directory first in the support path: C:\AV\AVWIN;C:\MAXAC;C:\R13\COM\SUPPORT; and so on.

Make sure **L***og File is checked*	Turns on logging of the text window to a file
In the **L***og File edit box, change the file name to* `C:\MAXAC\ACAD.LOG`	Redirects the log file
Choose OK	AutoCAD saves the changes to C:\MAXAC\ACAD.INI

Because we made the Preferences changes with the configuration directory (/C) set to C:\MAXAC, they only apply when you launch AutoCAD with the MaxAC Development shortcut. When you launch AutoCAD with the standard AutoCAD shortcut, it still uses your unchanged configuration files in C:\R13\WIN.

 Warning: Be careful not to add too many long directory names to the support path. AutoCAD allows up to 10 directories on the support path, but the entire path string should be shorter than about 255 characters (in our tests, the exact length varied, depending on the existing path). If you try to type a longer support path in the Preferences dialog, AutoCAD stops accepting characters. Be aware of this limitation when you create additional support directories. Keep the names short and the number of levels of nesting small.

If you absolutely must use more characters in the support path, you can edit the `ACAD=` line in the ACAD.INI file with a text editor and add directory names there, but then you lose the ability to control the support path from the Preferences dialog.

An alternative to multiple configuration files, each with a distinct AutoCAD support path, is the /S switch. If you add /S and a path to the AutoCAD launch command (that is, Target) in the program shortcut's Properties, AutoCAD uses the /S path instead of the one in Preferences.

The advantage of /S is that you have only one set of configuration files. When you want to make global changes to all of your AutoCAD configurations (for example, adding a new plotter driver), you only have to do it once.

The disadvantage is that your different AutoCAD configurations aren't completely segregated from one another. You might end up making a configuration change (for example, modifying a global system variable) that's desirable for one AutoCAD configuration but undesirable for another. In addition, your Target string in the Properties gets quite long, and you can't control the support directory path from the AutoCAD Preferences dialog.

Using Different AutoCAD Configurations in DOS

The method of launching DOS AutoCAD from a batch file hasn't changed in R13 DOS. We won't cover every detail of the R13 DOS batch file, but the following exercise gives you enough information to set up an AutoCAD configuration similar to the Windows configuration described earlier. The file you will create should be similar to the following listing (the [»] character indicates a line that had to be printed on two lines):

```
@echo off
C:
CD \MAXAC\DEVELOP
SET ACAD=C:\MAXAC;C:\R13\COM\SUPPORT;C:\R13\DOS\SUPPORT;  [»]
C:\R13\COM\FONTS
SET ACADCFG=C:\MAXAC
SET ACADDRV=C:\R13\DOS\DRV

C:\R13\DOS\ACAD %1 %2

SET ACAD=
SET ACADCFG=
SET ACADDRV=
CD \
```

Creating a Customization Environment

Note: If you're not familiar with using a text editor, read the "Selecting and Using Text Editors with AutoCAD" section of this chapter before editing MAXAC.BAT.

Warning: Don't try to put DOS and Windows (especially Windows 3.1) configuration files in the same directory. If you want to create custom configurations for DOS and Windows, set up two different configuration directories.

Exercise

Creating a Custom MaxAC Batch File for DOS AutoCAD

Skip this exercise if you're not using DOS AutoCAD.

The default R13 DOS batch file is called ACADR13.BAT. AutoCAD's installation program creates it in C:\ . Find ACADR13.BAT and make a copy of it called MAXAC.BAT. You can keep ACADR13.BAT and MAXAC.BAT in any directory that's on your DOS path (for example, C:\BAT).

Open MAXAC.BAT in a text editor (for example, EDIT.COM)

Add the following three lines to the beginning of the MAXAC.BAT:
```
@echo off [Enter]
C: [Enter]
CD \MAXAC\DEVELOP [Enter]
```

These lines turn off echoing of the batch file contents and change to the appropriate directory before AutoCAD starts.

Change the line:
```
SET ACADCFG=C:\R13\DOS
```
to:
```
SET ACADCFG=C:\MAXAC
```

The ACADCFG environment variable is DOS AutoCAD's equivalent to the configuration directory switch. Next, you'll add C:\MAXAC to the beginning of the ACAD environment variable (the [»] characters in the exercise listing indicates lines that had to be printed on two lines, but you should enter them on one line).

Maximizing AutoCAD® R13

Change:
SET ACAD=C:\R13\COM\SUPPORT;C:\R13\DOS;C:\R13\DOS\SUPPORT; ▷]
C:\R13\COM\FONTS
to:
SET ACAD=**C:\MAXAC;**C:\R13\COM\SUPPORT;C:\R13\DOS\SUPPORT; ▷]
C:\R13\COM\FONTS

The ACAD environment variable is DOS AutoCAD's equivalent to the setting accessed through **O**ptions, **P**references, **Su**pport in Windows.

You can remove C:\R13\DOS from the support path. It's part of AutoCAD's library search path anyway, because the DOS ACAD.EXE file resides there.

Add the following four lines to the end of the MAXAC.BAT:
SET ACAD= [Enter]
SET ACADCFG= [Enter]
SET ACADDRV= [Enter]
CD \ [Enter]

These commands execute after you exit AutoCAD and before you return to the DOS prompt. They clear the environment variable settings to prevent them from conflicting with other AutoCAD configurations, and then change to the root directory.

Your finished batch file should look similar to the one shown in the listing that precedes this exercise, with adjustments for any drive and directory differences on your system. Note that the SET ACAD support directory path must be on one line.

Save MAXAC.BAT and exit to DOS

Finally, you'll test the new batch file.

C:\> **MAXAC** [Enter]   Starts ACAD with the Maximizing AutoCAD configuration

The MAXAC batch file changes to the C:\MAXAC\DEVELOP directory and then launches AutoCAD using the configuration files in C:\MAXAC. When you exit AutoCAD, the batch file changes to the root directory.

Warning: If you see an Out of environment space error message when you run MAXAC.BAT (or if the settings don't appear to be working), your DOS environment size may not be large enough. To increase the DOS environment, add the line SHELL=C:\DOS\COMMAND.COM C:\DOS\ /E:1024 /P to your CONFIG.SYS file.

If you already have a similar line, increase the number after /E:, which tells DOS how many bytes of memory to reserve for the environment. You must reboot the computer after editing CONFIG.SYS in order to have the changes take effect. See your MS-DOS manual for more information about DOS environment size and CONFIG.SYS.

Selecting and Using Text Editors with AutoCAD

Most AutoCAD customization, including the exercises in this book, depends on the use of a text editor. A text editor is the Swiss army knife of customization tools. You use it to edit, view, print, and search a wide variety of AutoCAD support files. Your text editor must have three essential capabilities:

- It must create pure ASCII files.

- It must be able to merge one text file into another.

- It must not automatically word wrap (or at worst, you must be able to turn off word wrapping).

Selecting an Editor for Windows

The text editor you use for AutoCAD customization must be able to create standard ASCII files. Most text editors and word processors can do so, although word processors often require a couple of extra steps to save in ASCII format. A good text editor loads and scrolls text quickly and saves files in plain ASCII format without any extra steps.

Table 2.1 lists some text editors and their advantages and disadvantages.

EDIT.COM is available in DOS 5 and higher, and it works acceptably for AutoCAD customization. It loads quickly and can handle all but the largest menu files. EDIT.COM is a reasonable choice for DOS AutoCAD customization. It will work in Windows, but it's less than ideal because cutting and pasting and other Windows operations are either more work or not possible at all.

Note: EDIT.COM depends on two other files, QBASIC.EXE and EDIT.HLP. If you decide to copy EDIT.COM to a floppy, be sure to copy all three files.

LispPad (Figure 2.6) is a Windows-based editor developed by Tony Tanzillo and included on the MaxAC CD-ROM. LispPad is optimized for AutoLISP programming in AutoCAD, but it also works well for editing script, font, small menu, and other AutoCAD support files. We'll use it for the AutoLISP exercises in this book. Because of its 32KB file limit, LispPad doesn't work for editing larger menu files.

Maximizing AutoCAD® R13

Table 2.1
Text Editor Comparison

Editor	Platforms	Advantages	Disadvantages
EDIT.COM	DOS	Small and fast	Awkward in Windows
LispPad	Windows	ACAD optimized	32KB limit
Norton Editor	DOS	Matches parentheses	Older DOS program
Notepad	Windows	Small and fast	32KB limit
TextPad	Windows	Fast and full-featured and edits large files	Not optimized for LISP
WordPad	Win 95	Edits large files	Better as word processor
Write	Win 3.1/NT	Edits large files	Better as word processor

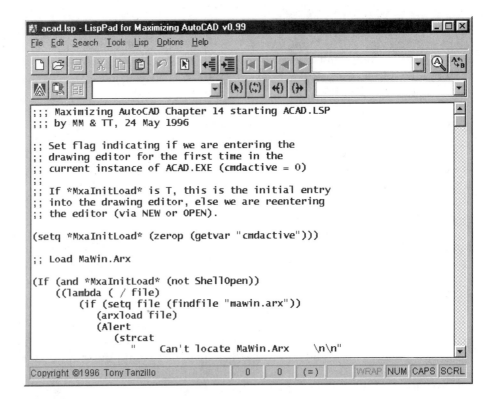

Figure 2.6
LispPad.

Creating a Customization Environment

NotePad is the Windows equivalent of EDIT.COM, but unfortunately NotePad is limited to 32KB files.

The Norton Editor is popular among long-time DOS users because it's very fast, not complicated to learn, and relatively efficient for DOS editing jobs.

TextPad is a popular shareware editor (Figure 2.7) with a 32-bit version for Windows 95 and NT and a 16-bit version for Windows 3.1. It's fast, handles large files easily, and includes features that make extensive editing tasks more efficient. The MaxAC CD-ROM includes the 32-bit and 16-bit shareware versions of TextPad. We'll use TextPad in most of this book's exercises.

Note: The publishers of TextPad, Helios Software, allow you to use TextPad for completing the exercises in this book, as part of or an extension to the shareware trial period. If you like TextPad and decide to use it for additional customization or other tasks, you must send in the modest registration fee. The simplest way to register is through CompuServe's SWREG (Shareware Registration) forum, with ID #3938. You can also register by fax or mail. See REGISTER.TXT in the TextPad directory after you install the program.

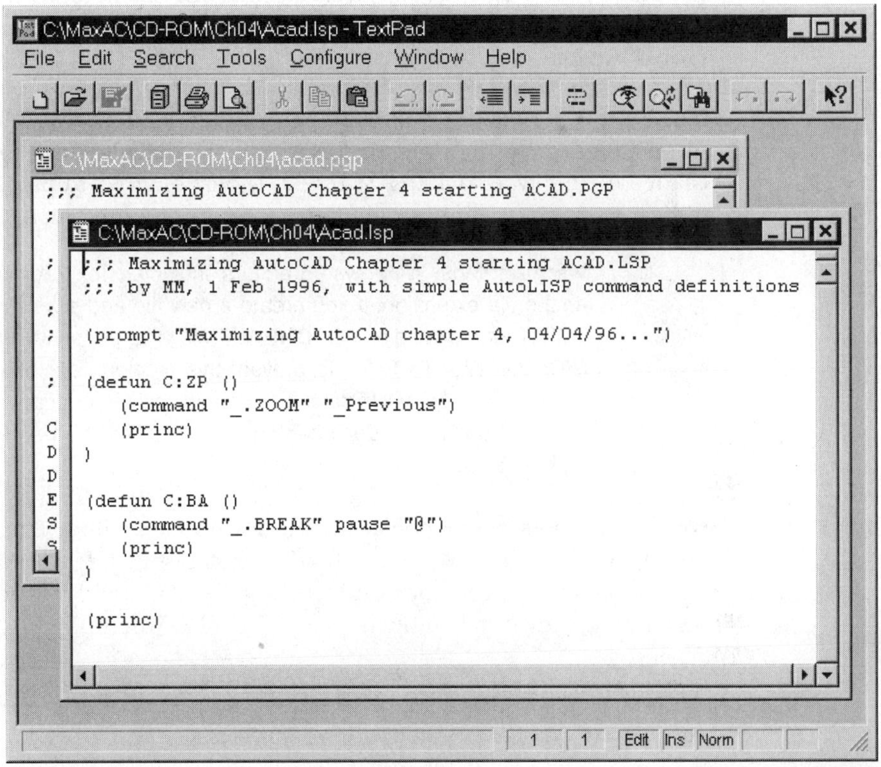

Figure 2.7
TextPad.

Windows Write is a no-frills word processor included with Windows 3.1 and NT. It can load large menu files and save them in ASCII format. Because Write is a word processor rather than a text editor, using it for customization is more of a hassle. WordPad is the Windows 95 replacement for Write. It adds even more word processing features (which were borrowed from Word, Microsoft's full-fledged word processor). WordPad shares Write's disadvantages for customization, but it's the one standard Windows 95 accessory that can edit large menus.

Word processors, including Windows Write, Word, WordPad, and WordPerfect, save their native documents in a binary format that won't work for AutoCAD support files. Fortunately, if you load an existing ASCII file, the better word processors remember the format and use that when saving. WordPad, the word processor that we'll use for editing larger menu files in this book, works that way.

When you load an ASCII file in Windows Write, the program asks whether you want to convert it to Write's native format (WRI). Always answer **N**o Conversion to keep the file in ASCII format when you save. When you load an ASCII file in WordPad, the program continues to save it in ASCII format unless you tell it to do otherwise.

If you create a new document in a word processor such as WordPad or Write, you must tell the program to save it in ASCII format. For example, in WordPad, you choose **F**ile, Save **A**s, and choose `Text Document - MS-DOS Format` in the Save as **t**ype drop-down list.

If you're not sure about your word processor, use **F**ile, Save **A**s to open a dialog box that shows what format the program is saving to by default. You might find that you need to change the file type back to ASCII text. (Some older word processors refer to ASCII text saving as non-document, programmer's, or unformatted mode.)

Warning: Most Windows 95 editors, including TextPad and NotePad, default to TXT as the file extension. If you create a new file and save it to a name that doesn't end in TXT (for example, ACADNEW.MNS), the editor will tack the TXT extension on (ACADNEW.MNS.TXT). To prevent this problem, enclose the file name in quotation marks (like "ACADNEW.MNS"). If you want, you can register other file types in Windows 95, so you don't need quotes to create them—see Windows Help for details on how to do so.

For the exercises in this book, you can use any text editor that meets the three criteria earlier in this section. We'll use TextPad and LispPad. If you want to use them too, see Appendix G for instructions on installing TextPad and LispPad on your system, if they were not already installed by the MaxAC CD-ROM setup.

Associating Registered File Types with Your Editor

All of the Windows operating systems allow you to associate file types (that is, extensions) with programs, so that when you double-click on a file of a particular type in Explorer or File Manager, Windows automatically loads the associated program. Use this feature to create associations for LSP (AutoLISP), MNL (menu LISP), MNS (Windows menu source), PGP (program parameters), and SCR (script) files.

Make sure you've installed TextPad and LispPad, as described in Appendix G. The exercise instructions are for these two editors, but you can substitute a different ASCII editor. If you are using Windows 95 or NT 4.0, use the following exercise; otherwise skip to the next exercise.

 Exercise

Associating Registered File Types with Your Editor In Windows 95 or NT 4.0

These instructions are for Windows 95 and NT 4.0 only.

Start Explorer

Choose **V**iew, **O**ptions, File Types	Opens the File Types associations dialog
In the Registered file **t**ypes *list, scroll down and select* AutoCAD Menu Source *(Figure 2.8, ①)*	The File type details section displays the extension MNS
Choose **E**dit	Opens the Edit File Type dialog
If **OPEN** *appears in the list of already defined* **A**ctions, *use the* **R**emove *button to remove it*	
Choose **N**ew	Opens the New Action dialog

Figure 2.8
Selecting a file type to associate.

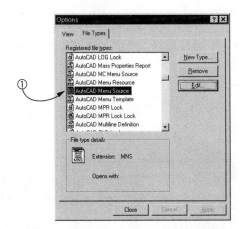

Figure 2.9
The action and application associated with the file type.

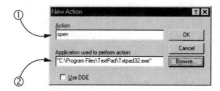

In the **A***ction edit box, type* **OPEN**   (Figure 2.9, ①)

Choose **B***rowse*   Opens the Open With dialog

Double-click on My Computer, *then locate and choose* TXTPAD32.EXE *in the TextPad directory*

Choose **O***pen*   Closes the Open With dialog

The New Action dialog should look similar to Figure 2.9.

Choose OK   Closes the New Action dialog

Choose **C***lose*   Closes the Edit File Type dialog

Repeat the previous steps for AutoCAD Program Parameters (PGP) and AutoCAD Script (SCR) file types.

To speed up the process, you can copy the string from the App**l**ication used to perform action box, ② in Figure 2.9 ("C:\PROGRAM FILES\TEXTPAD\TXTPAD32.EXE") to the Clipboard by selecting the entire string and pressing Ctrl+C. Then for subsequent file types, paste the string with Ctrl+V.

There are other, more efficient ways to create file type associations in Windows 95, but the long-winded way described in this exercise works on any system, no matter what the current associations.

You've created associations for TextPad; now do the same for LispPad and LSP and MNL files. The file types are AutoLISP Application Source (LSP) and AutoLISP Menu Source (MNL). The string to enter in the App**l**ication used to perform action box is
C:\MAXAC\TOOLS\LISPPAD\LISPPAD.EXE.

Finally, you should test the associations.

In Explorer, double-click on   TextPad opens the file
\MAXAC\BOOK\CH02\ACADMA.MNS

In Explorer, double-click on   LispPad opens the file
\MAXAC\BOOK\CH02\ACADMA.MNL

Creating a Customization Environment

As you can see, AutoCAD adds a huge number of file types to the Windows 95 Registry. You can associate TextPad or another editor with any of the text file types (for example, AutoCAD XREF Log (XLG) and AutoCAD Plot Configuration Parameters (PCP)). You also might want to associate TextPad with Text Documents (TXT).

If you are using Windows 3.1 or NT 3.51, use the following exercise.

Exercise

Associating Registered File Types with Your Editor in Windows 3.1 or NT 3.51

These instructions are for Windows 3.1 and NT 3.51 only.

Start File Manager

Choose File, Associate	Opens the Associate dialog
In the Files with Extension *edit box, type* **MNS**	
In the Associate With *list, select TextPad*	
Choose OK	Closes the Associate dialog

Repeat the previous steps for AutoCAD Program Parameters (PGP) and AutoCAD Script (SCR) file types.

You've created associations for TextPad; now do the same for LispPad and LSP and MNL files. The file types are LSP and MNL. Use the **B**rowse button to select the application C:\MAXAC\TOOLS\LISPPAD\LISPPAD.EXE.

Finally, you should test the associations.

In File Manager, double-click on \MAXAC\B\BOOKCH02\ACADMA.MNS	TextPad opens the file
In File Manager, double-click on \MAXAC\BOOK\CH02\ACADMA.MNL	

Selecting and Configuring an Editor for DOS

Because DOS is not a multitasking operating system, you must shell out of DOS AutoCAD in order to run a DOS text editor. EDIT.COM works fine for DOS AutoCAD customization, and the standard ACAD.PGP file includes an external command definition that allows you to launch EDIT.COM by typing **EDIT** at the command prompt.

If you want to use a DOS editor other than EDIT.COM, then you should add a line to \R13\COM\SUPPORT\ACAD.PGP for it. The standard ACAD.PGP line for EDIT.COM looks like this:

```
EDIT,EDIT,      0,File to edit: ,4
```

Make a copy of this line and in the copied line change the two instances of EDIT to the executable file name of your editor. For example, you would add the following line for the Norton Editor (NE.EXE):

```
NE,NE,    0,File to edit: ,4
```

The directory containing the program (NE.EXE in this example) must be included in the DOS path—otherwise DOS won't be able to locate the program file. Alternatively, you can include the path in the second parameter:

```
NE,C:\BIN\NE,    0,File to edit: ,4
```

Chapters 3 and 4 discuss the ACAD.PGP file and the syntax of external command definitions.

Testing Your Editor

If you're using an editor other than TextPad and LispPad, test its ability to produce ASCII files with the following exercise. You can skip the exercise if you'll be using only TextPad and LispPad.

 Exercise

Testing Your Text Editor

The exercise instructions are for TextPad, but most Windows text editors should work similarly.

Start TextPad and choose **F**ile, **N**ew

Enter at least a dozen lines of text, or open a
small text file on your disk

Choose **F**ile, **S**ave *and use the file name*
`"C:\MAXAC\DEVELOP\TEST.TXT"`

Creating a Customization Environment

The quotation marks aren't required in this case, because the file name ends in TXT. But it's a good idea to get in the habit of using the quotation marks so that you'll remember to add them when creating other type of files, such as LSP files.

Enter the MTEXT command and pick two points — The Edit MText dialog appears

Choose **Im**port — The Import Text File subdialog appears

Locate TEST.TXT *and choose* **O**pen

Choose **OK** *in the main Mtext dialog* — AutoCAD imports the text file into an MText paragraph

Your text editor is suitable if the paragraph displays text identical to what you typed in your editor (Figure 2.10, ①), with no extra control or other strange characters such as ^L or ?Í?àïŧ?á ② or smiling faces. If you see any such "garbage," particularly at the top or bottom of the text, then your text editor is not suitable or is not configured correctly for ASCII output.

Avoiding ACAD.LSP Conflict

If your system has an ACAD.LSP file on the AutoCAD library search path, AutoCAD automatically loads it. An existing ACAD.LSP file may contain functions that interfere

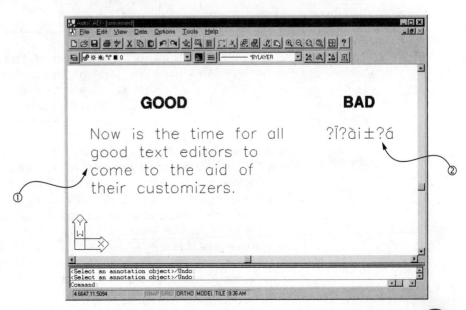

Figure 2.10
Good and bad results with a text editor.

with the book's setup. Play it safe and make a dummy ACAD.LSP file that is found and loaded instead of any existing file.

Exercise

Making a Dummy ACAD.LSP File

Start LispPad and choose File, New

Type the following line:
`(prompt "\nMaximizing AutoCAD chapter 2 04/01/96...\n")` [Enter]

Save the file as C:\MAXAC\DEVELOP\ACAD.LSP

Start a new drawing or open an existing one   AutoCAD loads ACAD.LSP

Flip to or scroll the text window

```
Loading C:\MAXAC\DEVELOP\acad.lsp...
Maximizing AutoCAD chapter 2 04/01/96...
loaded.Regenerating drawing.

AutoCAD Release 13 menu utilities loaded.

Command:
```

This dummy ACAD.LSP file loads instead of any other ACAD.LSP file on your system because the drawing directory (that is, the directory in which AutoCAD started) is the first directory in the library search path. See Figure 2.3 earlier in this chapter for an illustration of this fact and Chapter 4 for details on the library search path order. Notice the `Maximizing AutoCAD chapter 2 date...` prompt that displays.

Warning: Similarly to the ACAD.LSP file, any custom settings in the ACAD.ADS file and any ACAD.RX file may affect your AutoCAD environment. The default ACAD.ADS file has only one line: `acadapp`. Make sure the first ACAD.ADS file found on Auto CAD's search path contains this line, and that it does not contain other lines that load programs that might interfere with your customization efforts in this book. To be safe, you can put an ACAD.ADS file containing only the line `acadapp` in your MAXAC diirectory. There is no default ACAD.RX file. If there is a custom ACAD.RX file on your AutoCAD search path, you can prevent it from loading in this book's configuration by putting an empty ACAD.RX file in the MAXAC directory. Neither of these changes will affect your other AutoCAD configurations.

Creating a Customization Environment

Setting the Default Menu

The final configuration steps in this chapter are to tell AutoCAD to use the custom MaxAC menu as the default and to set a few global AutoCAD system variables, which get stored in the ACADNT.CFG file.

Note: If you're using DOS AutoCAD, substitute ACADMA.MNU for ACADMA.MNS in the following exercise (and in all exercises in this book).

Exercise

Setting the Default Menu and Global System Variables

Copy ACADMA.MNS and ACADMA.MNL from C:\MAXAC\BOOK\CH02 to C:\MAXAC\DEVELOP.

Continue in AutoCAD from the previous exercise.

Command: **CMDDIA** [Enter]

New value for CMDDIA <1>: *Make sure yours is set to 1*   Sets PLOT to use a dialog box

Command: **FILEDIA** [Enter]

New value for FILEDIA <1>: *Make sure yours is set to 1*   Sets the file commands to use a dialog box

Command: **PREFERENCES** [Enter]   Opens the Preferences dialog

Choose the **S**ystem *tab and make sure that the* **K**eystrokes *setting is* Menu **F**ile *(not* **A**utoCAD Classic*)*   Allows AutoCAD to read accelerator key definitions from the menu

Choose the **M**isc *tab and make sure that* **U**se Menu in Header *is turned off*   Tells AutoCAD to use ACADMA.MNS with all drawings

Command: **MENU** [Enter]   Opens the Select Menu File dialog

Choose OK

In the **D**irectories *list, navigate to* C:\MAXAC\DEVELOP *(Figure 2.11, ①)*	
In the **L**ist Files of Type *drop-down list, choose* *.MNS ②	The file list displays only MNS files
Substitute MNU for MNS in DOS.	
In the File **N**ame *list, double-click on* ACADMA.MNS ③	AutoCAD closes the dialog, compiles ACADMA.MNS to ACADMA.MNC and ACADMNA.MNR, and loads the MNC and MNR files

```
Menu loaded successfully.

MENUGROUP: ACAD
```

These settings ensure that AutoCAD behaves as the book's exercises expect. If you need more information on these settings and how they work, see the system variable table in Appendix A of the *AutoCAD Command Reference*.

Finishing Up

At the end of each chapter's exercises, you should wrap up your customization session by exiting AutoCAD, moving your newly created or modified files from C:\MAXAC\DE-VELOP to C:\MAXAC, and deleting test files from \MAXAC\DEVELOP. This procedure is equivalent to putting the "stamp of approval" on the finished custom files and cleaning the scratch paper off your desk.

Figure 2.11
Loading
ACADMA.MNS.

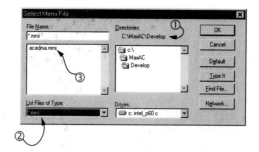

Creating a Customization Environment

In this chapter, the new customization files you created or copied are ACAD.LSP, ACADMA.MNS, and ACADMA.MNL, and the scratch files are TEST.TXT and TEST2.TXT. In addition, AutoCAD created ACADMA.MNC and ACADMA.MNR when it compiled ACADMA.MNS.

 Exercise

Finishing Up

Move ACAD.LSP, ACADMA.MNC, ACADMA.MNL, ACADMA.MNR, and ACADMA.MNS to C:\MAXAC.

Delete TEST.TXT and TEST2.TXT.

Use the MENU command to change the menu to C:\MAXAC\ACADMA.MNS.

Quit AutoCAD. You don't need to save changes.

 Warning: With the new R13 "always use the same menu file" feature (that is, <u>U</u>se Menu in Header) turned off, AutoCAD stores the path name with the menu file, instead of doing a library search. Thus, if you launch AutoCAD after moving the menu files to \MAXAC, AutoCAD will complain that it can't find \MAXAC\DEVELOP\ACADMA.MNS. To avoid this problem, we have you change the menu at the beginning and end of each chapter. The first exercise in each chapter of this book starts by changing the menu to \MAXAC\DEVELOP\ACADMA ,and the last exercise ends by changing it back to \MAXAC\ACADMA.MNS. These steps ensure that you can run AutoCAD with the MaxAC Development shortcut even when you're not working through the chapter exercises.

Conclusion

You can take some of the mystery out of how AutoCAD works by becoming familiar with the AutoCAD directory structure and configuration files. Organize your files and system by setting up a support directory search path that enables AutoCAD to find your application files efficiently. Keeping some directories separate (such as \MAXAC) saves you from interference woes with other projects.

If you've followed all the instructions in this chapter, your AutoCAD customization environment should be on common ground with this book's assumptions. Make sure that you're comfortable with creating and saving simple text files in TextPad and LispPad or another editor. You'll be using your editing program extensively in the chapters to come as you work with menus, scripts, and AutoLISP.

In the next chapter, you begin to perform real customization by adding keyboard macros and Windows toolbars to your system.

chapter 3

Creating Keyboard Macros and Toolbars

Keyboard macros and toolbars are two of the easiest ways to improve your AutoCAD productivity, by speeding up access to the AutoCAD commands you use most often. Under the topic of "keyboard macros," this chapter covers three different ways of creating keyboard shortcuts: ACAD.PGP command aliases, simple AutoLISP command definitions, and menu accelerator keys. The second part of the chapter shows you how to use and customize the new R13 for Windows toolbars.

Much of the material in this chapter is new with Release 13. ACAD.PGP command aliases haven't changed since Release 11, but accelerator keys add a new method for creating keyboard macros. The toolbar and toolbar customization capabilities of R13 are completely different than those of the limited toolbox in R12 for Windows. Note that all of the new R13 customization features covered in this chapter are unique to the Windows platforms (3.1, NT, and 95), and thus are unavailable in R13 for DOS.

Much of the information in this chapter applies to AutoCAD LT. You can create ACAD.PGP command aliases and define menu accelerator keys (but you cannot create simple AutoLISP command definitions). The procedures for creating Windows toolbars work for LT, but you won't be able to use some of the macros in the customization toolbar because they depend on AutoLISP.

About This Chapter

In this chapter, you will learn how to do the following:

- Create ACAD.PGP command aliases (that is, abbreviations) for commonly used AutoCAD commands
- Create simple AutoLISP command definitions for commonly used AutoCAD commands that require specifying command options
- Define menu accelerator keys for accessing commonly used AutoCAD commands with function keys and CTRL and SHIFT key combinations

- Learn how to manage the standard R13 toolbars
- Create toolbars containing favorite commands and customization tools

Warning: The exercises from this point forward assume that you've set up your system for customizing AutoCAD as described in Chapter 2. Please make sure that your system is configured according to the instructions in Chapter 2 before you proceed.

The resources you'll use in this chapter are:

- TextPad (or another ASCII text editor)
- ACAD.LSP, ACAD.PGP, ACADMA.MNS, ACADMA.MNL, MAWIN.ARX, MA_UTILS.LSP, and custom bitmap (BMP) icons from the MaxAC CD-ROM

In order to work through the exercises, you'll need a text editor. The book's exercises use the TextPad shareware editor and Tony Tanzillo's LispPad editor, which are included on the MaxAC CD-ROM. Any ASCII text editor capable of loading a fairly large menu file (200KB) will do, as described in Chapter 2.

Understanding Keyboard Macros

Many experienced AutoCAD users rely on the keyboard for rapid command and option input. Despite the increasing use of dialog boxes and menu improvements in AutoCAD's user interface, the keyboard remains the most direct, and in some cases the fastest, way to drive the program.

As is often the case with AutoCAD, you have a confusing array of options for defining keyboard macros. There isn't any single, simple way to record keyboard macros in AutoCAD, as there is with many other programs. Instead, AutoCAD provides three different ways of attaching command sequences to keystrokes. Each of these methods requires editing a text file and then reloading it in AutoCAD.

- **ACAD.PGP command aliases.** Command aliases are the easiest customization tool available in AutoCAD, and they work well for creating command abbreviations, such as L for the LINE command. The ACAD.PGP file also includes a section for defining external commands, which is especially useful with DOS AutoCAD.

- **Simple AutoLISP command definitions.** With a bit of AutoLISP, you can extend the command abbreviation idea so that you can add command options or pauses for user input.

Creating Keyboard Macros and Toolbars

- **Menu accelerator keys.** Accelerator keys, which are new to R13 for Windows, make it easy to assign command sequences to the function, CTRL, and SHIFT keys in combination with ordinary letter and number keys.

In the first part of this chapter, you learn how to use all three methods and how to decide among them when you want to create a keyboard macro.

Defining Keyboard Aliases in the ACAD.PGP File

By customizing the ACAD.PGP file, you can define abbreviations for AutoCAD commands, so that you can execute them with one or two keystrokes instead of entering long command names. The AutoCAD documentation calls this ACAD.PGP feature "command aliases." The standard R13 ACAD.PGP file (located in \R13\COM\SUPPORT) includes definitions for 16 commonly used AutoCAD commands, 14 dimensioning commands, and several other miscellaneous commands.

```
; Sample aliases for AutoCAD Commands
; These examples reflect the most frequently used commands.
; Each alias uses a small amount of memory, so don't go
; overboard on systems with tight memory.

 A,        *ARC
 C,        *CIRCLE
 CP,       *COPY
 DV,       *DVIEW
 E,        *ERASE
 L,        *LINE
 LA,       *LAYER
 LT,       *LINETYPE
 M,        *MOVE
 MS,       *MSPACE
 P,        *PAN
 PS,       *PSPACE
 PL,       *PLINE
 R,        *REDRAW
 T,        *MTEXT
 Z,        *ZOOM

 3DLINE,   *LINE

; Give Windows an AV command like DOS has
 AV,       *DSVIEWER
; Menu says "Exit", so make an alias
```

```
EXIT,    *QUIT

; easy access to _PKSER (serial number) system variable
SERIAL, *_PKSER
; Dimensioning Commands.

DIMALI,  *DIMALIGNED
DIMANG,  *DIMANGULAR
DIMBASE, *DIMBASELINE
DIMCONT, *DIMCONTINUE
DIMDIA,  *DIMDIAMETER
DIMED,   *DIMEDIT
DIMTED,  *DIMTEDIT
DIMLIN,  *DIMLINEAR
DIMORD,  *DIMORDINATE
DIMRAD,  *DIMRADIUS
DIMSTY,  *DIMSTYLE
DIMOVER, *DIMOVERRIDE
LEAD,    *LEADER
TOL,     *TOLERANCE
```

As you can see from the ACAD.PGP listing, the syntax for command aliases is simple. Each line defines one command alias and contains only two text strings (or *fields*) separated by a comma. Lines beginning with a semicolon are comments and are ignored by AutoCAD.

Look at the first command alias definition, which is for AutoCAD's ARC command. A is the *abbreviation*. The first field tells AutoCAD the alias you want to use for the command at the command prompt. ARC is the native AutoCAD *command name*. The second field names the AutoCAD command that is being abbreviated. You must place an asterisk (*) in front of the command name so that AutoCAD knows that this line defines a command alias, rather than an external command. (We discuss external commands later in the chapter.)

Note: Additional spaces in the command alias lines are for readability, and they don't have any influence. You can add as many or as few spaces as you want. Capitalization is unimportant as well. Thus `A,     *ARC` is the same as `a,*arc`.

You should define aliases in ACAD.PGP for the commands you use most frequently, and use the abbreviations that make the most sense to you. For example, people who use the COPY command more frequently than CIRCLE may want to redefine C as an alias for COPY and change CIRCLE to CI. Many important AutoCAD commands, such as STRETCH and BHATCH, lack command aliases in the standard ACAD.PGP file.

Creating Keyboard Macros and Toolbars

Warning: Where the ENTER key symbols (Enter) are shown in the following exercise, do not put a space between the input you type and the Enter.

In the following exercise, you start with the standard ACAD.PGP file, add a comment to the top of it, modify the existing aliases, and add some of your own.

Exercise

Adding Command Aliases to ACAD.PGP

Copy all files from \MAXAC\BOOK\CH03 to your \MAXAC\DEVELOP directory. Use the MENU command to change the current menu to \MAXAC\DEVELOP\ACADMA.MNS.

Open \MAXAC\DEVELOP\ACAD.PGP with your text editor.

Start AutoCAD using the MaxAC Development shortcut.

Add the following comment lines to the top of your new ACAD.PGP file, substituting your initials and the current date:

```
; Customized ACAD.PGP Enter
; by MM, 04/03/96, added preferred aliases Enter
```

Modify the COPY and CIRCLE alias lines in your ACAD.PGP file to match the following:

```
C,        *COPY Enter
CI,       *CIRCLE Enter
```

Add the following lines, or your own favorite aliases, to your ACAD.PGP file. Although AutoCAD doesn't care where you put the lines, you may want to arrange them alphabetically with the other aliases.

```
;Some Favorite Aliases Enter

AY,       *ARRAY Enter
BH,       *BHATCH Enter
CP,       *DDCHPROP Enter
```

Maximizing AutoCAD R13

```
DT,       *DTEXT  [Enter]
ED,       *DDEDIT [Enter]
I,        *INSERT [Enter]
LS,       *LIST   [Enter]
QS,       *QSAVE  [Enter]
S,        *STRETCH [Enter]
```

Save and close the file. Remain in AutoCAD for the next exercise.

Although you've modified the aliases in ACAD.PGP, your changes won't take effect until you reload the file in the next exercise.

Before you modify an AutoCAD support file, always add a comment to the top of the file. List your name, the date, and a brief description of the changes. Comments make it much easier to determine what version of a file you're using and how it's different from the standard version.

Reloading and Testing ACAD.PGP

AutoCAD rereads the ACAD.PGP file each time you open an existing drawing or create a new one. Thus, the changes you make to ACAD.PGP normally won't be active until you use the NEW or OPEN commands. Or, instead of using NEW or OPEN, you can force the reinitialization with the AutoCAD REINIT command (or RE-INIT system variable; note the dash). REINIT is a more efficient way to reload ACAD.PGP.

In the following exercise, you reload your modified ACAD.PGP and test your changes.

 Exercise

Reloading and Testing ACAD.PGP

Continue from the previous exercise, in the AutoCAD drawing editor.

Command: *Choose* Tools, Reinitialize *or enter*   Displays the Re-initialization dialog
REINIT   box (see Figure 3.1)

Turn on PGP File *in the* Device and File
Initialization *group*

Creating Keyboard Macros and Toolbars

Figure 3.1
Reloading ACAD.PGP with REINIT.

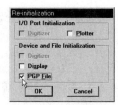

Choose OK AutoCAD closes the dialog box and reloads ACAD.PGP

Test each change and addition to ACAD.PGP. For example:

`Command: c `[Enter] Issues the COPY command

`COPY` Cancels the COPY command

`Select objects:` *Press ESC*

Repeat this test for the other changes and additions you made to ACAD.PGP.

Always test every modification that you make to AutoCAD support files. When you've made a lot of changes, it's tempting to spot-check just a few of them and assume that you did the rest properly. But if you test thoroughly, you'll catch lots of mistakes that aren't obvious when you're editing a support file. Thorough testing is especially important when you're distributing custom support files to other users.

 Note: If you write a program that modifies and needs to reload the PGP file, you can use the RE-INIT system variable. The RE-INIT system variable code values are: 1 digitizer I/O port, 2 plotter I/O port, 4 digitizer device, 8 video display device, and 16 ACAD.PGP file. Thus, the command sequence `RE-INIT 16` reloads the ACAD.PGP file. RE-INIT is a *bit-coded* system variable, which means that you can enter the sum of the codes corresponding to the items you want to reinitialize. For example, `RE-INIT 5` (= 1 + 4) attempts to reinitialize both the digitizer I/O port and the digitizer device itself. `RE-INIT 31` (= 1 + 2+ 4 + 8 + 16) reinitializes all of the options.

The MaxAC CD-ROM includes an ACAD.PGP file with a complete set of command aliases for R13. If your system has plenty of memory and you want to use this ACAD.PGP file, copy it from the \MAXAC\BOOK directory to \MAXAC\DEVELOP. If you're using an ACAD.PGP file that's already been customized (for example, by a third-party application), you'll need to combine the two files in order to ensure that you don't disable any of your existing custom command aliases or external commands.

Defining External Commands in the ACAD.PGP File

The ACAD.PGP command alias feature was added to AutoCAD in R11, but ACAD.PGP has a much longer history. Through Release 10, ACAD.PGP's only purpose was for defining external commands that allow you to run other programs by shelling out of AutoCAD. In fact, PGP is an acronym for ProGram Parameters (although long-time AutoCAD programmers refer to it as the PiGPen file, because some DOS users would throw every conceivable external command definition into it).

A *shell* is an operating environment under which you can run operating system commands and programs. Many programs, such as AutoCAD, enable you to suspend the current program temporarily, shell out to the operating system, and free memory to execute other commands or programs. This capability makes it possible to run programs, utilities, or operating-system commands without having to end AutoCAD or reload your drawing. External command definitions in the ACAD.PGP file define the link between AutoCAD commands and operating-system commands and programs.

Obviously, the ability to shell out and create external command definitions in ACAD.PGP is most useful in operating systems like DOS that don't directly allow multitasking. These features are less valuable in multitasking operating systems, including Windows, since you can simply load another program from the desktop or Program Manager without shelling out of AutoCAD.

On the other hand, even Windows users sometimes need to use or define external commands. Occasionally it's handy to shell out to the DOS prompt quickly from AutoCAD. Some third-party applications designed for DOS require ACAD.PGP external commands, and AutoCAD's own AUTOSPOOL plot spooling feature does its work by shelling out.

Tip: In Windows 95, you can execute a Windows application from DOS. Thus, if you define an external command definition for a Windows application in the PGP file, it enables you to launch that application from AutoCAD.

Using ACAD.PGP External Commands

The standard ACAD.PGP includes seven external command definitions.

```
; External Command format:
;    <Command name>,[<DOS request>],<Memory reserve>, [*]<Prompt>, [»]
<Return code>

; Examples of External Commands for DOS

CATALOG,DIR /W,0,File specification: ,0
DEL,DEL,      0,File to delete: ,4
DIR,DIR,      0,File specification: ,0
```

Creating Keyboard Macros and Toolbars

```
EDIT,EDIT,      0,File to edit: ,4
SH,,            0,*OS Command: ,4
SHELL,,         0,*OS Command: ,4
TYPE,TYPE,      0,File to list: ,0
```

An ACAD.PGP external command can either provide general-purpose access to the operating system command prompt or run a specific operating system command or program automatically. The SH and SHELL external command definitions (which are identical) are of the former kind; they simply provide a DOS prompt. The other external command definitions shown in the preceding listing run specific DOS commands or programs.

In the following exercise, you use two of the standard AutoCAD external commands and learn the difference between a partial (single command) and full (multiple command) shell.

Exercise

Using ACAD.PGP External Commands

Command: **SH** Enter — Prompts for an OS shell command

OS Command: **DIR** Enter — A directory listing scrolls by, and then AutoCAD returns to the command prompt

In Windows, and with some DOS display drivers, you probably didn't have a chance to read the directory listing before the command prompt returned. This time, press ENTER at the OS Command prompt to get a full shell.

Command: **SH** Enter

OS Command: Enter — The DOS prompt appears (in a window if you're running AutoCAD for Windows see Figure 3.2)

C:\MAXAC\DEVELOP: **DIR** Enter — A directory listing scrolls by, but this time the DOS prompt remains

C:\MAXAC\DEVELOP: **EXIT** Enter — Exits the DOS shell and returns to AutoCAD

Maximizing AutoCAD® R13

Figure 3.2
Full shell in Windows.

Now use the EDIT external command definition to launch EDIT.COM.

```
Command: EDIT Enter
```

```
File to edit: TEST.TXT Enter          EDIT.COM starts
```

Next, you will close EDIT.COM without making any changes.

Press ALT+F   EDIT.COM's File menu appears

*Choose E*x*it*   AutoCAD returns to the command
                                       prompt

When you used DIR in the first part of the preceding exercise, AutoCAD freed up memory, permitting the operating system to run the DIR command. Although the DIR command is not a true AutoCAD command, you can enter it as if it were because of the external command definition in the ACAD.PGP file.

When you used SH and pressed ENTER in the second part of the exercise, AutoCAD passed control to DOS, but remained in the background. In this *full shell*, you can execute as many commands as you like. Control remains with DOS until you type **EXIT** and press ENTER, which closes the shell and returns control to AutoCAD.

Warning: Be careful not to delete or move AutoCAD's temporary files when you're shelled out. Doing so can crash AutoCAD. AutoCAD's temporary files often contain dollar signs ($) in the file name. See Chapter 1 for more information about temporary file locations.

ACAD.PGP External Command Definition Syntax

As with command aliases, each external command definition in ACAD.PGP goes on its own line and consists of fields separated by commas. With external commands, though, there are five fields instead of just two. The five fields, together with an example, are:

Creating Keyboard Macros and Toolbars

```
AutoCAD command, OS command, memory reserve, user prompt,      return code
EDIT          , EDIT      , 0             , File to edit: ,4
```

The last two fields are optional, but most programmers include the commas and leave the fields blank when they're not needed.

Here's a description of the five fields.

- *AutoCAD command* (EDIT) is the name you type at the AutoCAD command prompt in order to run the external command. This command name cannot contain spaces because AutoCAD treats a space as an ENTER when you're typing a command.

- *OS command* (EDIT) is the command sequence that AutoCAD feeds to the operating system prompt. This field can include anything you normally would type at the DOS prompt, including spaces, command switches, and file paths. It can also be a batch file name. In many cases, the *AutoCAD command* and *OS command* fields will be the same, which simply means that the command you type at AutoCAD's command prompt is the same as what you normally type at the operating system prompt. In other cases, the two fields will be different to allow for additional operating system parameters or prevent conflicts between identically named AutoCAD and DOS commands. For example, if you wanted to create an external command definition for DOS's COPY command, you might use DOSCOPY as the *AutoCAD command* field to avoid conflicting with the AutoCAD COPY command. If the external command field is blank, as in the SH and SHELL lines, then AutoCAD prompts the user with OS command: and feeds the response to the operating system (or opens a full shell if the user presses ENTER alone). Note that you don't ever see the *OS command* field when you type the *AutoCAD command*. AutoCAD passes the *OS command* invisibly to the operating system.

- *memory reserve* (0) is a vestige of older AutoCAD versions in which you needed to specify the amount of memory that AutoCAD should try to release for the operating system to use. Memory reserve is ignored by all current AutoCAD versions, which free up as much memory as they can, regardless of this setting. You still need a number in this field for compatibility purposes, and zero is as good as any other number.

- *user prompt* (File to edit:) is the prompt that AutoCAD displays to the user. When you include this field, AutoCAD waits for input and feeds the user's response to *OS command* as a parameter. If an asterisk precedes the prompt, AutoCAD allows spaces in the input line and only accepts the input when the user presses ENTER. If no asterisk precedes the prompt, either ENTER or a space terminates the user's input. Typically, your user prompt should end with a colon and a space.

- *return code* (0) controls what AutoCAD does after the external command returns control to AutoCAD. In DOS AutoCAD with some display drivers, a value of 0 causes AutoCAD to return to the text screen, while 4 returns to whichever screen (text or graphics) was current when you launched the external command. In Windows, AutoCAD always returns to the screen that you left; the 0 value doesn't force the text window to pop up. (Return codes 1 and 2 are for applications that need to process DXB files. Such applications are rare now. For more information, see the *AutoCAD Customization Guide* section on ACAD.PGP.)

Sample External Command Definitions

If you're customizing ACAD.PGP for DOS AutoCAD, or you'd like to experiment with external commands, here are some samples that you can add to your custom ACAD.PGP.

```
; Maximizing AutoCAD External Commands for DOS
REN,RENAME,     0,*Oldname Newname: ,0
DWGS,DIR *.DWG,0,,0
DUP,COPY,       0,*SOURCEfile(s) TARGETfile(s): ,0
SHOW,MORE <,    0,File to list: ,0
```

Windows and some DOS display drivers cause external commands to close the text window before you can read the results of the command, as you probably saw in the DIR exercise earlier. To work around this problem, you can create a batch file with a pause at the end. For example, make a batch file called DWGSP.BAT that contains the following two lines and store it in a directory on the DOS path:

```
DIR *.DWG
PAUSE
```

Then add an external command definition line to ACAD.PGP:

```
DWGSP, DWGSP, 0,,4
```

With the batch file command PAUSE, DOS displays `Press any key to continue...` and waits for your key press before returning to AutoCAD.

Creating Simple AutoLISP Command Definitions

Although ACAD.PGP command aliases are useful and simple to create, they suffer from one major limitation: they allow only single commands without any command options. For example, you might be tempted to define an alias for ZOOM Previous:

```
ZP, *ZOOM Previous
```

Creating Keyboard Macros and Toolbars

If you do so, when it loads ACAD.PGP, AutoCAD will report: `the aliased command must be a single word`.

One way to get around this limitation is to create simple AutoLISP command definitions. A more thorough discussion of AutoLISP comes later, in Chapter 9, but in this section we'll show you examples of simple command definitions in AutoLISP and give just enough explanation to let you create your own. Don't worry about the details of AutoLISP for now; the code will be fully explained in Chapter 9.

We'll start by adding a command definition for ZOOM Previous, and then we'll discuss the AutoLISP syntax.

Exercise

Creating an AutoLISP Command Definition for ZOOM Previous

Open \MAXAC\DEVELOP\ACAD.LSP with your text editor.

Add a line at the top of the file similar to this one:
`(prompt "\nMaximizing AutoCAD Chapter 3 ACAD.LSP 04/01/96\n")`

The existing lines in this chapter's ACAD.LSP load MAWIN.ARX and MA_UTILS.LSP, which contain utility functions required later in the chapter.

Add the following four lines to the end of ACAD.LSP:
```
(defun C:ZP ()   [Enter]
   (command "_.ZOOM" "_Previous")   [Enter]
   (princ)   [Enter]
)   [Enter]
```

The syntax for this command definition is described at the end of this exercise. Check that you've typed all of the punctuation, including parentheses and quotation marks, correctly.

Save ACAD.LSP and return to AutoCAD

Command: `(load "ACAD.LSP")` [Enter]   Reloads ACAD.LSP

AutoLISP returns: `Maximizing AutoCAD Chapter 3 ACAD.LSP 04/01/96 C:ZP`

load is the function for loading an AutoLISP file. You'll learn more about it in Chapter 9.

Zoom in on an area of the drawing, so you can try your new ZP command.

```
Command: ZP Enter
```
Your ZP command issues ZOOM and the Previous option

```
_.ZOOM
All/Center/Dynamic/Extents/Left/
Previous/Vmax/Window/
<Scale(X/XP)>: _Previous
```
AutoCAD zooms back out

Note: If AutoCAD reports an error message when you load ACAD.LSP, then you probably mistyped something. Re-edit ACAD.LSP, check your typing, resave the file, and reload ACAD.LSP.

Let's look at each part of the command definition in ZP.

- (defun
  defun tells AutoLISP that you're defining a function. The open parenthesis matches the close parenthesis in the fourth line.

- C:ZP
  C: has nothing to do with your hard disk. It indicates that the function will be an AutoCAD-style command that you can type at the command prompt like any other AutoCAD command. ZP is the name of the newly defined command.

- ()
  The empty pair of parentheses holds arguments and local symbols in other kinds of AutoLISP function definitions. You'll learn more about arguments and symbols in Chapter 9. For simple command definitions, leave the parentheses empty.

- (command
  The AutoLISP command function feeds commands to the AutoCAD command prompt. The open parenthesis matches the close parenthesis at the end of the line.

- "_.ZOOM"
  "_.ZOOM" is the approved way of starting the ZOOM command from AutoLISP. The quotation marks are required and tell AutoLISP to treat _.ZOOM as a literal string. The underscore and period before the command name are optional in most cases but reflect good programming practice. The underscore tells foreign-language versions of AutoCAD

Creating Keyboard Macros and Toolbars

that the command name is in English. The period tells AutoCAD to use its native ZOOM command, even if you or another application happen to have redefined the ZOOM command to do something else.

- **"_Previous"**
  Now that the ZOOM command is active, `"_Previous"` chooses the Previous option. Again, the quotation marks are required and the underscore is optional. You don't use periods with command options, only with commands.

- **)**
  This close parenthesis matches the open parenthesis before `command`, and it tells AutoLISP that this command sequence is finished.

- **(princ)**
  `(princ)` is a magic function that tells AutoLISP to exit "quietly" when it executes the function. `(princ)` is optional, but if you omit it, you'll see a harmless **nil** on the command line each time you run ZP.

- **)**
  The final close parenthesis matches the open parenthesis before `defun`, and it tells AutoLISP that the function definition is complete.

This description of AutoLISP syntax is necessarily brief, but it should provide you with enough information to create additional simple command definitions. If you have any questions about the AutoLISP syntax, refer to Chapter 9. Let's add one more command definition: BA for BREAK @, which breaks a line in two at the point you pick. This example uses the AutoLISP **pause** symbol to pause for user input.

Exercise

Creating an AutoLISP Command Definition for BREAK @

First, go through the BREAK @ sequence in AutoCAD to remind yourself of the sequence.

Draw a line with the LINE command.

Command: **BREAK** [Enter]

Select object: *Pick a point on the line*

Enter second point (or F for first point): **@** [Enter]

Notice that the command sequence was to enter **BREAK**, then pick a point, then enter **@**. The trick in this macro is getting it to pause while you pick a point, then have it enter the @ automatically.

Return to ACAD.LSP in your text editor.

Add the following four lines to ACAD.LSP:
```
(defun C:BA () Enter
   (command "_.BREAK" pause "@") Enter
   (princ) Enter
) Enter
```

Add the following line at the end of ACAD.LSP:
```
(princ) Enter
```

This final `(princ)` makes ACAD.LSP load quietly, without returning a function name on the command line, as it did in the previous exercise.

Save and close ACAD.LSP and return to AutoCAD.

Command: **(load "ACAD.LSP")** Enter

AutoLISP returns:
Maximizing AutoCAD 04/01/96...

Command: **BA** Enter	Your BA command issues BREAK and then pauses for you to specify a point
_.BREAK Select object: *Pick a point on the line*	
Enter second point (or F for first point): @	Your command supplies the @
Command: *Pick one end of the line*	Only selects one part of the line; it is broken in two
Command: *Press ESC twice*	Cancels the selection and grips

You use the AutoLISP **pause** symbol with the **command** function to pause for user input. Note that pause is not a string, so you don't put quotation marks around it.

Creating Keyboard Macros and Toolbars

Note: As you saw in the exercises, your AutoLISP command definitions feed commands to the AutoCAD command prompt, and AutoCAD displays all of its normal prompts, even when AutoLISP is responding to them. In professional applications, letting all of these prompts scroll by usually is undesirable. In Chapter 13, we'll show you how to use the CMDECHO system variable to turn off the prompts while AutoLISP functions are active. When you're learning to customize AutoCAD, it's helpful to leave all of the prompts visible so that you can see what's going on.

Warning: These simple AutoLISP command definitions contain no error trapping. They will work fine for small-scale customization (for example, for yourself or a small office), but in some cases, they aren't robust enough for an application that you distribute more widely. In Chapter 9 and Part 5, you'll learn about error trapping and building more robust applications.

AutoLISP-based macros are versatile and easy to implement once you've mastered the basic format. The only challenge is getting the command sequence just right and then transcribing it to a (**command** ...) function. Always start by going through the command sequence in AutoCAD and noting the exact order of prompts and your responses. Remember to enclose command names and options in quotation marks and include a **pause** (without quotation marks) at each place where the macro should wait for the user to pick a point or answer a prompt. Refer to Chapter 9 for more information about using the (**command** function

Defining Accelerator Keys in the Menu File

Many DOS programs rely heavily on the function keys (F1 through F12) for command access. Some DOS AutoCAD users learned tricks for assigning function keys to AutoCAD commands with the ANSI.SYS driver or memory-resident keyboard macro programs. Although Windows is less reliant on function keys, savvy Window users learn function key shortcuts like ALT+F4 and F1. In addition, Windows programs are converging on a more or less standard set of keyboard shortcuts using the CTRL key.

R13 for Windows makes it easier to assign AutoCAD commands and options to various combinations of function, CTRL, SHIFT, and letter keystrokes. Autodesk calls this feature *accelerator keys*, and includes a new section in the menu file for it. Chapters 5 and 14 cover menu organization and syntax in detail, but this section will teach you enough to define your own accelerator keys.

Note: User-defined accelerator keys are available only in R13 for Windows. AutoCAD for DOS and other platforms don't support accelerator keys defined in the menu file.

Accelerator keys are defined in a short section of the standard ACAD.MNU file that begins with the line ***ACCELERATORS. The three asterisks indicate the beginning of a new section, and ACCELERATORS is the name of this particular section. The standard ACCELERATORS section looks like this:

```
***ACCELERATORS
[CONTROL+"L"]^O
[CONTROL+"R"]^V
ID_Undo          [CONTROL+"Z"]
ID_Cut           [CONTROL+"X"]
ID_Copy          [CONTROL+"C"]
ID_Paste         [CONTROL+"V"]
ID_Open          [CONTROL+"O"]
ID_Print         [CONTROL+"P"]
ID_New           [CONTROL+"N"]
ID_Save          [CONTROL+"S"]
```

These examples are not obvious at first. Each line defines one accelerator key. The part between the brackets indicates the keystroke sequence: [CONTROL+"L"] means CTRL+L (press and hold the CTRL key, and tap the L key). Thus the first two lines below ***ACCELERATORS make assignments to CTRL+L and CTRL+R, respectively.

When code follows the bracketed accelerator keystroke assignment, AutoCAD treats that code as a *menu macro*. What makes these particular menu macros confusing is that they correspond to old DOS AutoCAD control key assignments (^O = CTRL+O and ^V = CTRL+V). If you remember your old DOS AutoCAD control key assignments, you know that CTRL+O toggled Ortho mode on and off, and CTRL+V toggled to the next viewport. The problem in Windows is that CTRL+O usually means open a document, and CTRL+V usually means paste the clipboard. Autodesk chose to give in to the Windows conventions and assign new CTRL key assignments to the old Ortho and viewport toggles. Therefore, the first two accelerator key definitions assign new control keys to these AutoCAD functions. Then, when you press CTRL+L, Windows sends the CTRL+L to AutoCAD, which internally translates it to CTRL+O and toggles ortho on or off. In a moment, you'll create some less confusing accelerator key macros.

The remaining eight accelerator key definitions use an alternative syntax. They have no menu macro after the keystroke assignment, but instead they have something called a *name tag* before (for example, ID_Save). Name tags refer to menu macros elsewhere in the menu file. If you use the search function in your text editor to find ID_Save earlier in the menu file, you'll see that the ID_Save name tag before [CONTROL+"S"] refers to a menu macro in the ***POP1 menu section. The actual menu macro is simply _QSAVE (that is, do a quick save). In other words, when AutoCAD sees a name tag before a keystroke assignment in brackets, it looks for the same name tag elsewhere in the menu file and runs the menu macro it finds there. Don't worry about name tags yet—we'll cover

Creating Keyboard Macros and Toolbars

them in Chapter 13. These eight definitions assign the standard accelerators for standard Windows functions to the equivalent AutoCAD commands.

An equivalent way to write the last eight accelerator key definitions is:

```
[CONTROL+"Z"]_U
[CONTROL+"X"]'_CUTCLIP
[CONTROL+"C"]'_COPYCLIP
[CONTROL+"V"]'_PASTECLIP
[CONTROL+"O"]^c^c_OPEN
[CONTROL+"P"]^c^c_PLOT
[CONTROL+"N"]^c^c_NEW
[CONTROL+"S"]^c^c_QSAVE
```

You've already encountered the underscore in the previous section of this chapter. Remember that it tells foreign-language versions of AutoCAD to expect a command name or option in English. The single quotation mark (') tells AutoCAD to run a command transparently, without interrupting any other command that happens to be active at the time. Only some commands can execute transparently, though. The two CTRL+Cs (^c^c) before _OPEN, _PLOT, _NEW, and _QSAVE cancel any command that's active, because none of these commands can execute transparently.

Note: You might wonder why Autodesk went to the trouble of using name tags, rather than the simpler menu macro version that we listed. The standard AutoCAD menu contains several references (accelerator, pull-down, and screen menus) to each of these commands. Although it makes little difference for such simple macros, the programming philosophy is to only have one definition for a macro and to reference that definition rather than create duplicate definitions. The benefit of this approach in more complex macros is reduced maintenance—if the macro needs revision, it only has to be revised once.

Creating Accelerator Key Definitions

In the following exercise, you create some of your own accelerator key assignments.

In this exercise, as in all of the Windows menu customization exercises, you edit the ACADMA.MNS file. R13 for Windows employs a complex series of menu files (MNU, MNS, MNC, and MNR) that you'll learn about in Chapter 4. For now, all you need to know is that MNS is a file that AutoCAD creates from the MNU file and then writes to when you customize the toolbars with the Toolbars dialog box. Once AutoCAD creates the MNS file, you should edit only the MNS file. This protects the MNU file from inadvertent changes, enables you to use the unmodified MNU file to recreate the original MNS file if needed, and prevents inadvertently overwriting changes in the MNS file.

Maximizing AutoCAD® R13

Warning: There shouldn't be any files named ACADMA.MNU in your \MAXAC or \MAXAC\DEVELOP directory. If there are, delete them or move them out of the way onto a floppy disk.

Exercise

Creating Accelerator Key Definitions

Open \MAXAC\DEVELOP\ACADMA.MNS with your text editor.

Add this comment at the top of the file:

`// Maximizing AutoCAD custom menu by MM 04/01/96` [Enter]

Menu file comment lines begin with two forward slashes (//).

Use your text editor's search function to locate ***ACCELERATORS. The ***ACCELERATORS section is near the end of the file and should look like the 11 lines shown earlier in this section. Look for the [CONTROL+"S"] line.

Add the following lines after the [CONTROL+"S"] line:

```
// Maximizing AutoCAD custom accelerators [Enter]
["DELETE"]^c^c_ERASE [Enter]
["INSERT"]^c^c_DDINSERT [Enter]
[SHIFT+"F2"]_FROM [Enter]
[SHIFT+"F3"]_ENDP [Enter]
[SHIFT+"F4"]_MID [Enter]
[SHIFT+"F5"]_INT [Enter]
[SHIFT+"F6"]_APPINT [Enter]
[SHIFT+"F7"]_PER [Enter]
[SHIFT+"F8"]_NOD [Enter]
[SHIFT+"F9"]_INS [Enter]
[CONTROL+"F2"]_NEA [Enter]
[CONTROL+"F7"]_CENTER [Enter]
[CONTROL+"F8"]_QUA [Enter]
[CONTROL+"F9"]_TAN [Enter]
[CONTROL+SHIFT+"F2"]_NON [Enter]
```

Creating Keyboard Macros and Toolbars

Leave a blank line between your last accelerator key line and the ***HELPSTRINGS line.

Save ACADMA.MNS

Reloading and Testing a Menu File

Before you can test the new accelerator key definitions, you need to load (or reload) the ACADMA.MNS file.

 Exercise

Reloading a Menu File and Testing Accelerator Keys

Command: **MENU** [Enter]	The Select Menu File dialog box appears
In the **D***irectories list, navigate to* C:\MAXAC\DEVELOP *(see* ① *in Figure 3.3)*	
In the **L***ist Files of Type drop-down list, choose* *.MNS *at* ②	The file list displays only MNS files
In the File **N***ame list, double-click on* ACADMA.MNS *at* ③	AutoCAD closes the dialog box, compiles ACADMA.MNS to ACADMA.MNC and ACADMA.MNR, and loads the MNC and MNR files

```
Menu loaded successfully.
MENUGROUP: ACAD
```

If AutoCAD reports any error messages while compiling the menu file, correct the errors in ACADMA.MNS, resave it, and rerun the MENU command. Now test each of the accelerator key definitions.

Draw some lines and circles before proceeding.

Figure 3.3
Reloading
ACADMA.MNS.

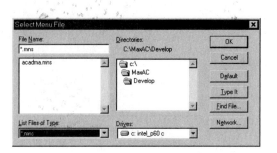

```
Command: Press the DEL (delete) key
```
Your new DELETE accelerator cancels any current command and issues the ERASE command

```
Command: *Cancel*
Command: _ERASE
Select objects: Press the CANCEL (CTRL+C
or ESC) key
```

```
Command: L Enter
```
The L alias issues the LINE command

```
LINE From point: Press SHIFT+F2
```
Your FROM accelerator issues the FROM input point modifier

```
_FROM Base point: Pick a point
```

```
<Offset>: Pick another point
```

```
To point: Finish the LINE command
```

Test the remaining accelerator key definitions that you added in the previous exercise.

Close ACADMA.MNS and exit your text editor

If you customized menus in previous AutoCAD releases, you'll notice that R13 for Windows menu compile times are much longer than you're used to. In part that's because R13 has more work to do, including creating the large MNR (menu resource) file containing the toolbar icons. In Chapters 13 and 14, you will learn to create small partial menu files that can be loaded more quickly and which share the menu space with other menus, such as the standard ACAD menu.

Creating Keyboard Macros and Toolbars

Accelerator Key Tips

As you can see, defining accelerator keys isn't complicated, especially if you ignore the name tag method and just use simple menu macros, as we did in our examples. In Chapters 5 and 14, you'll learn how to use name tags and create more complex menu macros. For now, remember these rules:

- Enclose all key names (letter, function, number pad, and arrow) *except* CONTROL and SHIFT in quotation marks. Valid key names are listed in Table 3.1.

- Put a plus sign between adjacent key names.

- Enclose the entire keystroke sequence in square brackets.

- Put a menu macro right after the close bracket, with no intervening spaces and no trailing spaces.

- End the accelerator with ^Z (the ^ and Z characters) if you want it to "type" the macro's contents at the command line but not enter it.

Tip: If you do not end a macro with a control or other special character, AutoCAD enters it as if you had pressed ENTER or the Spacebar. A ^Z suppresses this automatic entry. For example, to create an accelerator to preface any of the previous object snap accelerators with the QUICK modifier, you could define: [CONTROL+"Q"]_QUI,^Z [Enter] (no space after the ^Z). Then, pressing CRTL+Q would "type" _QUI, and wait for additional input, whereupon you could use any object snap's accelerator, such as SHIFT+F4 to enter _MID, resulting in _QUI,_MID being entered for a quick search for a midpoint.

Table 3.1	String	Description
Accelerator Key Names	"F1" through "F12"	F1 through F12 keys
	"NUMPAD0" through "NUMPAD9"	0 through 9 keys
	"LEFT"	Left arrow key
	"RIGHT"	Right arrow key
	"UP"	Up arrow key
	"DOWN"	Down arrow key
	"HOME"	Home key
	"PAGEUP"	PgUp key
	"PAGEDOWN"	PgDn key
	"END"	End key
	"INSERT"	Ins key
	"DELETE"	Del key
	"ESCAPE"	Esc key

Here are more ideas that you can use with accelerator keys:

- Assign F3, F11, and F12, which are unused by AutoCAD, to commands or options that you use frequently.

- To make AutoCAD for Windows work more like other Windows programs, assign F5 to REDRAW and CTRL+A to the selection option ALL. (But note that these keystroke combinations already have default assignments.)

Tip: Many key combinations are assigned to AutoCAD commands or options, either in the core program or through the default ACAD.MNU file. For example, CTRL+D toggles coordinate display on the status bar and CTRL+L toggles Ortho mode. Before you create an accelerator key definition, test the key combination in AutoCAD to see whether it's already assigned, and if so, whether you want to override the default assignment.

You can create accelerator definitions for single-letter keys. For example:

```
["L"]^c^c_LINE
```

defines an accelerator for the L key. Note that AutoCAD will run the LINE command as soon as you press the L key—it doesn't wait for you to press ENTER. This ENTERless approach makes for speedy keyboard macros, but it's different from the usual AutoCAD convention, in which ENTER or Spacebar signals AutoCAD to process the command. As a result, you and the other users of your customized menu probably will find single-letter accelerators confusing. You may want to reserve accelerator key definitions for those keys that traditionally respond immediately in AutoCAD or Windows (that is, function keys and CTRL and SHIFT key combinations).

Also, you can define accelerators for the number pad and arrow keys. See Table 3.1 for a list of available keys.

Warning: Some key combinations don't work in R13c4. In our testing, we weren't able to create accelerator key definitions for F10, CTRL+F4, CTRL+F6, CTRL+F10, CTRL+SHIFT+F4, CTRL+SHIFT+F6, CTRL+SHIFT+F10, and SHIFT+F10. Some of these key combinations have standard functions in Windows or Microsoft applications (although many of them don't do anything in AutoCAD).

Deciding among Aliases, AutoLISP, and Accelerators

With three different ways to create keyboard macros, you might wonder which ones to use. The strengths and weaknesses of each customization approach usually will suggest the best method.

Creating Keyboard Macros and Toolbars

- Use ACAD.PGP command aliases to create shortened names for AutoCAD commands when you don't need to specify a command option (for example, ZOOM). Command aliases are the easiest type of keyboard macro to define. They load quickly and use very little memory.

- Use simple AutoLISP command definitions to create shortened names for AutoCAD commands when you need to include command options or pauses for user input (for example, ZOOM Previous). AutoLISP command definitions are a little more complicated than ACAD.PGP command aliases, but they get around the "no command option" limitation of aliases.

- Use menu accelerator keys to assign menu macros to function, CTRL, and SHIFT key combinations. Menu accelerator keys are the one standard way in R13 for Windows to take advantage of these special keys.

If you're customizing AutoCAD for other users in your office, ideally you want everyone to agree on one set of keyboard macros so that you can maintain a single set of custom ACAD.PGP, ACAD.LSP, and menu files. After you modify these files, print out a list of macros and give it to each user (including yourself!).

AutoLISP command definitions override ACAD.PGP command alias and external command definitions. Thus, if you were to define T as an ACAD.PGP alias for MTEXT and also create an AutoLISP C:T command definition for DTEXT, AutoCAD would ignore the ACAD.PGP definition and run the DTEXT command when you typed **T**.

Menu file accelerator key definitions override both AutoLISP command definitions and ACAD.PGP definitions. Accelerator key definitions also override native AutoCAD key assignments. Thus if you create an accelerator definition for the F1 key, AutoCAD will use your definition instead of calling help.

Tip: ACAD.PGP command aliases aren't limited to AutoCAD core commands. You can define aliases for any command that you can type at the AutoCAD command prompt, including AutoLISP-defined C: commands and custom display driver commands.

Warning: Each alias, AutoLISP command definition, and accelerator key uses a small amount of memory that otherwise would be available to AutoCAD for storing drawing data and executing programs. If your system is short on memory, remove infrequently used keyboard macros from the support files.

Understanding Windows Toolbars

AutoCAD's toolbar capabilities are completely new in R13 for Windows. Although R12 for Windows included a toolbar and toolbox, they were limited and difficult to customize if you wanted to do anything more than change an existing button. R13 supports multiple customizable toolbars, toolbar docking along any side of the drawing area (see ① in Figure 3.4), flyouts (that is, cascading toolbars) ②, and ToolTips that help users decode cryptic icons ③. Even the two rows of icons at the top of the default R13 screen are defined as ordinary toolbars and can be moved around or docked elsewhere. These two toolbars are called Standard ④ and Object Properties ⑤.

In this section, we assume that you understand the basics of how to use toolbars, including flyouts and docking procedures. We show you how to show and hide toolbars and how to customize them with the Toolbars dialog box.

Note: See the "Windows Toolbars" section of Chapter 1 and Appendix A, "Customizing Toolbars," in the *AutoCAD Release 13 User's Guide* if you're not familiar with how to move, dock, and use flyouts in toolbars.

Figure 3.4
R13 for Windows toolbar features.

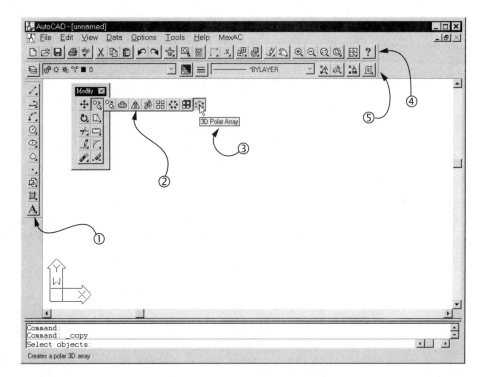

Creating Keyboard Macros and Toolbars

Showing and Hiding Toolbars

To show toolbars, you use the TOOLBAR command or the **T**ools, **T**oolbars pull-down menu (see Figure 3.5). To hide toolbars, you use the TOOLBAR command or the close button (or control menu in Windows 3.1) at the top of a floating toolbar. The only problem with using the TOOLBAR command is that you need to know the name of the toolbar. R13c4 for Windows includes definitions for about 50 toolbars (17 top-level toolbars and the rest flyouts)! Figure 3.6 shows the 17 top-level standard toolbars and Table 3.2 lists their default names.

Tip: Another way to show or hide a toolbar is with the Toolbars dialog box (right-click on any toolbar or type `TBCONFIG` and press ENTER). Select the toolbar name, choose the **P**roperties button, and check or uncheck the Hi**d**e box.

The "full name" of each toolbar includes the prefix `ACAD.TB_` (for example, `ACAD.TB_STANDARD_TOOLBAR`), but you can omit the prefix.

Figure 3.5
The Tools,
Toolbars menu.

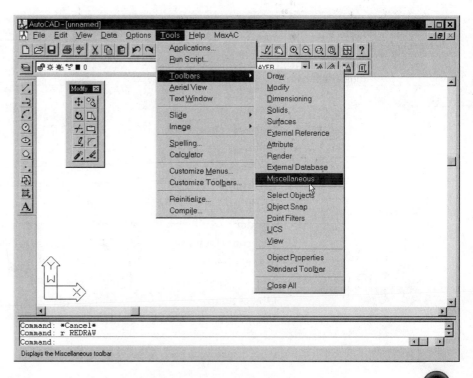

101

Figure 3.6
The 17 top level standard toolbars

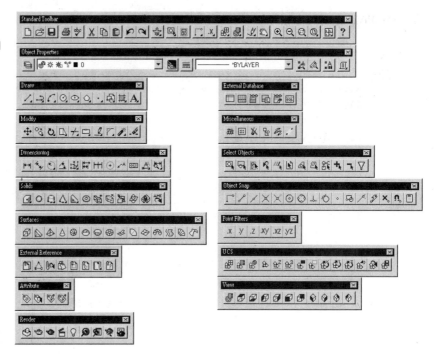

	Title Bar Name	Toolbar Name
Table 3.2 Seventeen Top-Level Standard Toolbar Names	Standard Toolbar	STANDARD_TOOLBAR
	Object Properties	OBJECT_PROPERTIES
	Draw	DRAW
	Modify	MODIFY
	Dimensioning	DIMENSIONING
	Solids	SOLIDS
	Surfaces	SURFACES
	External Reference	EXTERNAL_REFERENCE
	Attribute	ATTRIBUTE
	Render	RENDER
	External Database	EXTERNAL_DATABASE
	Miscellaneous	MISCELLANEOUS
	Select Objects	SELECT_OBJECTS
	Object Snap	OBJECT_SNAP
	Point Filters	POINT_FILTERS
	UCS	UCS
	View	VIEW

Creating Keyboard Macros and Toolbars

In the following exercise, you use several methods to show and hide toolbars.

 Exercise

Showing and Hiding Toolbars

Make sure that \MAXAC\DEVELOP\ACADMA.MNS is loaded.

`Command:` *Choose* **T**ools, **T**oolbars, **O**bject Snap	Shows the Object Snap toolbar
`Command:` **TOOLBAR** [Enter]	
`Toolbar name (or ALL):` **EXTERNAL_REFERENCE** [Enter]	
`Show/Hide/Left/Right/Top/Bottom/Float: <Show>:` [Enter]	Shows the External Reference toolbar
`Command:` **TOOLBAR** [Enter]	
`Toolbar name (or ALL):` **DRAW** [Enter]	
`Show/Hide/Left/Right/Top/Bottom/Float: <Show>:` **L** [Enter]	Specifies docking at the left side of the screen
`Position <0,0>:` [Enter]	Shows and docks the Draw toolbar

The **T**ools, **T**oolbars menu choices do the same thing as TOOLBAR Show. If the Modify toolbar isn't yet visible, use **T**ools, **T**oolbars, **M**odify to show it. Then dock it, as follows:

`Command:` *Click on and drag the Modify toolbar and dock it next to the Draw toolbar (see ① in Figure 3.7)*	Docks the toolbar—if it doesn't go where you want it, try again; sometimes positioning toolbars is tricky
`Command:` *Choose* **T**ools, **T**oolbars, **C**lose All	Hides all of the toolbars
`Command:` *Choose* **T**ools, **T**oolbars, **D**raw	Redisplays the Draw toolbar
`Command:` *Choose* **T**ools, **T**oolbars, **M**odify	Redisplays the Modify toolbar
`Command:` *Choose* **T**ools, **T**oolbars, Object **P**roperties	Redisplays the Object Properties toolbar
`Command:` *Choose* **T**ools, **T**oolbars, Standard Toolbar	Redisplays the Standard toolbar

Maximizing AutoCAD® R13

Figure 3.7
The four default toolbars, with Draw and Modify docked.

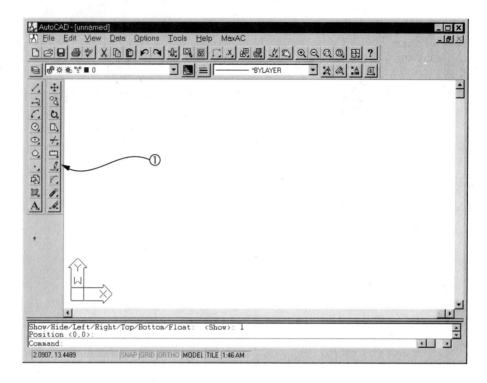

If you show a floating toolbar and later change to a lower resolution, you may find that the toolbar has moved off the screen. AutoCAD still thinks the toolbar is open, but you can't get to it. To move it back on screen, use the TOOLBAR command to dock the toolbar at the left side of the screen. From there, you can move it where you want it.

If you want to see all 50 of the toolbars, enter **TOOLBAR ALL SHOW** at the command prompt (see Figure 3.8). Then enter **TOOLBAR ALL HIDE**, or repeatedly use **U** (undo) until they go away.

Warning: Toolbars consume a large chunk of Windows resources (a small pool of memory allocated for menus, toolbars, and other graphical program elements). The resources consumption usually isn't a problem in Windows 95 or NT, but it can cause out-of-memory errors and other strange behavior in Windows 3.1. For example, you may notice erratic truncation or graying out of pull-down menus, and other display anomalies. You can remove memory-resident programs from CONFIG.SYS and AUTOEXEC.BAT and close other Windows applications to maximize Windows resources, but the fundamental problem is that R13 doesn't fit well with Windows 3.1.

Creating Keyboard Macros and Toolbars

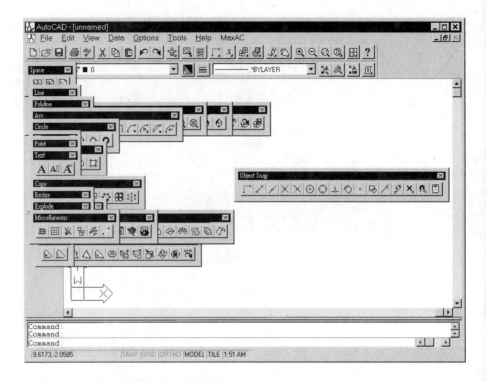

Figure 3.8
A flock of toolbars.

As you saw in the previous exercise, AutoCAD remembers changes you make to toolbar positions, even after you hide the toolbars. AutoCAD stores this information in the [toolbars] section of the ACAD.INI file.

Creating Your Own Toolbars

There are two ways to create your own R13 Windows toolbars: with the Toolbars dialog box (see Figure 3.9—TBCONFIG command) or by editing the menu (MNS) file. When you use the Toolbars dialog box, AutoCAD automatically makes any changes to the MNS file and recompiles it for you. Thus, the Toolbars dialog box method is easier, but the menu file method is more direct and gives you greater control. In this chapter, we cover the dialog box method. The menu file editing method is covered in Chapter 13.

Creating Toolbars with Existing Icons

Toolbar customization is easiest when you use the icons provided by Autodesk. The following exercise shows you how to create a new toolbar for your favorite commands.

Figure 3.9
The Toolbars dialog box.

Exercise

Creating a New Toolbar with Existing Icons

Start with the toolbar arrangement shown in Figure 3.7.

Command: *Right-click on any icon in any of the four visible toolbars*	Displays the Toolbars dialog box with the toolbar you clicked on selected

Two other ways to open the Toolbars dialog box are the TBCONFIG command and the Tools, Customize Toolbars pull-down menu choice.

Choose **N**ew	Displays the New Toolbar subdialog box

Type **Favorites** *in the* **T**oolbar Name *field (① in Figure 3.10) and leave* **M**enu Group *set to* ACAD ②

Choose OK	AutoCAD adds ACAD.Favorites to the Toolbars list and displays a new, blank toolbar titled Favorites
Choose **P**roperties	Displays the Toolbar Properties subdialog box

Figure 3.10
The New Toolbar subdialog box.

Creating Keyboard Macros and Toolbars

Figure 3.11
The Toolbar Properties subdialog box.

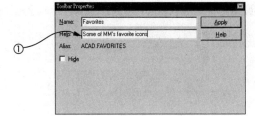

Type **Some of MM's favorite icons** *in the Help field (① in Figure 3.11)*	Defines the string that appears on the status bar when you point at the toolbar (you can use your initials instead of MM)
Choose **A**pply	Applies the changes

Notice the Hi**d**e check box, with which you can hide or show the toolbar. Leave the box unchecked so that the toolbar remains visible.

Close the Toolbar Properties dialog box with its close button or control menu (at the very top of the dialog)

Autodesk didn't include a Close button in the dialog, so you must close it with the Windows controls at the top of the dialog.

Choose **C**ustomize	Displays the Customize Toolbars subdialog box (see Figure 3.12)

There are 13 groups of icon categories in the Customize Toolbars subdialog box. Use the C**a**tegories drop-down list to select the different categories, and then view the available icons.

Click on any icon and drag it onto the new Favorites toolbar	AutoCAD adds the icon to the toolbar
Point at the new icon and pause for a moment	The ToolTip displays, indicating what the icon is for

Figure 3.12
The Customize Toolbars subdialog box.

107

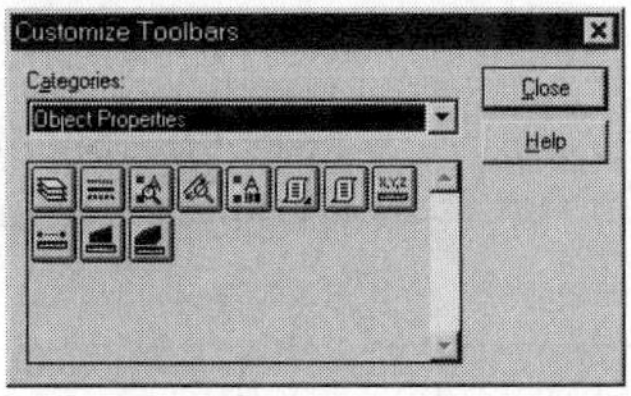

Figure 3.13
Sample Favorites toolbar .

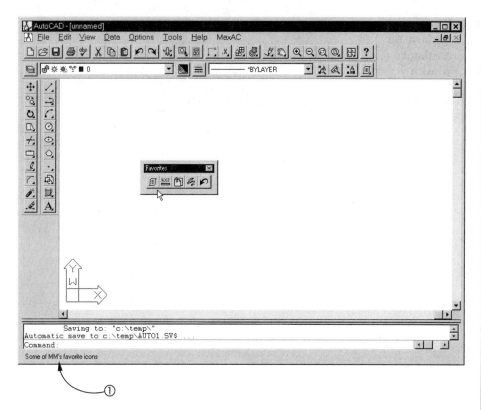

If you want to remove the icon from Favorites, drag it off the toolbar and release. To rearrange icons on the toolbar, simply drag them to a new position.

Add several other icons from various categories	Creates a toolbar similar to that shown in Figure 3.13

You'll also notice that you see a help string at the bottom of the screen, as shown at ① in Figure 3.13.

When you're finished adding icons, choose **C***lose*	Closes the Customize Toolbars subdialog box
Choose **C***lose in the main Toolbars dialog box*	AutoCAD modifies ACADMA.MNS and recompiles ACADMA.MNC and ACADMA.MNR
`Command:` *Test each of the icons in the new Favorites toolbar*	AutoCAD runs the command corresponding to the icon

Creating Keyboard Macros and Toolbars

This exercise demonstrates how you can create new toolbars containing arrangements of the standard AutoCAD icons and commands. In addition, you can modify the default toolbars that come with R13 by adding, removing, and moving icons in the same way that you did with your new Favorites toolbar.

Warning: The Toolbars customization dialog box writes changes to the MNS menu file. If you have an older version of the MNU file and later try to load it with the MENU command, AutoCAD will create a new MNS file from the old MNU file, thus overwriting your toolbar changes. That's why you always should edit and reload the MNS file, and hide the MNU file, in R13 Windows.

Creating Toolbars with New Icons

Creating a new toolbar with R13's stock of predefined icons is a good way to group your favorite AutoCAD commands onto one toolbar, but eventually you'll want to create custom icons and macros. You can do so from the Toolbars dialog box, although the process doesn't seem intuitive at first. In the next exercise, you begin creating a customization toolbar that you can use to speed up the process of editing and reloading support files like ACADMA.MNS, ACAD.LSP, and ACAD.PGP.

Creating a New Toolbar with New Icons

Command: *Right-click on the new Favorites toolbar*   Displays the Toolbars dialog box with ACAD.Favorites selected

If Large Buttons are turned on, turn them off, choose Close, and reopen the Toolbars dialog box so that you're working with small buttons.

Choose New *and type* **MaxAC** *in the* Toolbar Name *field*

Choose OK   AutoCAD adds ACAD.MaxAC to the Toolbars list and displays a new, blank toolbar titled MaxAC

Choose Properties *and type* **Maximizing AutoCAD customization icons** *in the* Help *field*

Choose Apply

109

Figure 3.14
Custom template icons.

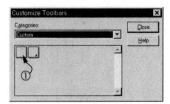

Close the Toolbar Properties subdialog box with its close button or control menu

Choose Customize

Choose Custom at the end of the Categories drop-down list	Displays blank templates for normal and flyout icons (see Figure 3.14)
Click on the blank (first) icon (① in Figure 3.14) and drag it onto the new MaxAC toolbar	Places a blank icon in the toolbar

Leave the Customize Toolbars subdialog box open.

Right-click on the new blank icon in MaxAC	Displays the Button Properties subdialog box (see Figure 3.15)
Type **Load Menu** *in the Name field (① in Figure 3.15)*	Defines the ToolTip
Type **Opens or reloads a menu file with the MENU command** *in the Help field ②*	Defines the status bar help string
Type **_Menu** *after ^C^C in the Macro field ③*	Defines what the button does

Figure 3.15
The Button Properties subdialog box.

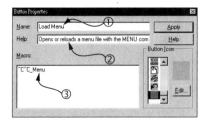

Creating Keyboard Macros and Toolbars

Figure 3.16
The Button Editor subdialog box.

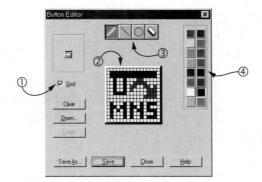

The **N**ame field defines the ToolTip, and the **H**elp field defines the status bar help string, both of which appear when you point to an icon and pause for a moment. The **M**acro field defines what the button does. The syntax for toolbar macros is the same as for accelerator key macros. Check that your entries are the same as in Figure 3.15.

In the next exercise, you will create a button icon using the Button Editor subdialog box (see Figure 3.16). The icon is represented by a 16x16 grid in the center of the Button Editor dialog box ②. Each small square represents one pixel in the icon. You create the icon by picking a tool at the top of the dialog box (pencil, line, circle, or erase) ③, picking a color at the right side of the dialog box ④, and then drawing on the button. Use the **U**ndo button to undo mistakes and the **C**lear button to start over.

Your icons don't have to be pretty in this exercise, as long as you can recognize them later. The preview box at the upper left of the Button Editor subdialog box shows how it will appear. Our version is shown in Figure 3.16. If you have trouble drawing pictures, just draw letters with the pencil and line tools. Picking a single grid square colors a single pixel. The default color is black; click on another color to select it. It takes a good deal of practice to draw 16x16 icons that are recognizable and look good.

 Exercise

Creating Button Icons

Continue from the previous exercise.

Choose **E**dit *from the Button Properties subdialog box*	Displays the Button Editor subdialog box
Turn on **G**rid *(① in Figure 3.16)*	Displays a pixel grid

Maximizing AutoCAD® R13

The first icon will be for the reloading of a menu with the MENU command.

Draw an icon for the MENU command

Choose **S**ave	AutoCAD saves the icon to a BMP file called ICONxxxx.BMP, where xxxx is an arbitrary number
Choose **C**lose	
Choose **A**pply	Applies the button changes
Close the Button Properties subdialog box with its close button or control menu	
Choose **C**lose *in the Customize Toolbars subdialog box*	
Choose **C**lose *in the main Toolbars dialog box*	
Test the new button	Runs the MENU command
Cancel the Select Menu File dialog box	

Tip: The Save **A**s button on the Button Editor subdialog box is for when you want to save to a name that you can recognize (for example, LOADMENU.BMP), so that you can open that icon later. Even if you use Save **A**s, AutoCAD still creates the ICONxxxx.BMP file and uses that in its menu. The Save **A**s file is for you to do what you want with; AutoCAD ignores it.

Drawing buttons is time-consuming. In the next exercise, you use existing bitmap (BMP) files as the basis for the remaining five buttons on the MaxAC toolbar.

Exercise

Adding More New Icons to Your Toolbar

Copy the 12 BMP files (MXLSPE16.BMP, MXLSPE32.BMP, and so on) from \MAXAC\BOOK\CH03 to your \MAXAC\DEVELOP directory.

Command: *Right-click on the new MaxAC toolbar*

Creating Keyboard Macros and Toolbars

Choose **C***ustomize*

Choose C*ustom at the end of the* C*ategories drop-down list*

Drag a blank icon onto the MaxAC toolbar

Leave the Customize Toolbars subdialog box open.

Right-click on the new icon in MaxAC

Type **Edit Current Menu** *in the* N*ame field*

Type **Edits the current menu file with TXTPAD32.EXE** *in the* He*l*p *field*

This help string assumes that you have configured TextPad as your editor. If not, substitute your editor's nmae for **TXTPAD32.EXE**. See Appendix G for information about TextPad.

Next, you will define the macro to start your text editor. Type the entire macro as one long line, without any returns (AutoCAD will wrap it around so that you can see the whole thing, but don't press ENTER). Type the spaces and punctuation carefully.

Delete the ^C^C *in the* **M***acro field and type:*
(MxaOpenFile (strcat (getvar "MENUNAME") ".MNS"))

Choose E*dit*

Choose O*pen*	Opens a file dialog box for selecting a BMP file
Choose MXMNSE16.BMP	
Choose O*pen in the file dialog box*	Imports the image from MXMNSE16.BMP
Choose S*ave*	Saves the button to ICONxxxx.BMP

Choose C*lose and then* A*pply*

Close the Button Properties dialog box

Optional: Now repeat the process of adding an icon to the MaxAC toolbar four more times: drag a blank button onto the toolbar, right-click on the new icon, edit the icon's properties in the Button Properties subdialog box, and import a button image in the Button Editor subdialog box. Table 3.3 lists all six of the icons, including the ones you've already done. See also the notes following the table. The finished toolbar is shown in Figure 3.17.

Figure 3.17
The completed MaxAC toolbar.

Choose **C**lose *in the Customize Toolbars subdialog box*

Choose **C**lose *in the main Toolbars dialog box*

Test the new buttons

Table 3.3
Button Properties for the MaxAC Toolbar

Name [BMP File]	Help	Macro
Edit Current Menu MXMNSE16	Edits the current menu file with TXTPAD32.EXE	(MxaOpenFile (strcat (getvar "MENUNAME") ".MNS"))
Load Menu MXMNSL16	Opens or reloads a menu file with the MENU command	^c^c_MENU
Edit ACAD.LSP MXLSPE16	Edits ACAD.LSP with TXTPAD32.EXE	(MxaOpenFile (findfile "ACAD.LSP"))
Load ACAD.LSP MXLSPL16	Reloads ACAD.LSP startup file	^c^c(load "ACAD.LSP")
Edit ACAD.PGP MXPGPE16	Edits ACAD.PGP with TXTPAD32.EXE	(MxaOpenFile (findfile "ACAD.PGP"))
Re-init ACAD.PGP MXPGPL16	Reinitializes ACAD.PGP with the RE-INIT system variable	^c^c_RE-INIT 16

1. Create a new button for each entry in Table 3.3. Enter the **N**ame, He**l**p, and **M**acro values into the fields in the Button Properties subdialog, as shown in Figure 3.15.

2. Type each value as one long string, without any returns. Type the spaces and punctuation in the **M**acro values carefully. Case isn't important.

3. The macros for editing files use the **MxaOpenFile** function that's defined in MA_UTILS.LSP, and which requires MAWIN.ARX. These two utility program files are loaded automatically by this chapter's ACAD.LSP file. **MxaOpenFile** opens a file using the application registered for that file's type, as if you had double-clicked on the file's name in Explorer or File Manager. Chapter 14 describes the functions in MAWIN.ARX. An alternative to **MxaOpenFile** is (**startapp** "*editor_path/editor_name*").

4. You'll learn more about the **startapp**, **strcat**, **getvar**, and **findfile** functions in Chapter 9. The macros in this table use these functions to ensure that your icons load the correct files into your editor.

Creating Keyboard Macros and Toolbars

If you don't want to add the four more buttons, you can skip them. The remaining exercises in the book don't depend on your using the MaxAC toolbar. Also, the menu files supplied on the CD-ROM for subsequent chapters include the finished MaxAC toolbar.

These macros use AutoLISP. You'll learn more about AutoLISP in Chapter 9.

As you saw during the exercise, some of the Toolbar subdialog boxes don't include Close buttons. To close these, use the close button or control box on the subdialog box's top bar.

Tip: In this exercise, we started with a blank template button. If you want to create a new button that looks similar to a standard AutoCAD button, start with one of the predefined buttons available in the Customize Toolbars box. When you modify and save the button, AutoCAD writes it to a new BMP file, so you don't have to worry about messing up the predefined buttons.

Note: When you create custom button bitmaps as you did in this exercise, AutoCAD needs the BMP file every time it compiles the menu. Thus you must make sure that the BMP files remain available to AutoCAD. You can do this by keeping them in the same directory with the menu file. If AutoCAD can't find a bitmap, it displays a yellow smiley face with sunglasses.

When you created a new toolbar and new icons in the previous exercise, you only created the 16-pixel version of the button. If you turn on Large Buttons in the Toolbars dialog box, AutoCAD will display blank buttons, because you haven't modified the 32-pixel buttons yet. If you prefer the large buttons, turn on Large Buttons, close and reopen the Toolbars dialog box, and then use the Button Editor to draw the 32-pixel buttons. (Or import the images from the CD-ROM BMP files whose names end with 32.) As Figure 3.18 shows, the Button Editor subdialog box looks a bit strange with 32-pixel buttons, but it does work. The menu files supplied on the CD-ROM for subsequent chapters include both 16-pixel and 32-pixel buttons.

Figure 3.18
The Button Editor subdialog box with 32-pixel buttons.

The Toolbars dialog box works fine for creating a few toolbars and icons, but it becomes tedious for more ambitious work. If your customization includes creating complete toolbar menu systems, you'll want to edit the menu file directly and use a full-featured icon editor. We cover these more advanced approaches, as well as creating flyout toolbars, in Chapter 13.

Finishing Up

In this chapter, you customized ACAD.PGP, ACAD.LSP, and ACADMA.MNS in your \MAXAC\DEVELOP directory. In addition, AutoCAD's Toolbar dialog box further modified ACADMA.MNS, created ICONxxx.BMP files, and compiled ACADMA.MNS into ACADMA.MNC and ACADMA.MNR. If you want to keep your work, exit AutoCAD and move these files from \MAXAC\DEVELOP to \MAXAC. If not, you can delete them and use the versions from the CD-ROM for subsequent chapters. In either case, use the MENU command to change the menu to C:\MAXAC\ACADMA.MNS before exiting AutoCAD.

Warning: The \MAXAC\BOOK\CHxx menu files for subsequent chapters won't contain the Favorites toolbar that you created in this chapter, however you now know how to add your own favorites whenever you want to do so.

Conclusion

Keyboard macros and toolbars are the "high road" to AutoCAD R13 customization. Without a lot of effort, you can use command aliases, simple AutoLISP command definitions, accelerator keys, and toolbars to streamline standard AutoCAD command input and begin creating your own custom macros. Even if you go no further with customization, the kinds of modifications covered in this chapter should make you more efficient in AutoCAD.

Chapter 4 begins the second part of the book, where you travel the customization "low road." You'll learn to write more sophisticated menu macros, discover other AutoCAD customization interfaces, and be introduced to two of AutoCAD's programming languages, AutoLISP and DIESEL. Part 2 begins with a detailed description of AutoCAD's support files and how you can customize them.

chapter 4

Support Files

In Part One of this book, you learned to optimize AutoCAD, establish an efficient customization environment, and customize AutoCAD with keyboard macros and toolbars. Part Two explores the essentials of AutoCAD customization in depth. You'll advance from the simple menu macros in Chapter 3 to powerful macros that take advantage of all the features of AutoCAD's menu macro language. You'll take this giant leap by developing custom pull-down menus that are versatile and well-suited to customization work. Then you'll learn to use the customization features that are independent of menus, including text, linetypes, hatch patterns, slides, scripts, PCP, and help files.

This chapter sets you off on the right foot with a guided tour of the AutoCAD support files and initialization sequence. Before you create and modify additional AutoCAD support files, you should understand the big picture. What kind of support files does AutoCAD use, what are they called, where are they located, and how does AutoCAD load them? What exactly happens behind the scenes when you launch AutoCAD or open a drawing? These details will help you control AutoCAD better and understand the consequences of your customization changes.

As you would expect, R13 introduces some new support files and initialization procedures. There are several new kinds of menu files in R13 for Windows. R13's new C and C++ language programming interface, ARX (Autodesk Runtime eXtension), necessitates some new files. The text and font changes, including support for TrueType fonts and new font mapping features, add other files. The AutoCAD initialization sequence hasn't changed dramatically since R12, but there are a few modifications that you'll need to be aware of if you're upgrading.

AutoCAD LTs support files and support file structure are a subset of AutoCAD's. You can use this chapter to get an overview of LT support files, but some of the specifics (especially about AutoLISP, ADS, and ARX programs) won't apply. Also, the locations and names of some support files are different and all LT support files are in the main LT program directory.

About This Chapter

The goals of this chapter are to explain the following:

- AutoCAD's standard support file structure and operation
- The AutoCAD initialization sequence
- Loading, listing, and unloading applications
- The support path and library search path
- Customization rules for operating system portability

The resources used in this chapter are:

- TextPad or another text editor

This chapter doesn't require any custom support files from the MaxAC CD-ROM, although you might want to use TextPad or another editor to browse some of the support files as we discuss them.

AutoCAD's Stock Support Files

You might think that Autodesk provides AutoCAD's array of customization options simply for the benefit of application developers and users, but Autodesk depends just as heavily on AutoCAD's customization interfaces as the rest of us do. AutoCAD has long since become too large to fit into one EXE file written by one group of programmers. As often as not, new features come in the form of AutoLISP, ADS, or ARX applications, menu macros, and other components that are external to the core of AutoCAD.

As a result, R13 comes with a large number of support files that you should know about, know how and when to modify, and know when to leave alone. You can think of these as AutoCAD's "in the box" customization. In this section, we'll review the most important support files. We'll highlight what's in each of these files, but if you're at your computer as you read through this chapter, you can also open the files in a text editor or viewer and peruse them yourself. This kind of exploration is a great way to learn how AutoCAD works and to learn about customization from examples.

Warning: If you choose to browse some of the support files in the R13 directory tree, do *not* modify any of them at this point. You might make undesirable changes to the way your standard AutoCAD configuration works. If you can't resist changing something, copy the file to \MAXAC\DEVELOP and edit it there.

Support Files

Release 13 Support Directories

Here is a listing of the most important R13 directories from a customization point of view. The ADS and ARX programming support directories aren't included because ADS and ARX programming are beyond the scope of this book. See Chapter 2 for a complete listing of R13 directories.

```
C:\
   R13\                        AutoCAD R13
      COM\                     Common directory for Windows and DOS
         FONTS\                AutoCAD, PostScript, and TrueType fonts
         SAMPLE\               Sample drawings and support files
         SUPPORT\              Common support files
      DOS\                     DOS program files
         DRV\                  DOS drivers
         SAMPLE\               DOS sample utilities
         SUPPORT\              DOS support files
      WIN\
         DRV\                  Windows drivers
         SAMPLE\               Windows sample application
         SUPPORT\              Windows support files
```

The COM branch includes fonts and support files that are common to both Windows and DOS versions of R13, and the \WIN\SUPPORT and \DOS\SUPPORT directories contain platform-specific support files. The SAMPLE directories in each branch include sample drawings and applications, some of which can be helpful as customization examples.

Release 13 Support Files

Table 4.1 lists the files and file types we discuss in this chapter, arranged alphabetically by file extension. It is based on a similar table in Chapter 1 of the *AutoCAD Customization Guide*.

Note: Not all of these files exist in the \R13 directory structure after you first install AutoCAD. For example, there are no standard ACAD.LSP, ACAD.RX, PCP, or SCR files. Some of the files you create later, and some you might never use.

Configuration Files: ACADNT.CFG, ACAD.CFG, ACAD.INI

Strictly speaking, the configuration files aren't support files, but you should keep track of them in the same way you keep track of support files. R13 uses a pair of configuration files. One of these files is called ACADNT.CFG in Windows 95 and NT, or ACAD.CFG in Windows 3.1 and DOS. We refer to this file as ACADNT.CFG in this book. ACADNT.CFG contains hardware configuration data and changes to default global system variable settings. The other configuration file, ACAD.INI, contains initialization information, such as AutoCAD directories, font mapping for the MTEXT editor, and toolbar

Table 4.1

R13 Support Files

File	Default Location	Description / Purpose
ACAD.ADS	\R13\COM\SUPPORT	List of ADS programs to load
ACAD.AHP	\R13\COM\SUPPORT	Platform-independent help file
progname.ARX	\R13\WIN or \R13\DOS	ARX program
ACAD.CFG	\R13\WIN or \R13\DOS	Windows 3.1 or DOS configuration file
ACADNT.CFG	\R13\WIN	Windows 95 or NT configuration file
dlgname.DCL	\R13\COM\SUPPORT	Dialog box definitions
ACADBTN.DLL	\R13\WIN	Bitmaps for toolbar buttons
progname.EXE	\R13\WIN	Windows ADS program
progname.EXP	\R13\DOS	DOS ADS program
fmapname.FMP	none	Font mapping for all AutoCAD fonts
ACAD.HDX	\R13\COM\SUPPORT	Help file index
ACAD.HLP	\R13\WIN\SUPPORT	Windows help file
ACAD.INI	\R13\WIN or \R13\DOS	Initialization file
ACAD.LIN	\R13\COM\SUPPORT	Linetype definitions
ACAD.LOG	\R13\WIN\	Log of text window
ACAD.LSP	none	User's AutoLISP to load
progname.LSP	\R13\COM\SUPPORT	AutoLISP program
ACADR13.LSP	\R13\COM\SUPPORT	AutoLISP functions for R13
ACAD.MLN	\R13\COM\SUPPORT	Multiline style definitions
ACAD.MNC	\R13\WIN\SUPPORT	Windows compiled menu
ACAD.MNL	\R13\COM\SUPPORT	AutoLISP functions for menu support

Support Files

Table 4.1 *(continued)*

R13 Support Files

File	Default Location	Description / Purpose
ACAD.MNR	\R13\WIN\SUPPORT	Windows compiled menu resources
ACAD.MNS	\R13\WIN\SUPPORT	Windows menu source
ACAD.MNU	\R13\WIN\SUPPORT	Windows menu template or "pre-source"
	\R13\DOS\SUPPORT	DOS menu source
ACAD.MNX	\R13\DOS\SUPPORT	DOS compiled menu
ACAD.PAT	\R13\COM\SUPPORT	Hatch pattern definitions
plotcfg.PCP	none	Plot configuration parameters
fontname.PFB	\R13\COM\FONTS	PostScript font
fontname.PFM	\R13\COM\FONTS	PostScript font metric
ACAD.PGP	\R13\COM\SUPPORT	Command aliases and external program macros
FONTMAP.PS	\R13\COM\FONTS	PostScript font mapping for PSIN
ACAD.PSF	\R13\COM\SUPPORT	PostScript font mapping, fills, and so on for PSOUT and PSFILL
ACAD.RX	none	List of ARX programs to load
scrname.SCR	none	AutoCAD command script
fontname.SHP	\ACADCOM\FONTS\	AutoCAD font or shape source SOURCE on CD-ROM
fontname.SHX	\R13\COM\FONTS	Compiled AutoCAD or PostScript font or shape
fontname.TTF	\R13\COM\FONTS	TrueType font
ACAD.UNT	\R13\COM\SUPPORT	Unit definition and conversion

locations. Chapters 1 and 2 include additional information about ACADNT.CFG and ACAD.INI.

Although R13 for Windows normally keeps its configuration information in the ACADNT.CFG and ACAD.INI files, there are alternative ways of controlling some of the settings contained in ACAD.INI. You can establish initial configuration values by setting DOS environment variables (for example, SET ACAD= and SET ACADCFG=). In addition, you can override ACAD.INI settings by specifying command-line switches in the AutoCAD program shortcut's properties (for example, /S and /C). For example, in Chapter 2, you used the /C switch to create a separate configuration for the book's exercises.

The AutoCAD documentation isn't very clear about which settings take precedence. These are the rules in R13c4:

- Command-line switches in the AutoCAD program shortcut's properties (for example, /S) override everything, but changes you make in the Preferences dialog still get written to ACAD.INI (and thus might not take effect if they duplicate a command-line switch setting).

- ACAD.INI settings are overridden by command-line switches but override any environment variable settings. If you start AutoCAD with a new configuration directory (that is, ACAD.INI doesn't yet exist), AutoCAD uses any environment variable settings as the defaults in Preferences and stores those settings in a new ACAD.INI file.

We recommend that you use the ACAD.INI file method to store all settings (other than the configuration directory). By using ACAD.INI, you keep the settings in one place and ensure that changes you make in Preferences take effect and are stored properly. The AutoCAD configuration directory setting isn't stored in ACAD.INI. The default configuration directory is \R13\WIN. You can leave it there for your standard AutoCAD configuration and use the /C switch to redirect it to other locations for alternate configurations, as you did in Chapter 2 to direct it to \MAXAC.

Warning: The R13 manuals mention an ACADENV.INI file. This file is supposed to be for storing settings that override settings of the same name in ACAD.INI. Despite what Autodesk's documentation says, ACADENV.INI doesn't have any effect. R13c4 simply ignores it.

ACAD.PGP

As you saw in Chapter 3, ACAD.PGP contains two kinds of keyboard macros: command aliases and external program definitions. Adding command aliases to ACAD.PGP is the easiest way to create abbreviations for AutoCAD commands.

ACADR13.LSP and ACAD.LSP

The ACADR13.LSP file contains AutoLISP utility functions that R13 requires and that your own applications can take advantage of. For example, the `autoload, autoxload, and autorxload` functions provide a convenient way to load some custom AutoLISP, ADS, and ARX programs. Chapter 14 discusses these functions and shows you how to use them. In most circumstances, you should avoid editing ACADR13.LSP. It is for AutoCAD's own library of AutoLISP functions, and if you change something, you might break AutoCAD.

ACAD.LSP is the place for AutoLISP functions that the customizer or user wants to have available in all AutoCAD sessions. It is like a user's version of ACADR13.LSP. As you saw in Chapter 3, ACAD.LSP is a convenient place to put simple AutoLISP command macros. Many chapters in this book demonstrate ways of modifying ACAD.LSP, but Chapter 9 discusses AutoLISP in detail.

Professional application developers should note that ACAD.LSP is primarily the *user's* customization file. If you're developing a professional application menu, you should keep ACAD.LSP changes to a minimum. Chapter 14 contains additional guidelines.

Menu Files: ACAD.MNC, MNR, MNS, MNU, MNX, MNL

Menu files were simpler before R13. There were only two files to worry about: ACAD.MNU and ACAD.MNX. ACAD.MNU was the ASCII source file that you edited in order to customize the menu. ACAD.MNX was the compiled form that AutoCAD created automatically whenever it detected that ACAD.MNU had been changed.

These conventions still apply in R13 DOS, but the picture has gotten more complicated in R13 Windows. Autodesk, in order to implement the new Windows toolbar features and avoid conflicts between Windows and DOS menus, created three new Windows menu file types: MNS, MNC, and MNR. Figure 4.1 shows the relationship among the files. The following discussion uses the default ACAD menu files (ACAD.MNU, ACAD.MNS, and

Figure 4.1
R13 Windows menu files.

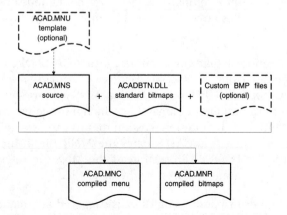

so on), but it applies equally to any set of menu files (ACADMA.MNU, ACADMA.MNS, and so on).

As in previous versions of AutoCAD, R13 for DOS compiles the ACAD.MNU (located in \R13\DOS\SUPPORT by default) into a binary ACAD.MNX file, and then uses the compiled version. R13 for Windows still uses the ACAD.MNU file if it finds one (the default Windows version is located in \R13\WIN\SUPPORT), but when AutoCAD compiles a menu for Windows, it creates a new source file called ACAD.MNS from ACAD.MNU. The ACAD.MNS file is in ASCII format and is quite similar to ACAD.MNU, but with most of the comments stripped out and a few other formatting changes in some cases. The important thing to realize is that the Toolbars dialog that you used in Chapter 3 to customize toolbars automatically modifies ACAD.MNS, not ACAD.MNU. If you make changes with the Toolbars dialog, ACAD.MNU becomes obsolete. If you later edit the now obsolete ACAD.MNU file, you'll have menu changes in two different files. Worse, if you use the MENU command to recompile the menu, AutoCAD will create a new ACAD.MNS file from your ACAD.MNU file, thus overwriting your toolbar customization (AutoCAD does display a warning dialog before overwriting an existing MNS file).

You can avoid the potential for disaster and most of the confusion by treating the MNS file as the only Windows menu source. If AutoCAD doesn't find the MNU file, it always uses the MNS file without complaining. You can think of the Windows ACAD.MNU file that comes with R13 as a "pre-source" menu file whose only purpose is to create the ACAD.MNS file and then be hidden away forever. Once you have an MNS file, move the corresponding MNU file where AutoCAD can't find it (for example, into a backup directory or onto a floppy). From that point forward, always edit the MNS file. This approach, in addition to being safer, avoids confusion between Windows and DOS menu source files. In other words, use ACAD.MNS as the Windows source file and ACAD.MNU as the DOS source file.

The compiled Windows menu occupies two files: ACAD.MNC and ACAD.MNR. ACAD.MNC is the analog of ACAD.MNX in DOS AutoCAD. ACAD.MNC contains a compiled binary version of the source file, and it defines menu organization, labels, and macros. The compiled toolbar bitmap images are contained in a separate file, ACAD.MNR (the R is for Resource, because bitmaps are one type of resource in Windows).

The bitmap image definitions that get compiled into ACAD.MNR don't come from ACAD.MNS; they come from a special Windows bitmap resource file called ACADBTN.DLL. A custom menu, like our ACADMA.MNS, can use bitmap images from ACADBTN.DLL or from its own menu bitmap file, which must be named *menuname*.DLL (for example, ACADMA.DLL). In addition, custom menus can use individual BMP files, as we did in Chapter 3. When AutoCAD creates the MNR file, it includes only those bitmap images that the ***TOOLBARS section of the MNS file references. AutoCAD doesnt include bitmaps that arent required by the MNS file.

Menu compilation, whether in Windows or DOS, is triggered automatically by a change in the MNS in Windows or MNU file in DOS. Compilation is also triggered by a missing MNC or MNX file. When you launch AutoCAD, the program checks whether the date of the MNS file (in Windows) or MNU file (in DOS) is newer than the MNC or MNX date. If the source file is newer, AutoCAD assumes that the compiled files are out of date and recompiles them from the source. Alternatively, you can force a menu recompilation with the MENU command. When you run the MENU command and select an MNC or MNS file (Windows) or MNX or MNU file (DOS), AutoCAD performs the same date comparison as when it launches. If the source file is newer, AutoCAD recompiles the menu. When you select an MNU file in Windows, AutoCAD warns you that compiling will cause any MNS file of the same name to be overwritten, and then creates new MNS, MNC, and MNR files, no matter what the file dates are.

Note: In previous versions, AutoCAD compared the dates of the source and compiled files every time you opened or created a drawing. In R13, as long as the Preference dialogs Use Menu in Header setting is turned off, AutoCAD only compares the dates when you launch the program. We describe the Use Menu in Header setting later in this chapter.

You might wonder why Autodesk bothered to include MNU files in the R13 Windows menu scheme, given the complications that they add. The original idea appears to have been portability. In theory, you can create a base MNU file that works in R13 DOS for Windows, and then add Windows-specific menu features by attaching a separate MNU/MNS file of a different name. In practice, this approach is difficult to use and of questionable practicality.

Table 4.2 summarizes and compares the menu files in R13 Windows and DOS. As you can see, Autodesk didn't make menu matters simple in R13.

The default AutoCAD menu files are all named ACAD, plus the appropriate menu file extension (ACAD.MNU, MNS, MNC, MNR, MNX). When you first install AutoCAD, whether for Windows or DOS, there will be an ACAD.MNU file in your \R13\WIN\SUP-

Table 4.2

R13 Menu File Comparison

File Type	Windows	DOS
Menu template or "pre-source"	*menuname*.MNU	not applicable
Menu source	*menuname*.MNS	*menuname*.MNU
Toolbar bitmaps	ACADBTN.DLL *menuname*.DLL *anyname*.BMP	not applicable
Compiled menu	*menuname*.MNC + *menuname*.MNR	*menuname*.MNX

PORT or \R13\DOS\SUPPORT directory. These files are not the same. The Windows and DOS menu features are very different now, and it's nearly impossible to maintain a single menu file for both platforms. After you first load and configure AutoCAD and enter the drawing editor, AutoCAD compiles the appropriate menu files (ACAD.MNS, ACAD.MNC, and ACAD.MNR in Windows, or ACAD.MNX in DOS). Again, remember to move the Windows ACAD.MNU file out of the way.

Note: In this book, we refer to the menu source file as *menuname*.MNS. If you're using DOS, always use *menuname*.MNU.

The Windows ACAD.MNU file contains abbreviated pull-down menus that are intended to complement the R13 toolbars. Autodesk also includes a menu file called ACADFULL.MNU, in the \WIN\SUPPORT directory, which contains the same, more complete pull-down menus as the DOS version. Users who prefer to turn off most of the toolbars and use the pull-downs as their primary means of command input should use ACADFULL.MNS instead of ACAD.MNS.

R13 for Windows supports another new feature called partial menu loading. No longer is it necessary to keep your entire menu in a single MNS file. You can break it into several smaller files, each of which might contain one or two pull-down menu pages, toolbars, or accelerator key macros, and load just the parts that you need. This feature is especially useful when you need to manage several applications. We cover partial menu loading in Chapters 13 and 14.

The last menu-related file that deserves mention is ACAD.MNL, the menu LISP file. ACAD.MNL contains AutoLISP functions that support the menu. The MNL file's name must match the MNS file's name. Thus when you use ACAD.MNS as the starting point of a custom menu, you should copy ACAD.MNS to *menuname*.MNS and ACAD.MNL to *menuname*.MNL. The MNL is where you should add AutoLISP functions that are required by your custom menu. It ensures that they are automatically loaded when the menu is loaded.

When you save a drawing, AutoCAD stores the name of the menu in an area of the DWG file called the *header*. But in R13, AutoCAD ignores this menu name if the `UseMenuHeader` setting in the `[General]` section of ACAD.INI is set to 0 (which is the default). Instead, AutoCAD retains the last menu that you used on your system with the current configuration. This menu's name is stored in the `MenuFile` setting in the `[AutoCAD]` section of ACADNT.CFG. This new feature prevents the old "can't find menu file" problems caused by different menus on different systems. If `UseMenuHeader` is set to 1, AutoCAD does use the menu name stored in the DWG file's header, as it did in R12. You control the `UseMenuHeader` setting with the **U**se Menu in Header check box located in the **M**isc page of the Preferences dialog

Support Files

As you saw in Chapters 2 and 3, you use the MENU command to change menus or compile a menu after making changes to the MNS file. Chapter 5 explores menu macros in depth, and Chapter 13 covers menu structure.

Programs: ADS, ARX, LSP, DCL

Many of the features that Autodesk added to R12 and R13 are not part of ACAD.EXE, but instead are separate ADS, ARX, or AutoLISP programs, some of which use custom dialog boxes.

AutoLISP is AutoCAD's subset of the LISP programming language. AutoLISP programs are stored in ASCII text files, which AutoCAD reads and interprets each time you run them. Although they may have any name and extension, by convention, AutoLISP files usually have an LSP extension (except for AutoLISP files associated with menus, which must end in MNL, as discussed in the previous section). Many of the newer dialog box commands (such as DDCHPROP, DDMODIFY, and DDRENAME) are AutoLISP programs in the \R13\COM\SUPPORT directory. You'll learn more about AutoLISP in Chapter 9.

ADS (the AutoCAD Development System) is AutoCAD's original method for writing C programs that work with AutoCAD. ADS programs are faster than AutoLISP programs at computationally intensive tasks and provide lower-level access to the operating system, but they're also more complicated to write. Some of the PostScript support and construction line features of R13, for example, are provided by ADS programs. C programmers write ADS programs and then compile them into binary files that AutoCAD can load directly, without having to interpret them each time. ADS programs must be compiled separately for each platform. ADS programs have an EXE extension in Windows and an EXP extension in DOS.

ARX (the AutoCAD Runtime eXtension) is a new method for writing C and C++ programs that work with AutoCAD. ARX programs communicate more directly than ADS programs with AutoCAD, so they can be even faster for very computationally intensive tasks such as rendering and solid modeling. The rendering and solid modeling modules of R13 are ARX programs. ARX programs have an ARX extension in Windows and DOS, although the file formats are different. Thus, if you install R13 for Windows and DOS, you'll find RENDER.ARX and SOLIDS.ARX files in both \R13\WIN and \R13\DOS.

Many AutoLISP, ADS, and ARX programs use dialog boxes to gather input. AutoCAD's platform-independent dialog box layout facility is called DCL (Dialog Control Language). Dialog boxes are defined in files with the extension DCL. For example, most of the AutoLISP DD programs in \COM\SUPPORT use a DCL file whose eight-character name is the same as the LSP file (such as DDCHPROP.DCL, DDMODIFY.DCL, and DDRENAME.DCL). In addition, AutoCAD uses quite a few dialog boxes (and dialog box subassemblies) that are defined in the ACAD.DCL file. Part Four covers the fundamentals of dialog boxes and ways you can create and use them in your customized system.

ACAD.ADS and ACAD.RX

ACADR13.LSP and ACAD.LSP provide places to put AutoLISP code that AutoCAD and you want to have loaded every time you enter the drawing editor. There's nothing completely equivalent for ADS and ARX programs, because there's no simple way to dump a bunch of binary files into one file for loading purposes. Instead, the ACAD.ADS and ACAD.RX files provide a simple and clean method for automatic loading of ADS and ARX programs, respectively. These ASCII files contain a list of the ADS and ARX programs, one program name per line, that you want to have loaded automatically.

The standard R13 ACAD.ADS lists only one ADS file: ACADAPP. This ADS file contains a variety of commands and functions that are vital to AutoCAD, including BHATCH, DDIM, and the **startapp** function that we used in Chapter 3. Note that the file name ACADAPP in ACAD.ADS contains no extension. AutoCAD is smart enough to load ACADAPP.EXE in Windows and ACADAPP.EXP in DOS.

Tip: A common support problem is the sudden disappearance of AutoCAD commands that used to work. Often this problem is the result of someone moving or modifying ACAD.ADS so that ACADAPP doesn't load. If the following commands and functions don't work, then ACADAPP isn't getting loaded:

AutoCAD commands:
    BHATCH
    BPOLY
    DDIM
    PSDRAG
    PSFILL
    PSIN

AutoLISP functions:
    `acad_colordlg`
    `acad_strlsort`
    `bherrs`
    `startapp`

There is no ACAD.RX in a standard R13 installation. Third-party applications or drivers may tell you to add their ARX program names to ACAD.RX or may add their names automatically during installation.

Although AutoCAD loads its own ADS and ARX programs automatically, you will sometimes want to load files explicitly for testing. An exercise later in this chapter

Support Files

demonstrates how you can load ADS and ARX programs at the AutoCAD command prompt.

Warning: If you add an entry to ACAD.ADS or ACAD.RX, put a return (press ENTER) after the last line. Some third-party application or driver installation programs fail to insert a return before appending to these files. The result can be two program names jammed together on one line, which will prevent both programs from loading.

Scripts: SCR Files

Script files are the AutoCAD equivalent of DOS batch files (but without any ability to branch). Scripts are an excellent tool for automating simple but repetitive tasks on one or more drawings, such as plotting and manipulating layers. They're also useful for creating AutoCAD slide shows. You create scripts in ASCII files with the extension SCR, and they are nothing more than a transcription of the keystrokes you would type at the AutoCAD command prompt. You load scripts with the SCRIPT command.

Chapter 8 shows you how to write scripts and use them for plotting and slide shows.

Definition Files: ACAD.LIN, MLN, PAT

AutoCAD's linetype, multiline style, and hatch pattern options come from ASCII files, the default names of which are ACAD.LIN, ACAD.MLN, and ACAD.PAT, respectively. You can add to these files or create your own files with LIN, MLN, and PAT extensions.

Use the Select Linetype (DDLTYPE) dialog's Load button to view and load linetypes in the Load or Reload Linetypes subdialog box (Figure 4.2). Use the Multiline Styles (MLSTYLE) dialog to load from and save to MLN files (Figure 4.3).

You'll learn how to create custom linetypes and hatch patterns in Chapter 7.

Figure 4.2
Viewing ACAD.LIN linetypes with DDLTYPE.

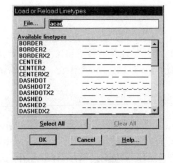

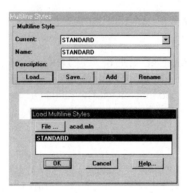

Figure 4.3
Loading multiline styles with MLSTYLE.

Warning: The multiline feature is not well-polished in R13, from either the user's or the customizer's point of view. In particular, the MLSTYLE dialog is unintuitive and limited in many ways. Multiline styles are stored in MLN files in a difficult-to-decipher DXF format. We don't cover multiline style customization in this book.

Fonts: PFB, PFM, SHP, SHX, TTF

R13 now supports three kinds of font files: AutoCAD's own SHX format, PostScript PFB format, and Windows TrueType TTF format. Even DOS AutoCAD can use TTF fonts. Chapter 6 covers most aspects of text customization.

AutoCAD's SHX fonts are the compiled form of ASCII SHP (shape) files. Although you don't normally see the SHP files, you can find them on the R13 CD-ROM. R13 SHP and SHX files can include Unicode character definitions. (Unicode is a 16-bit character encoding standard that allows for many characters and symbols in one font. See Chapter 6 for details.) Some SHP and SHX files are not fonts, but instead they define block-like symbols called shapes. Shapes aren't used much any more, but we'll discuss them briefly in Chapter 6.

PostScript fonts require two files: the main PFB font definition file and the PFM font metric file. You can compile PFB fonts to AutoCAD SHX format with the COMPILE command in order to make the loading and regenerating of drawings with PostScript text faster.

The Text Style dialog box (DDSTYLE), which is new with R13c4 (Figure 4.4), is Autodesk's attempt to make working with text fonts and styles easier. It's a convenient way to manage styles, but there are several problems with the dialog, which we discuss in Chapter 6.

Figure 4.4
The new R13c4 DDSTYLE dialog box.

Font Mapping Files: FMP

R13 includes several new font mapping options, including support for font mapping files. These files direct AutoCAD to substitute fonts when displaying and plotting text. You might use a font mapping file to substitute simpler fonts temporarily to improve performance or to enforce company standards. Autodesk suggests FMP as the extension for font mapping files, but you can use any extension you like. There is no default font mapping file, but see \R13\COM\SAMPLE\SAMPLE.FMP for an example. In order to use a font mapping file, you must set the FONTMAP system variable to the file's name.

PostScript Support Files: ACAD.PSF and FONTMAP.PS

Importing and exporting PostScript files requires mapping PostScript fonts to AutoCAD fonts and vice versa. AutoCAD uses the FONTMAP.PS and ACAD.PSF files to define font mapping for PostScript files going in and out, respectively. In other words, FONTMAP.PS defines the PostScript-to-AutoCAD font mapping for PSIN, and ACAD.PSF defines the AutoCAD-to-PostScript font mapping for PSOUT. ACAD.PSF also includes definitions for PostScript fills and some other information that controls how AutoCAD creates EPS files.

See Chapter 6 for more information about font mapping and PostScript.

ACAD.UNT

The AutoCAD units definition file, ACAD.UNT, contains definitions of six fundamental units and a large number of derived units. AutoLISP, ADS, and ARX programs can use these definitions to perform unit conversions. For example, AutoLISPs `cvunit` function performs units conversions by reading the values in ACAD.UNT. See *Maximizing AutoLISP R13* or Chapter 8 of the *AutoCAD Customization Guide* for details.

Figure 4.5
Save and restore PCP files with these buttons.

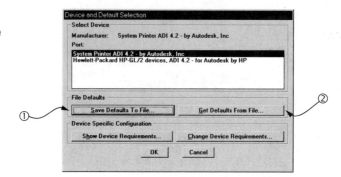

Plot Configuration Parameters: *.PCP

PCP files are a relatively straightforward way to save and restore plotter settings, including pen assignments. They also can be one part of automated plotting schemes, as we demonstrate in Chapter 15. You save and restore PCP files with the buttons in the File Defaults section of the Plot dialog's Device and Default Selection subdialog (Figure 4.5). PCP files are in ordinary ASCII format, so once you've saved one of them, you can edit it with a text editor.

Help Files: ACAD.AHP, HDX, and HLP

R13 can use two kinds of help files: the Windows-specific ACAD.HLP and the platform-independent ACAD.AHP. When you ask for help in R13 for Windows, AutoCAD normally opens ACAD.HLP with the standard Windows help engine (Figure 4.6). In DOS and on other platforms, or if a program requests help from an AHP file, AutoCAD displays platform-independent help in an AutoCAD dialog. Figure 4.7 shows the results of entering (`help "ACAD.AHP"`) at the AutoCAD command prompt.

Figure 4.6
Windows-specific help.

Support Files

Figure 4.7
Platform-independent help.

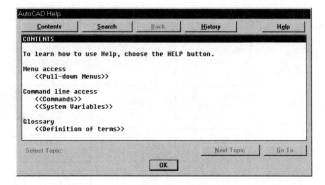

Chapter 14 discusses help files in more detail and presents techniques for creating custom help files.

ACAD.LOG

AutoCAD's log file stores the most recent contents of the text window. You can control the log file's name and location on the Environment page of the Preferences dialog.

Tip: If you have file locking turned on and you load two instances of AutoCAD from the same shortcut (that is, with the same configuration directory), AutoCAD reports that it is unable to lock ACAD.LOG. This error message appears because the first AutoCAD session opened ACAD.LOG and created an ACAD.LOK file to indicate that ACAD.LOG is in use. There is no harm done—you can simply choose OK to acknowledge the message—but the second session won't write its text window output to the ACAD.LOG file. If you need to run multiple AutoCAD sessions with multiple active log files, set up a separate shortcut and configuration directory and redirect the log file for each session, as we did in Chapter 2. If you are not on a network, you can avoid the problem by turning file locking off in the CONFIG command's Configure operating parameters settings. On a network, file locking is essential for file protection, but otherwise it is only a nuisance.

Understanding AutoCAD's Initialization Sequence

When you start AutoCAD and each time you open or create a new drawing, R13 performs an initialization sequence that includes loading some of the support files discussed in the previous section (Figure 4.8). You should understand this sequence so that you know how to control AutoCAD's initial working settings.

Figure 4.8
AutoCAD's initialization sequence.

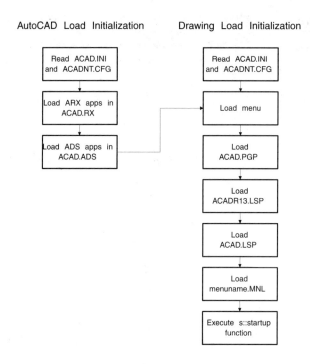

AutoCAD Load Initialization

Each time you launch AutoCAD from a Windows shortcut or DOS batch file, the program reads its configuration file, loads the ARX and ADS files that are listed in ACAD.RX and ACAD.ADS, and opens the appropriate menu file. The exact steps AutoCAD goes through to do so are:

1. Read configuration settings from ACAD.INI and ACADNT.CFG.

2. Load ARX files listed in ACAD.RX.

3. Load ADS files listed in ACAD.ADS (including, by default, ACADAPP.EXE).

AutoCAD loads these files once because ARX and ADS programs remain active throughout an AutoCAD session; unlike AutoLISP programs, they don't need to be reloaded each time you open a new drawing.

After completing these three steps, AutoCAD loads the prototype drawing and proceeds with drawing load initialization described next (skipping the first drawing initialization step, since its already been performed).

Support Files

Drawing Load Initialization

 After the AutoCAD load initialization, and again each time you open or start a new drawing, AutoCAD performs another series of initialization steps:

1. Reread configuration settings from ACAD.INI and ACADNT.CFG.
   Because you might have made changes to the configuration file settings with Preferences, AutoCAD rereads these settings at each drawing load.

2. Load the menu *menuname*.MNC + *menuname*.MNR (or *menuname*.MNX in DOS).
   As described earlier in this chapter, the Use Menu in Header setting in the Preferences dialog (Figure 4.9, ①) determines whether AutoCAD loads the last menu that you used on your system with the current configuration or the menu name in the DWG files header. AutoCAD checks for missing or outdated compiled menu files (MNC or MNX) and recompiles them from the source files (MNS or MNU) if necessary. If no compiled menu file is found, or if the source menu file is newer than the compiled menu file, AutoCAD recompiles the menu as described in the previous section. If Use Menu in Header is turned off, R13 doesn't bother reloading the menu.

3. Load command alias and external command definitions from ACAD.PGP.

4. Load AutoLISP code from ACADR13.LSP.
   R12 included an ACADR12.LSP file that was similar to ACADR13.LSP, but ACADR12.LSP was loaded explicitly from ACAD.MNL. By contrast, R13 loads ACADR13.LSP automatically.

5. Load AutoLISP code from ACAD.LSP.

6. Load AutoLISP code from *menuname*.MNL.

Figure 4.9
The Use Menu in Header setting.

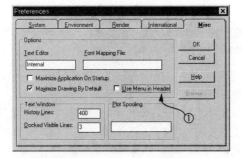

7. Execute any `s::startup` function defined in ACADR13.LSP, ACAD.LSP, or *menuname*.MNL.

Often you want to include initialization commands (that is, commands that get fed to the AutoCAD command prompt) in ACAD.LSP or *menuname*.MNL. For example, you might want to run the LAYER command to establish startup layer settings. The problem is that the drawing editor isn't yet "awake" when ACAD.LSP and *menuname*.MNL load. To get around this limitation, AutoCAD provides the special `s::startup` function name. If you define a function with this name in ACAD.LSP or *menuname*.MNL (or ACADR13.LSP), AutoCAD stores it in memory until the end of the initialization sequence, and then executes it right before giving control back to the user. You'll learn more about the `s::startup` function in Chapter 14.

 Tip: The error message `Command list interruption` when opening a drawing indicates that your ACADR13.LSP, ACAD.LSP, or *menuname*.MNL file is trying to execute the `command` function before the editor is awake. You need to put any such functions in an `s::startup` function.

Loading, Listing, and Unloading AutoLISP, ADS, and ARX Programs

R13 loads and manages most of its program files and dialog boxes automatically. Occasionally, you'll want to load some of these files, your own custom programs, or companion applications explicitly, and often you'll need to load your custom programs in order to test them. The user-friendly method is to use the APPLOAD command (**T**ools, **A**pplications), which brings up the Load AutoLISP, ADS, and ARX Files dialog box, as shown in Figure 4.10. APPLOAD.DFS saves your list of applications to a file called APPLOAD.DFS in the AutoCAD startup directory, so that you can load any of these applications again easily.

The command-line method for loading applications is to use the AutoLISP `load`, `xload`, and `arxload` functions. The `ads` and `arx` AutoLISP functions list the currently loaded ADS and ARX files, and the `xunload` and `arxunload` functions unload selected files.

Figure 4.10
The Load AutoLISP, ADS, and ARX Files (APPLOAD) dialog box.

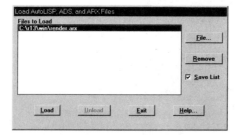

Support Files

There are no built-in functions for showing which AutoLISP files are loaded or for unloading files.

The following exercise demonstrates how to use APPLOAD and the AutoLISP functions.

Exercise

Loading, Listing, and Unloading Programs

Start AutoCAD.

First, you'll use APPLOAD to load an ARX program.

Choose **T**ools, **A**pplications, *or enter* **APPLOAD**	Opens the Load AutoLISP, ARX, and ARX Files dialog
Choose **F**ile	Opens the Select AutoLISP, ARX, and ARX File subdialog
Navigate to C:\R13\WIN	
In the **L**ist Files of Type *drop-down list, choose* *.AR**X**	Displays ARX files in the directory
Choose RENDER.ARX	
Choose OK	Adds RENDER.ARX to list of **F**iles to Load (Figure 4.10)
Choose **L**oad	Loads RENDER.ARX

```
Loading C:\r13\win\render.arx ...
File C:\r13\win\render.arx loaded.
```

Next, you'll use the AutoLISP functions to load AutoLISP, ADS, and ARX files. When you enter AutoLISP functions at the AutoCAD command prompt, you must always enclose them in parentheses.

Command: **(LOAD "DDMODIFY")** [Enter]

DDMODIFY loaded.

Command: **(XLOAD "ACADPS")** [Enter]

Maximizing AutoCAD® R13

LISP returns: `"ACADPS"`   Indicates that the file loaded successfully

Command: **(ARXLOAD "SOLIDS")** [Enter]

LISP returns: `"SOLIDS"`

Next, you'll list the currently loaded ADS and ARX applications.

Command: **(ADS)** [Enter]   Lists currently loaded ADS applications

LISP returns:
```
("C:\\R13\\WIN\\ACADPS.exe"
 "C:\\R13\\WIN\\geomcal.exe"
 "C:\\R13\\WIN\\acadapp.exe")
```

ACADAPP.EXE is loaded automatically when you start AutoCAD.

Command: **(ARX)** [Enter]   Lists currently loaded ARX applications

LISP returns:
```
("C:\\R13\\WIN\\SOLIDS.arx"
 "C:\\r13\\win\\render.arx")
```

Finally, you'll unload ADS and ARX applications.

Command: **(XUNLOAD "ACADPS")** [Enter]

LISP returns: `"ACADPS"`   Indicates that the file unloaded successfully

Command: **(ARXUNLOAD "SOLIDS")** [Enter]

LISP returns: `"SOLIDS"`

This exercise serves no useful purpose other than to demonstrate how you can use APPLOAD and the AutoLISP functions for loading, listing, and unloading programs. AutoCAD will load these particular files automatically as it needs them.

 Tip: If your system is low on memory, you can unload ADS or ARX applications in order to free up some memory for the drawing and other parts of AutoCAD.

The Support Path and Library Search Path

In Chapter 2, we discussed the *support path* setting stored in ACAD.INI, and we modified it to include our C:\MAXAC directory. Remember that the support path is a series of directories in which AutoCAD looks for its support files (menus, fonts, and AutoLISP, ADS, and ARX programs, and so on).

In fact, AutoCAD looks for support files in other places in addition to the support path that you specify. It looks in the following places and order:

1. The directory that was current when AutoCAD was launched first, which is sometimes called the *startup* or *current* directory (Figure 4.11, ①).

2. The directory containing the DWG file on which you're currently working ②.

3. The support path directories you specify ③.

4. The directory containing the AutoCAD executable program file (ACAD.EXE) ④.

The complete list of directories in which AutoCAD looks for support files is called the *library search path*. Thus the *support path* is a subset of the *library search path*. Each time AutoCAD needs to find a support file, it searches this entire path. If the path contains more than one version of the same file, AutoCAD uses the first one it comes to, searching in the order listed. For example, an ACAD.LSP file in the AutoCAD startup directory will always be used in lieu of an ACAD.LSP file anywhere else in the AutoCAD library search path.

 Tip: To find out what your current AutoCAD library search path is, enter a nonexistent block and file name at the block name prompt of the INSERT command. An AutoCAD Message dialog box will appear and display the library search path (Figure 4.11).

As Chapter 2 describes, you can set the support path in the S<u>u</u>pport edit box in the <u>E</u>nvironment page of the Preferences dialog box or by adding a /S setting in the AutoCAD program shortcut's properties. AutoCAD stores this value in the [GENERAL]

Figure 4.11
The AutoCAD library search path.

section of ACAD.INI, after the ACAD key word. The equivalent in DOS AutoCAD to the setting available in Windows through **O**ptions, **P**references, S**u**pport is the environment variable ACAD.

The default R13 support path comprises four directories:

- C:\R13\COM\SUPPORT
- C:\R13\WIN\SUPPORT
- C:\R13\WIN\TUTORIAL
- C:\R13\COM\FONTS

These directories except for \R13\WIN\TUTORIAL, which is superfluous, should remain on your AutoCAD support path. Add any custom support file directories to the front of the path, as described in Chapter 2, so that AutoCAD uses your custom files whenever there's any duplication with standard AutoCAD support files in the SUPPORT or FONTS directories. Watch out for AutoCAD's support path length limitation of about 255 characters (see Chapter 2).

Tip: A common support problem is using the wrong versions of AutoCAD support files because of duplicates in the library search path (for example, ACAD.LSP, ACAD.PGP, ACAD.ADS). Make sure that you understand and control the library search path.

Third-Party Application Issues

If you use a third-party application with AutoCAD, the application will complicate the support file picture further. Some applications require their own versions of standard AutoCAD support files such as ACAD.LSP and ACAD.ADS, or at least require appending to these files. Many applications add at least one directory to the AutoCAD support path. Most applications make use of application-specific support files that control layers, custom command defaults, and other application settings.

The big issues are avoiding standard AutoCAD support file conflicts and controlling application-specific support files. Before you install an application, try to get an idea of what files it will create and where. Check the application's installation documentation, and when in doubt, call the dealer or developer. It's prudent to back up your system before installing a major application, especially if you suspect that it might modify or overwrite existing files.

Applications aren't supposed to replace the standard AutoCAD support files with their own versions, but some of them still do. Sometimes they write these files (for example, ACAD.LSP, ACAD.PGP, ACAD.ADS) to their own application directories, and occasionally they go so far as to overwrite files in the \R13 directory tree. In such cases, you'll need to reconcile the application's version with the standard AutoCAD version or your own customized version.

Better-behaved applications load their own support files or append to, rather than replace, AutoCAD support files. But most application installation programs append to the files in the \R13 directory tree, not to any custom version that you might have elsewhere. Again, you'll need to reconcile the differences after installation. Software tools such as the Visual Diff shareware program can help with this task. Visual Diff is available in Library 2 of CompuServes WINSHARE forum (VISDIF.ZIP).

Note: Keep in mind the library search path order. If an application installs its own ACAD.LSP file, and you have a different ACAD.LSP file in your custom support directory, AutoCAD will use the one that's located earlier in the library search path.

Ideally, applications aren't supposed to write their program files to directories in the \R13 tree, but some do, especially small utilities and drivers. When possible, keep application files segregated from the AutoCAD program files so that you can manage and update each type separately.

Tip: If you use a large number of third-party utilities, create a single directory for them (for example, C:\ACUTILS) and add that directory to the search path.

Well-behaved applications should add no more than one directory to the support path, but some large applications add more. If the support path becomes too long, it is truncated and directories at the end of the path won't be included, which can cause errors. Applications should create a separate shortcut or batch file for launching AutoCAD with the application, and they should use separate configuration files (or the /S support path) switch to avoid conflicts with your standard AutoCAD configuration. Some application installation programs copy your ACADNT.CFG and ACAD.INI files from \R13\WIN in order to create a new configuration, as we did in Chapter 2. Others require you to perform these steps.

Application-specific support files usually don't cause any immediate AutoCAD conflicts, but managing them can be a headache, especially if you make modifications and then upgrade the application. Some applications look for their support files on the AutoCAD library search path, in which case you can manage these files the same way you manage AutoCAD's support files. Copy the standard support file to a custom support directory and modify the new version. Make sure that your custom directory comes earlier than the application's directory in the support path. Other applications look only in specific directories for their support files, in which case you're stuck with modifying the files there.

Keep track of which files you change, and back them up before installing an application upgrade.

Every application has a different method for storing its application-specific configuration settings, but look for small files with extensions like CFG, INI, and DBF. Note that R13 applications may store their own settings in separate sections of the ACAD.INI file.

Customizing for Portability

"Portability" means the ability to transfer programs easily among computers, operating systems, program versions, and human languages. There's no way to achieve complete portability, but a few good customization habits will minimize your portability problems.

- Don't edit or add to files in the \R13 directory tree. Copy files to a custom support directory and edit them there. Put your custom support directory at the front of the support path so that AutoCAD always finds your version before the R13 stock version.

- Prefix AutoCAD command names and options in menu macros, scripts, and AutoLISP programs with an underscore (_) for portability with foreign-language versions of AutoCAD, like: _ZOOM _Previous.

- Prefix AutoCAD command names (but not command options) with a period (.) when you need to ensure that your menu macro, script, or AutoLISP program uses the native AutoCAD command, and not any redefined version, like: ._ZOOM _Previous. The order of the period and underscore doesn't matter.

- Use real AutoCAD command names (for example, LAYER, COPY, and ZOOM) in menu macros, scripts, and AutoLISP programs. Don't use command aliases or other macro names (for example, LA, CP, and Z), because these are more likely to change than the AutoCAD command names.

- Avoid explicit (or *hard-coded*) paths when you reference support files. In your menu macros, AutoLISP programs, and other custom files, don't refer to other support files by their path (for example, C:\COM\FONTS\ROMANS.SHX). Instead, use just the file name (ROMANS.SHX) and let AutoCAD find the files on the library search path. This precaution will save endless headaches for you and other users when you decide to move files or rename directories. Many AutoCAD file selection dialogs strip the path when you select a file thats on the library search path, but some (notably DDSTYLE) dont. The Type It button, which is included on most file dialogs, is a good way to avoid explicit paths.

- Maintain the same program and support directory structure on all computers in your office whenever possible. If you customize or support AutoCAD for others in your office, your work will be much easier if all of the computers use the same directory structure. In cases where you can't maintain the same directory structure, set the support path on each computer to create the same library search path order.

- In an office with more than one AutoCAD user, it's easiest to maintain one set of customized support files for everyone (ACAD.LSP, ACAD.PGP, and so on). This approach helps enforce standards and makes the job of updating support files easier. In practice, though, the "one size fits all" approach isn't always desirable. Some users may have favorite command aliases or AutoLISP programs with which they can work more productively. In these cases, create a user custom support directory (for example, C:\ACUSER) and put it at the front of the support path. Tell users to put any user-specific custom support files there. Note that you may have to help them reconcile their versions of support files with the standard office versions.

- Read about compatibility issues in the AutoCAD documentation. Appendix D of the R13 *AutoCAD User's Guide* describes compatibility issues with R12 and earlier versions. This section alerts you to potential problems with your custom support files. Previous versions of the documentation also included a "Features to be Dropped in the Next Release" section, which helped you avoid future version compatibility problems. Unfortunately, this section was dropped from the R13 documentation.

Conclusion

Understanding support file structure and operation is vital to anyone who does much AutoCAD customization. Now that you have that understanding, you're ready to move on to Chapter 5, where you'll master menu macro techniques and learn to build custom pull-down menus.

chapter 5

Introduction to Menu Macros

A customized menu file is the centerpiece of every efficient AutoCAD system. AutoCAD's standard menu file, ACAD.MNS, is fine as a starting point, but you can make it much more efficient for your work by adding commands and macros and grouping together the choices that you use most often.

Although keyboard macros, such as those developed in Chapter 3, are a good way to speed up command access, they require users to memorize key sequences. Remembering one or two dozen frequently used macros isn't too difficult, but few of us can remember hundreds of macro names. Menu choices are an easier way to provide access to and organize many commands, especially for less knowledgeable AutoCAD users.

Your menu is also a place for custom menu macros that go beyond the standard AutoCAD commands. At its simplest, a menu macro is a blindingly fast and unerringly accurate typist. Macros help you increase productivity by reducing drawing steps. They can help you set drafting standards and control consistency. They enable you to add your own new commands to accomplish tasks that are tedious to do without menu customization. Figure 5.1 diagrams two of the example macros developed in this chapter.

 Despite the many changes to menu structure in Release 13 for Windows, such as the capability of loading multiple menus, there is little new about menu macro content in R13. AutoCAD's menu macro syntax remains essentially unchanged through the last few major releases. We do mention a few R13 macro tricks and traps in this chapter, but if you're already a menu macro expert, you can skim this chapter.

About This Chapter

If you complete this chapter, you will achieve the following goals:

- Understand enough about menu structure to create and edit pull-down menu sections

Maximizing AutoCAD® R13

- Learn the details of menu macro syntax and how to write macros
- Add a custom pull-down menu page

This chapter covers menu macros and their syntax in detail. It shows you how to write menu macros and use the full range of AutoCAD's macro syntax. Most of the material in this chapter is independent of any particular menu section (for example, pull-down, toolbar, side-screen, and so on)—you can use these macros in any section. We use pull-down menus in our examples because they are easy to create and customize, and they are portable across menu files and AutoCAD platforms. The other part of menu customization—structuring menus using different menu sections—is covered in Chapter 13.

The resources you'll use in this chapter are:

- TextPad (or another ASCII text editor)
- ACAD.LSP, ACAD.PGP, ACADMA.MNS, ACADMA.MNL, , MAWIN.ARX, MA_UTILS.LSP, ARROW.DWG, MXACH05.DWG, VALVE.DWG, custom bitmap (BMP) icons, and MXACH05.DWG from the MaxAC CD-ROM

Make sure that you have TextPad (or another editor capable of loading large menu files) configured as described in Chapter 2.

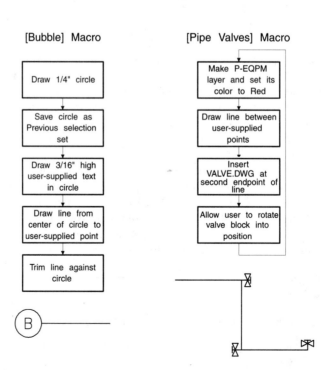

Figure 5.1
Pull-down menu macros.

Introduction to Menu Macros

Defining Macros and Menus

A *menu macro* (or *macro* for short) is a series of AutoCAD commands and parameters put together to perform a task. Macros can call AutoCAD commands, pause for input from the user, and draw or edit objects on the screen. Macros can even repeat automatically. In this chapter, you'll learn how to carry out the following macro techniques:

- String commands together to form a basic macro
- Put AutoCAD commands in a menu file
- Control the labels that appear in the menu to identify macros
- Make macros pause for input from a user
- Use special characters in a macro to control snap, grid, ortho, and other AutoCAD settings
- Set modes with control characters
- Use blocks and path names in macros
- Extend macros with polyline tricks
- Use object snaps in macros, and use the QUIck modifier to filter points
- Control layers with macros
- Repeat entire macros automatically by using the * modifier
- Use selection sets in macros

AutoCAD macros live in menu files. As you saw in Chapters 3 and 4, an AutoCAD menu is defined in an ASCII text file that lists the commands and macros for each of AutoCAD's menu interfaces (accelerator keys, pull-down, cursor, button, toolbar, image, side-screen, and tablet).

Macros and Menus in This Chapter

You create the following macros in this chapter:

- **Bubble** draws column line grids and bubbles with text labels inside the bubbles.
- **Leader** is a straight-line annotation leader with an arrow block.
- **Titlex Style** is a series of four macros that demonstrates how to call dialog boxes in macros.

- **Leader Arrow—>** is a single-object leader that uses a polyline to create a straight segment and arrow end.

- **Leader Arrow<—** starts with an arrow and can have unlimited curved and/or straight leader segments.

- **Pipe Valve** initializes layer controls, draws pipe flow lines, and puts in valve symbols.

- **Pipe Valves** is the same as Pipe Valve, except that it repeats.

- **Duplicate Layer** duplicates a set of objects on another layer.

Your working menu file for developing macros in this chapter is ACADMA.MNS from the \MAXAC\BOOK\CH05 directory.

Menu Structure Overview

Although we defer a full discussion of menu structure until Chapter 13, you should understand enough about menu structure to know what you're editing and where to put new items. AutoCAD menu source files are divided into *sections*, each of which starts with a label prefixed by three asterisks (***). If you open any MNS or MNU file and search repeatedly for three asterisks, you'll see most of the section names, as shown in Table 5.1 (see Chapter 13 for details).

Table 5.1
Menu Sections

Section Name	Defines
***POP1 through ***POP16	Pull-down menus
***POP0	Cursor menu
***AUX1 through ***AUX4	System pointing device buttons
***BUTTONS1 through ***BUTTONS4	Digitizer buttons
***IMAGE	Image (that is, icon) menus
***SCREEN	Side-screen menus
***TABLET1 through ***TABLET4	Digitizer tablet menus
***TOOLBARS	Windows toolbar menus
***ACCELERATORS	Windows keyboard accelerators
***HELPSTRINGS	Windows toolbar and pull-down menu status line help

Introduction to Menu Macros

Note: The standard AutoCAD menu files don't include all of these sections (for example, ***AUX3 and ***AUX4). Also, there is a ***MENUGROUP=ACAD label near the top of the menu that is not a menu section, but a menu group name. See Chapter 13 for details.

The standard AutoCAD pull-down menus (File, Edit, View, and so on) are defined in sections ***POP1 through ***POP7 in the R13 for Windows ACAD.MNS file. The R13 for Windows ACADFULL.MNS file and R13 for DOS ACAD.MNU files use three additional pull-down menu slots: ***POP8 through ***POP10. The remaining pull-down menus (***POP8 through ***POP16 in the standard Windows menu and ***POP11 through ***POP16 in the standard DOS menu) aren't defined. You can use one or more of these sections for custom menu macros, as we do in this book.

In this chapter, you'll add macros to a custom pull-down menu that we've already created as ***POP7 in ACADMA.MNS (or ***POP10 in the DOS menu ACADMA-D.MNU). We changed the Help menu from ***POP7 to ***POP8 (from ***POP10 to ***POP11 in the DOS menu) so that it remains as the right-most pull-down menu. Partial menu loading provides a more sophisticated method of inserting pull-down menus before the Help menu—see Chapters 13 and 14.

Writing a Simple Menu Macro

In this chapter's exercises, you edit the ACADMA.MNS file and add macros to the ***POP7 MaxAC pull-down menu section (***POP10 in DOS). To open ACADMA.MNS in your text editor, you can use the Edit Current Menu toolbar button you created in Chapter 3, or simply launch your editor and open ACADMA.MNS with the editor's File menu.

In the first exercise, you'll use the MXACH05.DWG drawing from the MaxAC CD-ROM as a prototype for your drawing. MXACH05.DWG is set up for architectural units and a scale of 1/8" = 1'-0" (1 = 96). It has snap set to 2 inches and grid set to 8 feet (that is, one plotted inch). Limits are set to correspond to a D-size plot sheet (288' feet, 192 feet). MIRRTEXT and REGENAUTO are turned off.

Exercise

Loading the Menu and Starting a Drawing

Copy all of the files from \MAXAC\BOOK\CH05 to your \MAXAC\DEVELOP directory. Open \MAXAC\DEVELOP\ACADMA.MNS with your text editor. (Use ACADMA-D.MNU with DOS AutoCAD.)

Add a comment similar to this one at the top of the file (use your initials in place of MM):
`// Maximizing AutoCAD custom menu by MM 04/01/96` Enter

Save ACADMA.MNS and start or return to AutoCAD.

Command: **MENU** Enter	Displays the Select Menu File dialog box
Select the C:\MAXAC\DEVELOP *directory, select* *.mns *in the* **L**ist files of Type *drop-down list, select the* ACADMA.MNS *file, then choose* OK	AutoCAD compiles and loads the menu

```
Menu loaded successfully. MENUGROUP: ACAD
Command:
AutoCAD Release 13 menu utilities
ACADMA.MNL loaded
```

If you encountered any problems using your text editor or loading the menu in AutoCAD, refer to Chapters 2 and 3 for help.

Next, you start a new drawing with the MXACH05 prototype.

Command: **NEW** Enter	Begins a new drawing; don't save changes to the drawing
Enter **TEST=MXACH05** *in the* New **D**rawing Name *edit box*	AutoCAD creates TEST.DWG, using MXACH05.DWG as the prototype

You can use the **P**rototype button and edit field instead of =MXACH05, but the =MXACH05 approach is faster.

In the next exercise, you'll create two simple menu items that issue the CIRCLE and TEXT commands, and then use them to draw a bubble symbol. The ACADMA.MNS menu file already contains a "dummy" ***POP7 section that you will modify.

Note: If you're editing the DOS menu (ACADMA-D.MNU), use ***POP10 wherever you read ***POP7 in this chapter.

Warning: When you edit menus, input characters exactly as the book shows them. Do not use tabs, invisible trailing spaces, or extra blank lines. When the exercises show the Enter symbol (for pressing ENTER) at the end of lines of code, do not type a space before you press ENTER.

Introduction to Menu Macros

 **Exercise**

Creating Two Simple Menu Items

Use your text editor's search function to locate ***POP7. The ***POP7 menu section should look like this:

```
***POP7
MA_MaxAC           [MaxAC]
MA_Message         [~Maximizing AutoCAD]
                   [~Chapter 5 Custom menu]
```

Delete the last two lines of this section, which we placed there just for informational purposes. Replace them with the following two lines, immediately after the MA_MaxAC [MaxAC] line.

```
MA_MaxAC           [MaxAC]
[Circle]_CIRCLE Enter
[Text]_TEXT Enter
```

Be careful not to add any spaces, including at the ends of the lines. Leave at least one blank line after [Text]_TEXT and before the ***TOOLBARS line.

Note that the strings before many of the menu labels in ACADMA.MNS (for example, MA_MaxAC before [MaxAC] and ID_About before [&About AutoCAD...] in the previous menu section are called *name tags*. They're new to R13 for Windows menus, and we'll discuss them in Chapter 13. Our menu macros in this chapter don't require tags.

Save ACADMA.MNS and return to AutoCAD.

Command: **MENU** Enter   Displays the Select Menu File
                                            dialog box

If you prefer, you can use the Load Menu toolbar button you created in Chapter 3 instead of entering MENU.

Choose OK *or press ENTER*   Reloads and recompiles the default
                                            menu, ACADMA

Before you test your new macros, zoom for a better view, as described in the next step.

Command: **ZOOM** Enter

All/Center/Dynamic/Extents/Left/Previous/Vmax
/Window/<Scale(X/XP)>: **0,0** Enter

```
Other corner: 32',16' [Enter]
```

From the pull-down menu, choose MaxAC, Circle — Issues CIRCLE just as if you entered it at the command prompt

```
Command: CIRCLE 3P/2P/TTR/<Center
point>:
```
Pick a point centered on the screen

```
Diameter/<Radius>: 24 [Enter]
```
Draws a circle (1/4" at 1:96 plot scale)

Choose MaxAC, Text — Issues the TEXT command

```
Command: TEXT Justify/Style/
<Start point>: M [Enter]
```
Specifies Middle justification

```
Middle point: @ [Enter]
```
Tells AutoCAD to reuse the last-entered point (@)

```
Height <default>: 18 [Enter]
```
Sets height for 3/16" at 1:96 plot scale

```
Rotation angle <0>: [Enter]
```
Defaults to 0

```
Text: A [Enter]
```
Draws the letter A in the bubble (see Figure 5.2)

As shown in Figure 5.2, a typical pull-down menu displays a list of choices in a menu "page" ① that appears (or *pulls down*) temporarily over the drawing area when you select its label ② from the menu bar. Typically, you see each menu item's label ③, as defined in the menu file, but you don't see the menu macro code in the pull-down menu. When you choose a menu item, AutoCAD feeds the item's macro to the command prompt. If the MENUECHO system variable is set to its default value of 0, you see the macro code echoed on the command line. This echoing is similar to what appears on the command line when you manually enter a sequence that performs the same function as the menu macro.

Note: When you have more experience writing menu macros, you'll want to suppress most of the command-line echoing in order to achieve a cleaner appearance when your macros run. For now, it's best to leave the echoing on so that you can see what's happening and debug problems more easily. Chapter 13 discusses MENUECHO and control of macro echoing.

Introduction to Menu Macros

Figure 5.2
The circle, with text.

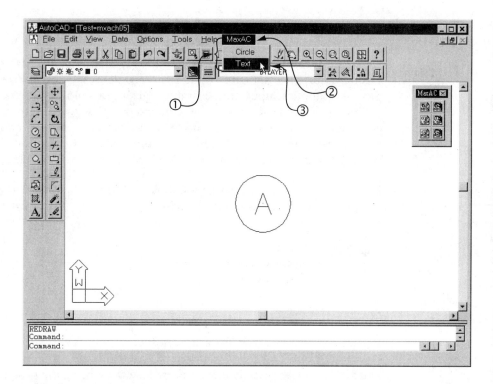

The menu item's label goes in square brackets (for example, [Circle]), and can be any string of characters that you choose. Keep menu labels reasonably short, but try to make them descriptive. Labels for single AutoCAD commands are easy, but deciding on names for more elaborate macros is more challenging. Use mixed case in the labels ([Circle] instead of [CIRCLE]), because it's easier for users to read and follows Autodesk's menu guidelines. Labels can include letters, numbers, and nearly any displayable character. A few characters have special meaning in labels; see Chapters 10 and 13 for details. True control characters and extended ASCII characters are simply ignored unless you are using an 8-bit international version of AutoCAD.

Most of the exercises in this chapter assume architectural units and a plot scale factor of 1/8" = 1'-0". The comments in the right column of the previous exercise indicate the equivalent plotted sizes. More flexible macros that work with other drawing scales require AutoLISP or DIESEL. We'll return to this topic in Chapter 14.

This first sample menu page is simple. Each item sends a single command to AutoCAD and terminates, leaving the user to do the rest. We use the TEXT command instead of DTEXT or MTEXT because TEXT is better suited to macros that place text and continue on with other tasks, as you'll see in subsequent exercises. DTEXT and MTEXT are for interactive use outside of macros.

Pausing Macros for User Input

To create macros that go beyond running a simple AutoCAD command, you string together multiple commands, options, and responses. Often macros need to pause while the user supplies input (for example, picking a point), and then continue on with additional automated options or commands. This process requires a special pause character, the backslash (\). The backslash makes AutoCAD pause for input, then resume execution of the macro with any commands or options that follow the backslash. This pause is the menu feature that makes interactive macros possible.

In the next exercise, you design a bubble macro that automates the two-step procedure you performed in the previous exercise.

 Exercise

Planning a Macro

Repeat the steps that you performed in the preceding exercise to draw the bubble with the CIRCLE and TEXT commands at the command prompt, including setting the text height. Write down each character of each step, ignoring the prompts. Use a blank space to represent each space you enter at the keyboard, or each time you press ENTER at the keyboard. Use a backslash to mark where you pick a point or enter input.

Your result should be the following line:

CIRCLE \24 TEXT M @ 18 0 \

Note the first space after CIRCLE, which represents pressing ENTER or the Spacebar to enter the CIRCLE command. The following backslash represents the pause while you pick the circle's center point. The final backslash represents the pause while you type the text character that you want to see in the circle.

When you write complex menu macros, your command syntax must be exact. Use the drawing editor interactively to test command sequences (as you did in the previous exercise) so that you know what options and input parameters are expected in your macros. Write your sequences down or use the AutoCAD log file to record keystroke sequences and paste them into your menu (removing all of the extra AutoCAD prompts, of course). Another option is to copy command sequences from the R13 for Windows text window: Press F2 if you need to see the entire text window, click and drag on the text you want to copy, right-click anywhere in the text window, and then choose **C**opy. When you perform this sequence, AutoCAD copies the selected text to the Windows

Introduction to Menu Macros

clipboard, from which you can paste it into your menu file. Whatever way you choose to record the command sequences, the idea is to make your mistakes in the drawing editor rather than in your text editor.

Next, you'll add a new Bubble macro to ACADMA.MNS using the sequence you just wrote down. This macro starts the CIRCLE command, pauses for location, draws a 24-inch circle, and waits for text before automatically putting the text into the center of the circle. The exercise shows additions to the existing menu in boldface type.

Warning: Take care to note the differences between l (the letter "ell") and 1 (the numeral 1), and 0 (zero) and O (the letter "oh") when writing macros.

Exercise

Writing a Macro with Pauses for Input

Add the bubble sequence to ACADMA.MNS below the Circle and Text lines:

```
MA_MaxAC             [MaxAC]
[Circle]_CIRCLE
[Text]_TEXT
[Bubble]_.CIRCLE \24 _.TEXT _M @ 18 0 \ [Enter]
```

Save ACADMA.MNS, return to AutoCAD, and compile and reload ACADMA.MNS with the MENU command, as described in the previous exercise.

From the MaxAC *pull-down menu, choose* Bubble	The macro issues the CIRCLE command and pauses
`Command: _.CIRCLE 3P/2P/TTR/` `<Center point>:` *Pick any point*	
`Diameter/<Radius> <2'-0">: 24`	Issues the radius and draws the bubble
`Command: TEXT Justify/Style/` `<Start point>: _M`	Issues the TEXT command and the Middle option
`Middle point: @`	Issues the @ to place text at the last point

155

Maximizing AutoCAD® R13

```
Height <1'-6">: 18              Issues the height

Rotation angle <0>: 0           Issues 0 rotation

Text: B Enter                   Draws the text in the bubble
```

Your screen should now show two circles with text (see Figure 5.3). The macro took only three steps (including picking the menu item) to draw the bubble, compared to the nine manual steps you used previously.

 Tip: If an error occurs while you're using a macro, use UNDO or U one or more times. Any error in a menu item crashes the rest of the item, but UNDO cleans it up.

A single backslash tells AutoCAD to wait for a single piece of input. In the circle-with-text macro, the first backslash pauses for the circle's center point. Without a backslash, AutoCAD would continue taking its input from the macro and pass the next item along to the command processor. It would read the 24 as the center point and get confused

Figure 5.3
Circles with text.

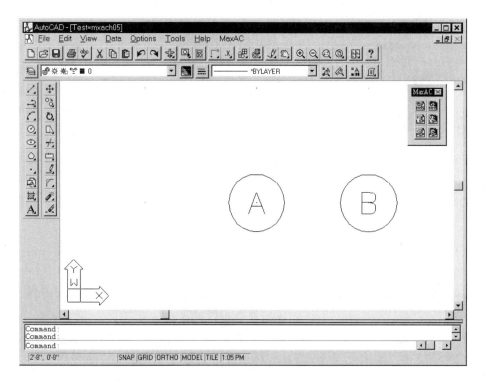

Introduction to Menu Macros

because 24 is not a coordinate point. The backslash at the end of the macro is not really needed because the TEXT command is waiting for input anyway. But adding the backslash reminds you of how the macro works.

Note: Each backslash pauses for exactly one item of input. Count your backslashes carefully. The menu macro resumes as soon as the backslash gets any form of input; it does not matter whether the input is correct. In other words, there's no built-in error trapping in menu macros.

Tip: The use of a backslash character to pause for input has some limitations. When the user wants to enter a point, angle, or distance, a backslash does not allow the user to pick an angle or distance with two points. To create routines that allow more input options, you need to use AutoLISP.

You may be wondering why you specified the text height and rotation in the macro when they were already set correctly in the current drawing. You never know what value was used last in the drawing. If modes and settings, such as style and text height, are important for your macro to function correctly, it is good practice to set them explicitly in the macro.

Ensuring Compatibility with Foreign Versions and Redefined Commands

You might also wonder when you should use only the underscore (foreign-language compatibility) before a command name (for example, _CIRCLE), and when you should also use the period (ignore command redefinition—for example, _.CIRCLE). There are no hard and fast rules, but a good rule of thumb is always to use the underscore and add the period when your macro assumes a specific sequence of command options. Thus, our Circle macro, like most of the standard ACAD.MNS macros, uses only the underscore in order to make the macro portable to foreign-language versions of AutoCAD. If you were to redefine the CIRCLE command on your system, this Circle macro would run the redefined version. Our Bubble macro answers CIRCLE and TEXT command option prompts, and thus the macro assumes that these commands operate in the standard AutoCAD way. For that reason, we used the period in addition to the underscore in the Bubble macro, just in case someone has redefined these commands to use a different prompt sequence.

If your macros never make it beyond users of English-language versions of AutoCAD who don't redefine commands, then these niceties aren't important. Nonetheless, it's prudent to develop good habits in the beginning. Chapter 14 discusses redefining commands with AutoLISP and UNDEFINE.

Using Special Characters in Macros

Backslashes and spaces are the two special characters that you use most frequently in macros. The following table lists all the special characters available for macros. The @ (last point) is not listed in this group because it does not share in the special treatment by the AutoCAD menu interpreter. In a menu, it works the same as it does when entered at the command line—simply returning the coordinates of the last point entered. Likewise, a period tells AutoCAD to use the original (not redefined) command, whether you enter it at the command line or with a macro. Table 5.2 lists the special characters for menu macros.

Table 5.2

Special Menu Characters

Character	Action in Macros
;	Issues ENTER
\	Pauses for input
+	Continues macro on next line
space	Issues a space (which usually acts like ENTER)
*	Autorepeats macro, or identifies a menu section or subsection
[ ]	Encloses label
^B	Turns snap on or off
^C	Issues one cancel
^D	Toggles through coordinate display modes
^E	Toggles through ISOPLANE modes
^G	Toggles grid on or off
^H	Issues a backspace
^I	Issues a Tab (which usually acts like ENTER)
^M	Issues a carriage return
^O	Toggles ortho on or off
^P	Toggles MENUECHO on or off
^Q	Toggles printer echo on or off
^T	Toggles tablet mode on or off
^V	Switches to the next viewport
^X	Deletes input in keyboard buffer
^Z	Suppresses automatic space at the end of a line
_	Tells AutoCAD to expect a command or keyword in English

Introduction to Menu Macros

In a macro, a space performs just the same as pressing the Spacebar at the keyboard. In most cases, this enters the previous input or causes the same results as pressing ENTER. AutoCAD acts as if there is a space at the end of a menu line unless the line ends with one of the special characters in Table 5.2. When AutoCAD encounters these special characters at the end of a menu line, it does not add an automatic space.

You represent control characters, such as CTRL+B, with two characters in a menu file, for example ^B. Enter a caret (^) followed by an uppercase letter, like ^B for CTRL+B. This coding avoids the conflicts that can occur if your text editor uses real control characters for its own purposes.

Note: The caret-plus-letter menu characters (for example, ^B and ^C) always function as described in Table 5.2, even when accelerator key definitions cause AutoCAD to do something different when you press CTRL key combinations on the keyboard.

Semicolons: A Special Case

Another important special character is the semicolon. Use it to add extension lines to the new bubble macro. The semicolon is the special character for the ENTER key. The new bubble macro in the following exercise uses more backslashes and the QUAdrant object snap mode. Again, the additions to the existing menu lines are shown in boldface.

Exercise

Using Semicolons and Object Snaps in a Macro

Return to editing ACADMA.MNS.

Add the boldface part to the end of the macro:

[Bubble]_.CIRCLE \24 _.TEXT _M @ 18 0 **_.LINE _QUA \\;** Enter

Save ACADMA.MNS, return to AutoCAD, and reload the menu.

From the MaxAC *pull-down menu, choose* Bubble	Issues the CIRCLE command and pauses
Command: _.CIRCLE 3P/2P/TTR/ <Middle point>: *Pick any point*	
Diameter/<Radius>: 24 Command: _.TEXT	Issues the radius and the TEXT command, then pauses
Text: CC Enter	Specifies text

Maximizing AutoCAD® R13

`Command: _.LINE From point: _QUA`	Issues the LINE command and the QUA object snap, then pauses
`of` *Pick the QUA point anywhere on circle*	Starts a line
`To point:` *Pick any point*	Draws the line
`To point:`	Issues a semicolon (ENTER) and ends the LINE command
`Command:`	

The two backslashes made the command pause for two input points after the LINE command; the last semicolon terminated it. You must supply one backslash for each entry that the user will make when the macro executes.

The last character is a semicolon. In most cases, you can use semicolons and spaces interchangeably in macros, just as you can either type a space from the keyboard or press ENTER. In macros, spaces have the same effect as pressing the Spacebar on the keyboard, and semicolons have the same effect as pressing the ENTER key. Unless there is a special character at the end of the line, AutoCAD executes the macro as if there is a trailing space. The preceding macro ends with a backslash, a special character, so you need to add the semicolon (or a space) to end the LINE command. In macros that need the final ENTER after a special character, always use a semicolon, which you can see, instead of an invisible training space.

Tip: Avoid using two spaces in a row for macros that require two ENTERs. You cannot easily count multiple spaces. Use semicolons, or use one space and then use semicolons for subsequent ENTERs.

Exercise

Spaces versus Returns in Text Macros

Return to editing ACADMA.MNS.

Add these two lines; note the `A1 LINE` versus the `A2;LINE`:

`[Bubble A1]_.CIRCLE \24 _.TEXT _M @ 18 0 A1 _.LINE _QUA \\;` [Enter]
`[Bubble A2]_.CIRCLE \24 _.TEXT _M @ 18 0 A2;_.LINE _QUA \\;` [Enter]

Save, return to AutoCAD, reload your menu and test both items. Which one works?

Introduction to Menu Macros

You might think that both lines should put their text label, A1 or A2, in their bubbles, but only the A2 item works. Because you must be able to type spaces in the middle of text strings, AutoCAD's menu interpreter treats them as true spaces in your text and does not enter the preceding input as it does in any other situation. If you want the macro to continue after text input, as in the A1 example, you need some way to tell AutoCAD that the string of text is complete. The way you do this is by using a semicolon to enter your text string.

Semicolons in Menu Text Strings

Except for spaces, all special characters act exactly the same whether they are in the middle of a text string or not. You can use a semicolon anywhere to issue ENTER. The sacrifice you make is that the only way to enter a real semicolon in a text string created by a menu is to use ASCII codes. The AutoCAD menu interpreter does not recognize ASCII codes, but the text string does.

You are probably familiar with AutoCAD's special characters in text, such as the %%u, which underscores a text string. The %% is AutoCAD's "escape" character for special text characters and formats. ASCII 59 is the ASCII code for the semicolon. If you embed a %%59 in a text string, you get a semicolon. For example, "Word%%59 item, stuff" becomes "Word; item, stuff." The %%*nn* format works for any character where *nn* is the decimal ASCII code number. AutoCAD also recognizes three special escape codes: %%d for the degrees symbol, %%p for the plus/minus symbol, and %%c for the diameter symbol. See Appendix F for a table of ASCII codes.

Note: U.S. versions of AutoCAD use 7-bit ASCII, but international versions use 8-bit ASCII codes. Prior to Release 12, you could use %%127, %%128, and %%129 for the degree, plus/minus, and diameter symbols, but Release 12 requires %%d, %%p, and %%c to create compatible drawings. You cannot use Unicode character codes (for example, `%\U+00B0`) in menu macros because AutoCAD's menu interpreter reads the backslash as a pause.

Calling and Avoiding Dialog Boxes in Macros

Macros can display dialog boxes, but they cannot provide predetermined input to dialog boxes. Do not use the DD*xxxxx* dialog box commands in macros unless you intend the user to provide all input. Some commands, such as PLOT, STYLE, and WBLOCK, have both dialog and older command prompt (non-dialog) forms for compatibility with previous AutoCAD releases. The CMDDIA system variable controls whether the PLOT command uses a dialog box, and FILEDIA controls whether commands that require a file name (for example, OPEN, SAVE, STYLE, and WBLOCK) use a dialog. Normally both system variables are set to 1 (on), but you can set them to 0 (off) to suppress dialogs during interactive use (that is, outside of menu macros).

When you include these dual personality commands in menu macros, AutoCAD automatically reverts to the command prompt form if any macro code or input follows the command name. In other words, during the macro, AutoCAD acts as if CMDDIA and FILEDIA are set to 0, even if they're currently set to 1. To plan such macros when typing the macro sequence at the command line, temporarily set the CMDDIA and FILEDIA system variables to 0 (off), so that you see the same AutoCAD prompts that the macro will "see." Remember to return CMDDIA and FILEDIA to 1 (on) when you're finished planning and testing the macro.

Dialog commands do not require a backslash for their input; they pause anyway until the user chooses OK or Cancel to close the dialog. If your macro requires the user to supply a file name and you want to allow dialog box selection of a file, follow the command name with a tilde (~) to display the file dialog box.

The next two exercises use the STYLE and DDSTYLE commands to demonstrate the subtleties of dialog box handling in menu macros.

 Exercise

Planning Macros with Commands that Use Dialog Boxes

First, you'll run through the STYLE command to remind yourself of the prompt sequence.

Command: **STYLE** [Enter]

Text style name (or ?) <STANDARD>: **TEST1** [Enter]

New style. *Choose* ROMANS.SHX *from the* File Name *list in the Select Font File dialog*   Because FILEDIA = 1, AutoCAD uses a dialog to prompt you for the font file (see Figure 5.4)

Height <0'-0">: [Enter]

Width factor <1.0000>: [Enter]

Obliquing angle <0>: [Enter]

Backwards? <N> [Enter]

Upside-down? <N> [Enter]

Vertical? <N> [Enter]

TEST1 is now the current text style.

Next, you'll repeat the previous sequence, but with FILEDIA turned off.

Command: **FILEDIA** [Enter]

Introduction to Menu Macros

Figure 5.4
STYLE command with dialog box.

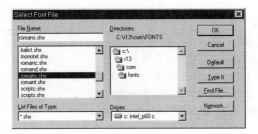

```
New value for FILEDIA <1>: 0 Enter

Command: STYLE Enter
```
Because FILEDIA = 0, AutoCAD uses the command line to prompt you for the font file

```
Text style name (or ?) <TITLE1>: TEST2 Enter
New style.
Font file <TXT>: ROMANS Enter
```

The remaining prompts are the same as in the first step. Use ENTER to accept the defaults for all six prompts. Lastly, you'll turn FILEDIA back on.

```
Command: FILEDIA Enter

New value for FILEDIA <0>: 1 Enter
```

In the previous exercise, the second sequence with FILEDIA = 0 mimics AutoCAD's behavior during the execution of a menu macro that includes code after the STYLE command name. In the next exercise, you'll write several variations on a macro that creates a new style.

 Exercise

Writing Macros with Commands that Use Dialog Boxes

Return to editing ACADMA.MNS. Delete the Bubble A1 and Bubble A2 macro lines and add the following five macros shown in bold directly below the Circle, Text, and Bubble lines:

```
[Bubble]_.CIRCLE \24 _.TEXT _M @ 18 0 \_.LINE _QUA \\;
[Title1 Style]_.STYLE TITLE1 [Enter]
[Title2 Style]_.STYLE TITLE2 \0 1 0 _No _No _No [Enter]
[Title3 Style]_.STYLE TITLE3 ~ 0 1 0 _No _No _No [Enter]
[Title4 Style]_.STYLE TITLE4 ROMAND 0 1 0 _No _No _No [Enter]
[DDStyle then MText]_DDSTYLE _MTEXT [Enter]
```

Save ACADMA.MNS, return to AutoCAD, and reload the menu. Next, you'll test each of the macros.

From the MaxAC *pull-down menu, choose* Title1 Style   Issues the STYLE command

```
Text style name (or ?)                  Issues the style name and opens
<TEST2>: TITLE1                         the Select Font File dialog
New style.
```

Although the STYLE command is being executed by a macro, there is no macro code after the part of the macro that precedes the dialog box (that is, the macro is finished), so AutoCAD displays the dialog box.

Choose ROMAND.SHX *from the* File Name *list in the Select Font File dialog*

Use ENTER to accept the defaults for the remaining six prompts.

From the MaxAC *pull-down menu, choose* Title2 Style   Issues the STYLE command

```
Command: STYLE Text style name (or ?)   Issues the style name and prompts
<TITLE1>: TITLE2                        for the font file on the command
New style.                              line

Font file <TXT>: ROMAND [Enter]

Height <0'-0">: 0                       Issues responses to the remaining
Width factor <1.0000>: 1                prompts
Obliquing angle <0>: 0
Backwards? <N> _No
Upside-down? <N> _No
Vertical? <N> _No
```

This time, the STYLE command suppressed the file dialog box because the macro includes code after the font file prompt.

From the pull-down menu, choose MaxAC, Title3 Style   Issues the STYLE command

Introduction to Menu Macros

`Command: STYLE Text style name (or ?) <TITLE2>: TITLE3 New style.`	Issues the style name and opens the Select Font File dialog

The tilde (~) in the Title3 Style macro forces AutoCAD to pop up the file dialog box, even though there is additional macro code.

Choose ROMAND.SHX *from the* File **N**ame *list in the* Select Font File *dialog*	Issues responses to the remaining prompts
From the pull-down menu, choose MaxAC, Title4 Style	Issues the STYLE command

`Command: STYLE Text style name (or ?) <TITLE3>: TITLE4`
`New style.`
`Font file <TXT>: ROMAND`
`Height <0'-0">: 0`
`Width factor <1.0000>: 1`
`Obliquing angle <0>: 0`
`Backwards? <N> _No`
`Upside-down? <N> _No`
`Vertical? <N> _No`
	Issues the STYLE command and answers all of the prompts

```
TITLE4 is now the current text style.
```

The Title4 Style macro includes the font file name, and thus is fully automatic.

 Warning: Autodesk added the DDSTYLE command in R13c4. Don't test the next macro if you're using an earlier version of AutoCAD.

From the MaxAC *pull-down menu, choose* DDStyle, *then* Mtext	Issues DDSTYLE and opens the Text Style dialog

```
Command: DDSTYLE
```

DDSTYLE in a menu macro always opens a dialog box because there is no command-line version of DDxxxx commands.

Choose TITLE1 *in the* Styles *list*	
Choose **C**lose	Makes TITLE1 the current style

> ```
> Command: MTEXT                              Issues the MTEXT command
> Attach/Rotation/Style/Height/
> Direction/<Insertion point>:
> ```
> *Complete the MTEXT command by selecting any two points and typing a sample title into the Edit MText dialog*
>
> If you see the error message `This font is not recognized by Windows. Select a font from the list to use while editing.`, choose Arial or any other uncomplicated Windows font. AutoCAD is asking for a font to use in the Edit MText dialog. The text will still appear with the ROMAND font in the drawing. See Chapter 6 for details about this error message.

Note that in the previous exercise's macros, we spelled out the responses to the prompts (_No instead of _N). The short form works fine, but as you string together more commands and options, macros become difficult to decipher. Using the long form of command options helps make the macro easier to read. In this book, we use uppercase letters for the minimum part of the option you would need to type at the command line and lowercase letters for the rest of the option.

Warning: Many third-party custom fonts for AutoCAD don't support vertical orientation. When you run the STYLE command and select one of these font files, AutoCAD doesn't show the final prompt (`Vertical? <N>`). Thus, the STYLE command can give a total of seven or eight prompts, depending on the font file. This variation can cause problems for menu macros. You need to use AutoLISP and the CMDACTIVE system variable to work around this problem.

Five of AutoCAD's newer dialog box commands whose names don't begin with DD (for example, BHATCH and MTEXT) have command-line versions that begin with a dash (-BHATCH and -MTEXT). These versions are handy for macros that supply all of the input and don't require the user's response. For example:

```
[Not for Constr. Note]_.-MTEXT _Height 18 \\Not for Construction!;;
```

The two backslashes pause for the user to select corners of the mtext box. The first semicolon enters the string, and the second one exits the -MTEXT command. The five dialog box commands with dash prefix command line versions are BHATCH, BOUNDARY, GROUP, HATCHEDIT, and MTEXT.

AutoCAD's conventions for when to display a dialog box can be a bit confusing at first, but once you master them, you'll have the flexibility to use dialog boxes or not in many of your macros. For example, to insert a symbol that the user selects from the file dialog box, you might use the following macro:

Introduction to Menu Macros

```
[Insert 1:1]_.INSERT ~ \1 1 0
```

(The backslash is for the insertion point, not the dialog box.)

Setting Modes with Control Characters, and Using Blocks in Macros

The next exercise shows you how to use control characters to set modes, such as snap and ortho. You will make a Leader macro that draws a two-segment polyline and inserts an arrow block at the end. The macro uses the ARROW block from the MaxAC CD-ROM, which is shown in Figure 5.5 for reference. It rotates the arrow to the midpoint of the line, using MIDpoint object snap. The Leader macro starts at the "tail" and ends at the arrow.

The three Leader macros in this chapter demonstrate a number of useful macro techniques. Although they duplicate to some degree the functionality of AutoCAD's LEADER command, there are reasons that you might prefer a macro similar to one of these for your drawing tasks. These macros are simpler to use without having to struggle with the subtleties of dimension styles and mtext. Many drafters find that drawing all of

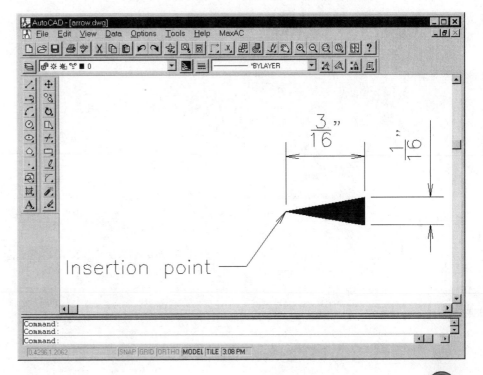

Figure 5.5
The pointing arrow block.

the text notations first and then drawing leaders to the text once all of the notes are positioned is a more efficient procedure in crowded drawings. In any event, these macros can be the springboard for others that automate your custom annotations.

Exercise

Making the Leader Macro

Continue in the previous drawing and erase everything, or start a new drawing using MXACH05.DWG as the prototype.

Return to editing ACAD.MNS. Add the following macro (as a single line) right after the previous ones:

[Leader]^C^C^C_.ORTHO _ON _.PLINE \\^O\ _.INSERT arrow @ 96 ; [»] _MID @ ^O [Enter]

Save ACADMA.MNS, return to AutoCAD, and reload the menu.

From the MaxAC *pull-down menu, choose* Leader

Command: _.ORTHO ON/OFF <Off>: _ON Command: _.PLINE	Issues ORTHO ON and PLINE, then pauses
From point: *Pick a point at ① for the shaft end of the leader (see Figure 5.6)*	
Current line-width is 0'-0"	
Arc/Close/Halfwidth/Length/Undo/Width/ <Endpoint of line>: *Pick a point at ②*	
Arc/Close/Halfwidth/Length/Undo/ Width/<Endpoint of line>: <Ortho off> *Pick a point at ③ for the arrowhead*	Issues CTRL+O to turn off ortho, then draws to the picked point
Arc/Close/Halfwidth/Length/Undo/ Width/<Endpoint of line>:	Issues a space to end PLINE
Command: _.INSERT Block name (or ?): arrow Insertion point: @ X scale factor <1> / Corner / XYZ: 96 Y scale factor (default=X):	Issues INSERT with ARROW block name at @ last point and 96 scale

Introduction to Menu Macros

```
Rotation angle <0>: _MID of @        Rotates angle to align with the line

Command: <Ortho on>                  Issues CTRL+O to end with ortho
                                     on
```

The ORTHO, PLINE, and BLOCK commands scroll by, and the macro ends with `<Ortho on>`, leaving a two-segment leader and arrow. Your screen should show the two-segment polyline leader (see Figure 5.6).

Several items in this Leader macro are new. The routine begins with three CTRL+Cs to cancel any currently active command. Autodesk uses two CTRL+Cs in its macros, which will cancel out of most commands. We recommend three CTRL+Cs because, in some deep dark recesses of AutoCAD and AutoLISP (for example, DIM HORizontal 'ZOOM), it takes three CTRL+Cs to get back to the command prompt. Whether you choose two or three CTRL+Cs, always use them to begin macros, unless the macro calls a transparent command, like ZOOM or LAYER. The CTRL+Cs ensure that your macro doesn't crash because the user forgot to finish the previous command.

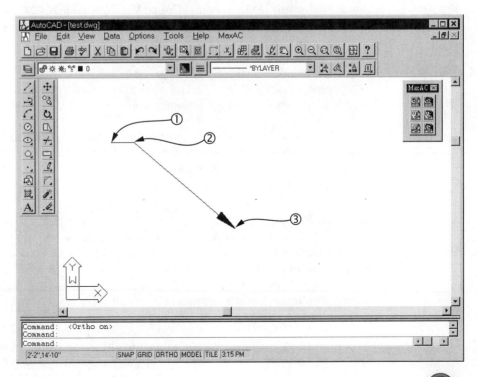

Figure 5.6
The results of the Leader macro.

On the other hand, omit the CTRL+Cs and use the apostrophe (') before macro commands that can run transparently. For example, the following macro zooms out and turns on all layers:

```
[Look at It All]'_.ZOOM 0.9X '_.LAYER _ON * ;
```

Because of the apostrophes, this macro will work in the middle of another command.

You probably are accustomed to using F8 or CTRL+O (in DOS) to toggle ortho. In menus, you can control such settings with commands, system variables, or control-characters, but not function keys. Because a setting may be already on or off, the Leader macro forces ortho on with the ORTHO command. After the first segment of the polyline, the ^O turns ortho off. Later another ^O turns ortho back on. To be certain of the initial ortho setting, the Leader macro initially used the ORTHO command instead of the CTRL+O toggle. Alternatively, you can use system variables in such cases.

Tip: The Leader macro leaves ortho on whether it was on or off before. It is better to check the current status of settings, make your changes, and then restore the original settings, but to do so in a macro requires AutoLISP. You learn to use AutoLISP to control settings in Chapter 9.

The Leader macro uses lowercase letters for the ARROW file name in order to maintain compatibility with UNIX systems. Unlike Windows and DOS, UNIX is sensitive to case in file names, and the UNIX convention is to use lowercase for user data files.

The Leader macro uses the @ symbol to recall the last point at the end of the polyline during the block insertion. The macro then feeds the X scale value of 96 (plot scale). The semicolon in `96 ;_MID` defaults the Y scale factor to equal the X. When you accept default values, show them with a semicolon.

Tip: You can make your macros more versatile by using AutoLISP to calculate scale. Chapter 14 discusses this topic, and the end of this chapter shows macros that demonstrate the technique.

Lastly, the Leader macro snaps the block rotation with `MID @`. Because the last point is at the end of the polyline, MIDpoint orients the arrow symbol to the angle of the line. This orientation is the reason we created the ARROW block pointing to the left with its insertion point at the arrow point.

Using Blocks with Path Names

In the previous exercise, the ARROW.DWG file required by the Leader macro was in the current directory. When the INSERT command doesn't find a block already defined in the current drawing, it looks in all the directories on the library search path for a drawing

Introduction to Menu Macros

of the same name. Because ARROW.DWG was in AutoCAD's current directory, the INSERT command found it. (See Chapter 3 for more about the library search path.)

You might want to write macros that use blocks or other files located in other specific directories outside of AutoCAD's library search path. In these cases, you must use forward slashes (/) as directory name separators instead of backslashes, because backslashes always represent pauses. For example, to insert \ODDBLOCK\WIDGET.DWG at a scale factor of 96 and rotation angle of 0, you might use the following macro:

```
[Widget]^C^C^C_.INSERT /oddblock/widget \96 ;0
```

Or, if you wanted to make sure that the macro always used the \ODDBLOCK directory on the C: drive, you could write:

```
[Widget]^C^C^C_.INSERT c:/oddblock/widget \96 ;0
```

If you're using path names that include spaces in Windows 95 or NT, enclose the entire file name in quotation marks:

```
[Widget]^C^C^C_.INSERT "/odd blocks/widget" \96 ;0
```

Warning: Avoid creating drawing file names with unusual characters (for example, spaces) when you plan to use those drawings as blocks or xrefs. Although R13c4 supports long file names and additional characters, it still requires that the block names stored in the drawing be composed only of letters, numbers and the three characters dollar sign ($), underscore (_), and dash (-). If your macro inserts a block containing other characters, AutoCAD will pop up the Substitute Block Name dialog every time the macro runs, as shown in Figure 5.7.

Note: In most cases, you should avoid menu macros that reference specific paths. The locations of blocks and other files on your system are likely to change over time, and when they do you'll have to change every path reference in your menu. This problem is even more common when your menus are used by other people on other systems. Instead of "hard coding" the path, keep most of your blocks and other support files organized in one or more directories and add those directories to the support path, as described in Chapter 2. Then when you reorganize your directory structure, all you need to change is one support path reference.

There are some legitimate uses for hard-coded path references. For example, if your company has a typical detail library comprising hundreds of details in a number of

Figure 5.7
The Substitute Block Name dialog.

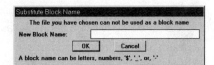

directories, you probably don't want to add all those directories to the support path. In that case, you might create a series of similar menu choices for inserting each detail:

```
[Concrete 01]^C^C^C_.INSERT /typdets/concrete/tdconc01 \1 ;0
[Concrete 02]^C^C^C_.INSERT /typdets/concrete/tdconc02 \1 ;0
...
[Steel 01   ]^C^C^C_.INSERT /typdets/steel/tdstel01 \1 ;0
[Steel 02]^C^C^C_.INSERT /typdets/steel/tdstel02 \1 ;0
...
```

If the company reorganizes its typical detail library, you can use your text editor's search-and-replace function to modify all of the paths quickly. Another alternative is to use AutoLISP to store and supply a path name to a macro. See Chapter 9 for more on AutoLISP.

Making Long Macros

As your sophistication with menu macros increases, you'll write macros longer than what will fit comfortably on one line. The Leader macro from the previous exercise is already getting uncomfortably close, and we aren't finished with it yet. Although AutoCAD doesn't mind long lines in menu files, you'll find them difficult to view, edit, and print. You can split a macro over multiple lines by ending each line with a plus sign. Let's split the Leader macro across two lines and give it a longer label to distinguish it from AutoCAD's LEADER command in this exercise.

Exercise

Splitting a Macro Across Multiple Lines

Return to editing ACAD.MNS. Modify the Leader macro so that it reads as follows. Changes are shown in bold.

```
[Leader Arrow]^C^C^C_.ORTHO _ON _.PLINE \\^O\ + [Enter]
_.INSERT arrow @ 96 ;_MID @ ^O [Enter]
```

Save ACADMA.MNS, return to AutoCAD, and reload the menu, then retest the macro.

Tip: Always retest macros after you make changes, no matter how insignificant the changes seem. It only takes one extra or missing space to crash a menu macro.

Introduction to Menu Macros

Look at the end of the first macro line. When AutoCAD sees a plus sign at the end of the line, it treats the next line as part of the same item. The most important thing to remember about the plus character is that you cannot put a space or semicolon after it.

Note: Although the ability to split a macro across multiple lines is handy, don't overdo it. Macros that get much longer than two or three lines become difficult to read and debug. If your macros get that complex, then they're good candidates for conversion to AutoLISP programs. See Chapter 9 for more about AutoLISP.

Polyline Macro Tricks

Now, try a few macro tricks. In the next exercise, you'll create a macro that draws an entire leader, including its arrow, as a single polyline. The advantage of this Leader Arrow—> macro is that you can select the resulting leader as a single object.

Exercise

Making a Single-Object Leader Arrow—> Macro

First, you'll draw the polyline leader interactively with AutoCAD, noting every character and pick so that you can transcribe them into a macro later.

Command: **ORTHO** [Enter]

ON/OFF <On>: **ON** [Enter]

Command: **PLINE** [Enter]

From point: *Pick a point*

Current line-width is 0'-0"

Arc/Close/Halfwidth/Length/Undo/Width/<Endpoint of line>: **W** [Enter]

Starting width <0'-0">: **0** [Enter]   Ensures a 0 width

Ending width <0'-0">: **0** [Enter]

Arc/Close/Halfwidth/Length/Undo/Width/<Endpoint of line>: *Pick a point*

Turn ortho off with F8 and pick another point.

You'll use ^O in the menu macro to toggle ortho, but you can't use CTRL+O at the keyboard to toggle ortho in R13 for Windows. As Chapter 3 demonstrated, an accelerator key defines CTRL+O as OPEN, in accordance with Windows conventions. Use F8 at the keyboard or double-click on the ORTHO indicator on the AutoCAD status bar.

```
Arc/Close/Halfwidth/Length/Undo/Width/
<Endpoint of line>: W [Enter]
```

```
Starting width <0'-0">: 0 [Enter]
```
Sets the width of arrow's head point

```
Ending width <0'-0">: 6 [Enter]
```
Sets the width of arrow's head base

```
Arc/Close/Halfwidth/Length/Undo/
Width/<Endpoint of line>: L [Enter]
```
Specifies the Length option

```
Length of line: -18 [Enter]
```
Specifies minus 18 length

```
Arc/Close/Halfwidth/Length/Undo/Width/
<Endpoint of line>: W [Enter]
```

```
Starting width <0'-6">: 0 [Enter]
```
Resets 0 width

```
Ending width <0'-6">: 0 [Enter]
```

```
Arc/Close/Halfwidth/Length/Undo/
Width/<Endpoint of line>: [Enter]
```
Exits PLINE

Collect your typed input and picks and string them together to make the Leader Arrow—> macro. Use a plus sign to split the macro across two lines. Add these two lines to ACADMA.MNS:

[Leader Arrow-->]^C^C^C_.ORTHO _ON _.PLINE _Width 0 0 \^O\+ [Enter]
_Width 0 6 _Length -18 _Width 0 0 ; [Enter]

Save ACADMA.MNS, return to AutoCAD, and reload the menu.

From the MaxAC *pull-down menu, choose* Leader Arrow—>

Executes macro

Pick points when Leader Arrow—> pauses, in the same sequence as you used to define it.

Select the leader to verify that it's a single polyline, including the arrowhead, as indicated by the highlighting in Figure 5.8.

Your screen should show a single-object leader. It looks like the earlier leader with the ARROW block. The Leader Arrow—> label uses the —> to graphically indicate the direction of the leader, as shown in Figure 5.8.

Introduction to Menu Macros

Figure 5.8
A single-object leader.

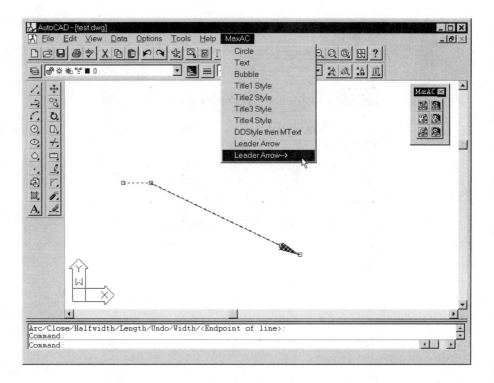

Remember that backslashes (as in _Width 0 0) and control characters, such as ^O, do not need a space or ENTER after them to execute. Because you cannot predict the incoming default polyline width, your polyline macros should always set the starting and ending widths explicitly, as in W 0 0. The _Width 0 6 sequence sets the width to draw the arrowhead.

The _Length parameter draws the 18-inch-long arrow segment by using a negative-length polyline. This trick folds the polyline arrowhead back upon the previous polyline segment with _Length -18. With this arrangement, you can easily pick the point for the arrowhead; otherwise, you would have to pick your last point exactly 18 units short of where you want the arrow to terminate.

Try defining one more leader macro, one that starts with the arrow instead of terminating with it. Label it with Leader Arrow<—. This macro uses a PEDIT trick to reset the last point to the start point of a polyline. With this arrangement, you can have the macro start with an arrow, but you must pick the second point twice. The macro also depends on the PLINE command's Length option, the angle of which defaults to the angle of the last segment of the current or previous polyline. This macro can draw as many line or arc polyline segments as you want. The macro is explained after the exercise.

Exercise

Making a Multisegment Leader Arrow<— Macro

Use ERASE to clean up your screen, and then add the following macro to ACADMA.MNS:

```
[Leader Arrow<-]^C^C^C_.PLINE \\;_.PEDIT _Last _Edit _Move @ +
_eXit ;_.ERASE _Last ;_.PLINE @ _Width 0 6 _Length 18 +
_Width 0 0 _Length 1 [Enter]
```

Save ACADMA.MNS, return to AutoCAD, and reload the menu.

From the MaxAC *pull-down menu, choose* Leader Executes the macro, starting with
Arrow<— issuing PLINE

Command: _.PLINE

From point: *Pick two points at ① and ② in Figure 5.9, keeping the cursor at the second point*

Many PLINE and PEDIT prompts scroll by, then:

Arc/Close/Halfwidth/Length/Undo/Width/
<Endpoint of line>: *Repick the ② point, and then several more points, such as ③ and ④*

Leaving snap on helps you select the same point twice.

Press ENTER to end the macro

Your screen should show a multisegment polyline leader, as shown in Figure 5.9. After the second point, you can enter an **A** at any Arc/Close/Halfwidth/Length/Undo/Width/<Endpoint of line>: prompt to draw arc segments, as shown at ⑤ in Figure 5.9.

_.PLINE \\; draws a single temporary polyline segment that establishes an angle for the arrow. The _.PEDIT _Last _Edit _Move @ is the key trick used in the macro. It does not actually move the vertex, but the @ resets the LASTPOINT system variable to the start of the polyline.

The _eXit ; exits the PEDIT command, then _.ERASE _Last ; erases the temporary polyline. The macro starts the real polyline with an arrow at the last point using _.PLINE @. The _Width 0 6 sets the arrow width, and the macro draws it with the length parameter using _Length 18. The _Width 0 0 sets width back to 0. The macro ends with _Length 1, a short segment, to keep the next segment from distorting the arrow. If you do not use a short straight-line segment, the macro miters the arrow

Introduction to Menu Macros

Figure 5.9
Multisegment polyline leader.

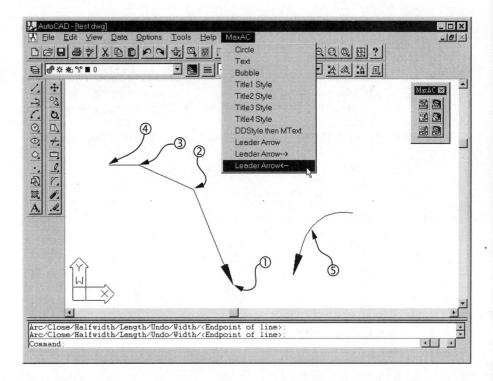

base relative to the angle of the next segment. The macro leaves you in the PLINE command. Repick the second point before moving the cursor, and draw as many line segments as you want.

Object Snaps in Macros

You can explore some more object snap features, making the Bubble macro require just two picks. SELECT, an AutoCAD command seldom used outside macros, also enhances the Bubble macro's features. SELECT does nothing except create a selection set, which can then be accessed with the Previous object selection option.

Exercise

Improving the Bubble Macro with the SELECT Command

Use ERASE to clean up your screen.

Edit the Bubble item in ACADMA.MNS so that it looks like the following two lines. The new code is shown in bold:

[Bubble]^C^C^C_.CIRCLE \24 **_.SELECT _Last ;**_.TEXT **_Middle** @ 18 0 + [Enter]

.LINE @ \;.ID _MID,_QUI @ _.TRIM _Previous ;_ENDP,_QUI @ ; [Enter]

Save ACADMA.MNS, return to AutoCAD, and reload the menu.

From the MaxAC pull-down menu, choose
Bubble

Command: _.CIRCLE 3P/2P/TTR/

<Center point>: *Pick a point*

Diameter/<Radius>: 24	Issues radius
Command: _.SELECT	Issues SELECT
Select objects: _Last 1 found.	Issues L to save the circle as the Previous selection set
Select objects:	Returns to end object selection
Command: _.TEXT	Issues the TEXT command, then _M @ 18 0^, then pauses at the Text: prompt
Text: **A3** [Enter]	Specifies the text
Command: LINE From point: @	Starts at the circle's center
To point: *Pick a point*	Specifies the other end of the line
To point:	Issues a semicolon to end the line
Command: ID Point: _MID,_QUI @ X = 130'-0" Y = 128'-0" Z = 0'-0"	Saves the line's midpoint as the LASTPOINT
Command: _.TRIM Select cutting edge(s)... Select objects: _Previous 1 found.	Issues TRIM and selects the circle SELECT saved as Previous
Select objects:	Issues a semicolon to end object selection

Introduction to Menu Macros

`Select object to trim: _ENDP,_QUI @`	Trims the end at the saved LASTPOINT
`Select object to trim:`	Issues a semicolon to end object selection

The object snap trick `ID MID,QUI@` uses the harmless ID command to reset the LASTPOINT system variable to the midpoint of the line, which is picked by @. QUIck mode makes the macro find the most recently entered nearby midpoint, rather than the closest midpoint. Because you just drew the line, you know it is the most recent. In this way, you can avoid having MID find the wrong line in dense areas of the drawing. QUIck also speeds up object snap. Use QUIck when you are using object snap to snap to an object that you are sure is more recent than its neighbors. This situation occurs frequently in macros. To control recent object selection, see the following "Object Sort Method Control" section.

To trim off the extra bit of line, `_.TRIM _Previous ;` selects the circle as a cutting edge, using the Previous option to select the circle that the SELECT command saved. Then `_ENDP, _QUI @` picks the line to trim. The macro uses ENDPoint to select the correct end, and @ to pick the last point that ID saved. Although both ends of any line are equidistant from its midpoint (your last point in the macro), AutoCAD almost always finds the first endpoint of the line if you snap with object snap ENDP from the midpoint of the line. The final semicolon exits from TRIM and ends the macro.

Tip: The macro uses ENDP (instead of END) to avoid accidentally ending the drawing.

Note: This bubble macro works nearly all the time. AutoCAD usually finds the first endpoint of a line, but occasionally it finds the second. As you can see, this macro pushed the limits of menu macros. A more reliable bubble macro would require AutoLISP for saving and recalling points.

Object Sort Method Control

Older versions of AutoCAD always sorted objects in their order of creation. Releases 12 and 13, however, use a more efficient oct-tree (octal tree) method of sorting objects. The sort order of the object database determines which objects are found first by object selection and object snaps, and the sort order also determines the display or output order of redraws, regenerations, MSLIDE, plotting, and PostScript output. Prior to R12, the sorting default found the most recently created objects during selection and object snaps, and it displayed or output objects in their order of creation. The oct-tree sorting method divides the drawing into rectangular areas and sorts, selects, displays, and outputs objects in this grid order. This type of sorting makes object selection and object snaps much faster because they search only the immediate area, not the whole drawing. However, with oct-tree sorting on, you cannot always be sure the most recent object will be found first.

For object selection or the QUIck mode to find the most recent object predictably, you need to turn oct-tree object sorting off for object selection and object snap. You can do this by making sure that the **O**bject Selection and Object **S**nap check boxes are checked in the Object Sort Method dialog box (choose Object Sort Method from the DDSELECT command's Object Selection Settings dialog box to display the Object Sort Method dialog box). You also can turn oct-tree sorting off with the SORTENTS system variable. SORTENTS is a bit-coded system variable; the values 1, 2, 4, 8, 16, 32, and 64 correspond to the seven check boxes in the Object Sort Method dialog box. A value turns oct-tree sorting off for that item. The sum of the values is the SORTENTS value. The default is 96 (32+64), which leaves oct-tree sorting on for all but Plotting and Postscript output.

To control the sort method properly would require AutoLISP to save and restore the settings, so the previous Bubble macro ignores this problem. See Chapter 9 for more on the use of AutoLISP to save and restore settings with the `getvar` and `setvar` functions.

Controlling Aperture in Macros

The APERTURE system variable controls what objects AutoCAD considers when you are using object snap. AutoCAD considers only those objects that cross the aperture box. You can do some interesting tricks by making the aperture extremely large or small. An aperture of one pixel is good for finding specific objects in crowded areas; an aperture of 50 (the maximum size) has a better chance of finding an object in sparse areas of a drawing. If your macros do change the APERTURE system variable in order to perform a specific task, they should reset it afterwards.

Controlling Layers with Macros

Thus far, all of our example macros have drawn objects on the layer that happens to be current when you run the macro. Presumably the user would change layers before running each macro. It's safer and more efficient to have your macros control layers automatically. This approach also helps ensure consistency and lets your macros put different objects (for example, text, bubble, and grid line) on different layers.

The Pipe Valve macro shows you how to control layer names, color, and linetype settings with a menu macro. Rather than assuming that a layer is created in a prototype drawing or by a setup routine, assume that it does not exist. Have the macro create the layer that

Introduction to Menu Macros

it uses. This type of macro control avoids problems if a user purges unused layers from a drawing.

The Pipe Valve macro uses several AutoCAD commands. It makes a layer, draws a line, then inserts a block named VALVE. The VALVE.DWG is on the MaxAC CD-ROM.

Exercise

The Pipe Valve Macro with Layer Setup

Continue in the previous drawing and erase everything, or start a new drawing using MXACH05.DWG as the prototype.

If you are unfamiliar with the exact syntax of AutoCAD's LAYER command, test and record the macro sequence in AutoCAD.

Edit ACADMA.MNS and add the following macro:

```
[Pipe Valve]^C^C^C_.LAYER _Make P-EQPM _Color _RED ;;+ Enter
_.LINE \\;_.INSERT valve @ 96 ;_DRAG Enter
```

Save ACADMA.MNS, return to AutoCAD, and reload the menu.

From the pull-down menu, choose Pipe Valve

`Command: _.LAYER ?/Make/Set/New/ON/OFF/Color/Ltype/Freeze/Thaw/LOck/Unlock: _Make New current layer <0>: P-EQPM ?/Make/Set/New/ON/OFF/Color/Ltype/Freeze/Thaw/LOck/Unlock: _Color Color: RED Layer name(s) for color 1 (red) <P-EQPM>:?/Make/Set/New/ON/OFF/Color/Ltype/Freeze/Thaw/LOck/Unlock: Command: _.LINE From point:` *Pick a point at ① in Figure 5.10*	Issues LAYER and makes a new layer named P-EQPM with the color red
`To point:` *Pick another point at ②*	Draws the pipe line
`To point:`	Ends LINE with a semicolon

```
Command: _.INSERT Block name          Issues the INSERT command
(or ?): VALVE Insertion point: @ X
scale factor <1> / Corner / XYZ: 96
Y scale factor (default=X):

 Rotation angle <0>: _DRAG             Drags rotation and inserts the valve
```
Pick a point at ③ to rotate the valve symbol into place

Use ZOOM to see the valve better (see Figure 5.10), then erase it and ZOOM Previous.

This is how the Pipe Valve macro works. First, `_.LAYER _Make P-EQPM _Color _RED ;;` sets the layer and its color. Instead of Set, the macro uses Make to set the layer current. The Make option creates the layer if the layer does not exist, and just sets it current if it does. In `_RED ;;` the space enters RED, the first semicolon defaults the current layer to RED, and the second semicolon exits the LAYER command. The plus sign continues the macro to the next line.

Next, `_.LINE \\;` draws a single pipe line when you pick two endpoints, and terminates the LINE command with the semicolon. `_.INSERT valve @ 96;` places the insert at

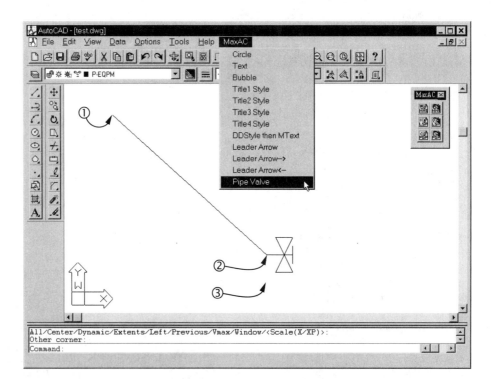

Figure 5.10
The pipe line and flow valve.

Introduction to Menu Macros

the last point with an X scale of 96 and with a semicolon defaulting the Y scale equal to X. Finally, _DRAG ensures that drag is on while you rotate the insert. The angle relative to which it drags makes it easy to align with an adjacent pipe section. Although drag's default is Auto, some users may have it turned off. If a macro depends on dragging to show the user orientation, you should explicitly drag to be sure.

Note: Professional applications shouldn't use explicit layer names in menus or AutoLISP programs. Instead, the application should include AutoLISP functions for creating and setting layers. These functions should look up layer names in a data file, so that programmers and users can modify the layer names easily in one place. These kinds of advanced techniques are covered in *Maximizing AutoLISP Release 13*.

Tip: Long macros, particularly ones that change settings or layers, may alter your AutoCAD settings if they fail due to an error or CTRL+C. An UNDO Mark at the beginning of macros enables an UNDO Back to restore the previous condition cleanly.

Repeating Macros and Multiple Commands

A little automation always calls for more. One limitation of menu macros is that, unlike AutoCAD commands, they don't repeat when you press ENTER. Why should you have to reselect a macro repeatedly from the screen if you want to use it several times? You can improve the Pipe Valve macro by making it repeat. One way to repeat is simply to add more macro code, as in the next exercise.

Exercise

Making the Pipe Valve Macro Repeat

Copy the Pipe Valve macro and modify the copy as shown in bold:

[5 Pipe Valve**s**]^C^C^C_.LAYER _Make P-EQPM _Color _RED ;;+ [Enter]

.LINE \\;.INSERT valve @ 96 ;_DRAG \+ [Enter]
.LINE **;**;.INSERT **;**@ 96 ;_DRAG \+ [Enter]
.LINE **;**;.INSERT **;**@ 96 ;_DRAG \+ [Enter]
.LINE **;**;.INSERT **;**@ 96 ;_DRAG \+ [Enter]
.LINE **;**;.INSERT **;**@ 96 ;_DRAG [Enter]

Figure 5.11
Five pipe lines and valves.

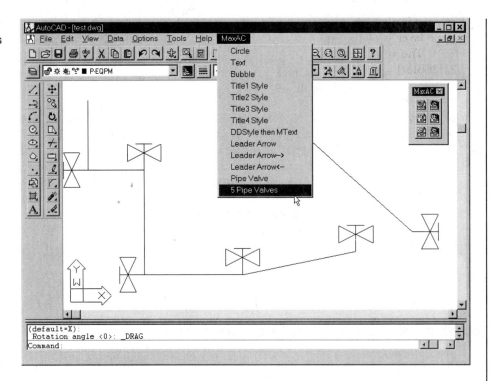

Save ACADMA.MNS, return to AutoCAD, and reload the menu.

From the MaxAC *pull-down menu, choose* 5 Pipe Valves

Pick points to draw five lines and insert five valves; it works as before, but draws five segments (see Figure 5.11).

You could fit this repeating macro on fewer lines, but this format makes it easier to see what it does. The original macro did not need a backslash after ending DRAG at the rotation prompt of the INSERT command. However, because each subsequent line starts with another LINE command, you have to fill the rotation prompt with a pause. The last line of an item never gets a plus sign continuation character, nor does the last line need a backslash.

Note the differences between the second line and the successive lines. The repetitive lines start with _.LINE ;\ instead of _.LINE \\, using a semicolon to default the new start point to the endpoint of the previous line. The macro is more efficient after the initial

Introduction to Menu Macros

line/valve set is placed because you do not need to repeat the layer setting or the block name. The layer is already set and the INSERT command has a default block from the first insert. The repetitive INSERT commands replace `valve` with a semicolon, to use the default block name.

If you want to draw fewer than five sections, you can always cut the macro short by pressing ESC.

Indefinite Repetitions

In most cases, repeating a macro's operation by copying the code several times isn't very elegant or flexible. You can make AutoCAD macros repeat indefinitely by putting an asterisk as the first character of the macro, right after the label's closing right bracket and before the first ^C.

In the next exercise, you make a new macro called Pipe Valves that repeats by using the asterisk feature.

Exercise

Making Commands and Macros Repeat Indefinitely

Erase the 5 Pipe Valves macro, then copy the previous Pipe Valve macro and edit it as shown.

[Pipe Valve**s**]*^C^C^C_.LAYER _Make P-EQPM _Color _RED ;;+ [Enter]

.LINE \\;.INSERT VALVE @ 96 ;_DRAG \ [Enter]

Save ACADMA.MNS, return to AutoCAD, and reload the menu.

From the MaxAC *pull-down menu, choose* Pipe Valves

The Pipe Valves macro works like the first Pipe Valve one, drawing a single segment and valve, but repeats until you cancel it with ESC. With each repetition, you can press ENTER to continue from the previous line or pick a point to start a new line.

The asterisk triggers the menu interpreter to repeat the macro in its entirety. The asterisk must be followed by at least one ^C or a ^X. Otherwise, it is interpreted as a character and interferes with the command or input that follows. AutoCAD would see *LAYER, not recognize it, and issue an error. This macro requires a backslash at the end to pause while you rotate the valve into place, before the macro returns to the beginning and repeats.

Note: In previous versions of AutoCAD, another option for repeating macros was the REPEAT command. REPEAT was discontinued in R13.

Using Selection Sets in Macros

You can use selection sets in macros to control how users select objects. In AutoCAD's normal verb/noun editing mode, edit and inquiry commands prompt for a set of objects before performing the actual operation. In macros, however, you use a specific number of backslashes to pause for selection. The macro must anticipate how the user will select objects. An explicit Window or Crossing selection requires three backslashes (the W or C and two points), Last requires one, and Fence, Cpoly, and WPoly require an indeterminate number. You need to predefine one backslash per input, but you have no way of knowing in advance how the user will want to select objects.

A solution exists: AutoCAD's SELECT command. SELECT is indispensable in macros because it has the unique feature of pausing the macro indefinitely, with only one backslash, until the entire selection process is completed. SELECT also recognizes a noun/verb selection set (with the PICKFIRST system variable turned on). Try an example, the Duplicate Layer macro, which copies a set of objects from one layer to another.

Exercise

Building Selection Sets in Advance

Edit ACADMA.MNS and add the following macro.

`[Duplicate Layer]^C^C^C_.SELECT \_.COPY _Previous ;0,0 ;+` [Enter]
`_.CHPROP _Previous ;_LAyer \;` [Enter]

Save ACADMA.MNS, return to AutoCAD, and reload the menu.

Your current layer should be P-EQPM, with red, continuous lines. Try Duplicate Layer on existing objects, or draw a few new ones to use.

From the MaxAC *pull-down menu, choose* Duplicate Layer

`Command: _.SELECT` [Enter]	Issues SELECT and pauses with a single backslash
`Select objects:` *Select several lines*	Creates the selection set
`Select objects:` [Enter]	Ends object selection

Introduction to Menu Macros

Command: _.COPY Select objects: _Previous 2 found Select objects: <Base point or displacement>/ Multiple: 0,0 Second point of displacement: Command: _.CHPROP Select objects: _Previous 2 found Select objects: Change what property (Color/LAyer/ LType/ltScale/Thickness) ? _LAyer	Issues COPY and CHPROP, which both reselect objects with Previous option
New layer <P-EQPM>: **P-EQPM-HIDD** [Enter]	Specifies a new layer for the objects

Change what property (Color/LAyer/
LType/ltScale/Thickness) ?

Turn off layer P-EQPM if necessary to see the lines that appear as blue and hidden on layer P-EQPM-HIDD.

Command: **PICKFIRST** [Enter]

New value for PICKFIRST <1>: *Make sure that it is set to 1*

Command: *Select some objects*	Highlights and selects the objects
Command: *From the* MaxAC *pull-down menu, choose* Duplicate Layer	Uses the pending noun/verb selection set instead of pausing for selection

Command: _.SELECT
Select objects: 8 found
Command: _.COPY
Select objects: P 8 found
Select objects:
<Base point or displacement>/Multiple: 0,0
Second point of displacement:
Command: _.CHPROP
Select objects: _Previous 8 found
Select objects:
Change what property (Color/LAyer/LType/
ltScale/Thickness) ? LAyer
New layer <P-EQPM>: 0 [Enter]

As you see in Duplicate Layer, SELECT does not do anything except create a selection set of objects enabling an indefinite number of picks. This capability is invaluable because you can use P (Previous) with a subsequent command, such as COPY, to reselect that selection set. The CHPROP command in the macro also uses Previous in Duplicate Layer, but technically it is reselecting COPY's set.

Notice the use of `0,0` ; to copy objects with 0,0 displacement. Many AutoCAD users are unaware that a response of ENTER to COPY's `Second point of displacement:` prompt causes it to take the first point as a relative displacement (0,0 in this case). The same ENTER response applies to MOVE and STRETCH.

Note: You always should use CHPROP instead of CHANGE when changing properties. CHANGE rejects many objects if they are 3D or if you are drawing in skewed UCSs.

Object Selection Settings and Menus

Releases 12 and 13 have several selection and editing options that were not available prior to R12. These options and the system variables that control them are listed here:

- Noun/Verb Selection (PICKFIRST system variable)
- Use Shift to Add (PICKADD system variable)
- Press and Drag (PICKDRAG system variable)
- Implied Windowing (PICKAUTO system variable)
- Grips (GRIPS system variable)

These features add flexibility to AutoCAD selection and editing, and they allow users the option of configuring AutoCAD to act more like other types of applications, such as drawing and illustration programs. You use the DDSELECT and DDGRIPS commands or the PICKxxxx and GRIPxxxx system variables to control these features. See Chapter 5 of

Figure 5.12
The Object Selection Settings dialog box.

Introduction to Menu Macros

the R13 *AutoCAD User's Guide* for more information. Figure 5.12 shows the Object Selection Settings (DDSELECT) dialog box.

Unfortunately, not all of these features work as you would expect in menu macros. The following list describes how these features interact with menus.

1. **Noun/Verb Selection.** Simple, singl-command macros such as `_.COPY` and more complicated macros that use SELECT, as in the previous exercise, use any pending selection set and skip normal verb/noun object selection. However, in macros that start with ^C and don't use SELECT, AutoCAD cancels any pending selection set and prompts for verb/noun object selection. Likewise, if any macro code follows a command that requires object selection, AutoCAD ignores the pending selection set.

2. **Use Shift to Add.** A macro that pauses with backslashes recognizes the Shift to Add feature. If a macro includes built-in selections, such as `_.COPY _Previous _Last`, the Previous and Last selections are added to the selection set without requiring you to hold down Shift.

3. **Press and Drag.** Macros work with the Press and Drag feature, so you can specify a window with a single pick instead of two picks by pressing the pick button at one corner, dragging to the other corner, then releasing the button. This sequence does fill two backslash pauses, just as it would with Press and Drag turned off. Thus macros work the same no matter what the current Press and Drag setting is. Also, because Implied Windowing does not work with macros containing backslashes, you must explicitly specify the Window or Crossing option to use Press and Drag.

4. **Implied Windowing.** A simple, single-command-only menu item, such as `_.COPY` or even `^C^C^C_.ERASE` works with Implied Windowing, which works like the AUto object selection option. Complex macros that use pauses, with or without SELECT, do not recognize Implied Windowing and require you to explicitly specify the Window or Crossing option. Unfortunately, even the AUto option doesn't work properly in macros. For example, in the macro fragment `_.SELECT _AUto`, the AUto object selection mode lasts only for the first selection. For subsequent selections, the user must again type Window, Crossing, or AUto.

5. **Grips.** Macros cancel grips, and thus ignore them.

Note: The PICKADD system variable can be confusing; when Use Shift to Add is on, PICKADD is 0 (off), not 1 (on). The rest of these system variables are 1 (on) when the selection settings they control are on.

Unfortunately, the way these features interact with menus makes it impossible to write complex macros that behave in the same manner as commands entered at the command prompt. The discrepancies create an undesirable and inconsistent environment for users who want to take advantage of AutoCAD's selection features. The best workaround is to use AutoLISP to define a function, and then call that function in the menu. For example, you could put the following lines in ACADMA.MNL:

```
(defun c:DUPLYR (/ SelSet)
   (setq SelSet (ssget))
   (command "_.COPY" "_Previous" "" "0,0" ""
            "_.CHPROP" "_Previous" "" "_LAyer" pause "")
)
```

Then the Duplicate Layer menu macro would become simply:

```
[Duplicate Layer]^C^C^CDUPLYR
```

This approach is completely compatible with the interactive object selection features. See Chapter 9 for more about AutoLISP and Chapter 14 for information on integrating AutoLISP with menus.

Selection Set Modes in Menus

All the selection modes—picking by point, Window, Last, Crossing, BOX, ALL, Fence, WPolygon, CPolygon, Add, Remove, Multiple, Previous, Undo, AUto, and SIngle—are available for use in menus. In fact, Last and Previous are particularly useful in macros; SIngle, BOX, and AUto are specifically intended for use in menus. The following list describes SIngle, BOX, and AUto:

- **SIngle.** SIngle stays in selection mode until one object or set is successfully picked.

- **BOX.** BOX acts like either Crossing or Window, depending on the order of the two points picked.

- **AUto.** AUto acts like BOX, unless the first point finds an object, which AUto then picks. AUto is the same as the default selection mode set by turning on Implied Windowing.

You can combine Single with other selection modes, like Crossing or Auto. You can also use it with SELECT if you want to force the creation of a Previous set with only a single selection. Before you assume that Single solves the problem of a missed pick ruining or aborting a menu macro, read on. Although Single ends object selection when an object or set is picked, it does not suspend the rest of a macro. Single works for simple macros, for example:

```
[Stretch1]*^C^C^C_.STRETCH _SIngle _Crossing
```

Single keeps you from having to press ENTER to end the selection process. However, if you miss your first selection and any additional commands or parameters follow, Single tries to use them as input. For example, this macro changes an object to yellow:

`[Change to Yellow]^C^C^C_.CHPROP _SIngle \ _Color 2 ;`

This macro works fine if you do not miss the object, but if you do miss, Single tries to use `_Color` as object selection input and crashes.

BOX is useful when your macros need the flexibility of selection by Window or Crossing. Remember to include two backslash pauses—one for each corner of the box. Pick left-to-right for a Window and right-to-left for a Crossing.

AUto is the explicit selection set option that's equivalent to Implied Windowing: it combines picking by point with the BOX selection set mode. AUto would be useful in macros if it worked correctly with SELECT. Unfortunately, `SELECT AUto \` in a macro only activates AUto for the first pick. Even when you're not using SELECT, AUto in macros can confuse users because of its ability to work with a single pick (that is, on an entity) or two picks (that is Window or Crossing). For example:

`[Change to Yellow] ^C^C^C_.CHPROP _AUto \\ _Color 2 ;`

This macro needs two backslashes for the Crossing and Window options of AUto. It works fine if you use it to select a Window or Crossing. However, if your first pick finds an object, AUto is satisfied, but the macro is still suspended by the second backslash. To avoid an error, you can press the pick button again at the same point and the macro reselects the same object, although this can be confusing to an unaware user.

Four of the selection modes were added in R12: ALL, WPoly, CPoly, and Fence. The following list describes them.

- **ALL.** The ALL mode selects all the objects in a drawing (although most commands ignore entities on frozen and locked layers).

- **WPoly.** The WPoly mode enables selection of all objects that fit in a free-form polygon window.

- **CPoly.** The CPoly mode enables selection of all objects that fit in or cross a freeform polygon window.

- **Fence.** The Fence mode selects objects that are cut by a multisegment fence line.

ALL is useful when you want a macro to operate on all objects in the drawing. For example, the following macro changes all entities to color bylayer:

`[All Color Bylayer]^C^C^C_.CHPROP ALL ;_Color BYLAYER ;`

You can use WPoly, CPoly, and Fence in macros, but in most cases you'll want to put them after SELECT, because they accept a variable number of picks. As is true with interactive use of these options, WPoly, CPoly, and Fence last for just one window polygon, crossing polygon, or fence. After that, the user returns to ordinary single object selection.

You should avoid writing macros that depend on a precise and rigid selection sequence, especially if others will use your custom menu. One of the advantages of AutoCAD is its wide range of selection options, and if your macros restrict these (or crash when they're used), you probably won't be increasing anyone's efficiency. Menu macros are great for automated drawing tasks, layer management, and simple editing tasks. But if you need flexible object selection that supports all of AutoCAD's options, AutoLISP is a better tool.

Finishing Up

This chapter's files include CH05MACR.TXT, which contains cleaned-up versions of the menu macros you have been creating in this chapter's exercises. The changed portions are shown in bold in the following listing. You can incorporate any of these macros in your own custom menus.

```
// Maximizing AutoCAD pull-down menu
***POP7
MA_MaxAC         [MaxAC]
[Circle          ]^C^C^C_CIRCLE
[Text            ]^C^C^C_TEXT
[Bubble          ]^C^C^C_.CIRCLE \(* 0.25 (getvar "DIMSCALE")) +
_.SELECT _Last ;_.TEXT _Middle @ (* 0.1875 (getvar "DIMSCALE")) 0 \+
_.LINE @ \;_.ID _MID,_QUI @ _.TRIM _Previous ;_ENDP,_QUI @ ;
[Title1 Style    ]_.STYLE TITLE1
[Title2 Style    ]_.STYLE TITLE2 \0 1 0 _No _No _No
[Title3 Style    ]_.STYLE TITLE3 ~ 0 1 0 _No _No _No
[Title4 Style    ]_.STYLE TITLE4 ROMAND 0 1 0 _No _No _No
[DDStyle then MText ]_DDSTYLE _MTEXT
[Leader Arrow->  ]^C^C^C_.ORTHO _ON _.PLINE \_Width 0 0 \^O\+
_Width 0 (* 0.0625 (getvar "DIMSCALE")) _Length +
(* -0.1875 (getvar "DIMSCALE")) _Width 0 0 ;
[Leader Arrow<-  ]^C^C^C_.PLINE \\;_.PEDIT _Last _Edit _Move @ +
_eXit ;_.ERASE _Last ;_.PLINE @ _Width 0 +
(* 0.0625 (getvar "DIMSCALE")) _Length (* 0.1875 (getvar "DIMSCALE"))
+
_Width 0 0 _Length (* 0.01 (getvar "DIMSCALE"))
[Pipe Valve]^C^C^C_.LAYER _Make P-EQPM _Color _RED ;;+
_.LINE \\;_.INSERT valve @ (getvar "DIMSCALE") ;_DRAG
[Pipe Valves     ]*^C^C^C_.LAYER _Make P-EQPM _Color _RED ;;+
_.LINE \\;_.INSERT VALVE @ (getvar "DIMSCALE") ;_DRAG \
[Duplicate Layer ]^C^C^C_.SELECT \_.COPY _Previous ;0,0 ;+
_.CHPROP _Previous ;_LAyer \;
```

Introduction to Menu Macros

All non-transparent menu macros should begin with ^C^C^C, as shown in this menu. The closing label brackets are aligned in order to make the labels and macros easier to read. The extra spaces have no effect on the macros because the spaces are inside the label brackets. These versions of the macros use the **getvar** AutoLISP function and DIMSCALE system variable to scale objects based on the current drawing's scale factor.

Conclusion

The macros you have created in this chapter are not perfect. You can make more efficient, flexible, consistent, and dependable macros with help from AutoLISP, but these macros show the AutoCAD command and macro syntax that you have at your disposal already.

It also helps to pay attention to the graphic structure of objects, their on-screen behavior, and the start points and endpoints that you can use as object snap points. The dependability of your macros will benefit. Play a little. It loosens up your imagination.

Look for tricks you have picked up while using AutoCAD, like the sequence _.LINE;^C that resets the LASTPOINT system variable to the end of the last line drawn. The _.LINE;^C sequence is great for inserting blocks at the end of a line. Look for ways to use object snaps as tools in macros, such as rotating block inserts to get correct angles. Remember, the SELECT command automatically pauses for a selection set. Use SELECT to make the set and then pass the set to other editing commands, such as COPY, by using the Previous option.

In the next three chapters, 6 through 8, we leave menus for a while to explore some of AutoCAD's other customization options, such as text, linetypes, hatch patterns, and scripts. If you're anxious to continue with menus, feel free to jump to Chapters 9 and 10, followed by 13 and 14. Chapters 9 and 10 will give you a foundation in AutoLISP and DIESEL, with which you can develop much more sophisticated menu macros. Chapter 13 discusses menu structure in detail, so that you can modify other menu interfaces such as cursor and image menus. Chapter 14 ties all of these concepts together.

chapter 6

Text Fonts and Styles

Text is a crucial component of almost every AutoCAD drawing. AutoCAD provides a range of font and style options for controlling the appearance of text in your drawings. In addition, you can extend the standard fonts with additional characters, or even create your own fonts (if you have the time and the patience!). In this chapter, you learn to control and customize text with AutoCAD's font and style features.

AutoCAD's text style options are easy to use, and they should be the basis of most of your text customization. Developing and using custom fonts is trickier, and it might or might not be worth the effort, depending on your needs. This chapter tells you how to customize fonts, but also suggests when it might be better not to.

Figure 6.1 shows how AutoCAD uses text fonts and styles. You use text-drawing commands (MTEXT, DTEXT, DDATTDEF, LEADER, DIMLINEAR, and so on) to create text objects, each of which inherits the properties of the text style assigned to it. You use the AutoCAD STYLE or DDSTYLE command to define a text style, which is stored in the DWG file's STYLE symbol table. The actual character definitions, however, are not stored in the drawing; for these, AutoCAD text styles reference font files, which can be native AutoCAD SHX fonts, PostScript PSF fonts, or TrueType TTF fonts. All three kinds of font files contain definitions of how to draw a variety of characters (letters, numbers, and special symbols).

 Text handling has never been one of AutoCAD's strong points, and Autodesk worked hard to improve it in Release 13. The results are mixed, and you should understand the strengths and weaknesses of R13's new text features, which we describe early in the chapter. The biggest change is a new multi-line "paragraph" text object (mtext), which automatically wraps text from line to line and supports stacked text and various formats and fonts within one mtext object. Other features include a spelling checker, the ability to use Windows TrueType fonts, a variety of font mapping (substitution) options, and AutoCAD SHX fonts that use Unicode rather than ASCII character mapping.

Maximizing AutoCAD® R13

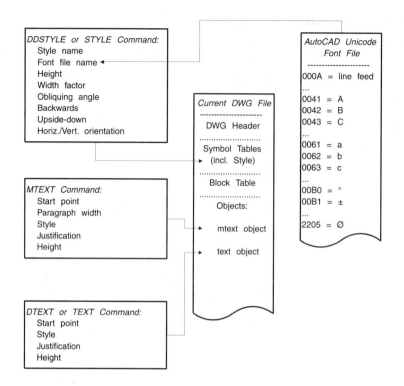

Figure 6.1
Text style and font organization.

 AutoCAD LT R3's text features are almost identical to those of R13, so most of the information in this chapter applies to LT. The Edit Mtext dialog box is simplified in LT and lacks character formatting and font selection features. You use the DDSTYLE command (Text Style dialog box) to select fonts.

This chapter assumes that you have a working knowledge of how to use the new text features, and especially MTEXT, in R13. If you need help with these features, read Chapter 8 of the R13 *AutoCAD User's Guide* and spend a few minutes experimenting with the MTEXT command.

About This Chapter

In this chapter, you'll do the following:

- Use the DDSTYLE command to make new styles and change style definitions

- Learn how and when to use PostScript and TrueType fonts

- Understand the text enhancements in R13

- Learn how to map fonts
- Add a custom angle character to the ROMANS font
- Create a custom center line character, fractions, and other custom characters in a big font file
- Learn about shapes, which can be used as symbols in complex linetypes

These are the tools you will use for the exercises in this chapter:

- TextPad (or another ASCII text editor)
- ALLFONTS.SCR, which creates styles for all of the R13 fonts
- ROMANSMA.SHP, which is a copy of the standard ROMANS.SHP AutoCAD font file
- BFSPECMA.TXT, which contains custom character definitions that you'll add to a big font file (BFSPECMA.SHP)
- ACAD.LSP, ACAD.PGP, ACADMA.MNS, ACADMA.MNL, MAWIN.ARX, MA_UTILS.LSP, and custom bitmap (BMP) icons from the MaxAC CD-ROM (these files are optional)

The text fonts you'll use in this chapter are:

- **ROMANSMA.SHP** is a font file to which you'll add a custom angle character.
- **BFSPECMA.SHP** is a big font file, containing a center line character, an angle character, subscript and superscript capabilities, and a set of fractional characters in 1/16 increments.

New Text Features in Release 13

For most users, the most significant improvement to text in R13 is the multi-line text object. The AutoCAD manuals refer to this new object as "paragraph text," but when you use the LIST command on a paragraph of text, you'll see that it's referred to as an *mtext* object (the older, "line text" object is simply a *text* object). The most appealing feature of mtext is that it handles word wrapping from line to line automatically. In addition, you can add formatting (such as bold, italics, different height, different font) to any sequence of characters, or even mix several different fonts in an mtext object. You can also stack fractions or other characters in mtext.

Figure 6.2
The Edit MText dialog and a sample of stacked text.

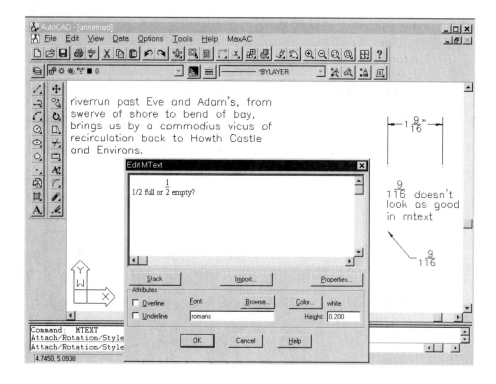

In R13 Windows, you create mtext objects with the Edit MText dialog box (MTEXT command) (Figure 6.2), or by dragging and dropping a text file into the AutoCAD drawing area. You can also use mtext in dimensions and leaders. To edit mtext in R13 DOS, the MTEXT command shells out to the MS-DOS EDIT.COM editor by default. You can specify a different text editor in DOS or Windows with the MTEXTED system variable. In R13c4, you can set MTEXTED to `:LISPED` in order to use a simple editing dialog that looks like DDEDIT. The :LISPED simple editing dialog is especially useful in DOS AutoCAD, since it allows editing of mtext objects without shelling out, but some Windows users also prefer the simple editing dialog to the Edit MText dialog.

 Tip: You can set MTEXTED to the names of two mtext editors—the default editor and one that appears when you pick the **F**ull Editor button in the :LISPED dialog. You separate the names of the two editors with a pound sign (#). For example, setting MTEXTED to `:LISPED#INTERNAL` causes R13c4 for Windows to use the simple editing dialog by default and the Edit MText dialog when you pick the **F**ull Editor button.

R13 also creates mtext automatically when you draw dimensions and leaders. Because the MTEXT command uses an interactive dialog box, you can't control it with a script, menu macro, or AutoLISP program. Instead, your programs can use the -MTEXT command line version, with the leading dash before the command name.

Stacked text is Autodesk's name for text that appears above or below the normal text baseline and/or smaller than normal. Numerical fractions are the most common application for stacked text. R13's fractions work reasonably well in dimensions, but they are difficult to use and control in mtext (see Figure 6.2). Also, you can't use stacked text or fractions at all in ordinary text objects.

The Text Style dialog (DDSTYLE command) is new with R13c4, and it puts a friendlier face on the old STYLE command. Unfortunately, it also adds some new problems, which we describe in the section on using text styles.

Previous releases of AutoCAD relied on a confusing array of 7-bit (that is, 128 character) and 8-bit (256 character) SHX fonts in order to support English and foreign-language versions of AutoCAD. R13 takes a big step toward eliminating this SHX Tower of Babel by moving to Unicode character mapping in its SHX fonts. The new Unicode SHX files include all of the characters for most of the languages that AutoCAD supports (AutoCAD still uses a different scheme for some Asian languages).

R13 text styles can use TrueType fonts, so any Windows user has a large collection of fonts for title blocks and other needs. Note, however, that regeneration and redraw performance is slower for TrueType (or PostScript) text than for SHX text, so you should use TrueType (and PostScript) fonts with discretion. When a drawing using TrueType fonts is opened or plotted on a system that does not support TrueType fonts (such as DOS), AutoCAD will substitute other fonts, using font mapping.

R13 adds three new kinds of font mapping to R12's font mapping (which was only for PostScript fonts). The new features are FONTALT, FONTMAP, and MTEXT editor font mapping. The FONTALT system variable stores the name of a font that R13 will substitute in any styles that reference missing font files, unless they have FONTMAP substitutions defined. The FONTMAP system variable points to a text file that defines substitutions on an individual font basis. MTEXT editor font mapping controls the way text of a particular font appears in the Edit dialog box. We discuss all of these font mapping options later in this chapter.

R13 also includes a spelling checker. It's arguable how useful a spelling checker is for most drafting work, given the large number of abbreviations, shortened words, and idiosyncratic spellings that are common in drafting practice (and unique to each discipline and office). Also, R13's spelling checker doesn't show you the location in the drawing of the words it flags, so it's difficult to tell what you're changing. Nonetheless, if you take the time to add your custom dictionary to the spelling checker's list of words, you might be able to reduce spelling mistakes.

The PostScript text features in R13 are the same as in R12, with one important addition. You can display and plot PostScript (and TrueType) font text with the outlines filled, even on non-PostScript plotters and printers. The TEXTFILL system variable turns filling on and off, and the TEXTQLTY system variable controls how accurately AutoCAD renders the fills. TEXTFILL = 1 (on) and higher values of TEXTQLTY cause slower performance.

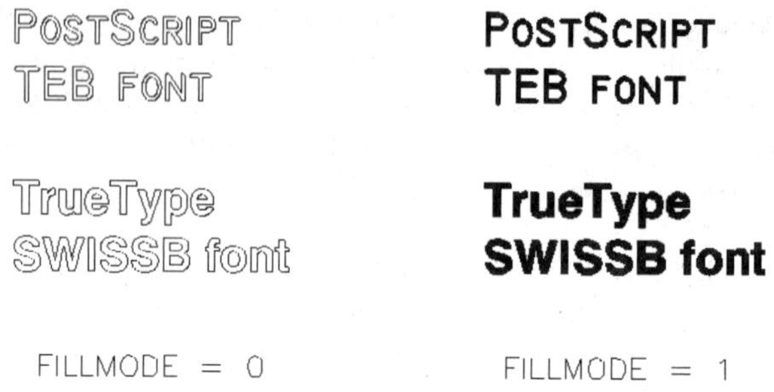

Figure 6.3
PostScript and TrueType fonts shown on screen, with fill options.

PostScript TEB font

TrueType SWISSB font

FILLMODE = 0

PostScript TEB font

TrueType SWISSB font

FILLMODE = 1

Figure 6.3 shows filled and unfilled text for both PostScript and TrueType fonts. The results would be similar on a plot.

Warning: R13 measures the height of PostScript text differently than did R12. R12 used the publishing convention in which "height" refers to the distance from the top of the tallest letter to the bottom of the lowest descender (for example, the bottom of a lowercase g). R13 uses the normal AutoCAD convention in which "height" means the height of a normal uppercase letter, such as A. Thus if you load an R12 drawing that contains PostScript text, the text will appear larger in R13, by a factor of approximately 1.45.

Finally, R13 enables you to use shapes (which are defined in a way similar to the way fonts are defined) as embedded symbols in complex linetypes (see Chapter 7).

Understanding Text Fonts and Styles

In AutoCAD, a *font* is a set of characters with a unified look to them, as shown in Figure 6.4. A font file defines both the characters that you can use and the default way each of them appears. An AutoCAD *style* comprises a font file name and additional characteristics such as height, width factor, and obliquing angle, altering the default appearance of the font characters. Every AutoCAD text object, whether it's free-standing text or mtext, part

Figure 6.4
Some fonts in AutoCAD.

AutoCAD's ugly TXT.SHX font

AutoCAD's ROMANS.SHX font

AutoCAD's ROMAND.SHX font

AutoCAD's ROMANC.SHX font

PostScript CIBT_____.PFB font

TrueType DUTCH.TTF font

Text Fonts and Styles

of a dimension, or a block attribute definition, is assigned a style, just as every object is assigned a layer.

AutoCAD's native fonts are defined in ASCII *shape* (SHP) files and then compiled into SHX files. Normally you don't need to worry about these details—you just select an SHX file from the \R13\COM\FONTS directory. But if you want to modify or create new native AutoCAD fonts, you must learn the format of SHP font source files and the procedure for compiling them into SHX font files. We describe the SHP format in detail later in this chapter.

AutoCAD can also use TrueType fonts and many Adobe Type 1 PostScript fonts. Both TrueType and PostScript fonts are based on more complex mathematical descriptions than the simple vector instructions in AutoCAD SHX fonts. TrueType fonts, which are the standard fonts used by Windows, are defined in binary TTF files. PostScript fonts are defined in binary PFB files, along with accompanying binary PFM (font metrics) files. AutoCAD's COMPILE command can convert PostScript fonts into native AutoCAD SHX fonts for faster performance. We don't cover modifying TrueType or PostScript fonts in this book. More advanced books on Windows and PostScript address these topics.

Note: PostScript fonts are resolution (and scale) independent, but converted PostScript SHX fonts only approximate the Postscript definition, with reduced accuracy at larger sizes.

Using AutoCAD Text Styles and Fonts

Taking advantage of AutoCAD styles and the variety of fonts that come with R13 is the simplest form of text customization. It's all that many users need to do in order to achieve the text characteristics they want in their drawings.

The DDSTYLE command's Text Style dialog box, which Autodesk added in R13c4, organizes all of the style options for you (Figure 6.5). You can use this dialog to select from among styles that are already defined, to define new styles, or to modify existing styles. Following are descriptions of the fields in the DDSTYLE dialog.

- Styles: Choose an existing style from the list to set it current. To create a new style, type its name and choose **N**ew. The new style inherits the properties of the previously current style. Use the Rena**m**e button to rename an existing style.

- Font: Enter the name of the font you want to use for the current style in the **F**ont File field. If you want to use a big font file along with the main font file, enter its name in the Bi**g**Font field. Big fonts are auxiliary font

Figure 6.5

The Text Style (DDSTYLE) dialog (new in R13c4).

files that work with the main font to provide additional characters. We discuss big fonts in more detail later in this chapter.

Warning: If you use the Browse buttons to select a font or big font file, AutoCAD always prefixes the font file's name with its path. DDSTYLE shouldn't add the path when the font is in a directory on the AutoCAD library search path (for example, \R13\COM\FONTS or the current DWG file's directory), but it does so anyway. This glitch will cause problems later if you send your drawing to someone whose drive letters and AutoCAD directories aren't exactly the same as yours, or if you rearrange your own directories. AutoCAD won't locate the font, because it will be looking only in the location where the font file was when you created the style, instead of in all the directories on the library search path.

To avoid problems, keep all of your fonts in directories on the library search path, don't prefix font file names with paths, and don't let DDSTYLE add the prefix. If you use the Browse buttons to select a font, strip the drive and path from the font file names in the **F**ont File and **B**igFont edit boxes. Alternatively, if you know the name, simply type it (without any drive or path) in the edit box.

Often you won't need to change any of the six options in the Effects section at the bottom of the dialog. If you do want to change something, it's likely to be one of the three options on the right:

- Hei**g**ht: If you want to create a fixed height style (that is, one that always uses the same height for text of that style), enter its height in this field. Make sure you enter the height in AutoCAD drawing units, not plotted units. If you want to be able to vary the height of the text, leave the style's height set to zero. Then you'll be prompted for the height each time you create new text with the current style. Usually it's best to use a height of zero, because AutoCAD then lets you change the height of existing text objects.

▶ **W**idth Factor: Use this field to control the width of characters in the style. 1.0 (normal width) is the default. A width factor less than 1.0 makes all characters narrower. A factor greater than 1.0 makes all characters wider.

▶ **O**blique Angle: Use this field to control the obliquing angle of text. Zero degrees means a normal, "upright" angle (neither left nor right). A positive angle causes text to lean to the right, and a negative angle to the left. A little slant goes a long way, so use small angles.

You'll seldom use the three check boxes:

▶ **U**pside-down: Turns characters upside-down.

▶ **B**ackwards: Creates text backwards, as if you were viewing it in a mirror.

▶ **V**ertical: Leaves each character in its normal orientation, but arranges them vertically, with each character below the previous one. The effect is like what you see on some book spines, where you don't have to turn your head to read the title.

The following exercise shows you all of R13's supplied fonts by defining a style for each. To automate the process of creating these styles, this chapter's exercise files from the MaxAC CD-ROM include the ALLFONTS.SCR script file. When the script finishes creating the styles, use DDSTYLE to take a look at them. (See Chapter 8 for more on script files.)

Warning: The ALLFONTS.SCR script assumes that all of R13's fonts are installed on your system. If you opted for a minimal installation, AutoCAD installed only the SHX font files and omitted the PostScript and TrueType fonts. In that case (or if any of the fonts are missing for other reasons), the script will stall when it tries to find the missing font file.

 Exercise

Looking at AutoCAD Fonts

Copy all of the files from \MAXAC\BOOK\CH06 to your \MAXAC\DEVELOP directory.

Maximizing AutoCAD® R13

Start AutoCAD, begin a new drawing named CH06A, and use the MENU command to make \MAXAC\DEVELOP\ACADMA.MNS the current menu.

Command: **SCRIPT** Enter   Opens the Select Script File dialog box

Select \MAXAC\DEVELOP\ALLFONTS.SCR, *then choose* OK   Defines new styles using the STYLE command (this may take a few minutes)

Command: **DDSTYLE** Enter   Opens the Text Style dialog box

AutoCAD displays the definition for the current style, which is the last style that ALLFONTS.SCR created. The list box at the left displays the defined styles. The Character Preview area shows a sample of some of the characters in the font.

Scroll toward the top of the Styles *list box and choose* AC-ROMANS   Displays sample text and current settings for AC-ROMANS

You can change the sample text by entering other letters in the box to the left of the **P**review button and then choosing **P**review.

Choose Char. Set   Displays the current font's first 256 characters in the Symbol Set subdialog (see Figure 6.6)

Choose OK   Closes the Symbol Set subdialog

Examine some other styles. When you are finished, choose Close.

Figure 6.6
DDSTYLE's Symbol Set subdialog.

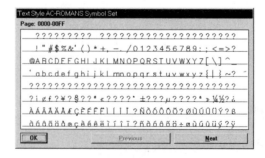

ALLFONTS.SCR created three groups of styles. The styles beginning with AC- use native AutoCAD SHX fonts, those beginning with PS- use PostScript PFB fonts, and those beginning with TT- use TrueType TTF fonts. You might've noticed while the script was running that the styles using PostScript and TrueType fonts took longer to define than did the styles using AutoCAD SHX fonts. That's because AutoCAD must compile PostScript and TrueType fonts in memory before it uses them in a drawing session, whereas SHX fonts are already compiled. If you haven't worked with PostScript and TrueType fonts before, create some text with several different styles so that you can compare quality and performance.

Choosing fonts for your AutoCAD drawings usually presents a trade-off between performance and good looks. Attractive fonts tend to be more complex and, especially when they're filled, make AutoCAD work harder. ROMANS.SHX is a good compromise between speed and attractiveness for many drawings. Use more elaborate fonts sparingly, especially if your drawings contain lots of text and you find yourself waiting a long time for text to regenerate, highlight, or redraw.

Native AutoCAD SHX Fonts and Unicode

Computer fonts keep track of individual characters by assigning them a coded number (for example, 65 = uppercase A, 66 = uppercase B, and so on). Operating systems and applications take care of converting the numbers to the characters you see on the screen.

In previous versions of AutoCAD, Autodesk supported different languages with a variety of 7-bit and 8-bit font files and different code pages. A 7-bit font uses a 7-bit number to represent each character. Because each bit can have one of two values (either 0 or 1), 7 bits can keep track of 128 distinct characters, because 2 to the seventh power = 128. The standard 7-bit character coding is called ASCII (the American Standard Code for Information Interchange). Appendix F includes a table of ASCII values.

The 128 characters in a 7-bit font are adequate for representing the English alphabet, Arabic numerals, and various additional control characters (for example, line feed). 128 characters are not enough for some other languages that use a lot of accented characters, so foreign-language versions of AutoCAD R12 used 8-bit fonts. An 8-bit font can keep track of 256 characters. Even 8 bits aren't enough to represent all of the possible characters in all of the languages that AutoCAD supports, so earlier versions of AutoCAD also had to specify a *code page*, which designated which set of 256 characters to use.

Note: Microsoft introduced code pages, or *character sets*, into MS-DOS as a way of dealing with foreign languages. A code page is simply a particular mapping of numbers to characters. In this context, "code" means the numbering code for characters, not program code. The appendices of most MS-DOS manuals contain listings of code pages.

As Windows, AutoCAD, and other software gets ported to more languages and grows more complex, this system of code pages has become increasingly inelegant for programmers and users. As a result, there is a movement away from ASCII and code pages to a new encoding scheme called Unicode. Unicode uses 16-bit numbers to represent characters, which enables it to keep track of 65,536 characters and thus to unify most languages and a large number of special symbols in a single encoding scheme. About 35,000 characters have been assigned numbers by the Unicode Consortium. The first 128 characters are the same as in ASCII. Windows NT uses Unicode as its character encoding scheme. Windows 95 still uses the older code page scheme, but it supports Unicode for network interoperability and some other operations.

With R13, Autodesk chose to move away from 7-bit and 8-bit fonts and toward Unicode with new SHX font files based on the Unicode scheme. The new font files (RO-MANS.SHX, TXT.SHX, and so on) don't include definitions for all 35,000 of the assigned Unicode characters. Instead, the R13 fonts define a small subset of about 300 characters, as shown in Appendix F. The Unicode fonts work with all R13 platforms, including DOS, Windows 95, and Windows 3.1. In other words, AutoCAD's support for Unicode SHX fonts is independent of the operating system.

You can see the characters supported by an R13 Unicode font by choosing the Char. Set button in the Edit Style (DDSTYLE) dialog box (Figure 6.6). The first page shows the first 256 Unicode characters (hexadecimal 0000 through 00FF, which equals decimal 0 through 255). Choose the **N**ext button once to see additional accented characters (Figure 6.7), and choose it three more times to see Cyrillic characters. Question marks indicate character positions (that is, 16-bit numeric codes) that aren't assigned any characters in the current font file.

Tip: Question marks also appear in your drawing's text when it includes references to undefined characters. Often the question marks are a sign that you're using a different font file than the one used by the creator of the drawing. Or, they can indicate that you entered a Unicode character string that doesn't match up with a defined character position. See the next section for details.

For many users, the switch from 7-bit and 8-bit SHX fonts to Unicode fonts won't have any obvious effect on how they use AutoCAD. The coding for the first 128 characters is

Figure 6.7
Some additional Unicode characters.

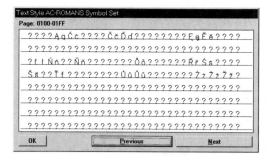

Text Fonts and Styles

the same as before, and R13c4 handles the changes for most other characters automatically. If you use accented or other special characters, look at the AutoCAD Unicode chart in Appendix F in order to see how those characters are coded now.

You might be interested in looking at the 34,700 or so Unicode characters that R13 doesn't include in its standard SHX file. Because of the huge number of Unicode characters, complete Unicode charts are a lot harder to come by than ASCII charts. The complete reference is *The Unicode Standard, Worldwide Character Encoding*, which Addison-Wesley publishes in two volumes. A few programming books that emphasize international and/or Windows NT software development (for example, Nadine Kano's *Developing International Software for Windows 95 and Windows NT*, published by Microsoft Press) document some of the Unicode characters.

Special Characters

Fonts in earlier releases of AutoCAD defined three special symbols: diameter, degree, and plus-or-minus, which you entered by typing **%%C**, **%%D**, and **%%P**, respectively. These codes still work in R13c4, but AutoCAD converts them internally to the equivalent Unicode character strings: \U+2205, \U+00B0, and \U+00B1, as shown in Table 6.1. Although there's no advantage to typing the longer codes for these three special symbols, you can type similar character strings in order to use the other Unicode characters and symbols that R13 supports. Find the Unicode character you want to use in Appendix F, and type its hexadecimal code, preceded by **\U+**. If you're editing in the Edit MText dialog box, put **%\U+** before the four-character code. For example, to use the less-than-or-equal-to sign, type **%\U+2264** in the Edit MText dialog, and **\U+2264** elsewhere.

Table 6.1
Special Symbols

Symbol	Description	%% Code (DTEXT TEXT)	Unicode string (MTEXT)	Unicode string
ø	Diameter	%%C	\U+2205	%\U+2205
°	Degree	%%D	\U+00B0	%\U+00B0
±	Plus-or-minus	%%P	\U+00B1	%\U+00B1

Warning: Previous versions of AutoCAD also allowed you to type %%129, %%127, and %%128 for diameter, degree, and plus-or-minus symbols. These codes don't work in R13.

Big Fonts

Big fonts are special AutoCAD SHX font files that can support up to 65,536 characters. Autodesk developed the big font scheme long before Unicode as a way of getting around the 256-character (8-bit) limit. This limitation is a serious problem for languages that comprise thousands of characters (for example, Japanese Kanji). R13 still uses big fonts for some Asian-language versions of AutoCAD. Big fonts are loaded along with another standard font file, and definitions in the big font can reference and combine characters from the standard font file.

Big fonts also can be useful for extending a Unicode, 7-bit, or 8-bit AutoCAD font file. For example, you can create a big font file that defines special symbols such as fractions or discipline-specific characters. When you enter the big font file's name in the **Big**Font field of the Text Style (DDSTYLE) dialog, these auxiliary characters then become available. The way you use the big font characters depends on the big font file. Each big font defines one or more *escape codes*, such as ~ or |, that tells AutoCAD "big font character follows." You type the escape code, followed by another key, in order to create the character. The choice of escape key and how other keys map to big font characters is up to the creator of the big font file. The big font customization part of this chapter discusses the details.

Tip: If you use the STYLE command instead of DDSTYLE, AutoCAD opens a Select Font File dialog and doesn't appear to let you specify a big font file. Pick the **T**ype It button to suppress the dialog, and then you can type both main font and big font file names at the command line. Separate the two names with a comma, like this: `ROMANS,BFSPECMA`.

Although there's no reason you couldn't define the same symbols in the main Unicode font file, big font files have the advantage of portability. You can define the characters once and use them with most AutoCAD font files, rather than defining the same characters over and over again in each "base" font file.

The important thing to remember is that, although a big font file uses the same SHX extension as a normal AutoCAD SHX file, the header information is different and AutoCAD knows the difference. You can't substitute a normal SHX file for a big font file, or vice versa. Also, there is no encoding standard for big font files, so you usually can't substitute one big font file for another.

PostScript and TrueType Fonts

Appendix C in the *AutoCAD User's Guide* lists the AutoCAD SHX, PostScript PFB, and TrueType TTF fonts that come with R13. The listings include samples of the SHX and PostScript fonts. The sample drawing \R13\COM\SAMPLE\TRUETYPE.DWG shows how the TrueType fonts look. Although TrueType fonts usually are associated with Windows, R13's support for TTF font files is independent of Windows. You can use TTF (and PFB) fonts with R13 DOS.

Text Fonts and Styles

 As we described earlier, in the "New Text Features in Release 13" section, you use the new TEXTFILL and TEXTQLTY system variables to control whether AutoCAD displays and plots PostScript and TrueType fonts as filled or open. For performance reasons, you'll usually want to leave TEXTFILL set to zero (off) while editing a drawing, and turn it on just before plotting.

If you use a PostScript font for anything more than testing, you should create a native AutoCAD SHX font out of it with the COMPILE command. Compiling an SHX version of the font will increase performance in drawings that reference the font. Also make sure that the style references just the font name without any extension (for example, CIBT instead of CIBT.PFB). When you omit the extension, AutoCAD uses an SHX file if it finds one, otherwise it uses a PFB or TTF file. By omitting the extension, you ensure that AutoCAD uses the faster compiled SHX file on your system, but you don't prevent it from using the PFB file on other systems where the users haven't compiled the PostScript font.

 Note: The PostScript font file names include underscore characters to make them all eight characters long (for example, CIBT____.PFB). You don't need to enter the underscores when specifying the file name in the STYLE command, but you must include them in the COMPILE command.

Unfortunately, R13 can't compile a TTF font, so TrueType font performance will always be slower than SHX font performance.

Beyond the compilation issues, PostScript and TrueType fonts usually are filled and tend to be more elaborate than AutoCAD's SHX fonts. If text regeneration, redraw, and selection performance are important in your drawings, you'll probably want to limit the use of PostScript and TrueType fonts to those places where you really need them (for example, in title blocks and for major titles).

 Warning: As mentioned earlier, a minimal installation of R13 omits the PostScript and TrueType fonts. This omission will cause problems if you transfer DWG files that use PostScript or TrueType fonts to a system with a minimal R13 installation. Advise the recipients of your drawings to make sure that they've installed all of the R13 fonts.

Custom Fonts

Although R13 includes a reasonable selection of fonts (especially with PostScript and TrueType additions), they don't cover everyone's needs. You can buy custom AutoCAD SHX fonts (see the Utilities section of *The AutoCAD Resource Guide*), and there are lots of additional PostScript and TrueType fonts available.

Windows includes a selection of TrueType fonts, and many Windows applications install additional ones. Look in your Windows font directory (for example, \WINDOWS\FONTS) to see what's available on your system. Adobe Type Manager and other PostScript-aware programs install PostScript fonts on your system.

Maximizing AutoCAD® R13

Tip: If you decide to use custom fonts, whether SHX, PostScript, or TrueType, include them in AutoCAD's library search path to make sure AutoCAD can find them. You can either add an existing directory (for example, C:\WINDOWS\FONTS) to the support path, as described in Chapter 2, or copy the font files to a special AutoCAD custom font directory (for example, C:\CUSTFONT) and add it to the support path. Although some users copy custom files into the standard AutoCAD font directory (\R13\COM\FONTS), we recommend against this practice because it makes upgrading AutoCAD more difficult.

Custom fonts can add pizzazz to drawings, but they also create headaches when you exchange drawings with others. If you don't send the custom fonts along with your DWG files, the recipient will receive an error message when loading the drawing. And if you do send the custom fonts along, there's a good chance that you'll be violating a license agreement. Most fonts are licensed for use on an individual computer (or, at most, a single site), in which case it's illegal to ship the custom SHX, PFB, or TTF files to others.

Because of the DWG portability and licensing issues, you might want to limit your use of custom fonts if you exchange drawings with others. If certain projects absolutely require custom fonts, you should attend to the licensing issues during the planning stages of the project.

Warning: Don't use TrueType fonts if you plan to SAVEASR12 and give your drawings to R12 users. R12 doesn't know how to use TTF font files.

Mapping Fonts

With all of these fonts—SHX, SHX big font, PFB, TTF, and custom variants of all four—the possibility for missing font files is large. R13's font-mapping features help you avoid the problem of missing fonts and control the fonts that AutoCAD uses.

The FONTALT system variable is a means of globally substituting one font for any fonts that are missing when you open a drawing. This feature eliminates the dreaded `can't open font file` error messages caused by drawings with missing custom fonts. However, you may not realize the substitution has occurred, and the alignment of the text with other objects may differ. To define the substitute font, you set FONTALT to the name of the font that you want R13 to use as an all-purpose substitute (for example, ROMANS). FONTALT is blank (.) by default, which means that R13 will warn you about missing fonts instead of substituting.

The FONTMAP system variable is a more precise way of substituting fonts, whether the fonts you want to substitute for are missing or not. To use FONTMAP, create a text file

Text Fonts and Styles

containing pairs of font names, with each pair on its own line and a semicolon separating the two names in each pair. For example:

```
TXT.SHX;ROMANS.SHX
ROMANC.SHX;DUTCH.TTF
HANDLETR.SHX;CIBT____.PFB
TOOFANCY.SHX,ROMAND.SHX
```

Save the file with an extension of FMP, and then set FONTMAP to the name of the file. AutoCAD stores the FONTMAP setting in the [Editor] section of the ACADNT.CFG file, with the key name `FontMappingFile`. See \R13\COM\SAMPLE\SAMPLE.FMP for an example, which maps all of the TrueType fonts to TXT.SHX.

Unfortunately, the FONTMAP mechanism is full of bugs in R13c4:

- AutoCAD issues no error messages when it can't find a FONTMAP file or can't read some of its entries. Thus there's no way to tell whether FONTMAP is working other than to look at the text in the drawing.

- The name of the font being mapped must be spelled identically in the FONTMAP file and in the drawing. If you create a style and assign it the font romans.shx in the drawing, then the FONTMAP file must use romans.shx (not ROMANS.SHX, romans.SHX, or romans). Similarly, if you receive a drawing whose font names include hard-coded paths (such as C:\ARCHFONTS\HANDLETR.SHX), AutoCAD won't perform a substitution unless the entry in the FONTMAP file matches the font name in the drawing exactly, including the path.

- If you edit the FONTMAP file while in a drawing and want to see the changes, immediately, you must set FONTMAP to nothing (by typing a period), regenerate the drawing, reset FONTMAP to the mapping file's name, and regenerate again.

- When you set the FONTMAP system variable, you must type the FMP extension, or else AutoCAD won't find the file.

- The **F**ont Mapping File edit box on the **M**isc page of the Preferences dialog doesn't work properly in versions earlier than R13c4a. Use the FONTMAP system variable instead.

A FONTMAP file entry takes precedence over FONTALT. In other words, if you open a drawing that includes a missing font file and the currently specified FONTMAP file includes a mapping for that font, AutoCAD uses it instead of any FONTALT font.

MTEXT editor font mapping is more an annoyance than a feature, and it's not described very clearly in the R13 manuals or readme files. The Windows text editor that appears by default when you create or edit mtext uses Windows screen fonts to display the text

Figure 6.8

MTEXT editor font mapping.

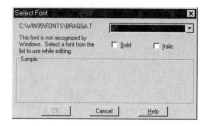

as you edit it. Unfortunately, when you first use MTEXT with a custom font, the editor isn't smart enough to know which Windows fonts to use. AutoCAD displays the error message `This font is not recognized by Windows`, as shown in Figure 6.8. In fact, Windows isn't the problem—the font isn't recognized by AutoCAD's MTEXT editor, even if it is a Windows font from the \WINDOWS\FONTS directory. You must select a Windows font for the Edit MText dialog to use when you edit AutoCAD text in the current font.

Fortunately, you'll only see this message once per font. Once you choose an MTEXT dialog display font, AutoCAD stores a mapping for it in the [MTE Fonts] section of ACAD.INI, and doesn't bother you again about it. Note that the font you choose only controls how the text appears when you're editing it in the MTEXT dialog. AutoCAD still uses the regular font file specified in the style definition when it displays the text in the drawing and plots it. Stick to simple fonts, such as Arial, for MTEXT editor font mapping because complex fonts slow down scrolling in the MTEXT dialog.

Tip: Back up your ACAD.INI file (or at least the [MTE Fonts] section) in case you need to restore your mappings.

PostScript Font Mapping

PostScript has its own way of defining and using fonts. Many PostScript output devices and programs include standard PostScript fonts like Times, Helvetica, and Palatino. If you want to exchange data between AutoCAD drawings and PostScript files or devices, you'll need to worry about mapping between AutoCAD font names and PostScript font names. The \R13\COM\FONTS\FONTMAP.PS file defines the PostScript-to-AutoCAD font mapping that occurs when you use PSIN to import a PostScript file into an AutoCAD drawing. The \R13\COM\SUPPORT\ACAD.PSF file includes a section that defines the AutoCAD-to-PostScript font mapping that occurs when you use PSOUT to create an encapsulated PostScript (EPS) file from an AutoCAD drawing. Both FONTMAP.PS and ACAD.PSF are ASCII files that you edit with any text editor.

The PostScript support features in AutoCAD (PSIN, PSOUT, PSFILL, and so on) are useful for people who work with publishing software, where PostScript is the accepted

Text Fonts and Styles

standard. But AutoCAD's PostScript support is quirky in places and not easy to master. If you don't have a compelling reason to use PostScript, you'll usually be better off avoiding it in AutoCAD. See the end of Chapter 7 for more warnings. You still can use PostScript PFB fonts, as described in the "PostScript and TrueType Fonts" section of this chapter, because printing the fonts in R13 doesn't require a PostScript device or exchanging PostScript data.

Stacked Text

R13's stacked text feature is a tantalizing step closer to being able to display true fractions and superscripts, but if you try to work with stacked text, you're likely to give up in frustration. As Figure 6.2 at the beginning of the chapter illustrates, AutoCAD can display reasonably good-looking fractions automatically in dimensions. Enhancements to the DIMUNIT and DIMTFAC system variables in R13c4 and earlier incremental versions give you some control over fractions in dimensions. The new DIMUNIT settings of 6 and 7 turn off stacked fractions in architectural and fractional units, respectively. DIMTFAC values less than 1.0 shrink the size of the numerator and denominator in fractions.

With mtext, fractions are a different story. You can create fractions and other stacked text with the **S**tack button, but instead of the fraction being centered on the line, the fraction denominators are bottom flush with the baseline of the other text. Also, there's no easy way to control the height of numerator and denominator or to create "slash" style fractions in which the numerator is offset slightly to the left from the denominator. AutoCAD provides no interface for entering fractions in ordinary text (that is, single line text) strings.

R13 does provide special codes for entering fractions in text and controlling their appearance in mtext. But the codes are devilishly complicated to type, and you can't use them in the Edit MText dialog box anyway (you must use another editor, such as EDIT.COM or the R13c4 :LISPED editor). Figure 6.9 shows some of the possibilities, along with the gibberish you must type to achieve them. If you want to try these examples

Figure 6.9
Magic codes for controlling stacked text.

Mtext Formatting Codes	Results
6\S7/8;	$6\frac{7}{8}$
\A1;6\S7/8;	$6\frac{7}{8}$
\A1;6{\H.75x;\S7/8;}	$6\frac{7}{8}$
6{\H.65x;\A2;7\A1;/\A0;8}	$6^{7}/_{8}$
3000 ft{\H.5x\A2;2} area	3000 ft² area

Table 6.2	Code	Meaning
Mtext Formatting Codes	`\Sn/d;`	Stack n over d with a horizontal bar between
	`\Sn^d;`	Stack n over d with no horizontal bar between
	`\A1;`	Center subsequent stacked text vertically
	`\A0;`	Align subsequent text bottom flush
	`\A2;`	Align subsequent text top flush
	`\H.75x;`	Make subsequent text 75% of current text height

in R13c4, temporarily set MTEXTED to **:LISPED#INTERNAL** in order to suppress the normal MTEXT dialog box. When you're finished experimenting, set MTEXTED to **INTERNAL** if you want to restore the MTEXT dialog.

Table 6.2 shows the mtext codes that are most useful for creating fractions. Chapter 8 in the *AutoCAD User's Guide* lists other mtext formatting codes. All codes begin with a backslash and end with a semicolon. Braces define the scope of the codes.

If you're willing to type mtext codes, Steve Johnson's shareware utility DDEDTEXT, which is included on the MaxAC CD-ROM, is a better choice than R13c4's :LISPED.

The big font customization section of this chapter demonstrates an alternative way to create fractions in AutoCAD text strings, using a big font file.

Customizing Text Font Files

The remainder of this chapter shows you how to modify and create AutoCAD SHX font files. If you're satisfied with R13's standard fonts (plus the many custom fonts that are available as commercial, shareware, and freeware products), you can skip or skim the rest of the chapter. But, if you need even a few special characters, or if you want to use symbols in complex linetypes, you may find it useful.

AutoCAD's native fonts are defined in SHP files as a series of vector-drawing instructions for each character. A *vector* comprises a length and direction. The vector-drawing instructions are analogous to what a plotter pen would do in order to draw the character ("draw a vector x units long in the y direction; pick up the pen and move it to new location m,n; put the pen down and draw another vector p units long in the q direction," and so on).

The ROMANS.SHP Font

To get a better feel for the structure and syntax of native AutoCAD font files, we'll explore the standard R13 ROMANS.SHP source file. A copy of this file, named ROMANSMA.SHP, is included in this chapter's exercise files. Later in the chapter, we'll modify

Text Fonts and Styles

ROMANSMA.SHP. If you want to look at R13's other SHP files, copy them from the \ACADCOM\FONTS\SOURCE directory on your R13c4 CD-ROM.

Warning: There's nothing to prevent you from modifying ROMANS.SHP rather than our ROMANSMA.SHP copy of it. Don't do it, though. If you modify a standard AutoCAD font and then load your drawings on another system that uses the original, unmodified font, your custom characters will display as question marks. Using a custom name sends a clear signal that you and recipients of your drawings must use a custom SHX file.

Also, don't forget to send your custom SHX files along with drawings. See the "Custom Fonts" section earlier in this chapter.

Open \MAXAC\DEVELOP\ROMANSMA.SHP in your text editor and look at the top lines:

```
;;; ROMANSMA.SHX: Maximizing AutoCAD copy of ROMANS.SHP
;;; for adding custom characters
;;; by MM 14 Mar 96
;;;
;;;    ROMANS.SHP - Extended Simplex Roman for UNICODE
;;;

*UNIFONT,6,Extended Simplex Roman for UNICODE
21,7,2,0,0,0
```

AutoCAD ignores the comment lines preceded by semicolons and the blank line. The first active line begins with an asterisk, and the word UNICODE indicates an AutoCAD Unicode-style font. (In older 7-bit and 8-bit SHP files, this field is 0 (zero) instead.) The 6 indicates that the next line contains six byte values, or fields, separated by commas. The remainder of the line describes the font.

The six fields in the next line are numbers that define certain global characteristics of the font. Each field is a number between 0 and 255 that's called a *specification byte*. The first byte, 21, is the *ascender code* and says that an uppercase character (or the ascender of a lowercase character) extends 21 units above the baseline. The second byte, 7, is the *descender code* and says that the descender of a lowercase character extends seven units below the baseline. The third byte, 2, is the *modes code* and says that text using this font may be drawn either in the normal horizontal manner or vertically, with each letter below the next, as on some book spines. (A modes code of 0 indicates horizontal only.) The fourth and fifth bytes, both 0, are supposed to be *encoding* and *type codes*, for specifying other kinds of encoding schemes and font embedding information. These codes are 0 in all of the R13 fonts. The sixth and last byte, 0, signals the end of this string of bytes. Note that older 7-bit and 8-bit SHP files don't have the fourth and fifth bytes.

Figure 6.10
Character ascender, descender, baseline, and line feed.

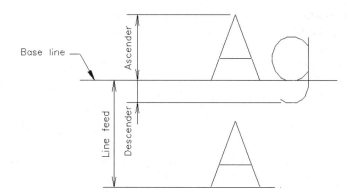

The ascender code also controls scale. In this font, 21 font vector units equal one drawing unit in the TEXT command. Most, but not all, fonts are the same in this respect. For example, the TXT.SHP codes are 6,2,2,0. Each of its characters extends as much as six units above and two units below the baseline. Six TXT font vector units equal one drawing unit. Figure 6.10 shows the measurements used in this font definition.

Before you can use the ROMANSMA.SHP shape file, AutoCAD must compile the shapes defined in the SHP file into an SHX file.

 Exercise

Compiling Shape Files

Begin a NEW drawing named CH06B. If you have not already done so, copy the ROMANSMA.SHP file from \MAXAC\CD-ROM\CH06 to \MAXAC\DEVELOP.

Command: **COMPILE** [Enter]   Opens the Select Shape or Font File dialog box

Select \MAXAC\DEVELOP\ROMANSMA.SHP
and choose OK

```
Compiling shape/font description file

Compilation successful.  Output file
C:\MaxAC\Develop\RomansMA.shx contains
14861 bytes.
```

Text Fonts and Styles

Next, you'll examine character height and spacing with the help of AutoCAD's snap and grid. Because the ROMANSMA font is defined with a character height of 21 units to one drawing unit, you'll use a text height of 21 in the drawing. Then, with a default snap setting of one, one snap increment equals one font unit. You'll set the grid to three so that seven grid increments total the 21 font-unit character height. This setup makes it easy to count units.

Exercise

Graphic Character Inquiry

Continue from the previous exercise, in the current drawing CH06B.

Set the grid at 3, the limits from 0,0 to 72,56, and then zoom All. Set the UCS Origin to 21,34, and the UCSICON to ORigin. Turn on ortho and snap.

Command: **DDSTYLE** [Enter]	Opens the Text Style dialog
Enter **ROMANSMA** *into the* **S**tyle *edit box and, choose the* **N**ew *button*	Creates ROMANSMA and makes it current
Enter **ROMANSMA** *into the* **F**ont File *edit box*	Changes the font file for the ROMANSMA style
Leave **H**eight *set to* 0.0000, **W**idth Factor *set to* 1.0, *and* **O**blique Angle *set to* 0	
Choose **A**pply, *then* **C**lose	Applies the changes and closes the Text Style dialog

Command: **DTEXT** [Enter]
Justify/Style/<Start point>: **0,0** [Enter]
Height <1.0000>: **21** [Enter]
Rotation angle <0>: [Enter]

The DTEXT box cursor is 21 units square.

Text: **Ag** [Enter]	Continues text to next line
Text: **A** [Enter]	Continues
Text: [Enter]	Accepts and exits

Each A is 21 units high, and the line-feed spacing is 34, which you can verify with the DIST command.

Command: **DIST** [Enter]

First point: *Pick the top of the upper A, at ① in Figure 6.11*

Second point: *Pick the top of the lower A ②*

Distance = 34.0000, Angle in XY Plane = 270, Angle from XY Plane = 0
Delta X = 0.0000, Delta Y = -34.0000, Delta Z = 0.0000
    Displays a distance of 34 units, the line-feed spacing

Command: **DIST** [Enter]

First point: *Pick the base of the upper A ③*

Second point: *Pick the top of the upper A ④*

Distance = 21.0000, Angle in XY Plane = 90, Angle from XY Plane = 0
Delta X = 0.0000, Delta Y = 21.0000, Delta Z = 0.0000
    Displays a distance of 21 units, the ascender value

Command: **DIST** [Enter]

First point: *Pick the baseline of the g ⑤*

Second point: *Pick the bottom of the g ⑥*

Figure 6.11
Graphic character inquiry.

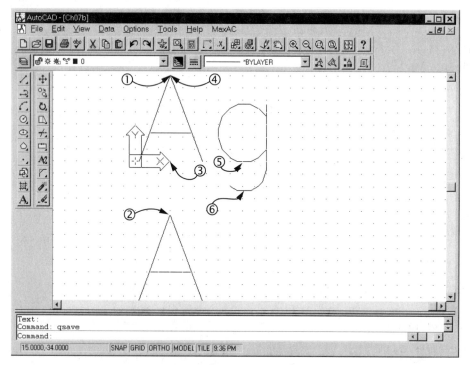

Text Fonts and Styles

`Distance = 7.0000, Angle in XY Plane = 270, Angle from XY Plane = 0 Delta X = 0.0000, Delta Y = -7.0000, Delta Z = 0.0000`	Displays a distance of 7 units, the descender value
`Command: DIST ⏎`	
`First point: 0,0 ⏎`	
`Second point:` *Pick the lower left corner of the upper A*	
`Distance = 1.0000, Angle in XY Plane = 0, Angle from XY Plane = 0 Delta X = 1.0000, Delta Y = 0.0000, Delta Z = 0.0000`	Displays a distance of 1 unit, the character-separation value
`Command: QSAVE ⏎`	

A Line-Feed Character Definition

The line-feed character is treated like any other character in the font file. AutoCAD uses the line-feed character to determine how much space to put between successive lines of text. Return to ROMANSMA.SHP in your text editor and look at the next two lines:

```
*0A,9,lf
2,8,(0,-34),14,8,(30,34),0
```

The asterisk introduces a new character definition. 0A is the hexadecimal number (= decimal 10) corresponding to the line-feed character in Unicode (and ASCII).

Font characters can be identified in SHP files with hexadecimal or decimal numbers. Hexadecimal numbers use decimal 16 instead of decimal 10 as their base, which means that in hexadecimal, you count 1, 2, 3, 4, 5, 6, 7, 8, 9, A, B, C, D, E, F, 10. Although hexadecimal numbers may seem awkward at first, they crop up often in computer programming, because 16 is a power of 2, and digital computers manipulate binary (0 and 1) values.

 Unicode character mapping uses hexadecimal numbers (0000 through FFFF), so it makes sense to use them in AutoCAD Unicode font files. Some older AutoCAD font files use decimal numbers instead. In SHP files, hexadecimal values must start with a leading zero (0A instead of A) so that AutoCAD can distinguish them from decimal values.

The Unicode 0A line-feed character code is followed by the number 9, which indicates that the following character definition line contains nine specification bytes, separated by commas.

The label lf (for LineFeed) is just a name to identify the character. You can choose any character names you want, but it's best to make them descriptive and keep them all lowercase for font files. If you use uppercase, AutoCAD stores the names in memory, which uses up additional system memory.

Note: When you define SHP files for shapes that get inserted with the SHAPE or LINETYPE commands (as described at the end of this chapter and in Chapter 7), you must make the character names all uppercase. AutoCAD then stores the descriptive name in memory so that you can reference it from the SHAPE command.

The second line of the code is the line-feed character's vector definition, and it consists of a sequence of vector drawing instructions, analogous to pen plotter instructions. The instructions include pen up, pen down, and draw a vector of a certain length at a certain angle. By convention, these codes usually are written in decimal form, although you can use hexadecimal numbers (with a leading zero) instead.

The 2 code is the pen-up code, and the line-feed character definition starts with it because line feeds are invisible.

The 8 code signifies that the next two codes are an X,Y displacement, so 8,(0,-34) means to move 0 units to the right and 34 units down. You verified this displacement previously when you measured the line feed with DIST. The parentheses simply improve readability, but they aren't required by the shape compiler. 8,0,-34 would work just as well.

The 14 code signifies that the next command is to be executed only if the style is configured for vertical ("book spine") orientation. If it is, the 8,(30,34) that follows moves the pen 34 units up and 30 units to the right. Supporting vertical orientation makes the design of new characters considerably more complicated, as you can tell by looking at the large number of 14 codes in ROMANSMA.SHP. We'll ignore vertical orientation issues in this chapter's exercises.

The final 0 signals the end of this character's vector definition.

Tip: By copying a font file and adjusting the line-feed definition, you can create a font whose line spacing coincides with the ruling lines commonly found in preprinted forms.

Note: When defining a complex character, you can split a character definition across two or more lines. Because the header line tells AutoCAD how many bytes to expect, you don't need to add any special indication that there's a line break, but you must keep all lines shorter than 128 characters.

Text Fonts and Styles

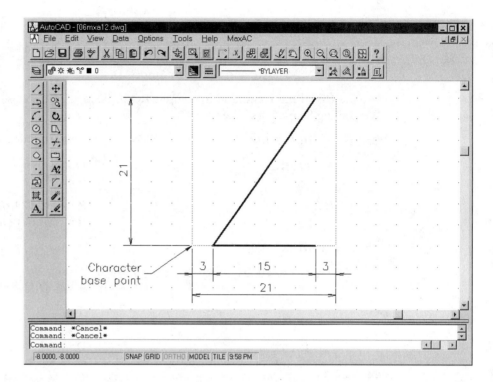

Figure 6.12
Designing the angle character.

Making an Angle Character

The next exercise shows you how to add a new character to ROMANSMA. We'll add Unicode character 2220 (hexadecimal), which is assigned to the angle sign. The easiest way to design a new character is to draw it with AutoCAD lines in a 21 x 21 box (that is, the ascender size for this font) and then measure the distances, as shown in Figure 6.12.

 Exercise

Defining the Angle Character

Continue from the previous exercise, in the CH06B drawing. Set the current color to one with good contrast.

If you want to, you can draw a 21 x 21 square and draw and measure a prototype of the angle character, as shown in Figure 6.12. Otherwise, just use the values shown in the figure.

Add the following lines to the end of ROMANSMA.SHP:

```
*02220,16,kangle Enter
2,8,(18,21),1,8,(-15,-21),8,(15,0),2,8,(3,0),0 Enter
```

Save ROMANSMA.SHP and return to AutoCAD.

Use the COMPILE command to compile ROMANSMA.SHP, as described in the earlier "Compiling Shape Files" exercise.

`Command: QSAVE Enter`	Saves the file so it can be reopened
`Command: OPEN Enter`	Opens the Open Drawing dialog box
Press ENTER	Reopens CH07B.DWG and reloads the revised ROMANSMA.SHX file

```
Command: DTEXT Enter
Justify/Style/<Start point>: 0,0 Enter
Height <21.0000>:
Rotation angle <0>:
Text: \U+2220 Enter                    Draws the angle character
Text: Enter
```

\U+ is the sign to AutoCAD that the following four characters are a Unicode hexadecimal number. See the "Special Characters" section earlier in this chapter.

```
Command: QSAVE Enter
```

You can use MTEXT instead of DTEXT to test new characters, but DTEXT is quicker. If you use MTEXT, remember to type the percent sign before the \U: `%\U+2220`.

Warning: AutoCAD reads the definitions from SHX files only when it loads a drawing. AutoCAD will not recognize the changes in the recompiled SHX until you reopen the drawing.

Note: When AutoCAD compiles an SHP file, it reports an error message if it detects any problems in the font definitions. The most common error is an incorrect byte count in the offending shape header. The error messages are `Shape exceeds specified length` or `Premature end-of-shape`. Sometimes the error messages can be misleading. If the cause is not obvious, look for the error in the line or definition just before or just after the line number indicated by the error message.

Text Fonts and Styles

The angle sign should look just like the one in Figure 6.12. Compare its shape definition to the line-feed definition. There is nothing new except the 1 (pen-down) code, which is described in Table 6.3.

As you have already learned, each asterisk marks the beginning of a new character in the SHP file. 02220 is the hexadecimal Unicode number for the font character, with a leading zero added to tell AutoCAD that this is a hexadecimal, rather than a decimal, number. 16 tells AutoCAD how many specification bytes of data to expect for the character definition. Table 6.3 shows the action for each item in the angle character definition.

Table 6.3
The Angle Character Definition Explained

Item	Explanation
Header:	
`*`	Begin new character
`02220`	Character code
`16`	Number of specification bytes in definition
`kangle`	Character name
Vector definition:	
`2,`	pen up
`8,(18,21),`	move pen 18 units right and 21 units up
`1,`	pen down
`8,(-15,-21),`	move pen 15 units left and 21 units down
`8,(15,0),`	move pen 15 units right
`2,`	pen up
`8,(3,0),`	move pen 3 units right (to beginning of next character)
`0`	end of this character shape

Font Definition Instruction Codes

The 0, 1, 2, and 8 codes are only 3 of 15 specification byte codes in AutoCAD's font-definition language. In addition, there are 16 vector codes, which we describe later. Table 6.4 shows the complete set of specification byte codes.

In the shape file, many of these codes determine how the *next* value is interpreted. For example, in the following line-feed character definition, the 8 causes the 0 and -34 to be read as X,Y displacements:

```
2,8,(0,-34),14,8,(30,34),0
```

Displacements are limited to a range from -128 to 127. If your character definition requires a series of displacements, rather than use the 8 specification byte over and over, you can use 9 to indicate that several X,Y pairs follow. (The 13 code works similarly for a series of arcs.) A 0,0 pair signals the end of the series. For instance, in our custom angle character, the following two definitions are equivalent:

```
2,8,(18,21),1,8,(-15,-21),8,(15,0),2,8,(3,0),0
2,8,(18,21),1,9,(-15,-21),(15,0),(0,0),2,8,(3,0),0
```

The subshape code, 7, allows you to make use of characters that are already defined in the SHP file. Subshapes are an efficient way to create new characters by combining existing ones, and we show you how to do that in the section of this chapter on big fonts.

Many of the other codes listed in Table 6.4 are for more advanced character definitions that require smooth arcs and finer control over vectors. The following paragraphs summarize some of the options, and the big fonts section shows some examples. In addition, you can examine how these codes are used in the R13 SHP files. Chapter 3 of the *AutoCAD Customization Guide* gives detailed syntax for all of the codes.

In the line-feed character definition, the 14 causes AutoCAD to ignore the 8 code that follows unless the style is configured for vertical ("book spine") orientation. Because the following 30 and 34 belong to the 8, AutoCAD ignores them as well when the style isn't vertical.

The divide and multiply (3 and 4) specification bytes allow you to subdivide the character's vector space more finely. The integer scale factor that follows the divide and multiply codes can range from 1 to 255. Multiply and divide scale factors are cumulative, so you can string them together to get the desired scale. Be sure to use the inverse (multiply versus divide) later in the character definition to restore the normal 1:1 scale.

You can push (save) up to four locations at one time. You pop (restore) these locations in the opposite order that they were pushed—last pushed, first popped. Be sure to pop everything that you push in a shape, or you will cause an error.

Note: The *AutoCAD Customization Guide* lists the specification byte codes as hexadecimal numbers 001 through 009 and 00A through 00E. The standard AutoCAD text SHP files use the corresponding 0 through 9 and 10 through 14 decimal values, however.

Any shape definition value can be in either decimal or hexadecimal form, but the convention used in the standard R13 shape files is decimal values for all X,Y and instruction codes, and hexadecimal values for vector length and direction codes. Just remember that hexadecimal values in SHP files always have a leading zero and decimal values never do, and you won't get confused between the two.

Text Fonts and Styles

Table 6.4
Font Specification Byte Codes

Dec.	Hex.	Description
0	000	End. Marks the end of a shape definition.
1	001	Pen down or draw on. Places the logical pen in a down position.
2	002	Pen up or draw off. Lifts the pen up.
3	003	Divide size control. Scales subsequent vector lengths by dividing by next number.
4	004	Multiply size control. Scales subsequent vector lengths by multiplying by next number.
5	005	Push. Saves the current location in memory.
6	006	Pop. Recalls the previously saved location.
7	007	Subshape. Draws the character specified by the next number.
8	008	X,Y displacement. Moves the pen the displacement specified by the next two numbers in file.
9	009	X,Y series. Followed by series of X,Y displacements. Ended by 0,0.
10	00A	Octant arc. Defined by next two numbers in file.
11	00B	Fractional arc. Defined by the next five numbers.
12	00C	Bulge-specified arc. Defined by following X,Y displacement and bulge.
13	00D	Arc series. Followed by series of X,Y, and bulge arcs. Ended by 0,0.
14	00E	Vertical flag. Process next command if vertical style.

Figure 6.13
Vector directions.

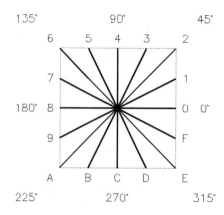

Font Vector Codes

An alternative to codes 8 or 9 followed by X,Y displacements is the set of vector length and direction codes. AutoCAD recognizes 16 predefined vector directions, 0 through F. Using this hexadecimal-based numbering system to define vector direction makes it easier to deal with 16 directions. Figure 6.13 shows the vector directions. The standard vector lengths also are hexadecimal, from 01 to a maximum of 0F (15 decimal). You can combine vector codes with size control (specification byte codes 3 and 4) to get greater lengths.

Vectors are defined by a 3-digit hexadecimal number. The first digit is 0, to indicate hexadecimal; the second digit is the distance; and the third digit is the direction code, as illustrated in Figure 6.13.

In Figure 6.13, notice that the vector endpoints in the diagram form a square, not a circle. All diagonal vectors are stretched to force their lengths to one vector unit in either the X or Y axis, and to one-half or one unit in the other axis. Therefore, the 1 vector is at 26.5651 degrees, not 22.5 degrees.

 Note: The standard AutoCAD text SHP files use a combination of X,Y displacements and vector length and direction codes. You can mix and match the two methods and use whichever is more convenient for the pen movement you need to define.

R13's 16-bit Unicode character numbers leave plenty of room for expansion in each SHP file. You can add other custom characters required by your discipline's or office's drafting conventions. A problem with this approach, though, is that you must add the characters to every SHP file you use. Because each font has a different look, adding your custom characters usually isn't a simple matter of copying the definitions between SHP files. Often you'll have to redesign the characters for each font. A simpler approach is to use a single big font file to supplement all of your main fonts.

Text Fonts and Styles

Tip: You can assign new characters to any 16-bit code that's not already used in the SHP file you're customizing (see Appendix F for a list of the codes that R13's SHP files use). But if you add custom characters that aren't part of the Unicode standard, it's best to use numbers from E000 to F8FF. In the Unicode standard, this 6,144 character region is set aside as the *private-use zone* for vendor-specific or user-designed characters.

Adding Big Font Files

As we described earlier in the chapter, Autodesk developed the big font file format before Unicode existed, as a way of making it possible to create SHX fonts with more than 256 characters. Many Asian-language versions of AutoCAD use big fonts to access the thousands of characters in those languages. But big fonts can also benefit fonts for languages like English. They can serve as supplements to main font files, in which capacity they add characters to the main font. This feature makes it possible to create a single big font file that adds custom characters to many main fonts.

Note: For written languages such as Kanji, which rely on composite characters made from many primitive characters, AutoCAD provides an *extended* big font file syntax. (Extended big fonts are discussed later in the chapter.)

A big font file can define up to 65635 characters. Each character is identified by a double ASCII code. Therefore, you type two keyboard keys to enter a single big font character at an AutoCAD text prompt. The first character acts an *escape code*, which tells AutoCAD to treat the next key as a big font character. For example, in the big font we'll develop in this section, you'll enter ~1 (a tilde followed by a 1) to print 1/16. In our big font, the tilde is the escape character and the 1 is the code for 1/16.

Big fonts are defined in SHP files and stored in compiled SHX font files that AutoCAD handles in a special way. The format for a big font SHP is nearly identical to a normal SHP, but the initial header is different. AutoCAD's special handling enables you to append your big font characters to any main (or *base*) SHX font. You can append big fonts without modifying the base font files.

The most efficient way to develop a big font is to reference characters in the base font with the 7 (subshape) specification byte. For example, you can direct your special big font character for 1/8th to use the 1, 8, and slash characters from AutoCAD's standard font files. By referencing existing characters, you cut down on your work, but you also ensure that the big font's special characters resemble the base font with which they're used. In other words, the custom characters will vary in appearance, depending on the base font to which the big font is attached. See Figure 6.14.

Figure 6.14
The same big font characters used with different base fonts.

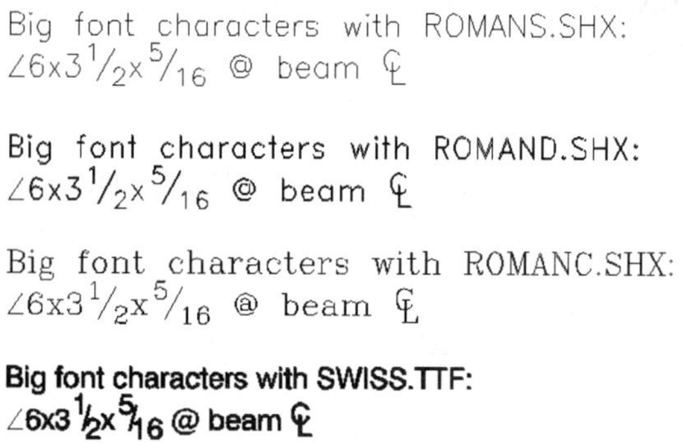

```
Big font characters with ROMANS.SHX:
∠6x3 1/2 x 5/16 @ beam ℄

Big font characters with ROMAND.SHX:
∠6x3 1/2 x 5/16 @ beam ℄

Big font characters with ROMANC.SHX:
∠6x3 1/2 x 5/16 @ beam ℄
```

Big font characters with SWISS.TTF:
∠6x3 1/2 x 5/16 @ beam ℄

Note: Big fonts are mainly for use with AutoCAD SHX base fonts. They will work with some TrueType fonts, but not with PostScript fonts.

Big Font Escape Characters

When you create a new big font file, you can choose any character you want for its escape character. In fact, if your big font contains a large number of custom characters (more than the number of available keys on the keyboard), you can define up to 256 different escape characters, each referencing a different 256-character big font character set. Escape characters should be characters that you use infrequently, but that are easily accessible on the keyboard. Although there are not many good ones to choose from, some choices are the following:

'   Reverse apostrophe

| Vertical bar or pipe symbol

~   Tilde

The big font file we develop in this chapter contains a small number of characters, so one escape character will be adequate. We'll use the tilde, which has a hexadecimal code of 07E (decimal 126). You'll designate the escape character in the header for the big font file, as shown following.

Creating a Big Font File

A big font file starts with a header that alerts AutoCAD to the special file format. Big font files are identified by a header line such as the following:

```
*BIGFONT 20,1,07E,07E
```

Text Fonts and Styles

The keyword BIGFONT tells AutoCAD that this is a big font SHP file, rather than a Unicode or older base font. Next come four numeric fields that tell AutoCAD how to treat the contents of the big font file.

The first number, 20, is the approximate number of character definitions in the file, to tell AutoCAD how much memory is needed to compile and store it. It does not have to be exact, but it should be within 10 percent of the actual number. If the number is too small, AutoCAD won't compile the file. If it's too large, the compiled SHX file will be larger than normal.

The second number, 1, tells AutoCAD how many ranges (that is, number pairs) of escape characters to expect in the rest of the line. Big fonts containing a large number of characters define ranges of escape characters. In our case, we only need one escape character, the tilde, so we have one range consisting of a single character (hexadecimal 07E, the tilde) at the "beginning" and "end" of the range.

Following the big font definition header line, the big font file still needs a font definition header, which is similar to the one you saw in ROMANSMA.SHP earlier. The codes for our big font are 21,7,2,21,0—ascender height, descender height, vertical/horizontal modes, character width, and 0 to end.

The combined header for the sample big font that you'll create, BFSPECMA.SHP, is:

```
*BIGFONT 20,1,126,126
*0,5,Specials - Maximizing AutoCAD
21,7,2,21,0
```

The ascender height and descender heights of a base font determine its compatibility with the big font. Both the font and big font must be close to the same size. Of the standard AutoCAD fonts, ROMANS, ROMAND, ROMANT, ROMANC, and ITALIC are all compatible with an ascender,descender size of 21,7, but TXT and MONOTXT have an incompatible size of 6,2.

The big font character definitions that follow the header use the same basic syntax as a regular font file.

Warning: Duplicate big font definitions override standard font characters. For example, if you redefine your zero (Unicode character 030, or decimal 48) in a big font file to put a slash through the zero, the slashed zero overrides the standard zero character of any base font that is loaded along with it. Make sure that you don't inadvertently redefine a character. See Appendix F for the codes that R13's standard fonts use.

Using Subshapes to Combine Characters

In addition to making new characters from scratch, you can also combine existing base font characters as subshapes to make new characters. Instruction code 7 (see Table 6.4)

Figure 6.15
Center line character diagram.

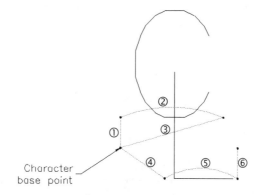

causes the subsequent code to be interpreted as a *subshape*, which is a reference to another character in the file. For example, 7,043 draws an uppercase C (hexadecimal code 043, which is equivalent to decimal 67) at the current pen position. Using subshapes is easier than duplicating definitions and makes a font file compact and easy to update.

Note: The 7 code takes on a special meaning for extended big font files. A zero following the 7 notifies AutoCAD that it is the special form of an extended big font character record. Extended big fonts are covered later in this chapter.

A Center Line Character

In the next two exercises, you'll combine the existing uppercase C and L characters to create a center line character, as shown in Figure 6.15. The center line font character has no new vectors. It simply references the C character, 7,043, and the L character, 7,04C, which are already defined in the base font file.

 Exercise

Designing the Center Line Character

Continue from the previous exercise in the CH06B drawing and erase everything. Zoom Window from -3,-9 to 27,33.

```
Command: DTEXT Enter
Justify/Style/<Start point>: 0,6         Sets the start point at 0,6
```

Text Fonts and Styles

```
Height <21.0000>: Enter
Rotation angle <0>: Enter           Accepts the defaults
Text: C Enter
Text: Pick the point 7,-6           Relocates the start of the next text
                                    line
Text: L Enter
Text: Enter                         Completes DTEXT
Command: QSAVE Enter
```

DTEXT is the perfect command for designing combined characters because it enables you to backspace and select multiple start points. The code for the center line character, shown following, is described in Table 6.5.

```
*32355,13,kcenterline~c
2,5,064,7,043,6,8,7,-6,7,04C,064,0
```

Table 6.5
The Center Line Character Definition Explained

Item	Explanation
Header:	
*	begin new character
32355,	character code (32256 + 99)
13,	number of specification bytes in definition
kcenterline~c	character name
Vector definition:	
2,	pen up
5,	save (push) current pen position on stack
064,	move up 6 in the 4 direction (① in Figure 6.15)
7,043,	draw subshape 043 (decimal 67), uppercase C ②
6,	move pen back (pop) to starting point (③)
8,7,-6,	move right 7 and down 6 from start point (④)
7,04C,	draw subshape 04C (decimal 76), uppercase L ⑤
064,	move back to baseline, 6 in 4 direction ⑥
0	end of this character shape

The character code calculation is a bit tricky, and it is based on decimal rather than hexadecimal numbers. To compute the number, you assign to a special character in a big font file, begin by taking the decimal (not hexadecimal) number of your escape code character—126 in this example—and multiplying it by the constant factor 256. This calculation yields 32256 for the character set. This number is called the *base offset*. If your big font file used other escape characters besides the tilde, each escape character would have its own base offset, which would be its decimal number times 256.

In our big font file, the tilde is the only escape character. Its base offset of 32256 means that it can access as many as 256 characters—decimal character numbers 32256 through 32512 out of the 65535 total. For each character in your set, add the base offset to the decimal value of the ASCII character that corresponds to the key you'll press after the tilde in order to generate this particular custom character. We've chosen to use ~c (tilde plus lowercase c) to generate the center line character. The decimal code for lowercase c is 99 (see Appendix F), so the code for this big font character is 32256 + 99, or 32355.

The push and pop codes, 5 and 6, allow us to calculate the pen movements for both C and L relative to the same character start position. Notice that there are no pen-down codes in the definition. The pen is automatically placed down at the start of each character definition, including subshapes. That is why most shapes start with a pen-up code (2) for their initial offset.

 Exercise

Creating the Center Line Character Definition

In your text editor, create a new big font file called \MAXAC\DEVELOP\BFSPECMA.SHP and add the following lines to it:

```
*BIGFONT 20,1,126,126 [Enter]
*0,5,Specials - Maximizing AutoCAD [Enter]
21,7,2,21,0 [Enter]
[Enter]
*32355,13,kcenterline~c [Enter]
2,5,064,7,043,6,8,7,-6,7,04C,064,0 [Enter]
```

Compile BFSPECMA.SHP as described in the "Compiling Shape Files" exercise earlier.

Use the STYLE command and create a new style called ROMANSBF, entering ROMANS,BFSPECMA as the font to specify ROMANS as the main font and BFSPECMA as the big font.

Text Fonts and Styles

Because AutoCAD wasn't using BFSPECMA.SHX before, you don't have to reload the drawing. Next, you'll test the new character.

```
Command: DTEXT Enter
```
```
Justify/Style/<Start point>: 27,0 Enter
```
Sets the start point at 27,0

```
Height <21.0000>: Enter
```

```
Rotation angle <0>: Enter
```
Accepts the defaults

```
Text: ~c Enter
```
Displays the center line character

```
Text: Enter
```
Completes DTEXT

Make sure you type a lowercase c. Big font character numbers (32355 in this case) refer to characters, not to letters, so they are case-sensitive.

The resulting character should look just like the C and L in the base font, ROMANS.

Note: You can use the 7 (subshape) specification byte in base (normal) SHP files as well as in big font files. If the normal SHP file is a Unicode font (that is, the file's header starts with `*UNICODE`), then each subshape counts as two bytes instead of just one. Therefore, for each Unicode character that you reference, you must add one to the byte count in the character's header. For example, our big font center line character's header is `*32355,13,kcenterline~c`. To convert it for use in a Unicode base font, you would change the big font character number to any Unicode character number and add two to the byte count of 13:

```
*0E000,15,kcenterline
2,5,064,7,043,6,8,7,-6,7,04C,064,0
```

E000 is the first number in the Unicode private-use zone, and since Unicode doesn't include a center line character, we chose a private-use character number. There are two subshape references: `7,043` and `7,04C`. Even though 043 and 04C look like single bytes, AutoCAD counts them as two bytes each in a Unicode SHP file. Thus the total number of bytes is 15. If you forget this rule, AutoCAD will report `Shape exceeds specified length` when you try to compile the SHP file.
  You would access this character in a main font by entering \U+E000.

Custom Fraction Characters

Another application for subshapes is to create a set of fractions. Although R13 supports fractions in the form of stacked text, there are severe usability problems with them, as described earlier in this chapter. We'll develop a set of big font fractions that are easier to use and that work in ordinary (line) text as well as in mtext.

For fractions, as for the center line character, you use characters that are already defined in the base font file, such as hexadecimal code 032 (decimal 49) for the numeral 1 and hex code 02F (decimal 47) for the forward slash character. Because the design and coding process is similar to what you did for the center line character, we'll describe only the 1/8 fraction:

```
*32306,29,oneeighth~2
2,0F4,3,06,4,05,020,7,031,3,05,4,06,0CB,7,02F,0CB,3,06,4,05,
7,038,3,05,4,06,094,0
```

Remember, you can split a character definition across two or more lines without any special indication that there's a line break, but be sure to keep all lines under 128 characters. Table 6.6 explains the 1/8 character codes and references Figure 6.16.

The fraction definition uses the 3 (divide) and 4 (multiply) scale-factor codes to scale the characters so that they are drawn slightly smaller than normal characters. The 3,06,4,05, sequence first divides the vector length by 6, then multiplies that length by 5, creating a vector length that is 5/6 the size of the normal vector. Only integers are acceptable scale input.

You should always reset the scale factors at the end of the shape. If you do not reset them, AutoCAD draws the next character with the last scale calculated.

Next, you'll add the 1/8 character to the ROMANSMA.SHP file and test it.

Figure 6.16
1/8 character diagram.

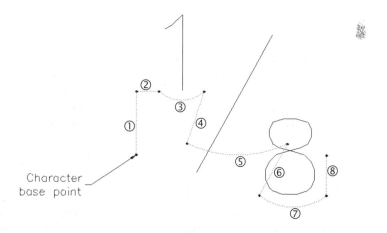

Text Fonts and Styles

Table 6.6
The 1/8 Character Definition Explained

Item	Explanation
Header:	
*	begin new character
32306,	character code (32256 + 50)
29,	number of specification bytes in definition
oneeighth~2	character name
Vector definition:	
2,	pen up
0F4,	move F units (15) in the 4 direction (① in Figure 6.16)
3,06,4,05,	divide by 6, multiply by 5: 5/6th scale for following strokes
020,	move 2 units in the 0 direction (②)
7,031,	draw subshape 031 (decimal 48), the numeral 1 (③)
3,05,4,06,	reverse the scale factor by 6/5 to normal
0CB,	move C units (12) in the B direction (④)
7,02F,	draw subshape 02F (decimal 47), the slash (⑤)
0CB,	move C units (12) in the B direction (⑥)
3,06,4,05,	scale by 5/6 again
7,038,	draw subshape 038 (decimal 56), the numeral 8 (⑦)
3,05,4,06,	reset the scale factor back to normal
094,	move 9 units in the 4 direction (⑧)
0	end of this character shape

Warning: If you have file locking turned on, AutoCAD won't let you recompile a big font file while it's in use in the current drawing. To avoid this problem, you can exit the current drawing by starting a new, blank one. Then return to your previous drawing to test. This back-and-forth sequence gets tedious quickly, so another approach is to turn off file locking (in the Configuration menu) while you're developing the big font. File locking is nothing but a nuisance unless you're on a network, but don't forget to turn it back on later if you share drawings over a network.

Exercise

Creating the 1/8 Fraction Character Definition

Continue from the previous exercise in the CH06B drawing, erasing everything.

Edit the BFSPECMA.SHP file, and add the 1/8 character definition to the end:

```
*32306,29,oneeighth~2 Enter
2,0F4,3,06,4,05,020,7,031,3,05,4,06,0CB,7,02F,0CB,3,06,4,05, Enter
7,038,3,05,4,06,094,0 Enter
```

Save CH06B.DWG and open a new, blank drawing. This procedure avoids AutoCAD's file locking error.

Recompile BFSPECMA.SHP, then reopen CH06B.DWG.

`Command: DTEXT Enter`	
`Justify/Style/<Start point>: 0,0 Enter`	Sets the start point at 0,0
`Height <21.0000>: Enter`	
`Rotation angle <0>: Enter`	Accepts the defaults
`Text: ~2 Enter`	Displays the 1/8 fraction character
`Text: Enter`	Completes DTEXT

Note: You can use big font characters, including fractions, with the MTEXT, DTEXT, and TEXT commands. One limitation of big font fractions, though, is that AutoCAD won't automatically use them in dimension text. You can let R13 use its native dimensions in fractions, or you can override the default dimension text and type a string that includes the big font fraction code (for example, `1'-1~2"` for "one foot, one and one-eighth inches"). If you type a dimension string, though, the text will no longer be associative with the actual distance of the dimension.

You can add other custom fractions to BFSPECMA.SHP in a similar way. As the number of custom characters grows, you need to come up with a sensible coding scheme, so that you and others can remember what to type after the tilde. We've chosen a scheme based

Table 6.7
BFSPECMA.SHP Big Font Character Map

Key	Fraction	Hex	Decimal	Big Font Number
~1	1/16	031	49	32305
~2	1/8 (2/16)	032	50	32306
~3	3/16	033	51	32307
~4	1/4 (4/16)	034	52	32308
~5	5/16	035	53	32309
~6	3/8 (6/16)	036	54	32310
~7	7/16	037	55	32311
~8	1/2 (8/16)	038	56	32312
~9	9/16	039	57	32313
~A	5/8 (10/16)	041	65	32321
~B	11/16	042	66	32322
~C	3/4 (12/16)	043	67	32323
~D	13/16	044	68	32324
~E	7/8 (14/16)	045	69	32325
~F	15/16	046	70	32326
~c	Center Line	063	99	32355
~<	angle	03C	60	32316
~S	Superscript	053	83	32339
~s	subscript	073	115	32371
~~	tilde	07E	126	32382

on sixteenths. Thus our 1/8 character's code is ~2, because 2/16 reduces to 1/8. Table 6.7 lists the other characters you'll add in a moment.

After ~9 (9/16), we run out of decimal numerals on the keyboard, so we continue to count in hexadecimal: ~A for 5/8, ~B for 11/16, and so on. Don't forget that these codes are case-sensitive. ~C is 3/4, but ~c is the center line character.

We've copied the angle character definition from **ROMANSMA.SHP** into the big font file and assigned it to ~<. In addition, we've added superscript and subscript codes. These are simple pen-up moves.

It's possible that you might want a real tilde to appear in a text string in your drawing. You can create a big font definition for it, just like for any other printed character. The hexadecimal code for the tilde is 07E (decimal 126), so its big font code is 32256+126, or 32382.

The complete set of big fonts is contained in BFSPECMA.TXT, which is included in the chapter's exercise files. It looks like this:

```
;BFSPECMA.SHX: Maximizing AutoCAD big font file
;with custom characters
;by MM 14 Mar 96

*BIGFONT 20,1,126,126
*0,5,Specials - Maximizing AutoCAD
21,7,2,21,0

;fractions
*32305,32,onesixteenth~1
2,0F4,020,3,06,4,05,020,7,031,3,05,4,06,0CB,7,02F,0CB,3,06,4,05,
7,031,7,036,3,05,4,06,094,0
*32306,29,oneeighth~2
2,0F4,3,06,4,05,020,7,031,3,05,4,06,0CB,7,02F,0CB,3,06,4,05,
7,038,3,05,4,06,094,0
*32307,32,threesixteenths~3
2,0F4,020,3,06,4,05,020,7,033,3,05,4,06,0CB,7,02F,0CB,3,06,4,05,
7,031,7,036,3,05,4,06,094,0
*32308,29,onequarter~4
2,0F4,3,06,4,05,020,7,031,3,05,4,06,0CB,7,02F,0CB,3,06,4,05,
7,034,3,05,4,06,094,0
*32309,32,fivesixteenths~5
2,0F4,020,3,06,4,05,020,7,035,3,05,4,06,0CB,7,02F,0CB,3,06,4,05,
7,031,7,036,3,05,4,06,094,0
*32310,28,threeeighths~6
2,0F4,3,06,4,05,7,033,3,05,4,06,0CB,7,02F,0CB,3,06,4,05,
7,038,3,05,4,06,094,0
*32311,32,sevensixteenths~7
2,0F4,020,3,06,4,05,020,7,037,3,05,4,06,0CB,7,02F,0CB,3,06,4,05,
7,031,7,036,3,05,4,06,094,0
*32312,29,onehalf~8
2,0F4,3,06,4,05,020,7,031,3,05,4,06,0CB,7,02F,0CB,3,06,4,05,
7,032,3,05,4,06,094,0
*32313,32,ninesixteenths~9
2,0F4,020,3,06,4,05,020,7,039,3,05,4,06,0CB,7,02F,0CB,3,06,4,05,
7,031,7,036,3,05,4,06,094,0
*32321,28,fiveeighths~A
2,0F4,3,06,4,05,7,035,3,05,4,06,0CB,7,02F,0CB,3,06,4,05,
7,038,3,05,4,06,094,0
*32322,34,elevsixteenths~B
2,0F4,020,3,06,4,05,020,7,031,7,031,3,05,4,06,0CB,7,02F,0CB,3,06,4,05,
```

Text Fonts and Styles

```
7,031,7,036,3,05,4,06,094,0
*32323,28,threequarters~C
2,0F4,3,06,4,05,7,033,3,05,4,06,0CB,7,02F,0CB,3,06,4,05,
7,034,3,05,4,06,094,0
*32324,34,thirteensixteenths~D
2,0F4,020,3,06,4,05,020,7,031,7,033,3,05,4,06,0CB,7,02F,0CB,3,06,4,05,
7,031,7,036,3,05,4,06,094,0
*32325,29,seveneighths~E
2,0F4,3,06,4,05,030,7,037,3,05,4,06,0CB,7,02F,0CB,3,06,4,05,
7,038,3,05,4,06,094,0
*32326,34,fifteensixteenths~F
2,0F4,020,3,06,4,05,020,7,031,7,035,3,05,4,06,0CB,7,02F,0CB,3,06,4,05,
7,031,7,036,3,05,4,06,094,0

;special symbols
*32355,13,kcenterline~c
2,5,064,7,043,6,8,7,-6,7,04C,064,0
*32316,16,kangle~<
2,8,(18,21),1,8,(-15,-21),8,(15,0),2,8,(3,0),0

;superscript and subscript
*32339,5,superscript~S
2,8,(0,18),0
*32371,5,subscript~s
2,8,(0,-18),0
```

In the following exercises, you'll copy and test all of the big font character definitions.

 Exercise

Using the Maximizing AutoCAD Big Fonts

In your text editor, erase all of the lines in BFSPECMA.SHP and copy all of the lines from \MAXAC\DEVELOP\BFSPECMA.TXT into BFSPECMA.SHP.

Erase everything in CH06B.DWG, then save CH06B.DWG and open a new, blank drawing.

Recompile BFSPECMA.SHP and reopen CH06B.DWG, then zoom Window from -3,-9 to 120,80.

Command: **DTEXT** [Enter]

Justify/Style/<Start point>: **0,69** [Enter]

`Height <21.0000>: 3` [Enter]	Sets the height to a more reasonable value
`Rotation angle <0>:` [Enter]	
`Text: ~<6x3~8x~5 @ ~c of beam` [Enter]	Adds a note with several special symbols
Text: *Pick the point 0,55*	Relocates the start of the next text line
`Text: ~1~2~3~4~5~6~7~8~9~A~B~C~D~E~F` [Enter] `Text: ~c ~< ~~` [Enter]	Tests all of the characters shown in Table 6.7
Text: *Pick the point 0,33*	Relocates the text start point
`Text: Normal ~SSuperscript ~sNormal ~sSubscript ~SNormal` [Enter]	Displays text with raised Superscript and sunken subscript
Text: *Pick the point 0,21*	Relocates the text start point
`Text: 3000 ft~S2~s area` [Enter]	Adds a note with a "squared" symbol
`Text:` [Enter]	Completes DTEXT

Figure 6.17 shows the results. With superscripts and subscripts, you must remember to reverse the operation when you're finished typing raised or lowered characters. Type ~S, raised characters, then ~s, or vice versa.

Note: When you type big font characters in the MTEXT dialog box, you won't see the characters immediately; you'll see the tilde codes. When you choose OK to accept the text, AutoCAD will convert the codes and show the big font characters properly in the drawing.

One problem with fractions and superscripts is that they bump into each other in adjacent lines. Your choices are to increase line spacing in the base font by customizing the line-feed character, decrease the size of the characters in the fractions by customizing the big font file, or move the lines of text slightly to provide more room. The last approach works only with ordinary line text, and not with mtext.

Text Fonts and Styles

Figure 6.17
Fractions, subscripts, and superscripts.

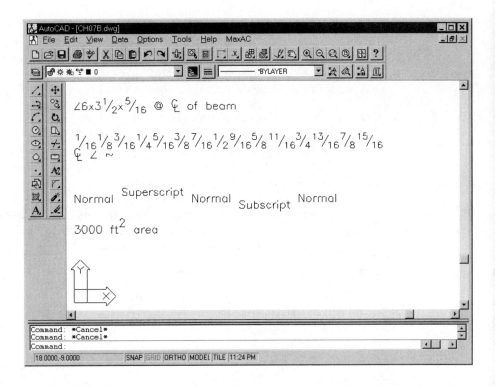

Note: As with any other font or shape, the SHX file that AutoCAD reads for the big font's definition is not stored inside the drawing. AutoCAD must locate the SHX file every time it opens the drawing. If you exchange drawings with others, don't forget to include all of the custom SHX files.

Extended Big Fonts

Ordinary big font definitions have one limitation: they don't let you reference characters within the big font file itself. The subshapes you reference with the 7 specification byte must be in the base font file, not the big font file.

AutoCAD provides a way around this limitation with *extended* big fonts. The SHP file format isn't any different; you just use an extended syntax of the 7 code. The extended big font syntax is especially efficient for developing some Asian fonts, which consist of a set of graphics primitives that are assembled many different ways into composite characters.

Extended Big Font Subshape Code Format

The extended code 7 syntax looks like this:

```
7,0,shape#,X_basepoint,Y+basepoint,width_scale,height_scale
```

The 0 tells AutoCAD that this 7 specification byte starts an extended code with six subsequent fields, instead of just one. The next field, shape#, is the number of the big font shape that you are referencing. Because big font character numbers are two-byte numbers, you must remember to add two bytes (not one) to the font definition character's byte count.

The remaining four fields tell AutoCAD where to put the shape and how large to make it. X_basepoint and Y_basepoint are always with respect to the character's origin, even if the logical pen happens to be at a different location. This convention makes it easy to draw several subshapes in a row without having to move the pen each time. width_scale and height_scale determine the scale of the subshape.

Note: When AutoCAD encounters a reference to the extended subshape, it first scales down the primitive character to a 1 x 1 matrix. The width and height scales are then used to arrive at the final size, relative to this 1 x 1 matrix.

Here's an example of a big font character that uses the extended syntax and that you could add to BFCUSTBF.SHP:

```
*32318,23,kangle2~>
7,0,32316,0,0,21,21,
5,2,8,(16,0),1,10,(12,001),2,6,8,(25,0),0
```

The character appears in Figure 6.18. The big font code for this character, 32318, corresponds to ~>. The second line references the big font angle character (number 32316) that we defined earlier. The pair of zeroes places the character in its normal position, and the pair of 21s scales it back up to the normal character size. The third line of the character definition draws a small arc inside the angle according to the numbered sequence in Figure 6.18, and demonstrates how to use the octant arc (10) specification byte.

Warning: Each time you use the 7,0 extended syntax, AutoCAD is supposed to restore the pen to the origin of the big font character. Because of an AutoCAD bug, it moves the pen four character units to the left each time. That's why the move to the beginning of the arc is (16,0) instead of (12,0) and the final move is (25,0) instead of (21,0). This may change, if this bug is fixed in future versions of AutoCAD.

Understanding Shape Files

Fonts aren't the only thing you can define with SHP files. You can also define *shapes* and insert them into your drawing. AutoCAD *shapes* are intended to be used as symbols, similar to primitive blocks, except that you define them with the same vector instructions as you use to create custom text characters. You create shape files the same way you create font

Figure 6.18
An extended big font character.

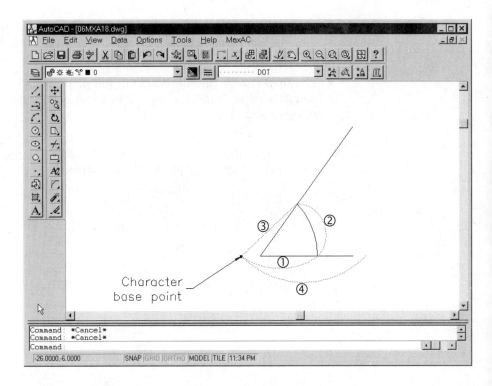

files, with two small differences. Once you've defined shapes in an SHP file and compiled them into an SHX file, you load the SHX definition into a drawing with the LOAD command and insert individual shapes with the SHAPE command. Chapter 7 shows an example of this procedure in the context of complex linetypes.

The main advantage of shapes over blocks is that they require less memory for AutoCAD to store. On the other hand, for the simple objects that lend themselves to shapes, the amounts of memory involved are small. Blocks are more flexible, more powerful, and much easier to define and use. Also, like custom text fonts, shapes require that the SHX file be available every time you load the drawing.

From the point of view of most AutoCAD users, shapes were just about extinct until R13 made them the basis of embedding symbols in complex linetypes (see Chapter 7). Although it would've been easier on users if R13 supported embedding blocks in linetypes, we're stuck with shapes for now.

Fortunately, there's really nothing new to learn once you're familiar with AutoCAD font customization. The specification byte syntax for shape files is exactly the same as for font files, explained earlier in this chapter. The only differences in the file formats are:

- Shape files don't have a file header. (Each shape has its own header, though, just as each character does in a font file.)

- The shape name (the third field in the shape's header) must be in uppercase so that AutoCAD stores it in memory and the SHAPE command recognizes it.

Here are the definitions of some sample shapes that R13 includes for demonstrating complex linetypes (\R13\COM\SUPPORT\LTYPESHP.SHP):

```
;;;
;;;   ltypeshp.SHP - shapes for complex linetypes
;;;

*130,6,TRACK1
014,002,01C,001,01C,0
*131,4,ZIG
012,02E,012,0
*132,6,BOX
014,020,02C,028,014,0
*133,4,CIRC1
10,1,-040,0
*134,10,BAT
025,10,2,-044,04B,10,2,044,025,0
```

The shapes are shown in Figure 6.19. As you can see, shape definitions tend to be simpler than character definitions.

In addition, the \R13\COM\SAMPLE directory contains ES.SHP and PC.SHP files with sample shapes for electrical schematics and printed circuit board drafting. Because you create shape files the same way you create font files, you already know how to do it.

Finishing Up

Use the MENU command to change the menu back to \MAXAC\ACADMA.MNS.

In this chapter, you created two custom font files (ROMANSMA.SHP/SHX and BFSPECMA.SHP/SHX). If you want to keep these, exit AutoCAD and move them from \MAXAC\DEVELOP to \MAXAC. Delete or move any other files from \MAXAC\DEVELOP.

Text Fonts and Styles

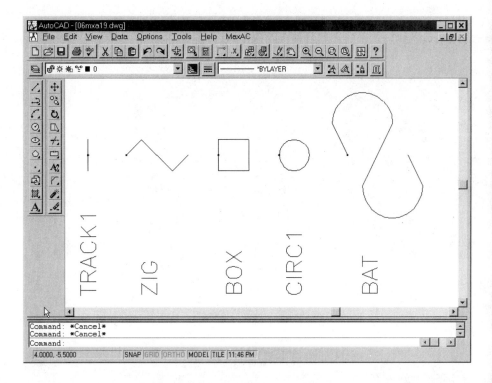

Figure 6.19
Sample shapes.

Conclusion

Choosing from R13's collection of SHX, PostScript, and TrueType fonts and understanding how to use them wisely are your primary text customization tasks. You can extend your font choices with third-party SHX, PFB, and TTF fonts, but remember that when AutoCAD opens a drawing, it must have access to all of the font files referenced by that drawing. Consider the drawing exchange issues before you adopt custom fonts. R13's font mapping features can help you keep the custom font situation under control.

Ambitious users can customize the standard AutoCAD font files to modify existing characters or add new ones. A big font file with subshape referencing usually is the easiest approach.

In the next chapter, you'll explore two other annotational features of AutoCAD—linetypes and hatching—and learn how to customize them.

chapter 7

Linetypes and Hatch Patterns

Continuous lines in various colors and line widths only go so far to convey the meaning of a drawing, whether on screen or on a plot. Most drawings also require special linetypes and hatch (or fill) patterns. As in manual drafting, linetypes and hatch patterns clarify what drawing components represent, make your drawings easier to read, and add visual appeal.

AutoCAD includes a useful selection of hatch patterns and dash-dot linetypes, but many users find that they need to add custom patterns and linetypes for their discipline or office. For example, although AutoCAD includes generic and ISO linetypes, it does not include linetypes meeting ANSI or common architectural standards. This chapter shows you how to customize and create linetypes and hatch patterns, and how to apply PostScript fills to closed polylines.

As Figure 7.1 shows, definitions for both linetypes and hatch patterns are stored in external files. AutoCAD copies linetype definitions to the drawing's LTYPE symbol table as you use new linetypes or when you load them with the DDLTYPE or LINETYPE command. Each time you use the BHATCH or HATCH command, AutoCAD reads the pattern definition from the external pattern file and adds a block or set of line objects to the drawing. When you use PSFILL, AutoCAD attaches extended entity data to the polyline describing the fill pattern.

Release 13 includes several welcome enhancements to AutoCAD linetypes and hatch patterns. You can now create, in addition to simple dash-dot linetypes, complex (symbolic) linetypes that contain embedded text characters, strings, or AutoCAD shapes. A new Select Linetype (DDLTYPE) dialog box makes using and managing linetypes easier. The new CELTSCALE system variable permits different linetype scaling for different objects. Associative hatching means that, in many cases, you can stretch a boundary and hatched area together, without having to rehatch.

Maximizing AutoCAD® R13

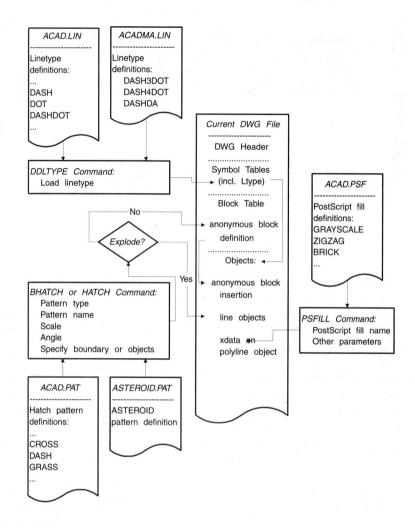

Figure 7.1
Linetype and hatch patterns in AutoCAD.

About This Chapter

In this chapter, you will learn how to do the following:

- Create simple and complex (symbolic) custom linetype definitions

- Control linetype scaling with global, object-specific, and paper space options

- Create simple and elaborate custom hatch pattern definitions

- Fill closed polylines with PostScript patterns

Linetypes and Hatch Patterns

These are the tools you will use for the exercises in this chapter:

- TextPad (or another ASCII text editor)

- AutoCAD's stock LTYPESHP.SHX and ACAD.PSF files

- ACADMA.LIN and ASTEROID.MA from the MaxAC CD-ROM

- ACAD.LSP, ACAD.PGP, ACADMA.MNS, ACADMA.MNL, MAWIN.ARX, MA_UTILS.LSP, and custom bitmap (BMP) icons from the MaxAC CD-ROM

- PATTERN.LSP, PATTERN.TXT, and CH07PAT.DWG (an optional hatch pattern generator with documentation and sample exercise drawing)

Linetypes, Hatch Patterns, and Programs in This Chapter

As you work through the exercises in this chapter, you will create and modify the following custom files:

- **ACADMA.LIN.** A custom linetype definition file.

- **CHECKP.PAT.** A hatch pattern definition representing checkered plates.

- **ASTEROID.PAT.** A hatch pattern definition representing asteroids. ASTEROID.MA is a completed version of this file.

This chapter assumes that you have a working knowledge of how to use linetypes and hatch patterns in R13. If you need help with these features, read the relevant parts of the R13 *AutoCAD User's Guide*: Chapter 2 (Creating Objects) and Chapter 6 (Using Layers, Colors, and Linetypes).

New Linetype and Hatching Features in Release 13

The most important change to R13 for customizing linetypes and hatching is the new complex linetype feature. You can now incorporate text characters and symbols in custom linetype definitions, as shown in Figure 7.2.

Other linetype and hatching changes don't affect customization directly, but they do improve the usability of custom linetypes and hatch patterns:

- Object linetype scaling. You still control global linetype scaling with the LTSCALE system variable, but AutoCAD now stores an additional linetype scaling factor with each object in the drawing. See Figure 7.2 for an example. This linetype scale is a new object property, similar to the layer, color, linetype, and thickness properties. The new CELTSCALE (Current

Figure 7.2
R13 linetype enhancements.

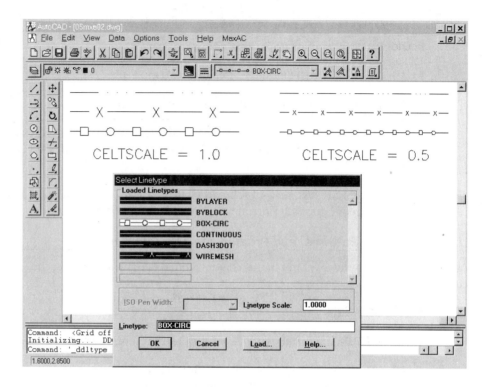

Entity LTSCALE) system variable (1.0 by default) controls the default object linetype scale applied to newly created objects. You can use DDCHPROP or CHPROP to alter the object linetype scale for existing objects.

- New Select Linetype (DDLTYPE) dialog box (see Figure 7.2). You can select linetypes and load them from LIN files with this dialog box. It's also available from the Layer Control (DDLMODES) dialog box. Note that the Linetype Scale edit box controls the linetype scale property (CELTSCALE) in R13c4a, not the overall linetype scale (LTSCALE).

- Associative hatching. In R13, hatching is associated with its boundary, and thus it stretches, moves, rotates, and scales with the boundary. Unfortunately, R13 still stores hatching as an anonymous block (a special kind of block definition that AutoCAD creates when you add hatching to the drawing), so large, complex hatches still bloat the drawing file and inhibit performance.

- Redesigned Boundary Hatch (BHATCH) dialog box (see Figure 7.3). It's arguable whether the new dialog is better than the R12 version. You can see and change all of the options without resorting to subdialogs, but se-

Linetypes and Hatch Patterns

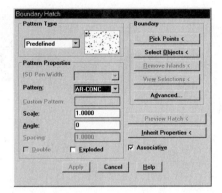

Figure 7.3
The R13 Boundary Hatch dialog box.

lecting a hatch pattern can be more difficult because you can see only one pattern icon at a time. However, the hatch performance in finding boundaries and islands is improved over R12's.

- HATCHEDIT command. With the new HATCHEDIT command, you can modify the properties (pattern, scale, and angle) of existing hatching in the drawing.

Simple Dash-Dot Linetypes

Previous versions of AutoCAD supported only one kind of linetype definition, which specifies a repeating series of dashes, spaces (or gaps), and dots. The R13 AutoCAD documentation refers to these dash-dot linetypes as *simple* linetypes, in order to distinguish them from the new complex linetypes that contain symbols—text or shapes.

Understanding Simple Linetypes

A simple linetype definition is a set of instructions that tells AutoCAD how to draw a sequence of dashes, spaces, and dots. Simple linetypes can have just dashes and spaces, just dots and spaces, or all three (dashes, dots, and spaces). In order to define a new linetype, you write a numeric pattern that describes the dash-dot sequence. The numeric dash-dot pattern you define tells AutoCAD how long each dash is, how much blank space to leave, and where to place a dot. Figure 7.4 shows three simple dash-dot patterns.

Here are the definitions of these three linetypes from the ACAD.LIN file:

```
*DASHED,__ __ __ __ __ __ __ __ __ __ __ __ __ __ __ __ __
A,.5,-.25
```

Figure 7.4
Simple dash-dot linetypes.

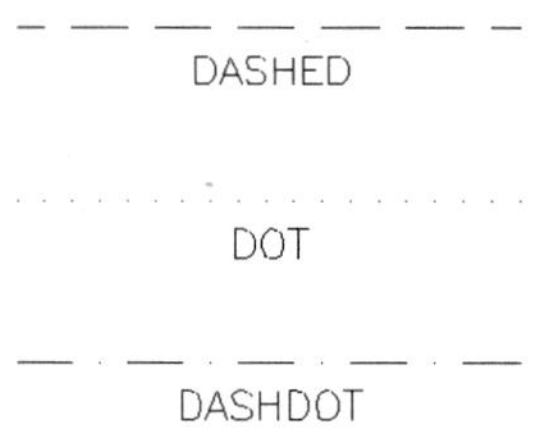

```
                    _ _ _ _ _ _
                      DASHED

                    . . . . . . . . . . . .
                          DOT

                    _ . _ . _ . _
                        DASHDOT
```

```
*DOT,. . . . . . . . . . . . . . . . . . . . . . . . . . . . . . . . . .
A,0,-.25
*DASHDOT,__ . __ . __ . __ . __ . __ . __ . __ . __ . __ . __ . __ . __
A,.5,-.25,0,-.25
```

Ignore for now the lines beginning with an asterisk (they are only identifying labels) and look at the sequence of numbers in each line that begins with A. Simple linetype definitions use the metaphor of a plotter pen that gets *raised and lowered* to draw the dashes, dots, and spaces. A positive number means "lower the pen and draw a segment of this length." A negative number means "raise the pen and leave a space of this length." A zero means "lower the pen and draw a dot." AutoCAD uses these imaginary pen-up/pen-down instructions to draw the linetype, both on the screen or on the plotter.

Creating Simple Linetypes

AutoCAD's standard linetypes are stored in the ASCII support file \R13\COM\SUP-PORT\ACAD.LIN. You can edit the ACAD.LIN file to add custom linetypes, or you can add them to new LIN files that you create. Alternatively, you can define simple custom linetypes "on the fly" with the AutoCAD LINETYPE command and store them in the file you choose. The LINETYPE method is handy for learning about and experimenting with new linetypes, so we'll use it for the first exercise.

 Note: Linetypes defined in your own LIN file are more portable, unaffected by updates to AutoCAD, and more compatible with other applications.

In the following exercise, you will use the LINETYPE command to create a simple linetype called DASH3DOT (see Figure 7.5). DASH3DOT is made of 1-unit long dashes and three dots, with each dot separated by a quarter-unit space. After AutoCAD verifies that the linetype definition does not exist, you are prompted to enter a description of the linetype's pattern. To describe the pattern, you'll type an ASCII facsimile of using underscores and periods. You'll save the linetype in the ACADMA.LIN file, which contains all of the standard AutoCAD Release 13 linetype definitions.

Figure 7.5
The DASH3DOT
linetype.

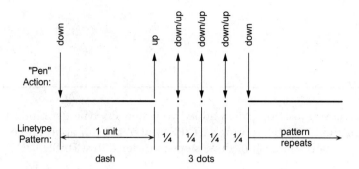

 Exercise

Creating a DASH3DOT Linetype

Copy all of the files from \MAXAC\BOOK\CH07 to your \MAXAC\DEVELOP directory.

Begin a new drawing named CH07A and use the MENU command to load the \MAXAC\DEVELOP\ACADMA.MNS menu.

Command: **LINETYPE** Enter

?/Create/Load/Set: **C** Enter

Name of linetype to create: Opens the Create or Append
DASH3DOT Enter   Linetype File dialog box

Choose \MAXAC\DEVELOP\ACADMA.LIN, then
choose OK

At the next prompt, you'll type a series of underscores, spaces, and periods that represent the look of the linetype. The descriptive text string only identifies it. It doesn't affect how AutoCAD draws the linetype.

Wait, checking if linetype already defined...

Descriptive text: ___ . . . ___ . . . ___ . . . ___ Enter

Enter pattern (on next line):
A,1.0,-0.25,0,-0.25,0,-0.25,0,-0.25 Enter

```
New definition written to file.

?/Create/Load/Set: [Enter]
```

The exercise you just completed added your new linetype definition to ACADMA.LIN, but it didn't add it to the drawing's LTYPE symbol table or apply it to objects. In the next exercise, you perform both of these steps in order to test DASH3DOT.

Using the New DASH3DOT Linetype

Continue in the previous drawing.

Command: *Choose the* Linetype *button on the Object Properties toolbar or enter* **DDLTYPE**   Opens the Select Linetype dialog

Choose L_ad   Opens the Load or Reload Linetypes subdialog

Make sure that F_ile *is set to* \MAXAC\DEVELOP\ACADMA.LIN

Scroll to the end of the Available linetypes *list and choose* DASH3DOT, *as shown in Figure 7.6*

Figure 7.6
The Load or Reload Linetypes subdialog box.

Linetypes and Hatch Patterns

Choose OK	Closes the Load or Reload Linetypes subdialog

Loading a linetype adds it to the current drawing's LTYPE symbol table, so that it can be applied to objects in the drawing.

Choose DASH3DOT *from the* Loaded Linetypes *list*	Selects DASH3DOT as the current linetype
Choose OK	Closes the Select Linetype dialog

Draw some lines and circles to see the effect of the change.

You can also use the Load and Set options of the LINETYPE command to load and set the current linetype and the LTSCALE command to set its scale, but the DDLTYPE command can do it all.

Making a linetype current tells AutoCAD to apply that linetype to objects that you draw from this point forward, regardless of what layers those objects are on. Many companies prefer not to apply linetypes to objects directly, but instead to assign linetypes to layers and leave object linetypes set to "by layer" (that is, inherit linetype from the layer's settings). In this chapter, we'll apply explicit ("by object") linetypes because it's easier when you're testing lots of linetypes.

The DASH3DOT linetype definition is stored in ACADMA.LIN and looks like this:

```
*DASH3DOT,____ . . . ____ . . . ____ . . . ____
A,1.0,-0.25,0,-0.25,0,-0.25,0,-0.25
```

The DASH3DOT definition is exactly what you entered in AutoCAD, plus an asterisk that indicates the start of a new pattern, and the A alignment code. The following list explains the linetype definition fields, each on a separate line:

```
*              Marker for a new linetype definition
DASH3DOT       The name of the linetype
____ . . .     A description to help you identify the linetype
A,             Alignment code
1.0,           One unit long dash
-0.25,         One quarter unit long pen-up space
0,             Zero length dot
```

```
        -0.25,          One quarter unit long pen-up space
        0,              Zero length dot
        -0.25,          One quarter unit long pen-up space
        0,              Zero length dot
        -0.25,          One quarter unit long pen-up space
```

In this example, we used 1.0 to define a 1-unit long dash. AutoCAD doesn't require the decimal point (1 would work as well as 1.0 here). Similarly, the leading zero in -0.25 is optional. We included the decimal point and trailing and leading zeroes in all cases so that all dash and space lengths are formatted consistently. This convention makes the linetype definition a bit easier to read, although it does increase the number of characters in the definition. According to the *AutoCAD Customization Manual*, linetype definitions should be limited to 80 character lines in the LIN file, but R13c4 appears to handle up to 280 characters. Even with a maximum of 80 characters, you're likely to run into the maximum of 12 dash, dot, and space fields first.

The numerical values defining the pen-up and pen-down motions are preceded by an alignment code. AutoCAD supports only one alignment code: A (presumably, Autodesk had plans to expand the alignment code offerings, but A has been the only valid code ever since AutoCAD first supported custom linetypes). AutoCAD automatically inserts this code for you when you define a linetype on the fly with the LINETYPE command. The alignment code is followed by a positive number for a pen-down dash, a negative number for a pen-up space, and a zero value for a dot. The fields are separated by commas.

The description field is optional, but you should include it so that you and others can identify the linetype later. Most people create an ASCII facsimile of the linetype using underscores, periods, and spaces, but you can use an ordinary text description instead of or in addition to these characters.

Note: You cannot begin a linetype with a negative number (for a space). If you try, the LINETYPE Load subcommand will report `The first dash/dot spec cannot be negative.` DDLTYPE's Load option won't complain, but it won't display the offending linetype.

You can make a linetype appear to begin with a space by starting the definition with a 0 (for a dot) and then a space, but you may not get equal spaces at the beginning and ending of lines drawn with such a linetype. AutoCAD doesn't handle linetypes that start with a dot-space-dash sequence properly.

Understanding the Linetype File Format

Although you can define custom linetypes on the fly with the LINETYPE command, it's easier to edit them directly in a linetype (LIN) file with your text editor. In the next exercise, you'll add two more linetypes directly to ACADMA.LIN. You'll also add comment lines (preceded by semicolons) to the beginning of the file.

Linetypes and Hatch Patterns

Exercise

Adding Linetypes to ACADMA.LIN

Open ACADMA.LIN in your text editor.

Take a moment to scroll through ACADMA.LIN and look at AutoCAD's linetype definitions. Appendix C in the *AutoCAD User's Guide* lists the standard linetypes.

Add the following comment lines to the top of the file:

; ACADMA.LIN: Custom Maximizing AutoCAD linetype file [Enter]
; by MM   6 Mar 1996 [Enter]

Go to the end of the file, where you'll see the DASH3DOT definition.

Copy the two lines defining DASH3DOT and make the changes shown in bold:

*DASH**4**DOT,____ .... ____ .... ____ .... ____
A,1.0,-0.25,0,-0.25,0,-0.25,0,-0.25,**0,-0.25**

Next, add one more linetype definition, called DASHDA. It's similar to AutoCAD's CENTER linetype, but with a shorter main dash.

*DASHDA,__ _ __ _ __ _ __ __ _ __ _ __ [Enter]
A,0.5,-.25,.25,-.25 [Enter]

Save ACADMA.LIN, return to AutoCAD in the previous drawing, and load and test each linetype.

When you're finished testing, use DDLTYPE to restore BYLAYER as the current linetype.

As you can see, there's not much to creating simple linetypes. As always, punctuation is important. Don't forget the asterisk, the A alignment code, or any of the commas. AutoCAD doesn't permit spaces within the linetype definition (although you can include spaces in the description). Begin comment lines with a semicolon.

Figure 7.7
Testing custom linetypes.

There aren't many new dash-dot linetypes waiting to be invented, once you get beyond the basic permutations of adding more dots and altering the dash lengths. You may find, however, that you need to create slightly altered copies of the standard AutoCAD linetypes to make them conform to industry standards. If you carefully measure their dashes and spaces, you will find that they don't meet either ANSI or most common architectural standards.

Setting Linetype Endpoint Alignment and Scale

You can apply linetype patterns to line, circle, arc, and 2D polyline objects. AutoCAD ignores linetypes for other object types (for example, text and 3D polylines). AutoCAD's linetypes apply a dash-dot pattern along the length or circumference of an object. AutoCAD automatically adjusts the beginning and ending portion of the line pattern to make it fit equally between the endpoints of line or arc segments. If the line length or circumference is too short, as is often the case with curve or spline-fit polylines, AutoCAD doesn't have room for a space and leaves the segment continuous. Polyline linetype generation, described later in this chapter, addresses this problem for polylines.

Tip: An alternative to AutoCAD's linetypes, especially if you want to apply dash-dot patterns to text or other objects for which AutoCAD doesn't support linetypes, is to use "hardware linetypes" that are built into your plotter or printer. If your output device is capable of applying its own linetypes, you can map specific colors to those hardware linetypes in the Plot Configuration dialog's Pen Assignments subdialog.

One disadvantage of this approach is that output is dependent on your plotter or printer. Lines drawn with plotter linetypes may not meet at corners. If you plot the drawing on a different device, you may not be able to reproduce the same linetypes. Another limitation is that you won't see the linetypes on the screen; they must instead be represented by color.

Now use your new DASH3DOT linetype to draw a few lines and circles, and see how AutoCAD aligns the linetype pattern to objects. To make it easier to see 1-unit relationships in linetypes, use an LTSCALE of 1 in the following exercise.

 Exercise

Linetype Alignment and Scale

Continue in the previous drawing, set snap to 0.25, and use ZOOM Left with a lower left corner point of 0,0, and a height of 6.

```
Command: LTSCALE Enter

New scale factor <1.0000>: Enter          Confirms a setting of 1
```

Linetypes and Hatch Patterns

Command: *Choose the* Layers *button on the Object Properties toolbar or enter* **DDLMODES**	Opens the Layer Control dialog
Create a new layer called TEST, *set its color to Red and its linetype to DASH3DOT, and make it the current layer*	

Make sure that the current linetype is BYLAYER (check the Object Properties toolbar or use DDLTYPE).

Next, you'll draw some lines and circles to test the new DASH3DOT linetype you have created.

Command: *Draw a line from 2,5 to the right 1 unit*	Displays as continuous
Command: *Draw a line from 2,4 to the right 2 units*	Displays one three-dot break with shortened ends
Command: *Draw a line from 2,3 to the right 3.5 units*	Displays one three-dot break with lengthened ends
Command: *Draw a line from 2,2 to the right 4 units*	Displays the correct line pattern twice
Command: *Draw a circle centered at 7,1 with a 0.25-unit radius*	Displays as continuous
Command: *Draw a circle centered at 7,2 with a 0.5-unit radius*	Displays with two three-dot breaks
Command: *Draw a circle centered at 7,4 with a 1-unit radius*	Displays with three three-dot breaks

Look at the set of lines in Figure 7.8 to see the way AutoCAD fits the linetypes. When the line segment or circle is too small, AutoCAD ignores the linetype and displays the object with a continuous line. "Too small" means that AutoCAD can't fit at least half of the first dashed segment (which we defined as one unit long) at the beginning and end of the line. Thus AutoCAD can't fit the DASH3DOT pattern within a 1-unit line. The 2-unit line leaves just enough room for AutoCAD to show the spaces and dots flanked by the dashes, shortened to one-half unit at each end of the line. With the 3.5-unit line, there isn't enough room for AutoCAD to insert a second set of dots and spaces, so it makes the ends longer (1.25 units) instead. The 4-unit long line does have enough room for a second set of dots and spaces, but once again, AutoCAD has to shorten the beginning and ending dashes.

Figure 7.8
AutoCAD's linetype endpoint alignment adjustments.

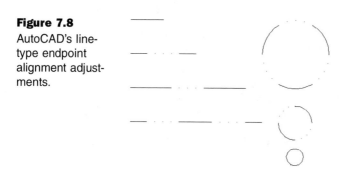

Now look at how AutoCAD applies the pattern to circles. The 0.25-radius circle (circumference 1.57 units) shows no linetype breaks. The 0.5-radius circle (circumference of 3.14 units) shows two sets of dots. The 1-unit radius circle (circumference of 6.28 units) shows three sets of dots at 120-degree intervals.

Scaling Linetype Patterns

The numbers defining linetype dash and space lengths (for example, 1.0 and -0.25 in DASH3DOT) represent drawing units. Thus, if you use decimal units in a drawing and your unit is a foot, a dash length of 1.0 will be one foot long. If your unit is a millimeter, the dash will be one millimeter long. If you use architectural units in a drawing, the dash will be one inch long.

AutoCAD multiplies the dash and space lengths by the current drawing's LTSCALE system variable setting in order to determine how long to draw dashes and spaces for all objects in the drawing. Thus, part of your drawing setup routine should be to set LTSCALE to an appropriate value, based on the current drawing's scale factor, just as you usually set DIMSCALE to the drawing's scale factor for dimensioning in model space. Some users set LTSCALE to equal the drawing scale factor (that is, the same as DIMSCALE). Most users find that this setting is too large, and instead use some fraction of the drawing scale factor. For example, 0.5 is a popular multiplier.

Thus, if you set up a 1=20 drawing where 1 represents one plotted inch, you should set DIMSCALE to 20 and LTSCALE to 0.5 times 20, or 10. If 1 represents one plotted millimeter, you normally need to multiply LTSCALE by an additional factor of 25.4 (the number of millimeters per inch), because AutoCAD's standard linetypes are defined for plotted inches. In a 1/8"=1'-0" drawing, you should set DIMSCALE to 96 (because one foot is 96 times 1/8") and LTSCALE to 0.5 times 96, or 48.

 LTSCALE controls the global linetype scale. When you need to alter the linetype scaling for individual objects (for example, a very short line for which AutoCAD wouldn't otherwise have room to show any spaces), apply an object linetype scale factor other than the default of 1.0. As described earlier in this chapter, you can control object linetype scale factors with the CELTSCALE system variable or DDLTYPE dialog (for new objects), or the DDCHPROP or CHPROP command (for existing objects). An object's linetype scale is cumulative with LTSCALE. Thus, if you draw a line with LTSCALE = 10 and CELTSCALE = 0.5, and the linetype definition specifies a dash length of 0.25, AutoCAD will make the dashes 10 times 0.5 times 0.25, or 1.25 units long.

Note: The AutoCAD standard linetype file ACAD.LIN, from which we created ACADMA.LIN, has three variations of each linetype: normal-scale, half-scale, and double-scale (for example, BORDER, BORDER2, and BORDERX2). These three versions were helpful before AutoCAD had paper space and object linetype scaling. You could use BORDER2 for short lines and BORDERX2 for long lines where you wanted longer dashes. Now you can achieve the same effect with greater control by using R13's object linetype scaling. Just use the main linetype definition (for example, BORDER), and apply object linetype scales of 0.5, 2, or any other values to the objects that require special linetype scaling.

Paper space can complicate the arithmetic further. An additional system variable, PSLTSCALE, controls whether AutoCAD performs special scaling in paper space viewports. Unlike LTSCALE and CELTSCALE, PSLTSCALE is a 0 or 1 (off or on) system variable. When PSLTSCALE is set to its default value of 0, AutoCAD doesn't do any special scaling of linetypes in paper space viewports. As long as all your viewports use the same scale and you've used the correct ZOOM XP scale factor, PSLTSCALE = 0 works well. If you have viewports showing model space geometry at different scales (for example, 1/8"=1'-0" plan and 1/2"=1'-0" plan detail), PSLTSCALE = 0 doesn't work well. Linetype dashes and spaces will be different lengths in the differently scaled viewports. To get around this problem, you need to set PSLTSCALE to 1 and LTSCALE to the value you would use for a full scale (1 = 1) drawing—for example, 0.5. The downside of PSLTSCALE is that you have to remember to change LTSCALE between model and paper space values every time you toggle the TILEMODE system variable.

In short, set LTSCALE to 0.5 times the drawing scale factor when you set up a new drawing. Use object linetype scaling for very short lines or other special purposes. Leave PSLTSCALE set to 0 unless you create differently scaled paper space viewports.

R13's CELTSCALE feature offers a new way of dealing with linetypes and differently scaled paper space viewports. You leave LTSCALE set to 1.0 (no matter what the drawing's scale factor is) and PSLTSCALE set to 0. You then apply object linetype scaling to each object based on the scale of the viewport it will appear in. For example, you would set CELTSCALE to 48 (0.5 x 96) for objects that will appear in a 1/8"=1'-0" viewport (ZOOM 1/96XP) and set CELTSCALE to 12 (0.5 x 24) for objects that will appear in a 1/2"=1'-0" viewport (ZOOM 1/24XP). The advantage of this method is that you can achieve paper space linetype scaling without the problems caused by setting PSLTSCALE to 1, which we described earlier. Note, however, that this method assumes each object will appear at one and only one scale.

Scaling ISO Linetypes

 The R13 ACAD.LIN file includes a new set of 14 linetypes that are intended to help users conform with the International Standards Organization (ISO) drafting standards. The first six ISO linetypes are listed here:

```
; dashed line
*ACAD_ISO02W100,__ __ __ __ __ __ __ __ __ __ __ __ __ __ __
A,12,-3
; dashed space line
*ACAD_ISO03W100,__     __     __     __     __     __     __     __
A,12,-18
; long dashed dotted line
*ACAD_ISO04W100,____ . ____ . ____ . ____ . ____ . ____ . ____
A,24,-3,.5,-3
; long dashed double dotted line
*ACAD_ISO05W100,____ .. ____ .. ____ .. ____ .. ____ .. ____
A,24,-3,.5,-3,.5,-3
; long dashed triplicate dotted line
*ACAD_ISO06W100,____ ... ____ ... ____ ... ____ ... ____ ... ____
A,24,-3,.5,-3,.5,-3,.5,-3
; dotted line
*ACAD_ISO07W100,. . . . . . . . . . . . . . . . . . . . .
A,.5,-3
```

Notice that the dash and space parameters are much larger than they are for the other ACAD.LIN linetypes (and for the custom linetypes that you created). ISO linetypes are designed to be used in metric drawings where sheet sizes are measured in millimeters rather than inches. Thus the linetype dash and space distances are in the neighborhood of 25 times larger (because 25.4 millimeters equal one inch). If you want to use the ISO linetypes in drawings whose sheet sizes are measured in inches, you'll need to divide your desired LTSCALE or CELTSCALE by 25.4.

R13 handles ISO linetypes a bit differently from other linetypes. The Select Linetype dialog (DDLTYPE command) lists ISO linetypes by their descriptive names (for example, Dashed or Dashed space) instead of the names in the ACAD.LIN file (ACAD_ISO02W100 or ACAD_ISO03W100). ISO linetype names are in mixed case, while non-ISO linetype names are in upper case. The drop-down linetype list on the toolbar, however, lists the ACAD.LIN names.

The Select Linetype dialog also includes an ISO Pen Width drop-down list, which you use in lieu of the Linetype Scale edit box to control CELTSCALE. (see Figure 7.2 early in this chapter). The ISO Pen Width setting is misleading to anyone who isn't familiar with the ISO standard. This setting does not control the plotted weights of the lines—you still must do that with the traditional color-to-lineweight mapping in the Plot Configuration dialog box. Instead, the setting changes the CELTSCALE system variable so that the dash and space lengths conform to the ISO correspondence between pen width and

Figure 7.9
Polyline linetype generation.

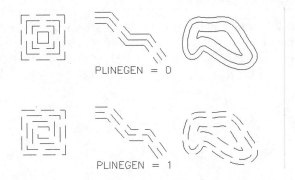

linetype scale. The default setting of 1.0 mm sets CELTSCALE to 1.0. A setting of 0.35 mm sets CELTSCALE to 0.35.

Using 2D Polyline Linetype Generation

You have two options when applying linetypes to 2D polylines (see Figure 7.9). By default, AutoCAD "justifies" linetypes for each segment of a polyline; it treats each polyline segment as if it were a separate line or arc, and doesn't allow the linetype pattern to flow through vertices. This approach works well for long lines and rectangles, but not for polylines comprising many short segments—especially splined or curve-fit polylines. In such cases, each polyline segment usually is so short that the entire polyline appears as a continuous line.

For curved polylines, you'll usually want to turn on AutoCAD's *polyline linetype generation* option. This option tells AutoCAD to ignore vertices when applying the linetype to the polyline. To turn on polyline linetype generation for subsequent polylines, set the PLINEGEN system variable to 1 (the default value of 0 means off). To change the polyline linetype generation setting for an existing polyline, use the PEDIT command's Ltype gen option. AutoCAD stores the polyline linetype generation setting with each polyline object, which allows you to turn it on for some polylines and off for others.

R13 Complex Dash-Dot Linetypes

 Thanks to R13's new *complex* (symbolic) linetype capability, you're no longer limited to dashes, dots, and spaces. You can embed individual text characters, entire text strings, or AutoCAD shapes from SHX files in your custom linetypes.

Understanding Complex Linetypes

Complex linetype definitions start with the same syntax as simple linetypes and add special parameters within square brackets to tell AutoCAD how to insert the text or shape within the line. Here are two examples from \R13\COM\SUPPORT\LTYPESHP.LIN, a sample file that comes with R13:

```
*GAS_LINE,----GAS----GAS----GAS----GAS
A,.5,-.2,["GAS",STANDARD,S=.1,R=0.0,X=-0.1,Y=-.05],-.25
*FENCELINE1,-----0-----0-----
A,.25,[CIRC1,LTYPESHP.SHX,S=.1],-.2,1
```

Figure 7.10 shows what these linetypes look like.

The syntax for an embedded text string is:

`["string",style,R=n,A=n,S=n,X=n,Y=n]`

The syntax for an embedded shape is:

`[shape,shape_file,R=n,A=n,S=n,X=n,Y=n]`

"string" is a literal text string of one or more characters enclosed in quotation marks, and *shape* is the name of a shape from *shape_file*, an AutoCAD SHX file. *shape* must exist in *shape_file*, or else AutoCAD won't let you use the linetype.

style is an AutoCAD text style name, and *shape_file* is an AutoCAD SHX shape file. If *style* doesn't already exist in the current drawing, AutoCAD won't let you use the linetype. If *shape_file* isn't in the library search path, AutoCAD complains and asks you to choose another SHX file. You can include a path with *shape_file*, but then AutoCAD won't recognize the linetype unless *shape_file* is in that exact location. It's better to omit the explicit path and keep the file on the library search path.

Warning: To use complex linetypes successfully, you must keep careful track of your text styles and shape files. If these files are missing, you'll cause lots of aggravation for yourself and other users. If you regularly use complex linetypes with text, automate the creation of any required text styles using menu macros, prototype drawings, or AutoLISP programs. If you use complex linetypes with shapes, make sure that the

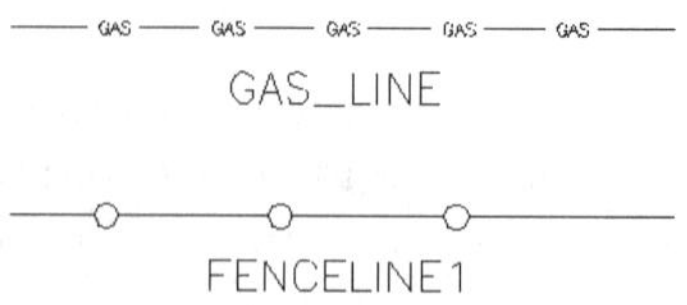

Figure 7.10
Symbolic linetypes.

SHX files accompany the drawings when you send them for plotting or to clients or consultants, just as you do with custom fonts and other custom shape files.

The five remaining fields, R=, A=, S=, X=, and Y=, are optional *transform* specifications. The `n` after each transform specification represents any number (which of course can be different for each transform).

R=`n` indicates a rotation of the text or shape relative to the current line segment direction. By default, R=0, which means that AutoCAD aligns the text or shape with the line segment. That's often the alignment you want; for example, it's used in the linetypes in Figure 7.10.

A=`n` indicates an absolute rotation of the text or shape with respect to the World Coordinate System X axis. Use A=0 when you want a text or shape always to appear horizontal, no matter what direction the line segments go. You can specify R= and A=, but not both. If you don't specify either one, AutoCAD uses R=0. R= and A= rotations are in degrees unless you add an R or G after the number to indicate radians or grads.

S=`n` specifies the scale factor for the text or shape. If you use a fixed height text style, then AutoCAD multiplies the height by `n`. If you use a variable (that is, zero) height style, then AutoCAD treats `n` as the absolute height. For shapes, the S= scale factor scales the shape up or down from its default scale factor of 1.0. In all cases, AutoCAD multiplies the S= scale factor by LTSCALE and CELTSCALE to determine the height or scale factor. Thus you should specify an S= value that's appropriate for a plot scale of 1:1 (full scale) with your normal LTSCALE (for example, 0.5). Then when you use the complex linetype in drawings with other scales and set LTSCALE appropriately for the scale of each drawing, the text or shapes will appear a consistent size on the plots.

X=`n` and Y=`n` are optional offsets from the current point in the linetype specification. By default, AutoCAD puts the lower left corner of the text string or insertion point of the shape at this current point; you specify offset values in order to move the string or shape relative to the current point. The offsets are measured parallel (for X=) and perpendicular (for Y=) to the current line segment direction, as though you had established a local coordinate system with the X axis pointing from the first endpoint to the second endpoint of the current linetype segment. Thus a positive X= offset moves the text or shape towards the second endpoint of the current linetype segment. A positive Y= offset moves the text or shape 90 degrees (counter-clockwise) from the positive X= direction. Both offsets are scaled by LTSCALE and CELTSCALE. These offsets allow you to position the text or shape more precisely. They are particularly important for text because AutoCAD doesn't center the text automatically.

Embedding Text in Linetypes

In the next three exercises, you create several complex linetypes with embedded text characters. The first exercise demonstrates an attempt to create a welded wire fabric (that is, wire mesh) linetype with embedded Xs whose height is 1/8 unit (or 0.125). The linetype will consist of a dash 1 unit long, a gap of 0.25 units, an X, and another gap of 0.25 units.

Exercise

Creating a Linetype with Embedded Text

Begin a new drawing named CH07B. You can discard or save the previous drawing.

Set LTSCALE to 0.5, snap to 0.125, and grid to 1.0, and use ZOOM Left with a lower left corner point of 0,0 and height of 5.

Open ACADMA.LIN in your text editor and add the following lines to the end:

```
*WIREMESH,--- X --- X --- X --- X --- Enter
A,1.0,-0.25,["X",STANDARD,S=0.125],-0.25 Enter
```

Save ACADMA.LIN and return to AutoCAD.

Command: *Use the DDLTYPE dialog to load and set current the new WIREMESH linetype*

Command: *Draw a line from 1,1 to 8,1, as shown in Figure 7.11 at ①*

As you can see from Figure 7.11, this first attempt isn't quite right. The Xs are left-aligned instead of middle-aligned. We need to specify X and Y offsets for the text in order to make it appear in the middle of the gap. Also, the Xs are only 1/16 (or 0.0625) tall, instead of 1/8, because AutoCAD multiplies the S=0.125 height by LTSCALE, which we've set to 1/2. Thus we need to change S to 0.25 in order to compensate.

Figure 7.11
Two versions of the WIREMESH linetype.

Linetypes and Hatch Patterns

Exercise

Adding Offsets to Embedded Text

Return to editing ACADMA.LIN and change the items in bold:

```
*WIREMESH,--- X --- X --- X --- X ---
A,1.0,-0.25,["X",STANDARD,S=0.25,X=-0.08,Y=-0.125],-0.25
```

Save ACADMA.LIN and reload and retest WIREMESH.

Command: *Draw a line from 1,2 to 8,2*   The improved linetype should look like that at ② in Figure 7.11

Command: **REGEN** Enter   Fixes the previous line you drew

S=0.25 times an LTSCALE factor of 0.5 results in our desired text height of 0.125. Y=-0.125 (half of S=0.25) centers the text vertically. X=-0.08 centers the text horizontally. We determined this value by zooming in and using the DIST command to measure the approximate width of an uppercase letter X and dividing that width in half.

You can use more than one character in an embedded text string. You can also use special Unicode and %% characters.

Exercise

More Embedded Text Linetypes

Return to editing ACADMA.LIN and add three new linetype definitions:

```
*PL,--- PL --- PL --- PL --- PL --- Enter
A,1.7,-0.4,["PL",STANDARD,S=0.2,X=-.16,Y=-.125],-0.4 Enter
*DIAM,--- o --- o --- o --- o --- Enter
A,2.0,-0.25,["%%C",STANDARD,S=0.25,X=-0.08,Y=-.125],-0.25 Enter
*HOUSE,--- ^ --- ^ --- ^ --- ^ --- Enter
A,2.0,-0.25,["\U+2302",STANDARD,S=0.5,X=-0.17,Y=-0.1],-0.25 Enter
```

Figure 7.12
More linetypes with embedded text.

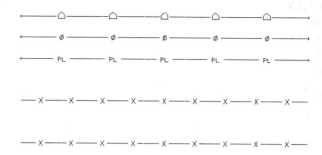

Save ACADMA.LIN, return to the previous drawing, and load and test each of the new linetypes. The results should look like Figure 7.12.

Creating complex linetypes with embedded text isn't difficult, but you'll often need to experiment with different scale and offset values in order to achieve the effect you want. Fortunately, testing new values is easy in Windows. Edit and save the LIN file, reload the linetype in AutoCAD, and regenerate the drawing to see the effects.

Tip: You can also use big font characters in complex linetypes. For example, assume that you've defined a style called NOTES whose fonts are ROMANS,BFSPECMA (ROMANS as the base font and BFSPECMA as the big font), and that BFSPECMA.SHX is a big font file in which ~c draws a centerline symbol (see Chapter 6). Then you can define a complex linetype like this:

```
*CL,--- CL --- CL --- CL --- CL ---
A,1.0,-0.25,["~c",NOTES,S=0.25,X=-.12,Y=-.125],-0.25
```

Embedding Shapes in Linetypes

The only difference between embedded shapes and embedded text is that you work with shape and shape file names instead of text strings and style names. In the next exercise, you experiment with shapes from the LTYPESHP.SHX shape file that comes with R13. LTYPESHP.SHX contains sample shapes that the LTYPESHP.LIN sample complex linetype file uses. The BOX, CIRCLE, and ZIG shapes are shown at the top of the drawing in Figure 7.13.

Linetypes and Hatch Patterns

Figure 7.13
Complex linetypes with shapes.

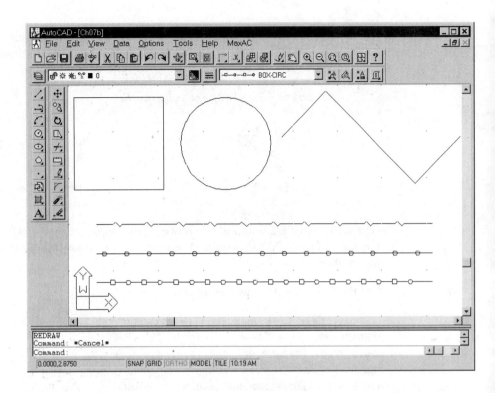

Creating Linetypes with Embedded Shape

Erase everything in CH07B.DWG.

Command: **LOAD** Enter   Opens the Select Shape File dialog

Select \R13\COM\SUPPORT\LTYPESHP.SHX *from the Select Shape File dialog*

The previous step loaded the shapes from LTYPESHP.SHX into the drawing, so that they can be used by the SHAPE command. You performed this step so that you can insert the shapes and see how they look before using them in complex linetypes. You don't need to load shapes before drawing a complex linetype that references them—you only have to ensure that the shape file is on AutoCAD's library search path.

Command: **SHAPE** Enter

`Shape name (or ?): BOX` [Enter]	The BOX shape appears at your crosshairs, similar to the way a block would
`Starting point:` *Pick any point as the insertion point for the shape*	
`Height <1.0000>:` [Enter]	Specifies 1.0 as the height for this shape
`Rotation angle <0>:` [Enter]	Specifies 0 as the rotation angle for this shape

Repeat the SHAPE command two more times and insert the CIRC1 and ZIG shapes. The top of Figure 7.13 shows the three shapes inserted with heights of 1.0 and rotation angles of zero.

Add the following linetype definitions to ACADMA.LIN:

```
;;NOTE: these linetypes require LTYPESHP.SHX [Enter]
*BREAKLINE,-----/\/------ [Enter]
A,0.5,[ZIG,LTYPESHP.SHX,S=0.1],-0.4,0.5 [Enter]
*BEADS,----0-----0---- [Enter]
A,0.5,[CIRC1,LTYPESHP.SHX,S=0.1],0.5 [Enter]
*BOX-CIRC,----[]-----0---- [Enter]
A,0.25,[BOX,LTYPESHP.SHX,S=0.1],-0.2,0.5,[CIRC1,ltypeshp.shx,S=0.1], [»]
-0.2,0.25 [Enter]
```

The last line in the exercise listing had to be printed on two lines, but you should enter it on a single line without the [»] continuation character.

Save ACADMA.LIN, then load and test each of the new linetypes. The results should look like Figure 7.13.

When you're finished testing, use DDLTYPE to restore BYLAYER as the current linetype.

Embedded shapes are often a bit easier to use than embedded text. If the shape has a reasonable insertion point, you don't need to specify offsets. You still need to work out an appropriate scale factor, though. And don't forget that the shape file (LTYPESHP.SHX in this case) must be located somewhere in AutoCAD's library search path every time you open drawings that reference these linetypes.

Note that the BEADS linetype doesn't have any gaps in the line—all segments are specified as positive, so the line runs through the shape. Also note that you can insert more than one shape, as shown in the BOX-CIRC linetype.

Linetypes and Hatch Patterns

Note: Polyline linetype generation also works for complex linetypes, although you won't see the results until you complete the polyline.

If you want to look at some other complex linetype examples, download Paul Richardson's file R13LT.ZIP from the ACAD forum libraries on CompuServe (the file currently is in Library 3). This file contains a number of useful and interesting custom linetypes that use embedded text and shapes.

Hatch Patterns

AutoCAD's hatch patterns extend the linetype idea into two dimensions. As you'll learn later in this section, AutoCAD hatch pattern definitions consist of families of linetype pattern definitions. The net effect is a two-dimensional pattern or fill inside a closed area. This section shows you how to create custom hatch patterns.

Tip: Many third-party applications include discipline-specific hatch patterns. In addition, you can buy collections of hatch patterns (see the Drafting Utilities section of *The AutoCAD Resource Guide*). These ready-made patterns can save you lots of time over developing your own custom versions, especially for complex patterns. On the other hand, developing your own patterns gives you the flexibility to create exactly what you want. It also gives you the freedom to distribute your pattern definitions to clients, fellow consultants, and friends without violating license agreements. (Like most software, third-party applications and commercial hatch pattern libraries usually limit their legal use to one computer or one site.)

Using Hatch Patterns

As with linetypes, AutoCAD includes a standard set of hatch patterns, which is stored in the ASCII support file \R13\COM\SUPPORT\ACAD.PAT. See Appendix C in the *AutoCAD User's Guide* for a listing and pictures of the standard hatch patterns. As with ACAD.LIN, you can add custom hatch patterns to ACAD.PAT. You can't create your own hatch pattern library (for example, ACADMA.PAT), but you can store individual custom hatch patterns in individual PAT files (for example, GRAVEL.PAT, PLYWOOD.PAT, and so on). Hatch patterns defined in your own PAT files are more portable, unaffected by updates to AutoCAD, and more compatible with other applications than if you put them in the ACAD.PAT file.

When you create hatching with the Boundary Hatch dialog box (Figure 7.14) or the HATCH command, you have the option of leaving it "whole" or exploding the parts of the hatching into individual line segments. When you leave the hatching whole, AutoCAD creates an *anonymous block*—a block whose name begins with an asterisk and usually

Figure 7.14

The R13 Boundary Hatch dialog box.

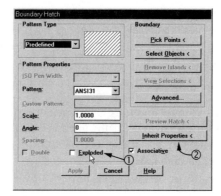

remains hidden (although you can see the names if you use the LIST command on a hatching block or use FILTER, Block Name, Select). Hatching blocks have names like *X0, *X1, and *X2. AutoCAD also attaches extended entity data to the block—data that defines the pattern, scale, angle, and other information about the hatching.

Tip: When you hatch identical areas, hatch one area and copy the hatching instead of hatching each area separately. Copied hatches reference the same block definition, thereby reducing the drawing's file size.

If you explode the hatching, either during creation (see ① in Figure 7.14) or later with the EXPLODE command, AutoCAD draws individual line objects. Usually you should avoid exploding hatching. Exploding obliterates the associativity of the hatching with its boundary, and leaves you with lots of little lines that can be difficult to erase or edit. It also does away with the extended entity data, which means that you can't use HATCHEDIT to revise the pattern or Inherit Properties ② to apply a similar pattern elsewhere.

When you explode hatching, the object count temporarily doubles in the drawing (one set of objects in the block definition and one set in the drawing). The next time you reopen the drawing, AutoCAD detects that the anonymous block is no longer referenced and automatically purges it.

Tip: Although you can create a solid filled area by using closely spaced hatch lines, this method makes your DWG files large and unwieldy. An alternative is to assemble filled objects such as wide polylines, donuts, and solids into the shape that you require. Align and overlay them in any combination that works. In 3D, however, solid fills that are not parallel to your current view do not display or plot as filled objects—AutoCAD shows and plots only the outlines of the areas. If you are using a PostScript output device, you can use PostScript fills instead of solids, as described at the end of this chapter.

Understanding Hatch Patterns

Hatch pattern definitions rely on the same plotter pen metaphor that you learned about in the linetype sections of this chapter. A hatch definition combines one or more parallel *families* of dash-dot lines and specifies an angle and offset spacing for the parallel lines in each family to create a two-dimensional pattern. AutoCAD applies these family characteristics to fill closed areas with the pattern of line families. Figure 7.15 shows all of these characteristics.

Here are three hatch pattern definitions from ACAD.PAT:

```
*DASH, Dashed lines
0,        0,0,              .125,.125,         .125,-.125

*CROSS, A series of crosses
0,        0,0,              .25,.25,           .125,-.375
90,       .0625,-.0625,     .25,.25,           .125,-.375
```

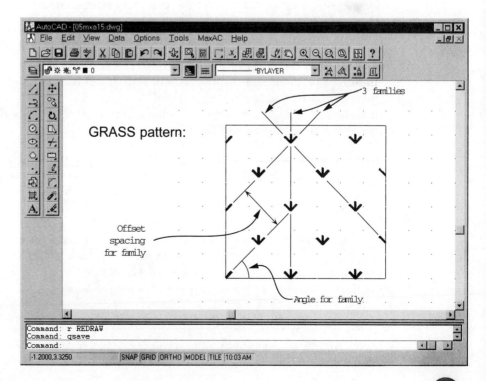

Figure 7.15
Hatch pattern family components.

Maximizing AutoCAD® R13

```
*GRASS, Grass area
90,      0,0,    .707106781,.707106781,    .1875,-1.226713563
45,      0,0,             0,1,             .1875,-.8125
135,     0,0,             0,1,             .1875,-.8125
```

As with linetypes, an asterisk introduces a new pattern definition and is followed by the pattern name and description. Each subsequent line of text defines one line family that AutoCAD will draw in order to fill in the hatched area. These lines use the format:

```
Angle,   X_origin,Y_origin,   X_offset,Y_offset,  dash_1[,dash_2, ...]
```

Pattern definitions, unlike linetype definitions, permit spaces in the definition. AutoCAD ignores blank lines and text following semicolons.

The `Angle` field specifies the default orientation of the line family (which AutoCAD rotates by the angle you specify when creating the hatch).

The next two fields, `X_origin` and `Y_origin`, control the imaginary starting point of the first line in the family. This point is not an AutoCAD coordinate, but rather a relative distance from the drawing's current SNAPBASE point (0,0 by default). To draw the first line in the family, AutoCAD "moves the pen" to this relative point, leaves the pen up, and moves it in the direction specified by `Angle` until it bumps into a hatching border line. At that point, AutoCAD "lowers the pen" and starts drawing. When it reaches another border, it "raises the pen" and stops drawing.

The `X_offset` and `Y_offset` fields specify the incremental distance between endpoints of successive lines in the family. The `X_offset` and `Y_offset` values are measured perpendicular to `Angle`. After AutoCAD is finished drawing the first line, it moves the pen to the point (`X_origin` + `X_offset` , `Y_origin` + `Y_offset`) from SNAPBASE and draws the second line. AutoCAD repeats this drawing and offsetting procedure until the area is filled.

The last set of fields—dash_1[, *dash_2,* ...]—are from one to six dash-dot linetype segment specifications, in the same format you used to create custom linetypes (positive for dashes, negative for spaces, and zero for dots). You can use any single positive number (for example, .25) to indicate a continuous line with no gaps.

 Tip: The smaller the offsets and dash lengths, the larger and slower the hatch will be. Don't make hatching definitions and hatch scale factors any smaller than they need to be. Also, draw your lines and dots so that as many lines as possible align with other lines—avoid excessive numbers of linetype families.

Creating a Simple Hatch Pattern

AutoCAD has no commands to help you create hatch patterns. You must edit ASCII files in order to create custom patterns. In the next exercise, you'll create a simple checkered

Figure 7.16
The checkered plate pattern.

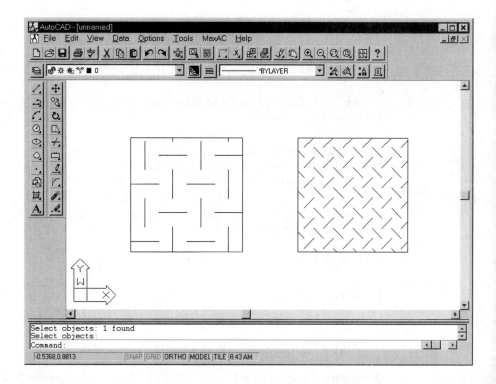

plate hatch pattern, as shown in Figure 7.16. After you define the pattern in your text editor, you'll hatch 1-unit squares with two different hatch rotation angles to see how the hatch pattern families relate to a drawing unit.

 **Exercise**

Making a Checkered Plate Pattern

Begin a new drawing named CH07C, then set snap to 0.1, and use ZOOM Center at 1.2,0.5 with a height of 2.

In your text editor, create a new CHECKP.PAT file in the \MAXAC\DEVELOP directory and enter the following lines:

```
*CHECKP, Checkered Plate Enter
0, 0,.09375, .25,.25, .25,-.25 Enter
90, .125,.21875, .25,.25, .25,-.25 Enter
```

Save CHECKP.PAT and return to AutoCAD.

Command: **RECTANG** [Enter]

First corner: *Draw a 1-unit square whose lower left corner is at the point 0,0*

Command: *Use the RECTANG command to draw another 1-unit square, but this time make 1.5,0 the lower left corner point*

Command: *Choose the* Hatch *button on the Draw toolbar or enter* **BHATCH**	Opens the Boundary Hatch dialog
Under Pattern Type, *choose* Custom	Specifies a custom PAT file
In the Custom Pattern *field, enter* **CHECKP**	Tells AutoCAD to look for CHECKP.PAT
Make sure the Scale *field is* **1.0**	Leaves the pattern full size
Make sure the Angle *field is* **0**	Leaves the pattern unrotated
Use Select Objects *to select the left square as the hatch boundary and press ENTER*	
Choose Preview Hatch	Shows what the hatching will look like
Choose Continue	
Choose Apply	Adds the hatching to the drawing and closes the Boundary Hatch dialog

The hatching you created should match the pattern in the left square in Figure 7.16. Notice how the horizontal dashes relate to the hatch pattern definition line 0, 0,.09375, .25,.25, .25,-.25. Each series of dashes is at zero degrees (first field in the family's definition). The first horizontal dash starts at 0,.09375 (second and third fields). The next series of horizontal dashes, directly above the first series, is offset .25,.25 (fourth and fifth fields). And each series of dashes comprises dash lengths of .25 separated by gaps of .25 (sixth and seventh fields). You can use the DIST command to verify these values, as well as the values for the vertical dashes.

Linetypes and Hatch Patterns

Command: *Repeat the BHATCH command, but change the* Scale *field to* **0.5** *and the* Angle *field to* **45**, *and hatch the right square*

Your screen should look like Figure 7.16.

Even though the pattern's linetype families use angles of 0 and 90 degrees, you can rotate the pattern to get the desired checkered effect. We defined the pattern using 0 and 90 degrees to avoid calculating line lengths at 45-degree angles.

In the previous exercise, we performed all the calculations for you without explaining how to figure out the origins and offsets. The exercises in the next section show you how to compute origins and offsets as you develop a more complex hatch pattern.

Creating a Complex Hatch Pattern

There is no difference in the syntax for more complex hatch patterns—just more work to figure out all of the proper numeric values. The task isn't difficult, but it can be tedious. It requires careful design and a lot of calculations—remember your trigonometry?

In the next six exercises, you'll develop a hatch pattern called ASTEROID, which is shown in Figure 7.17. Although the pattern is a bit fanciful, it demonstrates all of the techniques for developing a complex hatch pattern. Also, its components are useful as the basis for other hatch patterns, such as gravel or the reflection on glass.

If you're not concerned about creating your own complex hatch patterns, you can skim or skip this section. The file ASTEROID.MA that you copied into \MAXAC\DEVELOP contains the completed pattern, so you can look at the file and try the pattern if you just want to see the end result. (Rename it to ASTEROID.PAT before using it.)

Figure 7.17
The ASTEROID hatch pattern.

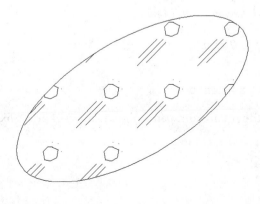

The method we'll use to develop ASTEROID relies on using a 1 x 1 unit box, drawn with RECTANG. You'll draw a facsimile of the pattern in that box, make a 4 x 4 array of the pattern, and then measure the objects with the LIST and DIST commands in order to determine the numeric values for the hatch pattern's line families. You can use any size box, but a 1 x 1 box makes some of the calculations simpler.

In the first exercise, you'll draw the boxes and pattern.

Exercise

Planning the ASTEROID Pattern

Begin a new drawing named CH07D, then set snap to 0.025, grid to 0.1, and use ZOOM Center at 0.5,0.5 with a height of 1.2.

Command: **RECTANG** [Enter]

First corner: *Draw a 1 x 1 square whose lower left point is 0,0*

Next, you'll use the POINT and LINE commands to draw the pattern with the coordinates in the next few steps. Use Figure 7.18 as your guide.

Command: **POINT** [Enter]

Point: *Draw points at 0.925,0.75, and 0.875,0.875, and 0.775,0.9*

Command: **LINE** [Enter]

From point: *Draw lines from 0.325,0.175 to 0.65,0.5, from 0.175,0.125 to 0.6,0.55, from 0.2,0.25 to 0.5,0.55, and then draw an irregular polygon from 0.7,0.825 to 0.775,0.825 to 0.875,0.775 to 0.85,0.65 to 0.75,0.60 to 0.65,0.625 to 0.625,0.75 and Close*

Zoom Center at 3,2 with a height of 5.

Next, you'll duplicate the square and pattern to make a 4x4 array, like that in Figure 7.18.

Command: **ARRAY** [Enter]

Linetypes and Hatch Patterns

Figure 7.18
Planning the ASTEROID pattern.

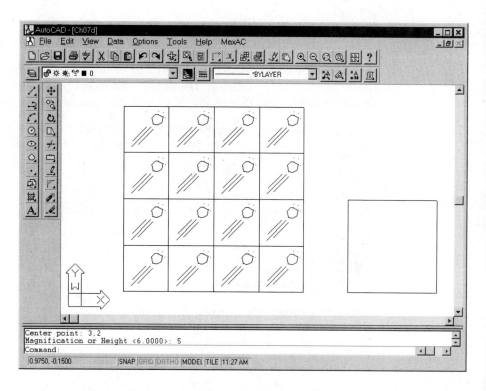

```
Select objects: ALL Enter

Select objects: Enter

Rectangular or Polar array (R/P) <R>:
Press ENTER to accept Rectangular

Number of rows (—) <1>: 4 Enter

Number of columns (||||) <1>: 4 Enter

Unit cell or distance between rows (—): 1 Enter

Distance between columns (||||): 1 Enter    Creates 15 copies
```

Next, you'll make a larger rectangle in which to test the pattern as you develop it.

```
Command: RECTANG Enter

First corner: Draw a 2 x 2 square whose lower
```
left point is 5,0

279

Maximizing AutoCAD® R13

Your screen should look like Figure 7.18.

To begin defining the pattern, you'll create a header with the pattern's name and description. After the header line, you'll write one definition line for each dot and line in the pattern. We'll start with the dots because they're the easiest. You can use the ID or LIST command to determine the coordinates of each point object.

Tip: If you have trouble seeing the point objects, set the PDMODE system variable to a value other than zero (for example, 32) and regenerate the drawing.

The first line family you will create is the dot shown in Figure 7.19.

 Exercise

Writing the Pattern Header and First Line Family

Turn the grid off to make the points easier to see.

Command: **LIST** Enter

Select objects: *Select the first point you drew, then press ENTER*   Lists the selected point's data

```
POINT        Layer: 0
Space: Model space
Handle = 1B8
at point, X=0.9250   Y=0.7500   Z=0.0000
```

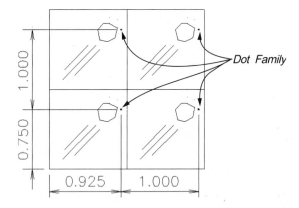

Figure 7.19
First line family—a dot.

Linetypes and Hatch Patterns

Note that the point's coordinates are .925,.75. In your text editor, create a new ASTEROID.PAT file in the \MAXAC\DEVELOP directory and enter the following lines:

```
; Maximizing AutoCAD complex hatch pattern example Enter
; by MM 7 Mar 1996 Enter
*ASTEROID, Asteroids hurtling through space Enter
0,  .925,.75,  0,1,   0,-1 Enter
```

Remain in your text editor for the next exercise.

The origins and offsets for this first line family are shown in Figure 7.19. Zero, the first entry, is the angle for this family of lines. Because dots don't have length, the angle doesn't really matter, but zero makes the other values easy to calculate. The point .925,.75 is the X,Y origin of the first dot, measuring from 0,0. The 0,1 is the X,Y offset to the next parallel member of its family.

An X offset of zero makes the next line's endpoint start perpendicular to the first. Any other value would stagger the starting origin for dashed and dotted lines. To create a family of dotted "lines" with one unit between each pair, you specify a Y offset of 1. The last two numbers, 0,-1, are the dash-dot specifications. The zero means put the pen down to draw a dot; the -1 means lift the pen and move it over one unit.

Add the other two dots in the same way. The only difference is the X,Y origin for each family. In order to determine the origins, you can use the LIST command again, or simply read the values from the first exercise in this section, when you drew the points.

 Exercise

Adding Dots and Testing the Pattern

Add these next two lines right after the previous ones you entered.

```
0,  .875,.875,  0,1,   0,-1 Enter
0,  .775,.9,    0,1,   0,-1 Enter
```

Save ASTEROID.PAT and return to AutoCAD.

Command: **BHATCH** Enter   Opens Boundary Hatch dialog box

Specify the following hatch parameters:
Under Pattern Type, *choose* Custom
In the Custom Pattern *field, enter* **ASTEROID**
In the Scale *field, enter* **1.0**
In the Angle *field, enter* **0**

Use Pick Points *or* Select Objects *to select the 2 x 2 square as the hatch boundary*

Choose Preview Hatch	Shows what the hatching will look like
Choose Cancel	Erases the temporary hatching

Warning: A common mistake when you're creating custom hatch patterns is to leave out commas. If your pattern doesn't look right, check carefully for missing commas between numbers.

The hatch pattern should be identical to the dots in the facsimile you drew. Next, you'll define the three parallel lines at 45 degrees. The origin and offset for the first line family are shown in Figure 7.20.

The origin for the first line family is 0.325,0.175. The offset distance from the beginning of the first line in the family to the beginning of the corresponding line in the square above is 0.7071,0.7071, when measured parallel and perpendicular to the first line. You might recognize .7071 as the cosine and sine of 45 degrees (the angle of this line family).

Figure 7.20
Diagonal line family.

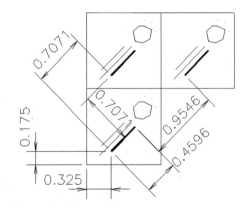

If trigonometry leaves you queasy, you can rotate the UCS so that it's parallel to the line (that is, rotate it 45 degrees about the Z axis) and use the DIST command to obtain the distances. Alternatively, leave the UCS set to World and measure the distance from the endpoint of the first line perpendicular to the next line in the family. Then reverse the process (endpoint of the second line in the family perpendicular to the first line).

The other numbers you need are the length of the line segment and the space or gap between segments in each family. The LIST command tells you that the first line is 0.4596 units long, and the DIST command tells you that the distance from its second endpoint to the first endpoint of the corresponding line in the upper right unit square is 0.9546 units (see Figure 7.20).

Tip: Note that 0.4596 plus 0.9546 is 1.4142, which is the square root of 2 and the length of a diagonal through the unit square. Thus for 45-degree lines, you really only need to know the line's length. Subtracting the length from 1.4142 gives you the gap.

 Exercise

Adding Diagonal Lines to ASTEROID

Use the LIST and DIST commands as described before the exercise to verify the origin, line length, and gap length. You can perform the same LIST and DIST procedures for the other two diagonal lines, or simply take our word for it and enter the values shown here.

Return to ASTEROID.PAT and add the next three lines to the end of the pattern definition.

```
45,  .325,.175,  0.7071,0.7071,  .4596,-.9546 Enter
45,  .175,.125,  0.7071,0.7071,  .6010,-.8132 Enter
45,  .2,.25,     0.7071,0.7071,  .4244,-.9898 Enter
```

Save ASTEROID.PAT and return to AutoCAD.

Test the revised pattern, as you did at the end of the previous exercise.

Your pattern should now include everything except the big boulder. Most of the lines representing it are at odd angles (that is, other than zero or 45 degrees), which complicates the process of determining the offsets and gaps. As you can see by comparing

Figure 7.21
Odd angle family.

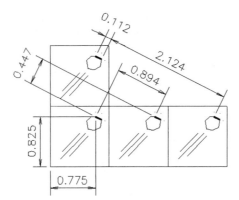

Figure 7.21 with Figure 7.20, odd angles cause the segments to repeat at longer intervals (2.12426458 units in this case) and move the lines in the family closer together in the Y direction (0.44721360 units here).

When you define hatch patterns with line families at odd angles, you need to increase AutoCAD's linear and angular precision and use the maximum number of digits (eight). If you round off to fewer digits, you'll find that the pattern segments "wander" and don't line up with one another.

In the next exercise, you'll create the first two sides of the boulder. See Figure 7.21.

The horizontal line at the top of the boulder is easy because its angle is zero. If you list the line, you'll see that its left endpoint is 0.7,0.825 and its length is 0.075. Subtracting 0.075 from 1.0 (the width of the unit square) leaves a gap of 0.925. The offset is 0,1 (parallel lines spaced 1 unit apart vertically).

 Exercise

Adding the First Boulder Lines to ASTEROID

Add this line to ASTEROID.PAT:

`0, .7,.825, 0,1, .075,-.925` [Enter]

Save and test ASTEROID.PAT.

Next, you'll add the line segment that's clockwise from the horizontal segment. Return to AutoCAD and refer to Figure 7.21.

Linetypes and Hatch Patterns

Command: **DDUNITS** `Enter`   Opens the Units Control dialog

Set the precision for both units and angles to eight decimal places, then choose OK

Command: **LIST** `Enter`

Select objects: *Pick the line running from ①*   Lists the line's data
to ② (see Figure 7.22) and press ENTER

```
LINE       Layer: 0
Space: Model space
Handle = 2BD
from point, X=0.77500000  Y=0.82500000  Z=0.00000000
to point, X=0.87500000  Y=0.77500000  Z=0.00000000
Length =0.11180340,  Angle in XY Plane = 333.43494882
Delta X =0.10000000, Delta Y = -0.05000000, Delta Z =0.00000000
```

The angle is 333.43494882 degrees. The start point is 0.775,0.825. The length of the line segment is 0.11180340.

Next, you'll measure the offsets and gap. Refer to Figure 7.22.

Command: **UCS** `Enter`

Origin/ZAxis/3point/OBject/View/
X/Y/Z/Prev/Restore/Save/Del/
?/<World>: **3** `Enter`

Origin point <0,0,0>: **0.775,0.825** `Enter`   Specifies the first endpoint of the line (① in Figure 7.22)

Point on positive portion of the X-axis: **0.875,0.775** `Enter`   Specifies the second endpoint of the line at ②

Point on positive-Y portion of the UCS XY plane: *Pick a point above the line, similar to ③*

Now the coordinate system should be aligned with the line. The nearest corresponding line in this family is located in the unit square directly to the right at ④.

Command: **DIST** `Enter`

First point: *Pick ①*

Figure 7.22
UCS pick points.

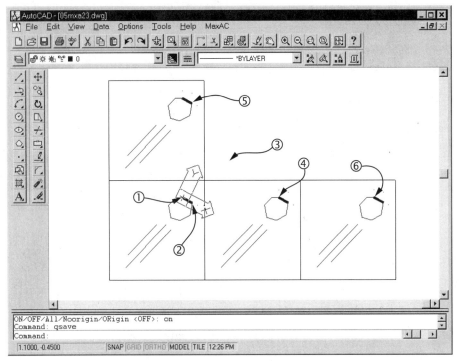

Second point: *Pick ④*   Measures and displays the distance
                                          and angle

```
Distance = 1.00000000,
Angle in XY Plane = 26.56505118,
Angle from XY Plane = 0.00000000
Delta X = 0.89442719,   Delta Y = 0.44721360,   Delta Z = 0.00000000
```

The offset for this family is 0.89442719,0.44721360. Because of the angle of the line at ①, there aren't any additional segments in line with it. Instead, you'll measure the distance from ⑤ to ⑥.

Use the DIST command to measure the distance from ⑤ to ⑥—the gap for this family is 2.12426458 units.

Command: **UCS** Enter

Linetypes and Hatch Patterns

```
Origin/ZAxis/3point/OBject/View/
X/Y/Z/Prev/Restore/Save/Del/
?/<World>: Enter
```
Returns to the World UCS

Now you have all the information to add the next family to ASTEROID.PAT. Add this line:

`333.43494882,   .775,.825,   .89442719,.44721360,   .11180340,-2.12426458` Enter

Save and test ASTEROID.PAT, as in the previous exercises.

As you can see, there are a lot of distances to determine for a family with an irregular angle. Also, it isn't always obvious where to measure the offset to. Look for the nearest adjacent line in the family perpendicular to the first segment. In order to determine the gap between segments, you sometimes have to zoom out and create the pattern in additional unit squares. Once you align the UCS with the line you're working on, your crosshairs should show you where to measure the gap.

In the next exercise, you finish the boulder and test the completed pattern.

 Exercise

Adding the Remaining Boulder Lines to ASTEROID

If you're a glutton for punishment, you can repeat the sequence in the previous exercise for each line. But we've already done the hard work for you. Add the remaining linetype families to ASTEROID.PAT:

```
258.69006753,   .875,.775,   -4.11843884,.19611614,   .12747549,-4.97154403  Enter
206.56505118,   .85,.65,     -.89442719,.44721360,    .11180340,-2.12426450  Enter
165.96375653,   .75,.6,,     -.97014250,-.24253562,   .10307764,-4.02002798  Enter
101.30993247,   .65,.625,    .98058068,-.19611614,    .12747549,-4.97154403  Enter
45,             .625,.75,    .7071,.7071,             .1061,-1,3081          Enter
```

Save and test ASTEROID.PAT one last time.

Command: **DDUNITS** Enter   Opens the Units Control dialog

Restore the units precision to four decimal places and the angle precision to no decimal places, then choose OK

Your screen should look like Figure 7.23 when you are finished.

The completed definition, which is in the ASTEROID.MA file from the MaxAC CD-ROM, is as follows:

```
;; Maximizing AutoCAD complex hatch pattern example
;; by MM 7 Mar 1996
*ASTEROID, Asteroids hurtling through space
;Dots:
0,              .925,.75,       0,1,                    0,-1
0,              .875,.875,      0,1,                    0,-1
0,              .775,.9,        0,1,                    0,-1
;Diagonal lines:
45,             .325,.175,      .7071,.7071,            .4596,-.9546
45,             .175,.125,      .7071,.7071,            .6010,-.8132
45,             .2,.25,         .7071,.7071,            .4244,-.9898
;Boulder:
0,              .7,.825,        0,1,                    .075,-.925
333.43494882,   .775,.825,      .89442719,.44721360,    .11180340,-2.12426458
258.69006753,   .875,.775,      -4.11843884,.19611614,  .12747549,-4.97154403
206.56505118,   .85,.65,        -.89442719,.44721360,   .11180340,-2.12426450
165.96375653,   .75,.6,         -.97014250,-.24253562,  .10307764,-4.02002798
101.30993247,   .65,.625,       .98058068,-.19611614,   .12747549,-4.97154403
45,             .625,.75,       .7071,.7071,            .1061,-1.3081
```

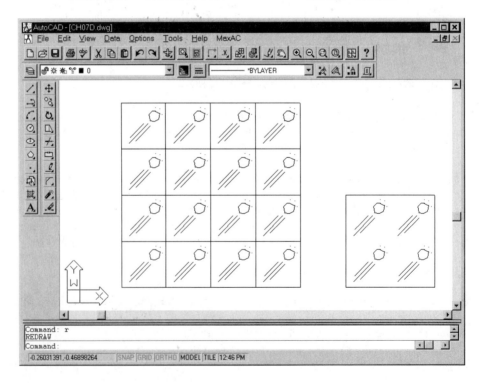

Figure 7.23
The completed ASTEROID pattern.

If you've been skimming this section and want to try the ASTEROID pattern, rename ASTEROID.MA to ASTEROID.PAT. Follow the BHATCH testing procedure in the exercise "Adding Dots and Testing the Pattern" at the beginning of this section.

Tip: Segregating custom hatch patterns in their own PAT files is a good way to keep them organized, but if you use a pattern frequently, it's more convenient to add it to ACAD.PAT. That way, you can select it from the Predefined pattern types rather than typing its name. If you decide to add to ACAD.PAT, copy it to your custom support file directory (for example, \MAXAC) and modify the copy.

If you decide to add custom hatch pattern definitions to ACAD.PAT, you should also add slides for them to ACAD.SLB. AutoCAD displays the slides in the Pattern Type area of the Boundary Hatch dialog. Chapter 8 covers slides and slide libraries.

This chapter's files include a custom hatch-pattern generator written in AutoLISP and called PATTERN.LSP. It automates the process of creating hatch patterns whose lines are limited to multiples of 45 degrees (0, 45, 90, and so on). PATTERN.TXT documents the program, and includes a short exercise that uses CH07PAT.DWG.

PostScript Fills

Among the PostScript options that Autodesk added to AutoCAD in R12 is the ability to define and use PostScript fill patterns. PostScript fill patterns are similar in concept to AutoCAD hatch patterns, but with the following advantages and limitations:

- PostScript fills can include color and grayscale (including black).

- You must use a PostScript plotter or printer. PostScript fills don't appear otherwise.

- You must use the PSOUT command instead of PLOT. PSOUT is more limited and has a more primitive interface than the PLOT command.

- PostScript fill patterns don't appear on the screen; they only appear when you use PSOUT.

- PostScript fills must be attached to closed polylines.

- Creating PostScript fills requires a knowledge of PostScript programming, but gives you all the flexibility of the PostScript language.

Warning: AutoCAD's PostScript support, including fills, is fraught with pitfalls for the unwary. If you don't have a compelling reason to use PostScript, you're probably better off ignoring it.

AutoCAD's PostScript fill support is made possible by the PSFILL and PSOUT commands and the \R13\COM\SUPPORT\ACAD.PSF support file. The R13 ACAD.PSF file includes 11 PostScript fill patterns, which are shown in Appendix C of the *AutoCAD User's Guide*.

The PSFILL command applies a PostScript fill pattern to the area inside a closed polyline. The prompts for PSFILL depend on the fill pattern. After you've added a PostScript fill with PSFILL, you must use the PSOUT command to generate an EPS (Encapsulated PostScript) file that contains the PostScript fill instructions. You then can copy this file to your PostScript printer or import it into another program that supports the PostScript language as defined in the "Red Book" (the *PostScript Language Reference Manual* written by Adobe Systems and published by Addison-Wesley).

Warning: Some illustration and publishing programs don't support the PostScript language as defined in the Red Book. Instead, they support the Adobe Illustrator Document Format Specification. Even though Adobe wrote both this document and the Red Book, and both types of files are called EPS files, they are not the same. Thus some programs won't accept EPS files created by AutoCAD's PSOUT.

The following exercise demonstrates how you can add a PostScript fill pattern definition to ACAD.PSF by modifying an existing pattern. Then, you'll apply it to a closed polyline.

Adding and Using a PostScript Fill Pattern

Copy ACAD.PSF from \R13\COM\SUPPORT\ACAD.PSF to \MAXAC\DEVELOP and open the copy of ACAD.PSF in your text editor.

Search for the string `%@Fill`, which marks the beginning of a PostScript fill pattern definition, then make a copy of these four lines:

```
%@Fill
/Grayscale %Grayscale,1, Grayscale=50
    {
        0 100 Rangefilter 100 div 1 exch sub setgray fill
    } bind def
```

Make the changes shown in bold to the copied lines:

Linetypes and Hatch Patterns

```
%@Fill
/Black %Black,1, Grayscale=100
    {
        0 100 Rangefilter 100 div 1 exch sub setgray fill
    } bind def
```

Save ACAD.PSF and return to AutoCAD, and then start a new drawing called CH07E.

Command: **PLINE** ⏎

From point: *Draw a closed polyline of any shape*

Command: **PSFILL** ⏎

Select polyline: *Select the new polyline*

PostScript fill pattern (. = none) <.>/?: **?** ⏎

Grayscale Black RGBcolor AIlogo Lineargray Radialgray Square Waffle Zigzag Stars Brick Specks

Note that your new Black fill pattern is included in the listing.

PostScript fill pattern (. = none) <.>/?: **BLACK** ⏎

Grayscale <100>: ⏎

Command: **PSOUT** ⏎   Opens the Create PostScript File dialog

Choose OK   Accepts default CH07E.EPS as the output file name

What to plot--Display, Extents, Limits, View, or Window <D>: ⏎

Include a screen preview image in the file? (None/EPSI/TIFF) <None>: ⏎

Most output devices don't want to see a screen preview image, but a screen preview is useful if you intend to insert the image in a word processing or graphics program that cannot display EPS images.

Size units (Inches or Millimeters) <Inches>: ⏎

```
Specify scale by entering:
Output Inches=Drawing Units or Fit or ? <Fit>: Enter

Standard values for output size

Size        Width        Height
A            8.00         10.50
B           10.00         16.00
C           16.00         21.00
D           21.00         33.00
E           33.00         43.00
F           28.00         40.00
G           11.00         90.00
H           28.00        143.00
J           34.00        176.00
K           40.00        143.00
A4           7.80         11.20
A3          10.70         15.60
A2          15.60         22.40
A1          22.40         32.20
A0          32.20         45.90
USER         7.50         10.50

Enter the Size or Width,Height          AutoCAD creates the EPS file
(in Inches): A Enter

Effective plotting area:  8.00 wide
by 4.55 high
```

Close your text editor

Figure 7.24
Black PostScript fill pattern.

If you have a PostScript output device, you can copy CH07E.EPS to it (for example, `COPY CH07E.EPS LPT1:`). The results should look similar to Figure 7.24.

Chapter 3 of the *AutoCAD Customization Guide* includes more detail about creating custom PostScript fill patterns, but you'll need some knowledge of PostScript programming in order to follow the details.

Finishing Up

Use the MENU command to change the menu back to \MAXAC\ACADMA.MNS.

In this chapter, you created custom linetype (ACADMA.LIN), hatch pattern (*.PAT), and ACAD.PSF files. If you want to keep your work, exit AutoCAD and move these files from \MAXAC\DEVELOP to \MAXAC. Delete or move any other files from \MAXAC\DEVELOP.

Conclusion

Linetypes, hatch patterns, and PostScript fills are AutoCAD's tools for adding clarity and visual appeal to your drawings. Although the standard linetypes and patterns are useful, you may need to create custom ones for your discipline or office using the techniques described in this chapter.

The next chapter describes customization tools that are useful to almost anyone who uses AutoCAD: slides, scripts, and PCP (plot configuration parameter) files.

chapter 8

Slides, Scripts, and PCP Files

This chapter explores a trio of related AutoCAD customization features: slides, script files, and plot configuration parameter (PCP) files. Together these tools provide a range of customization options, from slide shows to batch plotting.

Slides are AutoCAD's way of storing "snapshots" of the drawing area in separate SLD files. You can view individual slides later or link a series of them into a scripted slide show, as described in this chapter. You can also use slides to develop image tile menus, as described in Chapter 13.

Scripts are similar to menu macros (see Chapter 5), except that they reside in separate SCR files instead of in menus. Also, scripts aren't designed to pause for user input and don't support many of the special characters that are available to you in menu macros. Despite these limitations, scripts are excellent tools for automating repetitive command sequences and for batch processing multiple drawings automatically.

PCP files store most of AutoCAD's plot settings in a single ASCII file, so that you can restore those same settings easily in the future. Integrating PCP files and scripts is a good way to automate plotting.

Figure 8.1 shows this chapter's sequence for batch plotting with the DOS FOR command, a script, and a PCP file.

There is nothing new about slides, scripts, or PCP files in R13. However, the move to Windows presents a new challenge for batch processing multiple drawings, and we reveal new techniques for Windows in the scripts section of this chapter.

AutoCAD LT supports slides, scripts, and PCP files the same way that AutoCAD does, so most of the information in this chapter applies to LT. Ignore any references to AutoLISP programs, such as SHOWSLDS.LSP.

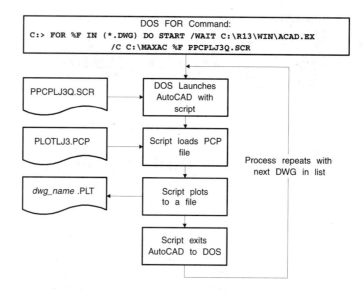

Figure 8.1
Batch plotting sequence.

About This Chapter

In this chapter, you'll learn how to do the following:

- Create and view slides
- Assemble slides into slide libraries
- Create a scripted slide show
- Develop scripts for plotting
- Learn how to configure devices with scripts
- Use a batch processing method to plot multiple drawings with a single command
- Learn how to use PCP files and integrate them into plotting scripts

These are the tools and files you will use for the exercises in this chapter:

- TextPad (or another ASCII text editor)
- TURBINE.DWG, a simple 3D drawing of a wind turbine for experimenting with slide shows
- MTURBSLD.SCR, which makes slides from TURBINE.DWG

- CFGLJ3.SCR, which demonstrates how to create a new plotter definition with a script

- CH08TSTA.DWG, CH08TSTB.DWG, and CH08TSTC.DWG, for demonstrating batch plotting

- PLOTLJ3.TXT, which contains the script sequence you'll use in PLOTLJ3.SCR

- PLOTLJ3.PCP, a sample plot configuration parameters file

- PPCPLJ3.SCR, a script that uses PPCPLJ3.PCP to set plot parameters before plotting

- AutoCAD's SLIDELIB.EXE program

- ACAD.LSP, ACAD.PGP, ACADMA.MNS, ACADMA.MNL, MAWIN.ARX, MA_UTILS.LSP, and custom bitmap (BMP) icons from the MaxAC CD-ROM

- SHOWSLDS.LSP (an optional program for viewing slides with indefinitely long pauses between them)

You'll create the following slides and scripts in this chapter:

- **TURB01.SLD** through **TURB03.SLD** are slides of TURBINE.DWG.

- **TURBINE.SLB** is a slide library containing TURB01.SLD through SEXT03.SLD.

- **TURBSHOW.SLB** is a slide library containing TURBSH01.SLD through TURBSH07.SLD.

- **PLOTLJ3.SCR** is a script that plots the current drawing to an HP-GL/2 LaserJet III file.

- **PLOTLJ3Q.SCR** is the same as PLOTLJ3.SCR, except that it quits AutoCAD after plotting.

Slides

Slides have been a popular tool for AutoCAD presentations since their early use in computer shows. Think of a slide as a "snapshot" of the current view in AutoCAD. When you make a slide, you create a file containing a simplified vector image that AutoCAD can redisplay quickly on the screen.

Maximizing AutoCAD® R13

Making and Viewing Slides with MSLIDE and VSLIDE

You can use slides for many purposes: to present a design to a client, as images for menu selections and custom dialog boxes (see Chapters 12 and 13), or to display diagrammatic help and other graphical information to users.

Tip: If you need to make traditional photographic slides or transparencies, you use means other than AutoCAD slides. For photographic slides, use AutoCAD's raster file "plotter" driver to plot drawing data, or a screen capture program to save images of the screen, including menus and dialog boxes. Then take the files to a service bureau for conversion to 35 mm slides. You'll need to coordinate the graphics format with the service bureau. The per-slide charge can be high, so check the costs before you start making dozens of slides. For transparencies, you can plot directly from AutoCAD onto blank overhead transparency stock. Office supply stores sell transparency stock designed for laser printers.

You use the MSLIDE command (Make SLIDE), which displays the Create Slide File dialog box (see Figure 8.2), to make a slide file, which has the extension SLD. MSLIDE captures only what you see in AutoCAD's drawing area. Thus, before you run the MSLIDE command, make sure that you've displayed the view you want and made any desired layer and color settings. MSLIDE doesn't record the image of the cursor, the UCS icon, or the

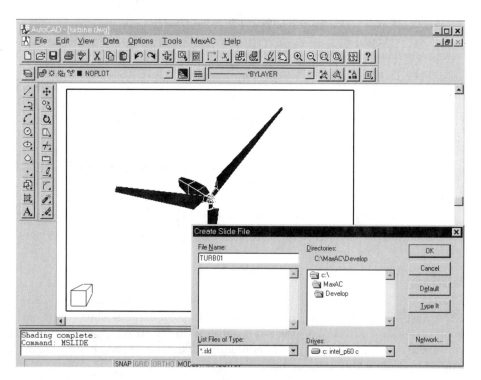

Figure 8.2
The Create Slide File (MSLIDE) dialog box.

grid, so if you want of any of these to appear in your slide, you'll have to replicate them by drawing them.

Tip: If, for other purposes, you need to save screen images that include more than just the drawing (for example, cursor, UCS icon, grid, menus or other parts of the AutoCAD program screen, or even the entire Windows desktop), use a screen capture program instead of MSLIDE. In Windows, you can press the PRINT SCREEN key to copy the entire screen to the Windows Clipboard. Press ALT+PRINT SCREEN to copy just the currently active Window. In either case, you can paste the image into Windows Paint or another program, and then save it in a graphics format such as BMP. If you do a lot of screen captures, utility programs such as Collage Complete (Inner Media, Hollis, NH, 603-465-2696) make the process more efficient and give you more flexibility.

In R13 DOS, the accelerated display driver by Vibrant Graphics includes a screen capture feature. Just press the PRINT SCREEN key (or SHIFT+PRINT SCREEN on some systems) to capture the entire screen to a BMP (Windows bitmap) file. The driver saves the first screen capture to \VIBSCRN.BMP, the second one to \VIB0001.BMP, the third one to \VIB0002.BMP, and so on.

AutoCAD does not store all of the drawing's data in the SLD file. MSLIDE stores a 2D vector representation of the drawing in its current state, using the current zoom area and screen resolution. You cannot edit slides in AutoCAD—if you want to change a slide, you must edit the drawing file that you used to create the slide and then create a new slide.

Note: Unlike a screen capture, a slide file is a vector file, not a bitmapped image. The size of a slide file depends on the number of vectors and how they relate to screen pixels. Circles, curves, and text are approximated with many short vectors. Fills, such as traces, solids, and filled polylines, are displayed as many closely spaced parallel vectors. Shaded images paint the screen with closely spaced horizontal vectors, which can cause slide files to be created with thousands of vectors, many having zero length. As a result, SLD files of shaded images can become quite large.

The VSLIDE command (View SLIDE) recalls the slide file and displays the image in the AutoCAD drawing area. When you use VSLIDE, it temporarily "paints over" your current drawing view (Figure 8.3). AutoCAD leaves your current drawing intact and active, but invisible. You use REDRAW to clear the slide and restore your drawing.

The only trick to making slides is setting up the views that you want to capture to the SLD files. In the following exercise, you make slides of shaded views of TURBINE.DWG, which is included with this chapter's exercise files.

Exercise

Making a Sequence of Slides

Copy all of the files from \MAXAC\BOOK\CH08 to your \MAXAC\DEVELOP directory.

Start AutoCAD, open TURBINE.DWG, and use the MENU command to make \MAXAC\DEVELOP\ACADMA.MNS the current menu.

Command: **UCSICON** Enter

ON/OFF/All/Noorigin/ORigin <OFF>:   Turns on the UCS icon
ON Enter

Note that TURBINE.DWG was saved in a perspective view. The DVIEW Camera option is set to approximately 35 degrees from the XY plane and -135 degrees within the XY plane.

Command: **SHADE** Enter   Shades the model

Regenerating drawing.
Shading complete.

Command: **MSLIDE** Enter   Opens the Create Slide File dialog
                                        (see Figure 8.2)

Enter **TURB01** *into the* File **N**ame *edit box and*   Saves the view as TURB01.SLD
choose OK

To create the next slide, you'll use DVIEW to rotate the camera 45 degrees clockwise around the turbine.

Command: **DVIEW** Enter

Select objects: Enter   Uses the default house icon for
                                        setting DVIEW parameters

CAmera/TArget/Distance/POints/PAn/   Accesses DVIEW's CAmera option
Zoom/TWist/CLip/Hide/Off/
Undo/<eXit>: **CA** Enter

Toggle angle in/Enter angle from XY   Leaves the first angle unchanged
plane <35.2644>: Enter

Slides, Scripts, and PCP Files

`Toggle angle from/Enter angle in XY plane from X axis <-135.00000>:` `-90` [Enter]	Rotates the camera 45 degrees clockwise
`CAmera/TArget/Distance/POints/ PAn/Zoom/Twist/CLip/Hide/Off/ Undo/<eXit>:` [Enter]	Exits the DVIEW command and generates the new view
`Regenerating drawing.`	
`Command:` **`SHADE`** [Enter]	Shades the new view
`Command:` **`MSLIDE`** [Enter]	
Enter **TURB02** into the File **N**ame edit box and choose OK	Saves the view as TURB02.SLD

Repeat the DVIEW, SHADE, MSLIDE sequence to make a third slide. This time, use an angle in the XY plane from X axis of -45 and a slide name of TURB03.

`Command:` **`REGEN`** [Enter]	Regenerates the unshaded view

This exercise showed how to create a series of slides in a 3D drawing, but you shouldn't overlook the possibilities for using slides of 2D drawings. You can use a series of slides as an on-screen "walk-through" of a set of plans or 2D details, without having to fumble with opening and zooming in drawings.

Tip: For better quality "shaded" images, you can use the RENDER, SAVEIMG, and REPLAY commands to create, save, and view rendered images. These features are independent of MSLIDE and VSLIDE, and do not create SLD files.

In the next exercise, you'll view the three slides you created.

Exercise

Viewing Slides

Continue from the previous exercise, in the current drawing TURBINE.

`Command:` **`VSLIDE`** [Enter]	Opens the Select Slide File dialog

Figure 8.3
Viewing a slide.

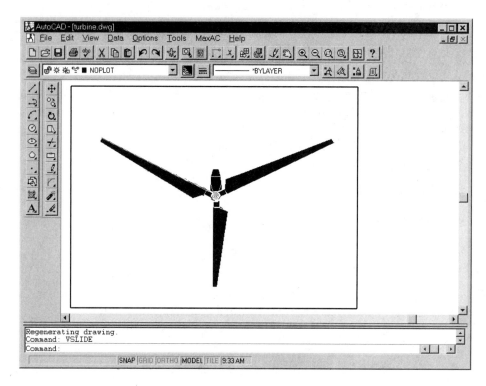

Double-click on TURB01.SLD	Displays the slide in the viewport (see Figure 8.3)
Repeat VSLIDE twice more to view TURB02 and TURB03	Displays the other slides (see Figure 8.4)

In these exercises, you created and viewed slides inside a paper space viewport, but you can use MSLIDE and VSLIDE with the crosshairs in paper space or with TILEMODE = 1 as well.

Using Slides in Model Space and Paper Space

In model space (TILEMODE = 1, or TILEMODE = 0 but crosshairs in a model space viewport), MSLIDE captures the image in the current viewport. If multiple viewports are displayed on the screen, the MSLIDE command makes a slide of the contents of the current viewport only.

In paper space (TILEMODE = 0 and crosshairs in paper space), the MSLIDE command captures the entire drawing area, including all viewports on the screen.

Slides, Scripts, and PCP Files

Figure 8.4
The SEXT03.SLD view of SEX-TANT.

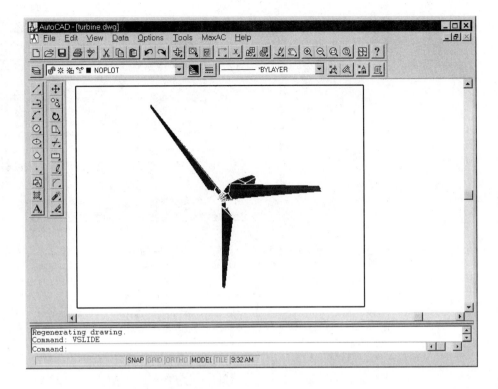

In model or paper space, VSLIDE displays a slide in the current viewport. The slide is scaled to fit the current viewport so that the image isn't distorted. Thus, depending on the relative aspect ratios of the viewports when you make and view a slide, you may see extra space at the top or right when you view the slide.

 Tip: You can use one viewport to display a slide for reference while you work in another viewport, or you can display multiple slides in multiple viewports.

Making Slides for Image Tile Menus and Dialog Boxes

When you prepare slides for an image tile menu or dialog box, you should center the image and leave a small blank area around it. Most slides look better when the images aren't jammed up against the borders of the slide display areas. Keep the images simple, because the time required for AutoCAD to display an image tile menu is proportional to the number of vectors in the slides. Users won't be inclined to use your image tile menus if displaying them causes a noticeable delay while AutoCAD draws lots of tiny, detailed slide images. To help speed up the process, AutoCAD represents solid-filled objects in image tile slides (for example, wide polylines and solids) with only their perimeters and omits the fills. Figure 8.5 shows good and bad examples of slides in image tile menus.

303

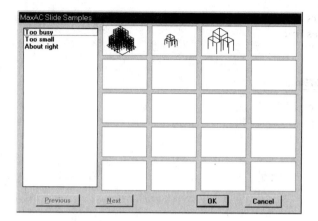

Figure 8.5
Good and bad image tile menu slides.

Another issue in image tile menu and dialog box slides is aspect ratio (that is, the ratio of the horizontal to the vertical length of the slide area). Image tile menu squares have an aspect ratio of 1.5:1 (1.5 units wide by 1 unit high). Dialog box image and image_button tile aspect ratios are up to the person who's creating the dialog box. In either case, your slides should have an aspect ratio that's similar to the rectangle in which they will be displayed. Otherwise, AutoCAD will add empty space at the top or right of the slide.

There are several ways to create a slide with a specific aspect ratio:

- Set TILEMODE to 0 and use MVIEW to create a viewport with the desired aspect ratio. Use MSPACE to enter the viewport and then arrange the view to your liking. Make the slide with the crosshairs in the viewport (they won't show in the slide, but they confirm that the correct viewport is current). This method works with all AutoCAD platforms, including DOS. It also is the easiest method to control precisely because you can specify exact coordinates for the viewport with the MVIEW command.

- Resize the non-maximized AutoCAD window so that the drawing area approximates the desired aspect ratio. You can put a ruler up to the screen to check. This method is simple but not precise, and it requires AutoCAD for Windows or another platform that uses a windowed interface.

- Unmaximize the drawing area (pick the unmaximize button that's underneath the AutoCAD program window's unmaximize button) and size it to approximate the desired aspect ratio. This method is similar to the previous one, except that you don't have to change the AutoCAD program window. On the other hand, AutoCAD regenerates the drawing every time you resize the drawing window, but it doesn't regenerate in R13c4 if you resize the program window.

Slides, Scripts, and PCP Files

Chapter 13 covers image tile menus and creating slides for them in more detail. Chapter 12 discusses slides in custom dialog boxes.

Using Slide Library Files

Slide files, whether for slide shows or image tile menus, have a way of multiplying quickly. *Slide libraries* give you a way to collect multiple SLD files into a single file for ease of portability and faster slide display performance. Slide libraries also reduce file congestion in your directories, so they make your other files easier to find. AutoCAD slide libraries are stored in SLB files.

For example, AutoCAD's standard ACAD.MNU file defines image tile menus that rely on slides in \R13\COM\SUPPORT\ACAD.SLB. The DDIM command uses slides in \R13\COM\SUPPORT\DDIM.SLB.

You can create your own slide libraries by using AutoCAD's SLIDELIB.EXE program, which is located in \R13\COM\SUPPORT. SLIDELIB.EXE is a primitive DOS prompt program, but its few options aren't difficult to use once you know a few DOS command-line tricks.

Tip: John Intorcio's shareware program SlideManager, which is included on the MaxAC CD-ROM, is a menu-driven alternative to SLIDELIB.EXE. SlideManager is a DOS program, but its menu interface makes it a bit easier to use than SLIDELIB.EXE, and it's much more flexible and powerful. With SlideManager, you can view and print individual slides and slides in libraries. One of its most valuable features is the ability to "explode" a slide library into individual SLD files.

In the next exercise, you'll make a slide library containing the slides from the previous exercise.

Exercise

Making a Slide Library and Viewing its Slides

Continue from the previous exercise, in the current drawing TURBINE.

```
Command: SH Enter

OS Command: Enter
```
Opens a DOS window (or shells out from DOS AutoCAD), with C:\MAXAC\DEVELOP as the current directory

First, you'll create a text file containing a list of SLD files.

`C:\MAXAC\DEVELOP>` **`DIR *.SLD /B /ON > TURBSLD.TXT`** [Enter]

This DOS DIR command string creates a Bare (/B) directory of slides Ordered by Name (/ON) and redirects it (>) to the text file SEXTSLD.LST.

`C:\MAXAC\DEVELOP>` **`TYPE TURBSLD.TXT`** [Enter]   Displays the list of SLD file names

```
TURB01.SLD
TURB02.SLD
TURB03.SLD
```

Next, you'll use the list of slide files as input to the SLIDELIB.EXE program. Type the command string as shown following, substituting the appropriate SLIDELIB.EXE path for your system.

`C:\MAXAC\DEVELOP>` **`\R13\COM\SUPPORT\SLIDELIB TURBINE < TURBSLD.TXT`** [Enter]

The Autodesk Slide library program now executes, creating a library file named TURBINE.SLB that contains the slides listed in the file TURBSLD.LST.

```
SLIDELIB 1.2   (3/8/89)
(C) Copyright 1987-1989,1994,1995 Autodesk, Inc.
     All Rights Reserved SLIDELIB 1.2   (3/8/89)
```

`C:\MAXAC\DEVELOP>` **`EXIT`** [Enter]   Closes the shell window

Finally, you'll view one of the slides from the slide library. Unfortunately, the Select Slide File dialog doesn't understand slide library syntax, so you must type the proper syntax at the command line.

Command: **VSLIDE** [Enter]

Choose the **T***ype It button*   Closes the Select Slide File dialog

```
Slide file <C:\MAXAC\DEVELOP\
TURBINE>:
```
TURBINE(TURB01) [Enter]   Displays the slide from the library

---

As you saw in the previous exercise, the syntax for referring to a slide in a slide library is *SLB_name(slide_name)*, with no spaces. Although this syntax is a bit cumbersome when you're using the VSLIDE command, more often you'll automate it in scripts and menu macros. In the scripts section of this chapter, you'll use the syntax to create a slide show from a slide library.

 Tip: When you create slides, make sure that you save all source drawings so that you can update your slides in the future. Similarly, when you make slide libraries, save the SLD files on a diskette or in a separate directory in case you need to rebuild the SLB file with additional SLD files later. If you need to alter an SLB file and you've lost the SLD files from which it was compiled, you can use SlideManager to re-create them.

Adding Hatching Slides to ACAD.SLB

The BHATCH command also relies on slides in ACAD.SLB in order to display images of hatch patterns. As discussed in Chapter 7, if you decide to add custom hatch pattern definitions to ACAD.PAT, you should also add slides for them to ACAD.SLB. AutoCAD displays the slides in the Pattern Type area of the Boundary Hatch dialog. Unfortunately, R13 doesn't make the process easy, because the SLD files that went into ACAD.SLB aren't included individually on the R13 CD-ROM, nor is any way of extracting them included with AutoCAD.

Following is the procedure you can use to add slides to ACAD.SLB:

1. Copy \R13\COM\SUPPORT\ACAD.SLB and SLIDELIB.EXE to a blank directory (for example, C:\SLB).

2. In DOS or from a DOS window, change to the C:\SLB directory.

3. Use the SlideManager program to extract all of the slides from C:\SLB\ACAD.SLB.

4. Copy the SLD files you want to add to ACAD.SLB into C:\SLB.

5. Create a template file by entering **DIR *.SLD /B > ACADSLB.TXT**.

6. Create a new ACAD.SLB file by entering **SLIDELIB ACAD.LIB < ACADSLB.TXT**.

7. Move the new ACAD.SLB into a directory on the library search path that AutoCAD searches before \R13\COM\SUPPORT (for example, \MAXAC or your custom support file directory, as described in Chapter 4). Alternatively, make a backup copy of the original \R13\COM\SUPPORT\ACAD.SLB and then copy your new version into \R13\COM\SUPPORT.

8. Save all of the SLD files in C:\SLD on a diskette or in a separate directory in case you need them later.

If you're adding slides for custom hatch patterns, make sure that the slide's name matches the hatch pattern's name exactly, otherwise BHATCH won't be able to find the slide. For

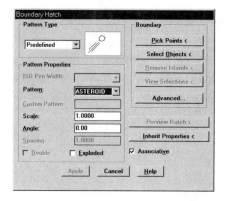

Figure 8.6
The Boundary Hatch dialog showing a slide added to ACAD.SLB.

example, if your hatch pattern is named ASTEROID and you've added it to ACAD.PAT, add a slide named ASTEROID.SLD to ACAD.SLB. Figure 8.6 shows the result.

Script Files

AutoCAD script files are akin to batch files for AutoCAD: you can use them to automate common AutoCAD command sequences. You can use scripts for establishing layer settings, plotting, and other repetitive tasks. Scripts are especially useful for batch-processing a group of drawings and creating slide shows.

Understanding Scripts

A *script* is an ASCII text file containing a sequence of AutoCAD commands. Script files have SCR extensions. Scripts resemble menu macros, with some important differences, which are described later. A script feeds commands and other input to AutoCAD, exactly as if it were being typed at the command line. You use the SCRIPT command to load and start a script.

A major difference between macros and scripts is that scripts cannot pause for input from the user (without an AutoLISP workaround described later). Thus scripts aren't appropriate for interactive programs—menu macros and AutoLISP are better suited for such tasks.

One big advantage of scripts is that, unlike menu macros and AutoLISP, they can continue to control AutoCAD even when you open or start new drawings. AutoCAD reinitializes AutoLISP and the menu when you change drawings, which means that AutoLISP and menu macros are confined to a single drawing session. Scripts, on the other hand, can skip in and out of various AutoCAD drawings. Because scripts can control AutoCAD from inside or outside the drawing, a script can instruct AutoCAD to start new drawings, edit existing drawings, plot drawings, or reconfigure AutoCAD.

Slides, Scripts, and PCP Files

Planning and Writing Scripts

To create a script, you should go through the command sequence you want to automate in AutoCAD. This process is just like planning a menu macro (see Chapter 5). Type everything at the keyboard (that is, no digitizer or menu picks) and write down each step. Don't use dialog box commands; use command-line versions instead (for example, LAYER instead of DDLMODES). Keep track of every ENTER.

Tip: R13's enhanced text Window is helpful for planning scripts. After you go through the command sequence, cut and paste in the text window to transfer the session to a text file. You'll need to delete a lot of prompt text and blank lines in order to make a script out of it, but you'll be less likely to forget an extra ENTER or command option.
Another useful tool for creating scripts is Tony Tanzillo's LispPad. You can use LispPad's **T**ools, Test script option to develop and debug scripts.

AutoCAD behaves a bit differently with some commands when a script is running, compared to when you're entering commands interactively. Most of the differences are for compatibility with older releases or to make some script tasks easier. The most important differences are:

- The file-handling commands (NEW, OPEN, SAVE, SAVEAS, and QUIT) don't display dialog boxes when a script is running. They prompt at the command line instead, so that the script can respond. Likewise for commands that normally pop up a file dialog box (such as STYLE, WBLOCK). If you set the FILEDIA system variable to 0 (off) and then run these commands interactively, you'll see the prompts that a script will "see."

- When you run QUIT in a script and there are unsaved changes in the drawing, the script must answer Yes (save changes) or No (discard changes). These answers aren't the same as the dialog choices on some platforms.

- When you run PLOT in a script, AutoCAD suppresses the plot dialog box and runs the old-fashioned command-line version of PLOT instead. If you set the CMDDIA system variable to 0 (off) before running PLOT interactively, you'll see the command-line version.

In other words, when you plan scripts that use file-handling commands or plot, you should temporarily set FILEDIA (file dialogs) and CMDDIA (plot dialog) to 0 before going through the command sequence. When you're finished, set FILEDIA and CMDDIA back to 1. When a script is running, AutoCAD acts as if both system variables were set to 0, no matter what their current settings are.

When you've written down the AutoCAD command sequence, enter each character into an ASCII file exactly as you typed it during the interactive AutoCAD session.

Here's an example of a script that sets layers and restores a view before plotting. It makes 0 the current layer and then turns off all layers containing the string NOPLOT or VIEWPORT. (The asterisks around the layer names are wildcards, and they ensure that AutoCAD turns off NOPLOT and VIEWPORT layers in xrefs and similarly named layers containing prefixes or suffixes.) The script then restores the named view SHEET.

```
;PREPLOT.SCR: Script to prepare drawing for plotting.
; Turns off NOPLOT and VIEWPORT layers
; and restores the SHEET view for plotting.
; by MM 8 April 1996
;
_.LAYER _Set 0 _OFf *NOPLOT*,*VIEWPORT*

_.VIEW _Restore SHEET
;end of script
```

AutoCAD ignores comment lines that begin with a semicolon. Unlike comments in AutoLISP files, comments in scripts cannot be on the same lines as commands or other input, but must be on separate lines that begin with semicolons. Each space or return in the SCR file acts like one press of the spacebar or ENTER key in AutoCAD. Keep careful track of these in the script file. Watch out for trailing spaces and returns (that is, spaces at the ends of lines or extra returns at the end of the file). When you need to add extra ENTERs, use blank lines instead of trailing spaces. This approach is shown in the sample script. You need to press ENTER one extra time to leave the LAYER command, and the script supplies this ENTER with a blank line instead of a trailing space. The blank line is easy to spot, whereas a trailing space is invisible in most editors.

Make sure that there's a return at the end of the last line in the SCR file, so that the last command gets entered into the AutoCAD command prompt.

Warning: Some text editors sometimes add one more return when you save the file. AutoCAD will see the extra return, and it can play havoc with your scripts. If a script doesn't seems to end correctly (for example, it leaves you in the middle of a command or does something one too many times), check carefully for trailing returns. Move to the very end of the file, and backspace over the extra returns.

As with menu macros, capitalization in scripts is up to you. We use the conventions described in Chapter 5: uppercase for what you would have to type at the command line and lowercase to fill out option names and make the script more readable. The underscore (indicating a command name or option in English) and period (telling AutoCAD to use the native command even if the command has been redefined) aren't required in most cases, but as with menu macros, using them is good practice.

Once you've created the SCR file, test it with the SCRIPT command in AutoCAD. If the script contains an error, it will stop and return you to the command prompt. Review the

Slides, Scripts, and PCP Files

text windows for clues about where the sequence went wrong. Then edit, save, and retest the SCR file.

You can stop a running script by pressing BACKSPACE or ESC. The script will complete its current command and then return you to the command prompt. You can run other commands and then enter **RESUME** to restart the script where you suspended it.

Creating Scripted Slide Shows

Scripts work well for automating everyday AutoCAD tasks, such as establishing layer settings and creating layers, text styles, and dimension styles. But scripts also have another, more specialized use: you can make a series of slides and then replay them with a script. You can control the time between slides and make the slide show repeat indefinitely, in a loop.

You use the DELAY command in scripts to create timed pauses as the script runs. Enter values for the DELAY command in milliseconds (1000 is one second), up to a maximum of 31768. You can put the RSCRIPT command at the end of a script file in order to make AutoCAD repeat the script over and over again, until you cancel it with BACKSPACE or ESC.

A slide show script usually is just a series of VSLIDE commands, separated by DELAYs. Add RSCRIPT to the end if you want the slide show to loop.

Warning: Chapter 6 of the *AutoCAD Customization Guide* describes a *preload* feature for slides. In the script, you load one slide with VSLIDE *slide1_name*, then put VSLIDE **slide2_name* (with the leading asterisk) before the DELAY line that defines the delay for *slide1_name*. You put VSLIDE after the DELAY line in order to display *slide2_name*. The VSLIDE **slide2_name* line causes AutoCAD to preload the slide into memory while the delay for *slide1_name* is happening, thereby avoiding the small delay caused by the VSLIDE command retrieving the slide name and image. On today's faster computers, the delay is imperceptible, so there's no reason to use preloading unless you're presenting a slide show on an old, slow computer.

Scripts also are good for creating a series of slides for a slide show. This chapter's files include a script called MTURBSLD.SCR that creates a series of seven shaded slides of TURBINE.DWG, rotating halfway around the model at 30-degree intervals:

```
_.DVIEW    _CAmera    180    _.SHADE _MSLIDE TURBSH01
_.DVIEW    _CAmera    -150   _.SHADE _MSLIDE TURBSH02
_.DVIEW    _CAmera    -120   _.SHADE _MSLIDE TURBSH03
_.DVIEW    _CAmera    -90    _.SHADE _MSLIDE TURBSH04
_.DVIEW    _CAmera    -60    _.SHADE _MSLIDE TURBSH05
_.DVIEW    _CAmera    -30    _.SHADE _MSLIDE TURBSH06
_.DVIEW    _CAmera    0      _.SHADE _MSLIDE TURBSH07
_.REGEN
```

Tip: If you're a stickler for order and neatness, you can use TABs instead of spaces to enter commands and options in your script files. AutoCAD interprets a TAB key the same as a spacebar, but using TAB will make the columns line up in a script such as the MTURBSLD.SCR file.

Note: Creating a series of rendered files in a script is trickier. The RENDER command displays a dialog box that requires interactive input, so it doesn't lend itself to scripts. Instead, you must use the `(c:render)` AutoLISP function. You need to do the same for REPLAY and `(c:replay)`. See Chapter 14 of the *AutoCAD Customization Guide* for details.

In the next exercise, you'll run the MTURBSLD.SCR to make the seven slides, compile them into a slide library, and then create and run a slide show script.

Exercise

Creating a Scripted Slide Show

Continue from the previous exercise, in the current drawing TURBINE.

Command: **SCRIPT** [Enter]

Choose and run the MTURBSLD *script*   Creates seven SLD files

The script will take a couple of minutes because of all the regenerations. Next, you'll create a slide library, as described in the previous exercise. Following are the commands you should type.

Command: **SH** [Enter]

OS Command: [Enter]   Opens a DOS window

C:\MAXAC\DEVELOP> `DIR TURBSH*.SLD /B /ON > TURBSHOW.TXT` [Enter]

C:\MAXAC\DEVELOP> `\R13\COM\SUPPORT\SLIDELIB TURBSHOW < TURBSHOW.TXT` [Enter]

C:\MAXAC\DEVELOP> **EXIT** [Enter]   Closes the shell window

Now you're ready to create the slide show script. Create a new file called VTURBSH.SCR and enter the following lines (cut and paste to do the job quickly):

```
_.VSLIDE TURBSHOW(TURBSH01)
_.DELAY 500
_.VSLIDE TURBSHOW(TURBSH02)
_.DELAY 500
_.VSLIDE TURBSHOW(TURBSH03)
_.DELAY 500
_.VSLIDE TURBSHOW(TURBSH04)
_.DELAY 500
_.VSLIDE TURBSHOW(TURBSH05)
_.DELAY 500
_.VSLIDE TURBSHOW(TURBSH06)
_.DELAY 500
_.VSLIDE TURBSHOW(TURBSH07)
_.DELAY 500
_.RSCRIPT
```

Command: **SCRIPT** Enter

Choose and run the VTURBSH *script* Shows the seven slides in sequence, and then repeats again from the top

Press ESC to stop the script

Command: **RESUME** Enter The script resumes from where you stopped it

Press ESC to stop the script again

Although you can use delays to control the pacing of slide shows, AutoCAD doesn't give you a simple way to control the delay between slides interactively, while the slide show is running. As with a slide projector, you'll sometimes want to stop on a particular slide to discuss it and then move on, or move more quickly through a few slides in order to save time. Of course, you can use ESC and RESUME to "pause" or the VSLIDE command interactively to load slides one at a time, but neither of these methods is elegant.

This chapter's files include SHOWSLDS.LSP, a simple AutoLISP program for viewing a series of slides with indefinitely long pauses between them. In order to use SHOWSLDS.LSP, you must have the SLD files (not a slide library) and a text file that lists the SLD names, one per line. You can create such a listing with the DOS DIR command and /B switch, just as you created a list of slides for SLIDELIB.EXE in the previous exercise: DIR TURBSH*.SLD /B /ON > TURBSHOW.TXT When you run

SHOWSLDS.LSP, it displays each slide in turn and prompts **Press any key to view next slide** after each one.

Tip: You can chain scripts together, end to end. For example, if you find yourself creating a lot of scripts to make and then view a series of scripts, you can link the make and view scripts together. At the end of the script that makes the slides, add a line that loads the script to view them (for example, `_.SCRIPT VTURBSH`). This line passes control to the other script, so if you call a second script from the middle of a first script, the rest of the first script won't get executed.

Pausing Scripts for Input

A limitation of scripts is that, unlike menu macros, they cannot pause for input without using a little trick. If your script includes a command that requires a response to a prompt, AutoCAD uses the next item in the script file as the input. If it isn't the right kind of input, AutoCAD reports an error and halts the script.

There is a way to work around this limitation. If a script stops because of an error, you can use the RESUME command to restart it. You can make use of this feature in an AutoLISP function that calls the **command** function with a nonexistent command name. The dummy command causes an error and thus halts the script, but the AutoLISP **command** function remains active. After the dummy command, supply whatever real **command** function arguments you want to, including **pause** or **getxxxx** functions for input. (See Chapter 9 for an explanation of these AutoLISP functions.) End the **command** function call with the RESUME command.

The following script uses this workaround to draw a circle. It starts a line at 1,1; pauses for the second point of the line; completes the line with the third point at 4,4; then draws another circle:

```
_.CIRCLE 2,2 1
(command "_.STOP" "_.LINE" '(1 1) pause '(4 4) "" "_.RESUME")
_.CIRCLE 3,3 1
```

This workaround is fine for short command sequences, but if your scripts need more extensive interactive control, you should consider writing the interactive portions as separate LSP files. Then you can chain together scripts and AutoLISP programs. Just remember that when an AutoLISP program calls a script (or vice versa), the calling program hands control to the new program and doesn't get it back. Thus, an AutoLISP program should run a script at the end of whatever it needs to do. The same is true with a script loading an AutoLISP program.

Plotting with Scripts

Automated plotting is one of the most common uses for scripts. A script can reduce the time that you spend plotting by eliminating the need to establish or verify settings in the

Slides, Scripts, and PCP Files

Plot Configuration dialog. Scripts can also help enforce plotting standards by ensuring that all of your plot settings are consistent each time you plot. And as you'll see later in this chapter, with some DOS tricks, you can batch plot drawings automatically.

Tip: It's also possible to plot from an AutoLISP program, and to redefine the PLOT command with AutoLISP. See Chapter 14 for details.

You must plan plotting scripts carefully. Each plotter driver can have a slightly different sequence of prompts, and as always, your scripts must respond to every prompt and in the right order. Thus, before you try to write a plot script, you should go through the plot sequence interactively (with CMDDIA and FILEDIA set to zero) and record each prompt.

Another plotting script issue is ensuring that you've selected the desired output device driver. AutoCAD provides two system variables for this purpose: PLOTTER and PLOTID. PLOTTER is a sequential number, starting at zero for the first plotter you configured. For instance, if you set PLOTTER to 2, AutoCAD will make the third plotter that you configured the current output device. You can see the order of your configured plotters in the Plot Configuration dialog's Device and Default Selection subdialog (Figure 8.7).

One problem with using the PLOTTER system variable is that the number and order of your configured output devices change as you reconfigure. This problem is even more severe when you manage multiple systems. The PLOTID system variable provides a more reliable method of selecting the output device. PLOTID corresponds to the description of the output device. If you enter the same device names when you configure all of the systems in your office, you'll ensure that PLOTID always selects the correct device, regardless of where it comes in the configured device list. Maintaining consistency of plotter descriptions still requires some effort, but it's a more practical approach than trying to maintain identical devices in an identical order for all your office's AutoCAD systems.

Figure 8.7
Configured plotters, in order.

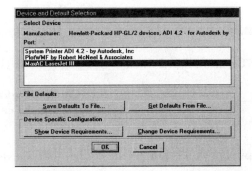

Figure 8.8
The plotter description corresponds to the PLOTID system variable.

Warning: If, during configuration, you don't type a description at the `Enter a description for this plotter:` prompt (that is, you just press ENTER), AutoCAD displays the default driver description in the Device and Default Selection subdialog (see Figure 8.7), but you won't be able to use this description with PLOTID to set the plotter. AutoCAD stores the description as a blank string ('"') and won't let you specify it with PLOTID. The only way to select such a plotter configuration from a script is with the PLOTTER number. To avoid this problem, always enter a short, descriptive name when you add a plotter configuration (see Figure 8.8).

In the next exercise, you'll create a new plotter configuration to use in subsequent exercises. We'll select the standard HP-GL/2 LaserJet III driver and call it **MaxAC LaserJet III**, which is the name we'll use to select it with PLOTID. (You can still perform the exercises, even if you don't have a LaserJet III, because we will plot to a file.)

Exercise

Adding a Plotter Configuration

Open the drawing CH08TSTA, and discard changes to TURBINE.DWG.

Command: **CONFIG** [Enter]   Starts the configuration program
                                    and displays the current
                                    configuration

We'll show prompts and input but won't show all the configuration text you will see.

Slides, Scripts, and PCP Files

Press RETURN to continue: [Enter]   Displays the Configuration menu

Enter selection <0>: **5** [Enter]   Specifies the Configure plotter choice and displays the Plotter Configuration menu

Enter selection, 0 to 4 <0>: **1** [Enter]   Specifies the Add a plotter configuration choice and displays the Available plotter list

At the next prompt, enter the number corresponding to the driver option named Hewlett-Packard HP-GL/2 devices, ADI 4.2 - for Autodesk by HP. That number is 11 on a standard R13c4 for Windows installation, but it can be different for different platforms and versions and on systems with additional drivers installed.

Select device number or ? to repeat list: **11** [Enter]   Specifies the HP-GL/2 driver and displays the Supported models list

Enter selection, 1 to 12 <1>: **11** [Enter]   Specifies the HP LaserJet III option

The driver displays a message telling you to use HPCONFIG to change HP-specific settings.

Press a key to continue [Enter]

Is your plotter connected to a <S>erial, or <P>arallel port? **P** [Enter]   Specifies parallel output

Enter parallel port name for plotter or . for none: **.** [Enter]   Specifies no output port, since we'll be plotting to a file

At this point, AutoCAD displays some default settings for the driver. We could change them now, but we'll wait until we plan the plot script to do so.

```
Plot will NOT be written to a
selected file
Sizes are in Inches and the style
is landscape
Plot origin is at (0.00,0.00)
Plotting area is 10.50 wide by 8.00
high (A size)
Plot is NOT rotated
Hidden lines will NOT be removed
Plot will be scaled to fit
available area
```

Maximizing AutoCAD® R13

`Do you want to change anything?` `(No/Yes/File) <N>:` [Enter]	Leaves settings at their defaults
`Enter a description for this` `plotter:` **`MaxAC LaserJet III`** [Enter]	Sets the PLOTID name (see Figure 8.8)

`Your current plotter is: Hewlett-Packard`
`HP-GL/2 devices, ADI 4.2 - for Autodesk by HP`
`Description: HP LaserJet III`	

`Enter selection, 0 to 4 <0>:` [Enter]	Exits the Plotter Configuration menu and returns to the Configuration menu
`Enter selection <0>:` [Enter]	Exits the Configuration menu
`Keep configuration changes? <Y>` [Enter]	Saves the new plotter configuration and returns to the drawing editor

Tip: If you need to add this same plotter configuration to many systems in your office, you can automate the process with a configuration script. The script would look like this (with prompts added as comments to make the script easier to read and change):

```
;CFGLJ3SCR: Add plotter configuration for LaserJet III
; by MM 8 April 1996
; NOTE: assumes that Hewlett-Packard HP-GL/2 driver is #11
.CONFIG
;Press RETURN to continue:

;Configuration menu - choose Configure plotter:
5
;Plotter Configuration Menu - choose Add a plotter configuration:
1
;Available plotters - choose Hewlett-Packard HP-GL/2:
11
;Supported models - choose HP LaserJet III:
11
;Press a key to continue:
;**Must press ENTER here - driver won't accept script input
;Is your plotter connected to a <S>erial, or <P>arallel port?
P
;Enter parallel port name for plotter or . for none <.>:
.
```

Slides, Scripts, and PCP Files

```
;Do you want to change anything? (No/Yes/File)
_No
;Enter a description for this plotter:
MaxAC LaserJet IIIb

;Plotter Configuration Menu - choose Exit
0
;Configuration Menu - choose Exit
0
;Keep configuration changes? <Y>
_Yes
```

One minor problem is that when the HP-G/2 driver displays its message about using HPCONFIG, it refuses to accept script input. You must press ENTER at the keyboard. After you press ENTER, the script resumes and runs to completion. You may encounter other prompts in other drivers that suffer from the same problem.

With the new plotter configuration in place, you're ready to plan a plotting script for it. The procedure is tedious the first time through, especially during the pen assignments configuration. Just remember that once you transcribe the sequence to a script, you'll never have to run through it again!

Exercise

Planning a Plotting Script

Continue from the previous exercise, in the current drawing CH08TSTA.

Command: **CMDDIA** Enter

New value for CMDDIA <1>: **0** Enter   Turns off the Plot Configuration
                                          dialog box

Command: **FILEDIA** Enter

New value for FILEDIA <1>: **0** Enter   Turns off File dialog boxes

Command: **PLOTID** Enter

New value for PLOTID   Ensures that the appropriate
<"MaxAC LaserJet III">:   plotter configuration is selected
MAXAC LASERJET III Enter

Maximizing AutoCAD® R13

`Command: PLOT `[Enter]	Starts the command-line plot sequence (because CMDDIA = 0)
`What to plot—Display, Extents, Limits, View, or Window <D>: L `[Enter]	Specifies drawings limits as the plot area

AutoCAD displays information about changing HP-specific settings (pen assignments, screened lines, date stamp, and so on) and lists the current AutoCAD plot settings.

`Do you want to change anything? (No/Yes/File/Save) <N>: Y `[Enter]	Continues with detailed plot settings

`Do you want to change plotters? <N> N `[Enter]

The HP driver displays another message about using HPCONFIG.

`Available options:`
`    0. Accept default HP-specific settings`
`Chosen option:, 0 to 0 <0>: 0 `[Enter]
	Accepts default HP settings

AutoCAD now displays the current, default pen assignments and gives you the option of changing them.

```
Pen widths are in Inches.
Object       Pen   Line   Pen      Object       Pen   Line   Pen
Color        No.   Type   Width    Color        No.   Type   Width
1 (red)       1     0     0.010    9             9     0     0.010
2 (yellow)    2     0     0.010    10            10    0     0.010
      ...and so on
```

`Do you want to change any of the above parameters? <N> Y `[Enter]	Begins pen assignment configuration

`Enter values, blank=Next, Cn=Color n, Sn=Show n, X=Exit`

```
Layer        Pen   Line   Line
Color        No.   Type   Width
1 (red)       1     0     0.010
```

When you perform pen assignment configuration from a script, it's a good idea to start the configuration for each color with C*x*, where *x* is the color. C*x* ensures that the script is configuring the right color.

Slides, Scripts, and PCP Files

Pen number <1>: **C1** [Enter] Pen number <1>: **1** [Enter] Line type <0>: **0** [Enter] Pen width <0.010>: **.003** [Enter]	Configures pen assignments for color 1 (red)

```
2 (yellow)      2      0   0.010
```
Pen number <2>: **C2** [Enter] Pen number <2>: **2** [Enter] Line type <0>: **0** [Enter] Pen width <0.010>: **.005** [Enter]	Configures pen assignments for color 2 (yellow)

We could continue in this vein and show the sequences up through color 16 (or even 255!), but you get the idea. It will be easier to configure pen assignments for all of the colors in the script, so let's continue on.

```
3 (green)      3      0   0.010
```
Pen number <3>: **X** [Enter]	Exits pen configuration
Write the plot to a file? <N> **Y** [Enter]	Specifies plotting to an HP-GL/2 plot file
Size units (Inches or Millimeters) <I>: **I** [Enter]	Specifies inches as the plot units
Plot origin in Inches <0.00,0.00>: **0,0** [Enter]	Specifies the plot origin

```
Standard values for plotting size
Size      Width      Height
A         10.50       8.00
A4        11.20       7.80
MAX       13.50      13.50
```

Enter the Size or Width,Height (in Inches) <A>: **A** [Enter]	Specifies the paper size
Rotate plot clockwise 0/90/180/270 degrees <0>: **0** [Enter]	Specifies plot rotation on the paper
Remove hidden lines? <N> **N** [Enter]	Specifies no hidden line removal
Specify scale by entering: Plotted Inches=Drawing Units or Fit or ? <F>: **F** [Enter]	Specifies plotting limits to fit on the sheet

```
Enter file name for plot           Accepts default plot file name (that
<C:\maxac\develop\ch08tsta.plt>: Enter    is, the drawing file name)

Effective plotting area:  10.50 wide by 7.88 high
Plot complete.
```

AutoCAD creates the plot file CH08TSTA.PLT.

In this case, many of the options were already set correctly (for example, PLOTID, plot rotation, paper size). However, we typed the values anyway because you can't rely on the settings being correct every time you run the script. You should make a habit of always setting values explicitly in your scripts—don't rely on default values being correct.

Although this script plotted to a file, you can use similar scripts to plot directly to a plotter, port, or network queue.

Tip: AutoCAD's plotter drivers allow you to map up to 255 colors to different lineweights, hardware linetypes, and pen definitions. Although this flexibility is admirable, you should limit your use of colors to a more reasonable number for most work, so that you don't have to worry about configuring all 255 colors. Many companies stick to the first 16 colors.

The one exception in this script is the plot file name. AutoCAD's default plot file name is the same as the drawing path and name, with a PLT extension. Often that's what you want (instead of the same PLT name for every drawing), so for that prompt, our script will simply press ENTER. If you plot to a network device, you might want to enter a specific plot "file" name (for example, the port or queue name) instead of accepting the default drawing name. See Chapter 1 for plotting suggestions.

Warning: If you've changed the default plot file name in the Configure operating parameters menu of the CONFIG program, AutoCAD will use that name, instead of the drawing name, as the default.

If you transcribe the plot sequence from the previous exercise into a script, it will look similar to this:

```
;PLOTLJ3.SCR: Plot to a file with MaxAC LaserJet III configuration
; by MM 8 April 1996
_.PLOTID MaxAC LaserJet III
_.PLOT
;What to plot
_Limits
;Do you want to change anything?
_Yes
```

Slides, Scripts, and PCP Files

```
;Do you want to change plotters?
_No
;0. Accept default HP-specific settings
0
;Do you want to change any of the above (pen) parameters?
_Yes
;Color Pen Linetype Width
C1 1 0 .003
C2 2 0 .005
C3 3 0 .007
C4 4 0 .009
C5 5 0 .011
C6 6 0 .013
C7 7 0 .015
C8 8 0 .017
C9 9 0 .019
C10 10 0 .021
C11 11 0 .023
C12 12 0 .025
C13 13 0 .027
C14 14 0 .029
C15 15 0 .031
C16 16 0 .033
X
;Write the plot to a file?
_Yes
;Size units (Inches or Millimeters)
_Inches
;Plot origin in Inches:
0,0
;Enter the Size or Width,Height (in Inches):
A
;Rotate plot clockwise 0/90/180/270 degrees:
0
;Remove hidden lines?
_No
;Specify scale - Plotted Inches=Drawing Units or Fit or ?:
_Fit
;Enter file name for plot: [use current DWG name]

;end of script
```

Tip: In this script, we're plotting to fit on an A-size (8.5" x 11") piece of paper. When you need to plot at a specific scale factor, you can use a little bit of AutoLISP to calculate the factor. Assuming that you've set DIMSCALE to your drawing scale factor and you want to plot full scale, substitute (/ 1 (getvar "DIMSCALE")) for _Fit. The expression (/ 1 (getvar "DIMSCALE")) returns the inverse of the drawing scale factor. See Chapter 9 for more about AutoLISP.

This version of the script includes most of the prompts as comments, but those are optional. In the next exercise, you'll create and test this plotting script.

Exercise

Creating and Testing a Plotting Script

Continue from the previous exercise, in the current drawing CH08TSTA.

Make a new, blank script file in your text editor called PLOTLJ3.SCR and create the script listed previously. You can either type the script sequence, copy it from AutoCAD's text window and edit it, or copy it from PLOTLJ3.TXT, which is included in this chapter's exercise files. In any case, save it as PLOTLJ3.SCR.

Use Explorer, File Manager, or the DOS DEL command to delete CH08TSTA.PLT. You need to delete this file because, if a plot file of the same name already exists, AutoCAD will ask whether to overwrite it, causing an error that will stop the script.

Next, you'll test the script.

Command: **SCRIPT** Enter	Starts the command-line version of the script command, because FILEDIA = 0
Script file <ch08tsta>: **PLOTLJ3** Enter	Launches the script, which answers all of the prompts and creates the plot file

Now that you've tested the script, finish up by resetting CMDDIA and FILEDIA each back to 1.

Tip: Many offices use different color-to-pen mapping conventions for different projects or other needs. You can create a plotting script for each convention.

Tip: You can enhance layer scripts by adding pre-plot or post-plot functions to them. For example, if you want to ensure that layers are set a particular way when you plot, add the appropriate layer commands to the beginning of the script (see PRE-PLOT.SCR earlier in this chapter for an example).

Slides, Scripts, and PCP Files

Developing a robust plotting script is not a trivial process, especially the first time you do it. You need to become intimately familiar with the plotter driver's prompts and AutoCAD's long-winded command-line plot configuration sequence. But the payoff in time saved is large if you plot very often. The payoff is even larger when you use scripts to batch plot drawings.

Batch Plotting Drawings

You can "batch plot" a group of drawings by combining the DOS FOR command with a script file. First, you create a script that ends with the AutoCAD QUIT command. Then you use the FOR command to launch AutoCAD with each drawing that matches a file specification that you provide. Your AutoCAD launch instructions include the script file that you want to run on each drawing.

FOR's syntax is:

```
FOR %F IN (filelist ...) DO command parameters ...
```

`command` can be any DOS command or any EXE, COM, or BAT file name, and it is executed once for each file name in `filelist`. `%F` is a variable name that assumes the name of the current file from `filelist` during each execution (that is, each pass through the FOR loop). You can use `%F` as a parameter to be fed to `command`.

The exact syntax you use with the FOR command for batch processing depends on your operating system. In Windows, you must use the START command with the /WAIT option to launch Windows AutoCAD. The syntax looks like this:

```
FOR %F IN (dwg_spec) DO START /WAIT C:\R13\WIN\ACAD.EXE %F scr_name
```

The START command launches a Windows program and the /WAIT option tells START to wait until the program exits and returns control before launching the program with the next file.

In DOS, you call a batch file with CALL. The DOS syntax is:

```
FOR %F IN (dwg_spec) DO CALL ACADR12.BAT %F scr_name
```

If you omit the CALL when calling a batch file, DOS doesn't return control to the FOR command, so the batch process only runs on the first file. If your AutoCAD batch file has a name other than `ACADR12.BAT`, substitute its name instead.

In both examples, you must substitute a drawing file specification for `dwg_spec` and your script file name for `scr_name`. The drawing file specification can include wildcards, lists of drawings, or both. For example, if you want to plot all drawings in the current directory, use `*.DWG` as the `dwg_spec` parameter. If you specify a list, separate the drawing names with spaces: (CH08TST1.DWG CH08TST2.DWG CH08TST3.DWG).

Maximizing AutoCAD® R13

Tip: If you just want to launch AutoCAD with a single drawing and automatically run a script on it, strip the FOR part of the preceding syntax. In Windows:

```
START /WAIT C:\R13\WIN\ACAD.EXE dwg_name scr_name
```

In DOS:

```
ACADR12.BAT dwg_name scr_name
```

The following exercise shows you how to convert PLOTLJ3.SCR into a batch plotting script and then run the batch process.

Exercise

Developing and Running a Batch Plotting Script

Continue from the previous exercise, or open the drawing CH08TSTA.

Return to your text editor and make a copy of PLOTLJ3.TXT by saving it as PLOTLJ3Q.SCR. The "Q" in the script file's name reminds you that this new script will quit AutoCAD at the end.

Make the changes shown in bold to the top of the script:

```
;PLOTLJ3Q.SCR: Plot to a file with MaxAC LaserJet III configuration
;QUITs at the end in order to work with batch processing [Enter]
```

Make the changes shown in bold to the end of the script (be sure to include the blank line):

```
;Enter file name for plot: [use current DWG name]

_.QUIT _Yes [Enter]
```

Delete the last line of the script:
```
;end of script
```

Because of a bug, AutoCAD ignores the QUIT command when there's a final comment line in the script file.

Save PLOTLJ3Q.SCR

Slides, Scripts, and PCP Files

Use Explorer, File Manager, or the DOS DEL command to delete CH08TSTA.PLT. This step ensures that the script doesn't stall because of an existing plot file.

Command: **SCRIPT** ⏎

Run PLOTLJ3.SCR The script plots and then exits AutoCAD

Next, you'll use the script to batch process three drawings, CH08TSTA.DWG, CH08TSTB.DWG, and CH08TSTC.DWG.

Use Explorer, File Manager, or the DOS DEL command to delete CH08TSTA.PLT once more.

If you're running Windows, open a DOS window and change to the \MAXAC\DEVELOP directory.

Type the appropriate FOR command sequence (all on one line). For Windows AutoCAD:

C:\MaxAC\Develop> **FOR %F IN (CH08TST?.DWG) DO START /WAIT C:\R13\WIN\ACAD.EXE /C C:\MAXAC %F PLOTJ3Q** ⏎

Notice the /C C:\MAXAC parameter, which ensures that AutoCAD launches with the appropriate configuration file, and thus the appropriate printer configurations.

For DOS AutoCAD, type (all on one line):

C:\MaxAC\Develop> **FOR %F IN (CH08TST?.DWG) DO CALL MAXAC.BAT %F PLOTLJ3Q** ⏎

Type the FOR command sequence in one long line, without pressing ENTER until you get to the end of the entire string. (The string may wrap on your screen, but don't add any extra ENTERs.)

The FOR command should launch AutoCAD, open the first drawing, plot it, quit, and then repeat the sequence for the other two drawings.

Although launching and closing AutoCAD each time may seem inefficient, remember that it enables you to batch plot while you are at lunch, home asleep, or attending to other work.

If you use the same FOR command string often or want to make it easy for less experienced users to run, you can put it in a DOS batch file. The only change you need to make is to double the percent signs. For example:

Maximizing AutoCAD® R13

```
@ECHO OFF
F1ECHO About to plot all drawings in the current directory
PAUSE
FOR %%F IN (*.DWG) DO START /WAIT C:\R13\WIN\ACAD.EXE /C C:\MAXAC %%F PLOTLJ3Q
```

Another trick is to use the START command's /MIN parameter to run AutoCAD minimized. Enter **START /?** in a DOS window in order to see the available options, which are different in the different Windows operating systems.

Stopping a FOR command batch processing run can be tricky, since you must cancel both the AutoCAD script and the DOS FOR loop. Here's the procedure:

1. Use ESC or BACKSPACE in AutoCAD to stop the script.

2. Enter **QUIT** at the AutoCAD command prompt.

3. Press CTRL+C quickly to stop the FOR command before it launches AutoCAD with the next file. Then answer Yes to abort the batch file.

Warning: The DOS FOR command combined with a script that saves the drawing sometimes causes DOS to process files more than once. Apparently DOS sees the newly saved file as a new file, even though it has the same name. Avoid saving the drawing in scripts for batch processing, save to another directory, or check the screen from time to time while the script is running and cancel when you notice that the same drawings are being loaded again.

Other Batch Processing Methods

Although scripts and the DOS FOR command work fine for batch processing, this method does require some technical skill. Also, the overhead of reloading AutoCAD for each drawing can be a problem if you're processing huge numbers of drawings. There are third-party utilities for batch processing that can make batch processing simpler, more powerful, and faster. Look in Library 10 (Utilities) of the CompuServe ACAD forum for shareware batch processing and script utilities. See also the Utilities section in *The AutoCAD Resource Guide*.

Another useful method is to load a script from a custom ACAD.LSP file. As long as AutoCAD uses that ACAD.LSP file, the script will run every time you open or start a new drawing. Make sure that you put (command "_.SCRIPT" *scr_name*) in the **s::startup** function (see Chapter 14 for information about **s::startup**).

You can, of course, create a single, long script with sections for processing multiple drawings. Just copy the section for one drawing multiple times and change the drawing names in each section. Usually, each section will start with the OPEN command and the drawing name. One advantage of this approach is that it eliminates the overhead of launching AutoCAD repeatedly. The disadvantage is that when you want to edit the script,

Slides, Scripts, and PCP Files

you have to make the same change to multiple sections. Your text editor's search and replace function can mitigate this disadvantage, though.

Warning: All batch processing methods require that all of the necessary support files (such as fonts, xrefs, and the script itself) for each drawing be located in the AutoCAD library search path. If any files are missing, the script will fail. Mike Dickason's shareware program SHOWREFS, which is included on the MaxAC CD-ROM, is helpful for identifying missing fonts and xrefs before you run a batch processing script.

Other Batch Processing Ideas

Batch processing with scripts isn't limited to plotting. Any repetitive procedure that you perform on a group of drawings is a candidate for batch processing with scripts.

Before submitting or archiving a set of DWG files, you might want to reset the menu name to ACAD and purge all unreferenced symbol table entries (block definitions, layers, text styles, and so on). The following script does the trick.

```
;PURGEALL.SCR: Script to purge entire DWG with WBLOCK *.
;QUITs at the end in order to work with batch processing
;Also changes menu name to ACAD.
; by MM 8 April 1996
_.MENU ACAD
_.WBLOCK (getvar "DWGNAME") Yes *
_.QUIT _Yes
```

You can add your own end-of-project processing commands (for example, redefine text styles, set layers, bind xrefs, and so on) to a script like this one.

Warning: Before you batch process a group of drawings, make copies of them in a separate directory or on your backup system, in case anything goes wrong.

Plot Configuration Parameters (PCP) Files

AutoCAD PCP files are ASCII files containing most of AutoCAD's plot settings: pen parameters, paper size, scale, plot file name, and so on. With PCP files, you can restore a previous set of plot parameters quickly and reliably. PCP files work well both for interactive use and in scripts. Although you don't have to use PCP files with scripts, they can simplify plotting scripts.

You create and load PCP files from the Plot Configuration's Device and Default Selection subdialog (see Figure 8.9). Use the Save Defaults to File button to save the current settings

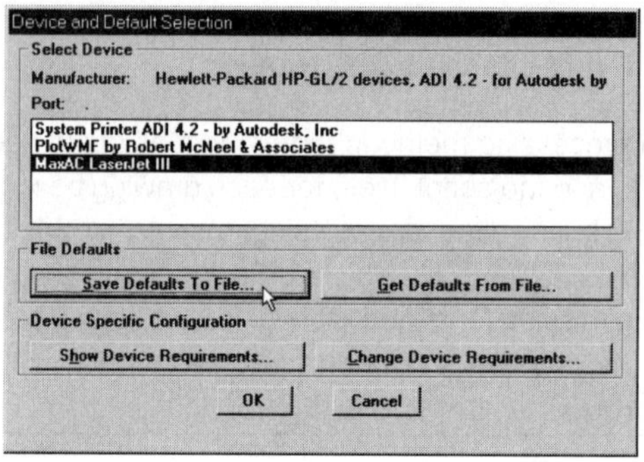

Figure 8.9
The Device and Default Selection dialog.

to a PCP file and the Get Defaults from File button to retrieve settings from a previously saved PCP file.

PCP files are specific to plotter configurations, so you usually want to restore the proper plotting device by choosing it in the Select Device list before getting defaults from a PCP file.

The one plot setting that isn't stored in a PCP file is what to plot (Display, Limits, Extents, Window, and so on), so make sure that you set this value, whether you're using a PCP file interactively or from a script. Also double-check the plot scale. It's easy to save a PCP file with one scale current and then forget to change it when you're plotting a drawing with all of the same settings except for plot scale factor.

Here's an example of the PCP file PLOTLJ3.PCP, which is included in the chapter's exercise files. You can view and edit the full file in your text editor:

```
;Created by AutoCAD on 03/24/1996 at 21:39
;From AutoCAD Drawing C:\MaxAC\Develop\ch08tsta
;For the driver: Hewlett-Packard HP-GL/2 devices, ADI 4.2 -
; for Autodesk by HP
;For the device: HP LaserJet III

VERSION = 1.0
UNITS = _I
ORIGIN = 0.00,0.00
SIZE = A
ROTATE = 0
HIDE = _N
PEN_WIDTH = 0.010000
SCALE = _F
PLOT_FILE = C:\MaxAC\Develop\ch08tsta
FILL_ADJUST = _N
```

```
OPTIMIZE_LEVEL = 1

BEGIN_COLOR = 1
   PEN_NUMBER = 1
   HW_LINETYPE = 0
   PEN_WEIGHT = 0.003000
END_COLOR

BEGIN_COLOR = 2
   PEN_NUMBER = 2
   HW_LINETYPE = 0
   PEN_WEIGHT = 0.005000
END_COLOR
```

...[and so on, up to color 255]...

AutoCAD ignores blank lines and lines that begin with a semicolon. The file consists of a header portion that defines the general plot parameters, followed by 255 five-line pen assignment blocks (one for each color).

One field that you might want to change is `PLOT_FILE =` in the header. As you can see, it defaults to the name of the drawing file that was current when you saved the PCP file (or to _NONE if you're plotting directly to a device and not to a plot file). If you leave this field set to a specific file name and then load the PCP file and plot from a script, AutoCAD will try to plot to the `PLOT_FILE =` name every time, no matter what the current drawing's name is. To prevent this problem, you should comment out this line by putting a semicolon at the beginning of it. Then AutoCAD will prompt `Enter file name for plot`, which the script can answer by pressing ENTER to accept the current drawing's name.

Plotting with Scripts and PCP Files

In order to load a PCP file from a script, you use the File option of the `Do you want to change anything?` prompt (see Figure 8.10).

Here's PPCPLJ3.SCR, a plot script that uses the PLOTLJ3.PCP file to do the same thing as PLOTLJ3.SCR earlier in this chapter:

```
;PPCPLJ3.SCR: Plot to a file with MaxAC LaserJet III configuration
;Uses PLOTLJ3.PCP to set plot configuration parameters
; by MM 8 April 1996
_.PLOTID MaxAC LaserJet III
_.PLOT
;What to plot
_Limits
;Do you want to change anything? (No/Yes/File/Save)
_File
```

Figure 8.10
Loading a PCP file with the command-line version of PLOT.

```
;Enter plot configuration file name:
PLOTLJ3.PCP
;Do you want to change anything? (No/Yes/File/Save)
_No
;Enter file name for plot: [use current DWG name]

;end of script
```

If you run PPCPLJ3.SCR, you'll notice that it runs considerably "cleaner" than PLOTPLJ.SCR. Most of the settings (including all of the pen settings) are set "silently" by the PCP file.

 Note: If your script refers to a Windows 95 or NT long file name containing spaces, such as A VERY LONG PCP FILE NAME INDEED.PCP, you must enclose the name in quotation marks.

Batch Plotting Summary

You have several choices for how to go about creating batch plotting routines. Here's an outline of one method, which will serve as a summary of this chapter's batch plotting techniques. A similar sequence would apply to other batch processing tasks.

1. Define a plotter configuration, if you don't have an appropriate one already. Write down the device name that you use, since you'll need it later for PLOTID.

2. Create and test plot settings interactively using the Plot Configuration dialog and a sample drawing. Save your desired settings to a PCP file.

3. Edit the PCP file and comment out the `PLOT_FILE = line`.

4. Set FILEDIA and CMDDIA to 0 in preparation for planning the script file.

5. Go through the plot sequence interactively, noting each step. Select the desired plotter definition first, and if necessary, establish layer settings.

6. Transcribe the interactive plot sequence into a script file by copying the AutoCAD text window and editing or commenting out the prompts and extra lines. Add `_.QUIT _Yes` to the end of the script.

7. Test the script on one drawing.

8. Correct any problems by editing the script file and test again until the script works properly.

9. In AutoCAD, set FILEDIA and CMDDIA back to 1.

10. From DOS, start the batch process with a FOR command string.

If you have any doubts, test the procedure on a small number of drawings first (for example, two or three). When you're confident that everything works, then you can batch process subdirectories with large numbers of drawings.

Finishing Up

You'll probably want to remove the plotter definition we added earlier in the chapter. Run the CONFIG command and choose Configure plotter, Delete a plotter configuration. Make sure you remove the right one (that is, MaxAC LaserJet III).

Use the MENU command to change the menu back to \MAXAC\ACADMA.MNS. In this chapter, you created slides, slide libraries, and scripts. If you want to keep these, exit AutoCAD and move them from \MAXAC\DEVELOP to \MAXAC. Delete or move any other files from \MAXAC\DEVELOP.

Conclusion

Slides and slide libraries are fun, but also useful additions to your customization toolbox. Together with scripts, slides can be an effective way to present your work to others. Slides also are the building blocks for other kinds of customization, as you'll see in Chapters 12 and 13.

Scripts are an excellent vehicle for AutoCAD customization, even for average users. They're relatively easy to create and use. Because they can operate independently of menus and other facets of your AutoCAD system, they lend themselves to on-the-fly customization. You can experiment and use them for quick-and-dirty tasks that don't merit permanent integration with your AutoCAD system. Scripts are especially useful for batch processing, because of their ability to work with multiple AutoCAD drawings.

In the next chapter, you are introduced to a full-blown programming language: AutoLISP. No matter what kinds of customization you choose to do in AutoCAD, AutoLISP will help you do it better, faster, and more elegantly.

chapter 9

Introduction to AutoLISP Programming

AutoLISP is the high-level macro programming language of AutoCAD. This easy-to-use language is a dialect of the LISP programming language, one of the oldest computer programming languages in existence. The name LISP is an acronym for **LIS**t **P**rocessing, which stems from the language's list-oriented nature, where *lists* are the primary means of representing and manipulating procedures and structured data.

AutoLISP provides you with a means of extending the AutoCAD drawing editor in useful and interesting ways. In simple uses, AutoLISP can serve as merely an intelligent typing assistant or programmable calculator. In more elaborate applications, AutoLISP can fully automate the creation of entire drawings and models from data obtained in a variety of ways, such as interactive user input or a database.

AutoLISP is seamlessly integrated into AutoCAD so that it operates transparently within the drawing editor. AutoLISP programs can instruct AutoCAD to perform commands in the same way you do with menus or the keyboard. AutoLISP programs can query and control the state of the AutoCAD drawing editor using AutoCAD system variables; access, examine, and modify the data of existing drawing objects; or even create new drawing objects with or without using AutoCAD drawing editor commands.

AutoLISP can add flexible automation to your customized system. Although many of the macros you've created so far are powerful and complex, they're basically just recorded keystrokes or command input that are played back when you select their associated menu item.

Representing the next tier in AutoCAD customization, AutoLISP adds intelligence and interactivity to your macros by making calculations and logical decisions based on user input and parameters derived from AutoCAD drawing objects. AutoLISP can also extend the AutoCAD command set. You can add new commands to AutoCAD that behave just like native commands do, but function in ways more specific to the type of work you use AutoCAD for. AutoLISP-based commands can prompt for input on the command line, present the user with options and defaults, and use dialog boxes, just as native AutoCAD commands do.

This chapter, although long and comprehensive, is merely an introduction to AutoLISP's full power. This book's companion volume, *Maximizing AutoLISP for AutoCAD R13*, is far more comprehensive, and teaches you how to write advanced AutoLISP applications that offer even greater power and control.

About This Chapter

Once you've completed this chapter, you should understand AutoLISP syntax and semantics, and you should be able to do the following:

- Create and call user-defined AutoLISP functions
- Add new commands to AutoCAD with AutoLISP
- Acquire user input and assign it to program variables
- Check for and reject errors in user input
- Query and alter the state of the AutoCAD drawing editor with AutoLISP
- Perform basic math, string, and list operations in AutoLISP
- Control AutoCAD commands from AutoLISP
- Write a basic error-handling function

The resources you'll use in this chapter are:

- Chapter 9 files from the MaxAC CD-ROM
- A text editor (LispPad, TextPad, DOS EDIT, or other)

Programs and Data in This Chapter

- **09ARRAY.LSP.** A sample program that creates helical arrays of objects, which demonstrates the use of option keywords in command input.
- **09COND.LSP.** Demonstrates the use of the AutoLISP **cond** function.
- **09HELIX.LSP.** A sample program that creates 3D helical coils, which demonstrates the use of the AutoLISP **repeat** function.
- **09INPUT.LSP.** A subroutine library that adds functions to simplify the use of default values in input prompts.
- **09MEAS.LSP.** A utility that adds a new object snap mode to AutoCAD,

Introduction to AutoLISP Programming

which computes a point on a line, a specified distance from one end. Demonstrates the use of several geometric functions including `polar`, `angle`, and `osnap`.

- **09MEAS1.LSP.** A modified version of 09MEAS.LSP. Demonstrates how to debug AutoLISP programs.

- **09MOFF.LSP.** Adds the MOFFSET command to AutoCAD, which offsets an object through multiple points. Demonstrates program iteration with the AutoLISP `while` function.

- **09RECT[1-4]LSP.** Adds a RECT command to AutoCAD, which draws a rectangle with options for specifying the rectangle's dimensions in terms of desired area and the length or width. Multiple versions are used in exercises, demonstrating the application of a wide range of AutoLISP functions.

- **09TSECT[1-2].LSP.** Adds the TUBESEC command to AutoCAD, which draws a hatched section through a tube or pipe. Several versions demonstrate input validation in AutoLISP commands.

- **MXA090[1-3].LSP.** Exercise support files that demonstrate local variables and dynamic variable scoping.

To make these files conveniently available in this chapter's exercises, copy the contents of the \MAXAC\BOOK\CH09 directory to your \MAXAC\DEVELOP directory.

Note: It's hard to get much benefit from just "a little AutoLISP." This chapter presents you with a robust, condensed dose of AutoLISP theory and application. For a more detailed examination of AutoLISP, look for this book's companion volume, *Maximizing AutoLISP for AutoCAD R13*. This companion volume covers AutoLISP in more gradual and easily digested doses than this book and also covers it much more completely.

Programming with AutoLISP

AutoLISP is very easy to learn and use. Various dialects of LISP are used as a *first language* in introductory computer science courses. What follows is a background summary of some general characteristics of LISP and AutoLISP.

Interpreted and Compiled Languages

There are two basic types of programming languages, *interpreted* and *compiled*. AutoLISP is an interpreted language. An interpreted language is one in which the *source code* that the programmer writes is executed directly by an *interpreter*. A compiled language is one in which source code is first translated into one or more intermediate forms, and then into *machine code*, which can be executed directly by the operating system and CPU. AutoCAD itself is an example of a program that is written in a compiled language.

A main advantage of interpreted languages is that no intermediate steps are required before you can execute programs written in them, as are required in compiled languages. You can interactively test or experiment with bits and pieces of a program independently of other parts, without the need to completely *compile* the entire program each time you want to test it.

Platform and Operating System Independence

Another advantage of interpreted languages like AutoLISP is *portability*. AutoLISP programs can execute in AutoCAD running on any supported platform, regardless of the host processor or operating system, provided the AutoLISP program does not use operating-system-specific extensions. For example, some built-in AutoLISP functions apply exclusively to the Windows environment and are not available in DOS or other non-Windows versions of AutoCAD. There may be some differences between platforms such as file-naming conventions, but those differences can be addressed within an AutoLISP program to achieve portability.

AutoCAD Version Independence

In addition to platform and operating system independence, AutoLISP programs that are written for any release or revision level of AutoCAD will, in most cases, run without modification on later releases of AutoCAD. AutoLISP was designed with backward compatibility in mind, ensuring that programs written for one version of AutoCAD will continue to run on future versions.

AutoLISP versus Mainstream Programming Languages

There are a number of significant differences between AutoLISP and other conventional programming languages in use today. The following discussion assumes some degree of experience in one of these languages (such as BASIC, Pascal, or C/C++). If you have no experience in any of these languages, you can skip this section.

- The most significant difference between AutoLISP and most other languages is the use of the *list* as a universal *container* for holding data. AutoLISP has no arrays, unions, structures, or records. In LISP, all complex data collections are represented and manipulated as lists.

- AutoLISP variables do not have declared types. The type of a LISP variable is determined dynamically, when it is assigned, so forward variable declarations like those used in C or Pascal, and special identifier naming conventions and DIM statements like those used in BASIC, are not required.

- AutoLISP does not require forward declarations for variables and functions (most languages require this mainly for the purpose of checking arguments and determining result types). Because AutoLISP variables are dynamically typed, there is no need for type checking arguments and re-

Introduction to AutoLISP Programming

Table 9.1
Language Comparison
AutoLISP versus Other Conventional Languages

Characteristic	AutoLISP	Basic	C/C++	Pascal
Declarative	No	No	Yes	Yes
Arrays	No	Yes	Yes	Yes
Variant types	Yes	Yes	No	Yes
Structured types	No	Yes	Yes	Yes
User types	No	Yes	Yes	Yes
Interpreted	Yes	Yes	No	No

sults. AutoLISP programs can determine the type of a variable at any time using a LISP function.

▶ Unlike BASIC, C, and Pascal, LISP has no *statements*, *keywords*, or *operators*. LISP is a function-oriented language, in which all operations are performed by function calls.

Table 9.1 summarizes the major differences between AutoLISP and other conventional languages.

Limitations of AutoLISP and Alternatives

AutoLISP has many advantages, but it also has some limitations and drawbacks. Because AutoLISP is interpreted and portable across hardware and operating systems, many functions available in other languages are not available in AutoLISP. Some of these include:

▶ Disk and directory operations

▶ Binary and random file I/O

▶ Access to Windows Application Programming Interfaces (API)

▶ Integrated Development Environment (IDE)

▶ Ability to call Dynamic Link Library (DLL) functions and operating system extensions

Some of these features that are not part of basic AutoLISP can be added to it using the AutoCAD Development System (ADS) or the AutoCAD Runtime Extension (ARX). With ADS or ARX, you can define new LISP functions that may be used within any AutoLISP program. These functions, called *external subroutines*, can do anything a standalone program can do. For example, AutoLISP cannot delete a file. The CD-ROM included with this book includes an AutoCAD Runtime Extension application that adds a number of extensions to AutoLISP, including one that can delete files. You'll learn more about AutoLISP functions and their use later in this chapter.

Prior to working through the exercises, please review the section in the Introduction of this book titled "Using The Exercises."

Fundamentals of AutoLISP

Before it was called AutoLISP, AutoCAD's high-level programming language was referred to as "Variables and Expressions." If you have little or no prior programming experience, this would be a better way to think of AutoLISP in your initial exposure to it.

Variables and Expressions

An AutoLISP program consists of *expressions*. Expressions can include variable references, literal values, and calls to functions. A function is a named procedure that can be passed zero or more values (referred to as arguments), and returns a result. AutoLISP functions are *called* by placing the name of the function, along with any arguments to be passed to it, in a space-delimited list surrounded by a pair of matching parentheses, as the conceptual example in Figure 9.1 illustrates.

Figure 9.1 shows a call to a LISP function whose name is **foo**. The **foo** function is passed two arguments; the real number 2.5, and the string "hello". AutoLISP has hundreds of *built-in functions* and also allows you to define your own functions, which you'll learn more about later in this chapter.

The following is a more complex example of an AutoLISP expression:

```
(+ 2 (* 2 3.5) 6)
```

This expression begins with a call to the built-in function **+** (addition) which returns the sum of its arguments. In this case, the **+** function is passed three arguments: the first is the integer 2; the second argument is itself a call to another function; and the third is the integer 6. The second argument is another function call, to the ***** (multiply) function,

Figure 9.1
Anatomy of an AutoLISP Expression

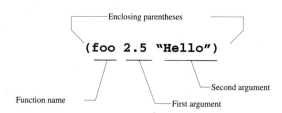

Introduction to AutoLISP Programming

which returns the product of its arguments. Both of these two examples consist entirely of literal values. All elements in a LISP expression must be separated by *at least* one space or *delimiter*. Delimiters are a special subset of characters used by AutoLISP to delimit elements of an expression. These characters are discussed in detail later in this chapter.

Prefix versus Infix Notation

Standard algebraic expressions use what is commonly called *infix* notation. With infix notation, every operand is separated from every other operand by an operator. The following is an example of an expression that uses infix notation:

```
4 + (6 * (12 - 5)) + 7 + 2 + (16 / 4) + (15 - 7)
```

In this expression, the parentheses are used to explicitly control the order of evaluation, when the intent is to alter that from the default implied by the *precedence* of each operator. AutoLISP uses *prefix* notation. In prefix notation, functions are used in place of operators, and function names (equivalent of an operator) precede the arguments (equivalent of operands). With prefix notation, there is no implied operator precedence because there are no operators, meaning that the order of evaluation is explicitly defined by the nesting of the expressions. When one complex expression contains another complex expression, the one most deeply nested is always evaluated *before* the expression that contains it. The AutoLISP equivalent of the preceding infix expression is shown here:

```
(+  4 (* 6 (- 12 5)) 7 2 (/ 16 4) (- 15 7))
```

Note that prefix notation is slightly more compact than infix notation when the number of arguments is great. For example, the following two expressions (the first using infix, and the second prefix notation) are precisely equivalent:

```
1 + 3 + 5 + 7 + 9              (+ 1 3 5 7 9)
```

In the following exercise, you will enter an expression that calls the AutoLISP **sqrt** function and passes it a number as an argument. The **sqrt** function will return the *square root* of that numeric argument.

Exercise

Using an AutoLISP Expression

Start AutoCAD and begin a new, unnamed drawing.

Command: **(sqrt 16.0)** [Enter]   Computes the square root of 16.0

Lisp returns: 4.0   Evaluates to a result of 4.0

```
Command: (sqrt 144.0) [Enter]         Computes the square root of 144.0

Lisp returns: 12.0                    Evaluates to a result of 12.0

Command:                              AutoCAD is ready for another
                                      command
```

Variables

Variables are the basic mechanism computer programs use to store and manipulate data. Variables are similar to *parameters* in a mathematical formula. For example, the formula of a circle's circumference is π * D, where D represents the diameter of *any* circle. Because this one formula may be applied to any circle to derive its circumference, the parameter D is said to be a *variable*, because it is replaced by an actual numeric value when the expression is evaluated for a specific case. Also, the expression that contains the variable D is valid for any value in its place. Put more simply, variables are a means of representing data in a program, where the actual data values to be represented are *unknown* when the program is written, and they are subject to change while the program is executing. Before a variable can be used in a program, it must be *assigned* an actual value when the program that contains it executes.

In AutoLISP programs, when a variable that has a value assigned to it appears in an expression, the variable is evaluated and is replaced with the assigned value, which is the result of the evaluation. AutoLISP uses *symbols* as variable identifiers, for example height, width, and your_name are all symbols that can be assigned values, and used to reference the assigned values in a program.

Symbols are sequences of alphanumeric characters that act as *unique identifiers* in a program. Symbols are described in greater detail later. AutoLISP also uses symbols for its function names. The preceding exercise and examples contained the symbols **sqrt**, **+**, and *****, all of which are the names of built-in AutoLISP functions.

Variable Assignment

The built-in AutoLISP **setq** function is used to assign values to symbols, allowing them to be used as variables.

```
setq

(setq sym expr [sym2 expr2]...)

Assigns each sym to the result of the following expr.
```

Introduction to AutoLISP Programming

AutoLISP provides functions that output values to the console display. The most basic of these is the **print** function, which simply prints its argument to the text console. You'll use this function in several exercises that follow.

In the following exercise, you will create and use a variable. The first expression assigns the numeric value 4.0 to the variable symbol **foo**. The second expression references the value assigned to **foo**, and multiplies it by 12.0.

Exercise

Using a Variable

Continue from the previous exercise, in the same drawing.

Command: **(setq foo 4.0)** [Enter]	Assigns the value 4.0 to foo
Lisp returns: 4.0	
Command: **(* foo 12.0)** [Enter]	Multiplies the value assigned to foo by 12.0
Lisp returns: 48.0	
Command: **(setq height 7.5)** [Enter]	Assigns the value 7.5 to height
Lisp returns: 7.5	
Command: **(setq area 32.0)** [Enter]	Assigns the value 32.0 to area
Lisp returns: 32.0	
Command: **(* height area)** [Enter]	Multiples the value assigned to height by the value assigned to area
Lisp returns: 240.0	Returns result of 32 multiplied by 7.4

When the expression (* foo 12.0) is evaluated, the value assigned to the symbol foo replaces that symbol in the expression, which becomes (* 48.0 12.0), yielding a result of 48.0. When the expression (* height area) is evaluated, the values assigned to `height` and `area` replace those symbols in the expression, which becomes (* 32.0 7.5), which yields a result of 240.

Expressions and Evaluation

The act of executing an AutoLISP program entails evaluating its expressions. When you perform simple math or algebraic calculations on paper or with a calculator, you are *evaluating* expressions. An expression is anything that can be evaluated to produce a *result*. Since every element in an AutoLISP program can be evaluated, everything is considered an expression. There are two basic types of LISP expressions, *simple* and *complex*. A complex expression contains other *subexpressions,* such as the expression (+22). All *lists* are complex LISP expressions. A simple expression is one that cannot be further decomposed into subexpressions. All simple LISP expressions are called *atoms*.

When an expression is evaluated, a *result* is always produced. The result of 24 is produced when the expression 6 * 4 is evaluated. When a complex expression is evaluated, each component or subexpression is evaluated first, in the order in which they appear (left to right). When each subexpression is evaluated, it is replaced by the result produced by the evaluation, until all subexpressions are evaluated, after which the complex expression that contains them is evaluated.

The following is an example of a complex LISP expression that contains two subexpressions. Each of the two subexpressions is also a complex expression that contains two simple expressions.

```
(sqrt (* base base) (* height height))
```

In case you don't recognize it, this expression computes the hypotenuse of a right-triangle, given the lengths of its base and height. When AutoLISP evaluates a complex expression, each subexpression is evaluated in a separate, discrete step. Each of these discrete steps in the evaluation produces an intermediate, partially evaluated form of the original expression, as Figure 9.2 demonstrates. The key to understanding the evaluation process lies in visualizing each of these intermediate forms. In the example shown in Figure 9.2, note that each complex expression is completely evaluated *after* all of the subexpressions they contain are first evaluated.

In the next exercise, you will use the AutoLISP **print** function to test the order of expression evaluation. The **print** function takes a single argument, evaluates it, displays the result on the text console, and returns that result to the calling function. Because the **print** function also returns its result, it can be used to display a value returned by an expression that is nested within another expression while still allowing the expression to perform its intended function. AutoLISP does not display the results of nested expression evaluation. Later in this chapter, you will learn to use **print** and other related AutoLISP functions to debug your programs.

Introduction to AutoLISP Programming

Figure 9.2
Understanding AutoLISP evaluation

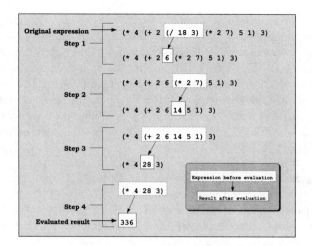

You will enter a modified version of the expression shown in the diagram in Figure 9.1, which contains calls to the **print** function, to display each subexpression's result *when* it is evaluated. The significance in this case is not the values displayed, but the *order* in which they're displayed. Remember that when entering the input shown in the example, you should not press ENTER until you reach the [Enter] symbol.

Exercise

Testing Order of Evaluation

Begin a new drawing or continue from the previous exercise.

Command: **(* 4 (print (+ 2 (print (/ 18 3))(print (* 2 7)) 5 1)) 3)** [Enter]	Prints expression results as they are evaluated
6 14 28 *Lisp returns:* 336	Result of each subexpression is printed in the order they are evaluated

Up to this point, all of the exercises required you to enter AutoLISP expressions at AutoCAD's command prompt, causing AutoCAD to display the result of evaluating them. At first, this may seem like a convenient replacement for a hand-held calculator, but in fact, there are even greater benefits to AutoCAD's ability to accept AutoLISP expressions directly on the command line. Not only can AutoLISP expressions be entered while

AutoCAD is at the command prompt, but they can be entered at almost any prompt where AutoCAD accepts input. When this happens, AutoLISP evaluates the expression, and AutoCAD accepts the result directly, as if it had been entered manually. If the result is numeric, AutoCAD receives it with no loss of numeric precision. Unlike using a handheld calculator to do calculations whose results must then be transferred to AutoCAD manually via the keyboard, AutoLISP calculations are accurate to 13 significant decimal places, and AutoCAD maintains this same degree of accuracy when it receives the result of an AutoLISP expression, regardless of the current units and precision settings.

In the following exercise, you will draw a circle, but rather than specifying the radius of the circle directly as requested by AutoCAD, you will express the radius in terms of the circle's area, by entering an AutoLISP expression that calculates the radius of a circle for a given area. In a practical application, you might create a circle in this way so that it can be used to subtract a desired area from a region.

You can obtain a circle's radius by dividing its area by π, and then taking the square root of the result. Recalling what you learned earlier about the order of expression evaluation, in AutoLISP you would express this equation as follows:

```
(sqrt (/ area PI))
```

In the exercise, you will replace **area** in the expression with a value of 120.0, which means "calculate the radius of a circle whose area is 120.0 square drawing units". You'll also be using the predefined variable **PI**, to which AutoLISP automatically assigns the value of π. After you've created the circle, you'll use the Object option of the AutoCAD AREA command to verify its area.

In the AREA command, you will make use of another built-in AutoLISP function called **entlast** to specify the circle. This function takes no arguments, and always returns the *entity name* of the last object in the database, which is also the one most recently added (in this case, the circle that you'll create). You'll see more use of **entlast** and entity names later in this chapter.

Exercise

Using AutoLISP to Calculate Input

Continue in the same drawing used in the previous exercise and zoom Center on the point 10,10 with a height of 30 units.

Command: **CIRCLE** [Enter]   Issues the CIRCLE command

Introduction to AutoLISP Programming

```
3P/2P/TTR/<Center point>: 10,10 [Enter]        Specifies a center of 10,10

Diameter/<Radius> <4>: (sqrt                   Calculates radius of a circle with an
(/ 120.0 pi)) [Enter]                          area of 120.0 square drawing units
```

AutoCAD draws the circle.

```
Command: AREA [Enter]                          Starts the AREA command

<First point>/Object/Add/Subtract:             Specifies Object option
O [Enter]

Select objects: (entlast) [Enter]              Specifies last object in database

Area = 120.0000, Circumference =               AutoCAD displays circle area
38.8325

Command:
```

Unlike the previous examples, you didn't see the result of the expression printed this time, because AutoCAD accepted that as input to the prompt for the circle's radius. So, not only can you use AutoLISP to perform complex calculations, but you can have the results of those calculations passed directly to AutoCAD with full numeric precision.

AutoLISP Syntax and Semantics

You've used a number of AutoLISP functions up to this point without knowing their exact format or *syntax*. Syntax is the formal set of rules that define the number, order, and data type of the arguments that each function accepts. This book uses the same syntax and notational conventions used in the *AutoCAD Release 13 Customization Guide* and related documentation to describe function syntax.

With all built-in AutoLISP functions, an argument can be either *required* or *optional*. Both required and optional arguments can also be *repeatable*. Some functions can also accept an argument, which can be one of several different data types, and in such cases, it is usually described in the text that accompanies each syntax description. Appendix A contains syntax descriptions of all functions available in AutoLISP. The following describes the conventions used to convey a function's syntax.

Required Arguments

A *required* argument is one that must be supplied in every call to a function. In syntax descriptions, a required argument identifier appears in **bold italic** type. The following is the syntax description for the **sqrt** function, which takes one required argument:

> **sqrt**
>
> **(sqrt *number*)**
>
> *Calculates and returns the square root of **number**.*

(sqrt *number*) looks like an example of the function's use, but is *not* an actual example of its use. It is called the function's *prototype*, since it resembles the calling form of the function as it would appear in source code.

Note that in this example prototype, ***number*** is an argument identifier. It is *not* a symbol. The specific identifier used in a prototype does not imply that variables passed to the function must be a symbol of the same name. The identifier that is used in a syntax description is one that generally describes the *data type* and/or *purpose* of an argument, which in this case means that the AutoLISP **sqrt** function requires a single argument which must be of a numeric type.

The notes that appear below the prototype are part of the description, and usually provide a literal description of what operation the function performs, what data it returns as a result, and any side effects it may have. In addition to these elements, a syntax description might also include more detailed descriptions for one or all of its arguments when necessary, as well as miscellaneous notes, and an actual example of its use.

Optional Arguments

An *optional* argument always appears within a pair of matching square brackets. Following is the syntax description for the AutoLISP **getint** function. This function prints its string argument if supplied, and waits for the user to enter an integer on the command line, which it returns. It accepts a single optional argument.

> **getint**
>
> **(getint [*prompt*])**
>
> *Displays the string **prompt** on the command line and waits for user to enter an integer, which is returned as the result.*

The square brackets surrounding the argument identifier in the prototype indicate that the argument is optional and can be omitted in calls to the function. The identifier ***prompt*** provides a vague indication of the argument's purpose, but the syntax description makes it clear that the argument must be a string which is used to "prompt" the user for input.

If a function prototype contains one or more required arguments, they *must* precede any optional arguments. If more than one optional argument appears in a prototype, each one can be supplied only if all preceding optional arguments are also supplied. Following is the syntax description for the AutoLISP **rtos** function, which converts real numbers to strings. This function accepts one required argument and two optional arguments.

---

rtos

(**rtos** *number* [*mode* [*precision*]])

Converts *number* to a string formatted in accordance with *mode* and *precision*.

The nested square brackets in this example prototype indicate that the *precision* argument can appear only if the optional *mode* argument is supplied.

Repeating Arguments

A *repeating* argument is one that can appear any number of times in each call to a function. A repeating argument can be optional or required. A repeating argument that's optional can have zero or more occurrences. A repeating argument that is required can have one or more occurrences. You've already used several functions that accept optional, repeating arguments, such as the basic math functions **+**, **-**, *****, and **/**. All of these functions accept any number of arguments (including none, in which case they return 0). There can be only one repeating argument in a function, and it must always be the *last* argument.

The next example is the prototype of the built-in AutoLISP **command** function, which passes its arguments to AutoCAD as command input. This function can take any number of arguments, including none (in which case AutoCAD behaves as if you pressed the ESC key to cancel the current command).

(**command** [*expr*] ...)

The ellipsis (**...**) that follows the argument identifier in the prototype indicates that there can be any number of occurrences of the *expr* argument. Because the argument in this example is optional, the prototype effectively states that there can be *zero or more* occurrences of the *expr* argument. If a function argument can be any data type, the argument identifier *expr* is often used.

These are all valid examples of calls to the **command** function:

```
(command "_.insert" "myfile" "0,0" 1.0 2.5 45.0)
(command)
(command "_.UNDO" "_.GROUP")
```

```
(command
   "_.OFFSET"
   12.0
   (list
      (entlast)
      (list 0.0 0.0 0.0)
   )
   "@25.0<0"
   ""
)
```

Formatting AutoLISP Code

Proper and consistent formatting of AutoLISP code is extremely important. It serves to make the structure of your code more explicit and easier to read. The formatting conventions that we use in this book are described here.

When an expression is too long to fit on a single line and must occupy multiple lines, as shown in the last expression in the preceding **command** example, the open and close parentheses of the expression appear in the same character column, and all elements of the expression are placed on a separate line and are indented three characters from the starting column of the parent expression. This type of indentation enables you to more easily find a matching parenthesis by looking in the same character column that contains its counterpart.

Tools that can find and/or select everything between pairs of matching structure delimiters including parentheses and square or curly braces are also available in the text and code editors included with this book.

AutoLISP Data Types

Up to this point in this chapter, all data you've used in the exercises is numeric data. AutoLISP programs can manipulate different types of data. Two of the most basic data types are numbers and strings. There are several different numeric data types, and other more specialized ones that you will encounter later in this chapter, but for now we'll focus on the basic data types. The type of a literal data element in a program's source code is determined when AutoLISP reads the source code. For example, when a number appears literally in a program's source code, if it contains a decimal point, then it is automatically bound as a real number. Conversely, if there's no decimal point and it is within the allowable range of integers, then it will be bound as an integer type.

Before writing and testing AutoLISP programs, you should be able to determine and recognize data types. Once you've learned to do this, you can usually determine the type of any data element or variable by displaying its value by using the **print** function. If you are unsure as to the type of a given data element, you can also use the built-in AutoLISP **type** function, which returns a symbol indicating the data type of its argument.

Introduction to AutoLISP Programming

> **type**
>
> (type *expr*)
>
> Returns a symbol indicating the data type of ***expr***.

The symbols returned by **type** for each of the most common data types, along with a sample of each type's printed format, are shown in Table 9.2.

Symbols

Symbols are sequences of alphanumeric characters, which are used as identifiers in AutoLISP programs to reference both variables and function names. Unlike identifiers in other programming languages, AutoLISP symbols are also data elements of the type *symbol*, and may be manipulated just like any other data type. Symbols are not case-sensitive, so you can refer to variables or functions via a symbol, regardless of case. For example, the symbol Foo and FOO is the same, and will always refer to same object. The first argument to the **setq** function is always a symbol.

When symbols appear in AutoLISP source, they must be separated from other source elements by at least one space or other LISP *delimiter* character: a space, period, double quote, single quote, open or close parenthesis, or semicolon:

" . ( ) ' ;

Table 9.2
Data Type Identifiers and Output Formats

AutoLISP Data Type	Symbol Returned by (TYPE) Function	Sample of Printed value
Integer	INT	35
Real number	REAL	21.31250
String	STR	"Thanks for all the fish"
List	LIST	(35.0 "A String" BAR)
Nil (empty list)	NIL	nil
Symbol	SYM	POINT1
Entity name	ENAME	<Entity name: 20c0568>
Selection set	PICKSET	<Selection set: 24>
File descriptor	FILE	<File: #2210f12>
Built-in function	SUBR	<Subr: #1bd045a>

These characters and the space are used to separate or delimit program source elements. For this same reason, these characters cannot appear in symbols.

Warning: AutoLISP symbols may contain alphabetic and numeric characters (such as `point_12`), but we advise you to avoid using leading numeric digits in symbols for two reasons. When reading source code, it takes AutoLISP longer to distinguish between numbers and symbols with leading numeric digits. Also, because numbers can be expressed literally in scientific notation, values that might look like symbols might be interpreted as real numbers. For example, `1e4` is interpreted as the number 10000.0 rather than a symbol, because AutoLISP recognizes numbers expressed in scientific units.

Integers

Integers are whole numbers and cannot have a fractional part. Integers are represented internally using 32 bits of information (or 4 bytes). The value of an integer can range from -2,147,483,647 to +2,147,483,648. However, integers passed between AutoCAD and AutoLISP (in either direction) are limited to a range of -32,768 to +32,767. When an integer appears literally in a program's source code, AutoLISP determines its type based on how it is formatted.

Literal integer expressions can have an optional leading + or - to indicate sign (+ by default). A literal integer may not be preceded or followed by a decimal point or fractional value. Integers cannot include commas or any other type of thousands separator.

Table 9.3 shows examples of valid and invalid literal integers.

Strings

Strings are sequences of zero or more ASCII characters that are treated literally by AutoLISP. A string may contain a virtually unlimited number of characters, restricted only by available memory, but within program source code, a literal string is limited to 132 characters in length. When a literal string appears in program source, it must be surrounded by double quotes, and should not span multiple lines of source code. If a string spans multiple lines of source code, the return(s) or linefeed(s) will become part

Table 9.3 Literal Integer Examples	Valid Integers	Invalid Integers
	-3265	335.
	53	-28.75
	+23472	0.0
	0	.25
	3676522	35,762,128

Introduction to AutoLISP Programming

Table 9.4
Literal String Examples

Valid Strings	Invalid Strings
"Hello"	FOO
" "	"oops
"D:\\ACAD\\WIN\\ACAD.LSP"	EX70"
"-50"	-45 degrees
"23,5,0"	"what's wrong with this?\"

of the string. There are also restrictions on the length of strings that can be passed between AutoCAD and AutoLISP, which are covered later in this chapter.

Warning: It is important to understand the difference between strings and symbols, which are not interchangeable. Symbols never appear within double quotes, cannot contain spaces or other LISP delimiters, and are used as identifiers of program elements such as variables and function names. Strings are data, generally used for messages, prompts, user-input, and AutoCAD command input.

There are no absolute restrictions on what characters can appear in a string, but the backslash (\) character has a special purpose in literal strings, affecting the interpretation of the character that immediately follows it, as you will see later in this chapter. Table 9.4 shows examples of valid and invalid literal strings.

Reals

Reals are numbers that can have fractional parts. Also known as floating-point numbers, they do not have finite range limits as integers do, but their accuracy is limited to 14 significant places. When AutoLISP displays real numbers on the command line, or via the **print**, **princ**, and **prin1** functions, they are rounded to 5 or 6 significant decimal places, but their internal accuracy remains at 14 places. Later in this chapter, you'll learn to format and display real numbers with greater control over precision.

When a real appears literally in program source code, AutoLISP determines its type based on how it is formatted. A literal real number must have a decimal point, and there must be at least one digit to the left of the decimal point. As with integers, a leading +/- is optional. A real number can also be expressed in scientific notation where the mantissa appears first, followed immediately by the letter "e", an optional +/- sign, and an optional exponent (default = 0). Commas or other thousands separators are not permitted. Table 9.5 shows examples of valid and invalid literal real numbers.

AutoLISP uses real numbers extensively for representing distances, angles, and other forms of geometric data. Coordinates are also composed of real numbers.

Table 9.5	Valid Reals	Invalid Reals
Literal Integer Examples	1.0	.005
	74369.3433	-5
	1.3e5	e-15
	1e-6	99
	-53.33334	(4.0)
	2.	35,70.25

Lists

There is one additional type of data that you've used already: the list. In all of the exercises so far, you entered *expressions*, each of which is surrounded by a pair of parentheses with several elements enclosed within them. These expressions are all *lists*. A list, as its name implies, is a container that can hold zero or more data elements of any type (including other lists). You can manipulate a list and all of its contents as if it were a single data element. The built-in AutoLISP **list** function is used to create lists in AutoLISP. The following is an example of its use:

Command: (setq mylist (list 1.0 "Hello" 3))
Lisp returns: (1.0 "Hello" 3)

Here is a slightly more complex example of a list:

Command: (setq yourlist (list 4.5 "2nd" (list 3.6 20) "4th" 99))
Lisp returns: (4.5 "2nd" (3.6 20) "4th" 99)

The first example list contains three elements. The second example list contains five elements, and the third element is another list containing two elements. Any list can contain other lists, in which case they are referred to as *nested* lists. There is no limit on the complexity or depth of nesting.

The most common use of lists in AutoLISP is to represent geometric coordinates or points. A three-dimensional point consists of three distance components or *ordinates*, which are the X, Y, and Z orthogonal axis offset distances of the point measured from the origin of their coordinate system (0, 0, 0). A 3D point is represented in AutoLISP as a list of three real numbers, where each real represents one of the point's three ordinate values.

There is one special list that is referred to as **nil**, or the *empty list*. This value is special because it represents *nothing* in an abstract sense. It is also the *default* value for all variables. Whenever a variable is created in AutoLISP, until it is explicitly assigned a value, its value is **nil**. AutoLISP uses **nil** to represent an empty list in source and printed output. When a list that contains no elements is printed, it appears as **nil** rather than a pair of matching parentheses with nothing between them. You'll learn more about **nil** and implicit program variables later in the chapter.

Entity Names

Entity names are pointer-like elements, which identify AutoCAD drawing objects such as lines, circles, and so on. AutoLISP provides a number of functions that accept or return entity names, to acquire and access drawing objects. Entity names provide a means to unambiguously reference drawing objects in AutoCAD commands and AutoLISP functions that manipulate entity data. The value of an entity name is temporary, and meaningful only within a single drawing editor session. The actual value of an entity name is significant only in terms of the ability to compare it to other entity names, to determine if two variables assigned to entity names reference the same drawing object. AutoCAD commands accept entity names whenever they expect the user to select objects. Entity names have no *external representation*. Hence, they cannot appear literally in source code.

Selection Sets

Selection sets are similar to entity names, but identify an array of AutoCAD objects instead of just one object. AutoLISP provides functions that acquire a selection set of objects from the user, using all of the same methods available in most native AutoCAD editing commands, such as Last, Window, Crossing, Previous, or by picking objects from the display. Whenever you see the familiar `Select objects:` message on AutoCAD's text console display, you are constructing a selection set. Selection sets also have no external representation, and cannot appear literally in AutoLISP source code.

File Descriptors

File descriptors or file "handles" provide a way to reference disk files that have been opened for input or output. AutoLISP provides basic functions for opening and closing files, and for reading and writing data to or from open files. File descriptors have no external representation and cannot appear literally in AutoLISP source code.

Using AutoLISP at the AutoCAD Command Prompt

In previous exercises, you entered AutoLISP expressions at the command prompt and at various input prompts. When you entered expressions while there was no command active, AutoCAD simply displayed their results and nothing else. In the following exercise you will enter an expression at the command prompt as before, but you will assign its result to a variable that you can then use in a subsequent command. You've seen several analogies between AutoLISP and a calculator so far. Another analogy between them is that AutoLISP variables are like the memory functions of a calculator, which allows you to save and recall values for later use.

Note that all AutoLISP symbols must be prefixed with an exclamation point when they're passed to AutoCAD on the command line (this is explained in detail following the next exercise).

In this exercise, you'll use the AutoLISP **getdist** function, which displays a message and waits for you to enter a distance. You can specify the distance using any method allowed by other AutoCAD commands (for example, picking two points, using object snap, point filters, and so on). **Getdist** *returns* the distance you enter as a real that represents a distance in drawing units, or **nil** if you press ENTER or SPACE. You'll use **setq** to assign the real number returned by **getdist** to an AutoLISP variable, and reference it as the displacement distance in the MOVE command. You'll also use the **polar** function to specify the move displacement and direction. The **polar** function computes a point at a specified distance and direction from a supplied basepoint. The direction is an angle in radians, so before calling **polar** you'll first convert a known angle in degrees into radians. **Polar** is described in more detail later in this chapter.

 Exercise

Using Variables on the Command Line

Continue from the previous exercise, or begin a new drawing. Zoom Center on the point 10,10 with a view height of 8 units, and draw a circle at 10,10 with a radius of 4 units.

Command: **(setq dist (getdist "Distance to move: "))** [Enter]	Obtains distance input and assigns the result to **dist**
Distance to move: 8.5	Specifies distance of 8.5 units
Lisp returns: 8.5	Echoes value assigned to **dist** variable
Command: **(setq ang (* PI (/ 30.0 180.0)))** [Enter]	Converts 30 degrees to radians and assigns result to **ang**
Lisp returns: 0.523599	Echoes 30 degrees converted to radians
Command: **(setq basept (list 0.0 0.0 0.0))** [Enter]	Assigns **basept** to the point 0, 0
Lisp returns: (0.0 0.0 0.0)	

Introduction to AutoLISP Programming

Command: **(setq newpt (polar basept ang dist))** [Enter]	Computes second point for MOVE displacement from 0,0 at angle of 30 deg. and distance of 8.5 units and assigns the point to **newpt**
Lisp returns: (7.36122 4.25 0.0)	Returns the point list assigned to **newpt**
Command: **MOVE** [Enter]	Starts MOVE command
Select objects: *Pick circle* 1 found	
Select objects: [Enter]	Terminates object selection

In the next two steps, note that AutoLISP symbols entered on the command line *must* be prefixed by an exclamation point:

Base point or displacement: **!basept** [Enter]	Specifies 0,0 as basepoint
Second point of displacement: **!newpt** [Enter]	Moves circle 8.5 units at 30 degrees
Command:	

In the preceding exercise, you had to prefix AutoLISP symbols with ! (the exclamation point) to use the AutoLISP variables as command input. When you enter an AutoLISP expression enclosed in parentheses, AutoCAD determines from the first character (the open parenthesis) that it's a LISP expression that must be evaluated. But, if you enter an AutoLISP symbol alone, AutoCAD doesn't assume it's a LISP symbol, since a symbol may consist of the same characters as an option keyword. (For example, entering the symbol midpoint at a point input prompt would be interpreted as the MIDpoint object snap keyword rather than as an AutoLISP symbol.) To make it possible to enter any AutoLISP variable on the command line, AutoCAD requires all AutoLISP symbols to be prefixed with an exclamation point.

There are several exceptions to the way AutoCAD handles AutoLISP expressions entered at the command prompt. When requesting *literal* text input (such as the value of a text object in the TEXT command), AutoCAD will not recognize AutoLISP expressions entered at the command prompt, if the TEXTEVAL system variable is set to 0. TEXTEVAL controls how AutoCAD handles AutoLISP expressions and symbols prefixed with !, when entered on the command line at a prompt where literal text is expected. By default,

TEXTEVAL is 0, which suppresses evaluation of AutoLISP expressions when entered as literal text input.

The following exercise demonstrates this. You'll perform the steps in the following exercise twice so you can observe the effects of changing the TEXTEVAL system variable. The first time through the exercise, TEXTEVAL should be set to 0. You'll verify this first by entering **TEXTEVAL** at the command prompt. If the current value is not zero, then change it to 0 before proceeding. After you complete the exercise, the value of TEXTEVAL should be 1. Erase the text you created, then work through the exercise a second time with TEXTEVAL set to 1.

 Exercise

Controlling Evaluation with TEXTEVAL

Begin a new drawing, or continue from the previous exercise and erase everything. Then, zoom Center on point 10,10 with a height of 30 units.

Command: **TEXTEVAL** [Enter]

New value for TEXTEVAL <0>: [Enter]	Verifies value of 0
Command: **(setq textdata "Hello World!")** [Enter]	Assigns string to symbol **textdata**

Lisp returns: "Hello World!"

Command: **TEXT** [Enter]	Starts the TEXT command
Justify/Style/<Start point>: **M** [Enter]	Makes text middle-aligned
Middle point: **10,10** [Enter]	Sets middle point of text to 10,10
Height <0.20>: **1** [Enter]	Sets text height to 1.0
Rotation angle <0>: **0** [Enter]	Sets rotation angle to 0d
Text: **!textdata** [Enter]	Tries to use value of **textdata**

AutoCAD displays !textdata in the drawing as a literal text object.

Command: **TEXTEVAL** [Enter]	Changes TEXTEVAL system variable

Introduction to AutoLISP Programming

New value for TEXTEVAL <0>: **1** [Enter]   Sets TEXTEVAL to 1

Erase the text and repeat the exercise steps with TEXTEVAL set to 1. This time, AutoCAD evaluates `textdata` and displays `Hello World!` as the value of the text.

The first time you worked through the exercise, the text you created contained the AutoLISP symbol **textdata** including the **!** prefix, rather than the string you had assigned to the symbol. After you changed the TEXTEVAL system variable to 1 and repeated the exercise, the text object then contained the string you assigned to the symbol, as you intended. This same rule also applies to AutoLISP symbols in menu macros.

A second exception for AutoLISP evaluation exists on the Windows 95 and Windows NT platforms. If you have *long filename* support enabled, any command that accepts a filename will also treat all input literally when the symbol is passed to it on the AutoCAD command line. The LONGFNAME system variable maintains the state of long filename support.

Working with Coordinates

In the next exercise, you'll use the **getpoint** and **getcorner** functions, which obtain coordinate input from the user. You'll use **setq** to assign the resulting points to a variable. You'll learn more about other **get*xxxx*** functions later in this chapter. When you are prompted for input by a **get*xxxx*** function, you can use any of the same input methods allowed by standard AutoCAD commands, such as entering input via the keyboard, picking with your pointing device, using AutoCAD's object snap and coordinate filter modifiers, and so on. You will also see how two of AutoLISP's **C*xxx*R** functions are used to access the elements of a list. These two **C*xxx*R** functions are described after the exercise.

 Exercise

Working with Point Lists

Continue from the previous exercise and erase everything, or begin a new drawing. Then zoom Center on point 10,10 with a height of 30 units.

Command: **(setq ll (getpoint**   Prompts for input (lowercase Ls
"Pick LL corner: ")) [Enter]   are not the number 11)

Pick LL corner: *Pick point 3,2*

Lisp returns: (3.0 2.0 0.0)	Stores point 3,2,0 as `ll`
Command: **(setq ur (getcorner ll "Pick UR corner: "))** [Enter]	Prompts with rubber-band box
Pick UR corner: *Pick point 9,6*	
Lisp returns: (9.0 6.0 0.0)	Stores point 9,6,0 as **ur**
Command: **(setq ul (list (car ll) (cadr ur)))** [Enter]	Stores x of `ll` (3.0) and y of `ur` (6.0) as new point list `ul`
Lisp returns: (3.0 6.0)	Point 3,6 assigned to `ul`
Command: **(setq lr (list (car ur) (cadr ll)))** [Enter]	Stores x of `ur` (9.0) and y of `ll` as new point list `lr`
Lisp returns: (9.0 2.0)	Point 9,2 assigned to `lr`

Leave the drawing in its current state for use in the next exercise.

You have stored four corners of a rectangle. The **getpoint** and **getcorner** functions prompt you for the points, and **setq** assigns the points to the variables `ll` and `ur` (lower left and upper right). The difference between **getpoint** and **getcorner** is that **getcorner** uses an existing point as the anchor of a rubber-band box, like AutoCAD's window object selection, and returns the point at the diagonal corner of the box when you press the pick button.

The **car** and **cadr** functions extract the first and second elements of the `ll` and `ur` points, and the **list** function combines the extracted coordinates into new upper left and lower right point lists, which are assigned to the UL and LR variables by **setq**. The *getxxxx*, **list**, **car**, and **cadr** functions are covered in more detail later in this chapter.

The COMMAND Function

So far you've used AutoLISP on AutoCAD's command line, both at the command prompt and at prompts within several AutoCAD commands. Next, you'll learn to pass data to AutoCAD with the **command** function, and to use the **setvar** function to change other settings. You'll use the four corner points that you assigned to the symbols `ll`, `lr`, `ul`, and `ur` in the last exercise to draw a rectangular polyline.

Introduction to AutoLISP Programming

The **command** function provides a way to execute AutoCAD commands in a scripted manner from an AutoLISP program, in a way that's similar to commands in menu macros and script files. The **command** function accepts any number of arguments, which can include AutoCAD commands (as case-insensitive strings) and any other type of input requested by AutoCAD on the command line. The **command** function is covered in greater detail later in this chapter.

The following example use of **command** draws a circle centered on the point 4, 7.25, 0. It passes all values that are requested by the CIRCLE command as arguments, in the same order they're acquired by AutoCAD on the command line.

```
(command "CIRCLE" (list 4.0 7.25 0.0) 12.5)
```
Lisp returns: `nil`

The first argument is a string, which is the name of the AutoCAD command to be started (CIRCLE). The second argument (a list) is the center point of the circle, and the last argument is the circle radius. The arguments must appear in *exactly* the same order in which they are requested by AutoCAD commands when they are used interactively.

The following example produces the very same result as the preceding example, except that the Diameter option keyword of the CIRCLE command is used to prompt for a circle diameter instead of a radius and the circle center point is passed as a *string* instead of a point list:

```
(command "CIRCLE" "4,7.25,0" "Diameter" 25.0)
```
Lisp returns: `nil`

AutoCAD System Variables

AutoCAD's system variables provide a means of querying and controlling the state of the AutoCAD drawing editor. You should already be familiar with system variables (from the perspective of the user) and the use of AutoCAD's SETVAR command, which allows you to change and view the current values of AutoCAD's system variables on the command line.

The **setvar** function is very similar to the AutoCAD SETVAR command. Its first argument is a string that is the name of an AutoCAD system variable (this string is *not* case-sensitive). The second argument is the new value for the specified system variable, which must have the same data type as the system variable. Read-only system variables cannot be changed using this function.

The **getvar** function accesses the existing value of a system variable. It takes one string argument, which is the system variable's name, and returns its current value.

> **setvar**
>
> **(setvar** *varname value***)**
>
> Sets an AutoCAD system variable to a supplied value.
>
> **getvar**
>
> **(getvar** *varname***)**
>
> Gets the current value of an AutoCAD system variable.

In the next exercise, you'll use the **command** function to draw a rectangular polyline, using the same points that you assigned to variables in the previous exercise. You'll also use the Close option of the PLINE command to make the polyline explicitly closed. Before drawing the polyline, you will first use the AutoLISP **setvar** function to set the current polyline width to 0.125 units. The current polyline width is maintained in the AutoCAD system variable PLINEWID.

You must continue in the same drawing because AutoLISP variables are not saved across different editing sessions.

 Exercise

Drawing a Polyline Rectangle with AutoLISP

Continue from the previous exercise and erase any objects remaining in the drawing, then zoom Center on the point 6,4 with a height of 12 units.

Command: **(setvar "PLINEWID" 0.125)** [Enter]   Sets current polyline width to .125 units

Lisp returns: 0.1250

Command: **(command "pline" ll lr ur ul "c")** [Enter]   Issues PLINE command and passes the four corner point variables and the close option keyword

Lisp returns: nil

Introduction to AutoLISP Programming

It should be clear at this point that, at a minimum, AutoLISP is a highly accurate calculator and blazingly fast typing assistant. Not only can an AutoLISP program perform calculations with high precision, but it can request various types of input that can be supplied using the same methods available in most AutoCAD commands. It can also store unlimited data, perform calculations on input, and pass the results directly to AutoCAD commands with no loss of precision. To take advantage of its capabilities, you can use AutoLISP in menu macros and in AutoLISP program files.

AutoLISP Program Files

Entering long AutoLISP expressions directly on the AutoCAD command line can be cumbersome and error-prone. User-defined functions entered on the command line are not saved automatically. They exist *only* in the current drawing editor session and are discarded when you exit AutoCAD or start a new editing session with the NEW or OPEN commands. Because user-defined functions are intended to be defined once and called many times, AutoLISP expressions and function definitions can be stored in ASCII text files, which can be *loaded* when needed. The effect of doing this is almost the same as entering each expression in the file on the AutoCAD command line manually, except that you do not see the result of the expressions as they are evaluated, except for the last one in the file. Rather than entering function definitions and expressions directly into AutoCAD's command line, you will store most of the AutoLISP programs you write in ASCII text files that can be loaded with the built-in AutoLISP **load** function.

AutoLISP program source files are simply ASCII files, which may contain any number of AutoLISP expressions in them. You can create these files using any standard text editor. You should avoid using word processors for creating LSP files unless you are very familiar with their file save options. Many word processors add non-ASCII formatting information to files, causing problems when loaded by AutoLISP. The standard file extension for AutoLISP source files is LSP, but in fact any extension can be used (although we generally recommend that you stick with LSP to avoid confusion).

---

load

(load *filespec* **[***onfailure* **[***verbose***]])**

Loads an AutoLISP file and evaluates its contents.

The **load** function loads an AutoLISP program source file and evaluates each of the expressions contained in it (as if each were typed on the command line). Each expression in the file is evaluated in order of appearance, and the result of evaluating the *last* expression in the file is returned by the call to **load**. The *filespec* argument must be a string that is the file specification of an AutoLISP program source file. If no extension is included in the file specification, the default extension LSP will be used. As with other types of files used by AutoCAD, a fully qualified file specification including drive and/or path can be included. If only a filename is supplied, AutoCAD will search for the file in each directory in the *library search path*, which is explained in detail in Chapter 4. If the specified file cannot be found and the optional **onfailure** argument is not supplied, an error condition occurs indicating that the file could not be loaded.

The optional **onfailure** argument is any valid LISP expression which, if supplied, is returned by the **load** function if it fails to find the specified file, and preempts the error condition of not being able to load the file.

Tip: The optional **verbose** argument of the **load** function, if supplied and non-nil, causes the contents of the AutoLISP file to be echoed to the AutoCAD text console as the file is loaded. This argument is *not documented* by Autodesk and is intended primarily as a debugging aid to find the general location of an extra right parenthesis or missing quote in a large AutoLISP file.

Warning: Undocumented AutoLISP features should be used only with extreme caution and good judgment because they are not supported in any way, and they are subject to omission from future releases or revisions of AutoCAD without warning.

In this book, you will only need to use **load** with the **filespec** argument. Both the **onfailure** and **verbose** arguments involve more advanced topics, which are covered in this book's companion volume, *Maximizing AutoLISP for AutoCAD Release 13*.

The following two examples load the standard Autodesk-supplied AutoLISP file AI_UTILS.LSP, which contains functions used by AutoCAD. The second example uses the optional **verbose** argument, which causes AutoLISP to display the contents of the file as it is loaded. Note that when the optional **verbose** argument is supplied, the optional **onfailure** argument must also be supplied as well.

Warning: Don't alter the AI_UTILS.LSP file! Doing so may cause some AutoCAD commands to malfunction.

```
Command: (LOAD "AI_UTILS.LSP")
Command: (LOAD "AI_UTILS.LSP" nil T)
```

Automatically Loading AutoLISP Programs

AutoCAD provides a means of automatically loading AutoLISP programs every time you enter the drawing editor. During initialization of the editor, AutoCAD will search its library path for a file named ACAD.LSP, and if the file is found, it will automatically load the file. ACAD.LSP provides a convenient way to have any number of useful functions

Introduction to AutoLISP Programming

loaded automatically whenever you enter the drawing editor. AutoCAD does not ship with an ACAD.LSP file, so if you want to make use of automatic loading, you must create this file, and place it in the library search path. If there are multiple copies of ACAD.LSP in several different locations on the library search path, AutoCAD will load only the *first* one it finds.

 Warning: Some third-party add-on applications for AutoCAD ship with an ACAD.LSP file. If you use such an application, do not add your own AutoLISP functions directly to this file—doing so can cause conflicts with the third-party application and make it much more difficult to upgrade the application in the future.

Although you can place frequently used programs and functions directly in ACAD.LSP, it is generally better to put groups of related functions in their own separate LSP files, and call the **load** function from ACAD.LSP to load those files whenever ACAD.LSP is loaded.

Loading separate files from ACAD.LSP has the same effect as having the contents of the loaded files appear directly in ACAD.LSP. The advantages to using separate LSP files for storing related functions are that it is easier to disable and enable the loading of a particular set of related functions that are stored in a single LSP file, and you can load additional LSP files *conditionally*, based on the result of a test that can be performed when ACAD.LSP is loaded.

When the following sample code fragment is placed in ACAD.LSP, it loads another LISP file called SETUP.LSP *only* when the current drawing is unnamed.

```
(if (= 0 (getvar "DWGTITLED"))    ;; If drawing is unnamed,
    (load "SETUP.LSP")            ;; then load SETUP.LSP
)
```

This code fragment checks to see if the value of the DWGTITLED system variable is 0, which indicates that the current drawing file is unnamed. If the drawing is unnamed, the file SETUP.LSP is loaded. SETUP.LSP might contain procedures that must be performed only when a new or unnamed drawing is opened.

Using MNL Files to Load AutoLISP Code

Another kind of AutoLISP program file is the *MNL* file (which stands for "MeNu Lisp"). These are AutoLISP files that are associated with an AutoCAD menu file and are loaded *automatically* when the associated MENU file is loaded. An MNL file has the exact same structure as a LSP file; the only difference is its extension. MNL files are associated with a menu (MNU or MNS) through a common filename, like ACAD.MNL and ACAD.MNS. When AutoCAD loads an MNU or MNS file, it will search for an MNL file having the same filename. If the MNL file is found, AutoCAD will load it just as if you had entered **(load "*MENUNAME*.MNL")** from the command line.

AutoCAD ships with the standard ACAD.MNU and ACAD.MNS files. It also includes a file called ACAD.MNL, which AutoCAD automatically loads immediately after loading ACAD.MNU/MNS. It will do this both at the start of an editing session and whenever ACAD.MNU/MNS is loaded explicitly through the use of the MENU or MENULOAD commands. You can also use MNL files with *partial* menus, which makes it easier to add new pull-down menus and new items to existing menus when the menu is loaded.

The MNL file is particularly useful for loading AutoLISP programs that a menu is dependent on. For example, if a custom menu invokes a custom command or LISP function, you must ensure that the code called by the menu's macros is loaded whenever the menu is loaded. You do not have to include the code a menu file depends on in the MNL file directly. Instead, you can call **load** from the MNL file to load any number of AutoLISP program files that are stored in other LSP files. This method is generally better than placing code directly in an MNL file, because it is easier to maintain and it makes it easier for several menus or partial menus to *share* the same programs or LSP files. In general, other than calls to the **load** function, the only AutoLISP code that should be placed in an MNL file is code that is *specific* to the associated menu, and not used by any other menu files.

The MNL file is also a good place to perform any initialization that may be required when a custom application with its own menu (or partial menu) is loaded for use. See this book's companion volume for more on using MNL files.

Tip: If you copy ACAD.MNU/MNS to another MNU/MNS filename, AutoCAD will not load ACAD.MNL when the renamed menu file is loaded. If you are creating a copy of ACAD.MNU/MNS to customize it, then you should also copy ACAD.MNL to have the same name as the customized version of ACAD.MNU/MNS. If you do not intend to modify ACAD.MNL, then rather than making a copy of ACAD.MNL, you can instead create a new MNL file and include a call to the AutoLISP **load** function, to load ACAD.MNL.

Loading ADS and ARX Programs.

In addition to the **load** function, AutoLISP also has functions that load ADS and ARX programs. **Xload** loads ADS programs, and **arxload** loads ARX programs. Each of these functions take the filename of the ADX or ARX program to be loaded (extension is not required), and uses the same method **load** uses to find and load the file.

Using the Maximizing AutoCAD Code Editor

Included on the MaxAC CD-ROM is a Windows-based AutoLISP editor called *LispPad for Maximizing AutoCAD*, which you can use to create and edit AutoLISP program files. We recommend you use LispPad or an equivalent with the exercises and examples in this book because it simplifies the process of editing, loading, and testing AutoLISP programs, allowing you to focus on the code you're writing rather than the distracting, tedious steps required to load and test it.

Introduction to AutoLISP Programming

With most normal editors that are not specifically designed to be used for AutoLISP code editing, you must edit code in the editor, save it to a file, switch to AutoCAD, and then call the **load** function to load the file. LispPad eliminates all of those steps by automatically loading the current selection or the entire contents of the editor buffer directly into AutoCAD with nothing more than a single keystroke or click of a button. LispPad can also simplify the creation, editing, testing, and debugging of menu macros, script, and DCL files, but these topics are not covered in this chapter.

In order to streamline the remainder of the exercises in this and other chapters, you will be briefly introduced to LispPad here, so you can use it for the following exercises, some of which require the use of a code or text editor.

LispPad operates in a manner that is largely consistent with Microsoft Notepad and most other Windows-based editors. It provides all of the same basic functionality as those tools, and also provides functions specifically designed for editing, testing, and debugging AutoLISP code. This introduction assumes that you are familiar with at least one or more of those tools.

Before you can use LispPad, you must install it from the CD. There's no installation program or other setup utility to complicate things, and you can simply copy the executable file (LISPPAD.EXE) from the CD-ROM to your hard disk and place it in whatever location you want, or create a shortcut to it. Refer to the instructions in Appendix G for specifics on installing LispPad on your computer.

You can refer to LispPad's documentation for more detailed information on all of its features and how to use them, but here we'll cover only the operations you'll need to know to open, edit, save, and load AutoLISP programs.

Figure 9.3 is an illustration of the LispPad window with prominent features and interface elements called out.

The following is a brief explanation of the elements annotated in Figure 9.3.

Figure 9.3
LispPad Editor interface elements.

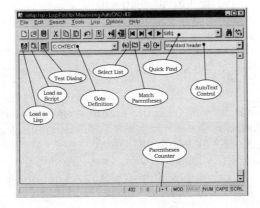

Load as Lisp. When you click on this button, the entire document or currently selected text is loaded into AutoCAD as AutoLISP code, and the AutoCAD window is brought to the top and given focus. If there is no selection the entire document is loaded. Using this button is the same as saving the contents of the editor to a file, switching to AutoCAD, and issuing (**load** "filespec") to load the saved file. You can also access this function by pressing CTRL+L.

Load as Script. When you click on this button, the entire document or current selection is sent directly to AutoCAD's command line as keyboard input. Doing this is similar to saving the document or selection to a file with an SCR extension and executing it using the AutoCAD SCRIPT command. This is useful for testing programs that request input, and for writing and testing script files on the fly. You can also access this function by pressing SHIFT+CTRL+ENTER.

Goto Definition. When you are editing AutoLISP files, this drop-down list displays the names of all user-defined functions in the current document and allows you to go to the definition of a function by selecting its name from the list.

Select List. This button will select a list or complex expression by extending the start and end of the selection highlight to the innermost pair of matching parentheses that encloses the caret. After selecting a list, you can then progressively select all of its enclosing parent lists by repeatedly clicking on the same button until you've selected a top-level list. You can also access this function by pressing ALT+L.

Find Matching Parentheses. This button will move the caret to the parenthesis that matches or balances with the one where the caret is. This makes it easier to find the start or end of a list or expression from either of its enclosing parentheses. You can click the button repeatedly to jump between the open and close parentheses surrounding a list or expression. You can also access this function by pressing CTRL+M.

Quick Find Control. This control consists of four buttons that resemble the controls on a VCR, and a combo box. Rather than having to display a dialog box to find text in the document, you can type it into the combo box, and then press ENTER to find the next occurrence of the search text. The buttons allow you to find the first, last, previous, and next occurrence of the search text respectively. The combo box holds a history of all text searched for in the current session, any of which can be restored by selecting it from the drop-down list. This function can also be accessed by pressing CTRL+F. The Find Next and Previous button functions can also be accessed by pressing ALT+N and ALT+P, respectively.

Parentheses Counter. This status bar pane dynamically displays the difference between the total number of open and closed parentheses in the current selection or the entire document when nothing is selected. When the parentheses in the selection or document are balanced, the contents of this pane will read (**=**), meaning that there are an equal number of open and closed parentheses. If the parentheses in the selection or document are not balanced, the value in this pane will begin with either an open or close

Introduction to AutoLISP Programming

parenthesis, a plus sign, and a number that represents the *difference* between the number of open and close parentheses in the document. For example, if there are 30 occurrences of "(" and 28 occurrences of ")" in the current document or selection, the contents of the pane will be "(+ 2". If there are five missing open parentheses in the selection or document, the contents of the pane will be ") + 5".

Send Line. You can send the contents of the editor line that the caret is currently on directly to AutoCAD's command line by choosing CTRL+ENTER or by selecting Send Line from the editor shortcut menu.

User-Defined Functions

All of the exercises you've worked through so far in this chapter required you to enter AutoLISP expressions directly into the command line. Although this is useful, it still requires a significant amount of effort. One way to reduce the typing involved is to store AutoLISP expressions in menu macros, so they can be executed easily when needed.

The AutoLISP language consists of hundreds of *functions* you can call to perform calculations or other operations. As you've already seen, you can call functions and pass data to them in the form of arguments. All of the functions you've used to this point are *built-in* functions AutoLISP provides to perform basic operations.

One of the most powerful aspects of the AutoLISP language is the ability to define and call your own functions. These are commonly called *user-defined functions*. Once you have defined a user-defined function, you can call it and pass it arguments just as you can with built-in functions. You can also have user-defined functions return values to a calling expression. User-defined functions allow you to define frequently used fragments of codes, and call them from one or more places in a program any number of times, without having to duplicate the code in more than one place. You can think of a user-defined function as the AutoLISP equivalent of an AutoCAD block definition, which can be referenced by any number of insertions of the block. In fact, there are a number of characteristics and advantages that are common to both. For example, when you redefine a block definition, all references to it in the current drawing will reflect the revision. Similarly, when a user-defined AutoLISP function is modified, all programs that call the function are affected.

`defun`

(`defun` *sym arglist expr*...)

Defines a new function named *sym*.

The **defun** function is used to create user-defined functions. The following is a simple example of a call to **defun** that defines a new user-defined function called **by-area**:

```
(defun by-area (area)
   (sqrt (/ area pi))
)
```

This function requires one argument, which it expects to be a real number representing the area of a circle. It computes and returns the radius of a circle having that area. You should recognize the expression **(sqrt (/ area pi))** from the previous exercise where you used it to draw a circle. In this case, the symbol **area** appears in place of the literal numeric value used in the previous exercise. Note that this symbol also appears between a pair of parentheses, identifying it as an argument, immediately following the name of the function in the definition.

Before examining the structure of a function definition, you'll use the **by-area** function in a variation of the exercise where you calculated the radius of a circle from its area. Before you can call a user-defined function, you must define it by calling **defun** and supply the new function's name, formal argument, and body as shown in the **by-area** example. Before drawing the circle, you'll first define the function.

Exercise

Defining and Calling a User-Defined Function

Start a new, unnamed drawing or continue in the drawing used in the previous exercise. Then zoom Center on the point 10,10 with a height of 30.

Command: **(defun by-area (area) (sqrt (/ area pi)))**	Defines the **by-area** function
Lisp returns: BY-AREA	**defun** returns name of function

Now, you'll test the new function by calling it.

Command: **(by-area 72.0)** [Enter]	Calls **by-area** function and passes value of 72.0 as argument
Lisp returns: 4.78731	Displays the value returned by **by-area**

Next, you'll use the function to compute the circle radius.

Introduction to AutoLISP Programming

`Command: `**`CIRCLE`**` [Enter]`	Starts the CIRCLE command
`3P/2P/TTR/<Center point>: `**`10,10`**` [Enter]`	Specifies a center of 10,10
`Diameter/<Radius> <12.50>:` **`(by-area 72.0)`**` [Enter]`	Calls **by-area** function to calculate radius of a circle having an area of 120.0 sq. drawing units
`Command: `**`AREA`**` [Enter]`	Starts the AREA command
`<First point>/Object/Add/Subtract:` `O [Enter]`	Specifies Object option
`Select objects: `**`LAST`**` [Enter]`	Specifies last visible object in drawing
`Area = 72.0000, Circumference = 30.0795`	AutoCAD displays circle area
`Command:`	

In this exercise, you first defined a new function called **by-area,** and called it to perform the same calculation made in the earlier exercise. The difference is that you've added a new function that can be used just like any other AutoLISP function, at any time in the editing session, from any AutoLISP program. User-defined functions have a clearly obvious advantage over entering complex AutoLISP expressions directly at AutoCAD input prompts: they reduce the complexity of AutoLISP expressions and the amount of typing required to use them, and along with that, mistakes and errors.

Warning: You should *never* assign functions or variable values to symbols that are the names of built-in AutoLISP functions or constants. Doing so will effectively disable the built-in function that was assigned to the symbol. Exercise caution when choosing symbol names for functions and variables. You can determine if a symbol is reserved by evaluating it on the command line, by entering it prefixed with an exclaimation point. If the result is anything other than **nil**, then the symbol is already in use, and you should not assign a variable or function to it. Common improperly used symbols are: **T**, type, pi, \, min, max, last, length, angle, distance, and chr.

Anatomy of a User-Defined Function

A function definition, defined with **defun,** contains several distinct components, as shown in Figure 9.4. The first argument is the name of the function to be defined. The second argument to **defun** is the *formal argument list* of the function that is being defined.

Figure 9.4

Anatomy of a function definition.

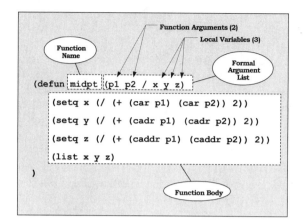

The remaining arguments are the expressions that comprise the defined function's body. There can be any number of expressions in the function body. When a function is called, it always returns the result of the last expression in its body to its caller.

The function name is the symbol that is used to reference the function when it is called. In the example in Figure 9.4, the name of the function is **midpt**.

The formal argument list is composed of two parts delimited by a forward slash. The first part contains the variable symbols that will be assigned to the values of the arguments passed to the function each time it is called. The formal argument list in the **midpt** function definition has two arguments that are assigned to the symbols **p1** and **p2**. The second part of the formal argument list, which appears to the right of the forward slash, contains symbols that are declared as *bound variables* within the function's environment. Bound variables are temporary variables in that the values assigned to them within a function's body are valid only when the body's expressions are evaluated. You'll learn more about bound and free variables later in this chapter. In the **midpt** function's formal argument list, there are three bound variables, which are the symbols **x**, **y**, and **z**.

The body of the function consists of the expressions that are evaluated when the defined function is called. In the **midpt** function, there are four expressions that comprise its body. One important point to remember is that when a function is defined by calling **defun**, the expressions passed that become the body of the function are *not* evaluated at that point. Those expressions are only evaluated when the function is actually called. Calling **defun** only creates the function. It does not execute the expression arguments.

The following few paragraphs give some examples of user-defined functions. The **delta** function requires four arguments, has no bound variables, and has one expression in its body. The **dimtext** function requires no arguments, has one bound variable, and has two expressions in its body. The **pspace?** function takes no arguments, has no bound variables, and contains one expression in its body.

Introduction to AutoLISP Programming

```
(defun delta (pt x y z)          ;; defines a function called delta
    (list
        (+ (car pt) x)
        (+ (cadr pt) y)
        (+ (caddr pt) z)
    )
)

(defun dimtextsize ( / h)        ;; defines a function called
    (setq h (getvar "DIMTEXT"))  ;; dimtextsize
    (* (getvar "dimscale") h)
)

(defun pspace? ()                ;; defines a function called
    (and (eq (getvar "TILEMODE") 0)  ;; pspace?
         (eq (getvar "CVPORT") 1)
    )
)
```

Most of the code listings and examples in this book are provided in electronic form on the accompanying CD, which eliminates the need to type them in. But the next exercise is an exception. It's not included on the CD, in order to encourage you to gain hands-on experience with the code editor by using it with the exercise. You will create an AutoLISP program file that will contain two user-defined functions. The following code listing is the entire contents of the file which you will enter into the LispPad editor, after which you'll load and test the functions in AutoCAD.

Note: Note that this exercise requires AutoCAD for Windows. If you're using DOS AutoCAD (which is not recommended for AutoLISP development), you can use the editor of your choice in place of LispPad, and use the AutoLISP load function to load the AutoLISP file where you see an instruction to press CTRL+L in LispPad.

Note that the first few lines in the code listing have semicolons at the start of each line. AutoLISP uses the semicolon as a means of including *comments* in your program files. AutoLISP treats anything that follows the first semicolon on each line as a comment, meaning that it ignores it.

You should enter the file exactly as it appears in the following listing, and with the same indentation and formatting. As you enter the code into the editor window, note how the parentheses counter pane changes to reflect parentheses balancing. After you've entered the entire definition for the first function **midpt,** the parentheses counter pane should read **(=)**, indicating that the parentheses up to that point are balanced. If the parentheses aren't balanced, then you should go back, examine what you've already entered to find the missing parentheses, and correct the error.

Maximizing AutoCAD® R13

Tip: When you are entering code into LispPad, it is best to check the parentheses counter pane after entering each function, since at those points, the number of open/close parentheses in the file should be equal. If the parentheses aren't balanced and you do not know where the missing one(s) are, you can select suspect portions of a file that should have balanced parentheses (for example, each function definition), and the parentheses counter will display the parentheses count of only the current selection.

 Exercise

Creating an AutoLISP Program File

Launch LispPad and enter the following code listing in its entirety.

```
; MIDPOINT.LSP   From Maximizing AutoCAD Release 13
; Implements MIDPT command and midpt function which
; finds a point halfway between two selected points.
;
; (midpt point1 point2)
;
; Returns the midpoint of an imaginary line
; between point1 and point2
(defun midpt (p1 p2 / x y z)
    (setq x (/ (+ (car p1) (car p2)) 2))      ;; Average of X ordinates
    (setq y (/ (+ (cadr p1) (cadr p2)) 2))    ;; Average of Y ordinates
    (setq z (/ (+ (caddr p1) (caddr p2)) 2))  ;; Average of Z ordinates
    (list x y z)
)
; (c:midpt)
;
; Prompts for two points and returns
; the midpoint of an imaginary line
; between them.
;
; Usage: 'MIDPT
(defun C:MIDPT ( / p1 p2)
    (setq p1 (getpoint "\n>>First point: "))   ;; Get first point
    (setq p2 (getpoint "\n>>Second point: "))  ;; Get second point
    (midpt p1 p2)                              ;; Compute and return
)                                              ;; the midpoint.
; END MIDPOINT.LSP
```

Choose **F**ile Save **A**s, and save the file under the name MIDPOINT.LSP

Leave LispPad active for the next part of the exercise.

Introduction to AutoLISP Programming

The preceding code listing contains two functions. The **midpt** function accepts two arguments, both of which must be point lists. It returns a point that is at the midpoint of an imaginary line between the two point arguments. The body of the **midpt** function calls three functions from the **CxxxR** family to extract the X, Y, and Z components from each point, and uses **setq** to assign values to three variables: **x**, **y**, and **z**. The second argument to each call to **setq** computes the average of the X, Y, and Z components of both points. The three resulting values are then passed to the **list** function, which returns them in a point list. All user-defined functions return the result produced by evaluating the *last* expression in their body, so the point list returned by the call to **list** (which is the last expression in the body of the function) becomes the result of the **midpt** function.

Next, you'll use LispPad (in Windows) to load the file entered in the first part of the exercise into AutoCAD. If you're working in DOS AutoCAD (not recommended), then use the AutoLISP **load** function to load the file instead, where you see the instruction to press CTRL+L in LispPad in the exercises.

Exercise

Loading an AutoLISP Program File

Launch AutoCAD, or continue in the current session after erasing everything.

In LispPad, press CTRL+L	Loads the contents of the editor into AutoCAD, and switches focus to AutoCAD
Lisp returns: C:MIDPT	Returns the result of the last expression in MIDPOINT.LSP

In the second part of the exercise, when you pressed CTRL+L (or you could choose File, Load) in LispPad, the AutoCAD main window was activated, and a message that resembles the following appeared on AutoCAD's command line:

```
Command: *Cancel*
Command: (load "d:/maxac/cdrom/lispbuf.$$$")
C:MIDPT
Command:
```

The call to the **load** function in this example was generated by LispPad. It loads a temporary file containing the editor or selection contents, which LispPad creates. The exact path may not be the same as what is shown here, depending on the name and location of the working directory.

Now you will test the **midpt** function.

 Exercise

Testing MIDPOINT.LSP

Continue from the previous exercise. Erase everything and zoom Center on the point 10,10 with a height of 10 units.

In LispPad, start a new file and enter the following on a single line:

`(command "LINE" "8,7" "8,12" "" "LINE" "12,7" "12,14" "")`	Defines code to draw two vertical lines
Press CTRL+L	Loads the line containing the cursor into AutoCAD and the two lines are drawn

In LispPad, delete the preceding expression and enter the following expressions formatted as shown:

`(setq p1`
`   (midpt (list 8.0 7.0 0.0)`
`          (list 12.0 7.0 0.0)`
`   )`
`)`
	Assigns midpoint between points 8,7 and 12,7 to **P1**

`(setq p2`
`   (midpt (list 8.0 12.0 0.0)`
`          (list 12.0 14.0 0.0)`
`   )`
`)`
	Assigns midpoint between points 8,12 and 12,14 to **P2**

Press CTRL+L	Loads the contents of the file and activates AutoCAD's window

In the AutoCAD window:

Command: **LINE** Enter	Starts LINE command
From point: **!P1** Enter	Specifies P1 as first point

Introduction to AutoLISP Programming

`To point: !P2 `[Enter]	Specifies P2 as second point
`To point: `[Enter]	AutoCAD draws a line between the two points
`Command: ERASE `[Enter]	Starts the ERASE command
`Select objects: LAST `[Enter] `1 found`	Selects middle line
`Select objects: `[Enter]	Ends object selection
`Command:`	

Leave the drawing as is for use in the next exercise.

In this exercise, you defined two new functions. The **midpt** function computes the midpoint between two points supplied as arguments. After the function is defined, it may be called any number of times in the current drawing editor session. You also saved the function's definition to the file MIDPOINT.LSP, which can be loaded in future editing sessions to use the functions it contains.

Unlike their built-in counterparts, user-defined functions may *not* have optional or repeating arguments. When a user-defined function is called, it must be passed one argument for each symbol in the first part of the formal argument list. If the number of arguments passed is not equal to the number of argument symbols in the function definition, an error condition occurs, and program execution stops. The following example illustrates this. The function **foo** is defined to accept two arguments (it also has one bound variable):

```
Command: (defun foo (a b / c)
(princ "\nHello from foo!"))
FOO
```

The following shows examples of calling **foo** with one, two, and three arguments.

```
Command: (foo 1)
error: incorrect number of arguments to a function
(FOO 1)
*Cancel*
Command: (foo 1 2)
Hello from foo!"\nHello from foo!"
Command: (foo 1 2 3)
error: incorrect number of arguments to a function
```

```
(FOO 1 2 3)
*Cancel*
Command:
```

Each user-defined AutoLISP function is like a separate program or *subroutine* that executes within another larger program. When a function is called, the following events occur:

1. Each argument to the called function is evaluated and replaced by its result. Each of those results is assigned to the variable symbol in the formal argument list that has the same position as the argument in the call to the function. In the **midpt** function you defined, the argument symbol **p1** is assigned to the result of the first argument passed in the call to the function, and **p2** is assigned to the result of the second argument passed in the calling expression.

2. The bound variables **x**, **y**, and **z** are all initially assigned the value nil. Bound variables are discussed in detail in the "Arguments, Bound Variables, and Rules of Scope" section of this chapter.

3. The expressions that comprise the body of the function are evaluated in the order they appear (in the case of **midpt**, there are four expressions in its body). The result produced by evaluating each expression in the function's body is *discarded* except for the result of the *last* expression, which is the value that is returned by the defined function when it is called.

4. The values of all arguments and bound variables are released; the function's environment is destroyed; and the function call returns the result of the last expression in its body.

Defining New AutoCAD Commands

The MIDPOINT code listing that you entered included two functions: **midpt** and **C:MIDPT**. The second function is special because, when loaded, it automatically adds a *new command* to AutoCAD, called MIDPT. Functions that are assigned to symbols that begin with the prefix "**C:**" automatically become new AutoCAD commands, whose name is the symbol the function is assigned to *without* the **C:** prefix.

For example, to add a new command called **TEST** to AutoCAD, you define a new AutoLISP function and assign it to the symbol **C:TEST**. Once you have defined a command function, you can execute it entering its name on the AutoCAD command line, like any AutoCAD command.

Introduction to AutoLISP Programming

Note that symbols and command names are not case-sensitive, but in this book we show the names of all commands in uppercase. The following example user-defined command will rotate the crosshairs (snap rotation angle) by a specified angle. This is useful when you only know the relative angle that you want to rotate the crosshairs, rather than the absolute value required by the SNAP command.

```
(defun C:SNAPROT ( / ang)
   (setq ang (getangle "\nRotation angle: "))  ; Get angle from user
   (setvar "SNAPANG"
      (+ (getvar "SNAPANG") ang)               ; Add supplied angle to
   )                                           ; current snap rotation
   (princ)                                     ; exit quietly
)
```

In the preceding user-defined function, the **getangle** function is called to obtain angle input. The result is assigned to the symbol **ang** and the **setvar** function is called to modify the value of the SNAPANG system variable, by adding the value supplied as input to its current value. The last expression in the body of **C:SNAPROT** is a call to **princ** with no arguments, which has the effect of returning a non-printing value, and helps make user-defined functions look more professional by suppressing the display of superfluous results on the command line when they finish. When you create a command function that is designed to be called from the command line, you should place a call to **princ** with no arguments, as the last expression in the function's body.

Calling Command Functions Transparently

User-defined commands can be called transparently by prefixing their names with a single quote, like AutoCAD commands. You can execute user-defined commands transparently only while native AutoCAD commands are active. You cannot execute a user-defined command while another user-defined command is active. Attempting to do so will generate a Can't reenter AutoLISP error.

The **C:MIDPT** command you created in the last exercise can be issued while the LINE command is active and waiting for an endpoint. When a user-defined command is used transparently, its result is accepted by AutoCAD as a response to the prompt where the command was issued.

The **C:MIDPT** command you created requests two points, and then passes them to the **midpt** function. In the following exercise, you'll test the **C:MIDPT** command within the LINE command.

Maximizing AutoCAD® R13

Exercise

Using the MIDPT Command

Continue in the current drawing from the previous exercise.

Command: **OSNAP** [Enter]	Starts the OSNAP command
Object snap modes: **ENDPOINT** [Enter]	Sets running osnap to ENDpoint
Command: **LINE** [Enter]	Starts the LINE command
From point: **'MIDPT** [Enter]	Issues MIDPT command transparently
>>First point: *Pick lower end of left line*	
>>Second point: *Pick lower end of right line*	Anchors rubber-band cursor to midpoint between two line endpoints
To point: **'MIDPT** [Enter]	Issues MIDPT command transparently
>>First point: *Pick upper end of left line*	
>>Second point: *Pick upper end of right line*	Line is drawn through midpoints of both existing line endpoints
To point: [Enter]	Exits LINE command
Command: **OSNAP** [Enter]	Starts OSNAP command
Object snap modes: **NONE** [Enter]	Restores running osnap to NONe
Command:	

You've just created and tested your first truly useful tool with AutoLISP. The new MIDPT command can be used like a running object snap modifier to compute a point halfway between two known points. You can add the function to your menu file and include its definition in its associated MNL file.

Arguments, Bound Variables, and Rules of Scope

In the last exercise, you defined a new function that accepts arguments and has bound variables. Bound variables are ones that exist only within the *environment* where the function's body expressions are evaluated. A function's environment is defined as the set of variables that are visible from the expressions in the function's body during evaluation. If a variable is not bound to a function's environment, then it is said to be *free*, with respect to the function. Bound variables are also referred to as *local* variables. All function arguments are also bound variables. A *global* variable is one that's not bound to *any* function's environment and is visible from all AutoLISP expressions. For example, all variables you create by calling **setq** on the AutoCAD command line are global variables, because they are not bound to any function.

The following exercise demonstrates free and bound variables, using the MXA0901.LSP file, from the MaxAC CD-ROM. Additional example functions are stored in the file MXA0902.LSP.

```
; MXA0901.LSP  Maximizing AutoCAD Release 13
; Exercise from Chapter 9
(defun foo (a / b)
   (princ "\n>>>> Entering foo.")    ;; Display message
   (setq b "value of b in (foo)")    ;; Assign value to b
   (setq c "value of c in (foo)")    ;; Assign value to c
   (princ "\n A = ")                 ;; Print value of a
   (prin1 a)
   (princ "\n B = ")                 ;; Print value of b
   (prin1 b)
   (princ "\n C = ")                 ;; Print value of c
   (prin1 C)
   (princ "\n>>> Leaving (foo).")    ;; Display message
   (princ)
)
; *** End listing
```

Exercise

Using Free and Bound Variables

Begin a new unnamed drawing, or continue in the drawing from the previous exercise. If you haven't already done so, copy all files from the MAXAC\BOOK\CH09 directory to your \MAXAC\DEVELOP directory.

Start LispPad and open MXA0901.LSP, press CTRL+L to load editor contents, then switch to the AutoCAD Text Window.

Command: **(setq a "Global value of a")** [Enter]   Sets global value of A

Lisp returns: "Global value of a"

Command: **(setq b "Global value of b")** [Enter]   Sets global value of B

Lisp returns: "Global value of b"

Command: **(setq c "Global value of c")** [Enter]   Sets global value of C

Lisp returns: "Global value of c"

Command: **(print a)** [Enter]   Prints global value of A
"Global value of a"

Lisp returns: "Global value of a"

Command: **(print b)** [Enter]   Prints global value of B
"Global value of b"

Lisp returns: "Global value of b"

Command: **(print c)** [Enter]   Prints global value of C
"Global value of c"

Lisp returns: "Global value of c"

Next, you call the **foo** function and see how it affects these values.

Command: **(foo "Argument to foo")** [Enter]   Calls **foo** function

```
>>> Entering foo.
A = "Argument to foo"
B = "value of b in (foo)"
C = "value of c in (foo)"
>>> Leaving (foo).
```

Command: **(print a)** [Enter]   Prints global value of A
"Global value of a"

Lisp returns: "Global value of a"

Introduction to AutoLISP Programming

```
Command: (print b) [Enter]                      Prints global value of B
"Global value of b"

Lisp returns: "Global value of b"

Command: (print c) [Enter]                      Prints global value of C
"value of c in (foo)"

Lisp returns: "value of c in (foo)"             Result of last call to print
                                                in foo

Command:
```

This exercise illustrates how bound or "local" variables have temporary values that exist only within the environment of the function they're bound to. In the exercise, the value of **B** has one value within **foo's** environment, and a different value outside of it both before and after **foo** is called. The same holds true for the variable **A**, which is an argument to **foo**. In fact, **A** behaves the same as the variable **B**, only its initial value inside of **foo** is the argument passed to **foo**.

In the function **foo,** the variable **C** is a *free* variable, since it does not appear in the formal argument list of foo's definition. For this reason, when a value is assigned to **C** within the body of **foo**, that value remains assigned to **C** after **foo** has returned.

AutoLISP variables use *dynamic scoping*. This means bound variables are *visible* from the body of any function that is called from the environment of the function to which a variable is bound. The following exercise demonstrates this. A variable that is bound to one function's body can be accessed and modified by another function when called from within the main function's body.

The following code is from the file MXA0903.LSP.

```
; MXA0903.LSP  Maximizing AutoCAD Release 13
(defun parent ( / a b)
   (setq a "A set in parent")       ;; assign value to a
   (setq b "B set in parent")       ;; assign value to b
   (child)                          ;; call child
   (print a)                        ;; print value of a
   (print b)                        ;; print value of b
)
(defun child ( / a)
   (setq a "A set in child")        ;; assign value to a
   (setq b "B set in child")        ;; assign value to b
```

)

; *** End MXA0903.LSP

 Exercise

Testing Variable Scope

Start AutoCAD with a new unnamed drawing, or continue in the current session.

Start LispPad, and open the file MXA0903.LSP, and press CTRL+L to load editor contents into AutoCAD, then switch to the AutoCAD Text Window.

Command: **(parent)** Enter   Calls **parent** function
"A set in parent"
"B set in child"

Lisp returns: "B set in child"

Command:

In this exercise, there are two user-defined functions, parent and child. Parent has variables A and B bound to its environment. Child has only A bound to its environment. When parent is called, it assigns values to both A and B, then it calls child, which assigns different values to A and B. After child returns control to parent, the values of A and B are printed to show that A has retained the value assigned to it in parent before child was called, and that B has changed from its original assignment, still having the value assigned to it in child.

Basic Programming Operations

This section provides a categorical summary of the various basic programming operations that are routinely performed by AutoLISP programs.

Math Operations

AutoLISP provides a number of built-in math and trigonometric functions, some of which you've already used or have seen used in examples. You've also learned the basic differences in the way math and other operations are expressed in AutoLISP's prefix notation, and in standard algebraic infix notation.

Introduction to AutoLISP Programming

Table 9.6
Basic AutoLISP Math Functions

Function	Description
(/ [number number]...)	*Division*
(* [number number]...)	*Multiplication*
(+ [number number]...)	*Addition*
(- [number number]...)	*Subtraction*
(1+ number)	*Increment by 1*
(1- number)	*Decrement by 1*
(abs number)	*Absolute value*
(exp number)	*Natural antilog*
(expt number power)	*Raises number to specified power*
(gcd [number]...)	*Greatest common denominator of arguments*
(log num)	*Natural log of argument*
(max [number number]...)	*Returns largest of its arguments*
(min [number number]...)	*Returns smallest of its argument*
(rem [number number]...)	*Remainder successive division of arguments*
(sqrt num)	*Square root*

Table 9.6 is a list of built-in AutoLISP math functions that you can use in your programs.

See Appendix A for more complete definitions of these functions.

All math functions whose prototypes contain **[number number]...** are functional equivalents of the basic binary math operators. They will all accept any number of arguments. If no argument is supplied, they return 0. If only one argument is supplied, these functions perform the specified operation on the argument, and the default value of 1 (for division and multiplication) or 0 (for addition and subtraction).

For example, **(+ 2)** is the same as **(+ 0 2)**, and **(/ 4.0)** is the same as **(/ 4.0 1)**.

The following two expressions (prefix and infix) are equivalent:

(+ 2 13 34 26 78)   2 + 13 + 34 + 26 + 78

In the following exercise, you will test some of AutoLISP's math functions.

Exercise

Using AutoLISP Math Functions

Continue in the previous drawing, or begin a new, unnamed drawing.

Command: **(+ 1 2.0)** [Enter]   Returns a real number

Lisp returns: 3.0

Command: **(/ 3 2)** [Enter]   Drops the .5 remainder and returns an integer because both arguments are integers (integer division)

Lisp returns: 1

Command: **(/ 3.0 2)** [Enter]   Keeps the .5 remainder and returns a real number because at least one argument is a real

Lisp returns: 1.5

Command: **(* 5 (- 7 2))** [Enter]   Evaluates the nested (- 7 2) expression first, then (* 5 ...)

Lisp returns: 25

Command: **(setq a 1)** [Enter]   Reassigns the value 1 to a variable

Lisp returns: 1

Command: **(+ (setq a (* a 3)) (+ a 2))** [Enter]   Assigns variable A before evaluating second expression

Lisp returns: 8

Command: **(+ (+ a 2) (setq a (* a 3)))** [Enter]   Uses variable A in (+ a 2) and (* a 3) then reassigns it

Lisp returns: 14

Introduction to AutoLISP Programming

Command: **!A** Enter

Lisp returns: 9

Note: Notice the important difference between the **(/ 3 2)** example and the **(/ 3.0 2.0)** example. They return different values. When AutoCAD does integer division, an integer is returned and the remainder value is dropped. With real division, the remainder is kept. Real division (or any other arithmetic operation) is performed when *any* of the arguments are real numbers. This difference illustrates why it is good practice to declare your variables properly (specify them initially with decimal points) when they are intended to be used as real values. Declare them as reals to ensure that calculation results return what you expect. This can be one of the most frustrating programming bugs to search for when your calculations keep coming out wrong.

String Operations

You've already seen strings used in a number of the preceding exercises and examples. You may have noticed several odd sequences at the start of some strings that start with a backslash, similar to the following:

(princ "\nHello!")

The backslash \ character in the string has special meaning, and it is called an escape sequence. When this character appears in a string, it signals that there is some special handling of the character that immediately follows it. AutoLISP strings can contain a number of special control characters that cause various effects. In this example, the lowercase n that follows the backslash causes the remainder of the string to appear on a new line when displayed, and this is commonly called the *newline* character. The following example demonstrates this:

Command: **(princ "\nLine 1\nLine 2\nLine 3\n")** Enter
Line 1
Line 2
Line 3
Lisp returns: "\nLine 1\nLine 2\nLine 3\n"
Command:

Note how the string was broken into several lines, one at each occurrence of the newline character. The built-in **princ** function is identical to the **print** function, except that it handles strings differently. Look at the following examples:

Command: **(setq mystring "Line 1\nLine 2\nLine 3\n")** Enter
Lisp returns: "Line 1\nLine 2\nLine 3\n"

Command: **(print mystring)** [Enter]
"Line 1\nLine 2\nLine 3\n" *Lisp returns:* "Line 1\nLine 2\nLine 3\n"

Command: **(princ mystring)** [Enter]
Line 1
Line 2
Line 3
Lisp returns: "Line 1\nLine 2\nLine 3\n"
Command:

Notice how the **print** function displays strings literally with surrounding double quotes, and does not interpret embedded control characters such as the newline, whereas the **princ** function interprets and expands control characters. In the preceding example, **princ** causes the string to be displayed on three separate lines because the newline control characters are interpreted. Also note that the string appears twice after the call to **print** and **princ**. This is because both of these functions return their first argument unaltered, so the first value that appears after each call is the value printed by the function, and the second value is what the function returns to its caller. Because the backslash is used as the escape character for all control codes, two backslashes are required in a literal string to produce a single backslash, as illustrated in the following example.

Command: (setq dwgname "d:\\acad\\com\\samples\\filter.dwg")
"d:\acad\com\samples\filter.dwg"

Note that each backslash in the result requires two backslashes in the literal string argument to **setq**. Failing to use two backslashes is a common error. The next example shows what happens when single backslashes are incorrectly used for path delimiters in the preceding example.

Command: (setq dwgname "d:\acad\com\samples\filter.dwg")
"d:acadcomsamplesfilter.dwg"

Warning: In menu macros, the backslash is used to pause the macro for input. When literal strings appear in menu macros, they must not contain backslash characters. AutoCAD permits the use of the forward slash (/) character in place of the backslash in menu macros for the purpose of including path delimiters in macros. Alternately, you can use ^M in a menu macro to obtain the same effect as "\n".

Table 9.7 lists control characters that can be used in AutoLISP strings.

Note: The control characters listed in Table 9.7 are case-sensitive. For example, to insert a newline break in a string, you must use "\n" with the "n" in lowercase. If you use an uppercase N as in "\N", this will not be interpreted as a control character.

Introduction to AutoLISP Programming

Table 9.7
String Control Characters

Code	Description
\\	\ (literal backslash) character
\"	" (literal double-quote) character
\e	Escape character
\n	Newline character
\r	Return character
\t	Tab character
nnn	Character whose octal code is *nnn*
\U+*xxxx*	Unicode character sequence
\M+n*xxxx*	Multibyte character sequence

AutoLISP has a number of built-in functions for manipulating strings, and converting strings to and from other data types. These functions are listed in Table 9.8.

See Appendix A for more complete definitions of these functions.

One of the most often used of these functions is **strcat**. It combines two or more strings together into a single string. It is extremely useful for generating prompt strings that are displayed on the command line to obtain input, as the following example shows. The first example demonstrates the general use of **strcat**, and the second shows how it is often used in AutoLISP programs.

```
(setq mystring (strcat "One," "Two," "Three"))
"One,Two,Three"

(setq dist 10.0)
(setq message
  (strcat
     "\nEnter flange width <"
     (rtos dist)
     ">: "
  )
)

(setq newdist (getdist message))
```

If the current linear units setting is decimal and the current precision is set to two places, the preceding code fragment causes the following input prompt to appear on the command line:

```
Enter flange width <10.00>:
```

Table 9.8
AutoLISP String-Handling and Conversion Functions

Function	Description
`(angtof angle [mode])`	Converts *string* expression of an angle to radians (a real number).
`(angtos angle [mode [precision]])`	ANGle TO STRing converts *angle* to a string.
`(ascii string)`	Returns the ASCII value of the first character of *string*.
`(atof string)`	Ascii TO Float converts *string* to a real number.
`(atoi string)`	Ascii TO Integer converts *string* to an integer.
`(chr integer)`	Converts *integer* to a string consisting of the ASCII character corresponding to integer (for ASCII values 032 through 127).
`(distof string [mode])`	Converts *string* expression of a distance to a real number.
`(itoa integer)`	Integer TO ASCII converts *integer* to an ASCII string representation of that integer.
`(rtos number [mode [precision]])`	Real TO string converts *number* to an ASCII string representation of that real number.
`(strcase string [which])`	Converts *string* to upper- or lowercase.
`(strcat [string]... )`	Returns the concatenation of one or more strings.
`(strlen [string]... )`	Returns the length of *string*.
`(substr string start [length])`	Returns the rest of *string*, beginning with the `start` (an integer) character position and continuing for `length` characters or to the end of the string, whichever comes first.
`(wcmatch string pattern)`	Returns **T** if *string* matches *pattern*.

Introduction to AutoLISP Programming

This format should be familiar, since it's used extensively by AutoCAD's own commands to signal that a default value (which appears between the angle brackets) may be accepted by pressing ENTER or SPACE. It is desirable to make your programs behave like this, making them consistent with the conventions used by AutoCAD's own built-in commands, since most of those who will use it are already familiar with these conventions.

In the preceding example, **strcat** is passed three arguments, two of which are literal strings, and one that is returned by the call to the **rtos** function. **Rtos** accepts a real number and converts it to a string representation of a distance, in a format that is determined by the current units and precision settings (which are maintained in the system variables LUNITS and LUPREC respectively). **Rtos** returns the string to its caller. This allows you to format distance and angular input in accordance with the current drawing editor's linear and angular unit format. You can also supply optional unit and precision arguments that override the current units and precision settings. In the following exercise, you'll use **rtos** to format some distance values.

Exercise

Converting Numbers to Distance Strings

Continue in the current drawing, or begin a new, unnamed drawing.

Command: **(setvar "LUNITS" 2)** [Enter]   Sets current linear units to 2 (decimal)

Lisp returns: 2

Command: **(setvar "LUPREC" 4)** [Enter]   Sets current linear unit display precision to 4

Lisp returns: 4

Command: **(rtos 14.25 4)** [Enter]   Converts 14.25 units to a distance string in arch. units format using current display precision setting in LUPREC

Lisp returns: "1'-2 1/4""

Command: **(rtos 2.5 2)** [Enter]   Converts 2.5 units to a distance string in decimal units format using current display precision setting in LUPREC

Lisp returns: `"2.5000"`

Command: **(setq a 326.3125)** [Enter]	Assigns 326.3125 to **A**

Lisp returns: `326.313`

Command: **(rtos a)** [Enter]	Converts value of **A** to a distance string in decimal units format with current display precision settings in LUPREC

Lisp returns: `"326.31"`

Command: **(rtos a 4)** [Enter]	Converts value of **A** to a distance string in arch. units format with current display precision settings in LUPREC
Command: **(rtos a 2 6)** [Enter]	Converts value of **A** to a distance string in decimal units format with 6 decimal places

There is one important thing you should note in this exercise. When you assigned the value 326.3125 to the symbol **A**, the value returned by **setq** on the command line was 326.313, the original value rounded to three places. Don't let what you see confuse you! When values are displayed at the command prompt, they are not displayed to full floating-point precision. Instead, they are rounded to between 3 and 6 decimal places, but that rounding only applies to display output, and the actual value remains unchanged, as is demonstrated by the expression that converted the value of **A** to a decimal string with six decimal places.

List Operations

You should already be acquainted with the list. Every complex expression that you've used so far is a list. A list can contain any number of elements of any type, including other lists. Unlike languages such as BASIC, C, or Pascal, AutoLISP does not have arrays. In AutoLISP, lists serve many of the same functions that arrays do in other languages. The **nth** function accesses a list element by position, similar to the way subscripts are used to access array elements in BASIC, C, or Pascal.

The coordinate data that you've worked with are also lists. You've already used the **list** function to create 3D coordinates, which AutoCAD accepts as input. AutoCAD also makes coordinate information available to your programs in this very same form. The following exercise shows how to access and modify coordinate data stored in AutoCAD system variables from AutoLISP.

Introduction to AutoLISP Programming

 Exercise

Accessing Point Lists in System Variables

Begin a new, unnamed drawing, and zoom Center on the point 6,4 with a height of 30 units.

Command: **(getvar "LIMMAX")** Enter

Lisp returns: (12.0 9)   Returns the upper right point of the limits

Command: **(type (getvar "LIMMAX"))** Enter   Determines the type of data returned by **getvar**

Lisp returns: LIST

Command: **(setq vctr (getvar "VIEWCTR"))** Enter   Assigns the current view center point to **vctr**

Lisp returns: (6.0 4.0 0.0)

Command: **(setvar "LASTPOINT" (list 2.0 3.0 0.0))** Enter   Sets the LASTPOINT system variable to 2,3,0

Lisp returns: (2.0 3.0 0.0)

Command: **CIRCLE** Enter   Starts the CIRCLE command

3P/2P/TTR/<Center point>: **@** Enter   Specifies value of LASTPOINT system variable as center point

Diameter/<Radius>: **3.0** Enter   Specifies radius of 3.0 units

Command:

In this exercise, you called the **getvar** function to obtain the value of the LIMMAX system variable, which stores the upper right coordinate of the current drawing limits. The value returned by **getvar** in this case was a list containing two reals, which represents a 2D coordinate (a 3D coordinate list contains three reals). In most cases,

AutoCAD accepts 2D coordinates where 3D coordinates are expected, and will fill in the Z component, which is obtained from the ELEVATION system variable.

You also used the **setvar** function to set the value of the LASTPOINT system variable to a 3D coordinate that was passed as an argument. The LASTPOINT variable maintains the point that is used when you use "@" at the keyboard to specify a relative coordinate.

AutoLISP provides functions to access and manipulate lists in many different ways. The following is a summary of these functions.

`list`

`(list [expression] ...)`

Returns a list containing the supplied **expression(s)**. *Calling* `list` *with no arguments returns* **nil**, *which is the same as* **()** *(the empty list).*

Because lists are containers that hold other elements, you need a way to access a list's contents by their position. **car** is shorthand for the first element of a list, **cadr** for the second element, and **caddr** for the third element.

`car`

`(car list)`

Returns the first element in `list`. `car` *is used to extract the X component of a point represented by a list.*

`cadr`

`(cadr list)`

Returns the second element in `list`. `cadr` *is used to extract the Y component of a point represented by a list.*

`caddr`

`(caddr list)`

Returns the third element in `list`. `caddr` *is used to extract the Z component of a point represented by a list.*

`cdr`

`(cdr list)`

Returns all but the first element in `list`.

Introduction to AutoLISP Programming

Creating Lists with the Quote Function

The **quote** function, which can be abbreviated as a single quotation mark ('), is also important. The single quotation mark form of the **quote** function is unique in that it is called without parentheses.

> **quote** or **'**
>
> **(quote** *expr***)** or **'***expr*
>
> *Returns* ***expr*** *without evaluation; the abbreviated* **'** *performs the same function.*

Unlike the **list** function, which evaluates its contents before forming a list, **quote** suppresses the evaluation of its expression(s). When it forms a list, it includes the list's contents literally, as shown in the following exercise. Look again at **list**, **car**, **cadr**, the similar **cdr**, and at the **quote** function. Enter the input at the command line to see how these functions work.

Exercise

Manipulating Lists

Begin a new drawing, or continue from the previous exercise.

Command: **(setq test (list 1 2 3.0))** [Enter]   Makes a list with three elements

Lisp returns: (1 2 3.0)

Command: **(car test)** [Enter]   Extracts first element

Lisp returns: 1

Command: **(cdr test)** [Enter]   Extracts the rest of the list: all but the first element

Lisp returns: (2 3.0)

Command: **(cadr test)** [Enter]   Extracts second element, the **car** of the **cdr**

Lisp returns: 2

Command: **(nth 2 test)** Enter   Extracts the third element

Lisp returns: 3.0

In the next expression, **list** evaluates Q, R, and S, which are all unassigned symbols (variables) with nil values, then it makes and returns a list of the three **nil**s.

Command: **(setq test (list q r s))** Enter

Lisp returns: (nil nil nil)

In the next expression, unlike **list**, **quote** forms a list without evaluating its arguments. It returns the unevaluated symbols as a list.

Command: **(setq test (quote (q r s)))** Enter

Lisp returns: (Q R S)

Command: **(setq test '(q r s))** Enter   Uses the abbreviated form of the **quote** function

Lisp returns: (Q R S)

Command: **!test**

Lisp returns: (Q R S)

Command: **!'test**

Lisp returns: TEST

Putting the List Functions to Work

You've already used the **list, car, and cadr** functions to calculate the four corner points of a rectangle from two diagonal points. In the next exercise, you'll define a new function that *encapsulates* those same calculations so they can be performed more easily in a program. The function is defined in the file 09RECT1.LSP, shown following.

```
;; 09RECT1.LSP From Maximizing AutoCAD Release 13
;;
;; (rectang corner1 corner3)
;;
;; Draws a closed rectangular polyline whose diagonal
```

Introduction to AutoLISP Programming

```
;; corners are the point arguments corner1 and corner3.
;;
;; Returns the entity name of the polyline.

(defun rectang (cor1 cor3 / cor2 cor4)
   (setq cor2 (list (car cor3)(cadr cor1)))    ; X of cor3 and Y of cor1
   (setq cor4 (list (car cor1)(cadr cor3)))    ; X of cor1 and Y of cor3
   (setvar "CMDECHO" 0)                        ; turn off CMDECHO first
   (command "_.PLINE" cor1 cor2 cor3 cor4 "_close")  ; draw polyline
   (entlast)                                   ; return last entity
)
;; END 09RECT.LSP
```

The **rectang** function takes two arguments and has two bound variables. It uses **list**, **car**, and **cadr** to extract the X and Y components of the two arguments, and transposes them to form the other two corners of the rectangle. The **setvar** function is called to set the value of the CMDECHO system variable to 0, causing AutoCAD to suppress the display of input and prompts on the command line. Next, the **command** function is called and passed the name of the PLINE command ("_.PLINE") to start that command, followed by the four corner points of the rectangle and the *close* option keyword (a string), which closes the polyline.

The point assigned to **cor2** is composed from the X component of the second point argument and the Y component of the first argument. The point assigned to **cor4** is composed from the X component of the first argument and the Y component of the second argument. Once the polyline is drawn, the **entlast** function is then called to get the entity name of the last object created (the new polyline), which becomes the result of **rectang**.

In the following exercise, you'll test the **rectang function.**

 Exercise

Testing the RECTANG Function

Begin a new unnamed drawing, or continue in the drawing from the previous exercise and erase everything. Then, zoom Center on the point 10,10 with a height of 30 units.

Start LispPad and open the file 09RECT1.LSP

Press CTRL+L to load editor contents into AutoCAD

Command: **(setq p1 (list 3.0 2.0 0.0))** [Enter]   Assigns **p1** to point 3,2,0

Lisp returns: (3.0 2.0 0.0)

Command: **(setq p2 (list 12.0 7.0 0.0))** [Enter]   Assigns **p2** to point 12,7,0

Lisp returns: (12.0 7.0 0.0)

Command: **(rectang p1 p2)** [Enter]   Calls **rectang** function, passing the points **p1** and **p2** as arguments

Lisp returns: <Entity name: 2110540>   Entity name of new polyline

Command:

The user-defined function **rectang** encapsulates all of the steps that are needed to draw a rectangular polyline given two diagonal corner points. It calculates the two opposite corner points; calls the command function to invoke the PLINE command; supplies the four points; passes the "close" option keyword to close the polyline; and returns the entity name of the polyline it creates to its caller. This same function can now be used in a more complex program that prompts for the corners, which makes it possible to draw the rectangle by interactively picking two points.

Note: In the preceding exercise AutoCAD displayed the prompts for input even when it is being supplied to by the **command** function. When using **command** in an AutoLISP program, you can suppress these prompts by setting the CMDECHO system variable to 0, as demonstrated in the preceding exercise. The CMDECHO system variable tells AutoCAD it should not display prompts during AutoLISP evaluation.

Other List Functions

AutoLISP has other functions to manipulate lists; among them are **caar**, **caddr**, **cadar**, and **nth**.

```
caar, caddr...

(c????r list)
```

Returns an element or list from **list**, *specified by the combination of a and d characters in the* **c????r** *expression, up to four levels deep. For example,* **caadr, cddr, cadar**, *and so on.*

Introduction to AutoLISP Programming

> **nth**
>
> (nth *index list*)
>
> *Returns the element specified by the integer* **index** *from* **list**. *Returns* **nil** *if the integer position exceeds the list length. A value of Zero returns the first element in the list. The last element in the list is returned by a value equal to the length of the list minus one.*

Warning: Be careful with functions that accept integer arguments that represent indices or positions. In some functions these arguments are *zero-based*, meaning that the value 0 references the first position or element (like arrays in other languages), while other functions take similar arguments that are *one-based*, meaning that the value 1 references the first element.

The **c???r** functions can be confusing. The easiest way to understand them is to view them as simply shorthand for a series of nested calls to **car** and **cdr**. In fact, all of the **c???r** functions can be decomposed into a series of calls to **car** and **cdr** that have the same result. Table 9.9 illustrates this. The bold **a** and **d** characters empahsize the relationship between the simple expressions consisting of each **c???r** function, and their complex equivalents using multiple calls to only **car** and **cdr**. In the table, note how equivalent **c???r** functions can be formed by taking the second character (**a** or **d**) from each **car** and **cdr** in the complex counterpart expression.

Other list manipulation functions include **last**, **reverse**, **length**, **append**, and **cons**. **Last** gives you the last element of a list. **Reverse** flips the order of the list. **Length** returns the number of elements in the list. **Append** takes any number of arguments, each a list, and merges them into a single list. **Cons** adds a new first element to a list or forms a special construction called a *dotted pair*. For more information on using these and the other list functions, see this book's companion volume, *Maximizing AutoLISP for AutoCAD Release 13*.

Geometric Functions

AutoCAD's primary function is to allow you to create and manipulate geometry. AutoLISP contains a number of useful functions for performing various types of geometric

	C???R Expression	Equivalent using only CAR and CDR
Table 9.9 Understanding C???R Functions	(c**a**a**r** *list*)	(car (car *list*))
	(c**d**a**r** *list*)	(cdr (car *list*))
	(c**a**d**a**r *list*)	(car (cdr (car *list*)))
	(c**d**d**d**r *list*)	(cdr (cdr (cdr *list*)))
	(c**a**a**d**a**r** *list*)	(car (car (cdr (car *list*))))

calculations. These functions can operate on coordinates, angles, and distances, and compute new angles, distances, and coordinates.

polar

(polar *point angle distance***)**

Computes and returns a point in the current UCS, which is a specified distance and direction from **point**. **angle** *is the angle (in radians) from point to the resulting point relative to the positive X axis of the current UCS.* **distance** *is a real number that specifies the distance from pt to the resulting point, measured in the XY plane of the current UCS.*

distance

(distance *point1 point2***)**

Computes and returns the 3D distance between **point1** *and* **point2**.

angle

(angle *point1 point2***)**

Computes and returns the 2D angle (in the UCS XY plane) of a line from **point1** *to* **point2** *(in radians, relative to the positive X-axis of the current UCS).*

inters

(inters *pt1 pt2 pt3 pt4* [*onseg*]**)**

Computes and returns the intersection of a line from **pt1** *to* **pt2** *and a second line from* **pt3** *to* **pt4**. *If the optional* **onseg** *argument is present and* **nil**, *the lines are considered to have infinite length and the intersection does not have to be on either of the segments.*

osnap

(osnap *pt mode***)**

Computes a point using AutoCAD's object snap facility by applying object snap to the point **pt** *using the modes specified by* **mode**. *The point must be visible in the current view. The* **mode** *argument is a string that must contain one more object snap mode option keywords separated by commas.*

Introduction to AutoLISP Programming

Warning: The **osnap** function requires its point argument to be *visible* in the current view. If it is not, **osnap** will *not* function and will return **nil**. **Osnap** does not work with all object snap modes that are available at the command line; for example, it can't be used with the apparent intersection mode of object snap (which requires more than one point).

The **polar** function returns a point that is a specified distance and direction from another point, like polar coordinate entry format on the command line. **Distance** and **angle** return the distance and angle respectively between two points. **Inters** returns the point where two lines intersect, or would intersect if they were of infinite length. **Osnap** computes a new point by applying AutoCAD's object snap feature to a supplied point.

Tip: Passing a negative distance argument to the **polar** function reverses the angle to the resulting point, which has the same effect as adding −180 degrees to the angle argument.

The meas function shown next is from the file 09MEAS.LSP. It can be used like an object snap modifier to compute a point that lies on a selected line, a specified distance from the end nearest to the point where it's picked. It uses the **entsel** function to select the line. **Entsel** returns a list containing the line's entity name and the point where it was picked. **Meas** uses the **osnap** function to find the line's midpoint and the endpoint nearest to the pick point. It also uses the **angle** function to find the angle of the line from the two points returned by **osnap**. Finally, the **polar** function is called to compose a new point that lies on the line, a specified distance from the selected end. The **polar** function requires an angle, which is produced by calling the **angle** function to compute the angle of an imaginary line between the two points, using the point nearest the pick point as the base. The resulting point is returned to the calling function, or if **meas** is called from an AutoCAD prompt, the resulting point is accepted by AutoCAD as input. You can use this function at any AutoCAD prompt where a point is expected.

```
(defun meas ( / line dist pickpt endpt midpt)
   (setq line  (entsel " from end of: ")       ;; select line
         dist  (getdist "\nDistance from end: ") ;; enter distance
         pickpt (cadr line)                     ;; get selection point
         endpt (osnap pickpt "ENDPOINT")        ;; find endpoint
         midpt (osnap pickpt "MIDPOINT")        ;; find midpoint
   )
   (polar endpt (angle endpt midpt) dist)       ;; compute new point
)
```

In the following exercise, you'll test the **meas** function by using it to set the center point of a new circle to a point that lies on a selected line, exactly 3.125 units from one end. Make sure that when you identify the line, you pick it *near* the end you want to measure from.

Testing the MEAS Function

Begin a new unnamed drawing, or continue in the drawing from the previous exercise after erasing everything. Then, zoom Center on the point 8,4 with a height of 12 units.

Launch LispPad and open the file 09MEAS.LSP.

Press CTRL+L	Loads editor contents into AutoCAD

In the AutoCAD Window:

Command: **LINE** [Enter]	Starts the LINE command
From point: **2,1** [Enter]	Starts line at 2,1
To point: **12,8** [Enter]	Ends line at 12,8
To point: [Enter]	Exits the LINE command
Command: **CIRCLE** [Enter]	Starts the CIRCLE command
3P/2P/TTR/<Center point>: **(meas)** [Enter]	Calls **meas** function
from end of: *Pick line near the lower left endpoint*	
Distance from end: **3.125** [Enter]	3.125 units from end of line
Diameter/<Radius>: **2** [Enter]	Specifies circle radius of 2.0
Command: **DIST** [Enter]	Starts the DIST command
First point: **END** [Enter]	Uses ENDpoint object snap
of *Pick line near the lower left endpoint*	
Second point: **CEN** [Enter]	Uses CENter object snap
of *Pick the circle*	

Introduction to AutoLISP Programming

```
Distance = 3.1250,   Angle in XY Plane = 35,   Angle from XY Plane = 0
Delta X = 2.5601,   Delta Y = 1.7921,    Delta Z = 0.0000

Command:
```

Like **midpt**, the **meas** function is another tool you'll find useful in situations where you might otherwise have to draw temporary construction geometry, or change the current UCS to obtain similar results. It combines the functionality of object snap and UCS to express a coordinate in terms of an existing object without the need to temporarily switch to the object's coordinate system.

Executing AutoCAD Commands

One of the most important aspects of AutoLISP is its seamless integration with AutoCAD, which allows it to *take control* of the drawing editor by issuing standard AutoCAD commands, just as you do from the keyboard and menus.

The AutoLISP COMMAND Function

You've already used the **command** function in some of the preceding exercises. Now you'll learn a few more details about this very essential function. The **command** function is the basic mechanism through which AutoLISP programs supply command input to AutoCAD, to control it and automate a series of complex operations.

---

command

(command [*expr*] ...)

*Evaluates each **expr** argument, and sends the result to AutoCAD as command input. Has the exact same effect as entering the input into the command line.*

This function accepts any number of arguments, including none. Arguments passed to **command** must be synchronized with AutoCAD's current input state. AutoCAD must be waiting for the value(s) passed to it by this function. Consider the following example.

```
(command "_.CIRCLE" (list 4.0 7.25 0.0) 14.25)
```

When this example sequence is executed via **command**, no command can be currently active, and AutoCAD must be at the command prompt waiting for a command to be entered. If this is not the case when this expression is evaluated, an error will occur that halts execution of the program.

 Note: Throughout this book, you will see references to pressing the ESC key in AutoCAD, which cancels the current command. However, if you are using DOS AutoCAD or AutoCAD for Windows with *AutoCAD Classic keystroke mapping* enabled, then you should substitute CTRL+C or CTRL+[for ESC. The setting for *AutoCAD Classic keystroke mapping* is on the System page of the Preferences Dialog in AutoCAD for Windows.

Calling **command** with *no arguments* has the same effect as issuing ESC from the keyboard, which *cancels* the currently active AutoCAD command and returns AutoCAD to its quiescent state, where it displays the command prompt. Depending on what command(s) are currently active, you may need to call **command** several times with no arguments to return AutoCAD to the command prompt. Passing a null string (*""*) argument to **command** has the same effect as pressing ENTER at the keyboard. The predefined AutoLISP symbol **pause** (which is assigned to the string "\") can be passed to **command** to allow you to supply input interactively, anywhere within a sequence of arguments. The use of **pause** is equivalent to using a backslash in a menu macro to pause for input. The following examples illustrate the use of the **command** function with no arguments, **pause**, and the null string.

```
(command "_.LINE" "12,34" (list 13.25 7.125) ""
         "_.LINE" "28,7" "@2<45" ""
)
```

This example uses **command** to draw two lines. Because the two lines do not share a common endpoint, they cannot be drawn with a single instance of the LINE command. To draw the second line, the LINE command must first be stopped and restarted. The null string ("") argument *stops* the LINE command just as pressing ENTER from the keyboard does. The LINE command is then restarted; the second line is drawn; and another "" is passed to stop it again, which causes AutoCAD to return to the command prompt and wait for another command. Note that values can be passed both as strings and as numeric data (or lists containing numeric data). Since the **command** function is just another way of entering AutoCAD commands, literal strings can be passed in place of numeric values. For example, you can pass a coordinate value as a list of three real numbers; a string containing numeric digits separated by commas; or as a string in polar notation, like that used in the preceding example.

```
(command "_.CIRCLE" pause "_Diameter" "24.75")
```

This example uses **command** to draw a circle, but *pauses* at the prompt for the circle's center point, allowing the user to supply it interactively. As soon as the user specifies the center point, the argument following the **pause** symbol is then passed to AutoCAD (the "Diameter" option keyword), followed by the string "24.75", which is the circle's diameter.

One reason for using **pause** in **command** is because it allows the user to drag objects (for example, within the INSERT command, pausing at the `Insertion point:` prompt will allow the user to drag the block into place).

Tip: The **pause** symbol can be used with the SELECT commmand to pause until object selection is completed. When used at most `Select Objects:` prompts, the **pause** symbol only pauses for a single selection (such as one pick, window, crossing box, and so on), but in the SELECT command, the **pause** symbol will repeatedly pause until the user has accepted the selection set and exited object selection mode.

```
(command "_.INSERT" "OLDBLOCK=D:\\DRAWINGS\\NEWBLOCK.DWG")
(command)
(command "_.REGEN")
```

This last **command** example starts the INSERT command and uses the *block=filename* option to redefine or replace an existing block called "OLDBLOCK" with the drawing file called "NEWBLOCK.DWG". When this form of INSERT is used to redefine a block, there's usually no new insertion of the block placed in the drawing, and no need to continue to supply input to INSERT, so it must be canceled. If you were using INSERT interactively, you could cancel the command by issuing ESC. The second call to the **command** function with no arguments does the very same thing from AutoLISP, canceling INSERT at the `Insertion point:` prompt. This method of canceling can be used to stop any AutoCAD command.

In all of the preceding example calls to **command**, AutoCAD commands are prefixed with both a period and underscore, and option keywords, such as "_Diameter", are prefixed with an underscore. The use of the leading period ensures that AutoCAD uses the *built-in* version of the command rather than one that may have replaced it using *command redefinition*. Command redefinition is covered in detail in this book's companion volume, *Maximizing AutoLISP for AutoCAD Release 13*.

The leading underscore that prefixes both main commands and option keywords allows the program to run on any foreign-language version of AutoCAD including localized versions that don't recognize English command names, option keywords, **ssget** mode strings, and standard color names. If you use the leading underscore, *any language version* of AutoCAD will accept the English equivalents.

Note: Underscores and periods that prefix command names can appear in any order. For example; "_.CIRCLE" and "._CIRCLE" are both accepted, but some commands must be invoked with a leading dash (such as -GROUP). In their case, the dash must *immediately precede* the command name. Hence, you can use `"._-GROUP"` or `"_.-GROUP"`, but you *cannot* use `"_-.GROUP"` or `"-._GROUP"`.

As demonstrated in most AutoLISP code that is included with this book, we strongly recommend the use of the leading period on all command names passed to the command function. And, if there is the possibility that a program will need to run on non-English-language versions of AutoCAD, then you should also use the leading underscore on all command names, option keywords, standard color names, and **ssget** mode strings.

Tip: As of the C4 release of AutoCAD Release 13, AutoLISP programs that issue commands to AutoCAD via the **command** function without a leading underscore will result in a warning dialog indicating that the command is *globalization unfriendly*. You can suppress this warning by setting the GLOBCHECK system variable to 0.

Warning: The leading underscore can be prefixed to command names, option keywords, standard color names, and the mode argument to the **ssget** function. Do not prefix the names of AutoCAD system variables with a leading underscore in calls to **command**, **setvar**, and **getvar**, since the latter are not language-independent.

Each argument to **command** must be a *single* command entry (just as everything entered on the AutoCAD command line up to the first terminating space or ENTER is one *command entry*). You *cannot* supply multiple command entries in a single string, like this:

(command "_.CHPROP _Last _Color _RED ")

The correct form of this **command** call is:

(command "_.CHPROP" "Last" "_Color" "_RED" "")

The **command** function *cannot* be used to supply input to AutoCAD dialog boxes. If an AutoLISP program passes a sequence of arguments to **command** that causes a dialog box to appear, the program's execution is suspended while the dialog box is active, leaving the user free to interact with the dialog. When the dialog box is dismissed, the AutoLISP program will then resume execution.

Note: Some AutoCAD commands such as PLOT, OPEN, SAVE, and so on use both dialog box and command-line interfaces. The values of the CMDDIA and FILEDIA system variables control which interface is used when they are invoked from the command line. When these commands are invoked from the AutoLISP **command** function, they *always* use the command-line interface. You can pass a string containing the tilde (~) character to **command** to force them to use their dialog box interface.

You can use the **command** function to execute built-in or *native* AutoCAD commands, *external commands* defined in ACAD.PGP, and commands defined by an AutoCAD Runtime Extension (ARX) application. You *cannot* execute command aliases defined in ACAD.PGP or AutoLISP- and ADS-based commands. The **command** function cannot execute AutoCAD commands transparently. For example, the following will generate an error:

(command "._'ZOOM" "_Previous")

Table 9.10
Angular Units in Command Function

Angular Units Setting	Value of AUPREC	Required angular Units
Decimal degrees	0	Degrees
Degrees/minutes/seconds	1	Degrees
Grads	2	Grads
Radians	3	Radians
Surveyor's units	4	Degrees

You cannot call AutoLISP **getxxxx** input functions such as **getdist**, **getpoint**, and so forth within a call to the **command** function, and you should not nest calls to the **command** function within each other. The following two examples generate errors.

```
(command ".CIRCLE" "23,4" (getdist "\nEnter radius: "))
(command "_.LINE" "2,2" "4,4"
        (command "_.CIRCLE" "4,4" "1")
        "3,8"
)
```

The first example has a call to the **getdist** function within a call to the **command** function, which results in AutoCAD rejecting the call to the **getdist** function and any remaining arguments of the **command** function. The second example contains a call to the **command** function within another call to it, which is unnecessary, and error-prone because doing so gets confusing. In this case, the LINE command is not properly terminated before the CIRCLE command is issued, causing an error. Also, because the **command** function returns **nil** and **nil** in a **command** function is interpreted as a cancel, a nested **command** function will cancel any pending AutoCAD command.

When numeric values representing angles are passed to AutoCAD by **command**, the units they must be expressed in vary and are dependent on the *current* angular units settings (the value of the AUNITS system variable). Table 9.10 shows the relationship between the units of numeric values representing angles and all angular units setting.

Conditional Branching and Program Logic

One of the most powerful aspects of all computer programming languages is their ability to make *decisions*. By performing tests on data or by examining conditions at the time of execution, a program can decide to execute an expression or not, or can select one of several expressions to execute. *Branching* is the act of selecting and following one of several different sets of instructions or *paths of execution*, where the selection of an execution path is determined by the outcome of a *test*. Tests can involve examining and comparing values that are obtained from the user, an object's data, state of the drawing editor, data in a file, and a wide range of other sources.

Before you can understand how AutoLISP programs can make logical decisions, you must understand how they represent and interpret *logic states*. A logical state is one that represents the concept of *true* or *false*. In AutoLISP, all values represent one of these two states. The special value **nil** represents a logical state of false. *All other values* represent a logical of true. That is, if the result of an expression is **nil** then it can be interpreted logically as meaning *false*. Conversely, any other result would be interpreted logically as meaning *true*.

Note: Many programming languages use numeric values to represent logic (where the value 0 usually represents false and all other numeric values represent true). AutoLISP does not use numbers to represent logical states. In AutoLISP, the symbol **nil** is the *only* value that represents *false*. *All other values* (including 0) represent *true*.

Many AutoLISP functions that perform tests return the symbol **T** (which stands for TRUE). There's nothing special about this value, and it always evaluates to *itself*. This symbol is returned by many functions that perform tests to indicate that the test was successful. Hence, while **T** is used in that context to convey a logical state of *true, any value* other than **nil** could be returned in place of **T**, and it would have the very same effect. This is because conditional functions like **if**, **cond**, and **while** actually compare the result of the test expression to **nil**. If the result is anything other than **nil**, it is interpreted as a logical *true* state.

The symbol **T** is predefined by AutoLISP as follows:

(setq T 'T)

The preceding expression assigns the symbol **T** to *itself*.

AutoLISP has several branching and control functions that enable your program to perform different actions depending on the outcome of a test or the *logical* true/false state of a value. The **if** function is the most basic of these.

---

if

(if *testexpr* *thenexpr* [*elseexpr*])

Evaluates **testexpr**. *If the result is non-nil(true),* **thenexpr** *is evaluated and the result is returned by* **if**, *and* **elseexpr** *(if present) is ignored. If* **testexpr** *is* **nil** *(false) and* **elseexpr** *is present,* **elseexpr** *is evaluated and its result is returned by* **if**. *If* **testexpr** *is* **nil** *and* **elseexpr** *is not present,* **if** *returns* **nil**.

The following is a simple example use of **if**. It will print a message only if the value of Z is equal to 6.

```
(if (= Z 6)
    (print "\nZ is equal to 6")    ; do this only if z = 6
)
```

The first argument to **if** is the *relational operator* = (which tests for equality). The **=** function returns **T** (true) if its arguments are equal, and **nil** (false) otherwise. If the value of Z is equal to 6, **thenexpr** (the call to the **print** function) is evaluated, and **if** returns the result of that evaluation. If Z is not equal to 6, the call to the **print** expression is not evaluated at all as if it were not there, and **if** returns **nil**.

The next example is similar, except that it also includes the optional **elseexpr** and will print one of two possible messages, depending on the value of the variable X in relation to number 3.

```
(if (> x 3)
    (print "\nX is greater than 3")        ; do this if x > 3
    (print "\nX is not greater than 3")    ; otherwise do this
)
```

The first argument to **if** is a call to the relational operator **>** (greater than), which returns **T** if its first argument is greater than its second argument, and **nil** otherwise. When the value of **x** is greater than 3, the call to **>** returns **T**; **thenexpr** (the second argument to **if**) is evaluated; and its result is returned by **if**. When this happens, **elseexpr** is skipped and does not get evaluated, as though it were not present. If the value of **x** is not greater than 3, the call to **>** returns **nil**; **thenexpr** is skipped; **elseexpr** is evaluated; and its result is returned by **if**.

Branching Tests

AutoLISP provides a number of functions that perform a test and return a result indicating the outcome of the test, which are commonly referred to as predicates and relational functions. These functions accept one or more arguments and return a result that answers a specific question relating to the argument(s). Logical functions are ones that perform logical operations on the result of one or more tests to produce a single logical value. The = (equal) and > (greater than) functions are two relational functions that were used in the preceding examples.

Predicates and Relational Functions

The following list summarizes the basic predicate and relational functions AutoLISP provides to perform tests on one or a series of values. The first two functions that follow can be used to perform equality comparisons on any type of data, including AutoCAD-specific types such as entity names, selection sets, and file descriptors.

eq

(eq *expr1 expr2***)**

eq *returns* **T** *if* **expr1** *is equal to* **expr2**, *otherwise it returns* **nil**. *The two arguments can be of any data type, but this function does not compare the contents of lists, and is primarily intended to compare atoms (non-lists).* **eq** *considers the real number 0.0 and the integer 0 to be equal.*

equal

(equal *expr1 expr2* **[***fuzz***])**

equal *returns* **T** *if* **expr1** *is equal to* **expr2**, *otherwise it returns* **nil**. *The first two arguments can be of any data type.* **equal** *produces the same result as* **eq** *for all atoms. Unlike* **eq**, **equal** *compares the contents of lists for equality. The optional* **fuzz** *argument must be a real number, and is used when comparing two real numbers for equality. The* **fuzz** *argument specifies the maximum difference between two real numbers that are considered equal.*

The following functions accept any number of arguments. Unless specifically noted otherwise, they should be used with string and numeric data types only. Although some of these functions will accept other data types as arguments, their result is defined *only* for strings and numbers.

=

(= *expr* **[***expr***]...)**

= *returns* **T** *if all* **expr** *arguments are equal, otherwise it returns* **nil**. *This function will compare symbols.*

/=

(/= *expr* **[***expr***]...)**

/= *returns* **T** *if its* **expr** *arguments are not equal, otherwise it returns* **nil**. *This function will compare symbols..*

Introduction to AutoLISP Programming

<

(< expr [expr]...)

< (less than) returns **T** if each **expr** argument is less than the arguments that follow, otherwise it returns **nil**.

<=

(<= expr [expr]...)

<= (not greater than) returns **T** if each **expr** argument is less than or equal to the arguments that follow, otherwise it returns **nil**.

>

(> expr [expr]...)

> (greater than) returns **T** if each **expr** argument is greater than the arguments that follow, otherwise it returns **nil**.

>=

(>= expr [expr]...)

>= (not less than) returns **T** if each **expr** argument is greater than or equal to the arguments that follow it, otherwise it returns **nil**.

Table 9.11 illustrates the use of the relational functions. It assumes the following variable assignments:

```
(setq X '(A B C) Y 1.5 Z '(A B C))
```

The **equal** function can be used to compare the *contents* of two lists, so it is very useful for comparing two coordinates for equality. If, for example, a program needs to determine whether two coordinates are the same or *coincident*, it can use **equal**, similar to the way it's used in the following example:

```
(setq p1 (getpoint "\nEnter first point: "))
(setq p2 (getpoint "\nEnter second point: "))
(if (equal p1 p2)
    (print "\nThe two points cannot be identical.")
    ...
)
```

Table 9.11

Relational Function Examples

Example	Read as	Returns
(< 2 y)	2 is less than Y—false	**nil**
(> 2 y 3)	2 is greater than Y or 3—false	**nil**
(<= 1.5 y)	1.5 is less than or equal to Y	**T**
(>= 2 y)	2 is greater than or equal to Y	**T**
(= 1.5 y)	1.5 is equal to Y	**T**
(equal 1.5 y)	1.5 evaluates to same as Y	**T**
(eq z x)	Z is identical to X—false	**nil**
(equal z x)	Z's contents are equal to X's contents	**T**
(/= 2 y)	2 is not equal to y	**T**

You can also use **equal** to determine if two points are within a specified orthogonal range using the optional *fuzz* argument. The following call to the **equal** function will return **T** only if the orthogonal distances between the X, Y, and Z ordinates of the two point arguments are all within 0.01 units apart.

```
(setq p1 (getpoint "\nEnter first point: "))
(setq p2 (getpoint "\nEnter second point: "))
(if (equal p1 p2 0.01)
    (princ "\nPoints are too close!")
)
```

Tip: Another way to compare two coordinates for equality is to use the **distance** function to determine if the distance between them is 0, or less than some specified tolerance. The distance function is described in greater detail later in this chapter. The following simple example code determines if the distance between the two points `p1` and `p2` are less than or equal to `fuzz` drawing units, which is assigned the value 0.005.

```
(setq fuzz 0.005)
(if (>= fuzz (distance p1 p2))
    (princ "\nClose enough!")
    (princ "\nPoints must be within 0.005 units")
)
```

Logical Functions

Logical functions are ones that operate on values *logically*, meaning they see all values as either **nil** (false) or non-nil (true). Logical functions can be used to operate on the result of one or more predicate or relational tests and produce a combined or logical union of their results, which is also referred to as a *connective*.

For example, if you want to perform an action only when the outcome of two tests are *both* true, you can use the logical function **and**, to operate on the result of each test logically, to produce a result that indicates whether they are both true. Similarly, if you wanted to perform an action only if the outcome of either one of two tests is true, you can use the logical function **or**.

AutoLISP provides the following logical functions.

---

and

(and *expr*...)

Evaluates each **expr** argument until it encounters one that evaluates to **nil**, or all arguments have been evaluated. Returns **T** if all arguments evaluate to non-nil, otherwise it returns **nil**.

or

(or *expr*...)

Evaluates each **expr** argument until it encounters one that does not evaluate to **nil**, or all arguments have been evaluated. Returns **T** if no argument evaluates to **nil**, and **nil** otherwise.

not

(not *expr*)

Returns the logical complement to its argument. If **expr** evaluates to **nil**, not returns **T**, otherwise it returns **nil**. Put more plainly, **not** converts true to false, and false to true.

All of these functions return either **nil** or **T** (where the symbol **T** is only significant in that it is a non-nil value). The **and** function returns **T** only when *all* of its arguments are non-nil, and **nil** otherwise. The **or** function returns **T** if at least one argument is non-nil, and **nil** otherwise. The **not** function returns **T** when its argument is **nil**, and **nil** otherwise. Both **and** and **or** perform *conditional evaluation*, which means that they do not always evaluate all of their arguments to return a result. For example, since the **and** function indicates if all of its arguments are non-nil, it can stop evaluating them as soon it encounters the first non-nil argument, since at that point the result is known, and further evaluation is pointless.

The following example illustrates how logical functions do not always evaluate all of their arguments. The bold portions of the expressions represent the parts that are evaluated, and the non-bold portions represent the parts that are not evaluated.

```
(setq a "I'm true"
      b nil
      c 0
      d 'foo
      e nil
)
```

(and a b c d)	*returns:* **nil**
(or b e c d)	*returns:* **T**
(and 1 "hello" c e)	*returns:* **nil**
(not c)	*returns:* **nil**
(and a c d)	*returns:* **T**
(not e)	*returns:* **T**
(or e c a b)	*returns:* **nil**

Tip: The preceding example also contains a new form of the **setq** function. In all prior examples and exercises, **setq** is called with two arguments. However, **setq** can perform multiple assignments in a single call, as shown in the preceding examples, with multiple pairs of arguments. For clarity, it is best to show each assignment on a separate line.

The PROGN Function

The **if** function accepts only one expression to evaluate when the test expression is true and one optional expression to evaluate when the test expression is false. There will be times when you'll need to perform an entire series of operations when a test succeeds or fails, all of which may not be easily coded as a single expression. The built-in **progn** function is used to enclose an entire *list* of AutoLISP expressions within a single construct, so they can all appear in places where only a single expression is accepted (such as the **thenexpr** and **elseexpr** arguments to **if**).

---

progn

(progn [*expr*]...)

Accepts any number of **expr** arguments and evaluates each in the order supplied. Returns the result of evaluating the last argument.

Introduction to AutoLISP Programming

Progn is commonly used with **if** to evaluate multiple expressions when the **if** test succeeds or fails.

The following code listing is from the included file 09TSECT1.LSP. The **C:TUBESEC** command requests a center point, outer diameter, and offset distance; draws two concentric circles; and hatches them using the current hatching style. It uses **if**, **and**, and **progn** to determine if the required input has been supplied and if not, it doesn't draw the geometry. If the command did not test to see if the input has been supplied, then an error would occur when it tried to pass the non-existent data to AutoCAD with the **command** function. TUBESEC uses **entlast** to save the entity names of both circles drawn to the variables **ent1** and **ent2** so it can supply them to the HATCH command.

TUBESEC also demonstrates how to save and restore the value of an AutoCAD system variable, when it needs to be changed within the program. Whenever your program needs to change a system variable, it should always exit with the value set to the value it had when the program started. The simplest way to save and restore system variables is to first call **getvar** to save the current value of a system variable to a local AutoLISP program variable; then call the **setvar** function to change the system variable as required by the program; and prior to exiting the program, call **setvar** again, passing it the value of the local AutoLISP variable that was assigned the original value of the system variable when the program started. TUBESEC saves the value of the HIGHLIGHT system variable in the variable **oh**. Then it changes the value of HIGHLIGHT to 0, and before it exits, it restores HIGHLIGHT to the value it had when the program started. The HIGHLIGHT system variable controls whether objects are highlighted during object selection. Also note that TUBESEC and all other programs in this chapter do not save and restore the current value of the CMDECHO system variable. This is mainly because all AutoLISP programs should explicitly turn this variable off before passing input to AutoCAD.

```
;; 09TSECT1.LSP   From Maximizing AutoCAD Release 13

(defun C:TUBESEC ( / odia thk cen ent1 ent2 oh)
   (setq odia (getdist "\nOuter diameter: "))
   (setq thk  (getdist "\nWall thickness: "))
   (setq cen  (getpoint "\nTube center point: "))
   (if (and odia thk cen)
       (progn
          (setvar "CMDECHO" 0)
          (setq oh (getvar "HIGHLIGHT"))
          (command "_.CIRCLE" cen "_d" odia)
          (setq ent1 (entlast))
          (command "_.CIRCLE" cen (- (/ odia 2) thk))
          (setq ent2 (entlast))
          (command "_.HATCH" "" "" "" ent1 ent2 "")
          (setvar "HIGHLIGHT" oh)
       )
```

```
            (princ "\nInvalid response(s), exiting.")
    )
    (princ)
)
```

This code calls **getdist** and **getpoint** to obtain a center point and two distances. If the user responds with a SPACE or ENTER to any of the prompts these functions display rather than entering the requested data, the **getxxxx** function will return **nil**. Remembering that **nil** is the value that represents *false* in a conditional test (and any other value represents *true*), the **and** function is passed the result of each **getxxxx** function (which is returned by **setq**) to determine if any of the input values is equal to **nil**. If none of the arguments to **and** is **nil**, **and** returns **T** which triggers a "true" condition, causing **if** to evaluate its second argument (the ***thenexpr*** argument), which is the **progn** that contains all of the expressions that do the work. This demonstrates how you can have several expressions appear where only one is allowed (by placing them all inside of a **progn**). In the case of a **T** result from **and**, the expression that follows the **progn** (the ***elseexpr*** argument), is not evaluated, and the **if** function returns the result of the last call to **command** within the **progn** (because the result of **progn** is the result of its last expression, and the result of **if** is the result of whatever argument it evaluates). If one of the three requested values is not supplied or **nil**, **and** returns **nil** to **if**, in which case the call to **progn** is skipped, and the expression that follows it is evaluated, which displays a message telling the user they didn't enter all of the required input.

Now you'll test the **C:TUBESEC** function. You'll go through the exercise twice. The first time you'll supply all required values and the section will be drawn. Then erase the section and issue the TUBESEC command a second time, but this time *do not* supply all of the values (instead press ENTER at the prompt), and the program displays an error message and does not create the section. This exercise serves to demonstrate how a program can make a decision and take different actions based on the outcome, and how programs should save and restore settings they need to change.

 Exercise

Testing the TUBESEC Command

Begin a new drawing, or continue from the previous exercise and erase everything. Then zoom Center on point 10,10 with a height of 15 units.

Start LispPad and open the file 09TSECT1.LSP

Press CTRL+L to load the file into AutoCAD

Introduction to AutoLISP Programming

In AutoCAD:

`Command: ` **`TUBESEC`** `[Enter]`	Starts TUBESEC command
`Outer diameter: ` **`4`** `[Enter]`	Specifies outer diameter of 4.0
`Wall thickness: ` **`.25`** `[Enter]`	Specifies wall thickness of 0.25
`Tube center point: ` **`10,10`** `[Enter]`	Specifies center point of 10,10; TUBESEC draws and hatches the circles

Erase everything, and use TUBESEC again, this time supplying a *null* response to the wall thickness prompt.

`Command: ` **`TUBESEC`** `[Enter]`	Starts TUBESEC command again
`Outer diameter: ` **`8`** `[Enter]`	Specifies outer diameter of 8.0
`Wall thickness: ` `[Enter]`	Deliberately fails to specify value
`Tube center point: ` **`10,12`** `[Enter]`	Specifies center point of 10,12
`Invalid response(s), exiting.`	TUBESEC displays an error message and exits without drawing anything
`Command:`	

The TUBESEC command demonstrates the basic use of conditional branching and logical functions to control execution and take different actions based on conditions that are tested when it runs. Specifically, TUBESEC uses conditional branching to determine if the input it requested was supplied, and aborts its execution if it was not.

The file 09TSECT2.LSP from the book's CD-ROM contains a modified version of the TUBESEC command. In the exercise where you replied to the `Wall thickness` prompt with nothing but ENTER, the program continued to prompt for more input, even though it *could not* complete its task because you failed to supply required input. When a program continues to execute beyond the point where it can determine that it can't successfully complete its task, the program is said to contain a *dead end* in its logic flow, which is something that should be avoided at all costs.

Try repeating the last exercise again using the version of the TUBESEC command from the file 09TSECT2.LSP. When you reach the `Wall thickness:` prompt, supply a null response and note that the program stops *immediately*, rather than continuing to prompt for additional input. Unlike 09TSECT1.LSP, this version does not contain a dead-end in its logic flow, and it demonstrates the correct way to obtain and validate command line input. Compare 09TSECT1.LSP with 09TSECT2.LSP, and try to determine precisely how the latter terminates as soon as it determines that it can't reach its objective. The 09TSECT2.LSP version also divides the TUBESEC command into two functions, the C:TUBESEC command function which gets the input and calls the TUBESEC function, and the TUBESEC function which does the calculations and issues the commands.

Conditional Program Branching with COND

The **cond** function is a multiconditional construct that works similar to **if**, but it accepts any number of associated tests and consequent or "then" expressions to evaluate when the test expression they are associated with succeeds.

---

cond

`(cond (testexpr [thenexpr]...)...)`

Evaluates only the first element of each list argument (called a clause) in the order they appear, until a non-nil result is encountered. When a list whose first element evaluates to non-nil is encountered, the remaining elements in that same list are evaluated in the order they appear and all subsequent lists are ignored. **cond** *returns the result of the last expression evaluated. If a clause contains only one expression that evaluates to non-nil, its result is returned by* **cond**.

The **cond** function accepts any number of *lists* as arguments. Each of these lists is called a *clause*. The first element of each clause is a test that determines if the remaining elements in the clause are to be evaluated. If the test expression in a clause fails (returns **nil**), the remaining elements in the list are not evaluated. The first clause whose test expression succeeds (returns anything except **nil**), all of the remaining elements in that same list are evaluated, and the result of the last one is returned by **cond**. If the clause contains only one expression (the test expression), and it evaluates to a non-nil value, then its result is returned by **cond**. All clauses that follow the one whose test succeeds are ignored.

The following function is from the file 09COND.LSP from this chapter's files on the MaxAC CD-ROM. Its sole purpose is to demonstrate more clearly how **cond** operates. When the function **condtest** is called and passed an argument of **3.5** (which is assigned to the symbol X in the function's body), the expressions that are shown in **bold** type are the *only* ones evaluated, and **cond** returns the result of the *last* expression in **bold** type.

```
;; 09COND.LSP   From Maximizing AutoCAD Release 13

(defun condtest (x)
   (cond
       (    (> 1 x)
            (princ "\nX is less than 1,")
            (princ " that's all clause 1 knows.\n")
       )
       (    (> 2 x)
            (princ "\nX is less than 2,")
            (princ " and not less than 1,")
            (princ " that's all clause 2 knows.\n")
       )
       (    (> 3 x)
            (princ "\nX is less than 3,")
            (princ " and not less than 2,")
            (princ " that's all clause 3 knows.\n")
       )
       (    (> 4 x)
            (princ "\nX is less than 4,")
            (princ " and not less than 3,")
            (princ " that's all clause 4 knows.\n")
       )

       (    (> 5 x))

       (T (princ "\nX is not less than 5,")
          (princ " that's all clause 6 knows.\n")
       )
   )
)
```

In this **condtest** function, **cond** is passed 6 lists (called *clauses*). The first element of each list acts as a *test condition* that determines whether the remaining elements in the same list are evaluated or not. In the theoretical example where X is given a value of 3.5, **cond** evaluates the first element of the first three lists, all of which evaluate to **nil** (because X is not less than 1, 2, or 3). When the first element of the *fourth* list is evaluated, it returns **T** (because X is less than 4), and **cond** then evaluates the remaining three elements in the fourth list, returns the result of the last of those, and ignores the remaining lists. The first element in the *last* clause is the symbol **T**. Since this symbol evaluates to itself and is a non-nil value, it serves as a *default* or "else" clause because the expressions that follow the symbol **T** are guaranteed to evaluate if the tests in *all* preceding clauses fail. This is a common method of preventing the logic flow from "falling through" the entire construct, and provides a means of controlling what happens when the tests in all other clauses in the construct fail.

In the following exercise, you'll call the **condtest** function and pass it several values to observe the result. Note that when you pass **condtest** a value of 4, the result is somewhat different. Study the **condtest** function closely and try to determine the reason.

 Exercise

Testing the COND Function

Launch AutoCAD with a new, unnamed drawing. Launch LispPad, open the file 09COND.LSP, and press CTRL+L to load it into AutoCAD.

In AutoCAD:

Command: **(condtest 0)** [Enter]

X is less than 1, that's all clause 1 knows.
Lisp returns: " that's all clause 1 knows.\n"

Command: **(condtest 3)** [Enter]

X is less than 4, but not less than 3, that's all clause 4 knows.
Lisp returns: " that's all clause 4 knows.\n"

Command: **(condtest 5)** [Enter]

X is not less than 5, that's all clause 6 knows.
Lisp returns: " that's all clause 6 knows.\n"

Command: **(condtest 7)** [Enter]

X is not less than 5, that's all clause 6 knows.
Lisp returns: " that's all clause 6 knows.\n"

Command: **(condtest 4)** [Enter]

Lisp returns: T

Command: **(condtest 1)** [Enter]

X is less than 2, and not less than 1, that's all clause 2 knows.
Lisp returns: " that's all clause 2 knows.\n"

Command:

Introduction to AutoLISP Programming

One important characteristic of **cond** is that it will only evaluate the expressions in the *first* clause whose test expression evaluates to non-nil. If more than one clause has a test expression that would evaluate to non-nil, only the first encountered will be evaluated, since **cond** always ignores any clauses that follow the one whose test succeeds.

```
(cond
   (   (> 2 z)
       (princ "\nZ is < 2")
   )
   (   (> 3 z)
       (princ "\nZ is < 3")
   )
   (   (> 1 z)
       (princ "\nZ is less than 1, ")
       (princ "but I won't tell anyone.")
   )
   (   (> 4 z)
       (princ  "\nZ is < 4)
   )
   (t (princ "Z must be 4 or more!"))
)
```

Study this expression closely, and try to determine why the expressions in the third clause in the call to **cond** (shown in **bold** type) will *never* be evaluated, no matter what value is assigned to z.

A common application of the **cond** function is to validate user input and stop a program when it does not have the required data to continue. The function in the following listing is from the file 09RECT2.LSP. The **C:RECT** command function serves as an *interface* for the **rectang** function you developed earlier in this chapter. **C:RECT** adds the RECT command to AutoCAD, allowing you to draw a simple rectangle by specifying two diagonal corner points.

```
;; 09RECT2.LSP From Maximizing AutoCAD Release 13
;;
;; C:RECTANG command.
;;
;; Requests two corner points and draws a rectangle
;; thru them.
;;
;; Uses the rectang function from 09RECT1.LSP to
;; draw the rectangle.

(defun C:RECT ( / pt1 pt2)
   (cond
```

```
      (   (not (setq pt1 (getpoint "\nFirst corner: ")))
          (princ "\nInvalid, must enter a point.")
      )
      (   (not (setq pt2 (getcorner pt1 "\nOther corner: ")))
          (princ "\nInvalid, must enter a point.")
      )
      (t (rectang pt1 pt2))
   )
   (princ)
)
```

In the **C:RECT** function, the **cond** has three clauses. The first two will trigger only if the result of either call to the **getpoint** contained in them is **nil** (meaning a point wasn't supplied in response to the **getpoint** prompt). If either of the first two clauses is triggered, the second expression in the triggered clause is evaluated (which are both calls to **princ** that display an error message), and the third clause, which draws the rectangle, is *not* evaluated.

The third clause has the symbol **T** as its test expression, causing it to be triggered if no preceding clauses succeed. The use of **not** in the first two clauses has the effect of inverting the logical values of the result returned by the **getpoint** calls, because the intent is to *trigger* the clause when the result of either **getpoint** is **nil**, rather than when it is non-nil (remember that a test expression in a clause is triggered when its first expression evaluates to a non-nil value). The general idea behind using **cond** in this manner is to validate user input as soon as it is entered, and to exit or abort the command immediately, as soon as a required value is not supplied or invalid.

Program Looping and Control Structures

In addition to branching, AutoLISP programs can also perform the same instruction (or series of instructions) repeatedly, any number of times. This is called *iteration*. A familiar example of iteration can be seen in most AutoCAD commands that allow you to select and edit multiple objects (such as MOVE, COPY, or ROTATE). In those commands, once you've entered all of the required parameters, AutoCAD begins to process the selected objects and modifies them one at a time until all of the objects have been processed. To do that, it uses the same set of instructions to modify each object, executing them repeatedly within a *loop,* or *processing* loop. Another example of iteration in AutoCAD is the LINE command, which repeatedly displays a To point: prompt, which allows you to draw any number of line segments.

AutoLISP provides a number of *iterative* process control constructs, which allow a program to perform a process repeatedly. A *process* is any set of instructions that are always executed together as if they were a single unit (or function). An *iteration* is the term that describes a *single* execution of a process that is performed repeatedly. So, *one iteration* means one trip through the loop.

Introduction to AutoLISP Programming

There are two basic types of iteration, *determinate* and *indeterminate*. In determinate iteration, the number of iterations or repetitions that are performed is known *before* the iterative process or looping begins. It is preferable to use determinate iteration when possible, since there's no need to perform a test on each iteration to determine if the process must be repeated again. With indeterminate iteration, the number of iterations cannot be determined before looping begins, which means that a test must be performed within each iteration to determine if another iteration should be performed.

For those familiar with BASIC, the FOR construct performs determinate iteration. Conversely, the BASIC WHILE construct performs indeterminate iteration.

AutoLISP provides several functions that control iteration.

The REPEAT Function

```
repeat

(repeat number [expr]...)
```
Evaluates each **expr** *the number of times specified by the integer* **number**.

The **repeat** function serves as AutoLISP's basic determinate looping construct. It accepts any number of **expr** arguments and repeatedly evaluates them in the order supplied. It does this a specific number of times, which is specified by its first argument. **repeat** returns the value of the *last* evaluation of the *last* **expr** argument. In the next exercise, you will use **repeat** to evaluate expressions multiple times.

Exercise

Using a REPEAT Loop in AutoLISP

Launch AutoCAD with a new, unnamed drawing.

```
Command: (repeat 5                          Evaluates the expression 5 times
(princ "\nDo some stuff"))  Enter

Do some stuff
Do some stuff
Do some stuff
Do some stuff
Do some stuff
```
Lisp returns: `"\nDo some stuff"`   Result of last evaluation

```
Command: (setq i 0) [Enter]              Result of last evaluation

Lisp returns: 0

Command: (repeat 6 (print                Prints the numerals 1 through 6
(setq i (1+ i)))) [Enter]

1
2
3
4
5
6
Lisp returns: 6                          Result of last evaluation

Command:
```

The file 09HELIX.LSP on the MaxAC CD-ROM contains the HELIX command, which relies on **repeat** to draw a flat helical coil or spring using either a polygon mesh or a 3D polyline. You can study this file and use it for generating helical forms as well. Figure 9.5 shows several objects created by the HELIX command.

The WHILE Function

while

(while test [expr]...)

while *first evaluates* **test**. *If the result is non-nil, each* **expr** *argument in the order supplied is evaluated. After the last* **expr** *argument is evaluated, the entire process is repeated until* **test** *evaluates to* **nil**. **while** *returns the result of the last evaluation of the last* **expr** *argument, or* **nil** *if there are no* **expr** *arguments.*

The **while** function serves as AutoLISP's basic indeterminate looping construct. It accepts any number of **expr** arguments (the body) and repeatedly evaluates them in the order supplied until the **test** expression evaluates to **nil**. The **test** expression is always evaluated *before* the body expressions, which means that it is possible that the body expressions will never be evaluated (in the case where the **test** expression returns **nil** the first time it is evaluated). In any case, the number of times the **test** expression is evaluated is *always* one greater than the number of times each body expression is evaluated.

Introduction to AutoLISP Programming

Figure 9.5
Objects created by the HELIX command from 09HELIX.LSP.

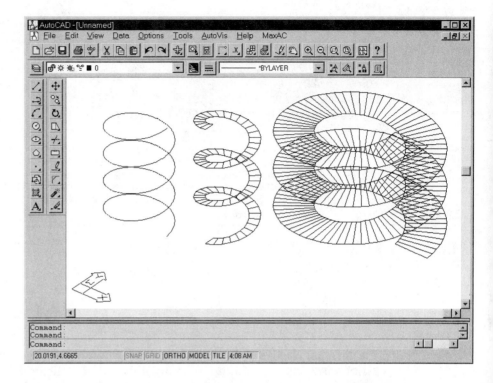

 Warning: The **while** function evaluates its body until the test expression evaluates to **nil**. There's no other way to exit a **while** loop (except through a program error, which will also terminate the program that contains the loop). When using **while**, you *must* ensure that at some point, the test expression will evaluate to **nil** or **while** will continue to loop infinitely. If this happens, your only option is to issue ESC or CTRL+C to cancel the program entirely.

The following program is from the file 09MOFF.LSP, which is included on the book's CD-ROM. It implements the **C:MOFFSET** command. It will offset an object through any number of points you pick in succession. It works similar to AutoCAD's OFFSET command, except that it repeatedly offsets the same object through different points, and does not require you to repeatedly pick the object to offset. MOFFSET is useful for creating offset curves that simulate curved multiline segments.

MOFFSET uses **while** to perform iteration. The test condition for **while** is the result of the call to **getpoint** (which is returned to **while** via **setq**). Just like **if**, the **while** function sees all non-nil values as *true*, so any value that is returned by **getpoint** other than **nil** will cause **while** to continue to iterate. When the response to the prompt from **getpoint** is ENTER or SPACE, **getpoint** returns **nil** to **while**, which causes

425

it to stop iterating. In the following code listing, the expressions that execute multiple times are shown in **bold** type.

```
;; 09MOFF.LSP from Maximizing AutoCAD Release 13

(defun C:MOFFSET ( / object pt)
  (setq object (entsel "\nSelect object to offset: "))
  (if object
    (progn
      (setvar "CMDECHO" 0)
      (command "_.OFFSET" "_T")                     ; start OFFSET command
      (while (setq pt (getpoint "\nThrough point: ")) ; while user
                                                     ;enters points,
        (command object pt)                          ; pass them to AutoCAD
      )
      (command "")                                   ; stop OFFSET command
    )
    (princ "\nInvalid, nothing selected.")
  )
  (princ)
)
```

The diagram in Figure 9.6 illustrates the relationship between the logic flow of the **C:MOFFSET** function and the structure of its body. It is helpful to sketch the logic flow and structure of major program elements in a similar fashion when developing programs in their conceptual stages, prior to coding.

The MOFFSET command illustrates one additional technique related to the use of the **while** function. Here is another way of writing the iterative part of MOFFSET, which demonstrates the improper use of **while**.

```
(setq pt (getpoint "\nThrough point: "))  ; Wrong! Do not initialize
(while pt                                  ; the test expression prior to
  (command object pt)                      ; entering the while loop.
  (setq pt (getpoint "\nThrough point: "))
)
```

When writing iterative constructs, you will be tempted to initialize the variable that is used to control the iteration (the first argument to **while**), before you enter the loop and pass just that variable as the first argument to **while** as shown in the preceding example. Don't do it! It is bad technique, because it requires the control variable to be initialized again within the body of the **while** loop, each time the loop executes. It is better to nest the initialization expression inside a call to **setq**, and pass that as the first argument to **while**, which works because **setq** returns the value that was assigned. The correct way to write this iterative construct is shown in the MOFFSET command.

Introduction to AutoLISP Programming

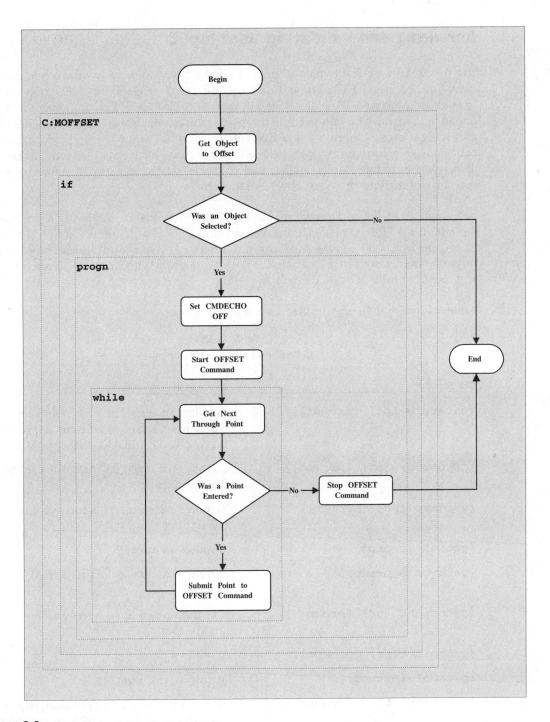

Figure 9.6
MOFFSET command program logic flow.

Acquiring and Validating User Input

You've seen and used a number of functions that *get* user input, most of which begin with (you guessed it) "GET". These functions are often referred to collectively as the "**get***xxxx*" functions. While the names of most AutoLISP input functions begin with get, not all of them do. The **entsel** function, for example, which you've seen used in several preceding exercises, doesn't follow that rule. There is an input function for each of the basic data types, including strings, reals, and integers, as well as ones for AutoCAD-specific data types, including points, angles, and distances. There are also a number of more specialized functions that we'll cover briefly here.

Just about all of the **get***xxxx* functions accept arguments for displaying a prompt on the command line. Some functions also accept optional arguments for things such as basepoints which allow *rubber-banding* input, where a "rubber-band" *displacement vector* or rectangle is anchored to a basepoint and whose opposite end or corner is anchored to and moves with the crosshair cursor.

Table 9.12 lists the basic AutoLISP user input functions.

Table 9.12
Command Line Input Functions

GET Functions	Explanation
`(getangle [pt] [prompt])`	Acquires an angle from two points or typed input, relative to the current settings for ANGBASE and ANGDIR
`(getcorner [pt] [prompt])`	Acquires the diagonal corner of a rubber-banded rectangle
`(getdist [pt] [prompt])`	Acquires a distance from two points or typed input
`(getint [prompt])`	Acquires an integer
`(getkword [prompt])`	Acquires one of a list of predefined keywords
`(getorient [pt] [prompt])`	Like **getangle** but ignores non-East base angle setting
`(getpoint [pt] [prompt])`	Acquires a point
`(getreal [prompt])`	Acquires a real number
`(getstring flag [prompt])`	Acquires text or a string (text can contain spaces). If FLAG is non-nil it accepts text, otherwise the first space terminates input.

Introduction to AutoLISP Programming

The optional **[pt]** argument for all the functions shown in Table 9.12 except **getcorner** is a point that is used as the *basepoint* or anchor for a dynamic rubber-band displacement vector. One end of the vector is anchored to the **pt** argument and the other is attached to and moves with the cursor. As you've already seen earlier in this chapter, the **pt** argument for **getcorner** is a point that is anchored to one corner of a rubber-band rectangle or *window* that is aligned with the display, and whose other corner is attached to and moves with the cursor.

The optional **[prompt]** argument for all functions is a string which is displayed on the command line, and used to inform the user that input is required. You've seen both of these arguments used extensively in this chapter.

INITGET Input Control

A major cause of errors in macros is unexpected or erroneous input. To control and restrict input, AutoLISP's **initget** function provides simple validation and filtering for most input functions, enabling you to specify what types of input are allowed or disallowed for their function calls. You do this by calling **initget** just prior to calling the input function which it is to control. For instance, you can tell **initget** to reject negative values, zero, or a null response, and the input functions that it controls will automatically display a generic error message and retry input until a suitable response is obtained. You can use **initget** in virtually all expressions that get user input, to eliminate the need to manually check and reject invalid input in an application. Unfortunately, **initget** isn't sophisticated enough to perform numeric or geometric range checking, which still must be done by other code in your programs.

The **initget** function also gives most input functions the ability to accept and return *option keywords* as well as *arbitrary input* in lieu of the primary type of input which each of those functions acquire.

---

initget

(initget *bits* [*string*])

Establishes options for **getxxxx** *and other input functions. The bits argument sets input filtering options, and the* **string** *argument sets keywords. Always returns* **nil**.

The **initget** function's control is initialized by passing integer bits (numbers) to the function. The bits and their resulting controls are shown in Table 9.13.

These individual **initget** control bits can be combined together using the + or the **logior** function to selectively enable one or more control states. For example, if you wanted a subsequent call to **getint** to reject negative, zero, and null input, you would

Table 9.13
INITGET Control Bits

INITGET Control Bit	Resulting Control Over Input
1	Not null. A value must be given.
2	Not zero. Numeric input cannot be zero.
4	Not negative. Input must not be less than zero.
8	No limits checking of points, regardless of LIMCHECK system variable.
16	(Not Used)
32	Shows a dashed rubber-band line.
64	2D distance. Ignores Z coordinate of 3D points with GETDIST.
128	Permits arbitrary keyboard input.

supply the value 7 to **initget** (which is the sum or *logical or* of the three control states to be enabled).

INITGET has a new localized and language-independent keyword feature, introduced in AutoCAD Release 13c2. The AutoLISP **initget** function can provide localized keywords. The following syntax for the keyword string allows input of the localized keyword while it returns the "language-independent" keyword:

```
"local1 local2 localn _indep1 indep2 indepn"
```

In this syntax, `local1` through `localn` are localized keywords, and `_indep1` through `indepn` are language-independent keywords. There must *always* be the same number of localized keywords as language-independent keywords, and the first language-independent keyword *must* be prefixed by an underscore.

For example:

```
(initget "Abc Def _Ghi Jkl")
(getkword "\nEnter an option (Abc/Def): ")
```

Entering "A" returns "Ghi" and entering "_J" returns "Jkl".

INITGET Bit Modes and Keywords

You must call **initget** before calling the input function that you want to control. **initget's** settings are not saved after the input function is evaluated, and they only apply to the next call to an input function. **initget** takes two arguments, an integer value between 1 and 255 and an optional string. The integer value establishes the type

Introduction to AutoLISP Programming

Table 9.14
Initget Flags and Keyword Support by Input Function

Input Function	Supports Option Keywords	No Null 1	No Zero 2	No Negative 4	No Limits 8	Uses Dashes 32	2D Distance 64	Arbitrary Input 128
getint	✓	✓	✓	✓				✓
getreal	✓	✓	✓	✓				✓
getdist	✓	✓	✓	✓		✓	✓	✓
getangle	✓	✓	✓			✓		✓
getorient	✓	✓	✓			✓		✓
getpoint	✓	✓			✓	✓		✓
getcorner	✓	✓			✓	✓		✓
getkword	✓	✓						✓

of input or point controls for the GET call. Table 9.14 shows which input functions are affected by which control bits.

In addition, the three entity-selection functions (**entsel**, **nentsel**, and **nentselp**) recognize option keywords, but not control bits. The **getstring** function is not affected by **initget** control bits and keywords.

Many AutoCAD commands accept option keywords. For example, the CIRCLE command first prompts for the center point and then with Diameter/<Radius>: to which you can enter **D** (for diameter) or a distance (the radius). Whether you enter **D**, **DIA**, **Dia**, or **Diameter**, AutoCAD interprets all of these as the Diameter option, and then prompts you for the circle diameter. **Initget** makes it possible for you to do the same in AutoLISP programs when they request input.

Warning: Avoid the use of option keywords that are also used by AutoCAD for things such as object snap modes or similar functions that are available at most input prompts. For example, if you initialize the option keyword "Middle" for use with **getpoint**, it will be accepted if you typed **M**, **MI**, or **MIDD**, but not **MID**, because the latter activates MIDPOINT object snap, which has precedence over **initget** keywords.

The optional string argument of **initget** passes a string of keywords to any of the **get**xxxx functions, with the exception of **getstring**. Keywords in an **initget** string argument permit the user to enter these words instead of the primary type of data that each particular **getxxxx** function acquires.

You can use two formats for specifying the keyword string in **initget**. In the first format, uppercase characters represent the minimum string that can be used to specify the keyword. For example, with a keyword string of **"Key WOrd LIST"**, entering **K**, **KE**, or **KEY** returns Key; entering **WO**, **WOR**, or **WORD** all return WOrd (making the first two characters in WOrd in the keyword list uppercase means that entering W alone is not enough); and LIST must be entered in its entirety. This format is the more commonly used and the easiest one. Keywords are always returned in the same case they are shown in the INITGET argument list, so entering **K**, **k**, **key**, or **KEY** in the preceding example returns Key, meaning that any subsequent comparison that is done must be case-sensitive.

In the second format, the full keyword must be capitalized, followed by a comma and the required portion. For example, **"KEY,K WORD,WO LIST"** is equivalent to **"Key WOrd LIST"** shown earlier, except only uppercase is returned by the second format. (The value returned by the first format can be forced to uppercase by the **strcase** function.)

initget is often used with the **getkword** function. **getkword** restricts input to the keyword list only. **Getkword** is covered in this book's companion volume, *Maximizing AutoLISP for AutoCAD R13*.

Try a few more **get** functions and an **initget**.

Exercise

Using GET Functions for Input

Continue from the previous exercise, or begin a new, unnamed drawing.

Command: **(getangle "Enter angle: ")** [Enter]   Prompts for input

Enter angle: **30** [Enter]

Lisp returns: 0.523599   Returns angle in radians, not degrees

In 360 degrees, there are $2 \times \pi$ radians. One degree = $180/\pi$. The AutoLISP symbol PI, used in the following is predefined to be equal to π.

Command: **(* (getangle "Enter angle: ")**   Converts input angle to
(/ 180 pi)) [Enter]   radians

Introduction to AutoLISP Programming

```
Enter angle: 30 [Enter]
```

Lisp returns: 30.0

```
Command: (setq pt1
(getpoint "Enter point: ")) [Enter]
```
Saves point

```
Enter point: Pick a point
```

Lisp returns: (2.0 2.0 0.0)

```
Command: (getangle pt1 "Enter angle: ")
[Enter]
```
Uses PT1 as base for rubber-banding

```
Enter angle: Pick a point at a 45-degree angle
```

Lisp returns: 0.785398

```
Command: (getangle pt1 "Enter angle: ")
[Enter]
```
Rubber-bands angle

```
Enter angle: 30 [Enter]
```

Lisp returns: 0.523599

```
Command: (initget 2) [Enter]
```
Disallows 0 input

Lisp returns: nil

```
Command: (getangle pt1 "Enter angle: ")
[Enter]
```

```
Enter angle: 0 [Enter]
```
```
Value must be nonzero.
```
Rejects and reprompts

```
Enter angle: 30 [Enter]
```
Accepts nonzero value

Lisp returns: 0.523599

```
Command: (getstring "Enter a word: ")
[Enter]
```

```
Enter word: Type This and press SPACE
```
Enters "This" when you press SPACE

Lisp returns: `"This"`

Command: **(getstring T "Enter words: ")** [Enter]   Allows spaces in input due to **T** flag

Enter sentence: **This is a sentence.** [Enter]   Enters input when you press Enter

Lisp returns: `"This is a sentence."`

Command: **(getcorner pt1 "Enter other corner: ")** [Enter]   Uses base point PT1 to rubber-band a rectangle

Enter other corner: *Pick a point*

Lisp returns: (5.4 4.8 0.0)

In these examples, the input automatically assumes the character of the data type requested. Responses that are not the requested data type are rejected. **getstring** accepts numbers, but converts them to string data, and **getreal** accepts integers, but converts them to floating-point reals.

Note: AutoLISP returns and internally uses angles in radians, not degrees. There are 2 × PI radians in 360 degrees. One degree = π/180.

Do not use **getreal** to obtain distance or angle input. It does not recognize all units formats, nor does it permit graphical input (picking two points to define a distance or angle). If you use **getdist** or **getangle** instead of **getreal**, you can enter a distance or angle by picking points with optional rubber-banding, or by typing values in any current linear or angular units format.

The file 09ARRAY.LSP from the companion CD-ROM contains the example HARRAY command, which produces helical arrays of objects. The HARRAY command uses **initget** flags and option keywords to provide a more flexible means of defining the parameters for the array.

The **getdist**, **getangle**, **getorient**, and **getpoint** functions can use an optional base point argument. **getcorner** must have a base point. The **getdist**, **getangle**, **getorient**, and **getpoint** functions rubber-band input when a base point is set or accept two points as input.

Introduction to AutoLISP Programming

Tip: If AutoCAD is requesting a point, distance, or angle, and displays a rubber-band cursor anchored to a reference point, or is dragging objects, you can use **getpoint**, **getdist** or **getangle** to move the point to which the rubber-band cursor is anchored before indicating the distance or angle, or disable object dragging, by entering using one of the following menu macros. The first macro disables the rubber-band displacement vector and dragging of objects for all forms of input. The second and third macros allow you to pick a new basepoint for 2-point distance and angle input respectively.

[No Drag/Basepoint] **(getpoint)**

[New Distance Basepoint] **(getdist)**

[New Angle Basepoint] **(* (getangle) (/ 180 pi))**

These sequences work well on the buttons or any other menu device; however, you cannot use them in response to input prompts from AutoLISP or ADS programs.

Selecting Objects

The **entsel** (ENTity SELection) function enables a program to get the entity name of a single object from the user by requiring them to select it from the display using the *pickbox*. This method of input is identical to that used by AutoCAD commands such as BREAK, FILLET, and TRIM. In addition to returning the entity name of the object that was picked, **entsel** also returns the point where the pickbox was located when the user pressed the pick button to select the object. **Entsel** returns both the entity name and the selection point in a list. You've seen **entsel** used in a number of examples earlier in this chapter.

---

entsel

(entsel [*prompt*])

Returns a list containing the entity name and the selection point used to select the object. The **prompt** *is optional; if not supplied, the standard object selection prompt* Select object: *is used.*

Many AutoCAD commands will accept entity names when they display a pickbox to allow an object to be selected visually. There are cases where both an entity name and a selection point are required. An example of the latter is the TRIM command, where the selection point is required because that is what determines what part(s) of the selected object is trimmed. In cases where a pick point is also required, the list returned by **entsel**, which contains both the entity name and the selection point is accepted as is, by AutoCAD commands.

When a selection point is not required or you are only interested in the entity name (the first element of the list returned by **entsel**), you can use the **car** function to retrieve the selected object's entity name from the list, like this:

(setq ename (car (entsel "\nPick something: ")))

Selection sets are nothing more than groups of entities. Most of AutoCAD's editing commands operate on selection sets. When an AutoCAD command displays the Select objects: prompt, you are constructing a selection set. AutoLISP's **ssget** function provides a way for your programs to use AutoCAD's object selection facility to obtain a group of objects from the user.

---

ssget

(ssget [*mode*] [*point1* [*point2*]] [*pointlist*] [*filterlist*])

Returns a selection set of entities. If **ssget** *fails, or the user does not select at least one object in response to the* Select objects: *prompt, it returns* **nil**. *All arguments are optional, but the appearance and value of the* **mode** *argument makes other arguments required, as explained following. The* **mode** *argument is a string that specifies options for non-interactive selection. With no arguments (or* **nil** *arguments),* **ssget** *uses AutoCAD's standard object selection to get a user selection The* **point1** *argument can appear by itself, to specify a single point which selects any object that passes through or near it, to within the current size of the pickbox. The* **filterlist** *argument specifies an object selection filter list that performs filtering of the resulting selection based on the criteria in the list.*

The **mode** argument must be a string specifying the type of non-interactive object selection to be performed. Table 9.15 describes each of these modes, their associated argument string, and the other arguments that are required for each possible value of the **mode** argument.

The **pointlist** argument is a list of points describing the vertices of a polygon or fence, which is used with the CP, WP, and F modes. **point1** and **point2** are points. When **point1** is used alone, it describes a single "pick point." The "C" and "W" modes both require **point1** and **point2**, which describe the diagonal corners of a rectangle.

Warning: The "L" (Last) mode and all other modes that accept point arguments are *view-dependent,* meaning they can select only objects that are entirely or partially *visible* in the current view. Objects that are not visible in the current view can only be selected using the "X" and "P" modes.

Introduction to AutoLISP Programming

Table 9.15
SSGET Modes

Mode	Returns	Additional Arguments Required
"P"	*Previous* selection set	None
"L"	*Last* visible object created	None
"I"	*Current* selection set	None
"X"	*All* objects	None
"W"	All objects within a *Window*	`point1, point2`
"C"	All objects within or *Crossing* a window	`point1, point2`
"WP"	All objects *Within* a *Polygon*	`pointlist`
"CP"	All objects within or *Crossing* a *Polygon*	`pointlist`
"F"	All objects crossing a *Fence*	`pointlist`

Formatting Input Prompts

When AutoCAD prompts for input on the command line, it frequently displays a *default* value that you can accept by pressing ENTER or SPACE. You can and should do the same in your own programs. Default values are displayed between pairs of angle braces (<>) at the end of a prompt string. The following example shows how to format a prompt string to show a default value:

```
(setq dist 4.0)
(setq msg (strcat "\nEnter length <" (rtos dist) ">: "))
(setq input (getdist msg))
```

With the current units and precision set to 2, the preceding code fragment produces the following prompt on the command line:

```
Enter length <4.00>:
```

The **rtos** function is used to convert the value of **dist** (a real) to a distance string formatted in accordance with the current unit and precision settings. To format the default value for an angle, you would use **angtos** instead of **rtos**.

Following the same convention used by AutoCAD's commands, when input prompts contain values that are displayed between angle braces, the displayed value can be accepted by just pressing SPACE or ENTER. When using defaults, you must determine if the response was SPACE or ENTER by testing the result of the **get***xxxx* input function with **if** or **cond**. It is generally more convenient to use **cond** for this. The following code fragment expands on the preceding one by adding the logic to test the result of the input function and substitute the default value for a null response.

```
(setq dist 4.0)
(setq msg (strcat "\nEnter length <" (rtos dist) ">: "))
(setq input (cond ((getdist msg)) (t dist)))
```

The value 4.0 is assigned to **dist**, which is the variable that holds the default value displayed between the angle braces in the input prompt. The call to **cond** returns the result of the call to **getdist** *only* if it is non-nil (in which case it is the distance that was entered). If the result of **getdist** is **nil** (meaning the user pressed SPACE or ENTER), then **cond** evaluates its second argument, which returns the value of **dist**. In any case, the value that is assigned to **input** is either the non-nil result of **getdist** or the value of **dist**.

It is often necessary to use default values when getting input. Therefore, it is better to define a function that encapsulates these operations, and call that instead of **getdist** (or another input function). Doing this will reduce the complexity of your code. The following is an example of a user-defined input function that can be called in place of **getangle** which formats the prompt with a default value and returns either the user-entered value, or the default value that was displayed in the prompt.

```
(defun mxa_getangle (msg default)
   (cond
      (  (getangle (strcat msg " <" (angtos default) ">: ")))
      (t default)
   )
)
```

Note that **mxa_getangle** uses **getangle** and **angtos** rather than **getdist** and **rtos** because it is used to get angle input in lieu of **getangle**. The file 09INPUT.LSP on the book's accompanying CD-ROM contains **mxa_getangle** along with other functions for getting different types of input with formatted prompts and handling of defaults. You can call these functions in your programs in lieu of their **get*xxxx*** counterparts. When using these functions, you cannot use the bit-1 flag of **initget**, because the functions accept a null response to signal the acceptance of the default value. The functions are described in Table 9.16.

The file 09RECT3.LSP uses many of the concepts introduced in the latter parts of this chapter, including the input defaults and option keywords and a number of the input functions introduced earlier. The C:RECT command is a further refinement of the previous version of the command of the same name. This revision uses **get*xxxx*** input functions and option keywords to permit you to specify the rectangle's size using discrete length and width entry or by specifying the desired *area* of the rectangle along with either its length or width. In the following exercise, you will test the various input options of the revised RECT command.

Introduction to AutoLISP Programming

Table 9.16
Functions from 09INPUT.LSP

Function	Description
`(mxa_getint prompt default)`	Obtains an integer with default
`(mxa_getreal prompt default)`	Obtains a real number with default
`(mxa_getdist prompt default)`	Obtains a distance with default
`(mxa_getangle prompt default)`	Obtains an angle with default
`(mxa_getbool prompt default)`	Obtains yes/no response with default

Exercise

Testing the RECT Command

Begin a new, unnamed drawing and zoom Center on the point 20, 20 with a height of 20 units.

Launch LispPad and open 09RECT3.LSP, then press CTRL+L to load file into AutoCAD.

The first use of RECT uses corner points as in earlier exercises.

Command: **RECT** [Enter]	Starts the RECT command
First corner: **13,16** [Enter]	Specifies one corner of rectangle
Length/Width/<Other corner>: **21,20** [Enter]	Specifies other corner of rectangle

RECT draws the rectangle through the specified corner points. Next, you'll specify the size of the rectangle using the discrete Width and Length options.

Command: **RECT** [Enter]	Starts the RECT command
First corner: **14,17** [Enter]	Specifies the lower left corner
Length/Width/<Other corner>: **LENGTH** [Enter]	Selects the LENGTH option

`Length: 4 `[Enter]	Specifies a length of 4
`Area/<Width>: 8 `[Enter]	Specifies a width of 8

RECT draws an 8.0 wide by 4.0 long rectangle. Next, you'll specify the size of the rectangle by supplying the width and area.

`Command: `**`RECT`** [Enter]	Starts the RECT command
`First corner: `**`14,20`** [Enter]	Specifies lower left corner
`Length/Width/<Other corner>:` **`WIDTH`** [Enter]	Selects the WIDTH option
`Width: 8 `[Enter]	Specifies a width of 8 units
`Area/<Length>: `**`AREA`** [Enter]	Selects the AREA option
`Area: 120 `[Enter]	Specifies an area of 120 sq. units

C:RECT draws a rectangle having an area 120 sq. units and a width of 8 units. Next, you'll use the AREA command to verify the area:

`Command: `**`AREA`** [Enter]	Starts AREA command
`<First point>/Object/Add/Subtract:` **`O`** [Enter]	Specifies the Object option
`Select objects: `**`LAST`** [Enter]	Selects the LAST object drawn
`Area = 120.0000, Perimeter = 46.0000`	
`Command:`	

Error Handling

While working with exercises in this book, you may have noticed cases where you canceled an AutoLISP function by pressing CTRL+C or ESC, while the program or function was waiting for you to supply input. Doing this generates an AutoLISP *error condition*. When an error occurs in an AutoLISP program, it causes AutoLISP to halt the program's execution and display an error message and *traceback*. The error message indicates the nature of the error, and the traceback shows the program expression where the error

Introduction to AutoLISP Programming

occurred, and the enclosing expressions. To view this first-hand, use the RECT command from 09RECT3.LSP. In the following exercise, where you see the instruction to press CTRL+C or ESC, you should only press CTRL+C if you are using DOS AutoCAD or have *AutoCAD Classic keystroke mapping* enabled.

Exercise

Viewing AutoLISP Errors

Begin a new, unnamed drawing, or continue in the previous drawing and erase everything.

Launch LispPad and open 09RECT3.LSP, and then press CTRL+L to load the file into AutoCAD.

Command: **RECT** Enter   Starts the RECT command

First corner: **9.25,12.5** Enter   Specifies first corner

At the following prompt, press CTRL+C or ESC.

Length/Width/<Other corner>:   Cancels current command
Cancel

error: Function cancelled
(GETCORNER P1 "\nLength/Width/<Other corner>: ")
(SETQ P2 (GETCORNER P1 "\nLength/Width/<Other corner>: "))
(C:RECT)
*Cancel*55

Command:

In this exercise, pressing CTRL+C or ESC caused an error that halted the execution of the RECT command at the prompt where it was waiting for you to enter the second corner or an option keyword. At that point, AutoLISP displayed the error message Function Cancelled. There are many different error conditions that can halt an AutoLISP program, and *Function Cancelled* is one of many different messages that can appear when an error condition occurs.

In addition to displaying the error message, AutoLISP also printed part of the C:RECT program itself on the text console. The first expression printed is the one that was currently being evaluated when the error occurred. The expressions that follow are its parent expressions, all of which are printed right up to the outermost expression, which in the preceding example is the C:RECT function. The expressions that were printed are sometimes called the *error traceback*. This traceback helps you find the exact location and nature of program errors, which can occur for many causes in addition to canceling the program. An error will also occur when you call a function with the wrong number of arguments, or wrong data type for one or more arguments. With a larger program the traceback can be quite long, sometimes consuming hundreds of lines on the text console. This is one reason why Microsoft Windows platform is the preferred environment for customization. On the DOS platform, since long error tracebacks will quickly scroll off your text screen, you must turn on the LOG file (use the LOGFILEON and LOGFILEOFF commands to turn logging to file on and off) and SHELL to the operating system to examine the file every time an error occurs. With AutoCAD running in a Windowed environment, you can scroll the text window to find the start of the error traceback. If the traceback is extremely long, you may need to increase the number of history lines that are displayed on the text screen, which can be set from the Misc tab of the Preferences dialog box.

Since it is common to cancel a program or command before it finishes, it is ill-advised to allow a traceback to appear on the display during normal use of a program, since it causes confusion and serves no purpose other than for debugging. To control this, you can use an *error handler*.

The *ERROR* Function

AutoLISP provides a means for a program to get control whenever an error occurs, so that it can handle the error in a more graceful and less generic way than AutoCAD does. In addition to suppressing the traceback, an AutoLISP program can perform other operations when an error occurs by using an *error handler*. For example, if you designed an AutoLISP program that needs to open a file, it must also ensure that the open file is *closed* before the program finishes. If an error occurs while the file is open, there needs to be a way to close the file. A program can arrange to perform any number of operations just before an error condition terminates it by installing an error handler. An error handler is a user-defined function that is always called when an error occurs, just before the program containing the error is halted.

AutoLISP defines the special symbol, ***error***, which can be assigned to the function that you want AutoLISP to call when an error occurs. As soon as an error condition occurs, AutoLISP checks to see if ***error*** is assigned to a function and if it is, it calls that function and passes it one argument, which is a string that contains the error message (the same message printed on the command line when there's no error handler installed). When an error handler is installed and an error occurs, AutoLISP also suppresses the traceback from being echoed to the command line. This solves the problem of having hundreds of lines of code scrolling up your text console every time an error occurs.

It is important to point out that ***error*** is *not* a built-in function that your program calls. Instead, AutoLISP expects you to define the ***error*** function, and AutoLISP calls

Introduction to AutoLISP Programming

the function when an error occurs—you do not call the function explicitly from your program. When a program that uses an error handler runs, it must *install* its error handler, which is necessary because there can only be *one* error handler installed at any given time. An error handler is installed by assigning it to the symbol ***error***, as shown following.

```
(defun your_error_handler (msg)
  (princ "\nHello from your_error_handler.")
  (princ (strcat "\nError: " msg))
  (setq *error* old_error)            ; Restore previous error handler
  (princ)
)

(defun C:YOURPROGRAM ( / old_error)

  (setq old_error *error*)            ; Save existing error handler

  (setq *error* your_error_handler)   ; Install new error handler

  (princ "\nHello from YOURPROGRAM!") ; Working part of program

  (setq *error* old_error)            ; Restore previous error handler

  (princ)
)
```

The **C:YOURPROGRAM** function first saves the value of an existing error handler by assigning the function currently assigned to ***error*** to the local variable **old_error**. This error handler is restored when the program finishes. Next, **C:YOURPROGRAM** installs its own error handler by assigning the function **your_error_handler** to the symbol ***error***. Before terminating, it restores the previous error handler that it saved at the start of the program. If an error occurs prior to reaching the point where the previous error handler is restored, the previous error handler is restored from the new error handler installed by **C:YOURPROGRAM**.

Notice that **C:YOURPROGRAM** does not call the **your_error_message** function, and it does not call the ***error*** function either. It simply assigns a function to the symbol ***error*** so that if an error occurs, that function will execute before AutoLISP shuts down your program. Within an error handler function, you can do any housekeeping or other actions that must be done before your program terminates. These are the most common tasks that are performed within an error handler.

- Examine the error message to determine the severity of the error and print the message if the severity warrants it.

- Restore the previous values of system variables that had been changed by the running program.

- Stop any active AutoCAD commands by calling **command** with no arguments.

- Erase any partially created objects and return the drawing to the state it was in when the command was issued.

- Close any open files.

- Restore any previous error handler that was installed when the program started.

- Call **princ** with no arguments to prevent the ***error*** function's result from appearing on the command line.

The **C:RECT** command saves and changes the value of the PLINEWID system variable to 0, because the polyline rectangle should always have a 0 width. Because **C:RECT** can't assume that the value of PLINEWID is 0 when the command is issued, it must do this in order to ensure that the polyline is drawn with a 0 width. After the polyline is drawn by calling the **rectang** function, the value PLINEWID had prior to issuing the **C:RECT** command is restored, but that will happen *only* if the program reaches that point without an error, which is by no means guaranteed.

In the next exercise, you'll use a revised version of the RECT command (from the 09RECT4.LSP file on the MaxAC CD-ROM) that can handle errors and terminate properly. Its error handler restores the value of the PLINEWID system variable to the value it had before the command was issued.

 Exercise

Using an Error Handler

Begin a new drawing, or continue from the previous exercise and erase everything. Zoom Center on point 20,20 with a height of 20 units.

Launch LispPad and open 09RECT4.LSP, and then issue CTRL+L to load file into AutoCAD.

`Command: `**`SETVAR`**` [Enter]`	Starts the SETVAR command
`Variable name or ?: `**`PLINEWID`**` [Enter]`	Specifies the PLINEWID sysvar
`New value for PLINEWID <0.0000>:` **`99`**` [Enter]`	Sets PLINEWID to 99

Introduction to AutoLISP Programming

Command: **RECT** [Enter]	Starts the RECT command
First corner: **12,17.5** [Enter]	Specifies first corner
Length/Width/<Other corner>: **W** [Enter]	Specifies Width option
Width: **7** [Enter]	Specifies width of 7.0
Area/<Length>: *Press CTRL+C or ESC*	Cancels command
Cancel	
Error: Function cancelled	Error handler prints message
Command: **PLINEWID** [Enter]	Starts SETVAR command
New value for PLINEWID <99.0000>: [Enter]	

In the exercise, you first set the PLINEWID system variable to 99. Then the RECT command was issued, which saves the current value of the PLINEWID system variable and sets it to 0. When you canceled the RECT command, the error handler restored the old value of PLINEWID. In the file 09RECT4.LSP, the function **recterr** is the error handler for the RECT command. You can use it as a template for adding error handlers to other commands.

Debugging AutoLISP Programs

Debugging is the process of isolating and detecting errors or *bugs* in an AutoLISP program. Some bugs will cause AutoLISP errors, and some will not, making the latter difficult to detect. When trying to find the exact source of a bug, there are several tools available to you. You can also create your own debugging tools. The most basic debugging tool available is the AutoLISP **print** function, which prints and returns its argument. Because **print** returns its argument in addition to displaying it, a call to **print** may be placed anywhere, to display the result of an expression at the point when it is evaluated, without interfering with the expression's primary function. Earlier in this chapter you used **print** to display in this manner to display the result of each subexpression in an enclosing complex expression, to understand the order of evaluation. The following example from 09MEAS1.LSP shows a modified version of the meas function from 09MEAS.LSP, which you used earlier in this chapter. In this version, calls to the **print** function have been spliced into the expressions that call the **osnap** function, to display their result. Note that since the **print** function returns the result of its argument, the result returned by the calls to **osnap** are still assigned to the variables endpt and midpt, via the result of the **print** function, just as they were prior to inserting the calls to **print**. The changes from the original version of the meas function are shown in bold type.

```
;; 09MEAS1.LSP From Maximizing AutoCAD R13

(defun meas ( / line dist pickpt endpt midpt)
   (setq line  (entsel " from end of: ")
         dist  (getdist "\nDistance from end: ")
         pickpt (cadr line)
         endpt  (print (osnap pickpt "ENDPOINT"))  ; print osnap result
         midpt  (print (osnap pickpt "MIDPOINT"))  ; print osnap result
   )    (polar endpt (angle endpt midpt) dist)
)
```

Try calling the modified version of **meas** to observe the result of the calls to the **osnap** function, and note how it displays the midpoint and endpoint of the selected line.

Another useful debugging tool provided by AutoLISP is *function tracing*. The **trace** and **untrace** functions provide a way to activate and deactivate function tracing for user-defined functions. When a user-defined function is being traced, AutoLISP will display a message on the text console when the function's body is entered, showing the values of the arguments it received, and a second message when that function exists, showing the result it returns to its caller.

---

trace

(trace *sym*...)

Activates tracing for the user-defined function(s) assigned to each **sym**, which must be an unquoted symbol. Returns the last **sym** argument.

untrace

(untrace *sym*...)

Deactivates tracing for the user-defined functions(s) assigned to each **sym**, which must be an unquoted syumbol. Returns the last **sym** argument.

In the following exercise, you will use function tracing to view calls to the **tubesec** function from 09TSECT2.LSP, which you used earlier in this chapter. The **tubesec** function, which is called by the C:TUBESEC command function, draws and hatches a cross section of a tube. This function takes three arguments, the center of the tube (a

Introduction to AutoLISP Programming

point); the outer diameter of the tube (a real number); and the wall thickness of the tube (also a real number). **Tubesec** returns a list that contains the entity names of the two circles and the hatch object it creates. The **tubesec** function from 09TSECT2.LSP is shown in the following listing.

```
(defun tubesec (cen odia thk / ent1 ent2)
    (setvar "CMDECHO" 0)                              ;; Turn off command echo
    (command "_.CIRCLE" cen "_diameter" odia)  ;; Draw outer circle
    (setq ent1 (entlast))                             ;; Save circle entity name
    (command "_.CIRCLE" cen (- (/ odia 2.0) thk))  ;; Draw inner cir-
cle
    (setq ent2 (entlast))                             ;; Save circle entity name
    (command "_.HATCH" "" "" "" ent1 ent2 "")  ;; Hatch circles
    (list ent1 ent2 (entlast))                        ;; Return entity names of
)                                                      ;; circles and hatching
```

Prior to issuing the TUBESEC command, you'll call **trace** to activate tracing for the **tubesec** function, and you'll deactivate function tracing by calling **untrace** after the command is finished.

Exercise

Tracing User-Defined Functions

Begin a new drawing, or continue from the previous exercise and erase everything. Then zoom Center on point 10,10 with a height of 15 units.

Start LispPad and open the file 09TSECT2.LSP.

Press CTRL+L to load the file into AutoCAD

In AutoCAD:

Command: **(trace tubesec)** [Enter]	Activates tracing for the **tubesec** function
Lisp returns: TUBESEC	
Command: **TUBESEC** [Enter]	Starts TUBESEC command
Outer diameter: **4** [Enter]	Specifies outer diameter of 4.0
Wall thickness: **.25** [Enter]	Specifies wall thickness of 0.25

```
Tube center point: 10,10 [Enter]          Specifies center point of 10,10;
                                          TUBESEC draws and hatches the
                                          circles

Entering TUBESEC: (10.0 10.0 0.0)         AutoLISP prints values of
4.0 0.25                                  arguments to the tubesec
                                          function upon entering its body

Result: (<Entity name: 21d0780>           AutoLISP prints result of
<Entity name: 21d0788>                    tubesec function upon exiting
<Entity name: 21d07a0>)                   its body

Command:
```

For much more powerful AutoLISP debugging tools, see this book's companion volume, *Maximizing AutoLISP for AutoCAD R13*.

Coding Practices and Standards

Now that you've been introduced to AutoLISP, you'll probably want to start writing your own programs to streamline your working environment. There are some general coding standards and practices that you should adopt and adhere to, that will help make your programs more readable and reduce the likelihood of conflicts with other AutoLISP programs or with AutoLISP itself.

Symbol Naming Conventions

One of the most important standards that you should define and strictly adhere to is in how you choose to name functions and variables. Remember that all of AutoLISP's built-in functions are assigned to *symbols* just as your own functions and variables are. This means that you will disable a built-in AutoLISP function if you assign a variable or function to the same symbol which built-in function is assigned to. *Be careful* when choosing variable names, since AutoLISP has a hundreds of functions whose symbols are *reserved*, and which should not be assigned to variables or user-defined functions. If you are unsure about the name that you want to use as a variable, you can easily determine if it is reserved by AutoLISP, by just entering the symbol at the command prompt, prefixed with an exclamation point. If the value that is returned is *not* **nil**, the symbol you entered is assigned to a value already, and should not be assigned to a program variable or user-defined function.

We also suggest that you examine the *AutoCAD Interoperability Guidelines* (AIG), published by Autodesk. A copy of this booklet is shipped with AutoCAD, and is also available in the AutoCAD (**GO ACAD**) forum on CompuServe as well as Autodesk's World Wide Web

Introduction to AutoLISP Programming

page at **http://www.autodesk.com**. The AIG discusses in greater detail AutoLISP symbol naming conventions, and other issues relating to avoiding conflicts with other third-party applications.

One of the most common mistakes made by novice programmers is to assign values to symbols such as **T** or **type**, both of which are reserved by AutoLISP. The **type** function was covered earlier in this chapter. Try the following exercise, and note what happens when you assign a value to a reserved symbol. After trying the exercise, you will have to exit and restart AutoCAD before doing anything else.

Exercise

Improper Use of Reserved Symbols

Begin a new unnamed drawing, or continue in the drawing from the previous exercise.

Command: **(type 300.0)** [Enter]   Gets the type of 300.0

Lisp returns: REAL

Command: **(setq type "ALBACORE")** [Enter]   Incorrectly assigns a value to the reserved symbol **type**

Lisp returns: "ALBACORE"

Command: **(type "A STRING")** [Enter]   Gets the type of "A STRING"

error: bad function
(TYPE "A STRING")
Cancel

This error occurs because the AutoLISP **type** function is no longer accessible

Command:

Exit the current editing session to correct the problem.

Comments and Annotations

Comments are a very important part of your programs and serve as the most basic form of documentation that describes the structure and function of its individual parts. When you assign a variable, you can add a comment indicating what logical value the variable represents in your program. If you adopt a consistent habit of commenting your code, you'll find it much easier to reacquaint yourself with it after not having seen it for several months.

You will see a number of files included in this book which are commented. You can also over-comment code, but doing so can cause the code itself to become more difficult to read, because it tends to get lost in the comments. One solution to this is to use an editor that supports syntax colorization, which displays different elements of a program file in different colors.

Finishing Up

If you want to keep any of the files you created or modified in this chapter, move them to the \MAXAC directory or another directory, then delete or move any remaining files out of \MAXAC\DEVELOP.

Conclusion

Here are some summary tips for integrating AutoLISP into your menu macros.

Use AutoLISP to create intelligent macros to replace repetitive menu macros. AutoLISP variables can eliminate menu selections that depend on drawing scale. When possible, calculate values to feed to AutoCAD commands, rather than pausing for user input. Add prompts to menu macros to clarify use. Coordinate input with backslashes. When you think about improving your macros with AutoLISP, use the examples as idea lists.

You will see more of AutoLISP and macros later in this book. With this book's companion volume, Maximizing AutoLISP for AutoCAD Release 13, you can extend this power, learning how to add more control.

The next chapter adds fuel to your programming fire with DIESEL, a string language developed specifically for menu macros and status line information.

chapter

Running on DIESEL

The DIESEL language is another AutoCAD application programming interface, which was first introduced in AutoCAD Release 12. DIESEL officially stands for **D**irect **I**nterpretively **E**valuated **S**tring **E**xpression **L**anguage (also known as Dumb Interpretively Evaluated Stupid Expression Language). The DIESEL acronym is also derived from the characteristic that all data is manipulated *efficiently* in string form.

For some purposes, the DIESEL language has advantages over AutoLISP because it provides a number of distinct capabilities not available to AutoLISP programs. Firstly, DIESEL expressions can appear in pull-down or cursor menu *labels*. When a DIESEL expression is defined in a menu item's label, it is evaluated every time the menu appears, and the results of the evaluation are displayed as the menu item's label. This enables you to create intelligent menu labels that can alter their contents and appearance dynamically, based on tests that can be performed by embedded DIESEL expressions.

Secondly, DIESEL expressions can also appear in menu macros. When a menu macro containing a DIESEL expression is executed, the DIESEL expression is evaluated and the resulting string replaces the expression and becomes part of the macro. This lets you add intelligence to menu macros without the use of AutoLISP.

Additionally, DIESEL expressions in both menu labels and macros can be evaluated *transparently*, independent of AutoLISP. This transparent evaluation solves the problem caused by AutoLISP's lack of reentrance. You probably have encountered the `Can't reenter AutoLISP` error message. This error occurs when you try to enter an AutoLISP expression in response to a prompt generated by another active AutoLISP (or ADS) program, which is not permitted. Hence, you cannot use an AutoLISP expression or variable as a response to an input prompt generated by an AutoLISP or ADS program. Because DIESEL works independently of AutoLISP and ADS, you *can* use DIESEL expressions to respond to and provide input to AutoLISP or ADS programs. DIESEL enables you to create macros that examine settings and return appropriate values, which are accepted by AutoLISP and ADS input prompts.

Lastly, DIESEL expressions can also appear in the MODEMACRO system variable, whose content is displayed in place of the standard AutoCAD status line in DOS AutoCAD, and in a panel of the status bar in AutoCAD for Windows. This enables you to add your own status items to the status line or bar.

Warning: DIESEL is not fully supported on all AutoCAD platforms. On the Sun SparcStation or Apple Macintosh, for example, you cannot use DIESEL in menu item labels. Unless you are using the DOS or Windows versions of AutoCAD, check your AutoCAD READxxx.TXT or HLP files or the *AutoCAD Interface, Installation, and Performance Guide* for compatibility details. If you are developing menus that will be used on various platforms, it would be prudent to avoid using DIESEL in menu item labels.

About This Chapter

This chapter shows you how to customize your AutoCAD system with DIESEL. After examining the format and syntax of the DIESEL language, you will learn to use DIESEL to:

- Create pull-down and popup cursor menu item labels whose contents can change dynamically.

- Disable, enable, check, and uncheck menu items conditionally, in response to current settings.

- Create intelligent menu macros that can automate responses to input solicited by any command, including those written in AutoLISP and ADS.

- Create custom dynamic status displays.

- Use DIESEL expressions in AutoLISP programs.

- Use DIESEL to add dynamic popup cursor menus to AutoLISP programs.

- Use DIESEL to dynamically manage and control popup cursor menus.

The resources you'll use in this chapter are:

- ACADMA.MNS, ACADMA.MNL, ACAD.PGP, ACAD.LSP, 10DIESEL.LSP, 10DMENU.LSP, 10MATCH.LSP, 10MODE1.LSP, 10MODE2.LSP, 10QHATCH.LSP, MAXAC.MNL, MAXAC.MNS, 10BLANK.MNS, 10BLANK.MNL, 10CURSOR.MNS, 10CURSOR.MNL, 10CURSOR.DWG, 10POPUPS.MNS, 10POPUPS.MNL, 10POPUPS.DWG, 10PSPACE.DWG, and DYNAMODE (see Appendix G).

Running on DIESEL

Menus, Macros, and Programs in This Chapter

The menus, macros, and AutoLISP programs in this chapter are:

- The DIESEL command from 10DIESEL.LSP is an AutoLISP program that allows you to interactively evaluate DIESEL expressions on the AutoCAD command line, for testing and instructional purposes.

- 10MODE1.LSP and 10MODE2.LSP contain custom DIESEL status lines and AutoLISP programs to manipulate them. These files add custom status lines to AutoCAD, and they also demonstrate techniques for adding command-specific extensions to custom status lines.

- The QHATCH command from 10QHATCH.LSP is a sample AutoLISP program that shows how to add custom status line extensions in AutoLISP programs.

- MAXAC.MNS is a partial menu file that adds the MaxAC pull-down menu and toolbar to the base menu. It also provides several example pull-down menus that support the examples and exercises.

- 10CURSOR.MNS is a partial menu file with popup cursor menus. This file is also used in Chapter 14. The file shows how how to use DIESEL to control the appearance of menus, and how to display information in pull-down and popup cursor menus.

- 10BLANK.MNS is a partial menu file that demonstrates techniques for adding dynamic popup cursor menus to AutoLISP programs.

- 10POPUPS.MNS is a partial menu that contains a toolbar, and it demonstrates the use of 10BLANK.MNS to create dynamic popup cursor menus.

DIESEL Mechanics

DIESEL is a macro string evaluation language consisting entirely of functions that manipulate data in string form and have no *side effects*. A side effect is any change that results from evaluating an expression or calling a function. Many, but not all, AutoLISP functions have side effects. An AutoLISP function that only returns a value and has no side effects is often called a *pure LISP function*. Examples of AutoLISP functions that have side effects, in addition to returning a value, are **print**, **setq**, and **redraw**. There is a side effect to calling each of these functions (**print** displays its argument; **setq** assigns a value to a variable; and **redraw** causes the current view to be redrawn). Examples of LISP functions that have no side effects are **+**, **strcase**, and **distance**. Each of these three functions returns a result, but none of them have any other effects on the environment or the AutoCAD drawing editor.

Like AutoLISP expressions, DIESEL expressions are evaluated primarily to produce a result. Calls to DIESEL functions can be nested as they are in LISP, so the result of one function can be used directly as an argument to another.

Differences between DIESEL and AutoLISP

There are a number of differences between DIESEL and AutoLISP. You should already be familiar with AutoLISP from reading Chapter 9, so we can draw some analogies that should help you understand DIESEL more easily.

The main difference between DIESEL and AutoLISP is that DIESEL consists entirely of *functions* that do not have side effects, whereas many AutoLISP functions have side effects and can make use of variables. For example, the primary side effect of the AutoLISP **setq** function is that it assigns values to variables; however, DIESEL has *no variables*, hence you cannot assign and reference variables.

Another important difference between these two languages is that DIESEL functions cannot accept more than nine arguments, whereas some AutoLISP functions can accept any number of arguments. Because DIESEL has no symbols or variables, there are no user-defined functions (which are just another kind of variable).

DIESEL Syntax versus AutoLISP Syntax

Both DIESEL and AutoLISP use parentheses to enclose a function call. The syntax of the two languages is strikingly similar with the following exceptions:

- A $ (dollar sign) precedes the opening parenthesis of every DIESEL function.

- A comma must be used to delimit arguments in a DIESEL function call.

- Because all data in DIESEL is string data, strings do not need to be delimited by double quotes as they are in AutoLISP, *unless* they contain a literal comma.

Here is the general syntax of a DIESEL expression:

```
$(func,arg1,arg2,...)
```

As in AutoLISP, DIESEL function calls are delimited by a pair of matching parentheses, and the first element that follows the open parenthesis is the name of the function. With a few exceptions, most DIESEL functions have names that are the same as the names of their functional AutoLISP counterparts. Again as in AutoLISP, function names in DIESEL are not case-sensitive, but all other data in a DIESEL expression is.

Running on DIESEL

The most notable difference between AutoLISP and DIESEL syntax is the comma delimiter in DIESEL function arguments. Two commas that appear consecutively with nothing between them are equivalent to the *null* or empty string. Because each argument in a DIESEL function call is delimited by the comma, spaces that appear between two consecutive commas are interpreted *literally* and will appear in the result.

Note: The syntax notational conventions of DIESEL function prototypes in this book are the same as those used for AutoLISP functions, except that all DIESEL functions have a dollar-sign ($) prefix added to distinguish them from like-named AutoLISP functions.

Here is an example of a DIESEL function call:

`$(substr,This is the first argument,13,5)`

This example's function call to the **$substr** function returns a 5-character long portion of the first argument (the string "This is the first argument"), starting with the 13th character, resulting in the word "first". Notice that spaces *do not* serve as argument separators or delimiters in DIESEL. In addition, note that because commas are used for delimiting arguments, the comma may not appear literally in a function argument unless the argument is surrounded by double quotes.

Here is another slightly more complex example of a DIESEL expression:

`$(rtos,$(getvar,ltscale),2,4)`

This example returns the value of the LTSCALE (LineType SCALE) system variable (a real number) formatted in decimal units with four places, like `"1.0000"`.

Here is one more example:

`$(GETVAR,LASTPOINT)`

The last example returns the value of the LASTPOINT system variable. The string returned is a 3D coordinate with the X, Y, and Z components delimited by commas, like `"10.00,9.75,0"`. This example serves to point out another important difference between DIESEL and AutoLISP, which is that DIESEL has no LIST data type. Instead, all complex data in DIESEL (such as coordinates) is represented in string form with individual elements delimited by *commas*.

Testing DIESEL Expressions

AutoLISP expressions can be evaluated by entering them directly into AutoCAD's command line, making interactive testing of code easy. DIESEL expressions can't be entered directly into AutoCAD's command line. But there are ways to make testing DIESEL expressions easier. The AutoLISP function in the following listing is from the file 10DIESEL.LSP and is included in this chapter's subdirectory on the accompanying

Maximizing AutoCAD® R13

CD-ROM. The function adds the DIESEL command to AutoCAD, which accepts DIESEL expressions entered on the command line for evaluation and displays the result. You will find this command a very convenient way to do ad hoc testing of DIESEL expressions or for general experimentation. To use it, load the 10DIESEL.LSP file, and enter **DIESEL** at the AutoCAD command prompt. The command displays a `Diesel>` prompt, at which you can enter any DIESEL expression, which is evaluated and printed to the command line. If you make an error and forget a closing parenthesis, DIESEL will print **nil**. To exit the command, press ENTER at the `Diesel>` prompt. The AutoLISP expression defining the DIESEL command is shown here:

```
(defun C:DIESEL ( / expr)
   (while (/= "" (setq expr (getstring t "\nDiesel> ")))
      (print (menucmd (strcat "M=" expr)))
   )
   (princ)
)
```

In the following exercise, you'll test several DIESEL functions, which have AutoLISP counterparts that you've already been introduced to. Make sure you have installed this chapter's code files from the CD-ROM or MAXAC\BOOK\CH10 directory to your working development directory

 Exercise

Testing DIESEL Functions with the C:DIESEL Command

Start AutoCAD and begin a new drawing named MA-DIESL, and zoom Center on the point 10,10 with a height of 10 units.

Command: **(load "10diesel")** [Enter]   Loads 10DIESEL.LSP

LISP returns: **10diesel.lsp loaded**

Command: **CIRCLE** [Enter]   Starts CIRCLE command

3P/2P/TTR/<center point>:   Specifies circle center point
3.5,7.25 [Enter]

Diameter/<radius>: **2** [Enter]   Specifies circle radius

Command: **'TEXTSCR** [Enter]   Displays text console

Running on DIESEL

`Command: `**`DIESEL`** [Enter]	Starts DIESEL command and displays the `Diesel>` prompt
`Diesel> `**`$(getvar,CIRCLERAD)`** [Enter]	Gets the value of CIRCLERAD system variable
DIESEL returns: `"2"`	
`Diesel> `**`$(getvar,DWGNAME)`** [Enter]	Gets value of DWGNAME system variable
DIESEL returns: `"D:\\MaxAC\\Develop\\MA-DIESL"`	The current drawing name and path (may be different on your system)
`Diesel>` **`$(rtos,$(getvar,CIRCLERAD),2,4)`** [Enter]	Gets value of CIRCLERAD system variable and formats it in decimal units with four places of precision
DIESEL returns: `"2.0000"`	
`Diesel> `**`$(getvar,LASTPOINT)`** [Enter]	Gets value of LASTPOINT system variable
DIESEL returns: `"3.5,7.25,0"`	
`Diesel>` **`$(index,1,$(getvar,LASTPOINT))`** [Enter]	Gets second element of the LASTPOINT system variable
DIESEL returns: `"7.25"`	
`Diesel> ` [Enter]	Exits DIESEL command

In this exercise, you made use of several DIESEL functions. The first was the **$getvar** function, which is identical to its AutoLISP counterpart (except that all results are in string form). You used **$getvar** to retrieve the value of the CIRCLERAD system variable (the radius of the most recently drawn circle); the DWGNAME system variable (the filespec of the current drawing); and the LASTPOINT system variable (the point referenced by @ in coordinate entry).

> **$getvar**
>
> **$(getvar,*varname*)**
>
> *Returns the value of the system variable* **varname**.

Note that the results displayed by the DIESEL command are surrounded by double quotes because they are strings. The double quotes surrounding the strings are not literally part of the result of the DIESEL expressions you entered—they are added by the AutoLISP **print** function in the **C:DIESEL** command. The **print** function is used to display strings in their *raw* or unformatted form, which makes it easier to identify leading or trailing spaces and other non-printing characters.

You also used the **$rtos** function to format the CIRCLERAD system variable as a decimal distance with four places of precision. Finally, you used the DIESEL **$index** function to extract the Y component of the LASTPOINT system variable. Since the DIESEL language doesn't have lists as AutoLISP does, complex data objects like 3D coordinates are represented in string form with commas separating each element or component. You can use the **$index** function to access the individual elements of a complex data object like coordinates as they are represented in DIESEL.

Most of the DIESEL functions you used so far will be examined in more detail later in this chapter.

Debugging DIESEL Macros

DIESEL macros usually consist of a single complex expression, with many other expressions nested within it. Because of their nested nature, DIESEL expressions can be difficult to debug as they increase in complexity. A single missing or extra comma is often a source of bugs and errors. Fortunately, AutoCAD provides a means of tracing a DIESEL expression's evaluation completely.

You can turn the MACROTRACE system variable on (set it to 1) to trace the evaluation of a DIESEL expression regardless of its complexity. When you set MACROTRACE to 1, the evaluation of all DIESEL expressions is *traced* (including those in menu macros and labels, and expressions on the status line). Tracing DIESEL expressions consists of displaying each DIESEL expression on the text console as it is evaluated, along with its result. To see how this works, try repeating the previous exercise after setting MACROTRACE to 1. For example, the trace narration of $(index,1,$(getvar,LASTPOINT)) is:

```
Eval: $(INDEX, 1, $(getvar,LASTPOINT))
Eval: $(GETVAR, lastpoint)
===>   3.5,7.25,0
===>   7.25
```

Running on DIESEL

Then set MACROTRACE back to 0 and reload the DIESEL command from 10DIE-SEL.LSP. See the last section of this chapter for more on MACROTRACE and debugging DIESEL macros.

Creating Custom Status Line Displays

The most basic application of DIESEL is also the one it was primarily designed for—a way to define a custom status line that can display anything that can be derived from the value of one or more AutoCAD system variables, operating system environment variables, and so on. In this book, we refer to the custom status line display area as the *status line* regardless of platform.

In DOS AutoCAD, the status line is located at the top of the graphics display (see Figure 10.1), extends from the extreme left side of the screen to the point where the coordinate display starts, and is displaced by the pull-down menu bar when you move the cursor onto it. Some display drivers let you configure the length of the status line. Other drivers automatically display the largest possible status line that will fit on the display, given the character font in use (which is also configurable). The DIESEL **$linelen** function returns the maximum length (in characters) of the status line in DOS AutoCAD.

On the Windows platform (as well as other windowed environments), the status line is a distinct panel that appears at the left end of the *status bar*, which is located at the bottom of the main AutoCAD window (see Figure 10.1). In AutoCAD for Windows, there is no default status line, and the panel that contains it is not visible if no custom status line is enabled.

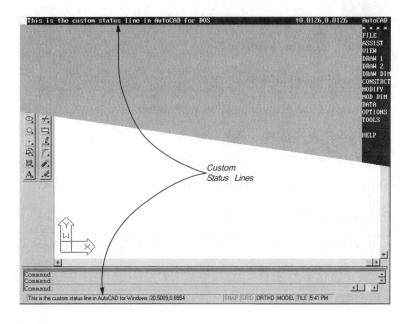

Figure 10.1
AutoCAD with custom status line.

Maximizing AutoCAD® R13

Much of the information that is displayed on the standard status line in DOS AutoCAD is displayed in separate panels on the status bar and in other places in Windows versions of AutoCAD. These settings include the current states of SNAP, GRID, ORTHO, and TILE modes, the current space (model or paper), and the time of day. In addition, the Object Properties toolbar displays the current layer, color, and linetype.

Warning: If you are using the Autodesk Mechanical Desktop, you cannot use a custom status line, since it is used by Mechanical Desktop to display the name of the current drawing.

The MODEMACRO system variable is a string variable. Unless it is set to " " (a null string), the status line displays its current contents. If you remember that DIESEL means Direct Interpretively Evaluated String Expression Language, you might guess what happens if the content of MODEMACRO is a DIESEL string expression—the DIESEL expression is evaluated and its result is displayed on the status line.

Simple Status Line Displays

The status line is also a good place to experiment with simple DIESEL expressions. In the next exercise, to see a customized status line, you'll just put some text in the MODEMACRO variable. Then you'll try displaying the current time in the status line using the **$getvar** function and the DATE system variable. Notice that, because everything in DIESEL is a string, you do not have to enclose system variable names such as DATE in double quotes, as you have to do in AutoLISP.

Exercise

Assigning Custom Status Lines

Start AutoCAD with a new, unnamed drawing.

Command: **MODEMACRO** [Enter]

New value for MODEMACRO, or . for   Displays "Have a nice day!" on the
none <" ">: **Have a nice day!** [Enter]   status line

Command: **MODEMACRO** [Enter]

New value for MODEMACRO, or . for   Displays the date in Julian time on
none <" ">: **$(getvar,DATE)** [Enter]   the status line

After you complete this exercise, the status line displays the current system date and time in Julian format, where time is represented as a real number, with the value 1.0 being equal to one day. This is not a friendly way to express a time or date, so DIESEL provides the **$edtime** function to format dates and times into more common formats.

Time and Date Format

> **$edtime$**
>
> (edtime, *time, picture*)
>
> Formats **time**, *a Julian date, into common date and time formats (year, month, day, hour, minute, second, and millisecond), as specified by* **picture**, *a string containing date and time formatting codes.*

The formatting codes for the **$edtime** function's *picture* argument are shown in Table 10.1.

When you use the **$edtime** function, you must specify at least one format code item. A typical format code used to express a date is shown in the following exercise. Try formatting the date by nesting the **$getvar** function in the **$edtime** function.

Table 10.1
DIESEL Time and Date Formatting Code

Format Codes	Example	Format Codes	Example
Day of Month		**Time of Day**	
D	2	H	4
DD	02	HH	04
Day of Week		MM	53
DDD	Thu	SS	17
DDDD	Thursday	MSEC	506
Month of Year		**Morning/Afternoon**	
M	9	AM/PM	AM
MO	09	am/pm	am
MON	Sep	A/P	A
MONTH	September	a/p	a
Year			
YY	92		
YYYY	1992		

Exercise

Creating a Custom Date Status Display

Command: **MODEMACRO** [Enter]

New value for MODEMACRO, or . for none <" ">$(getvar,DATE)">: **Today is $(edtime, $(getvar,DATE),DDDD MO/DD/YY)** [Enter]

Displays "Today is" followed by the day of the week and the date on status line

To remove the new status line and make AutoCAD use its built-in status line, set the value of the MODEMACRO system variable to a period with the SETVAR command, or to a null string with the AutoLISP **setvar** function.

You can intersperse other characters in the **$edtime** format string, such as spaces, dashes, slashes, colons, and even other words, but not commas. When you use the am/pm and AM/PM codes, the slash and entire code is required; otherwise, DIESEL would interpret the M in PM as the month code.

The previous MODEMACRO example demonstrates the basic use of DIESEL to define a simple custom status line. Defining a more complex status line gets a bit tedious at the command prompt.

Complex Status Line Displays

The best way to define complex status lines is to use the AutoLISP **setvar** function to assign a DIESEL string to the MODEMACRO system variable. To load such an expression automatically, you can include an MNL file.

The following listing, from the 10MODE1.LSP file from the accompanying CD-ROM, is a more complex MODEMACRO definition. This status line is mainly useful with DOS AutoCAD, because the AutoCAD for Windows interface already displays much of the same information using other means. Nevertheless, this status line demonstrates a good sampling of complex MODEMACRO definitions regardless of your AutoCAD platform.

You can use the AutoLISP **load** function to manually load 10MODE1.LSP at the command prompt, as you do in the next exercise. You can also use the APPLOAD command to load LSP files. To automate loading, you can add **(load "10mode1")** to the end of the default MNL file in the working directory that contains the chapter files, or to ACAD.LSP or another AutoLISP file automatically loaded at the start of each drawing editor session in your working configuration.

Running on DIESEL

Here is the MODEMACRO code defined in 10MODE1.LSP:

```
(setvar "MODEMACRO"                     ;assigns DIESEL macro to MODEMACRO
  (strcat                               ;formats MODEMACRO string

    ;; Displays current entity color
    "$(if,$(eq,$(substr,$(getvar,CECOLOR),1,1),B),"    ;if

        "$(substr,$(getvar,CECOLOR),1,3),"              ;then
        "$(if,$(<,$(getvar,CECOLOR),8),"                ;else,if
            "$(substr,RYGCBMW,$(getvar,CECOLOR),1),"    ;then
            "$(getvar,CECOLOR)"                         ;else
    ")) "                                               ;closes ifs

    ;; Displays current entity linetype
    "$(if,$(<,$(linelen),34),"          ;if less than 34
       ","                              ;then omit linetype
       "$(getvar,CELTYPE) "             ;else display it
    ")"

    ;; Displays current layer
    "$(substr,$(getvar,CLAYER),1,8) "   ; first 8 chars of layer name

    ;; Displays Ortho status
    "$(if,$(getvar,ORTHOMODE),O ,o )"   ;if 1, then O (on), else o (off)

    ;; Displays Snap status and non-uniform snap unit
    "$(if,$(getvar,SNAPMODE),"                          ;if 1
       "S:$(if,"                                        ;then S (on) and if
            "$(=,$(index,0,$(getvar,SNAPUNIT)),"        ;test X=Y snap
                "$(index,1,$(getvar,SNAPUNIT))),"
            "$(index,0,$(getvar,SNAPUNIT)),"            ;then display X
            "/="                                        ;else display /=
         ") "                                           ;close if
       ",s "                                            ;else s (off)
    ")"                                 ;close if

;;Displays tablet mode
    "$(if,$(getvar,TABMODE),T,t)"   ;if 1, then T (on), else t (off)
) )                                 ;close STRCAT and SETVAR
(princ)                             ;suppresses result when loaded.
```

As you can see from this listing, combining DIESEL and AutoLISP expressions can make them somewhat confusing and difficult to read, which is mainly a result of the similarities between AutoLISP and DIESEL syntax and semantics. When you combine AutoLISP and DIESEL, you must adhere to the syntax of both languages. Within the AutoLISP **strcat** expression, DIESEL expressions appear within pairs of double quotes because in terms of AutoLISP, they are just strings. Although you need not quote a DIESEL string when entering it at the MODEMACRO prompt, you must use quotes in AutoLISP to identify the AutoLISP data type as a string. Also remember that within the double quotes, every character in a DIESEL expression, even a space, is significant, and commas, which delimit DIESEL function arguments, are critical.

The entire MODEMACRO definition is enclosed in a single call to the AutoLISP **strcat** function. The **strcat** function combines its string arguments and returns them as a single string, which is then assigned to the symbol `modeline1`, which in turn is assigned to the MODEMACRO system variable using the **setvar** function. The most important thing you must remember when defining large, complex status lines is that the maximum string length that **setvar** can accept is 460 characters; longer strings will result in a `string too long` error.

Note: If an error occurs in a combined AutoLISP/DIESEL expression, you should count double quotes in pairs and make sure that all DIESEL strings and their parentheses are within pairs of quotes and that all AutoLISP expressions and their parentheses are outside of the quotes. You can use LispPad to count quotes in both DIESEL expressions as well as the AutoLISP code that constructs it. Misplaced or missing quotes and parentheses are likely to cause errors such as `string too long`, `invalid dotted pair`, `too few arguments`, `bad argument type`, `malformed list`, `bad list`, or `extra right paren`. Mismatched parentheses or missing commas within DIESEL expressions will not generate an AutoLISP error, but they will instead cause DIESEL evaluation errors when the DIESEL expression is evaluated.

Load and test the custom status line from 10MODE1.LSP in the following exercise, and then examine it more closely in the following sections.

Exercise

Displaying a Complex Status Line

Continue from the previous exercise or begin a new, unnamed drawing.

Command: **(load "10MODE1")** [Enter]   Loads MODEMACRO definition

The status line now displays BYL BYLAYER 0 o S:0.1 t (BYLAYER is omitted if your status display is limited to less than 34 characters).

The current status line display indicates the following:

BYL	First letter of current entity color, or BYL for BYLayer, or BYB for BYBlock.
BYLAYER	Current entity linetype name
0	Current layer name
o	Ortho: o for off; O for on

```
S:0.1        Snap: s for off; S: for on; includes snap unit if
             X,Y are equal
t            Tablet tracing mode: t for off; T for on
```

Now you'll change some settings in AutoCAD to observe how the custom status line behaves.

 Exercise

Testing a Complex Status Line

Command: **COLOR** [Enter]

New entity color <**YLAYER**>: **3** [Enter]   Status displays G (green)

Command: *Turn ortho on with CTRL+O or F8*   Status displays O

Command: *Turn snap off with CTRL+B or F9*   Status displays s

Use the LINETYPE Set option to set DASHED current. Use the LAYER Make option to make a layer named DOORS, and set its color to 1 (red). Use SNAP to set the snap increment to .125.

The status line now displays G DASHED DOORS O S:0.125 t (see Figure 10.2).

Some linetype names are nine or more characters long. If your display lacks sufficient status line width, this complex status line expression omits the linetype display. It does this by using the DIESEL **$linelen** function to check the display's capacity.

$linelen

$(linelen)

Returns the number of characters that can be displayed on the status line.

In 10MODE1.LSP, the value returned by **$linelen** is compared to 34, a number of characters that can be displayed by virtually all systems, by the **$<** (less than) function. The DIESEL less-than function works like its AutoLISP counterpart; see the Relational Operators section in Chapter 9 for details.

Figure 10.2
A complex custom status line.

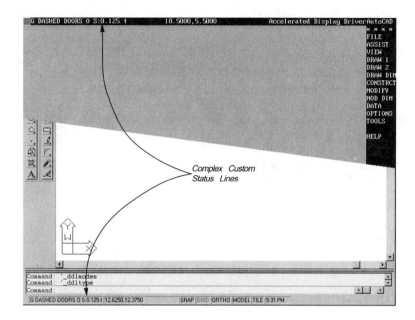

The DIESEL expression $(<,$(linelen),34) returns 1 if true (the maximum line length is less than 34) and 0 if not true. The returned value is used by the **$if** function to decide what to display.

Conditional Branching in DIESEL

You can use an if test to direct your DIESEL program to execute in a predictable order. The **$if** conditional function is your branching tool for controlling DIESEL.

$if

$(if,*testexpr*,*thenexpr*,*elseexpr*)

If ***testexpr*** *evaluates to true (non-zero),* ***thenexpr*** *is evaluated. If the test is false (equal to 0), the optional elseexpr is evaluated if supplied. The* **$if** *function returns the value of the evaluated expression or a null string.*

Another important distinction between AutoLISP and DIESEL is in how Boolean logic is represented. You'll recall from Chapter 9 that AutoLISP represents the logical value of false using the *symbol* nil and interprets all other values as representing the logical value of true. DIESEL, on the other hand, has no symbols. Hence, DIESEL uses numeric values to represent logical true/false states, where 0 represents false and all other values

Running on DIESEL

(non-zero) represent true. The use of numeric values as a means of representing true/false logic can be exploited in DIESEL, because AutoCAD system variables also use numeric values to represent logical true and false states. For example, the ORTHOMODE system variable indicates if ortho is on, by returning a value of 1, or 0 if ortho is off. Consider the following two examples, both of which do exactly the same thing. The first is written in AutoLISP and the second in DIESEL.

```
(if (eq 1 (getvar "ORTHOMODE")) "Ortho is ON" "Ortho is OFF")

$(if,$(getvar,ORTHOMODE),Ortho is ON,Ortho is OFF)
```

Notice that the DIESEL expression is the simpler of the two and requires no equality test to determine the condition. This is because in the DIESEL version, the numeric value returned by **$(getvar,ORTHOMODE)** is used *directly* as the test condition, and there is no need to compare it to the value 0, because in the case of DIESEL (and unlike AutoLISP), the value 0 represents false. The ORTHOMODE system variable can have one of two possible values (1 = on, 0 = off), but since any non-zero numeric value also represents true, the DIESEL **$if** function interprets all non-zero values the same (as representing true).

The following example indicates whether there are any running object snap modes active. It uses the OSMODE system variable to determine this. The *bit-encoded* OSMODE system variable can have many possible values, depending on what object snap modes are active. When there are no running object snap modes active, the value of OSMODE is 0. In this case, the example DIESEL code doesn't care what specific object snap modes are set, but only if there are any set. Because the **$if** function interprets all non-zero values as true, it can use the value of the OSMODE system variable directly, as the test expression.

```
$(if,$(getvar,OSMODE),Running object snap is active)
```

Later in this chapter, you'll learn to use DIESEL's bitwise logical operators with the OSMODE system variable to determine what specific object snap modes are currently active.

In the next example, two **$if** expressions are nested, to determine if the current space is paper space (which can only be the case if TILEMODE = 0 and CVPORT = 1).

```
$(if,$(eq,0,$(getvar,TILEMODE)),
     $(if,$(eq,1,$(getvar,CVPORT)),We must be in PSPACE)))
```

The *thenexpr* argument to the **$if** function is required, but it can be a null string, as in $(if,*testexpr*,,*elsexpr*). The presence of the second comma in this example defines the *thenexpr* argument, even though it is null and returns nothing. A null *thenexpr* argument is used in the 10MODE1.LSP MODEMACRO example's current entity linetype expression, shown combined into a single DIESEL expression in the following:

$(if,linelengthtest,,$(getvar,CELTYPE))

If linelengthtest (any valid test expression) returns 1 (the maximum line length is less than 34), *then* ,, displays nothing, *else* the returned $(getvar,CELTYPE) linetype name value is displayed. Notice the blank space before the closing parenthesis; it is part of the *else* argument to place a space before the next item in the 10MODE1.LSP MODEMACRO expression.

The Color Status

The task of formatting the value returned by $(getvar,CECOLOR) is a little more complex. The CECOLOR system variable returns either the string BYLAYER or BYBLOCK, or a color number. If the color number is 7 or less, it can be more easily recognized by its standard color name, or at least the first character of its name, such as Y for yellow (color 2). The colors are 1 Red, 2 Yellow, 3 Green, 4 Cyan, 5 Blue, 6 Magenta, and 7 White (which appears black on a display with a light background). The 10MODE1.LSP MODEMACRO entity color expression formats BYLAYER and BYBLOCK as BYL and BYB, colors 1-7 as the first letter of their names, and other colors as their color numbers. To do this, it uses two nested **$if** functions, as follows:

```
"$(if,$(eq,$(substr,$(getvar,CECOLOR),1,1),B),"      ;if
    "$(substr,$(getvar,CECOLOR),1,3),"                ;then
    "$(if,$(<,$(getvar,CECOLOR),8),"                  ;else,if
        "$(substr,RYGCBMW,$(getvar,CECOLOR),1),"      ;then
        "$(getvar,CECOLOR)"                           ;else
")) "                                                 ;closes ifs
```

The first test expression of the $(if,(eq,$(substr,$(getvar,CECOLOR),1,1),B), asks: "Is the first character of the color a B?" To check this, it uses the **$eq** and **$substr** functions. The **$eq** function returns 1 (true) if its two string arguments are the same, otherwise it returns 0 (false). The **$substr** function works exactly like its AutoLISP counterpart.

$substr

$(substr, *string*, *start*, [*length*])

Returns the portion of **string**, *beginning with the* **start** *(an integer) character position and continuing for* **length** *(an integer) characters. If* **length** *is not specified, the entire balance of* **string** *is returned. The first character position is referenced by a* **start** *value of 1.*

In this case in 10MODE1.LSP, $(substr,$(getvar,CECOLOR),1,1) returns the first character of the current color, then **$eq** compares it to B. The first **$if** determines whether the first character equals 'B'. The *then* expression is $(substr,$(getvar,CECOLOR),1,3), which uses **$substr** to extract the first three characters (BYB or BYL) of the value returned by $(getvar,CECOLOR). The *else* is a nested **$if** expression, which uses the **$<** (less than) function to decide if the color number is less than 8. The *then* expression uses **$substr** to determine which character of the string RYGCBMW to return for display. The color number returned by $(getvar,CECOLOR) is the start number for **$substr**. The *else* of this second **$if**, which returns the color number if 8 or greater, is simple: $(getvar,CECOLOR). Notice the space after the two closing parentheses of the **$if** expressions. It puts a space on the status line before the next item.

The only other complex expression in the 10MODE1.LSP MODEMACRO expression is the snap status expression. It displays an S: or s to indicate if snap is on or off. You can do this with a simple **$if** expression. But it also checks the snap unit setting and, only if the X,Y aspect ratio is equal, displays the snap unit. The SNAPUNIT system variable returns an X,Y value. A comma-delimited list, such as a point value, is to DIESEL as a list is to AutoLISP. To compare the X and Y components, you need to extract them, as explained in the following section.

Complex Data Manipulation

As you saw earlier in this chapter, DIESEL represents complex data as comma-delimited lists, such as "Red,Green,Blue". To retrieve one item of a list, you can use the **$index** function.

$index

($index,*position*,*list*)

Returns an element in a comma-delimited string. Returns the element specified by **position** *(an integer) in* **list**, *a string containing a comma-delimited list. The first item in the list is referenced by the position value 0.* **$index** *returns "" (a null string) if* **position** *is greater than the number of items in a list. A list can also contain null elements, which are represented by consecutive commas.*

Note how the **$index** function is used in the SNAP part of the status line macro:

```
"$(if,$(GETVAR,SNAPMODE),"              ;if 1
    "S:$(if,"                            ;then S (on) and if
        "$(=,$(index,0,$(getvar,SNAPUNIT)),"   ;test X=Y snap
            "$(index,1,$(getvar,SNAPUNIT))),"
        "$(index,0,$(getvar,SNAPUNIT)),"      ;then display X
        "/="                                  ;else display /=
    ")"                                      ;close if
```

```
        ",s "                                      ;else s (off)
")"                                                ;close if
```

SNAPMODE works like many system variables that maintain the ON/OFF state of a setting: 1 is on and 0 is off. The expression $(getvar,SNAPMODE) returns 1 if SNAPMODE is on and 0 if it is off. The first part of the snap expression is a simple **$if** to test if SNAPMODE is 1. The *else* expression is a simple ",s " which displays an s. The *then* expression is more complex, an S: to indicate that SNAP is ON, followed by another nested **$if** to check the SNAPUNIT system variable.

To check SNAPUNIT, **$if** uses **$=** and **$index** functions. The **$=** function returns 1 (true) if its two number arguments are the same, otherwise it returns 0 (false). **$index** extracts the X and Y components from SNAPUNIT (a 2D point), which **$getvar** returns as a list of comma-delimited numbers. For example, $(index,0,$(getvar, SNAPUNIT)) extracts the X component and $(index,1,$(getvar,SNAPUNIT)) extracts the Y component. If the components are equal, the *then* of the second **$if** expression extracts one of the components for display. Otherwise the *else*, "/=", is displayed.

Compared to this, the rest of the expressions in the 10MODE1.LSP MODEMACRO definition, such as for ortho, are simple **$if** expressions.

The 10MODE1.LSP MODEMACRO definition is about as complex as you can get in one chunk without encountering a `string too long` error. You can get around SETVAR's string length limitation by breaking the expression up into several special system variables.

DIESEL Data Storage and Retrieval

DIESEL has no variables. However, it can access AutoCAD system variables using the **$getvar** function, which allows you to pass information between AutoLISP programs and DIESEL expressions. AutoCAD has five undefined system variables for each of the three basic types of data: real, integer, and string. Any string up to 460 characters long can be stored in the five user-definable string system variables, USERS1 through USERS5. You can put multiple data elements in a string system variable by delimiting each with a comma and then using the DIESEL **$index** function to extract specific elements from the comma-delimited list. You can also store numbers. The five user-definable integer system variables, USERI1-USERI5, can store integers. The five user-definable real system variables, USERR1-USERR5, can hold real (decimal) numbers.

Warning: The USERI1-5 and USERR1-5 system variables (integer and real) are saved in the current drawing, but the USERS1-5 system variables are *not* saved with the drawing. Values assigned to the latter exist only within the drawing editor. Before using any user system variables, you should confirm that they are not used elsewhere in your programs, or in other third-party programs you are using.

Running on DIESEL

MODEMACRO String Length Limit

The MODEMACRO system variable can store up to 460 characters, and effectively limits the complexity of a status line DIESEL expression to what can be defined in that length. One way to overcome this is to use one or more of the five user-definable string system variables (USERS1-5) to store additional DIESEL expressions. When the DIESEL **$getvar** function returns a string, DIESEL does not attempt to evaluate it as an expression, and treats it as if it were a literal string in the DIESEL macro. The DIESEL **$eval** function can force DIESEL to evaluate any string, which is what permits you to store additional parts of a DIESEL macro in the USERS*n* system variables, and have them evaluated after being returned by **$getvar**. The **$eval** function will accept any string and evaluate it as a DIESEL expression.

$eval

$(eval,*expr*)

Evaluates ***expr*** (a string) as a DIESEL expression.

The following example shows how to use three user-defined string system variables to store a custom status line DIESEL macro. The system variables USERS1, USERS2, and USERS3 are each assigned portions of the total DIESEL macro that defines the custom status line, and their contents are each evaluated in sequence as DIESEL expressions using the **$eval** function after they are combined into a single string with the AutoLISP **strcat** function

```
(setvar "USERS1" "Part One")
(setvar "USERS2" "Part Two")
(setvar "USERS3" "Part Three")
(setvar "MODEMACRO"
   (strcat "$(eval,$(getvar,USERS1))  "
           "$(eval,$(getvar,USERS2))  "
           "$(eval,$(getvar,USERS3)) ")
   )
)
```

This will display `Part One   Part Two   Part Three` on the status line.

Another way to extend the string space available for DIESEL status line macros is by storing parts of them in operating system environment variables. There can be cases where USERS*n* system variables are needed for other purposes and are not available for use with a custom status line. In those cases, you can get the same functionality provided by USERS*n* system variables by using operating system environment variables, along with the DIESEL **$getenv** function to access them in a status line macro. The following variation on the last example uses the USERS1 and USERS2 system variables, along with an operating system environment variable named DIESEL, which is set in the AUTOEXEC.BAT file under DOS or Windows.

The following would be in the AUTOEXEC.BAT file:

```
SET DIESEL=Part Three
```

And the following would be in a LSP file.

```
(setvar "USERS1" "Part One")
(setvar "USERS2" "Part Two")

(setvar "MODEMACRO"
   (strcat "$(eval,$(getvar,USERS1)) "
           "$(eval,$(getvar,USERS2)) "
           "$(eval,$(getenv,DIESEL))"
   )
)
```

Again, this will display `Part One  Part Two  Part Three` on the status line.

The preceding example causes the value of the DIESEL environment variable to be treated as a literal part of the status line. You can also use the **$getenv** function to include other information from the operating system environment, such as a user's network login name or any other value that can be assigned to an environment variable.

$getenv

$(getenv, varname)

Returns the value of the operating system environment variable **varname**.

Custom Status Lines in AutoCAD for Windows

AutoCAD for Windows already provides much of the same information that's displayed by the custom status line definition in the preceding listing on its status bar and object properties toolbar, so the use of that status line with AutoCAD for Windows is somewhat redundant. But, you can use a custom status line in Windows to display other information that is not displayed elsewhere by AutoCAD.

Running on DIESEL

In this chapter's disk directory on the accompanying CD-ROM, you will find the file 10MODE2.LSP, which implements a more elaborate custom status line that is mostly useful for DOS AutoCAD, but which contains elements that can be adapted for use in AutoCAD for Windows versions. The major elements of the custom status line defined in 10MODE2.LSP are:

- La: *current layer*
- Lt: *current linetype* (displayed only if not BYLAYER)
- El: *current elevation* (displayed only if non-zero)
- Th: *current thickness* (displayed only if non-zero)
- Z: *Z component of last entered point*
- O/o (Ortho ON/OFF)
- S/s (Snap ON/OFF)
- T/t (Tablet ON/OFF)
- Vh: *height of current view in drawing units*
- Rot: *current snap rotation angle* (displayed only if non-zero)

The custom status lines included in this chapter's files are useful in their existing forms, but are mainly intended to serve as a starting point for developing your own custom status lines. Other useful information that can be displayed on the status line include the current running object snap modes, values from operating system environment variables (such as a user's login name), and command-specific variables such as the current text style and height.

10MODE2.LSP also provides functions that demonstrate AutoLISP programming techniques for controlling status lines dynamically. These routines can be called by your own AutoLISP programs to display extra information relevant to the currently active command. Included in the file are several variables that hold DIESEL macros that display information relating to specific commands, such as HATCH, TEXT, and dimension commands. You can add many other command-specific extensions to the default status line dynamically by calling the functions provided in 10MODE2.LSP.

The following listing is from the file 10QHATCH.LSP, which implements the QHATCH command. QHATCH is a simple example command that provides direct access to the internal point selection of the AutoCAD BHATCH command for quickly generating one or more hatches from an internal point that defines the boundary. The QHATCH command uses the functions in 10MODE2.LSP to dynamically alter the status line when the command executes. QHATCH adds the current hatch pattern, scale, and rotation to the default status line while the command is active.

```
;; 10QHATCH.LSP from Maximizing AutoCAD Release 13
;;
;; ADDS the QHATCH command to AutoCAD, to directly
;; access the BHATCH point selection routine
;;
;; Supports extended status line from 10MODE2.LSP

;; Status line extension for displaying hatch parameters:

(setq modestr:hatch
    (strcat
        "Pattern: $(getvar,HPNAME) "         ; pattern name
①       "Scale: $(rtos,$(getvar,HPSCALE)) "  ; scale
        "Angle: $(angtos,$(getvar,HPANG)) "  ; rotation
    )
)

(defun C:QHATCH ( / olderr point)       ; C:QHATCH command function
    (setq olderr *error*)               ; save existing error handler
    (defun *error* (s)                  ; define new error handler
        (setq *error* olderr)           ; restore old error handler
②   (addmodeline nil)                   ; restore default status line
    (repeat 3 (command))                ; cancel all active commands
    (princ)                             ; Return nothing
    )

    (setvar "CMDECHO" 0)                ; set CMDECHO off

③   (setmodeline modestr:hatch)         ; add hatch parameters
                                        ; to status line
    (while
        (setq point
④          (getpoint "\nInternal point: ") ) ; While user enters points
        (c:bhatch point)                ; submit points to (C:BATCH)
    )
⑤   (setmodeline nil)                            ; restore default status line

    (setq *error* olderr)                        ; restore old error handler
    (princ)                                      ; return nothing
)
(princ
    (strcat "\n" _FILE
        ".lsp from Maximizing AutoCAD Release 13 loaded."
))
(princ)

;; END 10QHATCH.LSP
```

Next, you will test the QHATCH command defined in 10QHATCH.LSP to observe how it dynamically alters the status line.

Exercise

Testing the QHATCH Command

Start AutoCAD with a new, unnamed drawing, and zoom Center on point 10,10 with a height of 10 units.

Command: **(load "10MODE2.LSP")** [Enter]	Loads 10MODE2.LSP
Command: **(load "10QHATCH")** [Enter]	Loads 10QHATCH.LSP

Note that the custom status line defined by 10mode2 is now active.

Command: **POLYGON** [Enter]	Starts POLYGON command
Number of sides <4>: **4** [Enter]	Specifies 4-sided square polygon
Edge/<Center of polygon>: **10,10** [Enter]	Specifies center point of 10,10
Inscribed in circle/Circumscribed about circle (I/C) <I>: **I** [Enter]	Specifies inscribed polygon
Radius of circle: **4** [Enter]	Specifies radius of 4 units
Command: **CIRCLE** [Enter]	Starts CIRCLE command
3P/2P/TTR/<Center point>: **8,8** [Enter]	Specifies center of 8,8
Diameter/<Radius>: **4** [Enter]	Specifies radius of 4 units
Command: **QHATCH** [Enter]	Starts QHATCH command

Note that the status line now includes the current hatch style, scale, and rotation.

Internal point: **9,9** [Enter]	Specifies internal point inside the circle and the square (see Figure 10.3)
Internal point: [Enter]	Terminates QHATCH command
Command:	Restores the status line to its default state

The code in 10QHATCH.LSP shown in the previous listing does several things to make the hatch parameters available, and to ensure that the default status line is restored when the command exits.

- When the file is loaded, the DIESEL macro that displays the hatch parameters is assigned to the global symbol `modestr:hatch` ①.

- When the **C:QHATCH** command executes, it first saves the existing error handler to a bound variable and defines a new error handler that is called in the event the command is canceled or some other error occurs ②.

- Then the CMDECHO system variable is turned off to suppress messages on the AutoCAD command line ③.

- Next, the user-defined AutoLISP function **addtmodeline** from 10MODE2.LSP is called and passed the symbol `modestr:hatch` ③. The **addmodeline** function encapsulates the task of adding the additional information to the default status line, and it also provides a way to restore the status line to its default state.

Figure 10.3
Custom status line during QHATCH command.

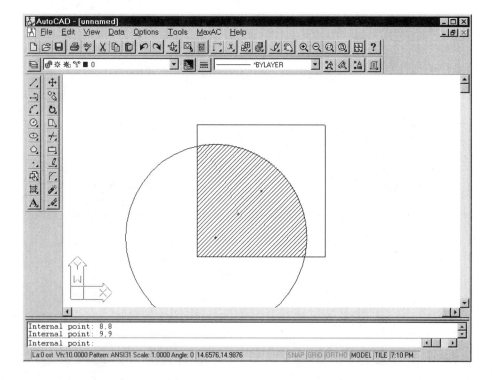

Running on DIESEL

- Next, a *while* loop is entered where the user is repeatedly prompted to enter an internal point that is submitted to the **C:BHATCH** function ④.

- When the user responds with SPACE or ENTER, the looping stops and the **addmodeline** function is called again ⑤, and is passed an argument of **nil**, which tells it to restore the status line to its default state.

The error handler that is defined by the **C:QHATCH** function also calls the **setmodeline** function to restore the default status line as well, and ensures that the default status line is restored in all cases. In addition to command-specific status displays for AutoCAD commands, you can also add information to a status line for custom programs and commands that you develop in AutoLISP, by passing the DIESEL macro that is to be added to the default status line to the **setmodeline** function in your program's main command function.

Dynamic Context-Sensitive Status Lines

The last exercise showed one way to add *context-sensitive* status lines to AutoCAD, which change with each command to display information relating to the currently active command. Although the technique used by 10MODE2.LSP is useful for custom commands defined in AutoLISP, it does not work as well for built-in AutoCAD commands, since they must be redefined to use context-sensitive status lines.

Included on this book's CD you will find a special version of DynaMode. DynaMode is a commercial utility that uses the AutoCAD Runtime Extension (ARX), which allows any command to have its own custom, context-sensitive status line. DynaMode adds functions to AutoLISP that allow you to easily define context-sensitive status lines for any built-in AutoCAD command, as well as commands that are defined in AutoLISP, ADS, or ARX.

Note: The version of DynaMode that's included on this book's CD-ROM enables you to define up to five context-sensitive status lines. The commercial version of DynaMode, which can be obtained directly from the authors, removes this restriction and allows you to define an unlimited number of context-sensitive status lines and associate each with a different AutoCAD command or group of commands. See Appendix G for more information.

Once a context-sensitive status line is defined and associated with one or more commands, the status line appears only when its associated command is active. A context-sensitive status line can be associated with one or several commands by the use of wildcards. You can define context-sensitive status lines using AutoLISP functions that are provided by DynaMode. In the following exercise, you'll load DynaMode and test several of the sample context-sensitive status lines that are included with it.

Exercise

Testing DynaMode

Start AutoCAD with a new, unnamed drawing, and zoom Center on point 10,10 with a height of 10 units. Install DynaMode from the MaxAC CD-ROM (see Appendix G) to the MaxAC directory.

Command: **(arxload "DYNAMODE.ARX")** [Enter]   Loads DYNAMODE.ARX

```
DynaMode 1.0 copyright ©1996 Tony Tanzillo all rights reserved
DynaMode 1.0 for Maximizing AutoCAD (Freeware Edition)
DynaMode context-sensitive mode line enabled.
"dynamode.arx"
DynaMode.DEF loaded
```

Command: **(SetDefaultModeLine "User: $(getvar,loginname)")** [Enter]   Sets the default status line to the value of the LOGINNAME system variable and displays your login name on the status line

Command: **PLINEWID** [Enter]   Starts SETVAR command

New value for PLINEWID <0.0000>: **.02** [Enter]   Specifies polyline width of 0.02

Command: **PLINE** [Enter]   Starts the PLINE command

Note that the status line has changed, and now displays the current polyline width.

From point: **7,6** [Enter]   Specifies first vertex of polyline

Current line-width is 0.0200

Arc/Close/Halfwidth/Length/Undo/Width
/<Endpoint of line>: **12,9** [Enter]

Arc/Close/Halfwidth/Length/Undo/Width
/<Endpoint of line>: **4,11** [Enter]

Arc/Close/Halfwidth/Length/Undo/Width
/<Endpoint of line>: **'ZOOM** [Enter]   Starts transparent ZOOM command

Note that the status line has changed again, and now displays the current view height.

```
>>Center/Dynamic/Left/Previous/         Increases magnification by 2x
Vmax/Window/<scale(X/XP)>: 2X Enter

Resuming PLINE command.
```

Note that the status line has changed again, and now displays the current polyline width.

```
Arc/Close/Halfwidth/Length/Undo/Width  Closes polyline
/<Endpoint of line>: C Enter

Command:
```

Note that the status line has changed again, and now displays the original default status line (the value of the LOGINNAME system variable).

As you can see from the preceding exercise, DynaMode makes the AutoCAD status line much more useful and enables you to take advantage of it with ease.

Controlling Pull-Down and Cursor Menu Labels with Diesel

Another application of DIESEL is within the labels of pull-down and cursor menus. DIESEL macros can be included in the labels of menu items to control their content and appearance dynamically. Both pull-down and cursor menus also possess a number of controllable display attributes such as check marks and disabling (or graying) labels as shown in Figure 10.4. By using DIESEL, you can dynamically apply these features to a menu item label every time the menu that contains it is displayed. You do this by including a DIESEL macro expression between the label brackets of a menu item.

You can also use DIESEL expressions to change the actual text of the menu label. DIESEL macros in menu labels work exactly the same as DIESEL macros used for displaying custom status lines. The important distinction lies in *when* a DIESEL macro is evaluated. DIESEL expressions that appear in a pull-down menu label are evaluated every time the menu becomes visible. The dynamic nature of DIESEL macro evaluation in menu labels provides for some interesting possibilities.

- You can limit the choices in a menu to only those available or applicable at the point when a menu appears, by disabling menu items conditionally based on the value of one or more system variables.

- You can have menu items whose labels reflect the state of a setting in the drawing editor, or a complex condition that can involve several settings.

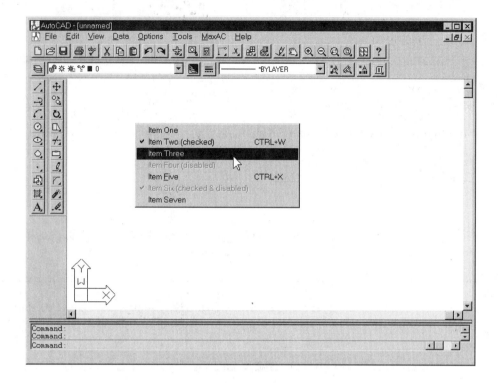

Figure 10.4
Cursor menu with checked and disabled items.

▶ You can create menu labels whose text can be established at the point when the menu appears, and which can be derived from system variables, operating system environment variables, or complex conditions involving a number of variables.

AutoCAD's menu macro language provides a number of special directives that alter the appearance of menu items. Chapter 13 covers menu structure and these directives in complete detail. In the following sections of this chapter, you'll see how you can use DIESEL to issue those directives to control a menu label's appearance dynamically.

Check Marks in Menu Labels

A user interface should strive to provide indications of the state of various settings in places where the user is likely to need that information. One way of doing this is to display check marks on menu items that represent and control an On/Off setting. The standard AutoCAD menu demonstrates a number of examples of how DIESEL can be used to determine the state of an on/off system variable, and display a check mark on the label of a pull-down menu item that controls that variable.

One way to control the state of check marks in menus is to use the AutoLISP **menucmd** function, which issues menu command processor directives like the ones that appear in menu macros. In menu macros, these directives start with a dollar sign; in **menucmd** expressions, the dollar sign is omitted.

---

menucmd

(menucmd *string*)

Provides access to AutoCAD command processor directives from AutoLISP.

For example, the following call to the **menucmd** function adds a check mark to the second item on the fifth pull-down menu:

```
(menucmd "P5.2=!.")
```

The number 5 in "P5.2" specifies the fifth pull-down menu. The number 2 specifies the second item on the specified menu. The directive **!.** is the instruction to add a check mark to the menu's label. The directive to remove or clear a check mark is simply a null string. The following example removes the same check mark added by the preceding example.

```
(menucmd "P5.2=")
```

AutoLISP programs can use **menucmd** to alter the display of menu items to reflect the current state of a setting such as an AutoCAD system variable, or one that is specific to a custom application. There are several problems relating to the use of **menucmd** to control menu items. First, the directives reference menus and the items on menus by their *position*. This will work reliably only if the positions of the menu and the items on it does not change. Because menus and their items can be added and removed, the numeric values that reference a specific menu item can change. What's worse, there is no way to determine if such a change has occurred.

Because of the potential for problems associated with using **menucmd** to reference menu items by position, we suggest you avoid using **menucmd** in that way. On the Windows platform, there is a better way to reference specific menu items. You can use the menu group name and name tag of a menu item with the **menucmd** function to reference a menu. The following pull-down menu item is from the View menu (***POP3) of the standard ACAD.MNU that ships with AutoCAD.

```
***MENUGROUP=ACAD

ID_Redall [Redraw &All]'_redrawall
```

The string `ID_Redall` is the menu item's name tag, which must be unique within the menu group (the menu group for ACAD.MNU is ACAD). The following call to **menucmd** will add a check mark to the Redraw **A**ll item on the **V**iew pull-down menu.

```
(menucmd "Gacad.Id_Redall=!.")
```

In this example, the string **Gacad** is the letter **G** prefixed to the name of the menu group that contains the item to be altered (acad), and **ID_Redall** is the name tag of the menu item. The menu group and name tag are always separated by a period. This latter method of referencing pull-down menu items is far more reliable than referencing them by numeric position, and it is the preferred way to alter the appearance of a menu item label from an AutoLISP or ADS program, in cases where the application has exclusive control of the underlying setting associated with the menu item.

There are cases where custom applications do not have exclusive control over a setting. For example, an AutoLISP program can determine and change the state of the ORTHOMODE system variable, and it can add a check mark to a pull-down menu item that reflects that state, but the menu that contains the item is available at any time, including when the AutoLISP program isn't running. Hence, it is impossible to use the **menucmd** function to maintain a check mark on a menu item that accurately reflects the state of ORTHOMODE any time the menu that contains it appears.

Setting Check Marks with DIESEL

The DIESEL language provides another more dynamic means of controlling menu item labels for AutoCAD-specific settings that can be changed by AutoCAD, or user-initiated actions. A DIESEL macro can be included in a pull-down or cursor menu label to control that menu label's appearance directly. Every time the menu appears, the DIESEL macro is evaluated, and its result becomes part of the menu label. The following is a simple example menu page that controls the display of check marks on several menu items.

```
**POP_EXAMPLE
[$(if,$(getvar,ORTHOMODE),!.)Ortho]^O
[$(if,$(getvar,GRIDMODE),!.)Grid]^G
[$(if,$(getvar,SNAPMODE),!.)Snap]^S
[$(if,$(getvar,TABMODE),!.)Tablet]^T
```

In each of the preceding menu items, the DIESEL **$if** function tests the result of **$getvar**, which is passed the name of a system variable. All of the system variables in the example are *toggles*, which can have only two possible values, 1 (on) or 0 (off). If a system variable has a value of 1, the second argument in each call to **$if** (the string **!.** which is the directive to add a check mark to the menu item) is returned and becomes part of the menu item label, and the menu item is given a check mark. If a call to

$getvar returns 0 (off), the parent $if expression returns a null string because there is no *else* argument to $if, and the menu item does not get checked.

Unlike the AutoLISP **menucmd** function, the DIESEL macros in the preceding menu example are evaluated every time the menu they are part of appears, which ensures that the menu items always reflect the current state of the system variables.

In addition to displaying the state of a toggle, check marks have other uses in pull-down and cursor menu labels. For example, a menu with several items can be used to represent the state of system variables that can have more than two possible values or states. An example of such a system variable is LUNITS, which maintains the linear units format settings for distance formatting as integer values ranging from 1 to 5. The range of possible values of the LUNITS system variable and their meanings are shown in Table 10.2.

One way to present the value of LUNITS to the user is by using a pull-down or cursor menu (or submenu) that contains one item for each of the five possible settings, with a check mark applied to the item that represents the current setting. In such cases, all of the menu items behave like a *radio group*, where only one of the items can have a check mark at any time. The advantage to using such an approach is that it allows the user to clearly see the current value of a setting within the context of all possible values for the setting, and also select a new value by making a single selection. The disadvantage is that it consumes a significant amount of precious menu real estate.

Figure 10.5 shows the Annotation and Text, Linear units submenu, which is a part of the **DDEMODES cursor menu page that is defined in the partial menu file 10CURSOR.MNS. *Partial menus,* as their name implies, are not complete menus, and are attached to the current *base* menu using the MENULOAD command. Partial menus are supported only on the Windows platform. Under Windows, this chapter's exercise and supporting files are set up to automatically load several partial menus. In DOS, the menu file ACADMA.MNU contains all of these menus combined. You will be seeing more of 10CURSOR.MNS later in this chapter. The Linear units submenu shown in Figure 10.5 contains five items, one for each of the five linear units settings. Because the current units setting is 2 (Decimal), that item is checked in the figure.

Table 10.2	Integer Value	Units Format
LUNITS System Variable Values	1	Scientific
	2	Decimal
	3	Engineering
	4	Architectural
	5	Fractional

Figure 10.5
The Linear units cursor submenu.

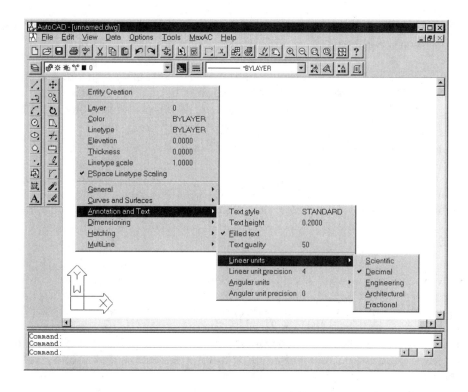

The following listing shows the menu source for the Linear units submenu:

```
[->&Linear units]
  [$(if,$(eq,1,$(getvar,LUNITS)),!.)&Scientific]'LUNITS;1
  [$(if,$(eq,2,$(getvar,LUNITS)),!.)&Decimal]'LUNITS;2
  [$(if,$(eq,3,$(getvar,LUNITS)),!.)&Engineering]'LUNITS;3
  [$(if,$(eq,4,$(getvar,LUNITS)),!.)&Architectural]'LUNITS;4
  [<-$(if,$(eq,5,$(getvar,LUNITS)),!.)&Fractional]'LUNITS;5
```

In each of these menu items, the **$if** function receives the result of a relational test using the **$eq** function. **$eq** is similar to its AutoLISP counterpart, except that like all logical values in DIESEL, it returns 1 (for true) or 0 (for false), rather than nil or non-nil as the AutoLISP **eq** function returns. In each expression, **$eq** compares the value of the LUNITS system variable to the value corresponding to the item's setting. After the DIESEL macros in all of the labels in this submenu are evaluated, the value of the LUNITS system variable is compared to every possible value the system variable can have. If any one of the tests succeeds, the second argument to **$if** becomes part of the menu label. In all cases, that argument is the directive to add a check mark to the menu item. If a test fails, then **$if** returns a null string, and that menu item is not checked. In this case, the test performed by one and only one of the DIESEL macros will always evaluate to true, and all others will evaluate to false, causing the appropriate menu item to have a check mark, and leaving all others without a check mark.

The preceding example demonstrates how menu labels are modified dynamically by DIESEL expressions to always reflect the current value of a setting. The example also shows how check marks can be used with multiple menu items to reflect the state of settings that can have more than two values. Try using the DDEMODES menu to change the current units settings from decimal to architectural, then to engineering (using the command line), and then back to decimal. When you do this, note that the check mark is properly synchronized to always reflect the current value of LUNITS, even when the variable is changed from the command line.

You can display the DDEMODES menu by placing the cursor over the graphics area of the AutoCAD window, holding down both the SHIFT and CTRL keys simultaneously, and clicking the same button on your pointing device that issues an ENTER when no keyboard modifiers are used (the first programmable button, which on a two- or three-button system mouse is the right button). In the following exercise, this button is referred to as *button 1*.

In DOS, the DDEMODES cursor menu is part of the base menu. Ignore the first instruction in the exercise, which loads 10CURSOR.MNS.

In Windows, before you can use the DDEMODES menu, you must first load the partial menu file 10CURSOR.MNS by choosing its name from the **M**enus submenu of the **M**axAC pull-down menu.

Exercise

Testing the Linear Units Cursor Submenu

Start a new, unnamed drawing using ACAD.DWG as the prototype.

Choose **M**axAC, **M**enus, 10CURSOR	Loads 10CURSOR.MNS and MNL
`10Cursor.mnl loaded.`	
`Command:` **LUNITS** Enter	Starts the SETVAR command for LUNITS
`New value for LUNITS <2>:` **2** Enter	Makes sure linear units are set to Decimal

Command: *With the cursor over the graphics area of the AutoCAD window, click CTRL+SHIFT+Button1*	The DDEMODES cursor menu appears
Choose **A**nnotation and Text, **L**inear units	The **A**nnotation and **T**ext, then **L**inear units submenus appear, with **D**ecimal checked
Choose **A**rchitectural	Sets linear units to **A**rchitectural

Move the crosshairs and see the architectural units in the coordinates display.

Command: *Click CTRL+SHIFT+Button1*	The DDEMODES cursor menu appears
Choose **A**nnotation and Text, **L**inear Units	The **A**nnotation and Text, then **L**inear units submenus appear, with **A**rchitectural checked
Press CANCEL 3 times	Dismisses the cursor menu

Next, you change the value of LUNITS from the command line:

Command: **LUNITS** [Enter]	Starts the SETVAR command for LUNITS
New value for LUNITS <4>: **3** [Enter]	Sets linear units to **E**ngineering
Command: *Click CTRL+SHIFT+Button1*	The DDEMODES cursor menu appears
Choose **A**nnotation and Text, **L**inear Units	The **A**nnotation and Text, then **L**inear units submenus appear, with **E**ngineering checked
Choose **D**ecimal	Sets linear units to Decimal
Command: **LUNITS** [Enter]	Starts the SETVAR command for LUNITS
New value for LUNITS <2>: *Press CANCEL*	Confirms linear units is set to Decimal

As the preceding exercises illustrate, DIESEL provides a powerful way to provide dynamic feedback in a user interface. The same methods shown in this popup cursor menu also apply to pull-down menus.

Using DIESEL in Menu Macros

In addition to controlling menu labels dynamically, DIESEL expressions can also appear in and control menu macros. DIESEL expressions in menu macros can issue commands and conditionally branch in the same way that DIESEL expressions in menu labels do. A DIESEL expression can also perform simple computations to derive input from the values of one or more system variables.

A DIESEL expression can be included in a menu macro by preceding it with the **$M=** directive. The **$M=** directive tells the menu interpreter that what follows is a DIESEL expression that requires evaluation by DIESEL *before* the expression is processed by the menu interpreter.

The following example menu item controls the BLIPMODE system variable. This system variable controls the display of blips.

```
[$(if,$(getvar,BLIPMODE),!.)Blips]'BLIPMODE;+
$M=$(xor,1,$(getvar,BLIPMODE))
```

The menu item in the preceding example uses DIESEL in both its label and in the menu macro. The DIESEL expression in the menu label adds a check mark to the item if the current value of BLIPMODE is 1 when the menu appears. The DIESEL expression in the menu macro will toggle the value of BLIPMODE between 1 and 0 when it is selected.

The menu macro begins with a transparent invocation of the SETVAR command (transparent because the system variable's name is prefixed with the apostrophe), followed by a semicolon, which issues an ENTER. The next part of the macro, which follows **$M=,** is the DIESEL expression. The **$xor** function is a logical operator that returns the bitwise complement of its arguments. Logical operators are discussed in more detail later in this chapter. The first argument to **$xor** is 1.

The second argument is a call to the **$getvar** function, which returns the current value of the BLIPMODE system variable. BLIPMODE can have one of two possible values, 0 (off) and 1 (on). When the value of BLIPMODE is 0, the result of **$xor** is 1, which becomes the new value of BLIPMODE. When the value of BLIPMODE is 1, the result of **$xor** is 0, which becomes the new value of BLIPMODE. Hence, the result of **$xor** is always the complement of its second argument, causing the value of the BLIPMODE system variable to be toggled between 1 and 0 each time the menu macro executes. You will see the DIESEL **$xor** function used often in this same way in many examples and exercises in this chapter.

The most important aspect of the preceding example macro lies in the fact that it can be used at *any time*, including when an AutoLISP program is running. As you already know, AutoLISP programs execute exclusively, and two different AutoLISP programs cannot execute simultaneously. If you try to run one AutoLISP program while another is running, it results in a `can't reenter AutoLISP` error. Hence, if the menu macro in this example were implemented in AutoLISP, it could not be used while another AutoLISP program is running. Since the menu item in this example uses DIESEL to compute the new value for BLIPMODE, it can be used at any time, including when an AutoLISP program is active. Because of the completely transparent nature of DIESEL, it is preferable to use it in menu macros wherever possible rather than AutoLISP, since there are no restrictions on when a menu macro that uses DIESEL can be used.

 Tip: You can put a plus continuation character (+) and begin a new line in the menu file anywhere in a DIESEL expression, except in a macro label. (You used to be able to put a plus in a macro label in Release 12, but you can't in R13c4). When it sees a plus at the end of a line, the AutoCAD menu interpreter ignores it and continues with the next line as if it were a single line of code. In menus with many long labels and macros, you can use the + in a consistent way to make the menu files more readable by placing them at the end of long labels, and starting the body of the menu macro on the following line.

DIESEL Expression Evaluation

DIESEL uses an *interpreter* to evaluate expressions. When AutoCAD's menu macro processor executes a menu macro, each item in the macro is read and passed to AutoCAD as input. When a DIESEL expression is encountered in a menu macro, it (along with the entire remainder of the menu macro) is passed to the DIESEL interpreter for evaluation before the remainder of the macro is executed. The result of the evaluation is then returned to the AutoCAD menu macro processor, where processing continues.

The full importance and effects of this description of DIESEL in menu macros may not be obvious. Study the description, noting the italics and bolding: "Each item in the macro is read and passed to AutoCAD as input. When a DIESEL expression is encountered in a menu macro, it (along with the entire *remainder* of the menu macro) is passed to the DIESEL interpreter for evaluation *before* the remainder of the macro is executed." This requires special formatting of a macro if the non-DIESEL remainder has any double quotes in it, because DIESEL evaluation strips quotes from the strings it returns.

To examine the full importance and effects of this, try experimenting with the test macros on the Eval test submenu of the MaxAc pull-down menu. The Eval test submenu contains two menu items that both draw a line and a circle, using the LASTPOINT system variable to place the circle. The difference in the formatting of quotes in the macros affects DIESEL's evaluation and controls whether the correct LASTPOINT value is used.

Running on DIESEL

Exercise

 Testing DIESEL Evaluation Order

Continue from the previous exercise or start a new unnamed drawing.

Command: *From the **M**axAC pull-down menu, choose **E**val test, Old**P**oint*

Command: ID Point: *Pick point 2,2*   Updates LASTPOINT

X = 2.0000   Y = 2.0000
Z = 0.0000

Command: LINE From point: 2,2,0   Enters LASTPOINT from DIESEL

To point: *Pick point 2,4*   Updates LASTPOINT

To point: Ends command

Command: CIRCLE 3P/2P/TTR/   Incorrectly draws circle at 2,2,0,
<Center point>: Diameter/<Radius>:   the ID command's LASTPOINT
1

Command: *From the **M**axAC pull-down menu, choose **E**val test, New**P**oint*

Command: ID Point: *Pick point 2,5*   Updates LASTPOINT

X = 2.0000   Y = 5.0000
Z = 0.0000

Command: LINE From point: 2,5,0   Enters LASTPOINT from DIESEL

To point: *Pick point 2,7*   Updates LASTPOINT

To point: Ends command

Command: CIRCLE 3P/2P/TTR/   Correctly draws circle at 2,7,0, the
<Center point>: 2,7,0 Diameter/   LINE command's LASTPOINT
<Radius> <1.0000>: 1

Command: *From the **M**axAC pull-down menu, choose **Q**uote test, Wrong*

```
Command: (prompt Wrong )              Fails
error: bad argument type
(PROMPT WRONG)
```

Command: *From the* **M**ax*AC pull-down menu,*
choose **Q**uote test

Choose Right

```
Command: (prompt "Right ") Right      Works OK
nil
```

To understand the behavior of DIESEL in these menu items, look at the code of the OldPoint and NewPoint menu items from the **M**axAC pull-down menu. Note the bold code and quotes:

[OldPoint]ID \LINE $M=$(getvar,LASTPOINT) \;+
CIRCLE $(getvar,LASTPOINT) 1
[NewPoint]ID \LINE $M=$(getvar,LASTPOINT) \;+
CIRCLE **"$M=**$(getvar,LASTPOINT)**"** 1

The OldPoint macro first uses ID to reset the LASTPOINT system variable, before DIESEL gets involved. Assume that LASTPOINT is set to 2,2,0 in this case. Then, the **$M=** sends the rest of the menu code to DIESEL for evaluation before execution. DIESEL substitutes the current LASTPOINT value (the 2,2,0 set by ID) for both cases of $(getvar,LASTPOINT) and sends LINE 2,2,0 \;CIRCLE 2,2,0 1 off for execution. That wasn't what was intended; you would expect the LASTPOINT used by CIRCLE to be the last point entered in the immediately preceding LINE command, not the earlier ID command.

The NewPoint macro demonstrates a subtle but important difference, which controls the order of DIESEL's evaluation. The only difference is that NewPoint puts the second $(getvar,LASTPOINT) in quotes, along with a second $M= code, such as **"$M=**$(getvar,LASTPOINT)**"**. This delays its evaluation. The NewPoint macro uses ID to reset LASTPOINT, before DIESEL gets involved. Assume that LASTPOINT is set to 2,5,0 in this case. Then, the $M= sends the rest of the menu code to DIESEL for evaluation before execution. DIESEL substitutes the current LASTPOINT value (the 2,5,0 set by ID) for the first case of $(getvar,LASTPOINT), but it only strips the quotes from the second case and doesn't evaluate what is between the quotes yet, so it sends LINE 2,2,0 \;CIRCLE $M=$(getvar,LASTPOINT) 1 off for execution. Only the LINE 2,2,0 \;CIRCLE part is executed, setting LASTPOINT to the point you pick at the backslash pause. Assume that you pick 2,7,0. Then, the second $M= code is encountered, sending the rest of the macro back to DIESEL for another evaluation. This time, DIESEL substitutes the current LASTPOINT value (the 2,7,0 set by LINE) for the

Running on DIESEL

second case of $(getvar,LASTPOINT), then sends the resulting 2,7,0 1 for execution, drawing the circle.

The fact that each DIESEL evaluation strips one level of quotes is also important to AutoLISP expressions that contain strings if they follow a DIESEL expression in a menu item. This fact is demonstrated by the behavior of the Wrong and Right items on the **Q**uote test submenu of the **M**axAC pull-down menu. Notice the number of quotation marks in the following listing:

```
[Wrong]$M=$(if,$(getvar,ORTHOMODE),,^O)(prompt "Wrong ")
[Right]$M=$(if,$(getvar,ORTHOMODE),,^O)(prompt """Right """)
```

When the $M= code is encountered in these macros, the rest of the macro code is sent to DIESEL for evaluation before execution. DIESEL strips off one level of quotes and returns the results for execution. Wrong returns (prompt Wrong), which is not a valid AutoLISP expression because the **prompt** function requires a string argument. Right returns (prompt "Right "), which is a valid AutoLISP **prompt** expression.

To sum it up, you must nest DIESEL expressions in "$M=..." when necessary to delay their evaluation until after the execution of other items in a menu macro, and you must nest quotes in AutoLISP expressions that follow DIESEL in menu macros, using """ to get a single quote (an extra two quotes for each level of DIESEL evaluation).

Tip: Carefully count spaces and semicolons in DIESEL expressions in menus. DIESEL returns all string characters, including spaces, literally. Some DIESEL expressions, such as $M=$(getvar,LASTPOINT), require a following space or semicolon to enter their returned values; other expressions, such as $(if,$(getvar,ORTHOMODE),,^O), do not. In some cases, you need to prevent the AutoCAD menu interpreter from adding an automatic space at the end of a macro that contains DIESEL code. To do so, end the line with the menu code for CTRL+Z, ^Z, unless it already ends with a special character such as a semicolon.

Calling DIESEL Expressions from AutoLISP

You can call any DIESEL expression from an AutoLISP expression. To evaluate a DIESEL expression by using AutoLISP, just pass the expression to the **menucmd** function, which invokes the menu interpreter to evaluate menu commands. In a menu macro, menu commands are the codes, such as $P2=pagename, $P2=*, $P3.2=!.~, or the $M=$(xor,1,$(GETVAR,blipmode)) expression in the previous menu macro example. $P*n* and $M are the menu command device codes with $P*n* addressing the POP*n* menu and $M calling the DIESEL expression that follows it. The AutoLISP **menucmd** function issues a menu command in the same manner as the menu interpreter, and the menu command argument syntax is the same except that the $ character is omitted from the menu command device code. The $ characters in the DIESEL expression itself are

Table 10.3
Menu Code and AutoLISP `menucmd` Function Equivalents

Menu Macro	AutoLISP
$P2=pagename	(menucmd "P2=pagename")
$P2=*	(menucmd "P2=*")
$P3.2=!.~	(menucmd "P3.2=!.~")
$M=$(getvar,ATTREQ)	(menucmd "M=$(getvar,ATTREQ)")

still required. Example menu code and AutoLISP **menucmd** function equivalents are shown in Table 10.3.

The following example calls the DIESEL **$edtime** function from AutoLISP, to format the value of the DATE system variable.

```
(setq date (getvar "DATE"))
(setq cdate
   (menucmd
      (strcat "M=$(edtime,"
              (rtos (getvar "DATE") 2 6)
              ", M/DD/YY)"
      )
   )
)
```

Because the syntax of DIESEL and AutoLISP can be difficult to read when the two are intermingled as they are in the preceding example, it is often useful to encapsulate specific or frequently used calls to DIESEL from AutoLISP in dedicated user-defined AutoLISP functions. A useful example of this can be seen in the file 10DIESEL.LSP from this chapter's directory on the accompanying CD-ROM, which adds the **edtime** and **index** functions to AutoLISP.

Both **edtime** and **index** access their DIESEL counterpart functions, and provide all of the functionality of the same to AutoLISP programs.

Tip: We encourage you to study the highly structured set of functions in 10DIESEL.LSP to see how they help simplify access to DIESEL from AutoLISP. Note that the functions in 10DIESEL.LSP handle the conversion of AutoLISP numeric data to strings and concatenation of them into the comma-delimited string format for you, eliminating the need to do that manually as you must when calling DIESEL functions directly via **menucmd**.

Now you'll load and test those functions.

Running on DIESEL

Exercise

Testing Functions from 10DIESEL.LSP

Start AutoCAD with a new, unnamed drawing or continue in the current drawing.

Command: **(load "10diesel")** [Enter]   Loads 10DIESEL.LSP

10diesel.lsp loaded

Command: **(setq today (getvar "DATE"))** [Enter]   Assigns the value of the DATE system variable to the symbol **today**

LISP *returns:* 2.45018e+006   Value of DATE system variable

Command: **(edtime today "M/DD/YY")** [Enter]   Calls the **edtime** function and passes it the value of the **today** variable and the **"M/DD/YY"** formatting template

LISP *returns:* "4/03/96"   The current date

Command: **(edtime today "H:MM AM/PM")** [Enter]   Calls the **edtime** function and passes the value of the **date** variable, and a formatting template

LISP *returns:* "8:59 AM"   The current time

Command: **(setq mylist "Zero,One,Two,Three,Four,Five")** [Enter]   Assigns a string of comma-delimited items to the variable **mylist**

LISP *returns:* "Zero,One,Two,Three,Four,Five"

Command: **(index 0 mylist)** [Enter]   Calls the **index** function to obtain the first element in **mylist**

LISP *returns:* "Zero"   First element of **mylist**

Command: **(index 3 mylist)** [Enter]   Calls the **index** function to obtain the fourth element in **mylist**

493

> LISP returns: "Three"
>
> Command:

The **edtime** function accepts a numeric value that represents a date in the same format that the DATE and TD*XXXXX* system variables return, along with a string formatting template. The conventions of the formatting template are exactly the same as that of the DIESEL **$edtime** function described earlier in this chapter. The **edtime** function returns the formatted time/date string just as its DIESEL counterpart does.

The **index** operates just like its DIESEL counterpart. **Index** accepts an integer and a string that contains a list of items delimited by commas, and returns the element of the string specified by the integer argument. The **index** function is extremely useful for extracting items from comma-delimited strings, since AutoLISP provides no direct way of doing that.

Disabling Menu Items with DIESEL

DIESEL macros can also be used to disable menu items conditionally. As shown in Figure 10.4 earlier in this chapter, when a POPn (pull-down or cursor) menu item is disabled, its menu label is grayed to indicate that the item is not available. Disabling a menu item aids the users by clearly showing which items on a menu are currently not available when they activate the menu.

There are many valid reasons why menus should be disabled. When a command is either not valid or not appropriate at any given time, the menu items that access it can and should be disabled to provide the user with a visual clue that the related command or operation is not available. For example, AutoCAD will not permit you to use the PAN and ZOOM commands if the current viewport contains a perspective view. The DIESEL macro in the following example Pan menu item's label will disable the item when the current viewport contains a perspective view.

```
[$(if,$(getvar,VIEWMODE),~)&Pan]'PAN
```

The VIEWMODE system variable returns 1 when the current viewport contains a perspective view, and 0 when it contains a non-perspective (orthographic) view. The **$if** function tests this system variable directly, and if it returns 1, **$if** returns its second argument, the tilde (~) character, which is the directive to disable the menu item. If VIEWMODE is 0, **$if** returns a null string and the menu item is enabled. Remember that this happens every time the menu appears, and because of this, the menu item will always be correctly disabled when the PAN command is not available.

Menu disabling can be linked to even more complex conditions as well. For example, the DVIEW command cannot be used when floating model space viewports are active and the current space is paper space. The TILEMODE system variable has a value of 0 when floating model space views are enabled, and if they are, the CVPORT system variable is 1 when paper space is the current space. The following example menu item issues the DVIEW command, and it disables itself when floating model space viewports are active and you are in paper space:

```
[$(if,$(and,$(eq,0,$(getvar,TILEMODE)),+
       (eq,1,$(getvar,CVPORT))),~)DVIEW]'DVIEW
```

The **$and** function is another logical operator. It accepts up to nine arguments and returns 1 when *all* of them evaluate to non-zero values. If at least one argument evaluates to 0, then **$and** returns 0. Remember that unlike the AutoLISP **and** function, which operates on symbolic (nil and non-nil) values, DIESEL represents Boolean logic with numeric values (zero=false, non-zero=true).

In the preceding example menu item, the values of the TILEMODE and CVPORT system variables are compared to the values 0 and 1 respectively. If TILEMODE is 0, and CVPORT is 1, **$and** returns 1, and **$if** returns its second argument, which is the directive to disable the menu item.

The following listing contains a modified version of the View, Zoom submenu from the standard AutoCAD menu file ACAD.MNU. The modified version appears as a submenu on the MaxAC pull-down menu for this chapter's exercises. This modified version of the Zoom submenu has DIESEL macros in its labels to properly disable the ZOOM command, as well as various ZOOM command options when they are not available. The first label appears on a single line in the menu file, but is broken into multiple lines and indented for clarity here. The menu name ID tags are also omitted for brevity. The DIESEL expressions that have been added are shown in bold.

```
[->$(if,$(or,$(getvar,VIEWMODE),+
         $(eq,$(getvar,CMDNAMES),DVIEW),+
         $(eq,$(getvar,CMDNAMES),DTEXT),+
         $(and,$(eq,0,$(getvar,TILEMODE)),+
              $(eq,1,$(getvar,CVPORT)),+
              $(getvar,CMDACTIVE))),~)&Zoom]
[&In]'_zoom;2x
[&Out]'_zoom;.5x
[--]
[&Window]'_zoom;_w
[$(if,$(getvar,CMDACTIVE),~)&All]_zoom;_all
[&Previous]'_zoom;_p
[&Scale]'_zoom
[&Dynamic]'_zoom;_d
```

```
[&Center]'_zoom;_c
[&Left]'_zoom;_l
[Li&mits]'_zoom;$M=$(getvar,LIMMIN);$(getvar,LIMMAX)
[$(if,$(getvar,CMDACTIVE),~)&Extents]_zoom;_e
[-$(if,$(and,$(eq,0,$(getvar,TILEMODE)),+
              $(eq,1,$(getvar,CVPORT)),~)&Vmax]'_zoom;_vmax
```

Note from the preceding listing that cascading submenu labels can also have DIESEL expressions in them, allowing you to disable the entire submenu. In the first label (which is the submenu's label), a somewhat lengthy DIESEL expression will disable the entire Zoom submenu when any one of the following conditions is true:

- The current view is *perspective*.

- The DVIEW or the DTEXT command is currently *active*.

- Any *command* is *active* while in *paper space*.

The subexpression starting with $(and,$(eq,0,$(getvar,TILEMODE))... determines if paper space is active *and* a command is active, using the CMDACTIVE system variable. The CMDACTIVE system variable maintains a bit-coded integer that is the sum of the following:

```
0    no command is active
1    a command is in use
2    a command and a transparent command are in use
4    a script is active
8    a dialog box is active
```

The **$or** function is another logical operator that accepts up to nine arguments and returns 1 (true) if *any* of its arguments is non-zero. If all arguments are 0, then **$or** returns 0 (false). The two expressions that start with $(eq,$(getvar,CMDNAMES)... determine if either of the two commands DVIEW and DTEXT is currently active using the CMDNAMES system variable. The CMDNAMES system variable contains the name(s) of all currently active commands, or the null string if none are active. The ZOOM command cannot be used while the DTEXT or DVIEW commands are active, so its menu item label is controlled by DIESEL in the Zoom submenu, as shown in Figure 10.6.

If the conditions for disabling the Zoom submenu fail and it is not disabled, then several of the submenu's items also have labels with DIESEL macros in them that determine if their respective options are currently available, and disables them if they're not. The ZOOM command Extents and All options are not available while a command is active, so DIESEL expressions in those two item's labels disable them if the CMDACTIVE system variable is non-zero. The ZOOM command's Vmax option is not available when in paper space, so that menu item's label has a DIESEL macro in it that disables it when in paper space. Notice that AutoCAD does not remove the Vmax option from the ZOOM

Figure 10.6
The Zoom sub-menu example.

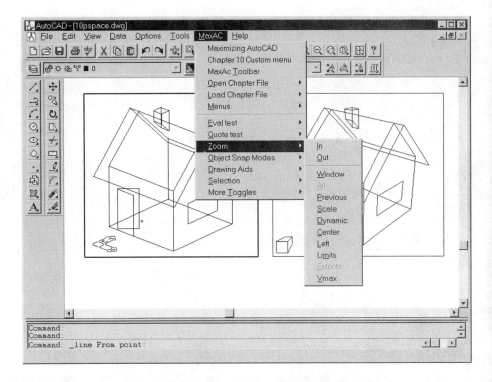

command's prompt when paper space is active, as it should (and does with other ZOOM options when they're not currently available).

Note: There are actually a number of other conditions under which the ZOOM command or some of its options are not available. It would be unreasonable or impossible to test every one of those in a menu label's DIESEL macro. In cases where you cannot test all conditions, choose those most likely to trigger the test conditions.

Now you'll test the **Z**oom submenu using the drawing file 10PSPACE from this chapter's exercise drawings. Note that the left viewport contains a non-perspective view, and the right viewport contains a perspective view.

Testing Menu Disabling with DIESEL Menu Label Macros

Open 10PSPACE.DWG from this chapter's exercise drawings.

TILEMODE is off, the current space is MODEL, and the left viewport is active.

Choose **M**axAC, **Z**oom	All ZOOM options are available
`Command: LINE` [Enter]	Starts LINE command
Choose **M**axAC, **Z**oom	The **E**xtents and **A**ll options are disabled
`From point:` *Press CANCEL*	Cancels command
`Command: DTEXT` [Enter]	Starts the DTEXT command
`Justify/Style/<Start point>:`	
Choose **M**axAC	The **Z**oom submenu is disabled
Press CANCEL	Cancels the DTEXT command
Click in the right viewport to make it active	Right viewport becomes current viewport
Choose **M**axAC	The **Z**oom submenu is disabled
`Command: PSPACE` [Enter]	Switches to paper space
Choose **M**axAC, **Z**oom	The **V**max option is disabled
`Command: LINE` [Enter]	Starts LINE command
`From point:` *Choose* **M**axAC	The Zoom submenu is disabled
`Command:` *Choose* **M**axAC, **Z**oom	All ZOOM options are available

As you can see from the preceding examples, DIESEL expressions provide a powerful means of adding visual clues about current settings to menus and limiting the choices on them to only those that are available or appropriate when a menu appears.

Bitwise Logical Operators

Earlier in this chapter, you saw the use of the three logical operators **$and**, **$or**, and **$xor**. In their most basic application, the first two of these functions are similar to AutoLISP's functions of the same names, except that since DIESEL uses numeric values

Running on DIESEL

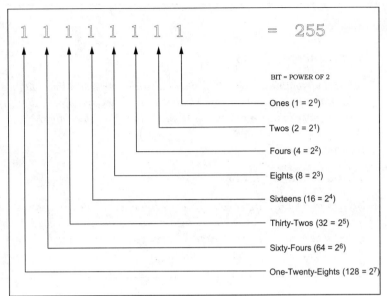

Figure 10.7
A bit-coded (binary) number.

to represent logic, the DIESEL versions operate on, and return, *numeric values*. What is not apparent from casual observation is their bitwise nature.

Bitwise operators operate upon binary (base 2 numbers) numbers. In a binary number, each digit can be 0 or 1, with a 1 in any digit representing the exponent of 2 corresponding to its position (see Figure 10.7).

A bitwise operator is one that treats each *binary digit* in a number as a distinct logical value or state, where any digit can be either on (1) or off (0) (or false or true respectively). Hence, a single numeric value whose value ranges from 0 to 255 when represented in binary notation requires eight digits, each of which can be either 1 or 0. That means an integer value between 0 and 255 can be used to represent up to eight distinct logical true/false (or on/off) states. Bitwise operators allow you to operate on these distinct logical states individually, by changing the corresponding bits.

$and

$(and, *integer*, ...)

Accepts up to nine arguments. Returns a binary-encoded integer with a 1 in each binary digit in which all arguments have 1 in the same binary digit. If not all arguments have 1 in the same binary digit, the result has a 0 in that same digit.

> **$or**
>
> **$(or,integer,...)**
>
> *Accepts up to nine arguments. Returns a binary-encoded integer with a 1 in each binary digit in which at least one of the arguments has 1 in the same binary digit. If all arguments have 0 in the same binary digit, the result has a 0 in that same digit.*
>
> **$(xor**
>
> **$(xor,integer,...)**
>
> *Accepts up to nine arguments. Returns a binary-encoded integer with a 1 in each digit in which the sum of the same binary digit of all arguments is an odd decimal value. If the sum of the same binary digit of all arguments is an even decimal value, the result is 0 in that same binary digit.*

The **$and** function is used in the following example, which contains two menu items from the **DDRMODES cursor menu page in the file 10CURSOR.MNS.

[$(if,$(and,1,$(getvar,PICKSTYLE)),!.)Group Selection]+
'PICKSTYLE $M=**$(xor,1,$(getvar,PICKSTYLE))**

[$(if,$(and,2,$(getvar,PICKSTYLE)),!.)Hatch Selection]+
'PICKSTYLE $M=**$(xor,2,$(getvar,PICKSTYLE))**

The preceding two menu items have DIESEL label macros that use the **$and** function to selectively *test* individual bits of a binary-encoded value. In this case, the value is the PICKSTYLE system variable. PICKSTYLE uses 2 bits of its value to represent two distinct logical states. That means it can have four possible values: 0, 1, 2, and 3. The right-most bit in PICKSTYLE controls group selection. If that bit is turned on (1), the expression $(and,**1**,$(getvar,PICKSTYLE)) returns 1. If that bit is turned off, the same expression returns 0. The second bit from the right in PICKSTYLE controls associative hatch object selection. If that bit is turned on (2), the DIESEL expression $(and,**2**,$(getvar,PICKSTYLE)) returns 2. If that bit is off, the same expression returns 0. In both of these expressions, the **$and** function is used to query the on/off state of individual bits in a binary-encoded value by specifying the bit that is to be queried as the first argument.

Similarly, the **$xor** function is used in DIESEL expressions in the menu macros to selectively *toggle* individual bits in the PICKSTYLE system variable. To toggle the right-most bit in PICKSTYLE, the expression $(xor,**1**,$(getvar,PICKSTYLE)) is used. If the right-most bit in PICKSTYLE is currently on, this expression returns a value with that same bit turned off (leaving all other bits unchanged). If the right-most bit of PICKSTYLE is currently off, the same expression returns a value with that same bit turned on (again, leaving all other bits unchanged). To toggle the second bit from the right in PICKSTYLE, the expression $(xor,2,$(getvar,PICKSTYLE)) is used. If the second bit from the right is currently ON, the result is a value with that same bit turned off, and with all other bits unchanged. If the second bit is currently off, the result of the same expression is a value with that same bit turned on, and with other bits unchanged. The menu macro changes the value of PICKSTLYE by setting the system variable to the result of the call to **$xor**. Notice that the first argument to **$xor** in the preceding expressions is a numeric value that specifies which bit is to be toggled.

There are a number of AutoCAD system variables that are binary-encoded integers. As with the PICKSTYLE system variable, in each of these values, each binary digit represents a separate or distinct logical state. AutoCAD's OSMODE system variable, which maintains the state of running object snap modes, is another example of a binary-encoded integer, where each binary digit in the value represents the ON/OFF state of a different running object snap mode. These values are shown in Table 10.4.

Note that each decimal value in the preceding table is represented in binary notation by a value that contains a single 1, each in a different position or digit. Bitwise operations consider each digit or *bit* in byte-encoded numbers as a *distinct logical state*, and they operate on them that way.

Table 10.4
The OSMODE System Variable

Decimal	Binary	Running Object Snap Mode
0	000000000000	None
1	000000000001	ENDpoint
2	000000000010	MIDpoint
4	000000000100	CENter
8	000000001000	NODe
16	000000010000	QUAdrant
32	000000100000	INTersection
64	000001000000	INSertion
128	000010000000	PERpendicular
256	000100000000	TANgent
512	001000000000	NEArest
1024	010000000000	QUIck
2048	100000000000	APParent intersection

Tip: Although you could use conditional and equality functions to test and change the state of system variables that are either on or off (bits-coded values of 1 or 0), it is safer to use bitwise logical operators. The advantage to using bitwise logical operators is that if, in a future version of AutoCAD, other bit-coded states are added to such a system variable, a bitwise macro will still work correctly, but a macro using conditional and equality functions will no longer work.

Note: There are a few system variables, such as DIMZIN, SURFTYPE LUNITS, CMLJUST, and the SOL*xxxxx* variables, which might at a glance appear to be bit-coded, but which are not.

The following are examples of logical operations on binary-encoded numbers. Values are shown in both decimal and binary notation side by side. Within this context, you can think of each binary digit as a distinct logical state where 1= true, and 0 = false.

Operation	Decimal	Binary	
	48	00110000	The **and** operation produces a value wherein each binary digit contains 1 only if *all* of the operands contain 1 in the same binary digit.
and	164	10100100	
and	36	00100100	
result	32	00100000	
	48	00110000	The **or** operation produces a value wherein each binary digit contains 1 if *any* operand contains 1 in the same binary digit.
or	164	10100100	
or	36	00100100	
result	180	10110100	
	164	10100100	The **xor** operation produces a value wherein each binary digit contains 1 if one of its two operands has a 1 in the same binary digit, and the other operand has 0 in that same digit.
xor	37	00100101	
result	129	10000001	

Another way to describe the **xor** operation is that it returns 1 in each binary digit, if the sum of the values in the same binary digit of all operands is *odd*. If the sum of a binary digit of all operands is *even*, the result has 0 in the same digit.

The popup cursor menu shown in the following listing is from the partial menu file 10CURSOR.MNS. It is the same as the cursor menu in the standard AutoCAD menu file

(ACAD.MNU), which provides access to immediate object snap modifiers, except this modified version adds cursor menu label macros that apply check marks to all current *running* object snap modes. The added check marks allow you to easily see which running object snap modes are currently active.

```
***POP0
[Osnap menu from 10CURSOR.MNS]
[From]_from
[--]
[$(if,$(and,$(getvar,OSMODE),1),!.)Endpoint]_endp
[$(if,$(and,$(getvar,OSMODE),2),!.)Midpoint]_mid
[$(if,$(and,$(getvar,OSMODE),32),!.)Intersection]_int
[$(if,$(and,$(getvar,OSMODE),2048),!.)Apparent Intersection]_appint
[$(if,$(and,$(getvar,OSMODE),4),!.)Center]_center
[$(if,$(and,$(getvar,OSMODE),16),!.)Quadrant]_qua
[$(if,$(and,$(getvar,OSMODE),128),!.)Perpendicular]perp
[$(if,$(and,$(getvar,OSMODE),256),!.)Tangent]_tan
[$(if,$(and,$(getvar,OSMODE),8),!.)Node]_node
[$(if,$(and,$(getvar,OSMODE),64),!.)Insertion]_ins
[$(if,$(and,$(getvar,OSMODE),512),!.)Nearest]_near
[$(if,$(and,$(getvar,OSMODE),1024),!.)Quick]_qui,^Z
[--]
[$(if,$(xor,1,$(getvar,OSMODE)),!.)None]_non
[--]
[.X].X
[.Y].Y
[.Z].Z
[.XZ].XZ
[.YZ].YZ
[.XY].XY
```

Note that each object snap menu item's label contains a DIESEL macro that passes the value of the OSMODE system variable along with the corresponding object snap mode's binary value to the **$and** function. For each item, if the running object snap mode is active, that mode's corresponding bit in the OSMODE system variable is 1 (ON), which causes the result of **$and** to be non-zero. If **$and** returns a non-zero value to **$if**, then it returns its second argument, an exclamation mark, which is the directive to add a check mark to the menu item. When there are one or more currently active running object snap modes enabled and this cursor menu is displayed, a check mark will appear on each of the current running object snap modes (see Figure 10.8).

Try using the cursor menu with several running object snap modes active. In this exercise, *Button 1* refers to the first programmable pointing device button (generally the right button on a mouse).

Figure 10.8
The customized object snap cursor menu from 10CURSOR.MNS.

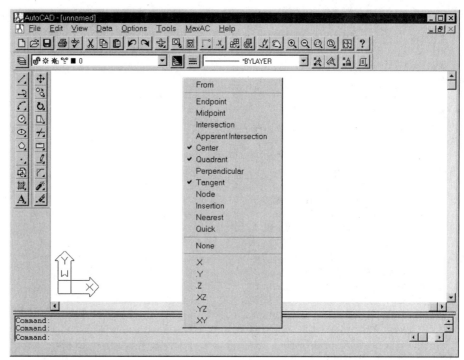

 Exercise

Testing the Enhanced Object Snap Cursor Menu

Start AutoCAD with a new, unnamed drawing. Choose MaxAC, Menus and make sure 10CURSOR is loaded.

Command: **OSNAP** [Enter]	Starts the OSNAP command
Object snap modes: **END,MID,PERP** [Enter]	Specifies Endpoint, Midpoint, and Perpendicular running object snap

Press SHIFT+Button1 and note which object snap modes are checked.

Command: **OSNAP** [Enter]	Starts the OSNAP command
Object snap modes: **CEN,TAN,QUA** [Enter]	Specifies Center, Tangent, and Quadrant running object snap modes

Press SHIFT+Button1 and note which object snap modes are checked.

```
Command: OSNAP [Enter]              Starts the OSNAP command

Object snap modes: NONE [Enter]     Specifies no running object snap
```

Press SHIFT+Button1 and note which object snap modes are checked.

In addition to the cursor menu in the preceding listing, the following listing shows several items from the **O**bject Snap Modes submenu of the **M**axAC pull-down menu for this chapter. The items in this menu also display check marks on the running object snap modes. In addition, it will toggle individual running osnap modes when they are selected.

```
[$(if,$(and,$(getvar,OSMODE),1),!.)Endpoint]+
'OSMODE $M=$(xor,1,$(getvar,OSMODE))
[$(if,$(and,$(getvar,OSMODE),16),!.)Quadrant]+
'OSMODE $M=$(xor,16,$(getvar,OSMODE))
[$(if,$(and,$(getvar,OSMODE),256),!.)Tangent]+
'OSMODE $M=$(xor,256,$(getvar,OSMODE))
[$(if,$(and,$(getvar,OSMODE),8),!.)Node]+
'OSMODE $M=$(xor,8,$(getvar,OSMODE))
[$(if,$(and,$(getvar,OSMODE),64),!.)Insertion]+
'OSMODE $M=$(xor,64,$(getvar,OSMODE))
```

Again, note how the **$xor** function is used to selectively toggle the corresponding bit of an individual running object snap mode, without disturbing the bits for all other modes. Try selecting items repeatedly from this menu, and note how each menu item toggles a different running object snap mode on and off, without affecting the state of other running osnap modes.

Creating Dynamic Menus with DIESEL

This section demonstrates even more creative uses for DIESEL in menus. Although AutoCAD commands are used here for demonstration purposes, the techniques that you will see in action can be applied to any custom application, limited only by your imagination. You will also learn about several AutoLISP tools and libraries included with this book's CD, which can be used as is for enhancing your application's user-interface.

Creating Dynamic Menu Captions with DIESEL

In addition to adding check marks and disabling (or *graying*) pull-down and cursor menu items, DIESEL macros can also generate menu caption text dynamically (a *caption* is the text that is displayed on a pull-down or cursor menu). The caption text can be derived from one or more AutoCAD system variables, environment variables, or for that matter, any value a DIESEL expression can compute and return.

The following menu item displays the name of the current text style in its caption.

`[$(eval,"")Text style\t$(getvar,TEXTSTYLE)]'textstyle`

In the preceding example, the menu label starts with a call to the DIESEL **$eval** function. The **$eval** function provides a way to coerce evaluation of a string as a DIESEL expression, when it would not otherwise be treated that way by the DIESEL interpreter.

$eval

$(eval,*string*)

Evaluates ***string*** as a DIESEL expression and returns the result. **$eval** takes one argument string, but it can contain a series of DIESEL expressions that are not separated by commas. Any literal text in the string is returned unchanged.

In the preceding text style example, **$eval** is passed a null string as an argument. This is done because, even though the menu item's caption starts with a literal string, the label *must* begin with a DIESEL macro, or subsequent DIESEL macros in the label will not be evaluated. The same menu label could also be written as follows.

`[$(eval,Text style\t$(getvar,TEXTSTYLE))]'TEXTSTYLE`

In the last example, the entire label is passed as the argument to **$eval**. You can write your own DIESEL macros either way, but you will find that the first method results in less complex expressions and reduces errors such as missing parentheses. Most DIESEL menu label macros you will see in this book and its accompanying files use the first method. You'll see more of the **$eval** function later in this chapter.

The two preceding examples both contain the literal string "Text Style" which is displayed in the menu's caption. Following the literal string is the sequence \t, which specifies the tab character. When a tab character (or \t) appears in a menu label, the text that follows it is left-justified in a second column, which will often contain the accelerator keystroke description for the menu item. After the \t sequence is a call to the DIESEL **$getvar** function, which is passed the TEXTSTYLE system variable name to retrieve the current text style name.

Note: When the \t sequence appears in menu labels in a MNU file, it will be replaced by an actual TAB character in the resulting MNS file that is generated by AutoCAD.

The next example is slightly more complex and shows how conditional branching functions can generate a menu item's caption with DIESEL. The example displays the current offset distance used by the OFFSET command, or the word "Through", if the current offset distance is less than 0. The OFFSETDIST system variable maintains the current offset distance, and uses a value of -1.0 to represent the "Through" option in the OFFSET command). Note that the example is broken into several lines and indented for readability in the following listing, but it must be on a single long line in the menu file.

```
[$(eval,"")Offsetdistance\t
    $(if,$(>,0,$(getvar,OFFSETDIST)),
        Through,
        $(rtos,$(getvar,OFFSETDIST)))]
```

The macro first compares the value of OFFSETDIST to 0. If the distance is less than 0, then **$>** (greater than) returns 1 to **$if**, and **$if** returns its second argument, the string "Through". If OFFSETDIST is not less than 0, **$>** returns 0 to **$if**, and **$if** returns its third argument, a call to the DIESEL **$rtos** function. The **$rtos** function converts the value of OFFSETDIST to a distance string, which becomes part of the menu caption.

The DIESEL **$rtos** function operates identically to its AutoLISP counterpart and converts distances to formatted strings using the current or a specified units and precision format. See the description for **rtos** in the chapter on AutoLISP.

The following example uses the **$index** function to display the current smooth surface type, used by the PEDIT command. The smooth surface type is maintained as an integer in the SURFTYPE system variable, which can have a value of 5 (Quadratic), 6 (Cubic), or 8 (Bezier). Again, this macro is shown here on several lines for clarity only.

```
[$(eval,"")Smooth surface type\t
    $(index,
        $(getvar,SURFTYPE),",,,,,Quadratic,Cubic,,Bezier")]'surftype
```

The **$index** function is passed a string surrounded by double quotes that contains the descriptive text of the three possible settings for the smooth surface type. The string contains commas, which requires it to be placed in double quotes (see the earlier section on DIESEL Macro Evaluation). The value of the SURFTYPE system variable is passed as the first argument to **$index**, which causes it to retrieve one of the eight elements of the comma-delimited list. The list is arranged so that the value of SURFTYPE can be used directly to specify the position of the desired element. The first through fourth, and the seventh, elements are all null strings. The value of SURFTYPE for a quadratic surface is 5. When this value is passed as the first argument to **$index**, it retrieves the *fifth* element in the comma-list, which is the string "Quadratic". This same technique can also be applied to other numerically coded system variables. Two examples are linear units (LUNITS), and multiline justification (CMLJUST).

Figure 10.9
The DDEMODES cursor menu.

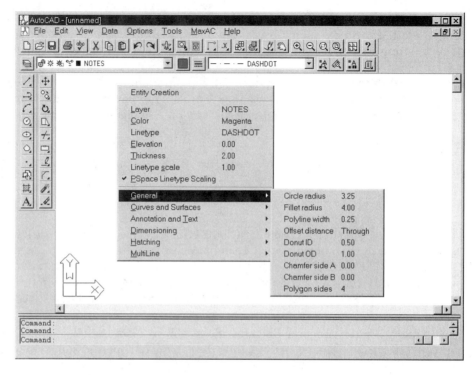

All three of the preceding examples are items from the DDEMODES cursor menu you worked with earlier in this chapter, which is shown in Figure 10.9. The menu is in the partial menu file 10CURSOR.MNS, which you can load by choosing its name from the **M**enus submenu of the **M**axAC pull-down menu for this chapter. The DDEMODES cursor menu contains many other DIESEL menu label macros that add drawing editor data to menu labels dynamically. After loading 10CURSOR.MNS, you can display the DDEMODES menu by holding down CTRL+SHIFT and pressing button #1 on your pointing device.

Creating Dynamic Menus with DIESEL and AutoLISP

It is useful to have menus whose contents can be defined dynamically at the time they are displayed. Using menus defined on the fly would serve to reduce the overhead resulting from an excessive number of static menus defined in a menu file. However, AutoCAD menus are static in nature. They are defined in MNU and MNS files and cannot be created or altered dynamically beyond what DIESEL macros provides for in labels. You cannot, for example, directly generate a pull-down or cursor menu dynamically under program control.

However, by combining AutoLISP, DIESEL and system variables, you can indirectly create dynamic POP*n* (pull-down or cursor) menus. One way to generate cursor menus dynamically from an AutoLISP program is by using a series of blank cursor menu *templates* that use DIESEL expressions in their labels to get each menu item's caption from a USERS*n* string system variable when the menu is displayed.

An AutoLISP program can easily set the value of one of these system variables to a comma-delimited string containing a list of menu captions, and then pop up the appropriate blank menu template containing the required number of items.

This scheme requires a separate blank menu template for each possible number of items that will appear on a menu. The following examples are two such dynamic cursor menu templates, which can be found in the partial menu file 10BLANK.MNS, in this chapter's directory. The file 10BLANK.MNS, along with the AutoLISP program in 10BLANK.MNL, implements a dynamic cursor menu that can contain between 2 and 15 items. You can add additional blank templates to 10BLANK.MNS to accommodate cursor menus with more than 15 items. (In DOS, the 10BLANK.MNS and MNL are integrated into ACADMA.MNU and MNL.)

```
// Example dynamic cursor menu
// templates from 10BLANK.MNS

// Used for menus with 4 items\
***POP4
[ ]
[$(index,0,$(getvar,USERS5))]$M=$(index,0,$(getvar,USERS5))
[$(index,1,$(getvar,USERS5))]$M=$(index,1,$(getvar,USERS5))
[$(index,2,$(getvar,USERS5))]$M=$(index,2,$(getvar,USERS5))
[$(index,3,$(getvar,USERS5))]$M=$(index,3,$(getvar,USERS5))

// Used for menus with 5 items
***POP5
[ ]
[$(index,0,$(getvar,USERS5))]$M=$(index,0,$(getvar,USERS5))
[$(index,1,$(getvar,USERS5))]$M=$(index,1,$(getvar,USERS5))
[$(index,2,$(getvar,USERS5))]$M=$(index,2,$(getvar,USERS5))
[$(index,3,$(getvar,USERS5))]$M=$(index,3,$(getvar,USERS5))
[$(index,4,$(getvar,USERS5))]$M=$(index,4,$(getvar,USERS5))
```

The preceding listing shows 2 of the 15 dynamic cursor menu templates from 10BLANK.MNS. Each template contains a different number of items, which provides for dynamic menus with a varying number of items. The AutoLISP file 10BLANK.MNL contains a number of functions that work with 10BLANK.MNS to allow AutoLISP programs to generate and activate popup cursor menus dynamically.

You can check or disable menu items by prefixing the appropriate menu directive to their caption text, and you can also assign accelerator keys (see Chapter 5). The following example code from the file 10DMENU.LSP in this chapter's directory defines the **C:POPTEST** function, which calls **PopUpMenu** to display a dynamic cursor menu with several items on it.

```
(defun C:POPTEST ( / input)
   (princ
      (strcat "\nEnter distance or "     ; Display input prompt
              "pick option from menu: "  ; before showing menu
      )
   )
   (PopUpMenu                            ; Display cursor menu
      (list "&Oranges"
            "!.&Apples"      ; This item is checked
            "&Pears"
            "~&Bananas"      ; This item is disabled \
            "&Grapes"
            "Pl&ums"
            "&Watermelon"
      )
   )
   (setq input (getdist))   ; Get user input after showing menu
   (cond
      (  (not input)
         (princ "\nNull response."))
      (  (eq (type input) 'str)
         (princ (strcat "\nYou chose: " input)))
      (t (princ (strcat "\nDistance: " (rtos input))))
   )
   (princ)
)
```

In the preceding example code, the **PopUpMenu** function is called to display a menu with five items (see Figure 10.10); one of the items is checked and another is disabled. The **PopUpMenu** function is defined in the file 10POPUP.MNL.

You can call this function from other AutoLISP programs to display cursor menus that a user can respond to when they are prompted for input. In order to use **PopUpMenu**, you must first attach the partial menu 10POPUPS.MNL by using the AutoCAD MENULOAD command, or by calling the **MxaMenuLoad** AutoLISP function defined in MA_UTILS.LSP.

In the next exercise, you'll test the **C:POPTEST** function.

Running on DIESEL

Figure 10.10
Dynamic cursor menu created by the PopUpMenu function.

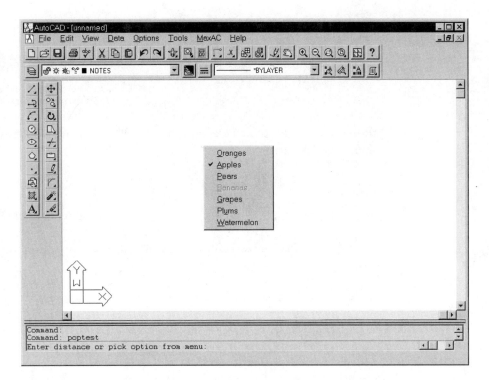

Testing the PopUpMenu Function

Launch AutoCAD with a new, unnamed drawing.

Command: **(load "10dmenu.lsp")** [Enter]	Loads 10DMENU.LSP, and in Windows, 10BLANK.MNS and 10BLANK.MNL
Menu loaded successfully. MENUGROUP: BLANK *LISP returns:* C:POPTEST1	(In Windows only)
Command: **POPTEST** [Enter]	Starts POPTEST command
Enter distance or pick option from menu: *Choose **P**ears from menu*	

511

```
You chose: Pears

Command: POPTEST [Enter]              Starts POPTEST command

Press CANCEL                           Dismisses cursor menu

Enter distance or pick option from     Specifies distance of 35.125 units
menu: 35.125 [Enter]

Distance: 35.1250

Command: POPTEST [Enter]              Starts POPTEST command
```

Press CANCEL to dismiss cursor menu and enter a menu option from the keyboard, as follows:

```
Enter distance or pick option from     Specifies APPLES
menu: APPLES [Enter]

You chose: Apples

Command:
```

Although the preceding example menu is fairly static and of little use in a practical sense, there are many interesting applications for dynamic popup menus. One is to display popup menus in response to pressing a toolbar button, which causes the button and menu to behave like a drop-down list or combo box.

Creating Toolbar Drop-Down List Menus with DIESEL and AutoLISP

The files 10POPUPS.MNS and 10POPUPS.MNL from this chapter's directory illustrate a more useful application of dynamic cursor menus. The menu file adds the POPUPS toolbar to AutoCAD for Windows, which contains three buttons. The first button sets the current text style by selecting it from a popup menu that appears when you press the button. You can use the same button to change the text style of any number of currently selected text objects to a new style selected from the menu. The second button displays a popup menu consisting of all defined dimension styles, and allows you to select one to become the new current dimension style. The third button allows you to change the hatch pattern of any number of selected hatch objects by picking the description of the style from a popup menu. The same button will also let you set the current hatch style as well.

Although 10POPUPS adds new flexibility to the AutoCAD for Windows interface, the AutoLISP code it uses goes a bit beyond the scope of this book. See the companion book *Maximizing AutoLISP for AutoCAD R13* for an explanation of how it works and an exercise demonstrating it.

If you want to try it out, first open the drawing file 10POPUPS.DWG from this chapter's directory, which contains an array of text and hatch objects along with several predefined text and dimension styles. Next choose 10popups from the Menus submenu of the MaxAC pull-down menu. A new toolbar named POPUPS will appear with three buttons. Try selecting some text at the command prompt. Then, click the Text Style button. A popup menu appears containing the names of all text styles defined in the drawing. Select a style to apply to the selected text. Repeat the process several times, and note how the button and menu work together to simplify editing the text style of multiple text objects, a task that usually requires the use of the DDCHPROP or CHANGE command.

Next, while at the command prompt, individually select two or more existing hatch objects (select only hatch objects, *do not* select their rectangular polyline boundaries). Then, press the Apply Hatch Pattern button from the POPUP toolbar. A popup menu will appear with the descriptions of several hatch patterns. Select a pattern and it will be applied to the selected hatch objects. Note that with both the text and hatch objects, after you change their styles, they remain selected for further editing, something that AutoCAD does only with grip editing commands. You can find out more about maintaining selections after editing and using these commands in this book's companion volume, *Maximizing AutoLISP for AutoCAD R13*.

Extending Custom Applications with DIESEL Power

One of DIESEL's primary advantages is its transparency. DIESEL expressions can be evaluated at any time, allowing them to be used in a wide variety of ways to make custom applications more flexible and dynamic. For example, if a custom application places user-supplied text in a drawing, you could pass the user-supplied strings through the DIESEL interpreter before placing them. This enables the user to include DIESEL expressions in the strings to, for example, dynamically specify data that can be derived from AutoCAD system variables.

A perfect example of how custom programs can use DIESEL in this way can be found in the demonstration version of the commercial application PlotStamp, which is included on the accompanying CD.

PlotStamp adds dynamic plot data stamping capability to AutoCAD R13 for Windows by allowing you to use DIESEL to dynamically specify what value an AutoCAD text object will have when the drawing that contains it is plotted. By using DIESEL to define the contents of the text, you can add the time and date when the drawing was plotted, along with other relevant information such as the drawing filename and user name. PlotStamp

Figure 10.11
The Edit PlotStamp dialog box.

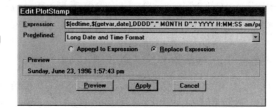

includes a custom dialog box (see Figure 10.11) for defining the stamp text, and you can select from a list of pre-defined DIESEL expressions for most common stamp data.

Note: The limited demonstration version of PlotStamp included on this book's CD adds a text prefix identifying itself to each stamp object. The standard edition of PlotStamp, which can be ordered directly from the authors, does not impose this limitation. See Appendix G for more information.

Warning: Because PlotStamp date and time data is updated at plot time, PlotStamp is not suitable for inserting static dates or times in text strings.

Tip: Take a look at the Predefined expressions drop-down list in the Edit PlotStamp dialog box, as they provide several examples of properly-coded, ready-to-use DIESEL expressions. You can copy them from the Expression edit box into your program code.

Debugging and DIESEL Error Messages

As your DIESEL expressions become more complex, you need to debug your macros. You can use two tools that help: tracing and error messages. You activate tracing by setting the MACROTRACE system variable to 1. Then, as each DIESEL expression is evaluated, the expression and its result is displayed on the text console. A typical trace narration is:

```
Eval: $(if,$(getvar,USERI1),=>UP,<=DOWN)
Eval: $(getvar,USERI1)
===>   0
===>   <=DOWN
```

Each level of the DIESEL expression is displayed from the outermost level to the innermost level. In this example, there are two levels of expression: a nested **$getvar** function call and a outer **$if** function. The 0 is the return value from the **$getvar** call. The <=DOWN is the return from the **$if** function.

Table 10.5
DIESEL Error Messages

DMessage	Meaning
$?	Syntax, missing parenthesis or quote.
$(func,??)	Incorrect arguments to a function.
$(func)??	Unknown function name; not a DIESEL function.
$(++)	Output string is too long. Truncated items.

The other debugging tool is a set of four error messages that report missing parentheses or quotes, incorrect arguments to a function, an unknown function name, or an output string that is too long, as shown in Table 10.5.

Finishing Up

If you want to keep any of the files you created or modified in this chapter, move them to the \MAXAC directory or another directory, and then delete or move any remaining files out of \MAXAC\DEVELOP.

Conclusion

In this chapter, you've seen DIESEL used in a variety of ways to provide a richer, more dynamic user interface. All of the working examples in this chapter make use of existing AutoCAD commands and functions to demonstrate underlying theory and technique. But in practice, you can adapt and build on these same techniques, to provide similar enhancements to the user interfaces of your own custom programs.

The DIESEL language provides some unique ways to customize your AutoCAD system. The format and syntax of DIESEL is similar to that of AutoLISP, making DIESEL easy to learn and use. With DIESEL, you can gray options to a command, disable commands, or provide check marks. You can control the menu labels and show labels dynamically updated. You can also customize the AutoCAD status line.

Although DIESEL is very dynamic, it is not as robust or comprehensive as AutoLISP. You will need AutoLISP to access entity data, deal with AutoCAD selection sets, or communicate with ADS programs. DIESEL is a string-based language for formatting and controlling the appearance and activation of pull-down and cursor menus, and for providing a limited capability to respond to AutoLISP expressions.

chapter

Introduction to Dialog Boxes

AutoCAD's user interface has been continuously evolving since the program's debut in 1982, with flexible programmability generally regarded as AutoCAD's hallmark. Many other programs developed *graphical user interfaces* (GUIs), but AutoCAD depended primarily on a *command-line interface* (CLI) to interact with the user. The command-line interface is not necessarily a bad thing. For experienced users, a CLI can be a more productive way to work. But as AutoCAD's instruction set continued to expand, this user interface became increasingly obtuse and difficult for new users. In the ninth major release of AutoCAD, the AUI (*Advanced User Interface*) was introduced. The AUI initially consisted of a relatively small number of dialog boxes, which provided a graphical means of controlling layer settings; object creation properties such as color, layer, and linetype; and the state of various drawing aids like SNAP, ORTHO, and so on. These first dialog boxes were mostly novelties, and they did little to improve the user interface or productivity other than providing a more convenient means of controlling layer settings and making the product seem more attractive to novices. More importantly, the AUI lacked a means of adding new dialog boxes for custom applications, and custom AutoLISP applications were still restricted to using the CLI to interact with the user.

In AutoCAD Release 12, Autodesk introduced a completely revamped user interface, which included toolbar buttons and numerous dialog boxes that both supplanted and expanded on the existing AUI-style dialogs. These dialog boxes were defined with a new *Dialog Control Language* (DCL), a non-procedural object description language that defines the appearance and functional characteristics of a programmable dialog box. You will be introduced to DCL later in this chapter and will learn more about it in Chapter 12. These two chapters provide a gentle introduction to DCL and how to control dialog boxes with AutoLISP. For complete coverage of DCL and driving dialog boxes with AutoLISP control, see the book *Maximizing AutoLISP for AutoCAD R13*.

> ### About This Chapter
>
> In this chapter, you will learn about predefined dialog boxes and the functions that access them. You will be introduced to Dialog Control Language and learn how to use it to create and modify custom dialog boxes.
>
> **The resources you will use in this chapter are:**
>
> - ACADMA.MNS, ACADMA.MNL, ACAD.PGP, ACAD.LSP, 11EX01-5.LSP, and LispPad.Exe
>
> **Programs and Functions in This Chapter**
>
> The programs and functions you will use in this chapter are:
>
> - DDLOAD.LSP, which shows how to use the file dialog box to acquire file specifications from the user
>
> - DDXPATH.LSP, which also uses the file dialog box to edit external reference paths
>
> - HELLO.DCL and 11EX02-5.DCL, which are exercise files
>
> All of the exercises in this chapter require AutoCAD R13 for Windows. Copy this chapter's code files from the CD-ROM or MAXAC\BOOK\CH11 directory to your working development directory.

Introduction

DCL-based dialog boxes are *user-programmable*, allowing you to create your own custom dialog boxes that can be controlled by AutoLISP and ADS programs. Most dialog boxes in AutoCAD Release 13's user interface are implemented in DCL and are controlled by AutoCAD, ADS, and AutoLISP programs. Many commands with names starting with the characters DD (such as DDCHPROP, DDMODIFY, and so on), are implemented in AutoLISP, and make use of programmable DCL-based dialog boxes. Figure 11.1 shows a programmable dialog box from the DDMODIFY command.

A distinguishing characteristic of dialog boxes implemented in DCL is that they are *platform-independent*. A properly designed dialog box written in DCL can be used on any platform that AutoCAD runs on without modification. When AutoCAD displays DCL dialog boxes, they automatically inherit visual and functional characteristics of native dialog boxes of the platform on which AutoCAD is running. DOS AutoCAD's dialog boxes are similar to Microsoft Windows dialog boxes. On DOS AutoCAD, all dialog boxes are implemented in DCL because DOS has no GUI. On all of the Windows platforms as well as all other GUI operating environments, AutoCAD uses both DCL and *native* dialogs.

Introduction to Dialog Boxes

Figure 11.1
The Modify Polyline dialog box in AutoCAD for Windows.

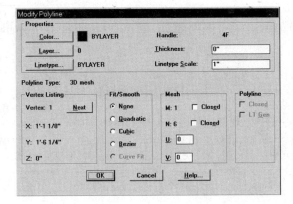

Native dialog boxes are created using an *application programming interface*, which is provided by the operating system vendor. Native dialogs cannot be controlled with AutoLISP, hence they are not available to AutoLISP applications.

Dialog Box Primer

Dialog boxes provide a richer, more robust way to interact with software applications. There are a number of reasons why dialog boxes are beneficial to the user. One is their non-linear nature. With AutoCAD's teletype-like command line, commands that issue multiple input prompts must do so in a predetermined and linear fashion. Input prompts are displayed sequentially, and you must respond to each prompt in order to advance to the next one. Take, for example, the AutoCAD CHANGE command. When a single object is selected to be changed, AutoCAD will issue a number of prompts that ask you for the new values of various properties of the selected object. You must respond to each prompt,

Figure 11.2
The Modify Polyline dialog box in AutoCAD for DOS.

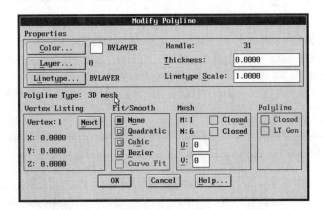

regardless of whether you want to change the property it is requesting or not. This means of interacting with AutoCAD can be tedious and error-prone, especially if one needs to change only a single value or determine its current state.

Another beneficial aspect of dialog boxes lies in the use of *graphical controls,* which eliminate extensive error-checking, rejection, and retrying for input, an inherent part of acquiring input on the command line. For example, a *list box* or *popup list* in a dialog box can be populated with the names of all layers defined in the drawing, from which you can select. Unlike typing layer names on the command line, choosing items visually from a list does not require you to remember or enter the exact name of a desired item because their names appear in the list, which also serves as the means of selection. Hence, the list box eliminates all possibility of input errors, such as entering the name of an unknown layer.

Graphical controls such as list boxes, popup lists, check boxes, and radio buttons greatly reduce the chances of making an invalid selection by eliminating the need to respond with arbitrary keyboard input, and by limiting input to valid responses. Hence, dialog box controls can greatly reduce the need to validate, reject, and reacquire mistyped or erroneous input, as you must must do on the command line.

If you've used AutoCAD or any Microsoft Windows-based application, you're already familiar with dialog boxes from the user's perspective. Dialog boxes can be positioned anywhere on the display. When a dialog box is visible and responding to keystrokes, that dialog box is said to have *focus.* A dialog box is said to have been *dismissed* when it is deactivated and removed from the display.

All dialog boxes fall into two basic functional classes, *modal* and *modeless.* A modal dialog box is one that has focus exclusively while it is active, and it restricts all user interaction with the associated application to occur within the dialog. In a windowed operating environment such as Microsoft Windows, each application has a *main* window. You cannot shift focus back to the application's main window while a modal dialog box is active in that application. Instead, you must interact with the application through the modal dialog and *dismiss* it before focus can return to the application's main window.

You dismiss a dialog either by clicking on the OK or Cancel button, or pressing the *accept* key (which is usually the ENTER key). A modal dialog box can also have one or more subdialog boxes, which may also be modal. When a modal subdialog box is active, you cannot shift focus to the parent dialog that displayed it until the subdialog has been dismissed. Most of the dialog boxes that AutoCAD uses are modal dialogs (and all DCL-based dialog boxes are modal dialog boxes, since DCL does not support modeless dialogs). When a modal DCL dialog box like that shown in Figure 11.1 is active, for example, you may not enter coordinates in the AutoCAD drawing view, type commands or options on the AutoCAD command line, or make selections from toolbars or pull-down menus. If you attempt such actions, AutoCAD complains by making a noise, and it prevents the active dialog box from losing focus. You must dismiss an active modal dialog box before AutoCAD allows you to continue working in the drawing view.

Introduction to Dialog Boxes

Figure 11.3
Modeless dialog boxes in AutoCAD for Windows.

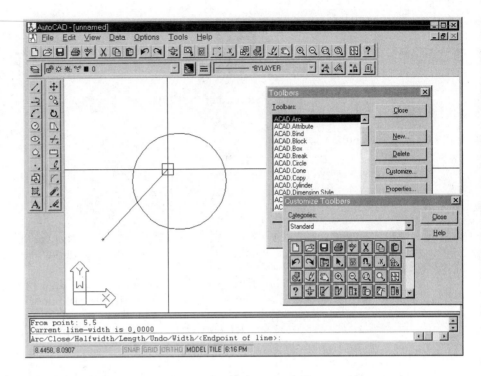

A modeless dialog box is one that behaves more like a separate application. It can remain visible and lose focus without having to be dismissed. A modeless dialog allows focus to repeatedly shift from itself to the parent or main application window that owns the dialog, or any other modeless dialog as well. The toolbars in AutoCAD for Windows are like modeless dialog boxes, since they remain visible when they do not have focus and they allow you to work in the drawing view or in other dialog boxes. AutoCAD for Windows uses several modeless dialog boxes. The Toolbars, Customize Toolbars, and Toolbar Properties dialog boxes (see Figure 11.3) are all modeless dialogs (they are also native Windows dialog boxes).

To better understand the behavior of modeless dialog boxes, try the following exercise with AutoCAD for Windows.

 Exercise

Using Modeless Dialog Boxes

Launch AutoCAD with a new, unnamed drawing and zoom Center on the point 10,10 with a height of 10 units.

Maximizing AutoCAD® R13

`Command: `**`PLINE`**` [Enter]`	Starts the PLINE command
`From point: `**`7.5,6`**` [Enter]`	Specifies **7.5,6** as start point
`Current line-width is 0.0000`	
`Arc/Close/Halfwidth/Length/Undo/` `Width/<Endpoint of line>: `**`@2<45`**` [Enter]`	Specifies second vertex point of polyline
`Arc/Close/Halfwidth/Length/Undo/` `Width/<Endpoint of line>: ` *Right-click on any toolbar button*	Displays the Toolbars dialog
In the Toolbars dialog, choose **C**ustomize	Displays the Customize Toolbars dialog

Arrange both dialog boxes so they do not obscure the drawing view, then click on the title of the AutoCAD window, and continue drawing the polyline.

`Arc/Close/Halfwidth/Length/Undo/` `Width/<Endpoint of line>: `**`@2<180`**` [Enter]`	Specifies third vertex point of polyline
`Arc/Close/Halfwidth/Length/Undo/` `Width/<Endpoint of line>: `**`CLOSE`**` [Enter]`	Closes the polyline
`Command: `**`CIRCLE`**` [Enter]`	Starts the CIRCLE command
`3P/2P/TTR/<Center point>: `**`8,10`**` [Enter]`	Sets center of circle to 8,10
In the Toolbars dialog, choose **C**lose	Closes the Toolbars and Customize Toolbars dialog boxes
`Diameter/<Radius>: `**`4`**` [Enter]`	Specifies a circle radius of 4 units
`Command:`	

Note that in the preceding exercise, the two modeless dialog boxes remained visible while you were interacting with AutoCAD's command line.

Common Dialog Boxes

AutoCAD provides several predefined dialog boxes that can be accessed by calling an AutoLISP function. Before you learn to design and program your own dialog boxes, you

Figure 11.4
The Alert Function Message dialog.

will first be introduced to the predefined ones that AutoCAD provides, which you can use in your programs.

The Message Dialog Box

The most basic of AutoCAD's predefined dialog boxes is the message box. Every AutoCAD user has seen or used the message box. You access the message box using the AutoLISP **alert** function. It is commonly used for very basic, generic, one-way communication, where no specific responses are required on the part of the user. The message box is particularly useful for notifying the users of an error or other exception, and to provide instructions on what they should do next, or explain why a given operation could not be completed.

The message box has only one button, which has the label OK. The **alert** function takes a single string as an argument, which is the message to be displayed in the dialog box. The user can dismiss the dialog by choosing the OK button, or by pressing either the ENTER or CANCEL key. You cannot determine precisely how the user dismissed this dialog, so you should not place any significance on it.

---

alert

(alert *message*)

Displays a message dialog box containing the string **message** *and an OK button, and waits for the user to press the button. This function always returns nil.*

The dialog displayed by the **alert** function can display multiple lines of text by including the newline (\n) character code in the *message* argument, which causes a new line to start at each occurrence of the character. The exact number of characters which this dialog can display is platform-, display-, and window-dependent. It is generally a good idea not to pass messages longer than 500 characters, or include long lines (over about 68 characters) that could require more room than is available on some displays.

Using the **alert** function is a good alternative to displaying prompts and messages on the command line, since the latter will generally not command a user's attention as well as a dialog box. At the other extreme, it can also be rude to constantly require the user to click on a button to acknowledge a message when there are many of them that must be displayed. Hence, overuse of the message box can also be undesirable.

Try testing the AutoLISP **alert** function by calling it to display some messages.

 Exercise

Testing the Alert Function

Continue from the previous exercise or start AutoCAD with a new, unnamed drawing.

Command: **(alert "Hello World!")** [Enter]   Displays the message dialog box

Choose OK *in the AutoCAD Message dialog box*   Dismisses the message dialog box

Lisp returns: nil

Command: **(alert "So long and thanks\nfor all the fish")** [Enter]   Displays the message dialog box with a two-line message (with apologies to Douglas Adams)

Choose OK *in the AutoCAD Message dialog box*   Dismisses the AutoCAD Message dialog box

Lisp returns: nil

Command: **(alert (strcat "The current drawing is " (getvar "DWGNAME")))** [Enter]   Displays the current drawing name in the message dialog box

Choose OK *in the AutoCAD Message dialog box*   Dismisses the AutoCAD Message dialog box

Lisp returns: nil

Command:

An Enhanced Message Dialog Box

As an alternative to the **alert** function, this book includes an enhanced message dialog box, which can display system icons along with a variety of predefined buttons with labels such as Yes, No, OK, Cancel, Retry, Abort, and Ignore. These tools are defined in an

Introduction to Dialog Boxes

Figure 11.5
The Enhanced Message dialog box from MAWIN.ARX.

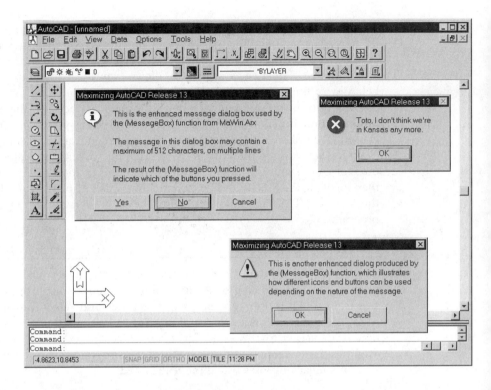

AutoCAD Runtime Extension (ARX) library of AutoLISP functions, in the file MAWIN.ARX on the accompanying CD-ROM. MAWIN.ARX provides a number of additional tools specifically for use in the Microsoft Windows environment.

You can access the enhanced message box from MAWIN.ARX using the **MessageBox** function. This function is available only when the MAWIN.ARX application is loaded, which can be accomplished using the APPLOAD command or AutoLISP **arxload** function. See Chapter 4 for instructions and examples on loading AutoCAD Runtime Extensions.

MessageBox

(MessageBox *title flags message***)**

An *MAWIN.ARX tool that displays an enhanced Windows message dialog box with a system icon. The* **title** *argument is a string that is used as the title of the dialog. The* **flags** *argument is a binary-encoded integer that specifies the type of icon used. The* **message** *argument is the message string, which is functionally the same as the* **alert** *function's argument of the same name. This function returns an integer indicating which button was chosen (where 1 represents the first or left-most button).*

525

The **MessageBox** function's ***flags*** argument provides numerous options, too many to list here. The included file MAWIN.TXT contains a complete description of the ***flags*** argument, which MAWIN.ARX automatically assigns to AutoLISP symbols for use with the argument. The ***flags*** argument allows extensive control over the appearance and behavior of the message dialog box, including the icon (and associated sound), which buttons appear on it, and which of the buttons acts as the default button. The **MessageBox** function returns an integer indicating which button on the dialog was chosen. You will see the **MessageBox** dialog used in example programs in this and other chapters. You can see the enhanced message box by starting AutoCAD with the starting directory set to your working directory (where this chapter's files have been copied), and then invoking the MSGBOXTEST command, which displays a sample message box. Study the MSGBOXTEST command to gain a better understanding of how to use it in your programs.

File Dialog Boxes

Another useful predefined dialog box AutoCAD makes available to your AutoLISP programs is the file dialog box. This dialog box allows your programs to display a dialog box from which a user can navigate directories and select files in them, just as many AutoCAD commands do. The AutoLISP **getfiled** function provides AutoLISP applications with access to the same file dialog box AutoCAD uses to obtain filenames of drawings, fonts, menus, and so on. AutoLISP programs can use this file selection dialog box for similar purposes by calling **getfiled**.

---

getfiled

(getfiled *title default ext flags*)

Displays a file dialog box from which the user can select a file. Returns the name or full path to the selected file, depending on the ***flags*** *argument, or* **nil** *if no selection was made.*

The **getfiled** function provides access to a file selection dialog box much like the one AutoCAD uses to obtain file specifications for drawings and other files it requests from the user. When **getfiled** is called, the file dialog box appears, and the user can use it to select a file. When the user dismisses the dialog box (by choosing the OK or Cancel button), **getfiled** returns the name of the selected file (and optionally the full path to the file, depending on the options specified in the arguments to the function, as described herein). If no file is selected or the Cancel button was pressed, **getfiled** returns nil. If the Type It button on the dialog is enabled, and the user chooses this button, **getfiled** will return a value of 1. See the following description of the ***flags*** argument for a more detailed description of the various options that can be enabled when calling **getfiled**. The file dialog box displayed by **getfiled** is shown in Figure 11.6.

Introduction to Dialog Boxes

Figure 11.6
The GETFILED file dialog box.

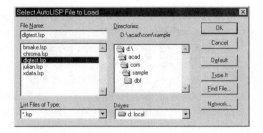

The **getfiled** function requires four arguments. The **title** argument must be a string, and it is used as the title or caption of the dialog box. In Figure 11.6, the caption of the dialog box is "Select an AutoLISP File to Load". The **title** argument can be a null string, but this value should never be used.

The **default** argument, also a string, specifies the default file specification, which can be accepted by choosing the OK button or pressing the accept key in the dialog box. The **default** argument may include a path, a filename, or both. If a path is supplied without a filename, the path should end with a slash or backslash. If a path is supplied, the initial directory of the file dialog box is the specified path, the path is displayed in the text label above the directory list in the dialog box, and the files in the specified path appear in the file list. If no path or a non-existent path is supplied, the path defaults to the current directory. If a filename is supplied in the argument, the **D**efault button on the file dialog box is visible and enabled. If the user chooses this button, the filename passed in the **default** argument appears in the file dialog's File **N**ame edit box. If an explicit filename is not supplied, the filename defaults to the asterisk wildcard. The **default** argument may include a file extension and the delimiting period. If an extension is supplied, it should match the extension supplied in the **ext** argument.

The **ext** argument is also a string, which specifies the required file extension. It can be any valid file extension, or the asterisk wildcard (*) to permit selection of a file with any extension or no extension. You may not include a delimiting period in the string. If **ext** is a null string (""), the extension defaults to the asterisk, and any extension (including none) is allowed. If an explicit extension is supplied, it appears in the File Type drop-down list in the dialog box, and the file selected by the user must have this extension. If the user specifies a file with an extension different from the one specified in the **ext** argument, an error message is displayed informing the user that the file type is incorrect.

Tip: The **ext** argument to **getfiled** may include *two* file extensions that are separated by a semicolon, as in **"dwg;dxf"** or **"mnu;mns"**. Only two extensions are permitted using this form. If two extensions are supplied, both appear in the File Type drop-down list in the dialog, and the first extension is always used as the default.

The **flags** argument is a binary-encoded integer that specifies functional behavior of the file dialog box. The **flags** argument has four significant bits, defined as follows:

Bit 0 (1)   Specifies if the dialog box is being used for a *Save* or *Open* operation. If this bit is not set (0), the dialog behaves like a common *Open* file dialog box. In Open mode, the user must select an existing file to be opened. If a nonexistent file is supplied, it is not accepted and an error message is displayed. Conversely, if this bit is set (1), the dialog behaves like a *Save* or *Save As* file dialog. In Save mode, the user can supply the name of a nonexistent file that is to be created and written to. If the user specifies the name of an existing file, a *Create File* dialog box appears, informing the user that the specified file already exists and, if it should be overwritten, requiring the user to confirm by clicking on a button.

Bit 1 (2)   If this bit is set (2), the **T**ype It button on the file dialog box is disabled. The **T**ype It button is provided to allow the user to supply the name of the file on the AutoCAD command line, rather than with the file dialog box. If **getfiled** is called while another dialog box is active, the **T**ype it button is always disabled regardless of the value of this bit. If you do not set this bit, you should check the result of **getfiled** to determine whether the user has chosen the **T**ype It button. If the user chooses this button, then **getfiled** returns the integer 1, and in that case you should obtain the name of the file on the command line using the **getstring** function.

Bit 2 (4)   If this bit is set (4), the file dialog box will accept files with *any* extension or files with no extension, regardless of the value of the ***ext*** argument or a file extension included in the ***default*** argument. This bit can be set if you want to allow files of two or more types to be selected.

Bit 3 (8)   If this bit is set (8) and bit 0 is *not* set (Open mode), and the user selects a file in a directory that is on the library search path, only the name of the file is returned. This is primarily intended for cases where the name of the selected file is to be stored in the drawing, and may be referenced in a subsequent session on a different system with a different directory structure. Storing *only* the name of the file, and searching for it on the library path, eliminates any dependence on a specific directory structure, so that a file having the same name can be located in the library path on any system, regardless of how file directories or folders are organized.

Using File Dialog Boxes in AutoLISP Programs

The **getfiled** function is an easy way to incorporate GUI elements to your programs without much effort. Later in this book, you will learn how to create your own custom dialog boxes in a way that allows them to be reusable, just as the file dialog box accessed via **getfiled** is.

Introduction to Dialog Boxes

The following code listing illustrates a simple AutoLISP application that uses **getfiled** to display a dialog box in which you can select an AutoLISP file to be loaded. This application produced the file dialog box shown in Figure 11.6.

```
;; 11FILED.LSP From Maximizing AutoCAD Release 13
;;
;; Adds the LSP command to AutoCAD which allows you
;; to select an AutoLISP file to be loaded using
;; the (getfiled) dialog box.
;;
;; Globals: *ld-file*   Stores name of last selected file

(defun C:LSP ( / file msg default)
  (setq msg "Select AutoLISP File to Load")
  (if *ld-file*
    (setq default *ld-file*)   ; use last file loaded
    (setq default "")          ; as default, or use ""
  )
  (setq file
    (getfiled
       msg             ; Dialog caption.
       default         ; Default to last file loaded.
       "lsp;mnl"       ; Allow .LSP or .MNL files
       2               ; and disable the Type-It button
    )
  )
  (if file
    (progn
      (load file)              ; If file was selected, load it.
      (princ

        (strcat
          "\nFile "             ; Tell user the file was loaded.
          file
          " successfully loaded."
        )
      )
      (setq *ld-file* file)    ; Remember file for use as
    )                          ; the default in next call.
  )
  (princ)
)

;; END 11FILED.LSP
```

Note that **getfiled** will accept a complete file specification, including a drive letter, path, filename, and extension, as the ***default*** argument, which makes it easy to use the file returned by each call to **getfiled** as the default file in the next call to the function. Each time **getfiled** is called, if the user selects a file, the LSP command stores it in the global variable *ld-file*, and uses that as the *default* file the next time LSP is invoked.

Also note that the LSP command sets bit 1 of the ***flags*** argument to **getfiled**, to disable the Type It button. Although this button is disabled in this case, because of the added complexity of handling command-line input, a well-designed program will allow the user to use the Type It button to specify the filename on the command line.

Note: If the user chooses Cancel or presses the CANCEL key to dismiss the **getfiled** dialog box, AutoCAD sets the value of the DIASTAT system variable to 0. The DIASTAT system variable indicates how the user dismissed the last active dialog box. If the user dismisses a dialog by pressing the OK button or ENTER, DIASTAT will be set to 1. You may use this system variable to determine if a dialog was canceled or accepted, immediately after the dialog is dismissed.

When using **getfiled**, you should consider several important points. In general, it is a good practice to design your custom programs so they behave in ways that are familiar to the user. Because you can expect the user to be familiar with AutoCAD and its built-in commands, designing your programs to resemble AutoCAD commands in terms of their appearance, feel, and behavior is a crucial aspect of good application design.

For example, consider that some AutoCAD commands allow you to choose between using a dialog box or the command line to specify the names of files. If you invoke the INSERT command from the command line, you can enter the tilde (~) character when you are prompted for the name of the block, which instructs AutoCAD to display a file dialog box that you can use to select a drawing to insert. When a file dialog box is active, you can often press the Type It button to dismiss the dialog and enter the name of the file on the command line. You can also disable file dialog boxes globally by setting the FILEDIA system variable to 0, which causes AutoCAD to always prompt for files on the command line, rather than with dialog boxes. In addition, AutoCAD automatically disables all dialog boxes when receiving input from AutoLISP programs or a running script file. One further consideration is in cases where one dialog box is already active and you are displaying the file dialog box. If the file dialog is displayed as a subdialog of another dialog box, then command-line input is not possible without dismissing both the file dialog box as well as the parent dialog, and while it is possible to provide for this, it is not advised. Hence, if a file dialog box is being displayed from another parent dialog, the Type It button should be disabled.

You can and should strive to design your own custom applications so they mimic, as closely as possible, AutoCAD's behavioral characteristics as described here. For example, if a program calls the **getfiled** function with the Type It button enabled, it should always check the result to see if the user chose that button (in which case, **getfiled** returns the integer 1). If the Type It button was chosen, you should then prompt the user for

Introduction to Dialog Boxes

the filename on the command line. Before you call **getfiled**, you should also check the value of the FILEDIA system variable, and if it is set to 0, also branch to a procedure that obtains the filename from the command line. If you want to allow your program to be driven by a script file, it can't use any kind of dialog box unconditionally, since scripts cannot supply input to them. To accommodate script files, you can check the value of the CMDACTIVE system variable before you display a dialog to see if a script is active and if one is, you can branch to a procedure that prompts for the filename on the command line. You should already be familiar with the CMDACTIVE system variable from Chapter 10, where it is used extensively with DIESEL expressions to control the appearance of menus. Bit 3 (decimal 4) in this variable is ON when a script file is running. The following example shows how to use this system variable to determine if a script file is active.

```
(if (eq 4 (logand 4 (getvar "CMDACTIVE")))
    (princ "A script file is running!")
    (princ "No script files are running.")
)
```

The **logand** function performs the same operation as the diesel **$and** function. It returns the *bit-wise logical and* of its arguments. If a script is active, the expression (logand 4 (getvar "CMDACTIVE")) returns 4. If no script is active, that same expression will return 0.

Along with 11FILED.LSP, this chapter's directory also includes a more advanced version of the LSP command shown in the previous listing, called DDLOAD, which can be found in the file 11DDLOAD.LSP. DDLOAD demonstrates many of the same concepts discussed here.

Unlike the more simple LSP command, the DDLOAD command supports command-line input in addition to dialog box input. When the command runs, it first checks the CMDACTIVE system variable to determine if a script file is currently active. The DDLOAD command also checks the FILEDIA system variable, which the user can set to 0 to disable file dialog boxes. If a script file is running or the user has disabled file dialogs, then DDLOAD prompts the user for the name of the file on the command line. Under Windows 95 and Windows NT, obtaining filename input on the command line is more complicated than it is under DOS or Windows 3.1 because *long filenames* may contain spaces, which normally terminate input to the **getstring** function if it is called without the optional *cr* prompt.

When obtaining a filename on the command line, the DDLOAD command provides for entry of long filenames by first checking the value of the LONGFNAME system variable to determine whether long filename support is enabled. If it is, DDLOAD adds the *cr* argument to the call it makes to the **getstring** function, which permits spaces in the response. If DDLOAD is run on DOS AutoCAD or Windows 3.1, the LONGFNAME system variable will be set to 0, and spaces will not be allowed in the response. The DDLOAD command also allows the user to press the **T**ype It button in the dialog box to enter the filename on the command line. DDLOAD serves to demonstrate how one can write custom AutoLISP programs that behave in a reasonably predictable manner that is consistent with many AutoCAD commands.

Selecting Drawing Files with a Dialog Box

The file dialog box displayed by **getfiled** has one special characteristic when it is used to select AutoCAD drawing files. When the file extension argument to **getfiled** is DWG, a drawing file *preview image* appears in the file dialog box and displays the preview image of the drawing file currently highlighted in the File list. Figure 11.7 shows the file dialog box with the drawing preview image.

The Library Search Path

Bit 3 (decimal 8) of the **flags** argument to **getfiled** is useful in cases where your program must obtain a filename that is to be stored in the drawing and referenced in a subsequent editing session (possibly on another computer). If you were to store the complete path to a file, then any subsequent reference to the file will require it to be in the exact same path. Since folder and directory names and structures are not the same on every system and a file may be located in a different path on other systems where a drawing might be edited, it is a bad practice to store the names of files with complete paths. Instead of storing an entire path, it is better to store only the name of a file, and search the library path for its actual location at the point when it is referenced. AutoCAD performs a library search for many kinds of files that are supplied to it in commands. For example, when you supply the name of a drawing file to the INSERT command, AutoCAD will search the entire library path for that file. Your programs can also search the library path for files by using the AutoLISP **findfile** function. This function takes the name of a file as an argument, and searches the entire AutoCAD library search path for an occurrence of the file. If the file is found on the library path, **findfile** returns the full path to the file.

Figure 11.7
The Select Drawing File dialog box.

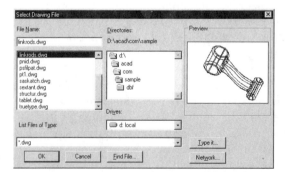

Introduction to Dialog Boxes

> `findfile`
>
> (`findfile` *string*)
>
> Searches for files on the library search path, or indicates whether a file exists anywhere on the file system. If **string** is a filename and extension without a path, the AutoCAD library path is searched for the existence of the file. If the file is found on the library search path, the complete file specification including the drive letter and path is returned. If **string** is a file specification with a path, the complete file specification is returned if the file exists, regardless of whether it is in the library search path or not. In all cases, if a file is not found, the result is nil.

Selecting Directories and Folders

There are times when it is necessary to request the name of a directory or folder from the user, rather than just a single file. Unfortunately, doing this with the **getfiled** file dialog box is nearly impossible or at least illogical. You can obtain the name of a directory with **getfiled** by asking the user to select a file in the directory, assuming there are files in the desired path, and then parsing it to remove the filename and extension. However, the Windows 95 and NT operating systems have a built-in dialog box with an Explorer-style Tree View control, that's designed for selecting folders, as shown in Figure 11.8. Basic AutoLISP cannot access system-level resources such as this dialog by itself. However, the **BrowseForFolder** function, another useful tool from the MAWIN.ARX file's collection of AutoCAD Runtime Extension tools included with this book, can. This function uses the Windows 95 and NT Browse for Folder dialog box to allow users to select a folder anywhere on their system or within their domain. With MAWIN.ARX loaded, you can call this function to obtain the name of a folder from the user.

The **BrowseForFolder** function is available under Windows 95 or Windows NT only. Do not attempt to use it under Windows 3.1. The function takes one string argument,

Figure 11.8
The Browse for Folder dialog box.

Maximizing AutoCAD® R13

which is a message displayed below the title bar of the dialog box. The function will return the name of the folder selected or nil if no folder was chosen. Note that in this context, the terms folder and directory are freely interchangeable.

Next, you'll test **BrowseForFolder**. A copy of the MAWIN.ARX runtime extension is included in this chapter's directory on the CD, which should be copied from the CD to your working exercise directory. MAWIN.ARX is loaded into AutoCAD automatically from the ACAD.LSP file from this chapter's files, when you start AutoCAD with the startup directory set to the one containing the files from this chapter's directory.

Exercise

Using the Browse For Folder Dialog Box

Continue from the previous exercise or start AutoCAD with a new, unnamed drawing.

Next, you verify that MAWIN.ARX is loaded:

Command: **(arx)** [Enter]	Displays loaded ARX programs
Lisp returns: ("c:\\maxac\\mawin.arx")	Returns the program name in a list
Command: **(BrowseForFolder "Test1")** [Enter]	Displays Browse for Folder dialog

Navigate to and select the AutoCAD support folder and choose OK

Lisp returns: "C:\\acad\\win\\support\\"	Returns the path to the AutoCAD support folder
Command: **(BrowseForFolder "Test2")** [Enter]	Displays Browse for Folder dialog
Choose Cancel	Cancels the dialog
Lisp returns: nil	
Command: **(BrowseForFolder "Test3")** [Enter]	Displays Browse for Folder dialog

Navigate to and select the Desktop and choose OK

Introduction to Dialog Boxes

Lisp returns: `"C:\\WIN95\\Desktop\\"`   Returns the path to the Desktop (your path may differ)

Command:

 Note: The `BrowseForFolder` function guarantees that all path strings it returns have a trailing backslash. This makes it easier to append a filename to the path.

The Color Dialog Box

Another common dialog box that AutoCAD provides access to from AutoLISP programs is the Color dialog box. AutoCAD uses this dialog in a number of instances, such as the DDEMODES and DDCHPROP commands. Note that DDCHPROP is written entirely in AutoLISP. The Color dialog box (see Figure 11.9) provides a means of selecting one of the 255 colors available in the AutoCAD Color Index (ACI). The ACI assigns integer values between 0 and 255 to the range of physical colors supported by AutoCAD. In addition to these physical colors, the Color dialog box provides two additional buttons that allow you to choose from the logical colors *bylayer* and *byblock*.

The Color dialog box is accessed through the AutoLISP **acad_colordlg** function. This function takes one or two arguments. The first is the index value of the color that is selected by default, and the second optional argument allows you to disable selection of the logical colors *bylayer* and *byblock*.

`acad_colordlg`

`(acad_colordlg color [flags])`

Displays the Color dialog box and allows the user to choose a color from it. Returns the index of the chosen color. The **color** *argument is an integer ranging from 0 to 256, and specifies the index of the color selected by default. The value 0 represents the logical color BYBLOCK, and 256 represents the logical color BYLAYER. If the* **flags** *argument is supplied and is nil, the BYBLOCK and BYLAYER logical color buttons are disabled. If the* **flags** *argument is not supplied, or is supplied and non-nil, the logical color buttons are enabled. If the logical color buttons are disabled, the* **color** *argument must be greater than 0 and less than 256, or an error occurs.*

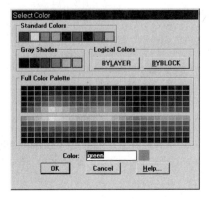

Figure 11.9
The AutoCAD Color dialog box

Using the Color dialog box is fairly simple. It requires a default color at minimum. It returns the index of the color selected, which can be used in AutoCAD commands that request object colors. There are cases where you might want to restrict input to only the physical colors, and prevent the user from selecting the logical *bylayer* and *byblock* colors.

Note: Always be sure that the **flags** argument is either not included or is non-nil when passing 0 or 256 as the first argument; otherwise, an error occurs, because the BYLAYER and BYBLOCK buttons *must* be enabled when either of these logical colors is the default.

You can pass **nil** as the second argument to **acad_colordlg** to *disable* the BYLAYER and BYBLOCK buttons on the Color dialog box. If you pass any other value as the second argument, it has the same effect as omitting the argument entirely.

Custom MTEXT Edit Dialog Box

AutoCAD allows you to specify the name of an AutoLISP program to be used to edit small amounts of MTEXT, in place of the default MTEXT editor, which does not exist on the DOS platform. The substitute dialog avoids the need for AutoCAD to open an external text editor for editing small amounts of multi-line text, such as those in dimensions. The MTEXTED system variable can be assigned the name of an AutoLISP function which is called and passed the text to be edited. This AutoLISP function can display a small dialog box in which you can edit the text. You'll see how to use this facility in the next chapter to create a custom text edit dialog.

Dialog Control Language

Up to this point, all of the dialog boxes that you've used are predefined by AutoCAD and Windows. Using these predefined dialogs is easy, since all you must do is call the LISP functions that display them, and use the results they return. AutoCAD and Windows do most of the work by controlling these dialogs for you internally.

Now you'll learn more about how dialog boxes are defined and controlled to respond dynamically to user actions. You'll also learn how to define custom dialogs that are more specific than those included with AutoCAD. You'll learn to use these dialogs for the more specialized applications that you'll create.

While you are interacting with a dialog box, many things are happening that are not readily apparent. Every time you click a button, enter text in an edit box, or select items in a list, the program that controls the dialog you're using is working to ensure that data you supply is valid and to dynamically update items in the dialog to reflect changes to other related or dependent items. For example, when you use the Selection Settings dialog to change the size of the pickbox, as you click on the slider to change the pickbox size, the program that manages the dialog box responds to the action and updates the preview image showing the current pickbox size. Similarly, in the Drawing Aids dialog box, when you choose the On toggle check box in the Isometric Snap/Grid cluster, the program that controls the dialog box changes the contents and enabled state of the X Spacing fields in the Snap and Grid clusters.

Although predefined dialogs are useful for obtaining generic data from the user, you will also have cases where they are unsuitable for more specific applications. AutoCAD allows you to define your own custom dialog boxes using Dialog Control Language, a non-procedural, object description language that serves as a non-graphical means of describing the visual and functional attributes of a dialog box. In conjunction with DCL, dialog boxes are controlled by AutoLISP or ADS programs. In this book, we present only AutoLISP control.

Dialog Control Language Basics

Unlike AutoLISP, Dialog Control Language is not a *procedural* language. A procedural language program contains expressions that are executed in a sequential fashion to produce side effects. DCL is a descriptive language in which you write expressions that describe the appearance and behavior of *interface objects*, and define their hierarchical relationship with other interface objects in a dialog box.

Interface objects include dialog boxes and all of the controls that appear within dialog boxes, such as buttons, toggles, list boxes, edit boxes, sliders, radio buttons, and so on. Figure 11.10 shows an example DCL dialog box that contains most of the interface objects supported by DCL.

In DCL terminology, interface objects such as those which appear in the dialog box in Figure 11.10 are referred to as *tiles*. In other programming environments such as Visual Basic, these same objects are generally referred to as *controls*. We will use the DCL terminology because it is used exclusively in AutoCAD's documentation. All objects that

Figure 11.10
A sample DCL dialog box.

appear in the dialog box in Figure 11.10 are tiles. In addition, the dialog box itself is a tile. There are many kinds of tiles, each varying in purpose and function.

Every kind of tile has a textual *representation*, which takes the form of a DCL expression. DCL expressions that comprise custom dialog boxes are stored in files with DCL extensions. These files are loaded by the programs that activate and control the dialogs. All tiles have properties that are referred to as *attributes*. A tile's attributes are what defines its appearance, behavior, and position within a dialog box. You'll learn more about tile attributes later in this chapter.

Subassemblies

Some kinds of tiles can act as *containers*, which hold one or more other tiles. When a tile contains other tiles, the contained tiles are referred to as *children* of the tile that contains them. A tile and its children are treated as a single unit, referred to as a *subassembly*. All dialog boxes are subassemblies. There are several examples of subassemblies in the dialog shown in Figure 11.10: the boxes that contain radio buttons and toggles are both subassemblies. The box that contains the slider is also a subassembly. Subassemblies can also be nested within other subassemblies to any reasonable depth.

Prototype Tiles

All of the tiles that appear in the dialog in the Figure 11.10 are predefined or *native* tiles, which are built into AutoCAD. In addition to the range of native tiles, Dialog Control Language allows you to define new kinds of tiles that are based on or *derived* from existing tiles. These derived tiles are referred to as *prototypes*. For those who are familiar with object-oriented programming (OOP), tiles are analogous to *classes*, whereas prototypes derived from other tiles or prototypes are analogous to *subclasses*. When a prototype is derived from an existing tile, the tile it is derived from is referred to as the prototype's

Introduction to Dialog Boxes

ancestor, and the new derived prototype is called a *descendant* of the ancestor. A prototype can have more than one ancestor. In OOP, this is analogous to *multiple inheritance.* When a new prototype is defined, it *inherits* the attributes of its ancestor(s). Once defined, a new prototype can then appear in a another prototype definition, subassembly, or dialog box, just as native tiles can. In this chapter from this point on, the term *tile* is meant to include references to built-in tiles, as well as prototypes derived from built-in tiles.

In this chapter, you will work mainly with native tiles. The topic of prototypes is a more advanced subject that will be examined more closely in the next chapter once you've learned the basics.

DCL Syntax and Semantics

The notational conventions used in this book to describe DCL syntax are the same as the conventions used for describing AutoLISP syntax. Most significantly, a pair of square brackets ([]) surrounding an item indicates the item is *optional,* and the ellipsis (...) indicates that any number of *additional* occurrences of the item preceding the ellipsis is allowed. Italicized items are descriptions that describe the items they represent; the specific strings used are not required in actual use, and should not be treated literally.

Elements in DCL expressions consist of *identifiers, numbers, strings,* and *delimiters.* Identifiers are strings of alphanumeric characters that are very similar to AutoLISP symbols. Identifiers may contain alphanumeric characters, numbers, and the underscore, but the first character in an identifier may not be a number. Also like AutoLISP symbols, identifiers must be separated from other elements by one or more delimiters. Delimiters in DCL serve the same purpose as their AutoLISP counterparts. DCL delimiters consist of the characters : ; { } = " and *space.*

Warning: One very important distinction between AutoLISP symbols and DCL identifiers is that identifiers in DCL are *always case-sensitive*. Failure to remember this convention can be the source of numerous errors. In fact, we suggest that you use all lowercase characters for all DCL identifiers exclusively, because doing so eliminates the need to remember exactly what case or capitalization you must use.

Numbers are simply that. They can be integers or floating-point numbers. Floating-point numbers must have at least one decimal place to the left of the decimal point. Strings are identical to their AutoLISP counterparts, and must be delimited by double quotes.

Note: Like AutoLISP, DCL allows *comments* in its source. Within a DCL file, anything that follows two slashes (//), up to the end of the line they appear on, is treated as a comment. Also, anything that appears between a matching pair of the sequences /* and */, is also treated as a comment. The latter form of commenting can also span multiple lines.

Tile Attributes

DCL tiles and subassemblies can have *attributes*. Attributes are properties that are associated with a tile. Attributes can be assigned values that define the behavior and appearance of a tile. Attributes are very much like *variables* in other programming languages. They are referenced by an identifier, and they can have values assigned to them in DCL source, using the assignment operator (=). There are both *predefined* and *user-defined* attributes. A predefined attribute is one that AutoCAD recognizes and associates with a specific meaning. For example, the width of an edit box can be defined by assigning a numeric value to the edit box's predefined width attribute. User-defined attributes are ones that mean nothing to AutoCAD or DCL, but can have meaning to an AutoLISP (or ADS) program that interacts with a dialog box. All of the attributes you will learn about and use in this chapter are predefined attributes. User-defined attributes are more commonly used with tile prototypes, which we'll examine in detail in the next chapter. The syntax for declaring a predefined or user-defined attribute and assigning a value to it is shown here.

```
attrib_name  = value;
```

Here, attrib_name is an identifier that is the name of this attribute. The actual sequence of characters used does not have to be unique within a DCL file, but *must* be unique within the tile whose definition it is part of. Value is the value that is assigned to the attribute, which can be another identifier, a string, or a number. There are several predefined identifiers built into DCL that have special meaning to AutoCAD. The most common of these are true and false, which represent those logical states. Others are left, centered, right, top, bottom, and so on. You should not use the names of predefined attributes as user-defined attributes. See the *AutoCAD Customization Guide* for a list of predefined attributes.

Tile References

The syntax for declaring a reference to a tile takes one of two basic forms. The first form is simply the name of the tile followed by a semicolon delimiter, as shown here.

```
tilename;
```

In the preceding example, tilename is the name of a tile (or defined tile prototype). This first form is the simplest one, and is used when the tile used in its basic form without modification. The second form of referencing a tile is used when there are instance attributes associated with the tile, or when the tile is a *cluster* that contains other tiles.

```
: tilename {
    [attribute = value;]...
    [tile]...
}
```

In the preceding example, *tilename* is the *name* of the tile. In this form, the name is *always* preceded by a colon. After the name is an open curly brace followed by one or more attribute assignments and/or one or more references to other tiles (which is permitted only if the tile is a descendant of a cluster tile). [*tile*]... represents one or more references to tiles, each of which can take either of the two forms of tile reference. There can be any number of tile references nested within a parent tile reference if the parent is a cluster type. Finally, a closing curly brace terminates the reference. Note that each attribute/value assignment pair as well as each abbreviated form of tile reference ends with a semicolon. Also note that there is *never* a semicolon after a closing curly brace, under any circumstance.

Because a dialog box is itself a kind of tile prototype, the syntax for defining a dialog box is similar to the syntax for defining a prototype, except there is also an ancestor, and it is always the predefined `dialog` tile, or a descendant of same.

```
tilename : dialog {
  [attribute = value;]...
  [tile;]...
}
```

In the preceding description, *tilename* is the name of the dialog box being defined, and is followed by a colon and the ancestor type, which must *always* be `dialog` or the name of a prototype that is a descendant of a dialog. Next is an open curly brace, and then the contents of the dialog, which can include any number of tile attributes and tiles. In addition, each tile reference in a dialog may contain other tile references. The dialog definition is terminated by a closing curly brace.

To better understand how DCL expressions are used to construct a dialog box, we'll examine a very simple dialog box definition in DCL.

```
// 11Hello.dcl from Maximizing AutoCAD R13

hello : dialog {                    // hello is a kind-of dialog

  label = "Hello";                  // The dialog's label attribute

  :text {                           // This is a text tile with
    label = "Here's some text";     // one attribute called value
  }

  :edit_box {                       // This is an edit_box tile
    label = "&New value:";          // with four attributes,
    key = "edit_1";                 // called label, key, value,
    value = "Edit me!";             // and edit_width
    edit_width = 12;
  }
```

Figure 11.11
The Hello dialog box from 11HELLO.DCL.

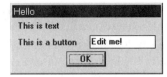

```
    ok_button;                    // An ok_button tile with
}                                 // no attributes
```

This Hello dialog box is shown in Figure 11.11.

Notice that the tiles in the definition of the Hello dialog box from the preceding listing are arranged in a single vertical column within the dialog box, in the same order they appear in the DCL source. By default, all tiles in a dialog will be arranged in a single column.

This dialog box definition has one attribute and contains three tiles (which are *children* of the dialog box). The dialog's label attribute is a predefined attribute (a string) whose value (which is the string "Hello") becomes the dialog box title. Most tile types have a label attribute which is the visible text associated with the tile. For example, a button tile's label attribute is the text that appears on the button, and a text tile's label attribute is the text itself. The label attribute also allows you to define a mnemonic character for a tile, by prefixing the mnemonic character with & (an ampersand). In the preceding dialog definition, the edit box label is **N**ew value, using the first letter M as the mnemonic. You should be careful to ensure that all mnemonics on a dialog box are unique.

The hello dialog definition also contains three child tiles: a text tile, an edit_box tile, and an ok_button tile. The text tile is *static text* that appears on the dialog box. The text tile also has one string attribute called label, which is the text that is displayed in the dialog box.

The edit_box tile follows the text tile. Edit boxes are rectangular areas containing text that can be edited like text in a text editor. AutoCAD's DCL edit_box is limited to a single line of text. The volume of text an edit box can hold is platform-dependent. The edit_box has four attributes: label, key, value, and edit_width. The label attribute of the edit_box is the descriptive text displayed to the left of the edit box. The key attribute is a predefined string attribute whose value is used by AutoLISP programs to reference the tile (you'll learn more about how AutoLISP is used to access and control DCL tiles later in this book). Because a tile's key attribute is the means of referencing the tile from AutoLISP, the value of this attribute *must be unique* within the DCL file that contains it. Hence, no two or more tiles in the same DCL file may have the

Introduction to Dialog Boxes

same value in their `key` attribute. The `edit_box` tile also has a `value` attribute. This attribute is also predefined and holds the data contents of each tile. All tiles have a `value` attribute, and how it is interpreted depends on the type of tile. For edit boxes, the `value` attribute is the text the user enters into the edit box. For a `toggle` tile, the `value` attribute is a string containing the number 1 or 0, which indicates if the toggle is checked or not. The `edit_width` attribute is another predefined attribute (a number) that defines the physical width of the edit box (12). The last tile in the dialog is a prototype called `ok_button`, derived from a `button` tile (hence, `ok_button` is a *descendant* of a `button` tile). The prototype definition for the `ok_button` is defined in BASE.DCL. This file contains global prototype definitions that are available to all dialog boxes including those used by AutoCAD. You'll learn more about prototypes and BASE.DCL in the next chapter.

Note: All dialogs must have at least an OK or Cancel button.

The *AutoCAD Customization Guide* contains a complete list of all native tile types and predefined attributes for each, which you can refer to for specifics about each predefined attribute and tile type.

DCL Learning and Development Tools

The process of creating a fully functional dialog box includes designing the dialog box in DCL, and writing the AutoLISP code that displays and controls the dialog box. To provide a gentle introduction to DCL, this chapter does not emphasize the AutoLISP programming concepts and techniques associated with managing dialog boxes, but instead focuses primarily on fundamental concepts of Dialog Control Language. For this purpose, you can use the LispPad code editor from this book's accompanying CD-ROM for working through the exercises in this chapter and for writing your own DCL code. All of the exercises in this chapter require AutoCAD R13 for Windows.

In addition to AutoLISP code editing, LispPad is also a useful tool for streamlining dialog box design in DCL. The LispPad editor can automatically instruct AutoCAD to display a dialog box defined in a DCL file that is currently open in the editor. You can repeatedly edit the source for a dialog in the editor, then press a single key to view the resulting dialog box on your display directly, without the need to first save the DCL file to disk and load it into AutoCAD (which involves significantly more steps than loading AutoLISP programs). With LispPad's dialog preview facility, developing DCL code is far simpler than doing the same task with a standard text editor, which also requires you to perform all the aforementioned steps each time you want to preview or test the dialog. More importantly, with a preview of the dialog you're designing only a keystroke away, you can focus on designing the dialog, without distraction.

Try using LispPad to modify and test the Hello dialog box from the previous listing. The Hello dialog definition is in the file 11HELLO.DCL from this chapter's directory, which you should copy to your working directory. It is assumed that you've already installed LispPad and are familiar with it from use in previous chapters.

 Exercise

Using LispPad to Design and Test Dialog Boxes

Make sure you have copied this chapter's code files from the CD-ROM or MAXAC\BOOK\CH11 directory to your working development directory.

Continue from the previous exercise, or start AutoCAD with a new, unnamed drawing. Launch LispPad, and open 11HELLO.DCL.

In the Goto Definition *drop-down list, choose* hello	Positions the cursor on the first line of the hello dialog box definition
Choose the Test dialog box button from the Lisp toolbar	The Hello dialog box appears (see the earlier Figure 11.11) and the Test dialog box button goes down

If the Hello dialog box is not visible, resize or minimize the LispPad window and/or drag the Hello dialog box to a location where both it and LispPad will remain visible.

Choose the Test dialog box button again from the Lisp toolbar, or press CTRL+L	The Hello dialog closes and the Test dialog box button goes up
Choose the Test dialog box button again from the Lisp toolbar	The Hello dialog reappears and the Test dialog box button goes down
Change the label attribute of the Hello dialog box from `"Hello"` *to* `"Hello World"`	
Press CTRL+L	The Hello dialog is updated with the modified label
Change the label attribute of the text *tile from* `"Here's some text"` *to* `"Maximizing AutoCAD"`	
Press CTRL+L	The Hello dialog is updated with the modified text value attribute

Introduction to Dialog Boxes

Insert the following line of DCL code exactly as shown, between the text tile and the edit_box *tile:*

```
: button {label = "Another
button!";}
```

Press CTRL+L	The Hello dialog is updated with the new button
Choose the Test dialog box button or press CTRL+L	The Hello dialog closes

Note: If you accidentally close LispPad before dismissing a dialog box that was previewed by LispPad, enter any invalid command on the AutoCAD command line to dismiss the dialog.

If you tried choosing one of the buttons on the dialog, then you noticed that it does not respond to your actions. Although the dialog preview is visible and some of its controls can be used, there is no code controlling it, hence it does nothing when you click on a button (such as the OK button), or edit text in the edit box. This is because the underlying AutoLISP code that controls the dialog is not present (a topic we'll take up in the next chapter). Once you've learned more about using AutoLISP to activate and manage dialog boxes, you'll come to appreciate LispPad's dialog preview to design dialog boxes.

Tip: While you are learning DCL, you'll endure your share of mistakes, as most DCL programmers do when creating or modifying dialog boxes. When this happens, the dialog will not appear, and AutoCAD will instead display a dialog that describes the error. If the error is severe, AutoCAD will output information about the error to a file called ACAD.DCE. You can open this file with another copy of LispPad or any other editor to examine the errors. When using LispPad for development, you can make the needed corrections to the DCL code and just press CTRL+L to update the dialog preview.

DCL Code Formatting

You've undoubtedly noticed that the DCL code in all of the listings in this book is very consistently formatted with indentation and matching colons and closing curly braces in the same column. For each tile reference that uses the complex form of reference, the children and attributes of the tile are indented two spaces from the parent or container reference. The closing curly brace that terminates a complex tile reference is always in the same column as the colon that marks the beginning of the same reference. Just as with AutoLISP, adoption of and strict adherence to formatting conventions are crucial

Tile Parametrics

DCL tiles are *parametric* in nature. They can automatically size themselves to fill the space they're contained in. Rather than having to explicitly define height, width, and locations of tiles, DCL tiles can automatically arrange themselves to fill the available space on a dialog box or another container tile. You can control the size of most tiles explicitly using the `width` and `height` attributes when needed. You can control how tiles align with respect to each other and to their container, and you can also inhibit automatic expansion of a tile to prevent it from filling the space it resides in as well.

Most DCL tiles posses a number of predefined attributes that allow you to control their size, location, and alignment. Some of these attributes are associated with all tiles (such as `height` and `width`), while others are specific to certain types of tiles. The AutoCAD Customization Guide describes these attributes and associated tiles.

The following code listing is from 11EX01.DCL from this chapter's directory. In addition to an OK button, the dialog contains a `text` tile, a `popup_list`, and two `button` tiles.

```
test : dialog {
  label = " Using Layout Attributes";
  : text {
①     label = "Exercise ";   // 1. Modify label attribute
  }
  : edit_box {
    label = "Value: ";
④                             // 4. insert edit_width attribute here
⑤                             // 5. insert fixed_width attribute here
  }
  : button {
    label = "Button 1";
②                             // 2. insert fixed_width attribute here
③                             // 3. insert alignment attribute here
  }
  : button {
    label = "Button 2";
  }
  ok_button;
}
```

In the next exercise, you'll change some attributes of the tiles in the test dialog from the preceding listing.

Introduction to Dialog Boxes

Exercise

Using Attributes to Control Tile Placement and Size

Continue in the current session or start AutoCAD with a new, unnamed drawing. Launch LispPad, and open the chapter file 11EX01.DCL.

In the Goto Definition *drop-down list, choose* test, *or move the cursor to the first line of the test dialog*	Editor scrolls to the test dialog
Press CTRL+L	The test dialog box appears

Arrange the dialog box and the LispPad window so both are visible.

Add the string **from Maximizing AutoCAD** *to the end of the text tile's label attribute (① in the previous listing)*	Changes the text tile's label to "Exercise from Maximizing AutoCAD"
Press CTRL+L	Updates the dialog box

Note that the width of the dialog increased to accommodate the longer label, and other tiles in the dialog have increased in width, to fill the additional space.

Insert the following attribute into the first button tile (Button 1) at ②.

`fixed_width = true;`	Prevents button from enlarging to fill the width of the dialog box
Press CTRL+L	Updates the dialog box

Note that Button 1 has decreased in width.

Insert the following attribute into the same Button 1 tile, at ③.

`alignment = centered;`	Causes the button to become centered in its container

Change the same alignment attribute's value from centered to right at ③.

`alignment = right;`	Causes the button to become right-justified in its container
Press CTRL+L	Updates the dialog box

547

Insert the following attribute into the edit_box tile at ④.

edit_width = 12;   Sets the edit box width to 12

Press CTRL+L   Updates the dialog box

Note that the width of the edit box has decreased and the edit control has moved to the right side of the dialog box.

Insert the following attribute into the edit_box tile at ⑤.

fixed_width = true;   Sets the edit box width to fixed

Press CTRL+L   Updates the dialog box

Note that edit control has moved back to the position it was in prior to adding the edit_width attribute in the previous step.

In LispPad, choose the Test dialog button   Closes the preview dialog

---

After you complete the exercise, the dialog box should appear as shown in Figure 11.12. If it does not, try repeating the exercise.

The previous exercise shows how to control the size of individual tiles using predefined attributes. Note the following points:

- Changing the value of the dialog's label attribute caused the dialog width to increase to accommodate the longer string.

- Adding the fixed_width = true attribute/value to the first button caused its width to decrease.

- Adding the alignment = right attribute/value to the first button caused it to move to the right side of the dialog box. Changing the same attribute's value to centered caused the button to become centered horizontally in the dialog box.

Figure 11.12
The test dialog box (as it should appear after completing the exercise).

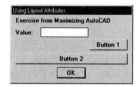

Introduction to Dialog Boxes

- Adding the edit_width = 12 attribute/value to the edit box caused the edit control to decrease in width and move to the right side of the dialog.

- Adding the fixed_width = true attribute/value to the edit box caused it to move to the left, near its label.

Note that assigning 12 to the edit_width attribute of the edit box had undesirable results, because it moved the edit control to the right, away from its label. This is because the edit control and its label are a single tile, and assigning an explicit width to the tile prevents the edit control from expanding to fill the width of its container (the dialog box), so the space between the edit control and its label increased instead. Setting the fixed_width attribute to true corrected this. You can use these and other predefined DCL attributes to fine-tune the appearance of your dialog boxes.

Using Spacers

Spacers are rectangular tiles that are completely invisible, and as their name implies, are used to add empty space to areas of a dialog. They are useful in cases where using the predefined layout attributes will not achieve the desired effect. The spacer tile's height and width attributes are not set by default, and if you use one, you should explicitly set the width or height as required. In addition to the spacer, BASE.DCL also defines two descendants of this tile called spacer_0 and spacer_1. The spacer_0 tile is a spacer that has no width or height. If two or more of these tiles appear in a cluster, assigning them a positive width or height causes all of them to uniformly expand to fill all excess horizontal or vertical space. The spacer_1 tile has a default width and height of 1, and can be used instead of the spacer, to eliminate the need to assign explicit height and width attributes of 1.

The next exercise shows how to use spacers to insert blank space into areas where it is often needed.

 Exercise

Using Spacer Tiles

You can continue in editing session from the last exercise, or launch AutoCAD with a new, unnamed drawing. Launch LispPad, and open the chapter file 11EX02.DCL.

In the Goto Definition *drop-down list, choose* test, *or move the cursor to the first line of the test dialog*	Editor scrolls to the test dialog definition

549

Press CTRL+L	The Using Spacers dialog appears (see the left dialog box in Figure 11.13)

Arrange the dialog box and the LispPad window so both are visible.

Note the extremely small amount of vertical space between the last edit box and the top of the OK and Cancel buttons.

Insert the following line after the last edit box, and before the `ok_cancel` *tile (on the* // Insert spacer_1; here *line):*	Places a space between the edit box and the OK and Cancel buttons

    `spacer_1;`

Press CTRL+L	Updates the dialog box (see the right dialog box in Figure 11.13)

Note that the height of the dialog has increased slightly, and there is now more space between the last edit box and the top of the OK and Cancel buttons.

Choose the Test Dialog button	Closes the preview dialog

Figure 11.13 shows the Using Spacers dialog before (left) and after (right) the addition of the spacer.

Clusters

Clusters are rectangular regions that may contain other tiles. Up to this point, all of the tiles you've seen, with the exception of the dialog box itself, are simple tiles, which cannot contain other tiles. You can use cluster tiles as containers to group other tiles. DCL provides a number of predefined tiles that are descendants of the native *cluster* tile. In this context, a cluster is any tile that is a descendant of the native cluster tile. The tiles in a cluster are referred to as its *children*. The basic cluster tiles are the `row` and `column` tiles. A `dialog` is also a cluster because it can contain other tiles. All clusters have a predefined `layout` attribute that defines how child tiles are laid out within the cluster. The layout of a cluster can be either horizontal or vertical. The layout of a `column` is vertical, and the layout of a `row` is horizontal. The layout of a `dialog` is vertical.

The `row` and `column` tiles are *invisible.* You can't see these tiles when they are part of a dialog box. These tiles are primarily intended to serve as invisible containers that are used to control the layout of a dialog box. With rows and columns, you can group related controls together and control their size, location, and alignment as if they were a single tile. For example, you may need to place a series of buttons on a dialog box and want

Figure 11.13
Using Spacers exercise.

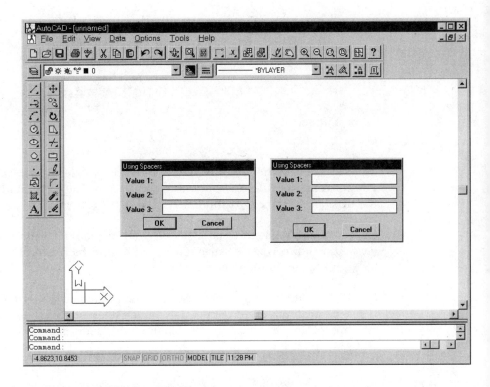

the buttons to be arranged in a `column` on the right side of the dialog. If you wanted the buttons to align vertically and have the same width, but do not want them to expand to the full width of the dialog or be affected by other controls that you will add afterwards, you could put the buttons inside a `column` tile. `row` and `column` tiles have the same syntax as other tile references, but unlike other tiles that cannot have children, rows and columns may also contain other tile references in addition to attributes. Here is a simple example of a dialog with a column and row.

```
example : dialog {
  : row {                         // start of row
    fixed_width = true;           // attribute of row
    : button {                    // Button inside row
      label = "Left button";
    }
    : button {                    // Button inside row
      label = "Right button";
    }
    : column {                    // start of column inside row
      : edit_box {                // edit box inside column
        label = "Upper";
```

```
            }
         : button {                    // button inside column
            label = "Lower";
         }
      }                                // end column
   }                                   // end row
   ok_button;                          // Ok button
}
```

The dialog box in the preceding listing contains two tiles: a `row`, and an OK button. The `row` contains three tiles: two buttons and a `column`. The `column` within the `row` contains two tiles: an edit box and a button.

Figure 11.14 shows a more complex dialog box containing a number of `row` and `column` tiles. Also shown is a tree diagram depicting the hierarchical relationships that exist between the tiles in the dialog.

In DCL, the tile references of those children are nested within the pairs of curly braces ({ }) associated with each row and column tile reference. Nesting columns and rows is an effective technique you can use to achieve more uniform and logical dialog layouts. The DCL source for the definition of the dialog in Figure 11.14 is shown here.

```
test : dialog {
   label = "Maximizing AutoCAD";
   : text {
      label = "Using Rows and Columns in Dialogs";
   }
   : row {
      : boxed_column {
         label = "Toggles";
         : toggle { label = "Toggle 1"; }
```

Figure 11.14
Row and column tiles.

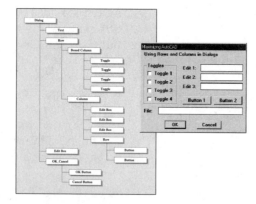

```
            : toggle { label = "Toggle 2";}
            : toggle { label = "Toggle 3";}
            : toggle { label = "Toggle 4";}
        }
        : column {
        : edit_box {
          label = "Edit 1:";
        }
        : edit_box {
          label = "Edit 2:";
        }
        : edit_box {
          label = "Edit 3:";
        }
        : row {
          : button {
             label = "Button 1";
          }
          : button {
             label = "Button 2";
          }
        }
    }
    : edit_box {
      label = "File: ";
    }
    ok_cancel;
}
```

In the next exercise, you'll add a descendant of a list_box, a row, and a column tile to the DCL dialog definition shown in the following listing. The dialog definition is from the chapter file 11EX03.DCL, which contains several versions of code demonstrating rows and columns.

```
/* 11EX03.DCL from Maximizing AutoCAD R13 */

// Topic: Rows and Columns

mxa_list_box : list_box {
   list =  "Item 1\nItem 2\nItem 3\nItem 4\nItem 5\nItem 6 [»]
\nItem 7\nItem 8";
}

test1 : dialog {
```

```
            label = "Maximizing AutoCAD - [Step 1]";

①      // Step 2: insert list_box, row, and column here

        : button {
          label = "Button 1";
        }
        : button {
          label = "Button 2";
        }
        : button {
          label = "Button 3";
        }

②      // Step 2: insert end row and column code here

    spacer_1;
    ok_cancel;
}
```

One line in this DCL listing had to be printed on two lines to fit on the page, but is a single line without the [»] continuation character in the DCL file.

Initially, the dialog contains only three buttons, arranged in the default manner (one column). Then, in the following exercise, code is added to display a list box and to arrange the dialog in two columns. Figure 11.15 shows the dialog as it should appear after each exercise step.

 Exercise

Working with Rows and Columns

Start AutoCAD with a new, unnamed drawing, or continue in the session from the last exercise. Launch LispPad, and open the chapter file 11EX03.DCL

In the Goto Definition *drop-down list, choose* test1, *or move the cursor to the first line of the test1 dialog definition*	Editor scrolls to the test1 dialog
Press CTRL+L	The test dialog box appears, showing only the three buttons in one column

Introduction to Dialog Boxes

Arrange the dialog box and the LispPad window so that you can see both.

Next, you try the test2 dialog definition. It adds the following code to test1, where indicated by ① and ② in the test1 listing preceding this exercise. This adds row, column and listbox definitions to the existing dialog box.

①
```
  : row {
    : mxa_list_box {
      height = 6;
    } // (end listbox)
    : column {
        alignment = top;
      fixed_height = true;
```
②
```
    }  // (end column)
  }  // (end row)
```

In the Goto Definition *drop-down list, choose* test2, *or move the cursor to the first line of the test2 dialog definition, then press CTRL+L*

Editor scrolls to the test2 dialog and updates the dialog preview to show two columns, the list box, and the buttons

Figure 11.15
The dialog from exercise 11EX03.DCL (at each step in the exercise).

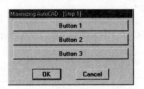

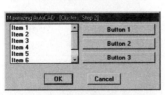

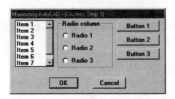

Move the cursor to the **: row { ** *line, and change the tile type from* **row** *to* **column**	Changes the row to a column
Press CTRL+L	
Undo the change made in the last step and change the **column** *back into a* **row**	Changes the column back to a row
Press CTRL+L	

Leave the dialog box, LispPad, and AutoCAD as is, for use in the next exercise.

Note that instead of using `list_box`, the dialog used `mxa_list_box`, which is a prototype descendant of a `list_box` which is defined in 11EXA03.DCL. You'll learn more about prototypes later in this and the next chapter.

Dialog boxes are easier to design and modify when tiles are arranged in a hierarchy of row and column clusters. In the exercise example, the hierarchy resulting from nesting tiles in rows and columns makes it relatively simple to add a column of radio buttons to the dialog. In the next exercise, you'll add a `boxed_radio_column` and three `radio_buttons` to the dialog resulting from the preceding exercise. The boxed radio column will be inserted to the left of the column containing the three buttons. The location of the added code is shown by the `// Step 3: insert boxed_radio_column here` comment in the test2 dialog definition in the 11EX03.DCL file.

Except for the portions shown in bold, the following code listing is the same as the test2 dialog definition from the previous exercise. The portions shown in bold define the radio column and buttons. The completed code is in the test3 dialog definition in 11EX03.DCL.

```
test3 : dialog {
  label = "Maximizing AutoCAD - [Clusters: Step 3]";

  : row {
    : mxa_list_box {
      height = 6;
    } // (end listbox)
    : boxed_radio_column {            // Begin boxed_radio_column
      label = "Radio column";
      : radio_button {
        label = "Radio 1";
      }
```

Introduction to Dialog Boxes

```
    : radio_button {
      label = "Radio 2";
    }
    : radio_button {
      label = "Radio 3";
    }
  }                          // End boxed_radio_column
  : column {
    alignment = top;
    fixed_height = true;
    : button {
      label = "Button 1";
    }
    : button {
      label = "Button 2";
    }
    : button {
      label = "Button 3";
    }
  }   // (end column)
 }   // (end row)
 spacer_1;
 ok_cancel;
}
```

Return to LispPad and try the test3 code, with the boxed_radio_column added.

 Exercise

Modifying Dialogs with Rows and Columns

Continue in LispPad, from the previous exercise.

In the Goto Definition *drop-down list, choose* test3, *or move the cursor to the first line of the test3 dialog definition, then press CTRL+L*	Editor scrolls to the test3 dialog and updates the dialog preview to show three columns: the list box, the radio column, and the buttons
Choose the Test dialog button	Closes the dialog box

Prototypes

So far, you've learned to create dialogs with structured elements, which help control the layout of dialogs and make revisions easier. If you structure your dialogs using clusters and their descendants, revisions are easier. You can, for example, change the location or order of all tiles in a cluster by moving the entire cluster in a dialog's DCL source. However, this requires moving potentially large blocks of code (a cluster and all of its contents), which can be error-prone and confusing. Now you'll learn to use another technique to make revisions to dialogs even easier. You'll also learn how to write reusable DCL code that can be shared by more than one dialog box.

Prototypes are formal definitions of a new type of tile that is based on one or more existing tile type(s), which include native tiles and other defined prototypes. As we discussed earlier in this chapter, a prototype is a descendant of the tiles it is based on, which are ancestors of the prototype. A new prototype tile inherits all of the characteristics and attributes of its ancestors. In this chapter, you've already used and defined tile prototypes. Each of the dialogs that you've worked with are, in fact, simply prototypes. For example, in the last exercise you worked with three dialogs called `test`, `test2`, and `test3`. Dialogs are also tiles, and each definition of a dialog is also a prototype. Each of these three aforementioned dialogs are tile prototypes which are descendants of their ancestor, the `dialog` tile type.

To recap from earlier in this chapter, the syntax of a tile prototype definition is precisely the same as the syntax for a dialog definition (simply because a dialog definition *is* a prototype).

```
typename : ancestor {
   [attribute = value;]...
   [tile;]...
}
```

Typename is the name of the new tile prototype that is being defined. The *ancestor* is the name of the ancestor tile type, which *typename* is derived from. The primary difference between a dialog prototype and a tile prototype, is that in the case of a dialog prototype definition, the *ancestor* type is always `dialog`. A dialog definition can reference only tiles in BASE.DCL. In a tile prototype definition, the *ancestor* type can be *any* tile type or even another prototype that has already been defined in the same way. Hence, you can create entire hierarchies or families of tile prototypes, where each descendant might be more specialized than its ancestors.

In addition to the `dialog` prototype, you've also used several other tile prototypes in the dialogs you've worked with so far in this chapter. The `ok_button` and `ok_cancel` tiles are both prototypes that are defined in BASE.DCL. Here is the definition of the `ok_button` tile.

Introduction to Dialog Boxes

```
ok_button : retirement_button {
  label = "  OK  ";
  key = "accept";
  is_default = true;
}
```

In this definition, the ancestor of `ok_button` is the `retirement_button` tile, which itself is a prototype with the following definition.

```
retirement_button : button {
  fixed_width = true;
  width = 8;
  alignment = centered;
}
```

Prototypes and Inheritance

The `retirement_button` is a tile that you rarely reference directly. Instead, you will use the `ok_button` tile, or the `ok_only` tile, which is another direct descendant of `ok_button`. The retirement button prototype tells AutoCAD that it is a button that is used to dismiss a dialog. AutoCAD requires every dialog to have at least one of these buttons or an error will occur. A prototype *inherits* all of the attributes of its ancestor. Understanding the concept of inheritance is the key to unlocking the true power of tile prototypes. When a new prototype tile is defined, the values of all the attributes of its ancestors, along with all of their ancestors, become the values of the attributes of the new tile prototype, as if they were included in its definition explicitly.

Look at the preceding definition of the `retirement_button`, which is the `ok_button`'s ancestor. Because the `ok_button` inherits the attributes of its ancestor, its definition is the *union* of its own attributes, *and* all of the attributes of `retirement_button`. As such, the exact same `ok_button` could be defined differently, as a direct descendant of the `button` tile with the addition of the `retirement_button`'s attributes, as shown here.

```
ok_button : button {
  fixed_width = true;          // inherited from retirement_button
  width = 8;                   // inherited from retirement_button
  alignment = centered;        // inherited from retirement_button
  label = "  OK  ";            // inherited from button
  key = "accept";              // inherited from button
  is_default = true;           // inherited from button
}
```

In fact, you could replace the existing definition of ok_button with this one, and it would result in the exact same tile. An important aspect that is not illustrated here is that when a new tile prototype is derived from an ancestor having an attribute that also appears in the definition of the derived prototype, the value of the prototype's attribute will *override* the value of the ancestor's attribute.

The following simple example demonstrates this. The listing shows one prototype called name which is derived from an edit_box, and another tile prototype called first_name, which is in turn derived from name. Note that both prototypes have a label attribute.

```
name : edit box {               // first_name is a kind-of edit_box
   width = 12;                  // which is 12 units wide, and
   label = "Your name:";        // has a label of "First name:"
}

first_name : name {             // first_name is a kind-of name
   label = "First name:";       // but with a different label.
}
```

The first thing that should be noted is that first_name has the same width as name, because first_name is derived from name, which has a width of 12. The definitions of these two tiles should make it obvious that the *only* difference between them is the value of their label attribute.

The ok_button prototype is an example of a simple prototype definition, which is simply a *derived subclass* of another button prototype. The ok_cancel prototype is a subassembly that contains other tiles. This second kind of prototype is a means of defining reusable dialog box components that may consist of anything that can be contained in a descendant of a cluster tile (such as a column or row).

The ancestor of ok_cancel is the row tile, which means that tiles within ok_cancel's definition will appear in a horizontal row, just as if they were children of row tile in the dialog box that references ok_cancel. You may recall from the last exercise that ok_cancel was referenced at the bottom of the test3 dialog definition, and that the resulting dialog box contained an OK button and a Cancel button, arranged in a row at the bottom of the dialog. Here is the definition of the ok_cancel tile prototype.

```
ok_cancel : row {
   fixed_width = true;          // don't expand to fill width of dialog
   alignment = centered;        // center contents
   ok_button;                   // an ok_button
   : spacer {                   // a spacer 2 units wide
      width = 2;
   }
   cancel_button;               // a cancel_button
}
```

Introduction to Dialog Boxes

As you can see, placing a reference to ok_cancel is simply another way of using all of its contents in a dialog. Prototypes can be used as a more structured means of defining many dialogs which contain subassemblies that are identical. Rather than having to include the contents of the ok_cancel tile in every dialog that requires both OK and Cancel buttons, you can simply reference ok_cancel instead.

You can use this same technique to define your own reusable components that can be referenced from any number of dialogs using a single identifier, just like the ok_button tile.

Unlike simple, non-cluster tile prototypes such as the ok_button tile, references to subassembly prototypes may *not* have additional attributes or tiles added to their references. All attributes and children of a subassembly prototype must be defined in the prototype definition. For example, if you try to add an attribute and/or child tile to a reference to the ok_cancel prototype as illustrated in the following example, AutoCAD will not include the children of the subassembly prototype in the resulting tile, and will replace them with the children in the reference.

```
: ok_cancel {
    width = 4;              // No good! The OK and Cancel buttons
    : button {              // will not be included in the tile,
      label = "Help!";      // only this button will appear.
    }
}
```

You also used another tile prototype in the last exercise, called mxa_list_box. The prototype for mxa_list_box, which appears at the top of 11EX03.DCL and is shown here.

```
mxa_list_box : list_box {
    list = "Item 1\nItem 2\nItem 3\nItem 4\nItem 5\nItem 6\nItem 7";
}
```

The mxa_list_box prototype is a descendant of the list_box tile. The only difference between it and a list_box is that mxa_list_box specifies the contents of the list box in its list attribute. The list attribute can be used to specify the initial contents of a list box or popup list in a DCL file. You can also populate list boxes and popup lists using AutoLISP while a dialog box is active (you will learn more about this in the next chapter). When you specify items using this attribute, each item is delimited by the newline character (\n) code, as shown in the previous listing. The list attribute is useful for specifying the contents of list boxes and popup lists when those contents never change. In the test3 dialog from the last exercise, you referenced mxa_list_box and the items shown in the list attribute appeared in the list box in the resulting dialog box.

Tile prototypes can be used to achieve a more structured and consistent design by using them to define more specialized versions of standard tiles. For example, it is sometimes necessary to require the user to supply or edit symbol names (blocks, layers, and linetypes) in edit boxes. Because all symbols are limited in length to 31 characters, it is necessary to limit the number of characters that can be entered into an edit box where a symbol will appear to that value. The `edit_limit` attribute can be used to specify the maximum number of characters that may appear in an edit box. Once the user enters the maximum number of characters into an edit box with this attribute set, no additional characters will be accepted.

The following prototype definition shows a prototype that defines a specialized version of an edit box. The prototype is given the name `symbol_edit_box`. The tile differs from the `edit_box` in that it limits the number of characters that can be entered to 31, and has a default width of 20 characters. The `symbol_edit_box` can be used wherever a symbol must be entered or edited in a dialog, eliminating the need to explicitly set the `edit_limit` attribute of a standard edit box every time it is used for the same purpose.

```
symbol_edit_box : edit_box {
  edit_limit = 31;           // Limits contents to 31 characters
  edit_width = 20;           // Sets visible width to 20 characters
}
```

The preceding prototype and the following dialog definition are from the chapter file 11EX04.DCL. The dialog uses the `symbol_edit_box` to obtain the name of a layer, linetype, and block.

```
test4 : dialog {
  label = "Symbol Entry";
  : symbol_edit_box {
    label = "Layer name: ";      // Edit a layer name
    key = "layer";
  }
  : symbol_edit_box {
    label = "Linetype name: ";   // Edit a linetype name
    key = "ltype";
  }
  : symbol_edit_box {
    label = "Block name: ";      // Edit a block name
    key = "block";
  }
  ok_cancel;                     // OK and Cancel buttons
}
```

Try using the test4 dialog and the `symbol_edit_box` prototype.

Introduction to Dialog Boxes

 Exercise

Using Tile Prototypes

Continue in the current editing session, or start with a new, unnamed drawing. Launch LispPad, and open the chapter file 11EX04.DCL.

In the Goto Definition *drop-down list, choose* test4,   Editor scrolls to the test5 dialog
or move the cursor to the first line of the test4 dialog
definition

Press CTRL+L   The test4 dialog box appears

Arrange the dialog box and the LispPad window so both are visible.

In the test dialog box, try to enter more than 31 characters in one of the edit boxes. It beeps and refuses additional characters.

Choose the Test dialog button   Closes the dialog

Structured Design Techniques

Now we'll examine a typical application of prototype subassemblies in the design of a dialog box. One advantage of prototype subassemblies is that they allow you to construct dialog boxes with groups of elements that can be rearranged with very little effort. The code listing that follows is from the chapter file 11EX05.DCL, which contains a revised version of the dialog box depicted in Figure 11.14 and its accompanying code listing. The following listing defines that same dialog, except that it uses several prototype subassemblies, which are also defined in the listing.

```
// Prototype: test5_toggles   Toggles subassembly

test5_toggles : boxed_column {
  label = "Toggles";
  : toggle { label = "Toggle 1"; }
  : toggle { label = "Toggle 2"; }
  : toggle { label = "Toggle 3"; }
  : toggle { label = "Toggle 4"; }
}
```

```
// Prototype: test5_edits   Edit boxes subassembly

test5_edits : column {
  : edit_box { label = "Edit 1:";}
  : edit_box { label = "Edit 2:";}
  : edit_box { label = "Edit 3:";}
}

// Prototype: test5_buttons   Buttons subassembly

test5_buttons : row {
  : button {label = "Button 1";}
  : button {label = "Button 2";}
}

// Prototype: test5_botedit   Bottom edit box

test5_botedit : edit_box {
  label = "File: ";
}

// Initial version of revised test3 dialog from ch. 11

test5 : dialog {
  label = "Maximizing AutoCAD";
  test5_text;                   // Text subassembly
  : row {
    test5_toggles;              // Toggle subassembly
    : column {
      test5_edits;              // Edit box subassembly
      test5_buttons;            // Buttons subassembly
    }
  }
  test5_botedit;                // Bottom edit subassembly
  ok_cancel;
}
```

Notice that the row of buttons, and the columns containing toggles and edit boxes that are children of the original test3 dialog, are defined outside of the dialog box as separate prototype definitions in this listing, and are included in the test5 dialog by reference. Now you'll make some changes to this version of the dialog, which can be done with far less effort than would be necessary to manually edit the test3 version of the dialog. Figure 11.16 shows how the dialog should look when it is updated after each revision.

Introduction to Dialog Boxes

Figure 11.16
The dialog from 11EX05.DCL (at each exercise step).

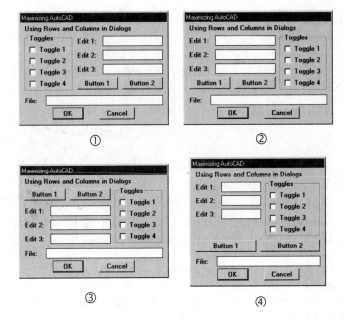

Exercise

Modifying Dialogs with Prototype Subassemblies

Continue in the current editing session, or start with a new, unnamed drawing. Launch LispPad, and open the chapter file 11EX05.DCL.

In the Goto Definition *drop-down list, choose* test5, *or move the cursor to the first line of the test5 dialog definition*	Editor scrolls to the test5 dialog
Press CTRL+L	The test4 dialog box appears (① in Figure 11.16)

Arrange the dialog box and the LispPad window so both are visible.

In the test5 dialog definition, move the `test5_toggles` line down after the closing curly brace of the column, as shown here:

Before revision
```
: row {
    test5_toggles;
    : column {
        test5_edits;
        test5_buttons;
    }
}
```

After revision
```
: row {
    : column {
        test5_edits;
        test5_buttons;
    }
    test5_toggles;
}
```

Press CTRL+L — The dialog box displays with the columns reversed ②

Within the column section of the test5 dialog definition, reverse the positions of the `test5_buttons` and `test5_edits` lines as shown here:

Before revision
```
: column {
    test5_edits;
    test5_buttons;
}
```

After revision
```
: column {
    test5_buttons;
    test5_edits;
}
```

Press CTRL+L — Updates the dialog box, with the position of the button row and edit boxes vertically reversed ③

Move the `test5_buttons` line from its previous position to where it is shown in bold below:

```
: row {
    : column {
        test5_edits;
    }
    test5_toggles;
}
test5_buttons;
test5_botedit;
```

Press CTRL+L — Updates the dialog, with the button row below the edit boxes and toggles, but the edit boxes are awkwardly spaced

Insert the fixed_height and alignment attributes in the column, as shown in bold below:

```
: row {
  : column {
    fixed_height = true;
    alignment = top;
    test5_edits;
  }
  test5_toggles;
}
```

Press CTRL+L   Updates the dialog ④

Choose the Test dialog button   Closes the dialog

It should be clear from the exercise that prototype subassemblies can make revisions and exploring design alternatives easier and less error-prone, since they allow you to manipulate entire clusters full of tiles with a single identifier.

Finishing Up

If you want to keep any of the files you created or modified in this chapter, move them to the \MAXAC directory or another directory, and then delete or move any remaining files out of \MAXAC\DEVELOP.

Conclusion

Adding graphical user interface elements such as custom dialogs to your applications is not easy. It requires a thorough understanding of DCL as well as AutoLISP. This chapter has provided only a gentle introduction to the former, which represents only one half of the equation. In addition to designing dialogs with DCL, you must also manage and control them with AutoLISP.

In the next chapter, you'll learn more about the use of DCL attributes, including user-defined attributes, and you will learn how to control dialog boxes with AutoLISP.

chapter

Designing Dialog Boxes

Chapter 11 introduced you to AutoCAD's programmable dialog boxes and Dialog Control Language. You learned the basics of DCL and how it is used to define the visual characteristics of dialog boxes. In this chapter, you'll explore dialog boxes further, and you will learn to control them with AutoLISP.

About This Chapter

Once you've completed this chapter, you should be able to do the following:

- Create a dialog box
- Write AutoLISP code to initialize a dialog box
- Assign and query values in dialog box tiles
- Control the enabled/disabled states of dialog box tiles
- Write callback functions to handle events in a dialog box
- Perform input validation on user-supplied data in a dialog box
- Manage and control list boxes

You should have already worked through Chapters 9 (Introduction to AutoLISP) and 11 (Introduction to Dialog Boxes).

The resource you will use in this chapter is:

- LispPad

Programs and Functions in this Chapter

- 12ATTSEL.LSP/DCL is a program that allows you to search for blocks by matching attribute values.

- 12HELLO.LSP/DCL and 12EX01-3.DCL are exercise files.

Copy this chapter's code files from the CD-ROM or MAXAC\BOOK\CH12 directory to your working development directory.

Introduction to Dialog Box Control

Adding a dialog-based interface to an AutoLISP application involves a number of tasks in addition to designing the dialog using DCL. You must also write AutoLISP code that initializes and activates the dialog box, and the AutoLISP code that controls the dialog and responds to user events that occur while the dialog is active.

The basic logic flow of the activities associated with the use of a dialog box in an AutoLISP program is shown in Figure 12.1.

Here is a summary of the steps shown in Figure 12.1:

1. Load the DCL file that contains the dialog box definition
2. Initialize the dialog box
3. Assign initial tile values enabled/disabled states
4. Associate action expressions with dialog tiles
5. Start the dialog event loop, allowing user to interact with dialog box
6. Handle user events while dialog is active, using action expressions
7. Close the dialog when user accepts or cancels the dialog box
8. Unload the dialog box

The first step in using a dialog box is to load the DCL file that contains the dialog's definition. Loading a dialog box is accomplished using the AutoLISP **load_dialog** function. This function takes the name of the DCL file to be loaded, and returns the *handle* of the loaded DCL file.

Designing Dialog Boxes

Figure 12.1
Dialog box process flow.

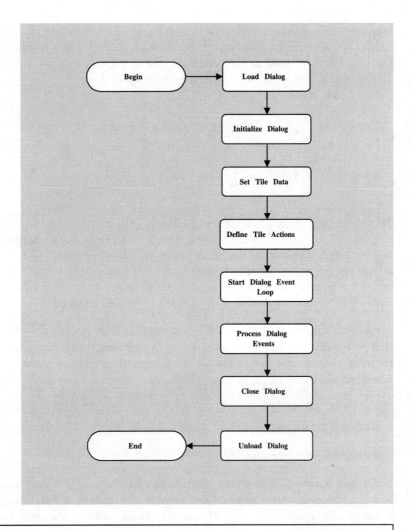

`load_dialog`

(`load_dialog` *dclfile*)

Loads a Dialog Control Language file into memory. The **`dclfile`** *argument, a string, is the file specification of the DCL file to load. If no file extension is included in* **`dclfile`**, *an extension of* DCL *is assumed. If no path is included in* **`dclfile`**, *the AutoCAD library path is searched for the specified file. If the DCL file is found and successfully loaded,* **`load_dialog`** *returns a positive integer, which is the handle of the loaded DCL file. This handle is used to refer to the loaded DCL file when calling the* **`start_dialog`** *and* **`unload_dialog`** *functions. If the specified DCL file is not found, or if an error occurs while attempting to load the file,* **`load_dialog`** *returns a non-positive integer that is less than 1.*

One important thing to remember when using **load_dialog** is that you *must* assign the result that it returns to a variable, because this result is the handle to the loaded DCL file, and must be used to reference the file in all related operations, such as starting a dialog box or unloading the DCL file when its dialogs are no longer needed. A typical call to **load_dialog** is shown here.

```
(setq hdcl (load_dialog "mydialog.DCL"))

(if (> 1 hdcl)
   (princ "\nCould not load DCL file.")
)
```

The preceding example use of **load_dialog** assigns the resulting integer handle of the loaded DCL file to the symbol **hdcl**. The second expression compares the result of the call to **load_dialog** to the value 1. If the handle is less than 1, the operation was unsuccessful, and a message is displayed indicating such. AutoLISP programs should never attempt to use a dialog box before verifying that the DCL file that contains the dialog's definition was successfully loaded by **load_dialog**.

A loaded DCL file occupies memory. AutoLISP also provides a way to *unload* a loaded DCL file, which is accomplished using the **unload_dialog** function. This function requires the handle of the loaded DCL file, that was returned by a previous call to **load_dialog,** which loaded the DCL file.

---

unload_dialog

(unload_dialog *handle*)

Unloads a currently loaded DCL file. The **handle** *argument, a positive integer greater than 0, is the handle of the loaded DCL file that is to be unloaded. This handle is obtained from the result of the previous call to* **load_dialog,** *which successfully loaded the DCL file. The* **unload_dialog** *function always returns* **nil**.

The following example use of **unload_dialog** unloads the DCL file loaded by the previous example use of **load_dialog**.

```
(unload_dialog hdcl)
```

Loading the DCL file is the first step that must be performed to display a dialog box. After a DCL file is successfully loaded, you must then create an instance of the dialog box you want to use. This is done using the **new_dialog** function. This function takes at minimum the name of the dialog box that you are creating an instance of, and the handle of the loaded DCL file containing the definition of that dialog box. **new_dialog** also accepts two additional arguments, both of which are optional.

`new_dialog`

`(new_dialog dlgname handle [action [position]])`

Creates a new instance of a dialog box. `dlgname`, *a string, is the name of the dialog box.* **handle**, *an integer, is the handle of the loaded DCL file containing the definition of the dialog box to be created. The optional* **action** *argument is a string, which is the default tile action expression for all tiles in the dialog box.* **position** *is a list of two integers, which specifies the screen coordinates of the upper left corner of the dialog box (where (***0 0***) is the upper left corner of the display). If the function succeeds, it returns* **T**, *otherwise it returns* **nil**.

The optional third and fourth arguments to **new_dialog** will be discussed in detail later in this chapter. The first argument is the name of the dialog box, which is the first identifier in the dialog's DCL definition. In the following simple example dialog definition, the name of the dialog is `hello`.

```
hello : dialog {                    // hello is the dialog name
   label = "Test dialog";
   :text {
      label = "Static text";        // Value displayed in dialog
      key = "text_1";               // key attribute
   }
   ok_button;
}
```

The second argument to **new_dialog** is a positive integer, which is the handle of the loaded DCL file that contains the definition of the dialog box. This is the same value that is returned by **load_dialog** when it is called to load the DCL file. You must call **load_dialog** to load the DCL file that contains the definition of the dialog you want to use *before* you can call **new_dialog** to create an instance of the dialog.

In the preceding chapter, you created and edited dialog boxes using LispPad, which instructs AutoCAD to display the dialogs for you without having to write any code at all. To add dialog boxes to your own programs, you must perform all of the needed steps required to display and control these dialog boxes with AutoLISP.

Table 12.1
Tile Access and Control Functions

Function	Description
(get_tile *key*)	Retrieves the current value of a tile
(set_tile *key value*)	Assigns a value to a tile
(mode_tile *key mode*)	Controls tile states

Dialog Box Tile Access and Control

In order to use a dialog box, you must access and control its tiles programmatically. These tasks include assigning values to tiles, accessing the current values of tiles, and controlling tile states such as enabled/disabled and focus. For example, in most cases, when a dialog box first appears, there will be default or current values assigned to some tiles. An edit box in a file dialog will often contain a default filename when the dialog appears, which you can accept by just pressing ENTER or choosing OK. The AutoLISP programs that you use to control your own custom dialog boxes can also assign or preset default values in tiles for the user, when the dialog first appears. In the last chapter, you learned how to assign values to tiles in DCL, by assigning the value to a tile's `value` attribute. AutoLISP provides several functions specifically for controlling and accessing tiles, which are summarized in Table 12.1.

Accessing Dialog Box Tile Data

The **get_tile** function provides a way to access the value of a tile. In the example dialog box definitions from the previous chapter, you saw the use of the predefined DCL tile attribute called *key*. This attribute is the means through which AutoLISP programs can reference individual dialog box tiles, to assign or retrieve their values and control their states. In order for a tile to be accessed from an AutoLISP program, it *must* have a key attribute. Furthermore, the key attribute of each tile in a DCL file must be *unique* within the file. If two or more tiles in the same DCL file have a key attribute with the same value, a *redefining key* DCL warning message occurs. DCL will permit you to have more than one tile with the same key value, but you will be unable to access the value of all tiles with the same key via AutoLISP. Hence, it is ill-advised to use duplicate tile keys.

---

get_tile

(get_tile *key*)

Retrieves the value of a dialog tile. The **key** *argument, a case-sensitive string, is the value of the key attribute for the tile whose value is to be retrieved. This function may be called only after a dialog box containing the tile has been initialized by calling the* **new_dialog**

> *function, and may only be used while the dialog that contains the tile is open and active. Once a dialog box is closed, its contents are destroyed and no longer accessible. Returns the value of the specified tile as a string if the tile exists. Returns* **nil** *if the specified tile does not have a value, or the tile does not exist in an active dialog box.*

The **get_tile** function can be used to access the current value of a tile, given the value of the tile's key attribute. If a tile has no key attribute, then its value cannot be accessed using **get_tile**. The result of **get_tile** is always a string if the specified tile exists. If no tile in the currently active dialog has a key value that's equal to the key argument, then **get_tile** returns **nil**. As pointed out in the preceding chapter, one very important thing to remember when referencing DCL identifiers from AutoLISP is that those identifiers are *case-sensitive,* meaning that the key argument to **get_tile** *must* appear in the same case as the key attribute value in the DCL file. Again, if you use lowercase identifiers exclusively, you will greatly reduce the number of errors resulting from mismatched case.

How the result of **get_tile** is interpreted depends on the type of tile. For example, the value of an edit box is the text the user enters into it (or the value assigned to the tile via the DCL value attribute, or the AutoLISP **set_tile** function). The value of a list box is the numerical index of the currently selected element(s) or a null string if no elements are selected. The value of a toggle is the string **"1"** if the toggle is checked, or **"0"** if the toggle is unchecked.

The complement of the **get_tile** function is **set_tile**, which allows you to assign tile values from AutoLISP.

> **set_tile**
>
> **(set_tile** *key value***)**
>
> *Assigns a value to a tile in an active dialog. The* **key** *argument, a case-sensitive string, is the value of the key attribute for the tile whose value is to be assigned. This function may be called only after a dialog box containing the tile has been initialized by calling the* **new_dialog** *function, and may only be used while the dialog that contains the tile is open and active. Once a dialog box is closed, its contents are destroyed and no longer accessible. Returns* **T** *if the specified tile exists and the assignment was made, and* **nil** *otherwise.*

The **set_tile** function does as its name implies, and assigns a value to a tile in an active dialog box. In the next exercise, you'll use the functions introduced up to this point in this chapter to load a DCL file, create a new instance of a dialog defined in the DCL file, and assign and retrieve values from tiles in the dialog box. The definition of this dialog from the chapter exercise file 12TILES.DCL is shown in the following code listing. The key attribute of each tile appears in boldface type.

```
// 12TILES.DCL From Maximizing AutoCAD R13

itemlist : tile {
  list = "Item 0\nItem 1\nItem 2\nItem 3\nItem 4\nItem 5\nItem6 [»]
\nItem 7";
}

maxacad :  dialog {                              // dialog name = maxacad

key = "maxac";                                   // dialog key = maxac
label = "Maximizing AutoCAD Release 13";         // dialog title

  : text {
    label = "Accessing Dialog Tile Data with AutoLISP";
    key = "text_1";
  }
  : popup_list : itemlist {                      // popup list
    label = "&Popup List:";
    key = "popup_1";                             // tile key
    edit_width = 28;
  }
  : row {
    : list_box : itemlist {                      // list box
      label = "&List box 1";
      height = 4;
      key = "listbox_1";                         // list box key
    }
    : boxed_radio_column {                       // radio column
      key = "radgroup";                          // key of radio column
      label = "Radio buttons";                   // 3 radio buttons
      : radio_button {
        label = "&Radio button 1";
        key = "rad_1";
      }
      : radio_button {
        label = "R&adio button 2";
        key = "rad_2";
      }
      : radio_button {
        label = "Ra&dio button 3";
        key = "rad_3";
      }
    }
  }
  : boxed_row {                                  // boxed row containing
```

```
      label = "Toggles";                  // 3 toggles with keys
    : toggle {
      label = "&Toggle 1";
      key = "tog_1";
    }
    : toggle {
      label = "T&oggle 2";
      key = "tog_2";
    }
    : toggle {
      label = "To&ggle 3";
      key = "tog_3";
    }
  }
  : slider {                              // A slider
    key = "slider_1";
    max_value = 100;
    big_increment = 10;
    small_increment = 1;
    value = 50;
  }
  : edit_box {                            // an edit box
    label = "&Edit box:";
    key = "edit_1";                       // key of edit box
    value = "Edit me!";
  }
  spacer;
  : row {                                 // a row with 3 buttons
    : button {                            // each with a key
      label = "&Button 1";
      key = "btn_1";
    }
    : button {
      label = "B&utton 2";
      key = "btn_2";
    }
    : button {
      label = "Butt&on 3";
      key = "btn_3";
    }
  }
  spacer;
  ok_cancel;   // Ok button key = "accept"
               // cancel button key = "cancel"
}
```

The fourth line in this DCL listing had to be printed on two lines to fit on the page, but is a single line without the [»] continuation character in the DCL file.

Figure 12.2

The Maximizing AutoCAD Release 13 (maxacad) dialog box.

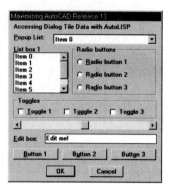

The maxacad dialog in the preceding code listing contains the most commonly used tiles. Each tile has a key, which allows it to be accessed from AutoLISP. Note that the **ok_cancel** tile reference, which is defined in BASE.DCL, contains an OK button, and a Cancel button. All DCL dialog boxes must have at least an OK or Cancel button, or an error will occur. Since the definitions of these two buttons do not appear in the dialog definition, their key attribute values are not obvious. The value of the OK button's key attribute is **accept**, and the value of the Cancel button's key attribute is **cancel**. The maxacad dialog from 12TILES.DCL appears in Figure 12.2.

 Next, you'll load and activate the dialog shown in the preceding code listing, and you'll access and change the values of several tiles from AutoLISP. After entering each call to **set_tile**, note the effects on the tiles in the dialog box. This exercise should be performed with AutoCAD for Windows.

Controlling Dialogs with AutoLISP

Make sure you have copied this chapter's code files from the CD-ROM or MAXAC\BOOK\CH12 directory to your working development directory.

Start AutoCAD with a new, unnamed drawing. Press F2 to make the AutoCAD Text window visible.

Command: **(setq testdlg (load_dialog ''12TILES''))** [Enter]   Loads 12TILES.DCL

Lisp returns: 1   Handle of loaded DCL file (actual value will vary)

Designing Dialog Boxes

Command: **(new_dialog "maxacad" testdlg)** [Enter]

Creates and displays new instance of dialog

Lisp returns: T

Arrange the AutoCAD text window and dialog box so both are visible.

Command: **(set_tile "text_1" "Hello World")** [Enter]

Sets value of text tile to "Hello World"

Lisp returns: T

Command: **(set_tile "radgroup" "rad_2")** [Enter]

Sets Radio Button 2 on (pushed)

Lisp returns: T

Command: **(set_tile "slider_1" "25")** [Enter]

Sets value of slider to 1/4 of its maximum

Lisp returns: T

Command: **(set_tile "edit_1" "Maximizing AutoCAD")** [Enter]

Sets value of edit box to "Maximizing AutoCAD"

Lisp returns: T

Choose Toggle 2

Checks Toggle 2

Command: **(set_tile "listbox_1" "3")** [Enter]

Selects fourth item in the list box

Lisp returns: T

In the Popup List, *choose* Item 4

Selects the fifth item in the popup list

In the Edit *box, select and delete the entire contents and then type* **MaxAcad**

Changes value in the edit box to MaxAcad

Command: **(get_tile "popup_1")** [Enter]

Retrieves the index of the selected item in the popup list

Lisp returns: "4"

Index of selected popup list item

```
Command: (get_tile "edit_1") [Enter]    Retrieves the value in the edit box
```
Lisp returns: `"MaxAcad"`   The contents of the edit box

```
Command: (get_tile "radgroup") [Enter]  Retrieves the selected radio button
```
Lisp returns: `"rad_2"`   Key of selected radio button

```
Command: (get_tile "slider_1") [Enter]  Retrieves the value of the slider
```
Lisp returns: `"25"`   Current value of the slider

```
Command: (get_tile "tog_3") [Enter]     Retrieves the value of Toggle 3
```
Lisp returns: `"0"`   The value of Toggle 3 (unchecked)

```
Command: (get_tile "listbox_1") [Enter] Retrieves the index of the selected
                                         item in the list box
```
Lisp returns: `"3"`   Index of selected item

```
Command: FOO [Enter]                     Causes unknown command error,
                                         and cancels and closes the active
                                         dialog box
```

```
Command: (unload_dialog testdlg) [Enter] Unloads 12TILES.DCL
```
Lisp returns: `nil`

As you saw in the exercise, the **set_tile** and **get_tile** functions provide a way to assign and retrieve tile values in an active dialog box. Later in this chapter, you will also learn to manipulate the contents of list boxes and popup lists using AutoLISP. How the value returned by **get_tile** is interpreted depends on the type of tile whose value is being retrieved. Table 12.2 lists DCL tiles that have a significant value along with the meaning of each tile's value. Note that the **radio_cluster** tile includes the **radio_column; boxed_radio_column; radio_row;** and the **boxed_radio_row**.

Controlling Tile States

AutoLISP programs can also manipulate other aspects of dialog tiles. For example, most tiles can be disabled, which prevents the user from changing their values. A well-designed dialog-based user interface will strive to make life easier for the user by disabling tiles whenever appropriate. For example, a dialog box may have required fields that must have values entered into them before the user can dismiss the dialog box, so a properly

Designing Dialog Boxes

Table 12.2
Tile Values

Tile	Value Description
`radio_cluster`	The value of the key property of the currently pushed radio button in the cluster, or a null string if no radio button in the cluster is pushed.
`dialog`	The title of the dialog box (same as a dialog's label property).
`edit_box`	The user-editable contents of the edit control.
`list_box, popup_list`	The indices of the selected item(s), delimited by spaces. The first item in the list has a value of "0". If nothing is selected, the value is a null string.
`radio_button`	The string "1" if the button is pressed, "0" otherwise.
`toggle`	The string "1" if the toggle is checked, "0" otherwise.
`slider`	A numeric value representing the position of the thumb button, where the `min_value` and `max_value` attributes are the values at the two extreme locations of the button.
`text`	The visible text (same as label property).

designed dialog box will disable the OK button until the user has satisfied all requirements. By disabling tiles when they should not be used or changed, it will serve to reduce the complexity of the error-checking.

Later in this chapter, you will see how you can use AutoLISP to disable controls on a dialog box dynamically whenever they should not be used. Next, you'll learn how to control the enabled, disabled, *focused*, and *selected* states of dialog tiles.

The AutoLISP **mode_tile** function provides a way to change various states of dialog tiles while a dialog box is active. These states include enabling and disabling tiles, selecting the contents of tiles, and setting focus to a tile. When a tile is disabled, its appearance changes to a "grayed" state, and it cannot get focus. When a tile has focus, depending on the type of tile, a *focus rectangle* appears around it, or a cursor appears in the tile, and all keystrokes are directed to the focused tile. When a tile's text or image contents are selected, the text or image is highlighted.

---

mode_tile

(mode_tile *key value*)

Controls various states of a dialog box tile. The **key** *argument, a case-sensitive string, is the value of the key attribute of the tile whose state is to be affected.* **value** *is an integer whose*

> range depends on the type of tile involved, and it specifies what state is to be changed. Can be used only while a dialog is visible and active. This function returns **T** if the operation succeeds, or **nil**.

Table 12.3

Mode_tile *values*

Value	Function	Applicable Tiles
0	Enable tile	All tiles
1	Disable tile	All tiles
2	Set focus to tile	All tiles
3	Select tile text	Edit boxes only
4	Select tile image	Image and Image buttons only

The value argument to **mode_tile** must be an integer ranging from 0 to 4. Each value specifies a different operation. Some operations apply only to certain types of tiles, while others may be used with any type of tile. Table 12.3 describes each of the values, and their functions.

In the next exercise, you'll use **mode_tile** to disable and enable some tiles in the same dialog used in the preceding exercise. Not all **mode_tile** options can be used from the AutoCAD command line. Some modes require the dialog box to be active and accepting input, which does not begin until you call **start_dialog** (which you'll learn more about later in the chapter). Also note that in the following exercise, the result of each call to **mode_tile** is omitted from the exercise listing for the sake of brevity. **Mode_tile** returns **T** if it succeeded or **nil** if it failed. In all cases in the exercise, it should return **T** if the expression was entered properly. Should an error occur or you want to restart the exercise, you can dismiss the visible dialog box at any time by simply entering any invalid input on the AutoCAD command line, such as an unknown command.

 Exercise

Controlling Tile States with mode_tile

Start AutoCAD with a new, unnamed drawing or continue from the previous exercise. Press F2 to make the AutoCAD Text window visible.

Designing Dialog Boxes

Command: **(setq testdlg (load_dialog "12TILES"))** [Enter]   Loads 12TILES.DCL

Lisp returns: 1   Handle of loaded DCL file (actual value will vary)

Command: **(new_dialog "maxacad" testdlg)** [Enter]   Creates and displays new instance of dialog

Lisp returns: T

Arrange the AutoCAD text window and the test dialog box so both are visible.

Command: **(mode_tile "listbox_1" 1)** [Enter]   Disables List box 1

Command: **(mode_tile "rad_2" 1)** [Enter]   Disables Radio Button 2

Command: **(mode_tile "edit_1" 1)** [Enter]   Disables Edit box

Command: **(mode_tile "text_1" 1)** [Enter]   Disables text tile

Command: **(mode_tile "btn_1" 1)** [Enter]   Disables Button 1

Command: **(mode_tile "listbox_1" 0)** [Enter]   Enables List box 1

Command: **(mode_tile "rad_2" 0)** [Enter]   Enables Radio Button 2

Command: **(mode_tile "edit_1" 0)** [Enter]   Enables Edit box

Command: **(mode_tile "text_1" 0)** [Enter]   Enables text tile

Command: **(mode_tile "btn_1" 0)** [Enter]   Enables Button 1

Command: **FOO** [Enter]   Unknown command error cancels and closes the active dialog box

Command: **(unload_dialog testdlg)** [Enter]   Unloads 12TILES.DCL

The design of the **mode_tile** function is a source of confusion. The value argument actually specifies one of five different functions which are too easily confused, and hence can become a source of frequent errors. Because these value arguments are somewhat cryptic, you may find it more convenient to create several user-defined functions that act as a more logical interface to **mode_tile**, each of which are dedicated to a different mode value. Rather than constantly looking up the meaning of each integer code, you can give these functions names that make the operation each performs more obvious. You can then call these functions in lieu of **mode_tile**, which will reduce the number of bugs that are likely to occur. As an added benefit, several of these functions will accept a single tile key argument (a string), or a single list containing any number of tile keys, and in the latter case, perform the operation on each tile whose key is passed in the list.

The user-defined helper functions in Table 12.4 are from the chapter file 12DCLSUB.LSP. Each function is dedicated to a different **mode_tile** operation, and has a name that makes the operation obvious.

The first two functions in the preceding table (whose argument is the symbol *keys*) will accept a single tile key argument (a string), or a *list* containing any number of tile keys. The following example shows how you can call the **tile_disable** function to disable all of the tiles that you disabled in the last exercise using **mode_tile**.

```
(tile_disable
  (list
     "listbox_1"   ; list box
     "rad_2"       ; radio button 2
     "edit_1"      ; edit box
     "text_1"      ; text tile
     "btn_1"       ; button 1
  )
)
```

In cases where you have a number of tiles whose states are always the same, and are manipulated together, you can place their key attribute values in a list, and assign it to an AutoLISP variable, so that you can reduce the enabling or disabling of all of the tiles to a single expression. The following example performs the same operation as the

Table 12.4
Helper Functions for Controlling Tile States
(from the file 12DCLSUB.LSP)

Function Prototype	Description	mode_tile Equivalent
`(tile_enable keys)`	Enables tiles	`(mode_tile key 0)`
`(tile_disable keys)`	Disables tiles	`(mode_tile key 1)`
`(tile_focus key)`	Sets focus to a tile	`(mode_tile key 2)`
`(tile_select key)`	Selects text in edit box	`(mode_tile key 3)`
`(tile_selimage key)`	Selects an image	`(mode_tile key 4)`

Designing Dialog Boxes

preceding example, except that it stores the key values of all the tiles in a list so they can be repeatedly passed to **tile_disable** and **tile_enable** more easily.

```
(setq tile_group1            ; Assign list of tiles
   (list                     ; to variable first, for
      "listbox_1"             ; for easier manipulation
      "rad_2"
      "edit_1"
      "text_1"
      "btn_1"
   )
)

(tile_disable tile_group1)   ; Disable tiles whose key values
                             ; are in the list tile_group1

(tile_enable tile_group1)    ; Enable the same group of tiles
```

It should be obvious from looking at this code that the use of these macro functions is more intuitive than the direct use of **mode_tile** with integer codes, since you can simply glance at your code and more easily understand what it does without the need to memorize the **mode_tile** integer codes.

Accessing Tile Attributes from AutoLISP

In Chapter 11, you were introduced to predefined tile attributes and you learned how to use them in DCL to control the visual characteristics of dialog box tiles. You can access the values of tile attributes from AutoLISP as well, using the **get_attr** function.

---

get_attr

(get_attr *key attrib***)**

Retrieves the DCL-file value of tile attributes in an active dialog box. **key**, *a case-sensitive string, is the value of the key attribute of the subject tile.* **attrib**, *a case-sensitive string, is the attribute whose value is to be retrieved. Returns the value of the specified tile attribute as a string, or a null string if the attribute or tile does not exist.*

The **get_attr** function is useful for accessing application-defined and predefined tile attributes, which can be assigned in a DCL file. One important aspect of **get_attr** is that it does not retrieve the actual run-time values of attributes (which can change while a dialog box is active, as a result of a user-action or calls to **set_tile** and **mode_tile**). The tile attribute values returned by **get_attr** are the values assigned to the attributes in the DCL file. Hence, when you use **get_attr** to retrieve the value attribute of a tile, it does *not* return the current value of the tile—it returns the value which was assigned to the tile's value attribute in the DCL file. You can use **get_attr** to retrieve both predefined and application-defined tile attributes. If you recall from the previous chapter, application-defined attributes are ones that have no specific meaning to AutoCAD and are intended primarily as a way to associate application-specific data with tiles in DCL.

One example of how application-defined attributes might be useful is to create new classes of tiles that impose some special restrictions on input or built-in default values. For example, the following DCL code listing defines a new kind of **image_button** tile called a **slide_image_button**, which has one application-defined attribute associated with it, called slide.

```
slide_image_button : image_button {
  slide = "maxacad.sld";           // name of slide file to be
}                                   // displayed in image button
```

The **slide_image_button** is a descendant of an **image_button** tile, with an added attribute called slide, which is assigned the name of the slide file to be displayed in the image button. The AutoLISP program that initializes the dialog containing this tile does not need to contain the name of the slide file it must use, because it can determine this by calling the **get_attr** function to get the slide file's name and display it in the image. Other tiles can be derived from this tile, with each having a different value in its slide attribute, demonstrating the concept of *encapsulation*.

In the next exercise, you'll test the **get_attr** function on several of the tiles from the maxacad dialog box used in the preceding exercises.

 Exercise

Accessing Tile Attribute from AutoLISP

Start AutoCAD with a new, unnamed drawing or continue from the previous exercise. Press F2 to make the AutoCAD Text window visible.

Command: **(setq testdlg (load_dialog**   Loads 12TILES.DCL
"12TILES")) [Enter]

Lisp returns: 1   Handle of loaded DCL file (actual
                                             value will vary)

Command: **(new_dialog "maxacad"**   Creates and displays new instance
testdlg) [Enter]   of dialog

Designing Dialog Boxes

Lisp returns: T

Arrange the AutoCAD text window and the test dialog box so both are visible.

Command: **(get_attr "listbox_1" "height")** [Enter]   Gets height attribute of the list box

Lisp returns: "4"

Command: **(get_attr "btn_2" "label")** [Enter]   Gets label attribute of Button 2

Lisp returns: "B&utton 2"

Command: **(get_attr "slider_1" "max_value")** [Enter]   Gets max_value attribute of slider

Lisp returns: "100"

Command: **(get_attr "popup_1" "list")** [Enter]   Gets contents of popup list

Lisp returns: "Item 0\nItem 1\nItem 2\nItem 3\nItem 4\nItem 5\nItem 6\nItem 7"

Command: **(get_attr "maxac" "label")** [Enter]   Gets label attribute of dialog box

Lisp returns: "Maximizing AutoCAD Release 13"

In the edit box, select and delete the entire contents and then type **MaxAcad**   Changes value in the edit box to MaxAcad

Command: **(get_attr "edit_1" "value")** [Enter]   Gets initial value of edit box

Lisp returns: "Edit me!"

Command: **FOO** [Enter]   Unknown command error cancels and closes the active dialog box

Command: **(unload_dialog testdlg)** [Enter]   Unloads 12TILES.DCL

Note that calling **get_attr** to retrieve the value attribute of the edit box, after you edited the initial value, did not result in the current value of the edit box. It is important to understand that the current values of tile attributes in an active dialog are not the values returned by the **get_attr** function. Instead, **get_attr** returns the initial values of tile attributes as they existed when the dialog was loaded (which are the same values assigned to the attributes in a DCL file). Since the values of some predefined tile attributes can change while a dialog box is active, you should never rely on **get_attr** always returning a tile attribute's current value.

Populating List Boxes and Popup Lists

The **set_tile** function can be used to set the value of tiles such as edit boxes, radio buttons, sliders, and toggles. Assigning values to list boxes and popup lists only selects items in them, but does *not* assign their contents. This discussion focuses on and refers to list boxes, but applies to both list boxes and popup lists.

To populate a list box with items, you must call the **start_list**, **add_list**, and **end_list** functions. The **start_list** and **end_list** functions are always called before and after **add_list** respectively. In between the calls to **start_list** and **end_list**, the **add_list** function can be called any number of times to add items to a list box. You can also use **add_list** to append items to a list box and change items in a list box. This discussion focuses mainly on adding items to a list box, which is the most common use of **add_list**. You will find more advanced techniques and tools for manipulating list boxes and popup lists in this book's companion volume, *Maximizing AutoLISP*.

The dialog boxes that you've been working with so far contained both list boxes and popup lists, with items in them. In these examples, placing items in the list boxes and popup lists was achieved by assigning the items to the DCL list attribute. However, more often than not the contents of a list box are dynamic, and you may not know what the contents are until it is displayed at run-time (for example, if you have to display a list of all layers in the drawing, the DCL list attribute would not work as the list contents varies with each drawing).

The **start_list** function takes the value of the list box or popup list's key attribute, and an optional index value. The index value is intended for changing one element in an already populated list box, and you will not use the index argument here. The **add_list** function takes a string which is to be added to the list box. You can call **add_list** any number of times between calls to **start_list** and **end_list**, but you must always call **start_list** before **add_list**, and you must call **end_list** after the last call to **add_list**. The following simple example adds three items to a list box whose key is "mylist".

```
(start_list "mylist")         ; begin list transaction
(add_list "Item one")         ;    add first item
(add_list "Item two")         ;    add second item
(add_list "Item three")       ;    add third item
(end_list)                    ; end list transaction
```

Designing Dialog Boxes

Since the sequence of events that are performed when you initialize a list box or popup list are very similar, it makes sense to define a subroutine that encapsulates the operation into a single function, which can be called and passed the key of the list box or popup list and a list of strings to load into the list.

The following example illustrates the use of **list_assign**, a user-defined function from the file 12DCLSUB.LSP. The **list_assign** function reduces the operation shown in the preceding example to a single function call.

```
(setq items
   (list "Item 1" "Item 2" "Item 3")      ; list of items
)

(list_assign "mylist" items)              ; adds all items in
                                          ; list to list box
```

In the file 12DCLSUB.LSP, you will also find similar functions for appending items to a list box and changing items in a list box, both of which eliminate the need to call **start_list** and **end_list**.

One important aspect of list boxes and popup lists is that the items that you place in them are read-only. You cannot access the strings added to a list box directly, so you must maintain a copy of the items in a list within your program if you need to find the text of an item in the list box given its positional index. The easiest way to do this is to first assign all of the items that you will add to a list box to an AutoLISP list, and then apply **add_list** to each element in the list within a **foreach** loop, or just use the **list_assign** function from 12DCLSUB.LSP. You will see how **foreach** is used for this purpose later in this chapter.

Next, you will use the list box functions just described to add some items to a list box in a sample dialog box. You will also load the chapter file 12DCLSUB.LSP and test the **list_assign** function as well.

 Exercise

Adding Items to a List Box

Start AutoCAD with a new, unnamed drawing or continue from the previous exercise. Press F2 to make the AutoCAD Text window visible.

Command: **(setq testdlg (load_dialog**   Loads 12TILES.DCL
"12TILES")) Enter

Lisp returns: 1   Handle of loaded DCL file (actual value will vary)

Command: **(new_dialog "testlist" testdlg)** [Enter]   Creates and displays new instance of dialog

Lisp returns: T

Arrange the AutoCAD text window and the test dialog box so both are visible.

Command: **(start_list "test")** [Enter]   Initiates list box transaction

Lisp returns: T

Command: **(add_list "First item in list")** [Enter]   Adds first item to list, but doesn't display it yet

Lisp returns: T

Command: **(add_list "Second item in list")** [Enter]   Adds second item to list

Lisp returns: T

Command: **(add_list "Third item in list")** [Enter]   Adds third item to list

Lisp returns: T

Command: **(end_list)** [Enter]   Ends list box transaction and displays list items

Lisp returns: nil

Next, use the list_assign function from 12DCLSUB.LSP to add items to the list box:

Command: **(load "12DCLSUB")** [Enter]   Loads 12DCLSUB.LSP

Lisp returns: STRTOINT

Command: **(setq items (list "Item one" "Item two " "Item three " "Item four))** [Enter]   Creates a list containing four strings

Lisp returns: ("Item one" "Item two" "Item three" "Item four")

Designing Dialog Boxes

Command: `(list_assign "test" items)` [Enter]	Adds the list of items to the list box
Lisp returns: `nil`	
Command: **FOO** [Enter]	Unknown command error cancels and closes the active dialog box
Command: `(unload_dialog testdlg)` [Enter]	Unloads 12TILES.DCL

Handling Dialog Events with Action Expressions

Up to this point, you've learned to load and initialize a dialog, and to access tiles before the user can interact with a dialog. After you've loaded and initialized a dialog box, you reach the point where you will allow the user to interact with the dialog. When the user interacts with a dialog box, your program must be able to determine what they do while the dialog is running. You can use action expressions for this purpose.

Event-Driven Programming

Before you undertake the next step in learning to create your own dialog boxes, you must first understand the abstract concept of the event-driven programming model. An event-driven program has a non-linear logic flow and has elements that are associated with and respond to events that occur in the computer. In an event-driven program, you specify what specific parts of the program are to be executed whenever an event occurs. Hence, you can associate specific events with different parts of your program, so that it can respond differently to each associated event. In this context, an event may be anything from pressing a key or moving a mouse to entering a string on the command line or clicking a button in a dialog box.

You already make use of event-driven programming to a limited extent. Each time you define a new command in AutoLISP by assigning a function to a symbol having the special prefix **C:**, you are telling AutoCAD to call the function assigned to the symbol when the user enters the name of the command (the portion of the symbol following the **C:** prefix). In this sense, the *event* is simply the act of the user entering the name of the command at the command prompt, and the program responsible for responding to the event is the AutoLISP function having the same name as the command with the added **C:** prefix.

In this scenario, the AutoLISP function that is called when the command is entered is referred to as an *event-handler*, because it is responsible for handling the event. What is significant about an event-handler is that it is usually invoked or called by another application, which in the case of a **C:XXXXX** function, is AutoCAD itself.

Dialog Box Events

AutoCAD is responsible for monitoring the events that occur when a custom DCL dialog box is active. But, since AutoCAD doesn't know exactly what should happen when one of these events occurs, it must notify the program that is responsible for controlling the dialog box each time an event in the dialog occurs. For example, if you choose a button on a dialog box, the AutoLISP program that controls the dialog box must respond to the event, and perform whatever actions are to occur when the button is clicked. In order for this to work, your program must instruct AutoCAD to *notify* it whenever the button is clicked. For this reason, AutoCAD permits your program to specify an expression that is to be evaluated, whenever an event occurs in a dialog box, by calling the **action_tile** function and passing it the name of a tile and a string that contains an AutoLISP expression. This AutoLISP expression is the *event handler* for events that occur within the associated tile.

---

action_tile

(action_tile *key action***)**

Associates a dialog box tile with an action expression. **key** *is a case-sensitive string that is the value of the key attribute of the tile to be associated with the action expression.* **action** *is a string that contains an AutoLISP expression that is to be evaluated whenever an event occurs in the associated tile. This function returns* **T** *if the tile specified by* **key** *exists, and* **nil** *otherwise.*

After you call **action_tile** and start the dialog, whenever an event occurs in the tile passed to **action_tile**, the expression passed in the second argument will be evaluated. In essence, **action_tile** is simply a way of telling AutoCAD what should happen whenever the user does something in a dialog box tile. These expressions are also referred to as *callbacks*.

You've been creating and displaying dialog boxes throughout this and the previous chapter, but up to this point, those dialogs have never become active. A DCL dialog box is active when it is processing events that occur in the dialog, and notifying the AutoLISP program that controls the dialog about those events as they occur.

Initializing a dialog box using **new_dialog** makes the dialog box visible, but does *not* make it active. In order to make a dialog box active, you must call the **start_dialog** function. When you call this function (which takes no arguments), your AutoLISP program's main thread of execution is suspended while the dialog box is active. The **start_dialog** function *does not return until the dialog is closed*. While the dialog box is active, your AutoLISP program can only get control when an event occurs in a tile that has an action expression associated with it.

Designing Dialog Boxes

After initializing a dialog box with **new_dialog**, and prior to making the dialog box active with **start_dialog**, you must set all tiles to their initial values, and register all action expressions with the dialog tiles. To recap from the start of this chapter, here is the sequence of events that you must perform prior to activating a dialog box:

1. Load the DCL file that contains the dialog box definition using the **load_dialog** function.

2. Initialize the dialog box by calling the **new_dialog** function.

3. Assign initial tile values and set initial tile states using **set_tile**, **mode_tile**, **start_list**, **add_list**, **end_list**, and so on.

4. Associate action expressions with dialog tiles using **action_tile**.

5. Call **start_dialog** to make the dialog active, thereby allowing the user to interact with it.

Action Expression Variables

When an action expression is triggered by an event in a dialog box, that expression needs to know something about the event that triggered it. Action expressions must be able to determine the reason for the event, they must be able to determine the key attribute value of the tile in which the event occurred, and they must be able to determine the value of the tile in which the event occurred as well. In order to make this information available to action expressions, AutoCAD automatically assigns it to special variables just before the action expression is evaluated. Table 12.5 describes these variables in detail.

The **$reason** variable is assigned to an integer that indicates the reason why the callback event occurred. The possible values for this variable are as follows.

Value	Callback Reason
1	A tile was selected.
2	An edit box has lost focus.
3	The user is dragging the thumb button of a slider.
4	The user *double-clicked* a list box item or a location in an image button.

Note: In this book, we use the $ character to distinguish DIESEL functions from AutoLISP functions, for example the **$getvar** DIESEL function versus the **getvar** AutoLISP function. Do not confuse this with the use of the $ character in the callback variables. It is only a coincidence that Autodesk chose to use the same character to denote the callback variables.

Table 12.5

Callback Variables

Symbol	Value
`$value`	The current value of the tile in which the event occurred
`$reason`	The reason for the event (description follows)
`$key`	The value of the key attribute of the tile in which the event occurred
`$data`	The value of the tile's data attribute
`$x`	The X component of the cursor location where the user clicked an image button (tile-specific)
`$y`	The Y component of the cursor location where the user clicked an image button (tile-specific)

The following code is from the chapter file 12HELLO.LSP. This file adds the HELLO command to AutoCAD, which implements a completely functional dialog box that is similar to the one used in the previous exercise, except it also has action expressions for most tiles. The action expressions of the tiles do nothing other than announce that a callback event occurred, and display a simple message box that shows the values of the special callback variables described in Table 12.5. In order for the message box to appear when a callback occurs, the **S**how Callbacks check box toggle at the bottom of the dialog box must be checked. This dialog also demonstrates techniques you can use to debug your own dialogs, or explore the callback mechanism in greater depth than can be done here. Note that references to all special callback variables are shown in bold type.

```
;; Implements a fully functional dialog box
;; that demonstrates action expressions in
;; action.

;; Hello dialog main function

(defun hello_dialog ( / hDcl lbitems show_cb)

  ;; Load dcl file.

  (setq hdcl (load_dialog "12HELLO"))

  ;; If the dcl file could not be found or loaded,
  ;; then display an error message and abort.

  (if (or (not hdcl) (> 1 hdcl))
      (fail "Can't load 12HELLO.DCL")
  )
```

```
(new_dialog "maxacad" hdcl)    ; Initialize and display dialog

;; Here is where you assign initial values
;; to dialog tiles, set all tile states, and
;; register action expressions.

;; First, load some items into the listbox:

(setq lbitems
   '( "Item 1" "Item 2" "Item 3" "Item 4"
      "Item 4" "Item 6" "Item 7" "Item 8"
   )
)

(start_list "listbox_1")   ; start transaction
(foreach item lbitems
   (add_list item)
)
(end_list)                 ; end transaction

;; Next put some text in the edit box

(set_tile "edit_1" "10")

;; Next check the second toggle

(set_tile "tog_2" "1")

;; Next push radio button 1 by setting
;; the boxed radio column value to the
;; key of the pushed radio button

(set_tile "radgroup" "rad_1")

;; Next, set the value of the slider to 1/4
;; of its full range (left = 0, right = 100)

(set_tile "slider_1" "25")

;; Now select the fourth item in the list box

(set_tile "listbox_1" "3")

;; Next, disable button 2

(mode_tile "btn_2" 1)      ; <- - - - - - - - - - - - - Step 3

;; Now, register action expressions for all tiles
;; except for the ok and cancel buttons. In this
;; demonstration the action expression is the same
;; for every tile, but in actual use, each action
;; expression will be different.

(action_tile "listbox_1" "(hello_action)" )
```

```
    (action_tile "rad_1"      "(hello_action)" )
    (action_tile "rad_2"      "(hello_action)" )
    (action_tile "rad_3"      "(hello_action)" )
    (action_tile "tog_1"      "(hello_action)" )   ; <- - - - - - Step 2
    (action_tile "tog_2"      "(hello_action)" )
    (action_tile "tog_3"      "(hello_action)" )
    (action_tile "slider_1"   "(hello_action)" )
    (action_tile "edit_1"     "(hello_action)" )
    (action_tile "btn_1"      "(hello_action)" )
    (action_tile "btn_2"      "(hello_action)" )
    (action_tile "btn_3"      "(hello_action)" )


    ;; After all tile values and states have been
    ;; initialized, and all action expressions are
    ;; registered via action_tile, its time to
    ;; allow the user to interact with the dialog
    ;; by calling (start_dialog). The start_dialog
    ;; function will not return until the dialog
    ;; box is closed.

    (setq diastat (start_dialog)) ; save result

    ;; The dialog has been closed, now unload
    ;; the DCL file

    (setq hdcl (unload_dialog hdcl))

    diastat                         ; return result of
                                    ; start_dialog
)

;; Shows callback data in Message box

(defun hello_action ( / tab)
    (setq tab "\t\t")
    (if (eq "1" (get_tile "show_cb")) ; do this only if the show
        (alert                        ; callback toggle is checked
            (strcat
                "Tile Callback Event Data\n"
                "\nReason:"      tab   (because $reason)
                "\nTile label:"  tab   (get_attr $key "label")
                "\nTile key:"    tab   $key
                "\nTile value:"  tab   $value
                "\nTile Data:"   tab   (if (eq $data "") "<none>" $data)
                "\nCursor X:"    tab   (itoa $x)
                "\nCursor Y:"    tab   (itoa $y)
            )
        )
    )
)

; <- - - - - - -Step 1: insert toggle_action function here
```

Designing Dialog Boxes

Notice that the **hello_action** function tests the value of a tile called show_cb (a toggle), and only displays the dialog box if the tile's value is "1" (which means the toggle is checked). This check box tile is at the bottom of the dialog box, and its label is **S**how Callbacks. Next, you'll test the dialog from 12HELLO.LSP and 12HELLO.DCL. Once you've completed the first part of the exercise, leave the drawing editor running for the second part.

Exercise

Testing Dialog Tile Callbacks

Start AutoCAD with a new, unnamed drawing.

Command: **(load "12HELLO")** [Enter]	Loads 12HELLO.LSP

Lisp returns: HELLO_ACTION

Command: **HELLO** [Enter]	Starts the HELLO command

In the Maximizing AutoCAD R13 dialog:

Choose **S**how Callbacks	Enables callback event display
Choose **B**utton 1	The AutoCAD Message dialog appears

Examine the message dialog and note the values it contains. These are the values of the special callback variables described earlier.

Choose OK	Closes the Message dialog
Choose **T**oggle 1	The AutoCAD Message dialog appears

In the Message dialog, note that the tile value is 1.

Choose OK

Choose **T**oggle 1 *again*	The AutoCAD Message dialog appears

In the Message dialog, note that the tile value is 0.

Try clicking on several other tiles in the dialog, and note the values that are displayed in the message box for each tile callback. This is the information that is available to the action expressions that are evaluated when a callback occurs. Once you've had enough, click OK to close the dialog box.

Choose OK   Closes the dialog

Command:

Leave everything as is, for use in the second part of the exercise.

The action expression for every tile in the dialog you used in this exercise is identical (in fact, a single function is being called by an identical action expression for every tile). In real applications, most tiles will have different action expressions, but there will be cases where several tiles that are similar in purpose or function may share a single action expression.

Note that the value passed as the second argument to **action_tile** is a *string* that contains an AutoLISP expression. A common mistake is to try to pass the actual

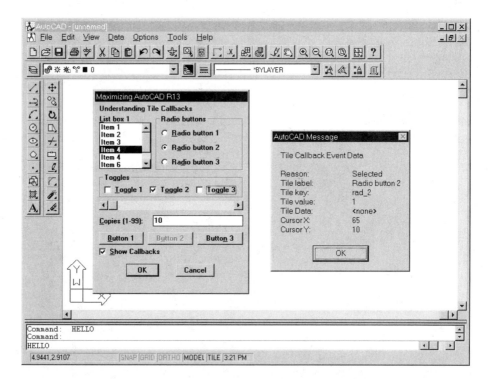

Figure 12.3
Displaying tile callback data.

Designing Dialog Boxes

expression as a list rather than as a string, which causes an error. Another common mistake often encountered when specifying action expressions that contain literal strings is the failure to use the \ (slash) character to *escape* the double quotes that surround the literal string within the expression. Here is an action expression that will cause an `incorrect number of arguments` error:

```
(action_tile "mytile" "(princ "Hello there!")" )
```

Count the number of double quotes in the action expression and note that there are four. AutoLISP requires strings to be *delimited* by double quotes, but in this case, the string itself must contain two double quotes, which requires them to be escaped with the \ (slash). Here is the proper way to pass the preceding action expression:

```
(action_tile "mytile" "(princ \"Hello there!\")" )
```

Next, you'll modify the code from 12HELLO.DCL to define the behavior of the three toggles on the dialog box. You will define a function called **toggle_action**, which will be called whenever one of the three toggles is chosen. The **toggle_action** function will disable or enable one of the three buttons in the row along the bottom of the dialog box, depending on which toggle was chosen and what its current value is (checked or unchecked). Toggle 1 will control the enabled state of Button 1, Toggle 2 will control the enabled state of Button 2, and Toggle 3 will control the enabled state of Button 3. When a toggle is checked, its associated button is *enabled*, and when a toggle is not checked, the associated button is *disabled*.

In addition to adding the new action expression, you will also modify the code that initializes the dialog box. There are several things that you must pay special attention to when modifying the initialization code. Mainly, you must synchronize the initial states of the three toggles and the three buttons properly. Since Buttons 1 and 3 are initially enabled, Toggles 1 and 3 must be checked when the dialog becomes active. And, since Button 2 is initially disabled, Toggle 2 must be initially unchecked. Next, you'll add the code for handling the callback events that occur when the user clicks one of the toggles. In this exercise and the 12HELLO.LSP file, the places where you must make changes are flagged with the comment "Step N" where N is a number.

 Exercise

Adding and Modifying Tile Callbacks

Start LispPad and open the file 12HELLO.LSP..

Before making any changes, choose Save **As**, and specify the file name 12EX01.LSP.

Step 1:

Go to the end of the file, and insert the following code:

```
(defun toggle_action () Enter
  (cond Enter
    (  (eq $key "tog_1") (setq btn_key "btn_1")) Enter
    (  (eq $key "tog_2") (setq btn_key "btn_2")) Enter
    (  (eq $key "tog_3") (setq btn_key "btn_3")) Enter
  ) Enter
  (mode_tile btn_key (if (eq "1" $value) 0 1)) Enter
) Enter
```

Step 2:

Navigate up to the following lines:

```
(action_tile "tog_1" "(hello_action)" )    ;; Step 2
(action_tile "tog_2" "(hello_action)" )    ;; Change hello_action's
(action_tile "tog_3" "(hello_action)" )    ;; to toggle_action's
```

In all three lines shown above, change hello_action to **toggle_action**.

Step 3:

Navigate up to the following line:

```
(mode_tile "btn_2" 1)    ;; Step 3
```

Insert the following three lines immediately after the line shown above:

```
(set_tile "tog_1" "1") Enter
(set_tile "tog_2" "0") Enter
(set_tile "tog_3" "1") Enter
```

With nothing selected, press CTRL+L Loads the entire contents of the editor into AutoCAD

If an error occurs when you load the file, the cause is most likely a missing or extra parenthesis or double quote. In this case, check your revisions, make any needed corrections, and reload the file by pressing CTRL+L again.

Command: **HELLO** Enter Starts the HELLO command

Note that when the dialog appears, **T**oggle 1 and **T**oggle 3 are checked, and **T**oggle 2 is unchecked. Also note that **B**utton 1 and **B**utton 3 are enabled, and Butto**n** 2 is disabled.

In the Maximizing AutoCAD R13 dialog, do the following:

Choose **T**oggle 1	Unchecks **T**oggle 1, disables **B**utton1
Choose **T**oggle 2	Checks **T**oggle 2, enables **B**utton 2
Choose **T**oggle 1	Checks **T**oggle 1, enables **B**utton 1

Try choosing each of the toggles several more times, noting how their checked/unchecked states are now linked to the enabled/disabled states of the three buttons.

In LispPad, issue **F**ile, **S**ave	Saves the contents of the editor

Close the dialog box, but leave AutoCAD and LispPad running for the next exercise.

The last two exercises show how to write AutoLISP code that responds to user actions in dialog boxes, and how that code may be used to *link* various states of two or more tiles in a dialog. Now you'll learn more about validating and handling input errors in dialog boxes.

Validating Edit Box Input and Handling Errors

Most of the input validation that you must do in a dialog box is limited to edit boxes, since they are the only kind of tiles that allow any degree of freedom to make a mistake. When the user shifts focus to an edit box and enters or changes its contents, no callback is triggered until the user exits the edit box and it loses focus. The user can exit an edit box in one of several ways:

1. Pressing the TAB key to advance to the next tab stop.

2. Pressing the ENTER or ACCEPT key to accept the dialog box.

3. Clicking the mouse on another tile to change focus to that tile.

In the next section, you will add a custom event handler for the edit box in the same dialog that you used in the last exercise. The edit box requires the user to enter a valid integer that is greater than 0 and less than 100. When an edit box loses focus, a callback occurs, which evaluates the action expression that was registered with the edit box using `action_tile`. This action expression will examine the value in the edit box, and determine if it is valid for the type and range of input required.

Validating Numerical Input

The process of determining if numerical input in an edit box is valid is fairly involved. Here are the steps.

1. Retrieve the contents of the edit box (a string) with **get_tile**.

2. Determine if the contents of the string in the edit box are a valid integer and if so, convert it to an integer data type.

3. Determine if the value is within the allowed range for input (in this case, greater than 0 and less than 100).

Since the value being requested is a number, and the contents of all edit boxes are initially strings, you must determine if the string can be converted to a number, and if so, perform the conversion. Once you've converted the string to a number, then you can perform additional validation, such as range-checking. If the value in the edit box is not valid (either not a number or a number that is not within the allowable range), then an error message is displayed at the bottom of the dialog box, in a text tile named **errtile**. The **errtile** is defined in BASE.DCL and is designed to provide a location for displaying error messages and other feedback on dialog boxes. This tile is the last element in the dialog definition in 12HELLO.DCL. Many of AutoCAD's dialogs make use of an error tile, which is always located at the very bottom of a dialog box below the location where the OK and Cancel buttons appear. You, also, should place your error tiles at the bottom of your dialog boxes. Figure 12.4 shows AutoCAD's DDCHPROP dialog with the message `Invalid thickness` in its error tile.

Figure 12.4
The error tile.

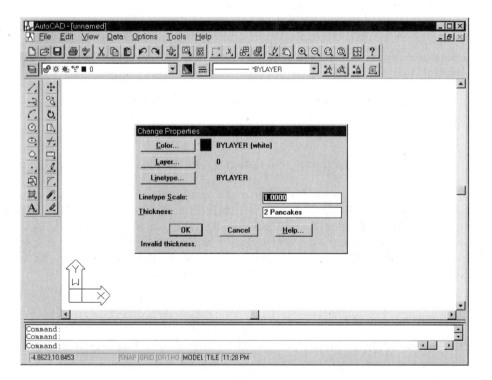

Designing Dialog Boxes

The first step in adding support for handling the edit box is writing the callback code. The function that you'll add to handle the edit box must determine if the contents of the edit box can be converted to an integer, and if it can, it must convert it. AutoLISP provides the **atoi** function for this purpose. One problem with using this function is that it will convert any string to an integer regardless of whether the string contains a valid integer value or not. Try using **atoi** in the following exercise.

Exercise

Converting Strings to Integers

Continue in the current editing session

Command: **(atoi "10")** [Enter]   Converts the string "10" to an integer

Lisp returns: 10

Command: **(atoi "33.33")** [Enter]   Converts the string "33.33" to an integer

Lisp returns: 33

Command: **(atoi "Hello")** [Enter]   Converts the string "Hello" to an integer

Lisp returns: 0

Command: **(atoi "-")** [Enter]   Converts the string "-" to an integer

Lisp returns: 0

Leave AutoCAD running for the next exercise.

As you can see from the preceding exercise, the **atoi** function does not reject values that are not valid integers. Instead, it simply returns the integer 0 in this case. This presents a problem, since you must be able to determine if the contents of an edit box are really a number or not. To validate a string that is to be converted to an integer, you can use the **StrToInt** function shown in the following listing. Unlike **atoi**, **StrToInt** first checks the contents of the sting it is given to determine if it contains a valid integer. If not, then **StrToInt** returns **nil,** which serves as an easy way to detect an invalid entry in an edit box that requires an integer.

```
(defun StrToInt (s)
   (if (and (eq (type s) 'str)
            (not (wcmatch s "*@*,#*-*,#*`.*")))
      (atoi s)
   )
)
```

StrToInt uses the AutoLISP **wcmatch** function to determine if the string contains a valid integer. The **wcmatch** function is a general-purpose wildcard pattern-matching function that allows you to easily analyze the contents of a string. **Wcmatch** is more generally used for determining if a given sequence of characters or *substring* exists within another string. In this case, the pattern supplied to **wcmatch** determines if any of the characters in the argument to **StrToInt** are not numeric digits. The pattern will permit only the first character in the string to be **-** (a minus sign), and also ignores any leading or trailing spaces. If any character in the string is not a numeric digit, then **wcmatch** will return **nil,** and **StrToInt** will fail. You can learn more about the **wcmatch** function and its many uses in this book's companion volume, *Maximizing AutoLISP for AutoCAD R13*.

Try repeating the previous exercise again, this time using **StrToInt** in place of **atoi**.

 Exercise

Converting Strings to Integers

Continue in the current editing session.

Command: **(load "12DCLSUB.LSP")** [Enter]   Loads the StrToInt function

Lisp returns: STRTOINT

Command: **(StrToInt "10")** [Enter]   Converts the string "10" to an integer

Lisp returns: 10

Command: **(StrToInt "33.33")** [Enter]   Tries to convert "33.33" to an integer

Lisp returns: nil

Command: **(StrToInt "Hello")** [Enter]   Tries to convert "Hello" to an integer

Designing Dialog Boxes

Lisp returns: `nil`

Leave AutoCAD running for the next exercise.

You will find **StrToInt** a useful tool for validating integer input in dialog edit boxes. You'll also see how to perform the same type of validation on real numbers including distances and angles, later in this chapter.

Handling Edit Box Tile Callbacks

In the following exercise, you'll learn how to handle events in edit boxes, validate, and accept or reject their contents.

Here is the code you'll add to 12EX01.LSP that will handle the edit box callback.

```
;; callback handler for edit box.

(defun edit_action (key / value errmsg)

   ; error message for this tile
   (setq errmsg "Copies must be an integer from 1 to 99")

   (set_tile "error" "")                ; Clear the error tile.
   (setq value (get_tile key))          ; Get contents of edit box.
   (setq copies (StrToInt value))       ; Convert contents to integer.
                                        ; If the tile doesn't contain
                                        ; a valid integer, StrToInt
                                        ; returns nil.

   (if (and copies                      ; If value in edit box is
           (> 100 copies 0)             ; an integer and is within
)                                       ; the required range, then
       copies                           ; return the integer value.
       (progn                           ; Otherwise display an error
         (set_tile "error" errmsg)      ; message and return nil.
         nil
       )
   )
)
```

The **edit_action** function will be called whenever the edit box loses focus. When this happens, **edit_action** accesses the text in the edit box using **get_tile**. The text could also be accessed using the **$value** callback variable, but doing so would not allow this function to be called from anywhere other than within the edit box callback. Also, as you will see later in the chapter, it is preferable to design callbacks in a way that does not make them dependent on the special callback variables, as you can then use the same function for several tiles of the same type.

Note the **edit_action** also accepts a single argument, key, which is the key of edit box. Again, this value can also be accessed in the **$key** special callback variable, but this is also undesirable, for the same reason that **$value** is not used.

Once **edit_action** retrieves the string from the edit box, it passes it to the **StrToInt** function, which attempts to convert the string to an integer. If that can't be done, the string in the edit box is invalid, and an error message is displayed in the error tile at the bottom of the dialog, and **edit_action** returns **nil**. If the edit box contents are successfully converted to an integer, **StrToInt** returns that integer, and it becomes the result of **edit_action**.

To associate the **edit_action** function with the edit box callback, you must modify the appropriate call to **action_tile**, in 12EX01.LSP, as shown here:

Before revision: (action_tile "edit_1" "(**hello_action**)")

After revision: (action_tile "edit_1" "(**edit_action $key**)")

Notice that in the revised call to **action_tile**, the action expression passes the value of the **$key** tile callback variable to the **edit_action** function. Passing **$key** as a parameter allows **edit_action** to be called from outside the edit box callback, as you will see later in this chapter.

Now you'll revise 12EX01.LSP to implement the edit box validation and error-handling code. AutoCAD should be running from the previous exercise, and the file 12EX01.LSP should be open in LispPad.

 Exercise

Handling Edit Boxes

In LispPad, with the file 12EX01.LSP open:

Choose File, Save As and save the file to 12EX02.LSP.

Move to the end of the file and insert the following code:

Designing Dialog Boxes

```
(defun edit_action (key / value errmsg) [Enter]
   (setq errmsg "Copies must be an integer from 1 to 99") [Enter]
   (set_tile "error" "") [Enter]
   (setq value (get_tile key)) [Enter]
   (setq copies (StrToInt value)) [Enter]
   (if (and copies (> 100 copies 0)) [Enter]
      copies [Enter]
      (progn (set_tile "error" errmsg) nil) [Enter]
 ) [Enter]
) [Enter]
```

Next, navigate up to the line shown here:

`(action_tile "edit_1" "(hello_action)" )`

Change the line shown above to the following:

`(action_tile "edit_1" "`**`(edit_action $key)`**`" )`

With nothing selected, press CTRL+L	Loads the entire contents of the editor into AutoCAD

If an error occurs when you load the file, check your revisions, make any needed corrections, and reload the file by choosing CTRL+L again.

Now test the revisions you've just made:

Command: **HELLO** [Enter]	Starts the HELLO command

The Maximizing AutoCAD R13 dialog box appears with the value 10 in the Copies edit box.

Delete the contents of the Copies edit box and in its place, type **HELLO**	Changes the value of the edit box
Choose Button 1	Causes edit box to lose focus

Note that an error message appears at the bottom of the dialog box, indicating that the value of the Copies field must be an integer from 1 to 99.

Delete the contents of the Copies edit box again and in its place, type **50**	Changes the value of the edit box
Choose Button 1	Causes edit box to lose focus

Note that the error message at the bottom of the dialog box is no longer visible.

Delete the contents of the **C***opies edit box again and in its place, type* **200**	Changes the value of the edit box
Choose **B**utton 1	Causes edit box to lose focus

Note that the same error message has reappeared at the bottom of the dialog box.

Choose OK	Closes the dialog box

Leave AutoCAD running for the next exercise.

One common pitfall that many dialog box developers fall into is forcing a user to correct an invalid entry in an edit box by shifting focus back to the edit box that contains the erroneous input when the error is detected. Doing this has the effect of trapping the user in the edit box without the ability to do anything else in the dialog box, other than cancel it. This trap should be avoided at all costs. However, you must prevent the user from accepting a dialog box if its contents do not make sense or contain errors. In the last exercise, note that you were able to choose OK and accept the dialog box with an invalid entry in the **C**opies edit box. Proper handling of the dialog would not allow this to happen. In order to prevent the user from accepting a dialog box while there is invalid input in one or more tiles, you can use one of these techniques:

1. You can disable the OK button to make it clear that the dialog's contents are not acceptable. This approach will prevent the user from accepting the dialog box when it contains invalid input, or when required input has not been supplied. This approach requires more code because you must check the state of every tile that may contain invalid content every time the state of any one of those tiles changes. If you use this technique, then you should ensure that the initial state of the dialog and the OK button's enabled state are synchronized. In other words, if the user cannot choose OK as soon as the dialog appears, then the OK button should be disabled initially, by calling `mode_tile` prior to activating the dialog.

2. You can leave the OK button enabled, add a callback for it, and perform final checking of the edit box from within the OK button's callback. This technique requires far less coding in the case of a dialog with multiple edit boxes.

 Note: The tiles in a dialog are destroyed when the dialog is accepted, which means that you must transfer their values to AutoLISP variables while the dialog is active, since you cannot access them once a dialog has been retired. Saving the final values of tiles is usually done in the OK button's callback.

Designing Dialog Boxes

In the next exercise, you'll add a callback for the OK button to the code for the dialog used in the previous exercise. This callback will prevent the user from accepting the dialog box when the content of the Copies edit box is not acceptable. Here is the code that will handle the OK button's callback:

```
(defun accept_action ()
   (if (not (edit_action "edit_1"))  ; edit_action returns nil if
                                     ; edit box contents are invalid
      (progn
         (mode_tile "edit_1" 3)      ; select edit box contents
         (mode_tile "edit_1" 2)      ; and set focus to edit box
      )
      (done_dialog 1)                ; otherwise we're done,
   )                                 ; close the dialog.
)
```

If you recall from the previous discussion concerning the rationale behind avoiding the direct use of the **$key** and **$value** tile callback variables in the **edit_action** function, the stated reason was because doing so would not allow **edit_action** to be called from any place other than the edit tile callback. Note that the **accept_action** function also calls the **edit_action** function to perform the final validation of the edit box to determine if the dialog can be accepted. If the contents of the edit box are not acceptable, **edit_action** returns **nil**, allowing the **accept_action** function to use that result to determine if it can close the dialog. The reason for avoiding the use of **$value** should be apparent here. When **accept_action** calls **edit_action**, it is doing so from within the OK button's callback, where the **$value** variable is not the contents of the edit box, it is the value of the OK button. This is why **get_tile** is used to access the contents of the edit box from within the **edit_action** function rather than **$value**. If the contents of the edit box were accessed through the **$value** variable, doing so would prevent the use of **edit_action** from any place other than the edit box callback, which would require the **accept_action** function to duplicate all of the error-checking that **edit_action** performs.

If the edit box contents are not valid, the **accept_action** function calls **mode_tile** twice; once to select the contents of the edit box, and a second time to set focus to the edit box. If the call to **edit_action** is anything but **nil**, that signals that the edit box contains an acceptable value, allowing the dialog to be accepted and closed. The **done_dialog** function is the means through which an AutoLISP program can close a dialog box.

---

done_dialog

(done_dialog [*status*])

Closes an active dialog box. This function must be called from a tile callback or directly in an action expression. **Status** *is an optional, non-negative integer. If* **status** *is supplied, it*

> is returned by the corresponding call to **start_dialog** that activated the dialog. If **status** is not supplied, the call to **start_dialog** that activated the dialog will return 1 if the dialog was accepted, or 0 if the dialog was canceled. **Done_dialog** returns the current location of the upper left corner of the dialog box in screen coordinates (left-hand coordinate system).

The **done_dialog** function is the only way to explicitly close an active dialog box. The status argument is designed to serve as a means of conveying how the dialog was exited (for example, accepted or canceled), and it is returned by the call to **start_dialog** which activated the dialog. If no status argument is supplied, then AutoCAD will substitute a value of 1 if the dialog was accepted, or 0 if the dialog was canceled. Generally, when you call **done_dialog** from either the OK or Cancel button callbacks, you should always pass a value of 1 or 0, to indicate whether the dialog was accepted or canceled, respectively. The status argument can also be used for the purpose of temporarily hiding a dialog box to obtain user input on the command line, but that is an advanced topic beyond the scope of this chapter. You can find out more about advanced dialog box handling in this book's companion volume.

Next, add the callback for the OK button to the 12EX02.LSP. You must also add a call to **action_tile** to register the callback with AutoCAD.

Exercise

Handling Edit Boxes

In LispPad, with the file 12EX02.LSP open:

Choose <u>F</u>ile, Save <u>A</u>s and save the file to 12EX03.LSP.

Move to the end of the file and insert the following code:

```
(defun accept_action () Enter
  (if (not (edit_action "edit_1")) Enter
      (progn Enter
         (mode_tile "edit_1" 3) Enter
         (mode_tile "edit_1" 2) Enter
      ) Enter
      (done_dialog 1) Enter
```

Designing Dialog Boxes

```
)  Enter
)  Enter
```

Next, navigate up to action expression for Button 3, shown here:

```
(action_tile "btn_3"      "(hello_action)" )
```

Just below the above line, add the following new line:

(action_tile "accept" "(accept_action)") Enter

Choose **F**ile, **S**ave	Saves contents of editor to 12EX03.LSP
With nothing selected, press CTRL+L	Loads the entire contents of the editor into AutoCAD

If an error occurs when you load the file, check your revisions, make any needed corrections, save the changes to the file and then reload it by choosing CTRL+L.

Switch to AutoCAD, and then test the revisions you've just made:

Command: **HELLO** Enter	Starts the HELLO command

The Maximizing AutoCAD R13 dialog box appears with the value **10** in the **C**opies edit box.

Delete the contents of the **C**opies *edit box and in its place, type* **50**	Changes the value of the edit box
Choose **B**utton 1	Causes edit box to lose focus
Delete the contents of the **C**opies *edit box again and in its place, type* **200**	Changes the value of the edit box
Choose OK	Returns focus to the edit box; the dialog box remains open

Note that when you choose the OK button with invalid input in the Copies edit box, the dialog does not close, but focus returns to the edit box and its contents are selected.

Choose Cancel	Closes the dialog box

Validating Distance and Angle Input

Because AutoCAD applications are graphics-intensive, applications frequently need to acquire point, angle, and distance input in dialog boxes. Many of AutoCAD's dialog boxes request these geometric units for a number of purposes. The techniques used to validate distance and angle input in edit boxes are similar to those used for integer values. The main difference is the procedure used to convert the string contents of the edit box to a numeric value. In the previous exercises, you made use of the user-defined **StrToInt** function from 12DCLSUB.LSP, which can validate and convert a string to an integer number. AutoLISP has built-in functions for applying this same operation to distance and angle data. The **distof** (Distance To Float), and **angtof** (Angle To Float) functions accept a string containing a distance or angle expression respectively, and convert it to a floating-point number. These two functions will also validate and reject invalid expressions as well.

---

distof

(distof *string* [*mode*])

Converts a string containing a valid distance expression to a real number. **String** *is a string that contains a distance expression in the format that is accepted by AutoCAD on the command line, depending on the current units setting.* **Mode** *is a positive, non-zero integer that species the units format that is accepted in* **string**. *If* **mode** *is omitted, it defaults to the current linear units setting stored in the LUNITS system variable. If* **string** *is a valid distance expression in accordance with* **mode**, **distof** *returns the distance as a real number, otherwise* **distof** *returns* **nil**.

The range of values accepted by the mode argument to **distof** are detailed in Table 12.6.

Table 12.6
Distof (Distance to Float) Unit Modes

Value	Example	Units Format
1	1.0e-03276	Scientific
2	44.75	Decimal
3	35'4.375"	Engineering (feet and decimal inches)
4	7'11-3/4"	Architectural (feet and fractional inches)
5	22 3/32	Fractional

Designing Dialog Boxes

You can use **distof** for the same purpose that **StrToInt** is used in edit box callback code. The following listing shows a modified version of the **edit_action** function from the previous exercise, which is designed to handle distance entry in an edit box. It requires the distance to be greater than 0.0.

```
(defun edit_action (key / len value errmsg)
   (setq errmsg "Length must be a positive, non-zero distance.")
   (set_tile "error" "")          ; Clear error tile.
   (setq value (get_tile key))    ; Get string from edit box.
   (setq len (distof value))      ; Try to convert string to real.
   (if (and len (> len 0.0))      ; If value is non-nil and > 0.0
       len                        ; then return it to the caller.
       (progn                     ; Otherwise, display an
           (set_tile "error" errmsg)  ; error message and return
           nil                    ; nil to the calling function.
       )
   )
)
```

The **angtof** function performs the same operation as **distof**, but for angular input. It also takes a string, and a mode specifying the accepted format.

---

angtof

(angtof *string* [*mode*])

Converts a string containing a valid angle expression to a real number. **String** *is a string that contains an angle expression in the format that is accepted by AutoCAD on the command line, depending on the current units setting.* **Mode** *is a positive, non-zero integer that species the units format that is accepted in* **string**. *If* **mode** *is omitted, it defaults to the current angular units setting stored in the AUNITS system variable. If* **string** *is a valid angle expression in accordance with* **mode**, **angtof** *returns the angle as a real number, otherwise* **angtof** *returns* **nil**.

The range of values accepted by the mode argument to **distof** are detailed in Table 12.7.

Angtof, like **distof**, can be used in the very same manner as the latter and **StrToInt**, to validate and convert the contents of edit boxes.

Handling List Boxes

List boxes and popup lists are useful in dialog box interfaces for the purpose of allowing the user to select from a list of predefined items. You've already learned to add items to

Table 12.7
Angtof (Angle to Float) Unit Modes

Value	Example	Units Format
1	45.00	Degrees
2	35d22'44"	Degrees/minutes/seconds
3	1.00g	Grads
4	2.25r	Radians
5	N 35d22'44" E	Surveyor's units

a list box using **start_list**, **add_list**, and **end_list**. List boxes can allow multiple items to be selected. Interpreting the value of non-multiple selection list boxes and popup lists is fairly simple. The value returned by **get_tile** and the **$value** callback attribute is a string that contains the index of the selected item in the list, or a null string if no item is selected. The index of the first item in a list box is 0.

In a multiple selection list box, the value of the list box is a string containing all of the selected items, each separated by a space. Converting this to data that can be easily accessed in AutoLISP is somewhat more complicated for multiple-select list boxes. To see how multiple selection list boxes represent multiple selections, you'll use a modified version of the dialog box used in previous exercises. To allow multiple selections in a list box, the `multiple_select` attribute must be set to true, as shown in the following listing.

In this exercise, you will not activate the dialog. Instead, you'll add some items to it manually, then you'll select several of those items, access the value of the list box, and convert it to a list. You'll also make use of the list-handling subroutines defined in 12DCLSUB.LSP, for convenience. The DCL definition of a modified version of the `testlist` dialog from 12HELLO.DCL is shown in the following code listing with the `multiple_select` list box attribute shown in bold face type.

```
testlist1 : dialog {
  key = "maxac";
  label = "Maximizing AutoCAD Release 13";
  : list_box {
     key = "test";
     label = "Adding Items to a List Box";
     multiple_select=true;
  }
  ok_cancel;
}
```

Designing Dialog Boxes

 Exercise

Handling Multiple-Selection List Boxes

Start AutoCAD with a new, unnamed drawing, press F1 to display the AutoCAD Text window, launch LispPad, and open the file 12HELLO.DCL.

In LispPad, choose testlist1 *from the* Goto Definition *drop-down list* — Scrolls to the definition of the `testlist1` dialog box

Press CTRL+L — Displays the dialog box

Arrange the AutoCAD Text Window and the dialog so both are visible, then switch to the AutoCAD Text window.

Command: **(load "12DCLSUB")** [Enter] — Loads the file 12DCLSUB.LSP

Lisp returns: STRTOINT

Command: **(setq contents (list "Item 0" "Item 1" "Item 2" "Item 3" "Item 4" "Item 5" "Item 6" "Item 7" "Item 8" "Item 9"))** [Enter] — Assign a list of 10 strings to contents

Lisp returns: ("Item 0" "Item 1" "Item 2" "Item 3" "Item 4" "Item 5" "Item 6" "Item 7" "Item 8" "Item 9")

Command: **(list_assign "test" contents)** [Enter] — Adds the list of 10 strings to the list box

Lisp returns: nil

In the list box, choose Item 0, Item 2, Item 3, Item 5 and Item 8

Next, get the value of the list box:

Command: **(setq items (get_tile "test"))** [Enter] — Assigns value of list box to items

Lisp returns: "0 2 3 5 8" — A string containing the indices of the selected items

Next, add an open and close parenthesis to the string:

Command: **(setq temp (strcat "(" items ")"))** [Enter]

Adds "(" and ")" to the string and assigns result to temp

Lisp returns: " (0 2 3 5 8) "

A string representation of a list

Next, convert the string to a list:

Command: **(setq selected (read temp))** [Enter]

Converts the string in temp to a list and assigns it to selected

Lisp returns: (0 2 3 5 8)

A list of integers assigned to selected

Next, print the selected elements from the original list contents:

Command: **(foreach item select (print (nth item contents)))** [Enter]

Retrieves and prints selected elements from original list contents

"Item 0"
"Item 2"
"Item 3"
"Item 5"
"Item 8"

Lisp returns: "Item 8"

In the last exercise, you called **list_assign** (from 12DCLSUB.LSP) to add ten items to a multiple-selection list box. Then you selected five items in the list and accessed their indices using **get_tile**, which returned a string containing the indices of the selected items. You then used **strcat** to add an open and close parenthesis to the string, giving it the same form that an AutoLISP list has. Then you used the AutoLISP **read** function to convert the string to an AutoLISP list, and finally, you used the indices in the converted list to access the corresponding elements in the original list that was used to populate the list box. This technique may be used reliably even if there are no items in the list selected. In that case, **get_tile** returns a null string, which results in " () " after the parenthesis are added. Since **nil** represents the empty list in AutoLISP, passing the string " () " to **read** produces **nil**.

Other techniques that you might need to perform on a multiple-selection list box include programmatically selecting or deselecting all items, and inverting the current selection (for example, deselecting items that are selected, and selecting items that are not

Designing Dialog Boxes

selected). Let's take a look at how to select and deselect all items in a multiple-selection list box. In order to perform these operations, you must know how many items are in the list box.

Putting Dialog Boxes to Work Finding Attribute Blocks

Included in this chapter's files are 12ATTSEL.LSP and 12ATTSEL.DCL. These two files add the ATTFIND command in AutoCAD, which uses a dialog box interface (Figure 12.5). ATTFIND allows you to select and locate block insertions by searching their attribute values. In addition to demonstrating many techniques related to multiple-selection list boxes, ATTFIND will prove to be a valuable tool for almost any AutoCAD user.

The ATTFIND command's dialog box uses a multiple-selection list box for selecting the blocks to be included in the search, along with two edit boxes for entering search patterns for the tag and value of the attribute that is to be searched for. When you choose the **S**earch button, ATTFIND begins searching through all insertions of the blocks that are selected in the list box, for attributes with tags that match the pattern in the **T**ag field, and having a value that matches the pattern in the **V**alue field. If the Stop at **F**irst Match toggle is checked, then ATTFIND stops searching as soon as it finds the first matching attribute, and optionally pans the display to put the attribute in the current view. If the

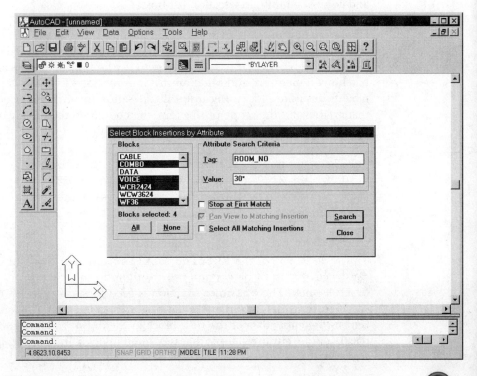

Figure 12.5
The ATTFIND dialog box.

aforementioned toggle is not checked, ATTFIND will search for all occurrences of matching attributes, and will optionally select and attach grips to them.

The code for ATTFIND demonstrates many common techniques used with multiple-selection list boxes, which include:

- Populating a list box with items from a drawing table
- Selecting and deselecting all items in a list box
- Controlling the enabled states of other tiles based on the number of selected items in a list box
- Adding buttons to select and deselect all items in a list box
- Using a text tile to reflect the number of items selected in a list box

Because of the complexity and advanced nature of the code used by ATTFIND for searching the drawing, that part of the code is not discussed here. You can learn more about these and other advanced topics related to ATTFIND, in this book's companion volume, *Maximizing AutoLISP for AutoCAD R13*.

Here, you'll learn how to perform the operations outlined in the preceding list to manage and control a multiple-selection list box.

Displaying drawing data such as layer, linetypes, and blocks in list boxes is a very common operation which many of AutoCAD's dialog boxes do. ATTFIND adds the names of all user-defined blocks that have attributes to the list box. It ignores any external references, hatching, dimension, and other anonymous blocks. ATTFIND iterates through the block table to find the blocks. Because this operation involves the use of advanced functions that are beyond the scope of this text, the code used to actually obtain the list of blocks that appears here is not explained in great detail.

```
(while (setq blk (tblnext "block" (not blk)))
   (if (eq 2 (logand 62 (cdr (assoc 70 blk))))
       (setq rslt (append rslt (list (cdr (assoc 2 blk))))) 
   )
)
```

This code is part of the **attfind_blklist** function from 12ATTSEL.LSP. The code constructs the list of blocks and returns them to the calling function, which adds them to the list box. The **tblnext** function is used to *iterate* through drawing table, which is specified in its first argument. Each time it is called with a second argument of **nil**, it returns the entity data of the next block in the drawing. The structure of a block's entity data is a more advanced topic that is covered in detail in *Maximizing AutoLISP for AutoCAD R13*.

Designing Dialog Boxes

Providing a means for the user to select all items in a list box is a necessity when the list box contains many items, since DCL list boxes have no extended-selection ability like that available in list boxes in native Windows dialogs (which allow you to select a range of list box items using the SHIFT key). For this reason, when a list box may contain a large number of items (more than a dozen), we strongly recommend that you provide the user with a convenient way to select and deselect all items in the list box using two buttons, one for each operation. The ATTFIND command does this, with the two buttons located below the list box, **All** and **None**. In order to select all items in a list box, you must construct a string of space-delimited indices of all the items, and assign it to the list box with **set_tile**. If the number of items in the list box is constant (that is, no items are deleted or added to it), then you can construct the string once, and use it whenever the user chooses the button that selects all items. Here is the code that constructs the string of indices (which is called once when the dialog is displayed and assigned to a variable), along with the **All** and **None** button callbacks, which select and deselect all items in the list box respectively. To create the value that is passed to select all items, the code loops a number of times equal to the number of items in the list box, which must be saved in a variable.

```
;; First, construct a string containing all indices
;; bcount is the number of items in the list box,
;; s holds the indices, i is the loop counter.

(setq s "0" i 0)                              ; s holds the indices
(repeat bcount                                ; bcount =
   (setq s (strcat s " " (setq i (1+ i))))    ; add numbers to string
)
(setq select_all_val s)                       ; save s for later

;; callback for the All button (select all)

(defun all_action ()
   (set_tile "blocks" select_all_val)
   (update_dialog)
)

;; callback for the None button (deselect all)

(defun none_action ()
   (set_tile "blocks" "")
   (update_dialog)
)
```

The `all_action` function calls **set_tile** and passes the string constructed by the loop as the value of the list box, causing all items to be selected. Deselecting all items in a list box is easy. You just call **set_tile** and pass a null string as the value.

Note that both of the callbacks call a function called **update_dialog**. This function is responsible for managing all buttons whose enabled/disabled states are dependent on the number of items selected in the list box. When there are no items selected in the list box, it makes no sense to choose the **N**one button, since it does nothing. And, when all items in the list box are selected, it makes no sense to choose the **A**ll button. So, when all items in the list box are selected, the **A**ll button is disabled, and when no items are selected, the **N**one button is disabled. In addition, since a block search requires *at least* one block, if none are selected a search cannot be performed, so **update_dialog** disables the **S**earch button when no blocks are selected.

The **update_dialog** function also performs several other tasks. One is to update the text below the list box that indicates how many items are selected. This is another convenience that should be provided for the user. It is not uncommon for a user to accept a dialog with a list box that has more items selected than they might know, because some may have scrolled out of view. The second additional task performed by **update_dialog** is to disable the **S**earch button if the **V**alue field is empty, since it is required for a search to take place.

As you can see, thoughtfully disabling buttons that can't be used, and adding text to show how many items are selected in list boxes, are two of many techniques that can result in a friendlier and less error-prone user interface. The code in 12ATTSEL.LSP avoids the use of the **mode_tile** function, and instead uses the macros defined in 12DCLSUB.LSP, and makes the code more explicit. Here is the definition of the **update_dialog** function.

```
(defun update_dialog ( / blocks)
   (setq blocks
      (read
         (strcat
            "(" (get_tile "blocks") ")"  ; Get the list of selected
         )                               ; items in Blocks list
      )                                  ; box and convert to list
   )
   (if blocks                            ; is at least 1 item selected?
      (progn                             ; Yes,
         (tile_enable "none")            ; enable the None button
         (if (eq (length blocks) bcount) ; All items selected?
            (tile_disable "all")         ; Yes: disable All button
            (tile_enable "all")          ; No: enable All button
         )
         (if (/= "" (get_tile "value"))  ; Is Value field empty?
            (tile_enable "search")       ; No, enable Search button
            (tile_disable "search")      ; Yes, disable Search button
         )
         (tile_focus "search")           ; Set focus to search button
         (set_tile "blkcnt"
            (strcat                      ; Update Blocks Selected: text
```

Designing Dialog Boxes

```
            "Blocks selected: "
            (print (itoa (length blocks)))
         )
      )
   )
   (progn                              ; Else no items are selected
      (tile_disable                    ; disable the Search and
         (list "search"                ; None buttons
               "none"
         )
      )
      (tile_enable "all")              ; Enable All button
      (set_tile "blkcnt"
         (get_attr "blkcnt" "value")   ; Set Blocks selected: to 0
      )
   )
  )
 )
)
```

Next, try the ATTFIND command with the supplied test drawing 12ATTSEL.DWG.

 Exercise

Testing the ATTFIND Command

Start AutoCAD and open the file 12ATTSEL.DWG.

Command: **(load "12ATTSEL")** [Enter]   Loads 12ATTSEL.LSP

12ATTSEL.LSP From Maximizing
AutoCAD R13. ATTFIND command loaded.

Command: ATTFIND   Starts the ATTFIND command

In the Select Block Insertions by Attribute dialog:

In the Blocks cluster list box, choose **TEST1** *and*   Selects TEST1 and TEST3 in list
TEST3   box

In the Blocks cluster, choose **A**ll   Selects all block names in list box

In the Blocks cluster list box, choose **TEST2** *and*   Deselects TEST2 and TEST4 in list
TEST4   box

Maximizing AutoCAD® R13

In the Blocks cluster, choose **N**one	Deselects all block names in list box
In the Blocks cluster, choose **A**ll	Selects all block names in list box
In the **V**alue *edit box, type* **2***	Specifies all block insertions with attribute values begining with the numeral 2
Click in the **T**ag *edit box, and type* ROOM	Enables the **S**earch button, and specifies the ROOM attribute tag
Choose **S**earch	Dismisses the dialog box
`Searching...` `Found and selected 16 insertions.`	Grips and selects 16 block insertions

 DCL is not notified of changes in an edit box until you exit the edit box. After you type a value in the **V**alue box, the **S**earch button is not enabled until you tab to it, or choose another item, such as the **T**ag edit box, or press ENTER.

Tools for Dialog Development

In addition to LispPad, the Maximizing AutoCAD R13 CD-ROM also includes a tool called DCL Spy (Figure 12.6), which is a learning and debugging aid that can help you understand and debug dialog box callback functions. DCL Spy provides the following functionality.

- Watch: A dynamic display showing the values of the special callback variables, which is updated each time a callback occurs.

- Log: A log display showing all callback events, and the values of the callback variables for each event.

- Attributes: An attribute page that shows each tile's design-time DCL attributes and their values.

- Filters: Allows you to filter tracing of callback events by tile, tile type, and callback reason.

DCL Spy allows you to trace dialog callbacks as they occur, and easily see the values of callback variables. It also helps in development of DCL by providing a dynamic display of each tile's DCL attributes, which is synchronized with callback events so that when a

Designing Dialog Boxes

Figure 12.6
DCL Spy for Maximizing AutoCAD.

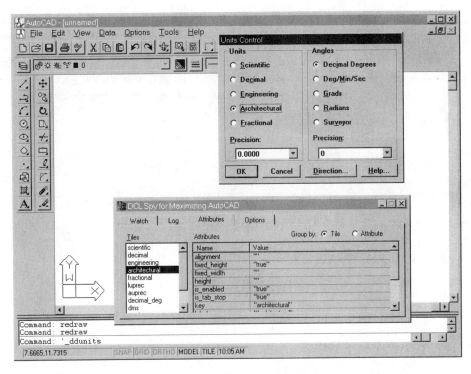

callback occurs, the attributes of the tile that generated the callback are displayed in a list.

See Appendix G for more information.

Finishing Up

If you want to keep any of the files you created or modified in this chapter, move them to the \MAXAC directory or another directory, and then delete or move any remaining files out of \MAXAC\DEVELOP.

Conclusion

After completing this chapter, you should be able to add a modest dialog box interface to an AutoLISP program. You should start with simple dialogs that do not have too many complex relationships between tiles. One thing you should avoid is creating a dialog for its own sake. Although dialog boxes do provide advantages in many cases, they do not make sense in all cases, and you should carefully consider the way you work and what your goals are when deciding to add a GUI front-end to a custom application.

In this book's companion volume, you will find more advanced DCL and dialog handling techniques, including how to write reusable dialog and callback code, hiding dialogs to get user input on the command line, and nesting dialog boxes.

chapter 13

Understanding Menu Structure

In Chapter 3, you learned how to create a custom toolbar and write simple macros for each button on it. In Chapter 5, you learned the ins and outs of writing more sophisticated macros, and you applied those skills to adding a pull-down menu page. In Chapter 10, you learned how to create dynamic pull-down and cursor menu labels using DIESEL. This chapter covers the remaining part of menu customization: menu structure. This chapter also introduces the partial menu loading feature, which is new to R13 Windows and receives further coverage in Chapter 14.

Although you can do extensive menu customization simply by adding more pull-down menu pages and toolbars, at some point you'll want a more sophisticated menu structure. For instance, you can use cascading pull-down menus and flyout toolbars to "compress" more options into smaller menu areas. In addition, you'll want to take advantage of AutoCAD's other menu interfaces: cursor popup menus, pointing device button menus, slide image menus, side-screen menus, and tablet menus. This chapter shows you how to develop more sophisticated menus and use the AutoCAD menu interfaces appropriately.

Figure 13.1 shows some of the menu interfaces and their corresponding sections in the menu source file.

Menu structure presents many new facets in R13, but most of the enhancements are unique to Windows AutoCAD. The changes include new menu file extensions, new menu sections (***TOOLBARS, ***ACCELERATORS, and ***HELPSTRINGS), menu groups and menu item ID name tags, and partial menu loading. See Chapter 4 for a description of the new Windows menu file extensions. The remaining new features are covered in this chapter.

R13 DOS menu structure is almost identical to that of R12 DOS. Because most of the new menu features aren't supported in R13 DOS, it's difficult to maintain a single menu file for DOS and Windows, especially if you want to take advantage of the new R13 menu features. Beginning with R13c4, you can define toolbars in UNIX versions of AutoCAD, but otherwise DOS and UNIX

Maximizing AutoCAD® R13

Figure 13.1
Menu interfaces and sections.

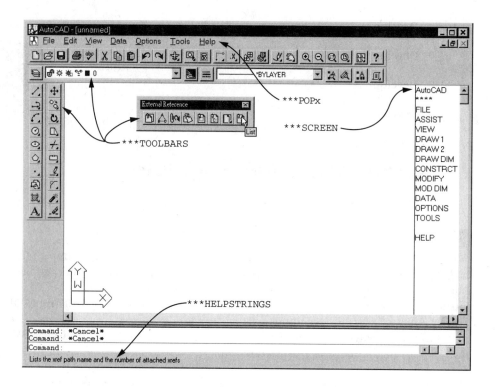

users lose out to Windows users in terms of flexibility of the R13 menu customization interface.

 The AutoCAD LT 3 menu structure (but not its content) is identical to that of R13 for Windows, so most of the information in this chapter applies to LT as well. This chapter includes some AutoLISP code in a few menu macros (in the Testing All Standard Menu Sections exercise, for example), so those macros won't work in LT becausae LT does not support AutoLISP.

About This Chapter

In this chapter, you will learn how to do the following:

- Create menu files and sections
- Use menu item labels
- Use menu groups and name tags
- Control command-line echoing of menu macros and AutoLISP code
- Create and control menu items in each type of menu section

Understanding Menu Structure

- Develop pointing device menus that interact with cursor menus
- Develop cascading pull-down menus
- Develop flyout toolbars
- Develop image tile menus
- Use two kinds of accelerator key syntax
- Define help strings to accompany pull-down, cursor, and toolbar items
- Develop side-screen menus
- Develop tablet menus
- Decide which menu interfaces to use
- Attach partial menus and use their menu sections

Because of the large number of AutoCAD menu interfaces and numerous new features in R13, this chapter covers a lot of ground. Many of the individual menu sections are treated independently, so that you can skip or quickly skim parts of the chapter that don't interest you (for example, screen and tablet menus).

This chapter assumes that you're familiar with the menu-related information in Chapters 3, 4, and 5. In particular, you should know how to use the Toolbars dialog box, understand the different types of menu files (MNS, MNC, MNR, MNU, and MNX), and know how to write menu macros and compile menus. You also should be acquainted with the portions of Chapter 8 that cover slides and Chapter 10 that cover using DIESEL in menu labels.

These are the tools and files you will use for the exercises in this chapter:

- TextPad (or another ASCII text editor)
- 13ALLSCT.MNS, which contains all of the standard AutoCAD menu sections
- 13ECHO.TXT, which contains a pull-down menu for testing MENUECHO and CMDECHO settings
- 13AUXCUR.TXT, which contains button and cursor menus you'll add to your test menu
- 13PCODES.TXT, which contains a pull-down menu that demonstrates cascading menus and other special pull-down label features
- 13TBSAMP.TXT, which contains a sample toolbars menu

- 13IMGFLO.TXT, which contains an image tile menu with block insertion macros for flow symbols

- FLOW*.DWG, the flow symbol blocks

- MSLDFLOW.LSP, which automatically creates slides for the flow symbol image tile menu

- FLOWGATC.SLD, a special slide file for the flow symbol image tile menu

- FLOWNAME.TXT, which contains the names of the flow symbol slides for SLIDELIB.EXE

- 13SCREEN.TXT, which contains a sample side-screen menu

- 13TABLET.TXT, which contains a sample tablet menu

- 13TABLET.DWG, a sample tablet overlay

- 13ATTACH.MNS, a partial menu

- AutoCAD's SLIDELIB.EXE program

- ACAD.LSP, ACAD.PGP, ACADMA.MNS, ACADMA.MNL, MAWIN.ARX, MA_UTILS.LSP, and custom bitmap (BMP) icons from the MaxAC CD-ROM

Menus and Support Files in This Chapter

You'll create the following menus and support files in this chapter:

- **13TEST.MNS** is a test menu that you'll create and then modify throughout the chapter.

- **BLANK.SLD** is a blank slide for the flow symbols menu.

- **FLOW*.SLD** are slides for the flow symbols menu.

- **FLOWSYMB.SLB** is the slide library used by the flow symbols menu.

- **13ATTACH.MNL** is a menu LISP file that accompanies the partial menu file 13ATTACH.MNS.

Understanding Menu Structure

Overview of New Menu Features in R13 Windows

As you learned in Chapter 4, R13 introduces three new types of menu files for Windows (MNS, MNC, and MNR). R13 Windows can create an MNS file from an MNU file, but in general, a Windows MNU file will not be the same as a DOS MNU file in R13. The MNS file is almost an exact copy of the MNU file, and it is the file that the Toolbars customization dialog modifies. It's safest to use the MNS file as your single Windows menu source file, so that you avoid making changes to two different ASCII files or accidentally overwriting toolbar modifications. Table 13.1 summarizes the menu file types. Refer to Chapter 4 if you need more information.

Note: The *AutoCAD Customization Guide* refers to the MNU file as the *template menu file*. Whether you call it a source file or a template file, the important thing to remember is that Windows creates and uses another menu source file with an MNS extension.

R13 also introduces three new menu sections, in order to support the new toolbar and accelerator key menu interfaces and the short help strings on the AutoCAD for Windows status bar. (You'll learn more about menu sections later in this chapter.) The new sections are named ***TOOLBARS, ***ACCELERATORS, and ***HELPSTRINGS. The ***TOOLBARS section contains all of the toolbar definitions in R13 Windows (and in UNIX, starting with R13c4). The ***ACCELERATORS section contains definitions of keyboard shortcuts, such as CTRL+S for SAVE. In Chapter 3, you learned how to add your own keyboard shortcuts to the ***ACCELERATORS menu section. The ***HELPSTRINGS section contains optional single-line descriptions corresponding to menu choices (see Figure 13.1).

Another new entry in Windows menu files is the menu *group name*. This name is preceded by three asterisks, and thus looks similar to a menu section name at first glance. Its syntax

Table 13.1
R13 Menu File Comparison

File Type	Windows	DOS
Menu template or "pre-source"	*menuname*.MNU	not applicable
Menu source	*menuname*.MNS	*menuname*.MNU not applicable
Toolbar bitmaps	ACADBTN.DLL *menuname*.DLL *anyname*.BMP	
Compiled menu: structure and macros bitmaps	*menuname*.MNC + *menuname*.MNR	*menuname*.MNX

is ***MENUGROUP=*group_name*. The group name is R13's way of keeping track of sections and menu items in different menu files, which is important for partial menu loading (described at the end of this chapter). Normally, you'll see the ***MENUGROUP=*group_name* label once in each Windows MNS or MNU file, near the beginning of the file. All menu items after a ***MENUGROUP=*group_name* label are assigned to that group. Thus, a single group name assigns all of the menu items in the file to one group. For instance, R13's standard ACAD.MNS file contains ***MENUGROUP=ACAD. It is possible to use more than one group name in a single MNS file, in order to divide the menu items into different groups.

Menu item *name tags* are a way of naming R13 Windows menu items so that they're easier to control from menu macros, AutoLISP, and ADS. The standard R13 **F**ile pull-down menu definition starts like this:

```
***POP1
ID_File          [&File]
ID_New           [&New...]         Ctrl+N]^C^C_new
ID_Open          [&Open...]        Ctrl+O]^C^C_open
```

The words ID_File, ID_New, and ID_Open are name tags for the **F**ile, **N**ew, and **O**pen menu items, respectively. Your menu macros and AutoLISP programs can use these name tags, together with the menu group name, to gray out or place a check mark next to a menu item. For example, the "full name" (menu group plus name tag) of the **O**pen menu item is ACAD.ID_Open, and an AutoLISP program could gray out the **O**pen menu choice with the expression (menucmd "GACAD.ID_Open=~"). This is more reliable than the old R12 method of addressing menu items by the menu number and item position. R13's help string feature also relies on the name tags in order to display the appropriate string in the status bar, as you'll see later.

Partial menu loading is a means of organizing menus into smaller, more manageable pieces and avoiding conflicts among application menus. Before R13, your AutoCAD menu had to be defined in a single, monolithic menu file. This requirement led to large menu files, long menu compilation times, menu inconsistencies among applications, and inelegant menu customization procedures. Most third-party applications simply replaced the stock ACAD.MNU (or your customized version) with their own. If you wanted to maintain custom menu sections, you often had to copy them into several different menu source files every time you made a change.

In R13, the menu file that loads when you open or create a new drawing (for example, ACAD.MNS) is called the *base menu*. The base menu is also the one that gets compiled and loaded when you use the MENU command. You can add to the base menu by attaching additional menu portions from other MNS files with the Menu Customization dialog box (MENULOAD command), either interactively or in an AutoLISP program (see Figure 13.2). Each MNS file (including the base menu) usually will contain a different menu group name, in order to avoid conflicts among the various menus.

Understanding Menu Structure

Figure 13.2
The Menu Customization dialog box.

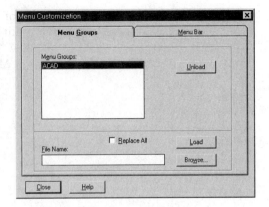

There is one small change that affects all platforms. Autodesk now calls the old *icon* menus by the new name of *image tile* menus, in order to minimize confusion with other meanings of the term "icon" in graphical user interfaces such as Windows. The menu file section that used to be called ***ICON is now called ***IMAGE. Although the section label ***ICON still works in R13, you should use ***IMAGE in order to avoid compatibility problems with future releases. Figure 13.3 shows one of the standard AutoCAD image tile menus.

Menu Structure

This part of the chapter is a guided, hands-on tour of menu file structure. You'll learn about the overall structure of a menu source file, and then create a menu from scratch in order to master menu sections, items, labels, name tags, and macros.

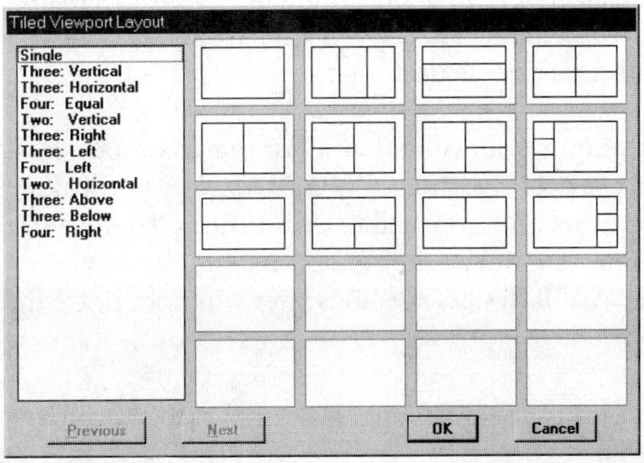

Figure 13.3
Icon menus are now called image tile menus.

Structure of the Windows MNS File

 An R13 Windows menu source file is organized conceptually like this:

```
// comment line

***MENUGROUP=group_name

***section1_name
title1_name_tag      [section1_title]
itemA_name_tag       [itemA_label]itemA_macro
itemB_name_tag       [itemB_label]itemB_macro
itemC_name_tag       [itemC_label]itemC_macro

***section2_name
title2_name_tag      [section2_title]
itemX_name_tag       [itemX_label]itemX_macro
itemY_name_tag       [itemY_label]itemY_macro
itemZ_name_tag       [itemZ_label]itemZ_macro
```

Comment lines can occur throughout the file, and they must be preceded by two forward slashes. The menu group name usually appears near the top of the file, so that all menu item name tags after it (generally all of them in the file) belong to that group. The bulk of the menu file comprises menu sections (described next), each of which may contain a title and a series of menu items. The menu section title appears first in the section, enclosed in square brackets, and may be preceded by a name tag. Each menu item has a label, which looks like a menu section title, optionally preceded by a name tag and followed by a menu macro.

In this example, the indentation of the menu item labels after the name tags is optional. You must put at least one space between the name tag and the label's open bracket, but AutoCAD doesn't care how many spaces you use. We've shown two levels of indentation in order to make the hierarchy clearer.

Not all menu sections require or even allow all of these components. For example, name tags don't serve any purpose in the ***BUTTONSx, ***AUXx, ***SCREEN, and ***TABLET menu sections and aren't allowed in the ***IMAGE section. Many sections don't use titles—the first line below the section name is an ordinary menu item and can include a menu macro. We'll discuss specifics later when we delve into the details of each section.

Some menu sections, notably ***TOOLBARS, ***IMAGE, and ***SCREEN, support submenus, which look like this:

Understanding Menu Structure

```
***section1_name
**submenu1_name
title1_name_tag          [submenu1_title]
itemA_name_tag              [itemA_label]itemA_macro
itemB_name_tag              [itemB_label]itemB_macro
itemC_name_tag              [itemC_label]itemC_macro

**submenu2_name
title2_name_tag          [submenu2_title]
itemM_name_tag              [itemM_label]itemM_macro
itemN_name_tag              [itemN_label]itemN_macro
itemO_name_tag              [itemO_label]itemO_macro
```

The submenus are reasonably sized groupings of menu items that can be displayed together, for example as a separate toolbar or side-screen menu page. Each submenu name starts with two asterisks (as opposed to three asterisks for a menu section or menu group).

Warning: Don't use duplicate submenu names, even in different menu sections. Duplicate submenu names confuse AutoCAD and usually lead to the wrong submenus being displayed in one or more sections.

Structure of the DOS MNU File

An R13 DOS menu source file is organized similarly, but without the menu group or name tags:

```
// comment line

***section1_name
     [menu_section1_title]
     [itemA_label]itemA_macro
     [itemB_label]itemB_macro
     [itemC_label]itemC_macro

***section2_name
     [menu_section2_title]
     [itemX_label]itemX_macro
     [itemY_label]itemY_macro
     [itemZ_label]itemZ_macro
```

As in the previous examples, indentation is optional. DOS menus can have submenus as well.

Menu Sections

AutoCAD recognizes 31 standard menu section names in DOS and Windows, plus three additional section names in Windows. Table 13.2 lists the 34 standard section names. (Remember that `***MENUGROUP=`*group_name* is not a menu section name, even though it starts with three asterisks.)

When AutoCAD loads a compiled menu file (MNC + MNR in Windows, or MNX in DOS), it automatically loads these standard sections into their corresponding interfaces and makes them available to the user (unless you use the MENULOAD command to attach a partial menu—see the end of this chapter for details). For example, the ***POP1 through ***POP16 menus are displayed on the pull-down menu bar, the ***POP0 menu appears when you click the pointing device button corresponding to the cursor menu, and AutoCAD stays on the lookout for any key combinations defined in ***ACCELERATORS. Some of the menu sections aren't available until you perform some other action. The ***IMAGE menus don't appear until another menu choice loads and displays an image tile menu, and you won't see the ***SCREEN menus at all if you leave the side-screen menu turned off in Windows AutoCAD.

Table 13.2
Menu Sections

Section Name	Defines
***POP1 through ***POP16	Pull-down menus
***POP0	Cursor menu
***AUX1 through ***AUX4	System pointing device buttons
***BUTTONS1 through ***BUTTONS4	Digitizer buttons
***IMAGE	Image (that is, icon) menus
***SCREEN	Side-screen menus
***TABLET1 through ***TABLET4	Digitizer tablet menus
***TOOLBARS	Windows toolbar menus
***ACCELERATORS	Windows keyboard accelerators
***HELPSTRINGS	Windows toolbar and pull-down menu status bar help

Understanding Menu Structure

In a menu section that contains submenus, AutoCAD automatically loads the first submenu that follows the section name. It doesn't immediately show the remaining submenus in the section. These are held in reserve until your menu or AutoLISP programs call them.

Note: ***POP16 is the highest numbered pull-down menu that AutoCAD displays automatically when you load a menu file. (You can display more than 16 pull-down menus with partial menu loading, but you risk overwhelming the user with too many menu choices.) In practice, few menus define all 16 standard pull-down menus. R13's ACAD.MNS file defines sections ***POP1 through ***POP7. The R13 for Windows ACADFULL.MNS file and R13 for DOS ACAD.MNU files use three additional pull-down menu slots: ***POP8 through ***POP10.

In the first exercise, you'll create a simple menu file with five sections from scratch, in order to see how menu files are structured.

Warning: If you're working through the exercises using R13 DOS, use the menu extension MNU instead of MNS throughout this chapter.

Exercise

Creating a New Menu File

Copy all of the files from \MAXAC\BOOK\CH13 to your \MAXAC\DEVELOP directory.

Start AutoCAD with a new, blank drawing.

Launch your text editor and create a new, blank text file. Enter the following lines of text, substituting your initials and the date in the comment line.

```
// 13TEST.MNS [Enter]
// 14 April 1996 by MM [Enter]
[Enter]
***MENUGROUP=13TEST [Enter]
[Enter]
***POP1 [Enter]
MxaPop1_T    [Pop1_Title] [Enter]
MxaPop1_1    [Pop1 Item1] [Enter]
[Enter]
```

```
***AUX1 [Enter]
[Aux1 Item1]^c^c^c_ARRAY [Enter]
[Enter]
***BUTTONS1 [Enter]
[Buttons1 Item1]^c^c^c_BREAK [Enter]
[Enter]
***SCREEN [Enter]
[Scr Itm1] [Enter]
[Scr Itm2] [Enter]
[Enter]
***TABLET1 [Enter]
[Tab1 Item1]^c^c^c_TRIM [Enter]
```

Omit the **MxaPop1_T** and **MxaPop1_1** name tags if you're using R13 DOS. (Also, the ***MENUGROUP=13TEST menu group line is superfluous in DOS AutoCAD, which sees it as an unknown section name and ignores it.) Save the menu file as \MAXAC\DEVELOP\13TEST.MNS. In some Windows text editors, you must enclose the new file name in quotation marks in order to ensure that the file is saved with an MNS extension instead of TXT.

Return to AutoCAD.

Command: **MENU** [Enter]	Opens the Select Menu File dialog box
Choose \MAXAC\DEVELOP\13TEST.MNS *and then* OK	Compiles and loads your new menu

A quicker alternative is to choose the **T**ype It button and then simply enter **13TEST** at the command prompt. Note that if you use the dialog box, AutoCAD doesn't search the support path, but if you use the **T**ype It button, it does.

Next, you'll turn on the side-screen menu (skip this step in DOS AutoCAD because the side-screen menu should already be on).

Command: **PREFERENCES** [Enter]	Opens the Preferences dialog box
Turn on Screen Men**u** *and choose* OK	Displays the screen menu area at the right side of the screen

Your screen should look similar to Figure 13.4. You can see the ***SCREEN menu and the title of the ***POP1 menu, but the remaining parts of your menu file are hidden. Next, you'll test the hidden sections.

Figure 13.4
The 13TEST.MNS menu.

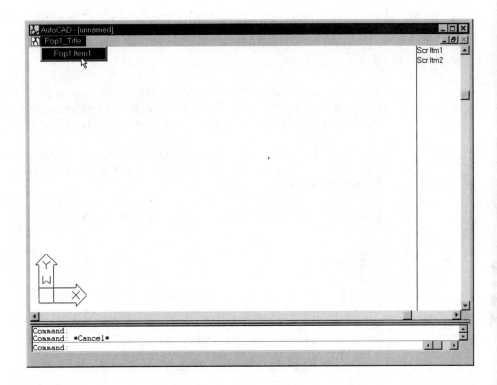

Command: *Click on the* POP1_Title *pull-down menu title*	Displays the Pop1 Item1 menu item (see Figure 13.4)
Command: *Click Button1 on your pointing device*	Runs the menu macro associated with either ***AUX1 or ***BUTTONS1

The location of Button1 (that is, the first programmable button, generally the first button after the pick button) depends on your pointing device. With a mouse, it's usually the right-most button. With a digitizer puck, it may have a number 2 on it. If you press each of the buttons in turn, you'll discover the correct one.

The menu issues either:
Command: _ARRAY
Select objects: *Press CANCEL*

or:
```
Command: _BREAK
Select object: Press CANCEL
```

If Button1 issued the ARRAY command, then your pointing device buttons are controlled by the ***AUX*x* menu sections. If Button1 issued the BREAK command instead, then your buttons are controlled by the ***BUTTONS*x* menu sections.

Try the next step if you have a digitizing tablet that's configured for at least one tablet menu area (refer to the TABLET CFG subcommand in the *AutoCAD Command Reference* for details on configuring it).

Command: *Pick the upper left menu square in tablet area 1*   Runs the menu macro associated with ***TABLET1

The menu issues:
```
Command: _TRIM
Select cutting edges: (Projmode = UCS, Edgemode = No extend)
Select objects: Press CANCEL
```

Remain in AutoCAD for the next exercise.

The point of this exercise was to see how AutoCAD assigned the menu sections to the various menu interfaces. We included the menu macros (ARRAY, BREAK, and TRIM) simply as mnemonics for the "invisible" sections (***AUX1, ***BUTTONS1, and ***TABLET1), to make it easy to tell which section AutoCAD used for a particular button or tablet menu square.

System pointing devices (for example, the Windows system mouse, or a digitizer with Wintab-compliant driver) are controlled by ***AUX*x* sections. Other devices (all DOS pointing devices, or a digitizer in Windows with an ADI driver) are controlled by ***BUTTONS*x* sections. In real menus, you'll usually want to make the ***AUX*x* and ***BUTTONS*x* sections identical to one another so that buttons behave the same no matter what kind of pointing device is active.

As you saw in the exercise, the syntax is similar for each of these five menu sections. The differences are:

- Some sections (***POP*x*, ***ICON, and ***TOOLBARS) require a title in brackets as the first line after the section name.

  - Some sections (for example, ***POP*x* and ***SCREEN) should have a bracketed label that defines what the user sees on the menu. Other sections (for example, ***ICON and ***TOOLBARS) require a "label" for specifying other information about the menu item, such as a slide file or button icon to display. Other sections (for example, ***AUX*x* and TABLET*x*) don't need a label because there's no way for AutoCAD to display it.

◗ Some sections can make use of name tags.

Table 13.3 summarizes these differences for all of the standard menu sections.

In this table, "Req'd" means that the menu definition element is required, and "Recm'd" means that it's recommended (you can get away without adding labels in many sections, but you'll want to include them so that you can control what appears to the user). "Optional" means that you don't have to include the menu definition element, depending on what else is in the menu and what you're trying to achieve. "Ignored" means that you can include the menu definition element, but AutoCAD doesn't do anything with it. Even when labels aren't required, they help you identify and keep track of the macros in the menu file. "No" means that the menu definition element doesn't apply to this section.

In the next exercise, you'll load and test a menu that defines all of the standard menu sections. The macros for all of the menus use the AutoLISP `princ` function to display a

Table 13.3
Menu Section Features

Section Name	Title	Label	Name Tag
***POP1 through ***POP16	Req'd	Recm'd	Optional
***POP0	Req'd, but ignored	Recm'd	Optional
***AUX1 through ***AUX4	No	Ignored	Ignored
***BUTTONS1 through ***BUTTONS4	No	Ignored	Ignored
***IMAGE	Req'd	Req'd	Not allowed
***SCREEN	No	Recm'd	Ignored
***TABLET1 through ***TABLET4	No	Ignored	Ignored
***TOOLBARS	Req'd	Req'd	Optional
***ACCELERATORS	No	Req'd	Optional
***HELPSTRINGS	No	Req'd	Req'd

message on the command line. The test menu (13ALLSCT.MNS) looks like this, with some repetitive sections removed:

```
// 13ALLSCT.MNS
// Shows all standard AutoCAD R13 Windows menu sections
// 14 April 1996 by MM

***MENUGROUP=13TEST

***AUX1
[Aux1 Item1]^p(progn (princ "Aux1 Item1") (princ))

**AUX2
[Aux2 Item1]^p(progn (princ "Aux2 Item1") (princ))

***AUX3
[Aux3 Item1]^p(progn (princ "Aux3 Item1") (princ))

***AUX4
[Aux4 Item1]^p(progn (princ "Aux4 Item1") (princ))

***POP0
MxaPop0_T    [P0T]
MxaPop0_1    [Pop0 Item1]^p(progn (princ "Pop0 Item1") (princ))

***POP1
MxaPop1_T    [P1T]
MxaPop1_1    [Pop1 Item1]^p(progn (princ "Pop1 Item1") (princ))
             [--]
MxaPop1_2    [Display Pop0 Cursor Menu]$p0=*
MxaPop1_3    [Display Image menu]$i=IMG_TEST $i=*

***POP2
MxaPop2_T    [P2T]
MxaPop2_1    [Pop2 Item1]^p(progn (princ "Pop2 Item1") (princ))

...

***TOOLBARS
**TB_TEST
MxaTb_T      [_Toolbar("MxaTB_T Toolbar Title", _Floating, _Show, 1, 101, 5)]
fMxaTb_1     [_Button("MxaTb_1", ICON_16_APPLOA, ICON_32_APPLOA) +
             ]^p(progn (princ "TB_TEST Toolbar Item1") (princ))
             [--]
MxaTb_2      [_Control(_Layer)]

***IMAGE
**IMG_TEST
[Image Menu Title]
[acad(box3d,Img Item1)]^p(progn (princ "Image Item1") (princ))

***ACCELERATORS
[CONTROL+"1"]^p(progn (princ "Accelerator for CTRL+1") (princ))
```

Understanding Menu Structure

```
***HELPSTRINGS
MxaPop0_1    [Sample Pop0 menu item #1]
MxaPop1_1    [Sample Pop1 menu item #1]
MxaPop2_1    [Sample Pop2 menu item #1]
...
MxaTb_1      [Sample Toolbar menu item #1]

***SCREEN
[Scr Itm1]^p(progn (princ "Screen Item1") (princ))
[Scr Itm2]^p(progn (princ "Screen Item2") (princ))

***TABLET1
[Tab1 Item1]^p(progn (princ "Tablet1 Item1") (princ))

***TABLET2
[Tab2 Item1]^p(progn (princ "Tablet2 Item1") (princ))

***TABLET3
[Tab3 Item1]^p(progn (princ "Tablet3 Item1") (princ))

***TABLET4
[Tab4 Item1]^p(progn (princ "Tablet4 Item1") (princ))
```

The ^p code at the start of the menu macros turns off echoing of the macro code on the command line, and is described in more detail later in the chapter. Encapsulating each **princ** expression with **progn** and following it by (princ) with no argument prevents AutoLISP from printing **nil** on the command line. In other words, much of the macro code simply makes for a cleaner command line. Figure 13.5 shows the loaded menu file.

 Exercise

Testing All Standard Menu Sections

Return to 13TEST.MNS in your text editor.

Erase all of the text in the file and import or copy the text from 13ALLSCT.MNS into 13TEST.MNS. (If you're using DOS AutoCAD, copy from 13ALLSCD.MNU.)

Save the menu file (as \MAXAC\DEVELOP\13TEST.MNS) and return to AutoCAD.

Command: **MENU** [Enter]	Opens the Select Menu File dialog box
Reload the 13TEST menu	Recompiles and displays the modified menu sections (see Figure 13.5)

Maximizing AutoCAD® R13

Figure 13.5
Sample menu with all standard sections.

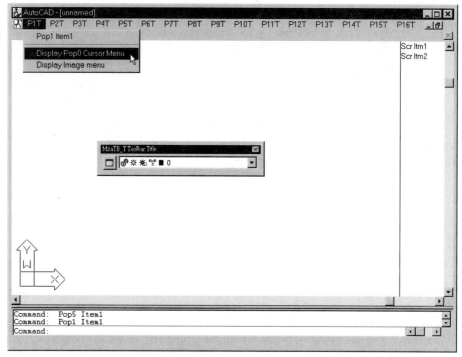

Test all of the menu sections, using the listing before this exercise as a guide. In particular, pick the Display Pop0 Cursor Menu and Display Image Menu items in the Pop1 pull-down menu (see Figure 13.5). Also notice the help strings that appear on the status bar when you rest the cursor on some of the pull-down menu items or the first toolbar icon.

Remain in AutoCAD for the next exercise.

This sample menu demonstrates the 34 standard menu sections that you can define in your custom menus. You can use this file as a template for creating new menus.

Using Submenus

In menu sections that contain submenus, AutoCAD automatically activates the first submenu defined in each section. In some cases (most commonly with ***SCREEN or ***ICON menus), you'll want to define additional submenus and activate them with menu macros or AutoLISP. In menu macros, you use special $ menu macro codes to call each type of page. Each menu interface, or "device," that supports submenus has a command code that sends menu pages to its particular device. Table 13.4 lists the codes.

Understanding Menu Structure

Table 13.4
Menu Device Codes

Code	Device
$S=	Screen menus
$P0=	Cursor menu
$P1= through $P16=	Pull-down menus
$B1= through $B4=	Button menus 1 through 4
$T1= through $T4=	Tablet areas 1 through 4
$I=	Icon menus
$A1= through $A4=	Aux menus 1 through 4

You can load a submenu to the device with the menu macro syntax:

$section=submenu

where *section* is one of the sections listed in Table 13.4, and *submenu* is the name of a submenu in the current section, without the two leading asterisks. For example, $S=scr_plot in a menu macro displays the ***SCREEN submenu named **scr_plot in the side-screen menu.

You can issue the $section=submenu call from a macro in any menu section. For example, it's common to load an icon menu with $i=*icon_submenu* $i=* in a pull-down menu item macro. The first part of the macro ($i=*icon_submenu*) assigns a particular icon submenu to the icon menu device, and the second part of the macro ($i=*) displays the icon menu. Similarly, it's common to attach $p0=* to ***AUX*x* and ***BUTTONS*x* macros. When the user pushes the button with this macro, $p0=* pops up the cursor menu (***POP0). The previous exercise showed examples of both of these techniques. You'll see more examples of using submenus throughout this chapter.

 Warning: Avoid loading submenus from one section into a different menu section. Submenus compiled for certain sections do not function properly if they're loaded into another section.

Flyout toolbars and cascading (child) pull-down menus are, in their own way, like submenus, with that structure defined in their menu item labels instead of with menu device codes.

Menu Item Labels

Menu item labels can serve different purposes, depending on the menu section:

▶ Labels define the item name that the user sees in the ***POP*x*, ***SCREEN, and ***HELPSTRINGS sections. Labels also serve this purpose in the ***IMAGE and ***TOOLBARS sections, although they do more in these sections. As you saw in Chapter 10, you can use DIESEL expressions in ***POP*x* labels to create dynamic labels that change depending on the current state of the AutoCAD system.

- Labels can include hierarchical (cascading) menu symbols in ***POP*x* menu sections. The two characters -> at the beginning of a pull-down or cursor label define the beginning of a cascading menu. The two characters <- at the beginning of a label define the end of a cascading menu.

- Labels define additional required information in the ***IMAGE, ***TOOLBARS, and ***ACCELERATORS sections. In the ***IMAGE and ***TOOLBARS sections, the label syntax includes fields that define the slide or bitmap image for AutoCAD to display. Also, when you define a flyout toolbar, the reference to the flyout goes in the "parent" toolbar item's label. In the ***ACCELERATORS section, the label defines the keystroke combination to which you're assigning the macro (for example, [CONTROL+"1"]). You'll learn more about the label syntax for these sections later in the chapter.

- Labels are optional documentation for menu items in the ***AUX*x*, ***BUTTONS*x*, and ***TABLET*x* sections. Because there is nothing for AutoCAD to display and no special information it needs besides the menu macro in these sections, AutoCAD ignores labels in square brackets. Thus, you can use them to keep track of menu item numbers or add comments to macros: [Aux1-1] or [T1-1 My Favorite AutoLISP Program]. (Without labels coding their positions, it's difficult to keep track of tablet menu items.)

Here are some tips and rules for good menu labels:

- Regardless of function, all menu item labels must be enclosed in square brackets (for example, [This is a Label]).

- Menu item labels can contain almost any displayable character, except for the few special symbols reserved for disabling and checking menu items, constructing cascading menus, and evaluating DIESEL expressions (~, !., ->, <-, and $ ()).

- AutoCAD doesn't impose any limitation on label length, but you should keep menu labels reasonably short for ease of reading and aesthetic reasons.

- Pull-down menu label text should use title capitalization—capitalize the first letter of all words, except for short words: [This is a Good Label].

- Don't use all uppercase, which is harder to read and makes it look as if your menus are shouting: [THIS IS AN ANNOYING LABEL].

Understanding Menu Structure

> Add an ellipsis ("...") to indicate that a pull-down menu item opens a dialog box: [Layers...]'_DDLMODES.

You can indent labels by adding leading spaces before the label opening brackets in order to emphasize a menu section's structure:

```
[Title]
  [Arc    ]^c^c^c_ARC
  [Circle]^c^c^c_CIRCLE
  [Line   ]^c^c^c_LINE
  [-]
  [Copy   ]^c^c^c_COPY
  [Erase  ]^c^c^c_ERASE
  [Move   ]^c^c^c_MOVE
```

AutoCAD ignores leading spaces when it compiles the menu, but they can make the menu source file more readable and thus easier to customize. The spaces show subordination and hierarchy in the menu file. This formatting is especially helpful in pull-down menus for identifying cascading (child) menus.

If you want blank spaces to appear on-screen as indents in the displayed menu item labels, place spaces inside the item label, as in [Label]. You can use spaces in side-screen menu labels to indent menu items on screen to organize them visually for the user.

Menu Item Name Tags

Name tags also serve several purposes:

In the ***POP*x* and ***TOOLBARS sections, name tags act as references that the ***HELPSTRINGS section items use. When you point to a pull-down, cursor, or toolbar menu item but haven't yet picked it, AutoCAD looks for an entry in the ***HELPSTRINGS section whose name tag matches the name tag of the menu entry you're pointing at. If it finds a match, AutoCAD displays the help string label (that is, the text in brackets) on the status line.

> In a similar way, ***ACCELERATOR section items can reference name tags in the ***POP*x* (but not ***TOOLBARS) sections. When an ***ACCELERATOR menu item's label is preceded by a name tag instead of followed by a macro, AutoCAD looks for a menu item in the ***POP*x* sections with the same name tag. If it finds a match, AutoCAD assigns the menu macro for that ***POP*x* menu item to the accelerator key combination. This feature lets you define a menu macro in one place and use it twice (once for the pull-down or cursor menu item and once for the accelerator key).

Maximizing AutoCAD® R13

▶ Name tags provide a way to control and query ***POP*x* menu items with the AutoLISP **menucmd** function. Chapter 10 demonstrated this feature.

The following exercise demonstrates the first two uses of name tags. This exercise applies only to R13 Windows—skip it if you're using R13 DOS.

Exercise

Using Name Tags

Return to 13TEST.MNS in your text editor. Make the changes to the ***POP1, ***TOOLBARS, ***ACCELERATORS, and ***HELPSTRINGS sections shown in bold:

```
***POP1
MxaPop1_T    [P1T]
MxaPop1_1    [Pop1 Item1](progn (princ "Pop1 Item1") (princ))
MxaDDim      [DDim Dialog...]'_DDIM [Enter]
             [--]

***TOOLBARS
**TB_TEST
MxaTb_T      [Toolbar("MxaTB_T Toolbar Title", _Floating, _Show, 1, 101, 5)]
MxaDDim      [_Button("Dimension Styles Dialog", ICON_16_DIMSTY, + [Enter]
             ICON_32_DIMSTY)]'_DDIM [Enter]
MxaTb_1      [_Button("MxaTb_1", ICON_16_APPLOA, ICON_32_APPLOA) +
             ](progn (princ "TB_TEST Toolbar Item1") (princ))
             [--]

***ACCELERATORS
[CONTROL+"1"](progn (princ "Accelerator for CTRL+1") (princ))
MxaDDim      [CONTROL+SHIFT+"D"] [Enter]

***HELPSTRINGS
MxaDDim      [Opens the DDIM Dimension Styles dialog box] [Enter]
MxaPop0_1    [Sample Pop0 menu item #1]
```

Save the menu file (as \MAXAC\DEVELOP\13TEST.MNS) and return to AutoCAD.

Use the MENU command to reload the 13TEST menu.

Test the changes by picking the new pull-down menu choice (① in Figure 13.6), picking the new toolbar icon (②), and pressing CTRL+SHIFT+D. Notice that AutoCAD displays the help string on the status bar when you point at the new pull-down or toolbar items (③).

Understanding Menu Structure

Figure 13.6
Menus that use name tags.

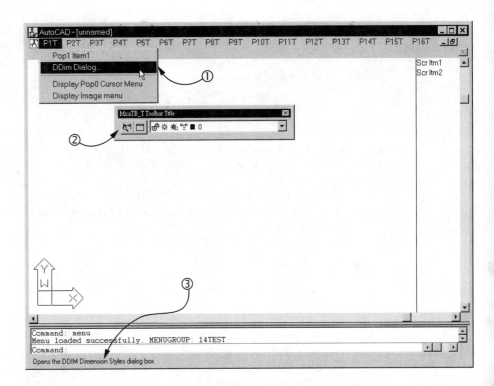

Remain in AutoCAD for the next exercise.

In this exercise, you used the same name tag—MxaDDim—four times. In ***POP1, the name tag creates a reference that's used by the ***HELPSTRINGS and ***ACCELERATORS entries. In ***TOOLBARS, the same name tag is only a reference for the ***HELPSTRINGS entry. The ***ACCELERATORS name tag tells AutoCAD to use the ***POP*x* menu macro ('_DDIM) with the corresponding name tag. AutoCAD displays the ***HELPSTRINGS entry whenever you point at either the ***POP1 or ***TABLET menu items. You could define different name tags and different help strings for the ***POP1 and ***TABLET menu items, but since both items refer to the same command, it's more efficient to use the same name tag and a single help string.

Name tags must start at the beginning of the line, and must be followed by at least one space or tab before the label's opening bracket:

```
Good_name_tag      [Label]
  Bad_name_tag_1   [Label]
Bad_name_tag_2[Label]
```

All of the name tags in the standard ACAD.MNU file that accompanies R13 Windows begin with the prefix `ID_`. You don't have to use a standard prefix—as you'll see later in the chapter, the menu group name prevents conflicts between identical name tags in different menus. But it's helpful to adopt some organized naming convention for name tags. In this book, we use the name tag prefix `Mxa`.

Menu Macros

Chapter 5 covered menu macros in detail, but here are a few reminders that concern how you enter macro text into the menu file:

- Start the menu macro directly after the close bracket (`"]"`) that ends the menu label. Also, remove trailing spaces from the end of the macro. AutoCAD treats every space in the macro as one press of the spacebar, which usually acts like ENTER.

- Where your macros do require multiple ENTERs, use semicolons instead of multiple spaces in order to make the extra ENTERs easier to see and count.

- If a macro becomes long enough that you want to continue it on the next line, add a plus sign at the (`"+"`) end of the first line. Don't indent the second line because the spaces will add lots of unwanted ENTERs to the macro!

- Every backslash in a menu macro acts as one pause for user input. If you need to specify path separators, use forward slashes instead.

Controlling Command-Line Echoing

As your menu macros become longer and incorporate AutoLISP expressions, AutoCAD's echoing of the macro code on the command line becomes more distracting. It also slows down your macros. You can use different values of the MENUECHO system variable and the special menu macro code ^P to control what gets echoed to the command line when users run your macros.

MENUECHO is a bit-coded variable. You set MENUECHO to the sum of one or more of the bit values shown in Table 13.5 to obtain the desired effect(s).

AutoCAD's default value of MENUECHO is 0, which causes all menu macro code to be echoed on the command line. You might think that the best interface would result from a setting of 3 (that is, 1 + 2), which suppresses menu macro code and system prompts. But turning off AutoCAD's system prompts can be confusing to the user if you create macros that issue a command and then pause for user input at one of the command's prompts. The user won't see the prompt and won't know what to enter. You can encounter

Understanding Menu Structure

Table 13.5
Effect of MENUECHO Settings

MENUECHO Bit Code	Effect
0	No suppression (the default)
1	Suppresses echoing of menu macro code (but the ^P macro code toggles echoing in an individual menu item)
2	Suppresses display of AutoCAD's system prompts during the menu macro
4	Prevents ^P code from toggling menu echoing
8	Echoes DIESEL input and output for debugging DIES

the same problem with the 1 bit turned on—the user sees a prompt, but doesn't see the name of the command that caused it, and therefore doesn't know what to do.

For most purposes, a MENUECHO setting of zero with liberal use of the ^P echoing toggle (described next) works well. An alternative is to set MENUECHO to 1 and use ^P when you want to toggle menu echoing on. The best choice depends on the types of macros you write most often.

You can mimic the effect of toggling the 1 bit (the suppress menu macro code echoing bit) by putting ^P in the menu macro. A ^P doesn't actually change the MENUECHO system variable, but it makes the macro behave as though the MENUECHO's 1 bit has been toggled. Another ^P in the same macro toggles the 1 bit back to its original behavior. When the macro ends, MENUECHO remains unaffected. In other words, the effect of a ^P lasts only until the next ^P or the end of the macro, whichever comes first.

Note: You used to be able to press CTRL+P at the keyboard during execution of a macro, unless MENUECHO's 4 bit was turned on. This was useful for debugging macros, but in R13 pressing CTRL+P no longer toggles echoing (even with keystrokes set to AutoCAD Classic to suppress the PLOT dialog).

In spite of the MENUECHO settings, commands and input issued by the AutoLISP **command** function still echo, even when the **command** function is issued within a menu macro. To suppress this echoing, you set the CMDECHO system variable to zero (1 is the default, for normal AutoLISP **command** function echoing).

In the next exercise, you'll experiment with menu and AutoLISP command echoing by using the following pull-down menu:

```
***POP12
[Echo]
[$(if,$(and,1,$(getvar,menuecho)),!.)MenuEcho 1 bit]+
'_MENUECHO $M=$(xor,1,$(getvar,menuecho))
```

```
  [$(if,$(and,2,$(getvar,menuecho)),!.)MenuEcho 2 bit]+
'_MENUECHO $M=$(xor,2,$(getvar,menuecho))
  [$(if,$(and,4,$(getvar,menuecho)),!.)MenuEcho 4 bit]+
'_MENUECHO $M=$(xor,4,$(getvar,menuecho))
[$(if,$(getvar,cmdecho),!.)CmdEcho]+
'_CMDECHO $M=$(xor,1,$(getvar,cmdecho))
  [--]
  [Layer Set 0]'_.LAYER _Set 0 ;
  [Layer Set 0 with ^P]^p'_.LAYER _Set 0 ;^p
  [Layer Set]'_.LAYER _Set \;
  [AutoLISP (command "LAYER")]+
^c^c^c(progn (command "_.LAYER" "_Set" "0" "") (princ))
  [AutoLISP (command "LAYER") with ^P]+
^c^c^c^p(progn (command "_.LAYER" "_Set" "0" "") (princ))
```

The first four menu items let you set and see the current values of MENUECHO and CMDECHO. (See Chapter 10 for more information about the use of DIESEL in these labels and macros.) The remaining menu items give you some macros to test with different MENUECHO and CMDECHO settings. Figure 13.7 shows the menu.

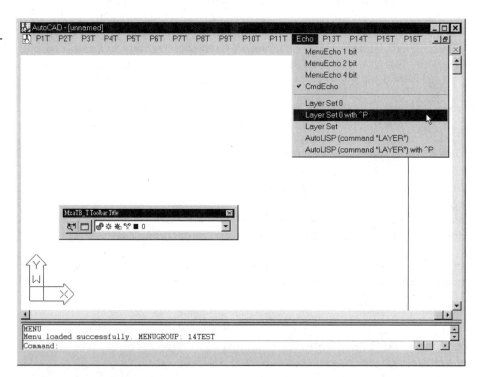

Figure 13.7
A menu for testing echoing.

Understanding Menu Structure

Exercise

Testing Menu and Command Echoing

Return to 13TEST.MNS in your text editor. Erase the existing ***POP12 menu section:

```
***POP12
MxaPop12_T    [P12T]
MxaPop12_1    [Pop12 Item1](progn (princ "Pop12 Item1") (princ))
```

In its place, add the new ***POP12 section listed just before this exercise. You can copy the section from 13ECHO.TXT.

Save the menu file (as \MAXAC\DEVELOP\13TEST.MNS), return to AutoCAD, and reload it with the MENU command.

Test each of the bottom four commands with a variety of MENUECHO and CMDECHO settings. Notice how the settings alter what you see on the command line.

Remain in AutoCAD for the next exercise.

You should decide on a MENUECHO setting (usually either 1 or 0) before creating a custom menu. You can use this ***POP*x* menu to experiment.

Note: If you decide to change your standard setting of MENUECHO from 0 to 1 (or vice versa), you'll probably need to add or remove ^P in many of your macros.

AutoCAD resets MENUECHO to zero and CMDECHO to 1 each time you open a drawing. If you want to use different settings, you'll need to establish them in ACAD.LSP or another initialization file. A *menuname*.MNL file is ideal for this purpose because it is automatically loaded when the menu of the same name is loaded.

Menu Section Details

So far we've covered the general aspects of menu structure. In this section, we delve into the details of using each kind of menu section.

Button and Auxiliary Menus

In AutoCAD, you can customize pointing device buttons by putting macros in the ***AUX*x* and ***BUTTONS*x* menu sections, where *x* is a number from 1 to 4. The number corresponds to button-clicks in combination with the SHIFT and CTRL keys. The ***AUX1 and ***BUTTONS1 sections define what happens when you click buttons without holding down SHIFT or CTRL.

As we described earlier in the chapter, Windows or UNIX system pointing devices (including a digitizer with Wintab-compliant driver) are controlled by ***AUX*x* sections. Other devices (including all DOS pointing devices) are controlled by ***BUTTONS*x* sections. In most menus, you should make the ***AUX*x* and ***BUTTONS*x* sections identical so that button behavior is consistent regardless of the pointing device. In this chapter, we often use ***AUX*x* to refer collectively to the ***AUX*x* and ***BUTTONS*x* sections.

Button assignments vary with the type of digitizer puck (that is, cursor) or mouse you use. The pick button usually is the left-most, upper-most, or lowest-numbered button on your pointing device. On a digitizer puck, it might be marked with a 1, a zero, or an asterisk, or might not be labeled at all. You can't customize the pick button's operation in the menu file—it always acts as the pick button. We call this the pick button in this book, and an instruction to just click or pick refers to this button.

Note: If you've used the Windows or AutoCAD driver configuration program to alter the pointing device button mapping (for example, swapping the left and right mouse buttons for a left-hander), then the button assignments will be different than what's described here.

The first button *after* the pick button is the first button that you can customize. On a mouse, this usually is the right mouse button. On a digitizer puck, it might be labeled with a 2 or a 1 (depending on whether the [pick button is labeled 1 or zero). Regardless of what it is marked with, we call this first configurable button "Button1". The next button (for example, third digitizer puck button or middle button of a 3-button mouse) is the second one that you can customize ("Button2"), and so on ("Button3", and so forth). In the diamond-shaped four-button layout that's common on many digitizer pucks, the pick button is at 12 o'clock, and they proceed counter-clockwise: Button1 at 9 o'clock, Button2 at 6 o'clock, and Button3 at 3 o'clock.

AutoCAD's standard ***AUX1 and ***BUTTONS1 sections look like this (the comments are not part of the actual menu file):

```
***AUX1         Section name
;               ENTER
$p0=*           Pop up the cursor menu
^C^C            Cancel
^B              Toggle snap
```

Understanding Menu Structure

```
^O        Toggle ortho
^G        Toggle grid display
^D        Toggle coordinates mode
^E        Toggle isoplane mode
^T        Toggle tablet mode
```

The first definition line after the section name (;) controls Button1. These definitions cover up through Button9. If you have a 12- or 16-button digitizer puck, then the remaining buttons are unassigned and don't do anything until you add more lines to the ***AUX*x* section.

Notice that the Button2 assignment ($p0=*) uses a device code macro in one menu section (***AUX1) to control another menu section (pop up the current ***POP0 menu). The remaining macros are simple. Button1 issues a single ENTER and Button3 issues two cancels. Button4 through Button9 toggle various modes with the old CTRL key special menu characters described in Chapter 5.

Because the number of buttons on most pointing devices is limited to either two or four, AutoCAD lets you define additional macros for button-clicks in combination with the SHIFT key, CTRL key, and both keys together. These combinations correspond to the ***AUX2, ***AUX3, and ***AUX4 sections, respectively. Table 13.6 summarizes the combinations.

Click indicates clicking one of the AutoCAD buttons. The instruction *click CTRL+Button1* means "hold down the CTRL key and click Button1." This action would run the first macro in the ***AUX3 menu section.

Note: R11 and previous versions recognized only a single button menu and single auxiliary pointing device menu: ***BUTTONS and ***AUX. R12 treated these sections as ***BUTTONS1 and ***AUX1 in order to maintain compatibility with older menus. R13 no longer recognizes ***BUTTONS and ***AUX.

Table 13.6
Key Combinations Used with ***AUXx and ***BUTTONSx Sections

Key Combination	Action
Click (default)	***AUX1 or ***BUTTONS1 definitions
SHIFT+click	***AUX2 or ***BUTTONS2 definitions
CTRL+click	***AUX3 or ***BUTTONS3 definitions
CTRL+SHIFT+click	***AUX4 or ***BUTTONS4 definitions

AutoCAD's standard ACAD.MNU makes little use of ***AUX2 through ***AUX4. The only assignment is for SHIFT+Button1:

***AUX2
$p0=*

This assignment makes the cursor menu available to users with 2-button mice. You can extend the usefulness of that second mouse button by adding ***AUX3 and ***AUX4 sections. You have even more options with a 3-button mouse or 4-button puck, but using a large number of button, SHIFT, and CTRL combinations requires a lot of mental and manual dexterity.

Although you can assign ordinary AutoCAD command macros to buttons (for example, ^c^c^c_LINE), it's more common to extend the usefulness of buttons by having them pop up different cursor menus. This approach is especially appropriate for 2-button mouse users, who have only four customizable button combinations: Button1, SHIFT+Button1, CTRL+Button1, and CTRL+SHIFT+Button1. In order to pop up different cursor menus, you create submenus in the ***POP0 section and define each ***AUX*x* macro so that it loads the desired ***POP0 submenu before popping up the cursor menu.

We cover the ***POP0 cursor menu in greater detail later in this chapter, and Chapters 10 and 14 include several sophisticated uses of cursor menus. Here you'll see how to call simple cursor menus from the ***AUX*x* sections. The file 13AUXCUR.TXT contains an example of this approach that you'll use in the next exercise. The ***AUX*x* and ***BUTTONS*x* sections look like this:

```
***AUX1
[1 Enter    ];
[2 Osnap Cursor]$P0=POP0 $P0=*
[3 Cancel   ]^C^C^C
[4 Snap     ]^B
[5 Ortho    ]^O
[6 Grid     ]^G
[7 Coords   ]^D
[8 Isoplane]^E
[9 Tablet ]^T


***BUTTONS1
[1 Enter    ];
[2 Osnap Cursor]$P0=POP0 $P0=*
[3 Cancel   ]^C^C^C
[4 Snap     ]^B
[5 Ortho    ]^O
```

```
[6 Grid    ]^G
[7 Coords  ]^D
[8 Isoplane]^E
[9 Tablet  ]^T


***AUX2
[Shift+1 Osnap Cursor]$P0=POP0 $P0=*

***BUTTONS2
[Shift+1 Osnap Cursor]$P0=POP0 $P0=*

***AUX3
[Ctrl+1 Draw Cursor]$P0=C-DRAW $P0=*

***BUTTONS3
[Ctrl+1 Draw Cursor]$P0=C-DRAW $P0=*

***AUX4
[Ctrl+Shift+1 Modify Cursor]$P0=C-MODIFY $P0=*

***BUTTONS4
[Ctrl+Shift+1 Modify Cursor]$P0=C-MODIFY $P0=*
```

AutoCAD ignores the optional button item labels. They serve as documentation to help you keep track of button numbers, SHIFT and CTRL key combinations, and button functions.

The ***AUX1 and ***AUX2 button macros function identically to those in the standard AutoCAD ACAD menu. The only addition is $P0=POP0 for Button2 and SHIFT+Button1. This submenu loading command ensures that AutoCAD reloads the default ***POP0 cursor menu before displaying the cursor menu on the screen with $P0=*. If you omitted the $P0=POP0, then Button2 and SHIFT+Button1 would display the most recently used cursor menu, which might not be the Osnap menu.

The ***AUX3 and ***AUX4 macros load custom Draw and Modify ***POP0 cursor submenus, and then display the cursor menu. Figure 13.8 shows the Draw cursor menu.

Here is the custom ***POP0 section:

```
***POP0
**C-OSNAP
[Osnap]
[From]_from
[--]
```

Maximizing AutoCAD® R13

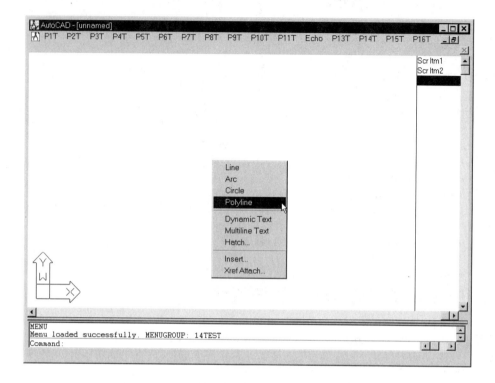

Figure 13.8
The custom Draw cursor menu.

```
[Endpoint]_endp
[Midpoint]_mid
[Intersection]_int
[Apparent Intersection]_appint
[Center]_center
[Quadrant]_qua
[Perpendicular]_per
[Tangent]_tan
[Node]_nod
[Insertion]_ins
[Nearest]_nea
[Quick,]_qui,^Z
[--]
[None]_non
[--]
[.X].X
[.Y].Y
[.Z].Z
[.XZ].XZ
[.YZ].YZ
```

Understanding Menu Structure

```
[.XY].XY

**C-DRAW
[Draw]
[Line]^c^c^c_LINE
[Arc]^c^c^c_ARC
[Circle]^c^c^c_CIRCLE
[Polyline]^c^c^c_PLINE
[--]
[Dynamic Text]^c^c^c_DTEXT
[Multiline Text]^c^c^c_MTEXT
[Hatch...]^c^c^c_BHATCH
[--]
[Insert...]^c^c^c_DDINSERT
[Xref Attach...]^c^c^c_.XREF _Attach ~

**C-MODIFY
[Modify]
[Erase]^c^c^c_ERASE
[Oops]^c^c^c_OOPS
[--]
[Copy]^c^c^c_COPY
[Move]^c^c^c_MOVE
[Rotate]^c^c^c_ROTATE
[Stretch]^c^c^c_STRETCH
[--]
[Break]^c^c^c_BREAK
[Extend]^c^c^c_EXTEND
[Fillet]^c^c^c_FILLET
[Trim]^c^c^c_TRIM
[--]
[Change Properties...]^c^c^c_DDCHPROP
[Modify...]^c^c^c_DDMODIFY
[Polyline Edit]^c^c^c_PEDIT
```

The default first submenu of the POP0 cursor menu, **C-OSNAP, contains the standard AutoCAD cursor menu, which includes a set of object snaps and point filters. Although the default submenu is labeled with **C-OSNAP, it need not be; the $P0=POP0 command loads the default submenu to the device regardless of name. The ***POP0 section contains two additional, custom submenus: **C-DRAW and **C-MODIFY. Figures 13.8 and 13.9 show these two custom menus.

In the following exercise, you'll import these ***AUX*x*, ***BUTTONS*x*, and ***POP0 menu sections from 13AUXCUR.TXT into your test menu and experiment with them.

Figure 13.9
The custom Modify cursor menu.

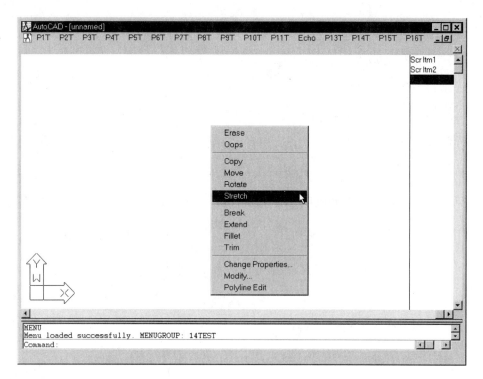

 Exercise

Coordinating Button and Cursor Menus

Return to 13TEST.MNS in your text editor. Erase all of the existing ***AUX*x* and ***BUTTONS*x* sections and the ***POP0 section (which are grouped together at the beginning of the file).

In their place, add the new sections listed before this exercise. You can copy the sections from 13AUXCUR.TXT.

Save the menu file (as \MAXAC\DEVELOP\13TEST.MNS), return to AutoCAD, and reload it with the MENU command.

Next, you'll test the buttons and associated cursor menus.

Command: *Click CTRL+Button1, then choose* Line   Displays the Draw cursor menu (Figure 13.8) and issues _LINE

Understanding Menu Structure

```
_LINE From point: Pick three or four points
```

To point: *Click SHIFT+Button1, then choose* Midpoint	Displays the Osnap cursor menu and issues _mid

```
_mid of Pick middle of first line
```

To point: *Click CTRL+SHIFT+Button1, then choose* Copy	Displays the Modify cursor menu (Figure 13.9) cancels LINE and issues _COPY

```
Command: _COPY
```

```
Select objects: Select some objects and copy
them in the usual way
```

Experiment with other commands on the Draw and Modify cursor menus.

Remain in AutoCAD for the next exercise.

Our example ***AUX*x*, ***BUTTONS*x*, and ***POP0 menus give you an idea of how you can streamline entry of the commands you use most often. Other ideas include creating a custom display cursor menu for zooming and panning, and adding dimensioning commands to the Draw cursor menu. Experiment with these menu sections to find out what kinds of customization offer the most efficiency gains for your work. If your pointing device has more than two buttons, you can add more cursor menus or use some of the buttons for your most common object snaps or editing commands.

Warning: Many AutoCAD users are accustomed to the default Button1, Button2, Button3, and SHIFT+Button1 assignments (ENTER, cursor menu, cancel, and cursor menu). You should avoid making drastic changes to these buttons in menus that others will use. On the other hand, the CTRL and CTRL+SHIFT buttons assignments are excellent for customization, since AutoCAD's default menu doesn't define these combinations.

Pull-Down and Cursor Menus

In Chapter 5, you learned how to add menu items to a custom pull-down menu page. Pull-down menus are probably the most commonly customized of AutoCAD's menu interfaces. They're easy to add or modify, and they provide for longer, more descriptive labels than does the side-screen menu. They're also flexible. Autodesk made significant improvements to pull-down menu customization in both R12 and R13.

When you use the MENU command to load a base menu, AutoCAD automatically displays up to 16 pull-down menus (***POP1 through ***POP16) on the menu bar and loads one cursor menu (***POP0) into memory. The ***POP*x* name comes from the way these menus "pop" up or down on the screen, over the drawing area. Figure 13.10 shows the 16 standard pull-down menu positions.

Warning: In DOS, the maximum number of menu title characters that AutoCAD can display in the pull-down menu bar depends on your screen resolution and display driver. A combination of long menu titles and large number of pull-down menus doesn't leave enough room for all of the title text at lower display resolutions, in which case AutoCAD truncates the titles in order to make them fit. Keep your menu titles short and avoid using more than about 10 pull-downs if your menus might be used at 640x480 resolution in DOS AutoCAD.

In Windows, AutoCAD wraps the menu titles onto a second line, which doubles the height of the pull-down menu bar. This approach avoids the truncation problem but consumes more of the drawing area.

As you saw in Chapter 5, your pull-down menu sections don't have to be consecutive. For example, you can define ***POP1, ***POP2, and ***POP16, without defining the intervening ***POP*x* sections. AutoCAD will display ***POP16 right next to ***POP2, instead of leaving a large gap. Nor do they have to appear in order on the menu bar; you can use the MENULOAD command to display them in any order.

Figure 13.10
The 16 standard pull-down menu positions.

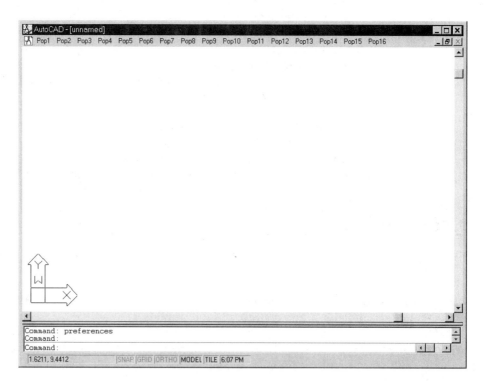

Understanding Menu Structure

Pull-Down Menu Titles and Item Labels

The first line in a pull-down menu section (or submenu section) defines a title rather than an executable menu item. For example, here's the beginning of the standard R13 Windows ***POP1 menu (see Figure 13.11):

```
***POP1
ID_File          [&File]
ID_New           [&New...      Ctrl+N]^C^C_new
ID_Open          [&Open...     Ctrl+O]^C^C_open
ID_Save          [&Save Ctrl+S]^C^C_qsave
ID_Saveas        [Save &As...]^C^C_saveas
ID_SavR12        [Save R12 DW&G...]^C^C_saveasr12
```

The first line contains only a name tag and label, which defines the menu's title on the menu bar (**File**—see ① in Figure 13.11). There's no menu macro after [&File] because the function of a ***POP*x* title is predefined—it displays the rest of the pull-down menu.

The optional ampersand (&) before the F in **F**ile tells AutoCAD to make the letter following it a shortcut key for accessing the menu. Thus, ALT+F opens this **F**ile menu. The same syntax works for items in the body of the menu, but the user must have the menu displayed (that is, pulled down) in order to use

Figure 13.11
The standard AutoCAD File menu.

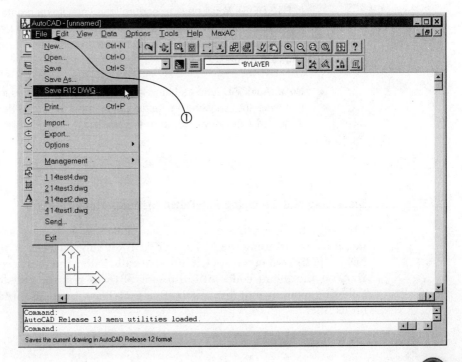

the shortcut. In other words, ALT+F, ALT+A (or simply ALT+F, A) opens the File menu and runs the Save As macro in this menu. The *AutoCAD Customization Guide* calls these shortcuts *accelerator keys*, but note that they have nothing to do with the ***ACCELERATORS menu section.

Warning: DOS AutoCAD doesn't support ALT+letter shortcut keys. If you include ampersands in DOS menu labels, AutoCAD will complain when it tries to compile the menu.

You can add as many items to a pull-down section as you want, but you should avoid extremely long menus. They're difficult for users to scan quickly, and therefore they are cumbersome to use. In addition, all display drivers place limits on the number of items that AutoCAD can display in a single pull-down menu. The maximum for DOS AutoCAD's VGA driver is 21 items, which is a good practical maximum even if you're using a higher resolution and Windows. If you need more than 21 items on a pull-down menu, use cascading menus as described later to group items and "collapse" them under parent items.

Warning: In R12, you could break a long pull-down menu label (for example, one containing a complex DIESEL expression) into more than one line in the same way that you can break a macro into more than one line: by adding a plus sign to indicate continuation. This technique no longer works in R13 Windows (although it does work in R12 DOS). You'll have to endure long lines if you create elaborate, DIESEL-laden menu labels in R13 Windows. You can, however, break toolbar labels by putting the + continuation character after a comma. The Toolbar Menus section of this chapter shows examples.

Note: AutoCAD ignores blank labels ([]) and blank lines in ***POP*x* menu sections. It moves the succeeding items up in the menu. Previous versions of AutoCAD truncated a pull-down menu at a blank line.

Displaying and Swapping Pull-Down Menus Automatically

Normally the user activates a particular pull-down menu by picking on its title on the menu bar, or by using the F10 key or ALT key shortcut. Occasionally you might want to make a pull-down menu appear automatically, with a menu macro or AutoLISP program. You use the menu codes $P1 through $P16 to control pull-down menu pages. The following menu macro displays the first pull-down menu, as though the user had picked on its title:

```
[Show POP1]$P1=*
```

This AutoLISP expression does the same thing:

(menucmd "P1=*")

An alternate AutoLISP syntax in R13 Windows uses the menu group and name tag instead of position, and uses a vertical bar (|) instead of an asterisk:

(menucmd "Gacad.ID_File=|")

The name tag approach is more reliable, because if a partial menu is loaded, you can't be sure of the position number of each pull-down menu.

It's also possible to define multiple submenus in a ***POP*x* section and swap them into that menu slot as needed. Assuming that ***POP12 contains a submenu called **MYTOOLS, the following menu macro and pair of AutoLISP expressions both load the submenu into the ***POP12 menu slot and then display it:

[Swap and Show POP12]$P12=MYTOOLS $P12=*
(menucmd "P12=MYTOOLS") (menucmd "P12=*")

Assuming that the **MYTOOLS menu had the name tag ID_ToolMenu and was included in the ACAD.MNS file (that is, was part of the ACAD menu group), then you could use this new syntax:

[Swap and Show POP12]$P12=acad.ID_ToolMenu $P12=*

(menucmd "P12=acad.ID_ToolMenu") (menucmd "P12=*")

You should avoid swapping pull-down submenus frequently, since it tends to confuse users. Also, Autodesk warns that the ability to swap pull-down menus may be removed in a future AutoCAD version, since it violates Microsoft's Windows interface guidelines.

Warning: If you intend to use R13's partial menu loading feature (described at the end of this chapter), or any third-party applications that rely on partial menu loading, you should use the $P*x*=*group_name*.*menu_name* syntax. With partial menu loading, you don't know for sure which pull-down menus occupy which ***POP*x* slot in the pull-down menu bar. The new syntax gives you a position-independent method of referencing each pull-down page. If you omit the *group_name*, AutoCAD looks for *menu_name* in the base menu.

The Cursor Menu

The cursor menu has all the features of a pull-down menu with one difference; it pops up on the screen at or near the current position of the cursor. The cursor menu is defined in the special ***POP0 section. Like the other ***POP*x* sections, ***POP0 must have a title line right after the ***POP0 line, but AutoCAD doesn't display this title on the cursor menu. It's there in the menu file only for consistency with the other ***POP*x* sections.

Because there's no title to pick in order to display the cursor menu, you must display it with the $P0=* menu macro code or (menucmd "P0=*") AutoLISP expression. As you learned earlier in the chapter, the most common method is to assign $P0=* to a pointing device button, such as SHIFT+Button1.

Tip: Although the ***POP0 section is the one reserved for the cursor menu, and the one that AutoCAD loads automatically, you can load submenus from other ***POP*x* sections and display them on the cursor menu. You might want to use this feature for a submenu that could appear as either a pull-down menu or a cursor menu in your application. If you don't have this need, it's less confusing to keep all of your cursor submenus in the ***POP0 section.

Chapter 10 and the Button and Auxiliary Menus section of this chapter demonstrate several ways to customize the cursor menu.

Special Codes in Pull-Down and Cursor Menus

You can take advantage of a variety of special ***POP*x* label codes in order to make your pull-down and cursor menus more usable and responsive. The options include disabling (graying out) menu items, placing check marks next to them, and defining cascading menus (also called child, hierarchical, or walking menus). Cascading menus pop out of the side of their parent menus to show nested sets of menu items.

Table 13.7 shows the ***POP*x* label codes that you can use in R13 Windows.

An item label containing two hyphens displays a horizontal separator line for the full width of the pull-down or cursor menu. AutoCAD doesn't let the user pick a separator line, and it ignores any macro code following the [--] label.

You use codes made up of hyphens and angle brackets (-> and <-) to make menus cascade in and out. A -> code, such as [->*title*], defines the cascading menu's title. The title displays a small solid triangle to indicate that it opens a cascading menu (for example, see the Options and Management menu items in Figure 13.11, earlier in this section). The cascading menu appears to the right of the title or, if space is limited, over the title. As in pull-down menu bar titles, AutoCAD ignores any code following the label of the cascading menu title.

You can disable or put a check mark next to a menu item (or both) with the tilde code (~) and the exclamation mark plus period code (!.). Normally you won't want to permanently disable or put a check mark next to a menu item. Instead, you use these codes in DIESEL or AutoLISP expressions to disable and check menu labels dynamically, based on system variable settings or other AutoCAD conditions. Chapter 10 contains all of the details.

The [^*icon_name*^] code lets you use bitmap icons instead of label text in pull-down menus (the icons must be contained in a DLL file whose name matches the menu file's name). This feature is a leftover from before AutoCAD supported toolbars. The

Understanding Menu Structure

Table 13.7
Special ***POPx Menu Item Label Codes

Code	Format	Description
-	[-]	Creates horizontal line
&	[la&bel]	Defines ALT+letter shortcut key (method 1)
/c	[/clabel]	Defines ALT+letter shortcut key (method 2)
\t	[label\tmore_lbl]	Puts a tab in between parts of the label text
->	[->title]	Begins cascade (child) menu
<-	[<-label]	Ends cascade menu (repeat for nested menus: <<-)
~	[~label]	Disables (grays out) menu item (see Chapter 10)
!.	[!.label]	Puts check mark before label (see Chapter 10)
$(	[$(expression)]	DIESEL language expression (see Chapter 10)
^^	[^icon_name^]	Displays BMP icon instead of label text

combination of icons in a pull-down menu results in a terrible interface, and you should avoid it (see Figure 13.12). The icons take up lots of menu space, and putting them on a pull-down menu corresponds to no recognized user interface convention. In addition, icons eliminate one of the major advantages of pull-down menus: the ability to use descriptive text in menu item labels. If you want to use bitmap icons, put them in toolbars, not pull-down menus.

 R13 DOS supports all of these special label codes except for &, /c, and ^icon_name^.

The following sample Windows pull-down menu, which is contained in 13PCODES.TXT, demonstrates many of the special label codes:
```
***POP13
MxaCodes    [Codes]
MxaGrayed   [~Grayed Out Item]
MxaChecked  [!.Checked Item]
            [--]
MxaSCasc    [->Single Level Cascade]
MxaSCascA      [Item A - Checks D](menucmd "G13test.MxaSCascD=!.")
MxaSCascB      [Item B - Grays D](menucmd "G13test.MxaSCascD=~")
MxaSCascC      [Item C - Checks and Grays D]+
(menucmd "G13test.MxaSCascD=!.") (menucmd "G13test.MxaSCascD=~")
```

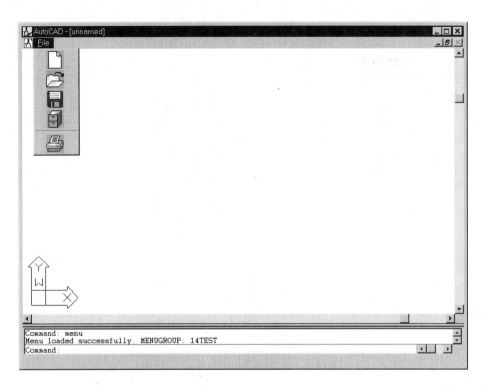

Figure 13.12
Pull-down menus and icons—a bad combination.

```
MxaSCascD      [Item D]
MxaSCascZ      [<-Last Item - Clears D](menucmd "G13test.MxaSCascD=")
               [--]
MxaMCasc       [->Multi-Level Cascade]
MxaMCasc1A       [Level 1 Item A]
MxaMCasc1B       [Level 1 Item B]
MxaMCasc2T       [->Level 2 Title]
MxaMCasc2A         [Level 2 Item A]
MxaMCasc2B         [Level 2 Item B]
MxaMCasc3T         [->Level 3 Title]
MxaMCasc3A           [Level 3 Item A]
MxaMCasc3B           [Level 3 Item B]
MxaMCasc3Z           [<-Last Level 3 Item]
MxaMCasc2B         [Level 2 Item C]
MxaMCasc2Z         [<-<-Last Level 2 and 1 Item]
               [--]
MxaMore        [More Items]
MxaFinal       [Final Item]
```

Try the following exercise to see these codes in action. In particular, notice the [<-<-Last Level 2 and 1 Item] code in the listing and how it terminates both levels of the cascading menus.

Understanding Menu Structure

Exercise

Looking at Pull-Down or Cursor Codes

Return to 13TEST.MNS in your text editor.

Delete menu sections ***POP2 through ***POP11 and ***POP13 to make room for three levels of cascading menu (but leaving the ***POP12 Echo menu in place).

Add the new ***POP13 section listed before this exercise. You can copy the section from 13PCODES.TXT. If you are using R13 DOS, delete all of the name tags (MxaCodes, MxaGrayed, etc.).

Save the menu file (as \MAXAC\DEVELOP\13TEST.MNS), return to AutoCAD, and reload it with the MENU command.

Command: *From the pull-down menu bar, choose* Codes — Displays the new ***POP13 pull-down menu

The first four item labels (see Figure 13.13) are a grayed-out item (from the tilde in the label), a checked item (from !. in the label), a horizontal separator line (from [-]), and a cascading menu call (from a -> code).

Command: *Choose* Single Level Cascade — Opens submenu

The checking and graying macros in this cascading menu reference name tags, so they won't work in R13 DOS.

Command: *Choose* Multi-Level Cascade, *then* Level 2 Title, *then* Level 3 Title — Opens first-level, second-level, and third-level cascading menus (see Figure 13.13)

Command: *Choose* Level 3 Item B — Closes the menu and does nothing (because there is no macro after the label)

Remain in AutoCAD for the next exercise.

The [->Single Level Cascade] item label is an example of the simplest kind of cascading menu. The -> code at the beginning of the label causes AutoCAD to treat the item as the title of a cascading menu. Notice that this menu item includes no macro—the

Figure 13.13
Multi-level cascaded menus.

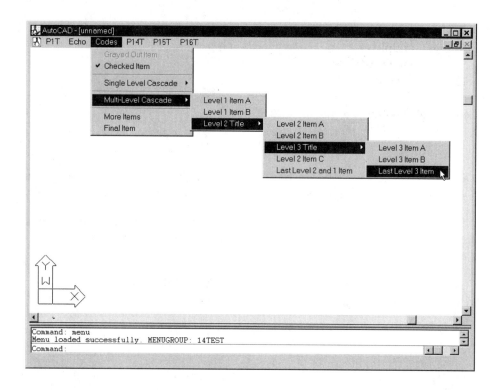

only action that AutoCAD allows when you click on a cascading menu title is to open the cascading menu.

The opening and closing behavior of cascading menus varies slightly in different operating systems. In all operating systems, the cascading submenu appears when you select the title by clicking on it with the pointing device button. If you move the cursor close to the triangle at the right side of the menu, or if you select the title by clicking the pointing device button, the cascading submenu appears. In Windows 95, pointing at the cascading menu's label and pausing for a moment also opens the submenu, unless you have tweaked your Windows 95 settings to alter this behavior. In DOS, moving the cursor close to the triangle at the right side of the menu opens the submenu. Also in DOS, the clicking method keeps the submenu displayed, even if you move the cursor away from it and point at another item in the parent menu.

The [<-Last Item - Clears B] label complements the opening -> code with its <- closing code. You must match each opening cascade -> code with a corresponding cascade <- code, or else the menu structure won't look the way you expect it to. If you nest a cascading menu inside a cascading menu, as in the Level 3 example, then the last item in the nested menu must close both levels with a double code: [<-<-Last Level 2 and 1 Item].

Understanding Menu Structure

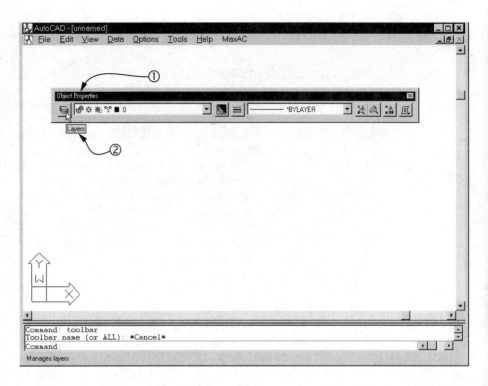

Figure 13.15
The standard R13 Object Properties toolbar.

```
                [--]
                [Control(_Linetype)]
                [--]
ID_Objcre       [Button("Object Creation", ICON_16_OBJCRE, +
ICON_32_OBJCRE)]'_ddemodes
ID_Mlstyle      [Button("Multiline Style", ICON_16_MSTYLE, +
ICON_32_MSTYLE)]^C^C_mlstyle
                [--]
ID_Ddmodi       [Button("Properties", ICON_16_MODIFY, +
ICON_32_MODIFY)]^C^Cai_propchk
                [--]
ID_TbInq        [Flyout("Inquiry", ICON_16_LIST, ICON_32_LIST, _OtherIcon, +
                 ACAD.TB_INQUIRY)]
```

**TB_OBJECT_PROPERTIES is the submenu name of this particular toolbar. You can use it to load the toolbar with the TOOLBAR command, as described in Chapter 3. You also use this name to reference the toolbar in other menu items, for example when you want to use the toolbar as a flyout in another toolbar.

Toolbar Titles

As in pull-down menus, the first line after the submenu name defines a title:

```
ID_TbObjpro    [Toolbar("Object Properties", _Top, _Show, 0, 100, 1)]
```

The key word _Toolbar indicates that this is a title definition. The label syntax is:

Table 13.8
Toolbar Title Label Syntax

Field	Description
`title`	The title that the user sees above the toolbar (① in Figure 13.15)
`orientation`	Default orientation of the toolbar (use the key words `Floating`, `Top`, `Bottom`, `Left`, or `Right`)
`show/hide`	Whether to show or hide the toolbar by default (use the key words `Show` or `Hide`)
`horiz_pos`	Default horizontal position of toolbar, measured in pixels from the left edge of the screen
`vert_pos`	Default vertical position of toolbar, measured in pixels from the top edge of the screen
`#_rows`	Default number of rows to display (AutoCAD determines the number of columns from the number of buttons)

`[Toolbar(title, orientation, show/hide, horiz_pos, vert_pos, #_rows)]`

Table 13.8 defines the five fields.

All but the first field are defaults that AutoCAD uses in the absence of similar settings in the [toolbars] section of ACAD.INI. When you load a new menu or reload and recompile an old one, AutoCAD copies these field values from the menu file to ACAD.INI and uses the ACAD.INI values from then on. When you reposition a toolbar, AutoCAD stores the new values in ACAD.INI and uses them the next time you launch AutoCAD with the same ACAD.INI file.

Buttons

The next line in the Object Properties toolbar defines an ordinary button:

`ID_Layers [Button("Layers", ICON_16_LAYERS, ICON_32_LAYERS)]'_ddlmodes`

The label syntax is shown following and defined in Table 13.9:

`[Button(tool_tip, 16_pixel_bitmap, 32_pixel_bitmap)]menu_macro`

The bitmaps can come from a DLL resource file or an individual BMP file, as described in the next section of this chapter.

The [--] label after the first button definition inserts a small space between buttons, and thus acts like a separator line in pull-down menus. You use [--] to group buttons visually.

Flyouts

The flyout button syntax is similar to that of an ordinary toolbar button, but it references another toolbar submenu. The last item in the Object Properties toolbar is a flyout button:

Understanding Menu Structure

Table 13.9
Button Label Syntax

Field	Description
`tool_tip`	The ToolTip that the user sees pop up when the cursor pauses on the button (② in Figure 13.15)
`16_pixel_bitmap`	The 16-pixel icon to display on the button's face when the user has **L**arge Buttons turned off in the Toolbars dialog
`32_pixel_bitmap`	The 32-pixel icon to display on the button's face when the user has **L**arge Buttons turned on in the Toolbars dialog
`menu_macro`	An ordinary menu macro

```
ID_TbInq  [Flyout("Inquiry", ICON_16_LIST, ICON_32_LIST, _OtherIcon, [»]
          ACAD.TB_INQUIRY)]
```

The label syntax is shown below and defined in Table 13.10:

```
[Flyout(tool_tip, 16_pixel_bitmap, 32_pixel_bitmap, own/other, [»]
subtoolbar)] [menu_macro]
```

These menu items had to be printed on two lines to fit on the page, but are on single lines without the [»] continuation characters in the menu file.

The `own/other` field determines the default for what happens each time the user picks an icon on the flyout menu. If `own/other` is `OwnIcon`, then the parent toolbar's icon always remains visible. It acts like a pull-down menu item that opens a cascading menu. If `own/other` is `OtherIcon`, then the flyout icon that the user selects "floats" to the top and displays in the parent toolbar. The advantage to this behavior is that it usually makes flyout toolbars more efficient to use, since the user's most recent choices are visible and accessible with a single click. The disadvantage is that it makes the interface inconsistent. The icon on the parent varies (as does its ToolTip), so unless the user is familiar with all of the icons on the flyout, it can make it hard to keep track of and find flyout tools. All of R13's standard toolbars use `OtherIcon`.

Table 13.10
Flyout Button Label Syntax

Field	Description
`tool_tip`	Same as Table 13.9
`16_pixel_bitmap`	Same as Table 13.9
`32_pixel_bitmap`	Same as Table 13.9
`own/other`	Whether to show own icon always or switch to flyout menu's icon when user picks a flyout menu option (use the key words `OwnIcon` or `OtherIcon`)
`subtoolbar`	Submenu name of the toolbar to display as a flyout
`[menu_macro]`	An ordinary menu macro (optional)

Tip: Putting a check in the Sho<u>w</u> This Button's Icon setting in the Flyout Properties dialog box (right-click twice on a flyout icon) sets any individual flyout to the `OwnIcon` mode. There's a subtle bug to be aware of, though. If you selected a flyout icon while `OtherIcon` was active (before you opened the Flyout Properties dialog), AutoCAD will restore the parent toolbar's icon. If you pick the icon, AutoCAD will run the macro corresponding to the old flyout button, not the currently visible button. To avoid confusion, you should restore the default (parent) button before turning on Sho<u>w</u> This Button's Icon.

You do not add a menu macro after a `_Flyout` button label (if you do, AutoCAD reports a `Menu Syntax Error`). Instead, the icon inherits the macro from the first button defined in the flyout toolbar (that is, the toolbar referenced by the *subtoolbar* field). In the Object Properties example, the Inquiry flyout button references ACAD.TB_INQUIRY, which looks like this:

```
**TB_INQUIRY
**INQUIRY
ID_TbInq     [_Toolbar("Inquiry", _Floating, _Hide, 10, 340, 1)]
ID_List      [_Button("List", ICON_16_LIST, ICON_32_LIST)]^C^C_list
ID_Id        [_Button("Locate Point", ICON_16_ID, ICON_32_ID)]'_id
ID_Dist      [_Button("Distance", ICON_16_DIST, ICON_32_DIST)]'_dist
ID_Area      [_Button("Area", ICON_16_AREA, ICON_32_AREA)]^C^C_area
ID_Masspr    [_Button("Mass Properties", ICON_16_MASSPR, [»]
ICON_32_MASSPR)]^C^C_massprop
```

The mass properties menu item had to be printed on two lines to fit on the page, but is on a single line without the [»] continuation character in the menu file.

When the user picks the Object Properties, Inquiry button, AutoCAD runs the Inquiry, List macro (`^C^C_list`) because the Object Properties, Inquiry button doesn't have a macro of its own. The icons (`ICON_16_LIST`, `ICON_32_LIST`) are defined to be the same for both buttons so that this behavior makes sense to the user. The first time the user opens the flyout and chooses a different button, that button "floats" to the top. Its image and macro replace the Object Properties, Inquiry button (until the user reloads the menu or selects a different tool from the flyout). For example, in Figure 13.16, picking the Distance button ① will cause it to replace the Object Properties, Inquiry button ②.

Controls

Toolbar control menu items are the easiest of all to use. The Object Properties toolbar has all three of the controls that you can use:

`[_Control(_Layer)]` Displays the standard AutoCAD layer control drop-down list box (③ in Figure 13.16)

`[_Control(_Color)]` Runs the DDCOLOR command (④

`[_Control(_Linetype)]` Displays the linetype control drop-down list ⑤

Understanding Menu Structure

Figure 13.16
With OtherIcon, flyout icons replace parent toolbar icons.

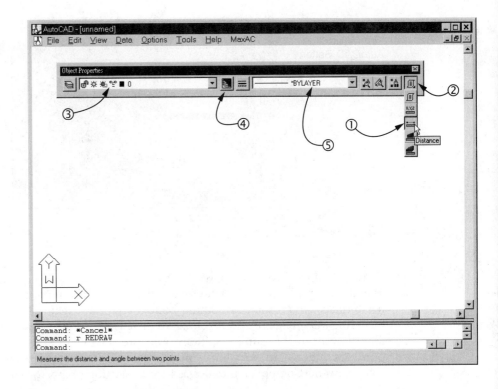

The syntax is simply:

[_Control(*control_name*)]

You don't use menu macros with controls; the controls and DDCOLOR dialog do all of the work themselves.

Icons and DLLs

Once you've mastered the four varieties of toolbar label syntax, the other task is finding or creating bitmap icons. There are two basic approaches:

- Use Autodesk's icons, which are defined in \R13\WIN\ACADBTN.DLL.

- Create your own BMP icons using the Toolbars dialog or a third-party icon editing tool.

The first approach is the easiest, but it limits you to the icons that Autodesk supplied with R13. ACADBTN.DLL contains 16-pixel and 32-pixel versions of almost 400 icons, so you have quite a few choices. The Maximizing AutoCAD Button Image Explorer, which is included on the MaxAC CD-ROM, lets you browse through the icons (see Figure 13.17). You also can extract icons as BMP files for editing.

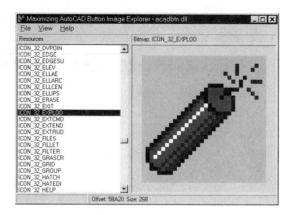

Figure 13.17
The Maximizing AutoCAD Button Image Explorer.

Chapter 3 showed you how to use the AutoCAD Toolbar dialog's Button Editor subdialog to edit bitmap files. It's adequate for small tasks, but if you intend to create more than a few bitmaps, you'll want something better. There are many shareware bitmap editors, including Microangelo™ (available as MUANGL.ZIP in the CompuServe ZDSTUF forum library or from the Impact Software web site on the Internet at http://www.impactsoft.com). In addition, you can find lots of freeware and shareware bitmap collections on CompuServe to use as is or as starting points for your own icons.

As you saw in Chapter 3, AutoCAD can use individual BMP files as toolbar bitmaps. You enter their names, including the BMP extension, as the `16_pixel_bitmap` and `32_pixel_bitmap` label fields and make sure that the BMP files reside in a directory on AutoCAD's library search path. The number of BMP files can get out of hand quickly, though, and it's better to compile them into a single DLL (dynamic link library) resource file. The name of the bitmap resource DLL must match the name of the menu file.

Tools for compiling BMPs into a resource DLL are more difficult to come by than bitmap editors. Borland's Resource Workshop, which comes with Delphi, and Microsoft's AppStudio, which comes with Visual C++, are two such tools.

Developing a Custom Toolbar

The file 13TBSAMP.TXT contains a short sample toolbar section that you can experiment with. It demonstrates how to define flyout toolbars, using the MaxAC customization toolbar from Chapter 3 and a Draw toolbar containing common AutoCAD drawing commands:

```
***TOOLBARS
**TB_SAMPLE
MxaTb_Main  [Toolbar("Sample", _Floating, _Show, 1, 101, 5)]
ID_Draw     [Flyout("Draw", ICON_16_LINE, ICON_32_LINE, _OtherIcon, +
```

Understanding Menu Structure

```
13TEST.TB_DRAW)]
            [--]
MXA_Custom    [_Flyout("Customize", MxMnsE16.bmp, MxMnsE32.bmp, +
_OtherIcon, 13TEST.TB_CUSTOMIZE)]

**TB_DRAW
ID_TbDraw      [_Toolbar("Draw", _Left, _Hide, 0, 0, 11)]
ID_Line        [_Button("Line", ICON_16_LINE, ICON_32_LINE)]^C^C_line
ID_Arc3pt      [_Button("Arc 3 Points", ICON_16_ARC3PT, +
 ICON_32_ARC3PT)]^C^C_arc
ID_Cirrad      [_Button("Circle Center Radius", ICON_16_CIRRAD, +
 ICON_32_CIRRAD)]^C^C_circle
ID_Pline       [_Button("Polyline", ICON_16_PLINE, ICON_32_PLINE)][»]
^C^C_pline
ID_Dtext       [_Button("Dtext", ICON_16_DTEXT, ICON_32_DTEXT)][»]
^C^C_dtext
ID_Mtext       [_Button("Text", ICON_16_MTEXT, ICON_32_MTEXT)][»]
^C^C_mtext
ID_Bhatch      [_Button("Hatch", ICON_16_BHATCH, ICON_32_BHATCH)][»]
^C^C_bhatch
ID_Dinser      [_Button("Insert Block", ICON_16_DINSER, +
 ICON_32_DINSER)]^C^C_ddinsert
ID_XreAtt      [_Button("Attach", ICON_16_XREATT, ICON_32_XREATT)][»]
^C^C_xref _a

**TB_CUSTOMIZE
MXA_MaxAC      [_Toolbar("MaxAC", _Floating, _Hide, 512, 256, 3)]
MXA_MnsEdit    [_Button("Edit Current Menu", MxMnsE16.bmp, [»]
MxMnsE32.bmp)]+
^C^C^P(MxaOpenFile (strcat (getvar "MENUNAME") ".MNS"))
MXA_MnsLoad    [_Button("Load Menu", MxMnsL16.bmp, +
 MxMnsL32.bmp)]^C^C^P_MENU
MXA_LspEdit    [_Button("Edit ACAD.LSP", MxLspE16.bmp, +
 MxLspE32.bmp)]^C^C^P(MxaOpenFile (findfile "ACAD.LSP"))
MXA_LspLoad    [_Button("Load ACAD.LSP", MxLspL16.bmp, +
 MxLspL32.bmp)]^C^C^P(load "ACAD.LSP")
MXA_PgpEdit    [_Button("Edit ACAD.PGP", MxPgpE16.bmp, +
 MxPgpE32.bmp)]^C^C^P(MxaOpenFile (findfile "ACAD.PGP"))
MXA_PgpLoad    [_Button("Re-init ACAD.PGP", MxPgpL16.bmp, +
 MxPgpL32.bmp)]^C^C_RE-INIT 16
```

Several of these menu items had to be printed on two lines to fit on the page, but are on single lines without the [»] continuation characters in the menu file.

The Draw toolbar is taken from various toolbar button definitions in the standard R13 menu. Figure 13.18 shows the toolbar with the Draw flyout. In the next exercise, you'll add these toolbars to 13TEST.MNS and test them. This exercise applies only to R13 Windows—skip it if you're using R13 DOS.

Exercise

Creating a Toolbar with Flyouts

Return to 13TEST.MNS in your text editor.

Delete the ***TOOLBARS section and add the new section listed before this exercise. You can copy the section from 13TBSAMP.TXT.

Save the menu file (as \MAXAC\DEVELOP\13TEST.MNS), return to AutoCAD, and reload it with the MENU command. AutoCAD should display the toolbar automatically (see Figure 13.18).

Command: *Point at the Line icon*	The ToolTip "Line" displays
Command: *Click on the Line icon*	Starts the LINE command
From point: *Draw several line segments*	
Command: *Click on the Line icon and hold down your pointing device button*	Displays the Draw menu as a flyout
Command: *While holding down your pointing device button, move the cursor down to the Polyline icon and click on it*	Starts the PLINE command and replaces the icon in the Sample toolbar with the Polyline icon

You can experiment with other buttons in both flyout menus.

Remain in AutoCAD for the next exercise.

This exercise demonstrated the structure of toolbars and flyouts in the menu file and how you can take existing toolbars and reorganize them with flyouts. You can use this technique to make AutoCAD's standard toolbars more efficient for your work. Similarly, you can create your own new toolbars, flyouts, and buttons. See Chapter 3 for exercises on using AutoCAD's Button Editor dialog box to create and load toolbar bitmaps. And if you have the appropriate bitmap editing and resource DLL compiling tools, you can use them to make and compile your own bitmaps for your custom application toolbars. Compiled bitmaps are more efficient, both in disk space and loading performance.

Understanding Menu Structure

Figure 13.18
A sample toolbar with flyouts (with Large Buttons turned on).

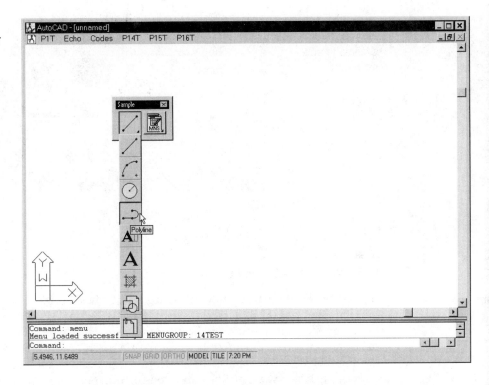

Image Tile Menus

Image tile menus aren't new in R13, but their name is. They were called icon menus in previous releases, but "icon" has come to mean something different in Windows, so Autodesk renamed this menu section. What used to be ***ICON is now ***IMAGE. R13 will still recognize a menu section called ***ICON, but you should switch to the new ***IMAGE section name in all of your custom menus for future compatibility.

Image tile menus are simply a series of AutoCAD slides (SLD or SLB files) with menu macros attached to them. The slide images help users identify and select desired commands (see Figure 13.19). Image tile menus resemble dialog boxes, but are much more limited (and easier to customize) than the programmable dialog boxes that you learned about in Part Four.

Tip: The standard R13 Windows ACAD.MNU file doesn't use any of the ***IMAGE submenus (although it does have them defined in the menu file). To display the image tile menu shown in Figure 13.19, load ACAD.MNU or ACADMA.MNS and enter **(menucmd "I=IMAGE_3DOBJECTS")**, then **(menucmd "I=*")** at the command prompt.

Figure 13.19
AutoCAD's 3D Objects image tile menu.

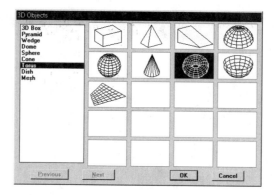

Making Image Tile Menu Slides

In Chapter 8, you learned how to create AutoCAD slide (SLD) files and compile them into slide libraries (SLB files) with the SLIDELIB.EXE program. When you prepare slides for image tile menus, you should center the image and leave a small border around each slide. Keep the images simple because the time required to display a slide depends on the number of vectors that AutoCAD must draw. Slides automatically show solid-filled objects, such as wide polylines and solids, with only their perimeters to simplify the images. You should trace over complex parts or symbols to simplify them. The following steps make suitable slides:

1. Insert the symbol into an empty drawing.

2. Zoom Extents to fill the screen.

3. Determine the center point of the image.

4. Zoom with the Center option to the center point.

5. Zoom out 0.9x to create a small empty border around the symbol.

6. Trace complex parts or symbols, then erase the original and redraw the screen. If any solid (filled) entities must show as filled, turn fill off, regenerate, use SELECT to select them, trace their perimeter lines, use ERASE P to erase the original, and fill them with a user-defined hatch pattern of lines spaced at about 10 pixels on center (see the following tip).

7. Make a slide with MSLIDE, using the same name as the symbol's block name.

Tip: To determine the distance of 10 pixels, zoom Extents in a drawing containing only the symbol to fill, turn snap off, issue the DIST command, press HOME, press ENTER, press the right-arrow key ten times, and then press ENTER.

If the symbols require no editing of the sort described in step 6 of the preceding list, you can automate the process with AutoLISP. The MaxAC CD-ROM includes the program MSLDFLOW.LSP file, for this purpose. You can edit it and replace the FLOW* symbol file names with others to adapt it for making other sets of slides.

```
(foreach caname
  '("FLOWAGIT" "FLOWARRO" "FLOWCHKV" "FLOWDOT" "FLOWDRAN"
    "FLOWGATO" "FLOWINC" "FLOWMOTR" "FLOWPUMP" "FLOWRED"
    "FLOWSAMP" "FLOWSHOW" "FLOWSMVL" "FLOWXMTR"
  )
  (command "_.ERASE" "_ALL" ""
           "_.INSERT" caname "0,0,0" 1 1 0 "_.ZOOM" "_E"
           "_.ZOOM" "_C"
           (mapcar '+ (getvar "extmin") (mapcar '* '(0.5 0.5 0.5)
           (mapcar '- (getvar "extmax") (getvar "extmin"))))
           ".9x" "_.MSLIDE" caname "_.ERASE" "_ALL" "" "_.VSLIDE" caname
  )
)
(command "_.REDRAW")
(princ)
```

Don't worry about the unfamiliar AutoLISP functions. Chapter 9 introduced you to the basics of AutoLISP and this book's companion volume, *Maximizing AutoLISP for AutoCAD R13*, covers more advanced functions like **mapcar**.

The FLOWGATC (closed gate valve) symbol was omitted from the symbols listed in the MSLDFLOW.LSP file because it needs to appear filled to distinguish it from the FLOWGATO (open gate valve). The FLOWGATC.SLD file is already included in this chapter's exercise files. You'll run MSLDFLOW.LSP in the next exercise to create the rest of the flow symbol slides. Then you'll run SLIDELIB.EXE to create the slide library file. Review Chapter 8 if you need help with SLIDELIB.EXE. You'll use the following file listing (in the ASCII file FLOWNAME.TXT) to tell SLIDELIB.EXE which SLD files to compile. Notice the BLANK slide name in the listing; the flow symbol image tile menu formatting requires a blank slide to display a blank image tile.

```
BLANK
FLOWAGIT
FLOWARRO
FLOWCHKV
FLOWDOT
FLOWDRAN
```

```
FLOWGATC
FLOWGATO
FLOWINC
FLOWMOTR
FLOWPUMP
FLOWRED
FLOWSAMP
FLOWSHOW
FLOWSMVL
FLOWXMTR
```

 Exercise

Making Slides and Slide Libraries

Begin a new, blank drawing, discarding changes to the old one.

Command: **MSLIDE** [Enter]   Opens the Create Slide File dialog box

Enter **BLANK** in the File **N**ame edit box, and choose OK   Creates BLANK.SLD

Command: **(load "MSLDFLOW.LSP")** [Enter]   Executes the AutoLISP code in MSLDIMAG.LSP

ERASE, ZOOM, MSLIDE, and VSLIDE commands flash by as the slide files are created.

Next, you'll create the slide library file. Substitute your support path for \R13\COM\SUPPORT if it's different.

Command: **SHELL** [Enter]

OS Command: [Enter]   Opens a DOS window (or shells out from DOS AutoCAD), with C:\MAXAC\DEVELOP as the current directory

C:\MAXAC\DEVELOP> **\R13\COM\SUPPORT\SLIDELIB FLOWSYMB < FLOWNAME.TXT** [Enter]

```
SLIDELIB 1.2   (3/8/89)                        Creates FLOWSYMB.SLB
(C) Copyright 1987-1989,1994,1995
Autodesk, Inc.
      All Rights Reserved SLIDELIB 1.2   (3/8/89)
```

Understanding Menu Structure

If you see any error messages, refer to Chapter 8 for help with SLIDELIB.EXE.

C:\MAXAC\DEVELOP> **EXIT** Enter   Closes the shell window and
                                            returns to AutoCAD

Remain in AutoCAD for the next exercise.

Now that you have the slide library, you're ready to create the image tile menu.

Designing Image Tile Menus

Image tile menus work similarly to the multiple cursor menus that you saw earlier in this chapter. The ***IMAGE menu section can contain multiple submenus, each of which starts with **submenu_name. You use the $I=submenu_name macro code or (menucmd "i=submenu_name") AutoLISP expression to make a particular image tile submenu current. Then you use $I=* or (menucmd "i=*") to display it on the screen.

Tip: A common mistake is to forget to load an image tile submenu into memory with $i=submenu_name before displaying it with $i=*. When you make this mistake, you'll see a blank image tile menu with strange characters in the title bar. You must explicitly load an image tile submenu before displaying it, unlike most other menu sections. AutoCAD doesn't automatically load the first submenu.

Image tile menu item labels require a special syntax to control the item name that appears in the list at the left side of the image tile menu and to reference the image tiles from slide files or a slide library. In this respect, the image tile labels are similar to toolbar labels.

As in the ***POPx and ***TOOLBARS section, the first label in an image tile submenu is its title, which is displayed in the title bar of the image tile menu (see ① in Figure 13.20). The remaining menu item labels supply AutoCAD with the names of the slides to show in the image tile boxes on the screen. The list at the left side of the image tile menu (②) displays these slide names unless you specify alternate label text to display. Each label corresponds to one image tile box on the screen. Several different label syntax formats are available, depending on the information you want to display. You can use slides from an SLB slide library file or from individual SLD slide files. The formats are as follows:

[*slidename*]
[*slidename,labeltext*]
[*libraryfilename(slidename)*]
[*libraryfilename(slidename,labeltext)*]
[*labeltext*]

Figure 13.20
AutoCAD's Spline Fit Variables image tile menu.

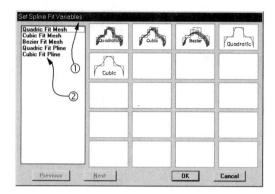

In these formats, `slidename` is the name of an SLD slide file or a slide in a slide library file, `libraryfilename` is the name of a SLB slide library file, and `labeltext` is the alternate text label to display in the listing at the left side of the image tile menu.

The standard AutoCAD image tile "dialog box" displays a listing of the slide names or label text to the left of the picture images. AutoCAD evaluates these item labels as the submenu is displayed on the screen. If you follow the slide name with a comma and some alternate text, AutoCAD displays the alternate text instead of the slide name in the side listing. Thus, you can present the user with more descriptive names than just the slide names. You tell AutoCAD that the slide is from a slide library by placing the slide name (and alternate label text, if any) inside parentheses preceded by the slide library file name outside the parentheses.

Warning: According to the *AutoCAD Customization Guide*, you're supposed to be able to use the following syntax as well:

[blank]
[`labeltext`]

[blank] is supposed to display a blank slide and a separator line in the list on the left. In R13c4, it displays the message `No icon!` in the image tile and the word blank in the list. [`labeltext`] with a leading space is supposed to display a blank slide and `labeltext` in the list. Instead, it too displays `No icon!` in the image tile. If you want a blank slide instead of the `No icon!` message, create a slide called BLANK and reference it in the usual way, as we do in this section.

AutoCAD displays a maximum of 20 slides at a time in a 4-column-by-5-row layout. Slides are laid out horizontally, from left to right, then from top to bottom. If there are more than 20 items in the submenu, AutoCAD activates the **N**ext and **P**revious buttons and automatically handles displaying additional "pages" of icons.

UNDERSTANDING MENU STRUCTURE

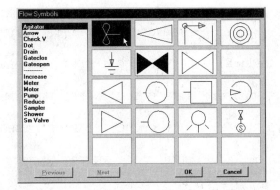

Figure 13.21
The flow symbol image tile menu.

In the next exercise, you'll add the following image menu section and pull-down menu item for displaying the flow symbols menu (see Figure 13.21). These menu sections are contained in 13IMGFLO.TXT:

```
***POP14
[Symbols]
[Flow Symbols]$I=I-FLOWSYMB $I=*

***IMAGE
**I-FLOWSYMB
[Flow Symbols]
[flowsymb(FLOWAGIT,Agitator)  ]^P^C^C^C_INSERT FLOWAGIT \1 ;[»]
\_COPY _L ;_M @
[flowsymb(FLOWARRO,Arrow)     ]^P^C^C^C_INSERT FLOWARRO \1 ;[»]
\_COPY _L ;_M @
[flowsymb(FLOWCHKV,Check V)   ]^P^C^C^C_INSERT FLOWCHKV \1 ;[»]
\_COPY _L ;_M @
[flowsymb(FLOWDOT,Dot)        ]^P^C^C^C_INSERT FLOWDOT \1 ;[»]
\_COPY _L ;_M @
[flowsymb(FLOWDRAN,Drain)     ]^P^C^C^C_INSERT FLOWDRAN \1 ;[»]
\_COPY _L ;_M @
[flowsymb(FLOWGATC,Gateclos)  ]^P^C^C^C_INSERT FLOWGATC \1 ;[»]
\_COPY _L ;_M @
[flowsymb(FLOWGATO,Gateopen)  ]^P^C^C^C_INSERT FLOWGATO \1 ;[»]
\_COPY _L ;_M @
[flowsymb(blank,-----------)  ]$i=*
[flowsymb(FLOWINC,Increase)   ]^P^C^C^C_INSERT FLOWINC \1 ;[»]
\_COPY _L ;_M @
[flowsymb(FLOWXMTR,Meter)     ]^P^C^C^C_INSERT FLOWXMTR \1 ;[»]
\_COPY _L ;_M @
[flowsymb(FLOWMOTR,Motor)     ]^P^C^C^C_INSERT FLOWMOTR \1 ;[»]
```

```
\_COPY _L ;_M @
[flowsymb(FLOWPUMP,Pump)      ]^P^C^C^C_INSERT FLOWPUMP \1 ;[»]
\_COPY _L ;_M @
[flowsymb(FLOWRED,Reduce)     ]^P^C^C^C_INSERT FLOWRED \1 ;[»]
\_COPY _L ;_M @
[flowsymb(FLOWSAMP,Sampler)   ]^P^C^C^C_INSERT FLOWSAMP \1 ;[»]
\_COPY _L ;_M @
[flowsymb(FLOWSHOW,Shower)    ]^P^C^C^C_INSERT FLOWSHOW \1 ;[»]
\_COPY _L ;_M @
[flowsymb(FLOWSMVL,Sm Valve)  ]^P^C^C^C_INSERT FLOWSMVL \1 ;[»]
\_COPY _L ;_M @
```

These menu items had to be printed on two lines to fit on the page, but are on single lines without the [»] continuation characters in the menu file.

A macro following an image tile menu item label works just like any other menu macro. When you select an image tile from this example menu, the macro inserts a symbol, prompting for the position and rotation, and then issues COPY Multiple to duplicate the symbol. The [flowsymb(blank,- - - - - - - - -)] items display a horizontal break in the listing. If the user selects the blank line, its macro issues a $I=* to redisplay the same image menu page.

Exercises

Adding an Image Tile Menu of Flow Symbols

Return to 13TEST.MNS in your text editor.

Delete the ***IMAGE and ***POP14 sections and add the new sections listed before this exercise. You can copy both sections from 13IMGFLO.TXT.

Save the menu file (as \MAXAC\DEVELOP\13TEST.MNS), return to AutoCAD, and reload it with the MENU command.

Command: *Choose* Symbols, *then* Flow Symbols	Displays the Flow Symbols image tile menu (Figure 13.21)
From the image tile menu, double-click on Agitator *or the top left image tile*	Closes the image tile menu and begins insertion of the flow agitator

Complete the insertion, pick a couple of points for COPY Multiple, then press ENTER to exit.

Remain in AutoCAD for the next exercise.

Understanding Menu Structure

Figure 13.22
The Point Style dialog box.

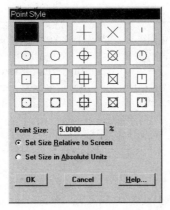

Using Image Tile Menus versus Dialog Boxes

AutoCAD also supports user-definable dialog boxes, which can contain *image buttons* (also called icons, picture boxes, or glyphs). These buttons can look and function much like image tile menus, or they can be quite different and have more flexibility. For two examples, see Figures 13.22 and 13.23.

Image tile menus are much easier to create than dialog boxes. The main advantage of dialog boxes is that you can combine slide images with the many other dialog box interface features. However, dialog boxes require strong AutoLISP, ADS, or ARX-language programming skills and a detailed understanding of AutoCAD's Dialogue Control Language. See Part Four for an introduction to custom dialog boxes, and *Maximizing AutoLISP for AutoCAD R13* for in-depth treatment of custom dialog box development.

Figure 13.23
The Viewpoint Presets dialog box.

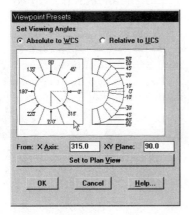

Accelerator Menus

  Although you might not think of keyboard shortcuts as a type of menu interface, the menu file is where Autodesk chose to support them. The advantage of defining keyboard shortcuts in the menu file is that doing so makes it easy for them to reference existing POP*x* menu macros by using their name tags.

 Accelerator keys are another feature that DOS AutoCAD doesn't support.

In Chapter 3, you learned how to define these accelerator keys:

```
***ACCELERATORS
// Maximizing AutoCAD custom accelerators
["DELETE"]^c^c_ERASE
["INSERT"]^c^c_DDINSERT
[SHIFT+"F2"]_FROM
[SHIFT+"F3"]_ENDP
[SHIFT+"F4"]_MID
[SHIFT+"F5"]_INT
[SHIFT+"F6"]_APPINT
[SHIFT+"F7"]_PER
[SHIFT+"F8"]_NOD
[SHIFT+"F9"]_INS
[CONTROL+"F2"]_NEA
[CONTROL+"F7"]_CENTER
[CONTROL+"F8"]_QUA
[CONTROL+"F9"]_TAN
[CONTROL+SHIFT+"F2"]_NON
```

Notice that each item's label (in square brackets) defines the key combination to which you're assigning a macro. This convention is different from most other menu sections, in which labels either define what the user sees on the menu or are ignored by AutoCAD. Also notice that the key names are spelled out (CONTROL instead of CTRL). Remember that key names other than CONTROL and SHIFT must be enclosed in quotation marks (for example, "DELETE", "F2", and "L"). Chapter 3 includes a list of accelerator keys that AutoCAD supports and suggestions for how to implement accelerators.

The syntax used in the previous examples is:

[key_combination]menu_macro

As you saw in the name tag section earlier in this chapter, an alternative syntax substitutes a name tag for the menu macro:

Understanding Menu Structure

name_tag   [key_combination]

The *name_tag* must match a name tag in one of the ***POPx menu sections. In this case, no macro is defined in the accelerator key item, but instead the accelerator key "inherits" the menu macro for the ***POP*x* item corresponding to the *name_tag*. If *name_tag* doesn't match a ***POP*x* name tag, then the accelerator key won't do anything.

In the next exercise, you'll add several more accelerators using both the macro and the name tag syntax. You'll also use some AutoLISP expressions to make more interesting and powerful menu macros.

Exercise

Adding Accelerator Key Definitions

Return to 13TEST.MNS in your text editor. Make the changes to the ***POP1 and ***ACCELERATORS sections shown in bold:

```
***POP1
MxaPop1_T    [P1T]
MxaPop1_1    [Pop1 Item1](progn (princ "Pop1 Item1") (princ))
MxaDDim      [DDim Dialog...]'_DDIM
MxaZ09       [Zoom 0.9x]^P'_.ZOOM 0.9X [Enter]
MxaZ11       [Zoom 1.1x]^P'_.ZOOM 1.1X [Enter]
             [--]

***ACCELERATORS
[CONTROL+"1"](progn (princ "Accelerator for CTRL+1") (princ))
MxaDDim      [CONTROL+SHIFT+"D"]
MxaZ09       ["PAGEUP"] [Enter]
MxaZ11       ["PAGEDOWN"] [Enter]
["HOME"]^P'_ZOOM (getvar "LIMMIN") (getvar "LIMMAX") [Enter]
["END"]^P'_ZOOM (getvar "EXTMIN") (getvar "EXTMAX") [Enter]
```

Save the menu file (as \MAXAC\DEVELOP\13TEST.MNS) and return to AutoCAD, and reload it with the MENU command.

Test the changes by drawing some objects and then pressing the PAGE UP, PAGE DOWN, HOME, and END keys.

Remain in AutoCAD for the next exercise.

The first two new accelerator key definitions (PAGE UP and PAGE DOWN) use the name tag syntax to reference pull-down menu macros. The remaining two definitions (HOME and END) use their own macros instead of name tags. These macros use the AutoLISP **getvar** function to extract system variable values that are fed to the ZOOM command. The HOME macro zooms to the drawing limits, as defined by the LIMMIN and LIMMAX system variables. The END macro zooms to AutoCAD's currently calculated drawing extents, as defined by the EXTMIN and EXTMAX system variables.

You may have noticed that AutoCAD displayed the ***ACCELERATORS section menu macro code on the command line, even though the macros start with ^P to toggle off macro code echoing. Unfortunately R13's accelerator key feature doesn't observe ^P or MENUECHO, even when the accelerator definition references a pull-down macro containing ^P.

Another limitation of the ***ACCELERATORS section is that you can't use the $*section*= codes to control menus or embed DIESEL expressions in macros. Otherwise, a better way to define the HOME and END macros in other menu sections would be:

```
[Zoom Limits ]^P'_ZOOM $m=$(getvar,LIMMIN) $m=$(getvar,LIMMAX)
[Zoom Extents]^P'_ZOOM $m=$(getvar,EXTMIN) $m=$(getvar,EXTMAX)
```

Normally, DIESEL is better than AutoLISP for such macros because it allows you to run the macros (and thus ZOOM) even when another AutoLISP program is active. Unfortunately, you can't use this technique in the ***ACCELERATORS section, because R13 doesn't evaluate the $m= DIESEL calls properly. This limitation applies even when the accelerator definition references a pull-down macro.

Help Strings

As you saw in the Menu Item Name Tags section of this chapter, the ***HELPSTRINGS section of the R13 Windows menu defines help strings that appear on the status bar when you point at toolbar icons or pull-down items (see ① in Figure 13.24). The syntax is simple:

```
name_tag     [help_string]
```

The *name_tag* must be identical to one or more name tags in the ***POP*x* or ***TOOLBARS sections (or both). AutoCAD then displays the *help_string* text on the status bar when you point at these pull-down menu items or tools.

Side-Screen Menus

The ***SCREEN menu section defines AutoCAD's original "screen" menu—the one that appears on the right side of the screen (and which is turned off by default in Windows AutoCAD). We call it the *side-screen menu* to distinguish it from the other menus that appear on the screen (pull-down, cursor, toolbar, and icon).

Understanding Menu Structure

Figure 13.24
A help string.

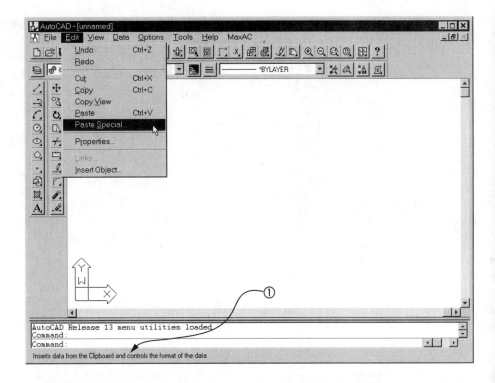

The side-screen menu is rapidly becoming obsolete in Windows AutoCAD. Pull-down menus, cursor menus, and toolbars match the Windows interface paradigm better and offer much greater flexibility. But side-screen menus still have their place, especially in DOS AutoCAD. One advantage is that a side-screen menu stays put. You don't have to keep opening it or wade through cascading pull-downs repeatedly in order to pick the same menu item. Another advantage is that side-screen menus can be context-sensitive, so that they automatically display a menu appropriate to the current command.

 Tip: If you supply fully developed Windows pull-down, cursor, and toolbar menus, then users of your custom menu may not require the side-screen menu. If that's the case, then you can delete the entire, long ***SCREEN section from the menu file. The resulting shorter menu file will load and compile more quickly and won't consume as much of your Windows resources. We've deleted the ***SCREEN sections from most of the Windows sample menu files on the CD-ROM in order to improve menu compile times.

Side-screen menus, like toolbars, make heavy use of submenus. You can think of the side-screen as being divided into "pages" that get swapped into the side-screen menu area, in the same way that we swapped submenus into the ***POP0 cursor menu earlier in the chapter.

Maximizing AutoCAD® R13

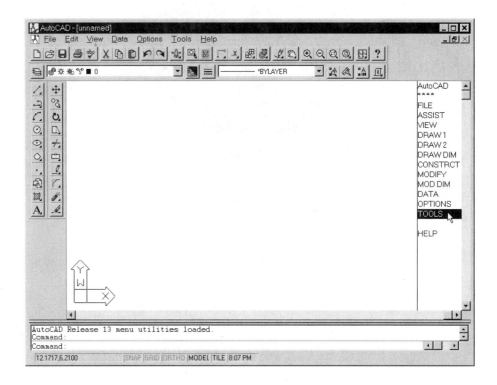

Figure 13.25
AutoCAD's main side-screen menu.

Here's the beginning of the ***SCREEN menu section in the standard R13 ACAD.MNU file, including the blank lines that determine where AutoCAD displays each menu item (see Figure 13.25):

```
***SCREEN
**S
[AutoCAD ]^C^C^P(ai_rootmenus) ^P
[* * * * ]$S=OSNAP
[FILE     ]$S=01_FILE
[ASSIST   ]$S=02_ASSIST
[VIEW     ]$S=03_VIEW
[DRAW 1   ]$S=04_DRAW1
[DRAW 2   ]$S=05_DRAW2
[DRAW DIM]$S=06_DRAWDIM
[CONSTRCT]$S=07_CONSTRUCT
[MODIFY   ]$S=08_MODIFY
[MOD DIM ]$S=09_MODDIM
[DATA     ]$S=10_DATA
[OPTIONS ]$S=11_OPTIONS
```

Understanding Menu Structure

```
[TOOLS     ]$S=12_TOOLS

[HELP      ]$S=13_HELP




[SERVICE]$S=service
[LAST    ]$S=
```

The name of the main side-screen menu is S, as indicated by the **S after ***SCREEN. After that comes a series of menu choices, all of which swap other menus into the side-screen menu area with the $S=submenu_name menu macro code. (The number at the beginning of each submenu name is Autodesk's way of keeping track of the submenus; side-screen submenus need not begin with a number.) The one new code is $S=, corresponding to the LAST menu item. This menu code tells AutoCAD to redisplay the most recently displayed side-screen menu.

The structure of a side-screen menu, defined by which menu pages are called from what other pages, is usually organized in a hierarchical tree, much like a directory tree except that any item on any page can display any other page. When you are designing a complex menu, it helps to draw a diagram of the tree.

Unlike in most other menu sections, AutoCAD treats blank lines in the ***SCREEN section literally: it leaves a blank line on the side-screen menu. For example, the S side-screen menu includes eight blank lines after HELP, which makes the entire menu 26 lines long. The use of blank lines enables a short menu page to "cover" a longer page that it is replacing on the screen.

Warning: In previous versions of AutoCAD, you could use a blank label ([]) in place of a blank line. R13c4 reports Menu Syntax Error when you include blank labels in the ***SCREEN section. However, you can use a label with nothing but a space in it ([ ]).

The standard AutoCAD ***SCREEN menu section defines a very large branching tree of menus. As an example, the TOOLS menu item on the main side-screen menu page calls this menu:

```
**12_TOOLS 3
[Appload:]^C^C_appload
[Script: ]^C^C_script
```

```
[TextScr:]'_textscr

[EXT DBMS]$S=122_EXTDBMS

[Hide:   ]^C^C_hide
[SHADE   ]$S=shade
[RENDER  ]$S=123_RENDER
[Vslide: ]^C^C_vslide
[Mslide: ]^C^C_mslide
[SaveImg:]^C^C_saveimg
[Replay: ]^C^C_replay

[Spell:  ]^C^C_spell
[GeomCal:]'_cal

[Menu:   ]^C^C_menu
[Reinit: ]^C^C_reinit
[Compile:]^C^C_compile
```

Many of these menu items run commands, but a few of them issue $S=*submenu_name* calls to other submenus (and so on, *ad nauseam*). Notice the colons at the end of the labels corresponding to ordinary command names. The colon is an AutoCAD side-screen menu convention to identify items that issue commands. The colon isn't required, but you might want to use it in custom side-screen menus for consistency. Also notice that all of the labels are eight characters wide (filled out with spaces where necessary). Eight is the maximum number of characters that AutoCAD displays in the side-screen menu with many DOS display drivers. If you use more than eight characters, the extra characters are truncated.

Overlaying Side-Screen Menus

If a side-screen menu page has fewer items than the page it is replacing on screen, some of the items on the previous page remain visible and active. This behavior enables you to overlay partial pages and combine pages. You can control the position of the overlaying page by using an offset code. For example, in the preceding submenu (and many ***SCREEN submenus), a vertical *offset* of 3 appears after the submenu name:

```
**12_TOOLS 3
```

The integer 3 tells AutoCAD to start this submenu at the third position from the top, instead of the first position. As a result, the first two choices from the main side-screen menu—AutoCAD and * * *—remain visible and active. You can use this offset code to partially overlay menus in menu sections other than ***SCREEN, but its use is rare outside of the ***SCREEN section.

Understanding Menu Structure

The exercise file 13SCREEN.TXT contains a sample ***SCREEN menu section that demonstrates side-screen menu paging, use of a vertical offset to preserve part of the current submenu, and adding custom commands:

```
***SCREEN
**S_MXA
[AutoCAD ]$S=S_MXA
[* * * * ]$S=OSNAP
[DRAW    ]$S=MXA_DRAW


[MnuEdit:]^c^cEDIT (strcat (getvar "MENUNAME") ".MNU")
[Menu:   ]^C^C_MENU

[LspEdit:]^C^CEDIT (findfile "ACAD.LSP")
[LspLoad:]^C^C(load "ACAD.LSP")

[PGPEdit:]^C^CEDIT (findfile "ACAD.PGP")
[PGPLoad:]^C^C_RE-INIT 16
```



```
[LAST    ]$S=

**MXA_DRAW 3
[Line:   ]^C^C_line
[Arc:    ]^C^C_arc
[Circle: ]^C^C_circle
[Pline:  ]^C^C_pline

[Dtext:  ]^C^C_dtext
[Mtext:  ]^C^C_mtext
[Bhatch: ]^C^C_bhatch

[DDinsrt:]^C^C_ddinsert
```

Maximizing AutoCAD® R13

Figure 13.26
A sample custom side-screen menu.

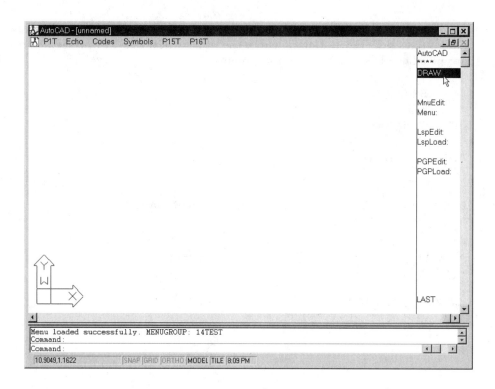

```
[Xref A:  ]^C^C_xref _a

**OSNAP 3
[Osnap:   ]'_osnap

[DDosnap  ]'_ddosnap

[App Int  ]_appint

[CENter   ]_cen^z
[ENDpoint ]_endp^z
[INSert   ]_ins^z
[INTersec ]_int^z
[MIDpoint ]_mid^z
[NEArest  ]_nea^z
[NODe     ]_nod^z
[PERpend  ]_per^z
[QUAdrant ]_qua^z
[TANgent  ]_tan^z
[NONE     ]_non^z
```

Understanding Menu Structure

```
Quick,     ]_qui,^z

[FROM      ]_from
```

  The file editing items on this screen menu (`MnuEdit:`, `LspEdit:`, and `PgpEdit:`) use the ACAD.PGP external command definition for the DOS EDIT.COM editor, so that they work properly in DOS AutoCAD. In R13 Windows, you could substitute macros from the Toolbar Menus section earlier; for example, `(MxaOpenFile (findfile "ACAD.LSP")` instead of `EDIT (findfile "ACAD.LSP")`. The `MxaOpenFile` function opens the file with the Windows application corresponding to its extension (for example, TextPad or LispPad).

In the next exercise, you'll experiment with this sample side-screen menu.

Exercise

Adding a Custom Side-Screen Menu

Return to 13TEST.MNS in your text editor.

Delete the ***SCREEN section and add the new section listed before this exercise. You can copy this section from 13SCREEN.TXT.

If you're using R13 Windows, change the line:

```
[MnuEdit:]^c^cEDIT (strcat (getvar "MENUNAME") ".MNU")
```

to:

```
[MnsEdit:]^c^cEDIT (strcat (getvar "MENUNAME") ".MNS")
```

Save the menu file (as \MAXAC\DEVELOP\13TEST.MNS), return to AutoCAD, and reload it with the MENU command.

`Command:` *On the side-screen menu, choose* DRAW, *then* Line: Displays the Draw menu and issues the LINE command

`From point:` *Draw several segments*

`To point:` Enter   Ends the LINE command

Command: *Choose* LAST   Returns to the main menu

You can experiment with other side-screen menu options if you like.

Remain in AutoCAD for the next exercise.

Context-Sensitive Side-Screen Menu Paging

If the ***SCREEN menu section contains a submenu name that matches the name of an AutoCAD command, then AutoCAD displays that submenu when you issue the command. This context-sensitive side-screen menu paging works no matter how you issue the command. For example, here's the side-screen menu that AutoCAD displays when you issue the PLINE command:

```
**PLINE 3
[Pline:   ]^C^C_pline

[Arc     ]_a
[Close   ]_c
[Halfwid ]_h
[Length  ]_l
[Undo    ]_u
[Width   ]_w

[Angle   ]_a
[Center  ]_ce
[Close   ]_cl
[Direct'n]_d \
[Line    ]_l
[Radius  ]_r
[2nd PT  ]_s \

[Pedit   ]^C^C_pedit
```

The context-sensitive paging feature is controlled by the MENUCTL system variable. The default is 1 (on—enabled); set it to 0 if you want to disable it.

 Tip: Chapter 14 shows you how to create context-sensitive cursor menus that mimic the MENUCTL = 1 behavior of side-screen menus.

Understanding Menu Structure

If you decide to develop custom side-screen menus, study and experiment with the ***SCREEN section in R13's ACAD.MNU. Long-time DOS users are likely to be familiar with its structure and operational conventions, so you can use them as guidelines for your own efforts.

Tablet Menus

The remaining menu sections—***TABLET1 through ***TABLET4—control up to four digitizer tablet menu areas. If you don't have a digitizer, you can skim or skip this section.

Like the side-screen menu, tablet menus are waning in popularity as more AutoCAD users move to Windows. Some users are giving up digitizers in favor of mice. A digitizer offers other advantages besides tablet menus, including absolute positioning (a point on the tablet always maps to the same point on the screen), higher accuracy, and additional pointing device buttons. AutoCAD even supports rubber-sheeting to correct distortion when using the digitizer for tracing. But even users who appreciate these features are being driven towards toolbars, pull-down menus, and cursor menus for command input, and you can use a digitizer for tracing and graphic input without using it for tablet menus.

Tip: You can remove the ***TABLET section from your menu file if users of your custom menu don't use digitizer tablets. The ***TABLET section isn't quite as large as the ***SCREEN section, but removing it will shorten menu compile and load times.

Nonetheless, tablet menus still appeal to many AutoCAD users, especially DOS users. The standard AutoCAD tablet areas are well-known to many DOS users. Custom tablet menus are a convenient place to put large numbers of symbols (blocks, and so on) because users can see all of them at a glance.

The standard R13 menu file uses all four tablet areas, which are arranged around the screen pointing area as shown in Figure 13.27. You can see the individual tablet square assignments by looking at the tablet overlay that comes with AutoCAD or opening the sample drawing \R13\COM\SAMPLE\TABLET.DWG.

There's nothing sacred about this arrangement of the four areas, but it's been the standard one for many AutoCAD releases, so you probably should use it as the basis for any tablet customization that you do. The ***TABLET1 area at the top of the digitizer isn't used by AutoCAD, so it's a good place for custom tablet menu items.

If you do have a tablet, make a copy of Figure 13.27 or plot 13TABLET.DWG at 1:1 scale and tape it on your tablet anywhere outside the screen pointing area. You'll reconfigure one tablet menu area in the next exercise. The reconfiguration affects only the configuration files you copied into the \MAXAC directory. It does not change your normal AutoCAD setup. (If you were going to use this ***TABLET1 area in a real custom menu, you would integrate it with AutoCAD's other ***TABLET*x* areas and maintain the layout shown in Figure 13.27.)

Figure 13.27
The standard AutoCAD tablet menu areas.

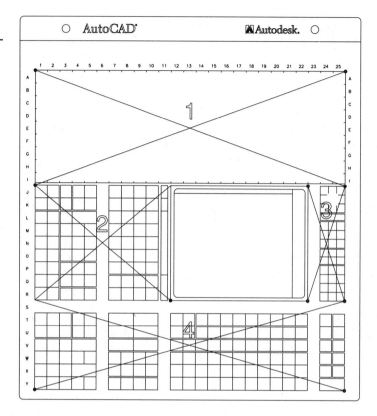

 Tip: You can copy, modify, and plot the TABLET.DWG to create your own custom tablet overlay.

The exercise file 13TABLET.TXT contains a ***TABLET1 section, shown in the following listing.

```
***TABLET1
**T-FLOWSYMB
[1A AGITATOR  ]^P^C^C^C_INSERT FLOWAGIT \1 ;\_COPY _L ;_M @
[2A ARROW     ]^P^C^C^C_INSERT FLOWARRO \1 ;\_COPY _L ;_M @
[3A CHECK V   ]^P^C^C^C_INSERT FLOWCHKV \1 ;\_COPY _L ;_M @
[4A DOT       ]^P^C^C^C_INSERT FLOWDOT  \1 ;\_COPY _L ;_M @
[5A DRAIN     ]^P^C^C^C_INSERT FLOWDRAN \1 ;\_COPY _L ;_M @
[6A ]
[1B GATECLOS  ]^P^C^C^C_INSERT FLOWGATC \1 ;\_COPY _L ;_M @
[2B GATEOPEN  ]^P^C^C^C_INSERT FLOWGATO \1 ;\_COPY _L ;_M @
```

Understanding Menu Structure

```
[3B INCREASE    ]^P^C^C^C_INSERT FLOWINC \1 ;\_COPY _L ;_M @
[4B METER       ]^P^C^C^C_INSERT FLOWXMTR \1 ;\_COPY _L ;_M @
[5B MOTOR       ]^P^C^C^C_INSERT FLOWMOTR \1 ;\_COPY _L ;_M @
[6B ]
[1C PUMP        ]^P^C^C^C_INSERT FLOWPUMP \1 ;\_COPY _L ;_M @
[2C REDUCER     ]^P^C^C^C_INSERT FLOWRED \1 ;\_COPY _L ;_M @
[3C SAMPLER     ]^P^C^C^C_INSERT FLOWSAMP \1 ;\_COPY _L ;_M @
[4C SHOWER      ]^P^C^C^C_INSERT FLOWSHOW \1 ;\_COPY _L ;_M @
[5C SM VALVE    ]^P^C^C^C_INSERT FLOWSMVL \1 ;\_COPY _L ;_M @
[6C ]
```

This menu section relabels and reorganizes the flow symbol macros that you used in the image tile menu exercise to fit the 6 x 6 tablet configuration defined in the following exercise. Notice that the items in the menu are ordered sequentially across the first row, then the second row, and so on. The first five boxes across each of the top three rows contain symbols. The numbers and letters in the labels correspond to the rows and columns in the tablet menu template layout. Box 5C has the macro for the block FLOWSMVL. The exercise instruction "*Choose* [5C SM VALVE]" means to pick the fifth column, third row. Because the location of macros on the tablet depends on the location of the menu items in the MNS file, you should always number and label items, even blank ones.

 Exercise

Adding a Custom Tablet Menu

Return to 13TEST.MNS in your text editor.

Delete the ***TABLET1 section and add the new section listed before this exercise. You can copy this section from 13TABLET.TXT.

Save the menu file (as \MAXAC\DEVELOP\13TEST.MNS), return to AutoCAD, and reload it with the MENU command.

Make sure that you've attached the tablet menu overlay shown in Figure 13.28 to your digitizer, leaving room for a screen pointing area.

```
Command: TABLET Enter
```

```
Option (ON/OFF/CAL/CFG): CFG Enter        Starts tablet area configuration
```

```
Enter number of tablet menus              Specifies only one tablet menu area
desired (0-4) <4>: 1 Enter
```

Figure 13.28
A small tablet menu template.

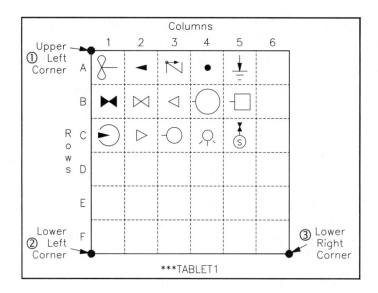

Digitize upper left corner of menu area 1: *Pick ① in Figure 13.28*

Digitize lower left corner of menu area 1: *Pick ②*

Digitize lower right corner of menu area 1: *Pick ③*

Enter the number of columns for menu area 1: **6** Enter

Enter the number of rows for menu area 1: **6** Enter

Do you want to respecify the screen pointing area? <N> *Press ENTER (or enter* **Y** *if you need to respecify it)*

If you are using a WinTab tablet driver, you will see the following prompt. For more information on the Floating Screen *pointing area and WinTab drivers, see the AutoCAD Installation Guide for Windows.*

Do you want to respecify the Floating Screen pointing area? <N> Enter

The flow symbols are now installed on the tablet.

Command: *From the tablet menu,, choose* [5C SM VALVE]   Executes the macro and issues _INSERT FLOWSMVL

Understanding Menu Structure

```
Command: INSERT Block name (or ?): FLOWSMVL
```

Pick points to insert it and then try some others.

Close 13TEST.MNS in your text editor.

Remain in AutoCAD for the next exercise.

As you can see, the ***TABLET*x* areas are simple to customize. But be aware that tablet menu customization requires keeping two custom components in synch: the menu source file and the paper tablet overlay.

Combining Menus by Loading Partial Menus

  Thus far in this chapter, and throughout most of the book, we've kept our custom menu in one MNS file. R13 Windows includes a new *partial menu loading* capability with which you can organize custom menu components in additional MNS files and load them "on top of" a standard base menu such as ACAD.MNS. You got a glimpse of this approach in Chapter 10, where the cursor and pull-down menus were attached as partial menus, but all of the details were handled transparently by that chapter's support files.

Partial menus are an excellent way to develop custom menus because they let you keep common menu elements in a single, standard base menu file and attach custom menu elements as needed. Partial menus also allow you to use the same customization with different base menus, which is especially useful when you develop applications for users who have their own base menus.

Be aware, though, that partial menus require an additional level of sophistication on your part. AutoCAD won't load your partial menus and MNL files automatically, as it does with a single, monolithic custom menu file. You must use ACAD.LSP or other initialization files to make sure that your partial menus are attached and displayed properly. We demonstrate these techniques in the remaining chapter.

In this chapter we introduce partial menu loading and show you how to attach partial ***TOOLBARS, ***POP*x*, and ***AUX*x* menus. Chapter 14 covers more advanced partial menu loading techniques.

 DOS AutoCAD doesn't support partial menu loading. You must continue to keep your entire DOS menu in one MNU file.

Partial menu loading introduces the concept of a *base menu* (before R13, all AutoCAD had was a base menu). The base menu is the one you load with the MENU command. AutoCAD stores its name in the MENUNAME system variable, which corresponds to the `MenuFile` entry in the ACADNT.CFG file. AutoCAD automatically loads this base menu file every time you launch the program (unless the Preferences dialog's Use Menu in Header setting is turned on, in which case AutoCAD loads the base menu whose name is stored in the DWG file).

The MENULOAD Command

You use the new MENULOAD command to attach a partial menu file to the base menu. The interactive version (the Menu Customization dialog box) is shown in Figure 13.29, but usually you'll write AutoLISP expressions that run the command-line version of MENULOAD and do the work of loading partial menus without user intervention. You can use the Menu Customization dialog box to experiment with partial menus. If you want to use the command-line version of MENULOAD interactively, set the FILEDIA system variable to zero first.

When you attach a partial menu with MENULOAD, AutoCAD automatically loads its toolbars, but displays nothing else. You can use the Menu Customization dialog's Menu Bar tab to add ***POP*x* menu sections and subsections to the pull-down menu bar (or remove them from the menu bar), as shown in Figure 13.30. This approach is fine for some simple experimentation, but it lacks many features. The dialog doesn't let you control other sections (for example, ***AUX or ***ICON). Also, it is cumbersome to add a pull-down menu to the end of the list (that is, to the right of all current pull-down menus), or to reorder the list. To do so, you must choreograph inserting and removing menus with the Insert and Remove buttons while highlighting the appropriate pull-down menus in the Menus and Menu Bar lists. When you choose the Insert button, the currently selected menu in the Menus list is placed above the currently selected menu in the Menu Bar list.

In the next exercise, you'll load ACADMA.MNS as your base menu and attach a stripped-down version of the menu you've been working on in this chapter as a partial

Figure 13.29
The new Menu Customization dialog (MENULOAD command).

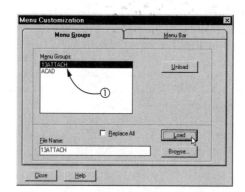

Figure 13.30
Adding and removing pull-down menus with the Menu Customization dialog.

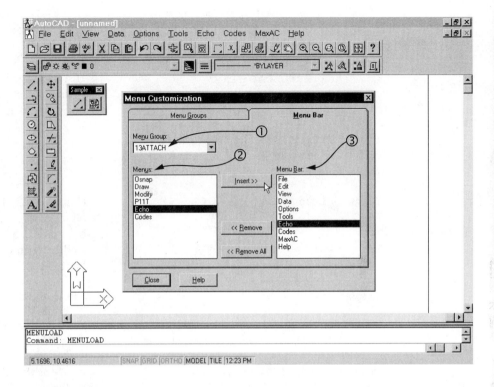

menu. The partial menu is called 13ATTACH.MNS, and it contains the ***AUX*x*, ****BUTTONS*x*, ***POP*x*, and ***TOOLBARS sections from 13TEST.MNS. Its menu group name is 13ATTACH.

Note: The ability to do partial loading of some menu sections wasn't added until R13c3. If you have problems getting partial menus to work, make sure that you have the current R13c*x* release.

Exercise

Attaching a Partial Menu

Return to the AutoCAD drawing editor.

If the side-screen menu is still on, turn off the Screen Men**u** check box in the Preferences dialog box.

Use the MENU command to load \MAXAC\DEVELOP\ACADMA.MNS, replacing all of the 13TEST.MNS menus.

Command: **MENULOAD** [Enter]	Displays the Menu Customization dialog, with only ACAD in the currently loaded Menu Groups list
Enter **13ATTACH** *in the* File Name *edit box, or use the* Browse *button to select* 13ATTACH.MNS *from the file dialog box*	
Choose the Load *button*	Compiles 13ATTACH.MNS, attaches its menu sections to the base menu, and adds 13ATTACH to the Menu Groups list (see ① in Figure 13.29)
Choose the Close *button*	Closes the Select Menu File dialog

Notice that the pull-down menus didn't change. Toolbars are the only partial menu elements that AutoCAD displays automatically.

Command: **MENULOAD** [Enter]	Reopens the Menu Customization dialog
Choose the Menu Bar *tab*	Displays the dialog tab for adding menus to the pull-down menu bar
Make sure that 13ATTACH *is selected, in the* Menu Group *drop-down list (see ① in Figure 13.30)*	Controls the Menus ② available for adding to the Menu Bar list ③

Notice that the Menus list includes ***POP0 submenus (Osnap, Draw, and Modify). AutoCAD doesn't distinguish between ***POP0 and the other ***POP*x* sections here.

Choose MaxAC *in the* Menu Bar *list*	Defines the pull-down page before which you'll add new pages
Choose Codes *in the* Menus *list*	Defines the pull-down you'll add
Choose the Insert *button*	Adds Codes to the Menu Bar list and to the pull-down menu bar
Choose Echo *in the* Menus *list and add it in the same way*	Adds Echo to the Menu Bar list and to the pull-down menu bar

Understanding Menu Structure

Your screen should look like Figure 13.30.

*Choose the **C**lose button*   Closes the Select Menu File dialog

Remain in AutoCAD for the next exercise.

This exercise showed you how to load a partial menu and manipulate some of its menu sections. Of course, you wouldn't want to have to go through the same procedure every time you launched AutoCAD in order to use a partial menu.

Partial Menu Loading with AutoLISP

You use the **command** function to load a partial menu and the **menucmd** function to manipulate its sections. The following series of AutoLISP expressions will perform the same partial menu loading steps that you performed in the previous exercise:

```
(command "_.MENULOAD" "13ATTACH")
(menucmd "P7=+13attach.pop2")
(menucmd "P8=+13attach.pop3")
```

The syntax for adding a menu to the pull-down bar is:

```
(menucmd "Px=+group_name.menu_name")
```

where *menu_name* can be a menu section name (for example, pop12), subsection name, or name tag (for example, MxaMyPop). AutoCAD inserts the new menu page at pull-down position POP*x*, moving any existing menu pages to the right. If the *menu_name* is already displayed on the pull-down menu bar, AutoCAD doesn't move it.

In order to remove a menu from the menu bar, you use the following syntax:

```
(menucmd "Px=-")
```

For example, this expression removes any menu page that's currently at pull-down menu location 10:

```
(menucmd "P10=-")
```

Note: In professional application custom menus, you should avoid referencing menus solely by pull-down menu bar position, since you don't know for sure what other menus might've been added by the user or another application. Chapter 14 discusses more robust techniques.

In the next exercise, you'll use AutoLISP expressions to manipulate the 13ATTACH partial menu. AutoCAD retains your partial menu configuration throughout a drawing session, including when you start new or open existing drawings. When you exit and restart AutoCAD, it loads only the base menu (ACADMA.MNS in this case). Instead of restarting AutoCAD, you'll use the Menu Configuration dialog's Unload button.

Exercise

Controlling a Partial Menu with AutoLISP

Return to the AutoCAD drawing editor.

Command: **MENULOAD** [Enter]

Choose 13ATTACH *in the* Menu Groups *edit box*

Choose the Unload *button*	Unattaches 13ATTACH.MNS and removes all of its menu sections from the pull-down menu bar
Choose the Close *button*	Closes the Select Menu File dialog
Command: **(command "_.MENULOAD" "13ATTACH")** [Enter]	Reattaches 13ATTACH.MNS to the base menu and redisplays the Sample toolbar
Command: **(menucmd "P7=+13attach.pop2")** [Enter]	Adds Echo to the pull-down menu bar
Command: **(menucmd "P8=+13attach.pop3")** [Enter]	Adds Codes to the pull-down menu bar

Next, you'll automate this procedure using ACAD.LSP and 13ATTACH.MNL.

Open \MAXAC\DEVELOP\ACAD.LSP and add the following lines near the end of the file, just before the final (princ):

(defun s::startup () [Enter]
   (command "_.MENULOAD" "13ATTACH") ↵
   (princ) ↵
) [Enter]

Understanding Menu Structure

See Chapter 14 for more about the `s::startup` function.

Create a new text file and add these lines:

```
;; 13ATTACH.MNL [Enter]
;; by MM   14 Apr 1996 [Enter]
(menucmd "P7=+13attach.pop2") [Enter]
(menucmd "P8=+13attach.pop3") [Enter]
(menucmd "A1=13attach.aux1") [Enter]
(menucmd "A2=13attach.aux2") [Enter]
(menucmd "A3=13attach.aux3") [Enter]
(menucmd "A4=13attach.aux4") [Enter]
(menucmd "B1=13attach.buttons1") [Enter]
(menucmd "B2=13attach.buttons2") [Enter]
(menucmd "B3=13attach.buttons3") [Enter]
(menucmd "B4=13attach.buttons4") [Enter]
(princ "\n13ATTACH.MNL loaded.""") [Enter]
(princ) [Enter]
```

Save the file as \MAXAC\DEVELOP\13ATTACH.MNL. This file activates the ***AUX*x* and ***BUTTONS*x* menus that you developed earlier.

Exit and restart AutoCAD. AutoCAD loads ACADMA.MNS as the base menu, attaches 13ATTACH.MNS, and loads 13ATTACH.MNL.

Test the menus and button assignments (CTRL+Button1 and CTRL+SHIFT+Button1). They should pop up the custom cursor menus from earlier in this chapter.

Remain in AutoCAD until you've completed the Finishing Up steps at the end of the chapter.

Your pull-down menu bar should look just as it did in Figure 13.30, at the end of the previous exercise, but now all of the partial menu procedures are automatic.

Partial Menus and MNL Files

When you attach a partial menu file, AutoCAD looks for an MNL (Menu LISP) file of the same name and, if it finds one, loads it. This convention is the same as for base menu files. Thus the MNL file is a good place to put **menucmd** expressions that activate sections of your partial menu.

Warning: There's a subtle bug in R13c4 that you'll encounter if you attach more than one partial menu at a time. On AutoCAD startup, only the MNL file associated with the last partially loaded menu loads. MNL files associated with any other partially loaded menus don't load until you reinitialize the drawing editor (by starting a new drawing or opening an existing one). This bug won't affect you if you attach only one partial menu, since it will be the last one. See Chapter 14 for workarounds.

Deciding Which Menu Interfaces to Use

The design of an effective menu system depends on your application. Here are some tips for using the different kinds of AutoCAD menus.

Pull-down menus are good for single-pick menu actions because they disappear after each pick. Pull-down menus also are good for giving users descriptive text in a menu item label. Cascading menus help you organize and group menu options. The dynamic label options available by using DIESEL with pull-downs allow you to develop a more sophisticated and responsive interface.

Button menus use the mouse or digitizer puck buttons. The ability to have user-defined macros at your fingertips is one of AutoCAD's best features. Buttons are perfect for items that affect the picking of points and objects, because picking naturally uses the cursor. Coordinating buttons and the SHIFT and CTRL keys with several cursor menus offers the most dynamic and interactive menuing system. Because they are related to pull-downs, cursor menus also offer cascading submenus and label formatting features.

Toolbars are new in R13 but are rapidly gaining favor, in part because of their popularity in other Windows programs. Flyout toolbars provide for a more streamlined toolbar interface without using up too much precious screen space. Bear in mind that small icons often aren't as descriptive as pull-down menu labels. ToolTips and help strings will help users become acquainted with your custom toolbar icons, but careful icon design and good toolbar organization are important, too.

Image tile menus are excellent graphic reminders of your symbols, and thus are well-suited to block libraries and other collections of graphical objects. Image tile menus are less static and more flexible than tablet menus, but they take a moment to appear on the screen.

The side-screen menu always remains visible (if you have it turned on), and that's both its strength and its weakness. You can use the context-sensitive paging feature to display menus appropriate to AutoCAD commands, or you can turn off MENUCTL and have side-screen menus stay put when you select them.

Tablet menus can divide your digitizer into four rectangular areas, each containing a different set of commands. Tablet menus enable you to you use templates to organize and display a large number of macros. Tablet menus are fairly static and require that you edit both the menu file and the tablet overlay.

Don't forget the keyboard when you design your interface. Chapter 3 covered ACAD.PGP command aliases, simple AutoLISP command shortcuts, and accelerator keys defined in the menu file. You can assign one- or two-character abbreviations to commands and custom programs. Once you learn the abbreviations, you can access any command with quick one-hand typing, while using the other hand for mouse or digitizer input.

Finishing Up

Use the MENU command to change the menu back to \MAXAC\ACADMA.MNS.

In this chapter, you created the 13TEST.MNS custom menu and associated support files (slide library, block DWG files, MNL file, and modified ACAD.LSP file). If you want to keep any of this work, move it to the \MAXAC directory, then delete or move any remaining files out of \MAXAC\DEVELOP. If you turned on the side-screen menu in Windows, you'll probably want to make sure that you turned it back off with the Preferences dialog.

Conclusion

Menu sections and submenus help you organize and integrate commands and macros in a logical way. You can use the submenu structure to inform users of program settings and to control your program's operation. Pages of symbols make accessing and inserting them into the drawing faster and easier for the user. Through the design of your menu pages, you can group macros by application task instead of AutoCAD function. These techniques make drawing straightforward and efficient. A menu system designed with specific pages for the pull-down, cursor, toolbar, image tile, button, side-screen, and tablet menus takes best advantage of the different features offered by each device.

Chapter 14 ties together the techniques you've studied throughout this book. You'll learn how to integrate DIESEL, AutoLISP, and menus into unified application functions, and you'll pick up some additional customization techniques and tricks along the way.

chapter 14

Creating an Integrated Professional Menu System

In the preceding chapters of this book, you learned how to use most of AutoCAD's customization interfaces, but mostly in isolation from one another. If you want to develop advanced menus and custom functions, you'll often need to use two or more customization interfaces. This chapter shows you how to integrate your customization, especially menus and AutoLISP, into more sophisticated application components.

You've already learned some useful integration techniques:

- Chapter 3 showed how to create keyboard macros with simple AutoLISP command definitions.

- Chapter 7 demonstrated techniques for incorporating text characters and shapes in custom linetypes.

- Chapter 8 included a scripted slide show that combined scripts and slides. Chapter 8 also showed how to automate plotting with scripts, PCP files, and operating system commands.

- Chapter 10 covered how to use DIESEL expressions in menu labels and menu macros.

- Chapter 13 showed how to build image tile menus from slides.

This chapter covers additional integration techniques that will help you develop more efficient and useful menus. The material in this chapter applies whether you're developing a menu for your own use only or a menu system that will be part of a professional application and used by thousands of people. But how you approach menu development, and thus how you use this chapter's material, will depend on your goals. For a "personal" menu, you can use any of AutoCAD's support files as you see fit. For a more widely distributed professional menu, you need to be more circumspect. Your custom components should be self-contained and shouldn't interfere with files and settings that users might want to customize on their own systems.

Maximizing AutoCAD® R13

Note: Throughout this chapter, we use "personal menu" to describe a custom menu that you develop for your own use and "professional menu" to describe a custom menu that you develop for wide distribution. Your situation may fall in between these extremes, but you can use the personal and professional menu suggestions to fashion a reasonable in-between development strategy.

Figure 14.1 diagrams the components of a sample integrated menu system.

The most important new R13 feature in this chapter was introduced in Chapter 13: partial menus. Here we explore some additional partial menu loading techniques and discuss some of the problems with this feature. Remember that R13 supports partial menus in Windows only.

This chapter also covers help files, which are different in R13. AutoCAD for Windows now supports the standard Windows help file format, which allows for a more powerful and friendly help system. In addition, all AutoCAD platforms use a revised platform-independent AutoCAD help file format. The revised format is closer to Windows help, so the concepts you learn for developing either kind of custom AutoCAD help files will apply to the other kind as well.

Most of the techniques described in this chapter require AutoLISP, and therefore won't work in AutoCAD LT. One exception is the Creating a Context-Sensitive Cursor Menu section, which LT users should find just as

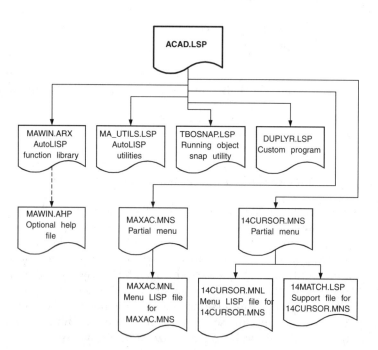

Figure 14.1
Menus and associated support files.

Creating an Integrated Professional Menu System

useful as AutoCAD users will. Also, there is no ACLT.AHP help file in AutoCAD LT, nor any way to call custom AutoCAD help files, other than Windows-specific HLP files (sometimes called WinHelp).

About This Chapter

In this chapter, you will learn how to do the following:

- Integrate DIESEL, AutoLISP, and menus into unified application functions
- Develop a menu LISP (MNL) file that supports your custom menu
- Employ more sophisticated partial menu loading techniques
- Use `autoload` with AutoLISP programs to automate program loading
- Control scale in menu macros and AutoLISP programs
- Create a context-sensitive cursor menu
- Improve the usefulness of the object snap cursor menu and toolbar with an AutoLISP program
- Use function libraries provided by ADS and ARX programs
- Develop application-specific help files

These are the tools and files you will use for the exercises in this chapter:

- TextPad (or another ASCII text editor)
- DUPLYR.LSP, a program to duplicate objects on a different layer
- TBOSNAP.LSP, a program to add running object snap functionality to cursor and toolbar menus
- MAXAC.MNS and MAXAC.MNL, a partial menu and menu LISP file with MaxAC toolbar and pull-down menus
- 14CURSOR.MNS and 14CURSOR.MNL, a partial menu and menu LISP file for context-sensitive cursor menus
- 14MATCH.LSP, a support file for 14CURSOR.MNS
- 14CURSOR.DWG, a chapter exercise drawing for context-sensitive menus
- MAWIN.ARX, an ARX file containing useful AutoLISP functions
- MAWIN.LSP, an AutoLISP support file for MAWIN.ARX

- MAWIN.AHP, a sample platform-independent help file

- ACAD.LSP, ACAD.PGP, ACADMA.MNS, ACADMA.MNL, MA_UTILS.LSP, and custom bitmap (BMP) icons from the MaxAC CD-ROM

Application Initialization

As you saw in Chapter 4, AutoCAD's initialization procedure is complex. But it does give you a variety of ways to ensure that your application and all of its components are loaded properly.

The ACAD.LSP File

AutoCAD loads the ACAD.LSP file (assuming that one exists on the library search path) every time the user launches AutoCAD or changes to a different drawing. Thus ACAD.LSP is a good place for AutoLISP functions that you want to have loaded always, regardless of what menus are loaded at any given time.

In Chapter 3, you created a simple ACAD.LSP file with command definitions and a load message. If you're creating a personal custom menu, you can add the following kinds of things to ACAD.LSP:

- Simple command definition macros, as described in Chapter 3

- AutoLISP utility programs that you use often

- A general-purpose error handler function (**error**—see Chapter 9)

- Startup settings

You should avoid adding too much to ACAD.LSP, though. An inordinately long ACAD.LSP file adds to the load time for every drawing. When inexperienced users find out about ACAD.LSP, they sometimes add dozens of AutoLISP programs they've collected, whether or not they need those programs in every drawing session. There are more efficient methods to load AutoLISP programs automatically when theyre needed, as described later in this chapter.

Another disadvantage of a long ACAD.LSP file with many different kinds of functions is that it's difficult to debug and maintain. If you do want to have some functions loaded at all times, you can keep ACAD.LSP more orderly by putting major functions or groups of related functions in separate LSP files.

Creating an Integrated Professional Menu System

For example, instead of this:

```
;;;ACAD.LSP file:
(defun c:SOMEUTL1 ()
   lots of AutoLISP code here
)
(defun c: SOMEUTL2 ()
   lots of AutoLISP code here
)
(defun c: SOMEUTL3 ()
   lots of AutoLISP code here
)
;;;end of ACAD.LSP
```

You could break ACAD.LSP into four files like this:

```
;;;ACAD.LSP file:
(load "SOMEUTL1.LSP")
(load "SOMEUTL2.LSP")
(load "SOMEUTL3.LSP")
;;;end of ACAD.LSP

;;;SOMEUTL1.LSP file:
(defun c:SOMEUTL1 ()
   lots of AutoLISP code here
)
;;;end of SOMEUTL1.LSP

;;;SOMEUTL2.LSP file:
(defun c:SOMEUTL2 ()
   lots of AutoLISP code here
)
;;;end of SOMEUTL2.LSP

;;;SOMEUTL3.LSP file:
(defun c:SOMEUTL3 ()
   lots of AutoLISP code here
)
;;;end of SOMEUTL3.LSP
```

This approach allows you to see at a glance what gets loaded from ACAD.LSP, and it makes your AutoCAD initialization procedure more modular. You can develop and debug each of the individual SOMEUTLx programs separately. You can turn the loading of any of them off temporarily by commenting out a single **load** line in ACAD.LSP.

If you're developing a professional application menu, you should keep ACAD.LSP changes to a minimum. ACAD.LSP is primarily the *user's* customization file, and filling it up with dozens of lines of code is bad behavior for a professional application. Of course, replacing ACAD.LSP is an even worse offense. At most, your application should add a line or two of code that loads the application:

```
;;;ACAD.LSP file:
user's AutoLISP code here
(load "CUSTMAPP.LSP")   ;load your custom application
;;;end of ACAD.LSP
```

Keep in mind that loading your application from ACAD.LSP causes it to load always, whether the user wants it loaded or not. Depending on the nature of your application, you might want to hold off on loading most of its components until the user enters one of the application's commands or otherwise indicates that the application is required.

If your professional application does need to include a line in ACAD.LSP, then you need to make sure that it gets added properly. A commercial application should include an installation program that searches for and/or prompts the user for ACAD.LSP and appends to it automatically. It should also give the user the option of making the changes manually, and inform the user of the required changes. Keep in mind that some users may have more than one ACAD.LSP file in their hard disks.

Warning: Another bad application initialization habit is making changes that alter the DBMOD system variable. DBMOD (database modification) is a read-only system variable that indicates whether the drawing has been changed since the last save (or open), and if so, how. As long as DBMOD is zero, there are no unsaved changes in the drawi"ng, and AutoCAD will let the user exit (to open another drawing, for example) without prompting to save changes. If DBMOD is other than zero, then there are unsaved changes and AutoCAD will ask the user whether to save them. (See Appendix E for a description of DBMOD values.) Applications don't have to modify objects in the drawing in order to trigger DBMOD; changing a system variable, modifying layer settings, or zooming will trigger DBMOD also.

Some ill-behaved applications make these kinds of changes every time they initialize. The consequence is that AutoCAD always prompts the user to save changes, even when the user did nothing more than open the drawing. This alteration to AutoCAD's standard file-handling behavior annoys users and deprives them of an important cue for deciding whether or not to save their work.

Note: AutoCAD is a highly user-customizable program, and interaction or conflicts between various custom and commercial applications may cause problems or impose limitations. Therefore, it is extremely important that commercial applications avoid unnecessary modifications to the user's AutoCAD environment and system, and that they document the modifications they do make.

Creating an Integrated Professional Menu System

The `s::startup` Function

The drawing editor isn't yet fully "awake" when AutoCAD loads the ACAD.LSP file: AutoCAD's command processor isn't ready to accept input until after ACAD.LSP has loaded. As a result, you can't use the **command** function in the body of ACAD.LSP (outside of a function definition). AutoCAD reports a `Command list interruption` error if you do. For example, the following expressions in ACAD.LSP all cause errors:

```
;;;Turn off blips:
(command "_.BLIPMODE" "_OFF")
;;;Zoom to limits:
(command "_.ZOOM" (getvar "LIMMIN") (getvar "LIMMAX"))
;;;Load application menu:
(command "_.MENULOAD" "MYAPP")
```

The following use of **command** in ACAD.LSP is acceptable because it's inside a function definition, and AutoCAD doesn't execute the contents of the function until the user enters **ZL**:

```
(defun c:ZL ()
   (command "_.ZOOM" (getvar "LIMMIN") (getvar "LIMMAX"))
   (princ)
)
```

For those situations where you do need to use the **command** function in ACAD.LSP, AutoCAD provides the **s::startup** function name. **s::startup** is a "magic" name. When you define an **s::startup** function, AutoCAD stores the function in memory and then executes it at the end of the drawing initialization sequence, when the command processor is active, but before the user gets control. Thus the examples that caused a `Command list interruption` error will work if you wrap them in an **s::startup** function:

```
(defun s::startup ()
   ;;Turn off blips:
   (command "_.BLIPMODE" "_OFF")
   ;;Zoom to limits:
   (command "_.ZOOM" (getvar "LIMMIN") (getvar "LIMMAX"))
   ;;Load application menu:
   (command "_.MENULOAD" "MYAPP")
   (princ)
)
```

Note that ZOOM Limits and BLIPMODE trigger DBMOD, so you wouldn't want to use them in a professional application.

Maximizing AutoCAD® R13

Tip: An alternative to `(command "_.BLIPMODE" ...)` is to use the **setvar** function:

`(setvar "BLIPMODE" 0)`

You can use **setvar** anywhere in ACAD.LSP; it doesn't have to be inside the **s::startup** function. The same caveat applies to `(command sysvar_name ...)` and **setvar**: changing the system variable triggers DBMOD.

Although it's common to define **s::startup** in ACAD.LSP, the other option is to define it in a menu LISP (MNL) file (see the next section). An MNL file is automatically loaded when a corresponding menu is loaded. The MNL option is useful for menu-based applications that need to ensure a particular initialization sequence each time the user changes drawings.

There can be only one function definition for **s::startup**, so all applications need to be careful not to replace any **s::startup** function that might be defined in ACAD.LSP or another MNL file. Your application should check whether the symbol **s::startup** is **null**, and if not, append to it. For example:

```
(if (null s::startup)
   ;;no s::startup defined yet, so define a new one:
   (defun s::startup ()
      (command "_.MENULOAD" "MYAPP")
      (princ)
   )
   ;;s::startup already defined, so append to it:
   (setq s::startup
      (append s::startup (list
         '(command "_.MENULOAD" "MYAPP")
         '(princ)
      ))
   )
)
```

Later in this chapter you'll see examples of using ACAD.LSP, MNL files, and **s::startup** to support custom menus.

Supporting Menus with MNL Files

ACAD.LSP is the place for AutoLISP expressions that you want to have loaded in every drawing session, but menu LISP (MNL) files are the place for menu-specific AutoLISP support code. When AutoCAD loads a base or partial menu (for example, ACAD.MNS or ACAD.MNU), it searches the library search path for a file with the same name and an

Creating an Integrated Professional Menu System

MNL extension (ACAD.MNL in this example). If AutoCAD finds the matching MNL file, it loads just as though you'd entered **(load "ACAD.MNL")** at the command prompt.

There is no difference in syntax between the expressions in an ordinary LSP file and those in an MNL file. Everything you learned about AutoLISP in Chapter 9 applies to MNL files. The only differences are the file-name extensions and the purposes of the files. LSP is for general-purpose AutoLISP files, while MNL is for AutoLISP files that support menus.

Warning: Because of a bug in R13c4, if you launch AutoCAD in a minimized state, the MNL file associated with the base menu isn't loaded automatically. If you regularly launch AutoCAD minimized, you can add (or ai_tiledvp_chk (load "ACAD.MNL")) to the **s::startup** function in your ACAD.LSP file in order to ensure that the MNL file gets loaded. **ai_tiledvp_chk** is the first function defined in ACAD.MNL. If it's null, that means that ACAD.MNL didn't load it, so the **or** expression will load it explicitly.

This workaround doesn't completely fix the bug, because AutoCAD will load the menu and MNL files in reverse order when you launch the program minimized (MNL then menu, instead of menu then MNL). If you modify ACAD.MNL to include **menucmd** partial menu loading instructions, they won't have any effect when you first launch the program minimized. In order to fix *this* bug, put the expression (while (not (menucmd "P1.1=?"))) before (or ai_tiledvp_chk (load "ACAD.MNL")) in the **s::startup** function:

```
(while (not (menucmd "P1.1=?")))
(or ai_tiledvp_chk (load "ACAD.MNL"))
```

The empty **while** loop forces AutoCAD to wait until the user restores the AutoCAD program window before loading ACAD.MNL. This delay fixes the load order problem.

MNL "Helper" Functions

When Autodesk introduced MNL files in R12, they were intended as a way to clean up and simplify menu files. Users and application developers were creating increasingly lengthy menu macros, which were hard to debug and made for unwieldy menu files. MNL files provide a place for "helper" functions that then can be referenced in the menu file. Because the MNL files are ordinary AutoLISP, they are easier to develop, read, and debug than long strings of menu macro code. AutoLISP functions also provide greater functionality and flexibility than menu macros.

For example, in Chapter 5, you created the following macro:

```
[Duplicate Layer]^C^C^C_.SELECT \_.COPY _Previous ;0,0 ;+
_.CHPROP _Previous ;_LAyer \;
```

To convert this macro to a function in an MNL file, you would define a function with **defun**:

```
(defun c:DUPLYR (/ SelSet)
   (setq SelSet (ssget))     ;get selection set from user
   ;;copy selected entities with 0,0 displacement:
   (command "_.COPY" "_Previous" "" "0,0" ""
   ;;change originally selected entities to a different layer:
           "_.CHPROP" "_Previous" "" "_LAyer" pause "")
   (princ)
)
```

Then you would change the Duplicate Layer menu macro so that it called the new function:

```
[Duplicate Layer]^C^C^CDUPLYR
```

This approach has several advantages. You can reference the **c:DUPLYR** function in more than one place in the menu file, so that users can start it from different menu interfaces. You can comment your function easily. And you can extend the function without it becoming unwieldy. This last advantage is especially compelling when you want to add error-trapping. It's difficult to include much error-trapping in menu macros, but AutoLISP removes that restriction. For example, we could rewrite the **c:DUPLYR** function like this:

```
(defun c:DUPLYR (/ SelSet NewLayer)
   (princ "\nDuplicate objects on a different layer...")
   (setq SelSet (ssget))      ;get selection set from user
   (cond (SelSet              ;if there are entities in selection set
      ;;get target layer name:
      (setq NewLayer (getstring "\nLayer for copied objects: "))
      (cond ((/= "" NewLayer)     ;if user supplied a layer name
         ;;create layer if it doesn't exist:
         (if (null (tblsearch "LAYER" NewLayer))
            (command "_.LAYER" "_New" NewLayer "")
         )
         ;;copy selected entities with 0,0 displacement:
         (command "_.COPY" "_Previous" "" "0,0" ""
         ;;change originally selected entities to a different layer:
                 "_.CHPROP" "_Previous" "" "_LAyer" NewLayer "")
      ))
   ))
   (princ)
)
```

The revised function provides helpful prompts for the user and exits gracefully if the user doesn't select entities or supply a layer name.

ACAD.MNL

R13c4's standard ACAD.MNL file includes 19 functions, which are listed in Table 14.1. Notice that most of the function names begin with the prefix **ai_** (Autodesk Inc.). This convention helps avoid function name conflicts with other applications. You should use a similar naming scheme in your own MNL files.

ACAD.MNL performs the following tasks, in addition to defining the functions listed in Table 14.1:

- Restores any alternate tablet menus from ACADNT.CFG
- Sets the MENUCTL system variable to 1 (see Chapter 13)
- Loads the AutoVision menu if AutoVision is installed (Windows only)

Using MNL Files and ACAD.LSP with Partial Menus

MNL files and ACAD.LSP take on added importance with R13's partial menus. As you saw at the end of Chapter 13, partial menus allow you to append menu sections to the currently loaded base menu. This modularity makes partial menus especially appropriate for custom application menus. You can add custom menu components without replacing the user's preferred menus. Partial menus also are appropriate for personal menu customization. Because they are smaller and more self contained, partial menus are easier to edit and debug, quicker to compile, and easier to integrate with AutoCAD updates.

One complication of partial menus is that you must instruct AutoCAD to load and display them each time the user launches AutoCAD—they don't persist across AutoCAD sessions in the way that base menus do. Chapter 13 showed the basic techniques: first load the MNS file with (command "_.MENULOAD" menuname), then display the desired sections with a series of **menucmd** function calls. The **menucmd** lines usually belong in the menu file's associated MNL file. The (command "_.MENULOAD" menuname) line goes in the **s::startup** function in ACAD.LSP if you want the partial menu to be loaded in every drawing session. Alternatively, you can define a command that allows the user to load the partial menu when it's needed. The next section of this chapter demonstrates these techniques in greater detail.

MNL Files vs. ACAD.LSP

In many cases, you have the choice of putting application code in an MNL file or in ACAD.LSP. Here are some guidelines to help you decide:

- Professional applications should keep ACAD.LSP additions to a minimum, as described earlier in this section.

Table 14.1
ACAD.MNL Functions

Function	Purpose
`ai_tiledvp_chk`	Displays an icon menu if TILEMODE = 1
`ai_tiledvp`	Creates tiled viewport layouts
`ai_tab1`	Swaps alternative menu into tablet menu area 1
`ai_tab2`	Swaps alternative menu into tablet menu area 2
`ai_tab3`	Swaps alternative menu into tablet menu area 3
`ai_tab4`	Swaps alternative menu into tablet menu area 4
`ai_popmenucfg`	Retrieves menu configuration data from ACADNT.CFG
`ai_putmenucfg`	Stores menu configuration data in ACADNT.CFG
`*merr*`	Restores original error handler
`*merrmsg*`	Displays an error message and restores original error handler
`c:rectang`	Draws a polyline rectangle (see Figure 14.2)
`ai_showedge_alert`	Displays alert dialog about invisible edges
`ai_hideedge_alert`	Displays alert dialog about invisible edges
`ai_rootmenus`	Displays root side-screen menus
`ai_fms`	Changes to "floating model space" (model space in a paper space viewport)
`c:ai_propchk`	Runs DDCHPROP or DDMODIFY
`ai_onoff`	Toggles a value from 0 to 1 or vice versa
`ai_dim_cen`	Centers dimension text
`set_alt_tabs`	Sets alternate tablet menus based on ACADNT.CFG settings

Creating an Integrated Professional Menu System

Figure 14.2
Example boxes drawn with the RECTANG command, defined in ACAD.MNL.

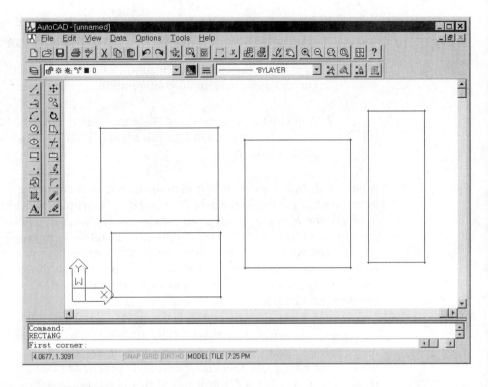

- Menu-related code usually belongs in an MNL file.

- Code that you use independently of any particular menu (for example, keyboard macros or general-purpose AutoLISP utilities) usually belongs in ACAD.LSP or in a file loaded by ACAD.LSP.

Partial Menu Loading Techniques

Chapter 13 demonstrated a simple method of attaching a partial menu automatically:

1. Include (command "_.MENULOAD" menu_file_name) in the **s::startup** function in ACAD.LSP.

2. Add a **menucmd** expression to the menu's MNL file for each menu component that you want to activate, where the syntax for **menucmd** is:

    (menucmd "Px=+group_name.menu_name")

This method works fine in some situations, but it has two disadvantages, especially for professional applications:

Maximizing AutoCAD® R13

- It assumes that you know where to insert your partial menu's pull-down menu pages (that is, that you know the current pull-down menu configuration). If your application is used with different base menus and perhaps with other partial menus, you can't depend on knowing the current pull-down menu configuration.

- AutoCAD reports an error (`A menu with that MENUGROUP name already exists`) when you try to MENULOAD a partial menu that's already loaded.

A more controlled way to add pull-down menus is to count the existing ones and then insert yours at the location where you want it. (Unfortunately, R13 doesn't give you a way to determine what position a given menu is in, but you can count them.) This approach gives you a way to ensure that your application's pull-down menu(s) get added in a known position relative to all of the existing menus. For example, you can add your pull-downs after all of the existing ones, or before the last one (which should be the Help menu and should remain as the right-most pull-down menu in order to conform with Windows interface standards).

Two functions included in MA_UTILS.LSP accomplish the tasks of attaching a partial menu only if it hasn't been attached already and inserting a pull-down menu page before the current final page: **MxaMenuLoad** and **MxaInsertMenu**. The AutoLISP code for these functions is beyond the scope of this book, but you can use the functions easily without understanding the internal details.

In the first exercise, you'll try the **MxaMenuLoad** and **MxaInsertMenu** functions interactively. The syntax for these functions is described after the exercise.

Warning: Because the exercises in this section deal with partial menu loading, they apply to R13 Windows only. If you're using R13 DOS, skip the partial menu loading exercises.

 Exercise

Attaching a Partial Menu with MxaMenuLoad and MxaInsertMenu

Copy all of the files from \MAXAC\BOOK\CH14 to your \MAXAC\DEVELOP directory.

Start AutoCAD with a new, blank drawing. Notice the files that are loaded automatically:

Creating an Integrated Professional Menu System

```
Menu loaded successfully. MENUGROUP: ACAD
Loading C:\MaxAC\Develop\acad.lsp...
Ma_Utils.Lsp loaded.
ACAD.LSP loaded.
...
AutoCAD Release 13 menu utilities
AcadMa.mnl loaded.
```

MxaMenuLoad and **MxaInsertMenu** are defined in Ma_Utils.Lsp, which was loaded from ACAD.LSP.

Command: **(MXAMENULOAD "MAXAC" "MAXAC")** [Enter]   Attaches the menu file MAXAC.MNS with the menu group MAXAC

```
Menu loaded successfully. MENUGROUP: MAXAC
T
```
AutoCAD reports success and **MxaMenuLoad** returns **T**

```
Command:
MaxAc.mnl loaded.
```

Command: **(MXAINSERTMENU "MAXAC.POP1")** [Enter]   Displays the POP1 menu from the MAXAC menu group before the Help menu

7   **MxaInsertMenu** returns the location where it inserted the new pull-down

Command: *Open the new menu (Figure 14.3)*

Notice what happens when you try to reattach MAXAC.MNS with the AutoCAD MENULOAD command:

Command: **(COMMAND "MENULOAD" "MAXAC")** [Enter]   AutoCAD tries to attach MAXAC.MNS

```
Unable to load menu: MAXAC
A menu with that MENUGROUP name already exists.
nil
```

Enter name of menu file to load: *Press CANCEL*

The **MxaMenuLoad** function doesn't have this problem:

Command: **(MXAMENULOAD "MAXAC" "MAXAC")** [Enter]

Figure 14.3
The pull-down menu from MAXAC.MNS.

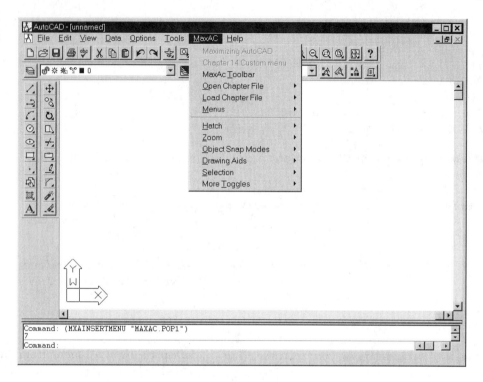

    " "

MxaMenuLoad returns an empty string, which indicates that the menu was loaded already

Remain in AutoCAD for the next exercise.

The syntax for **MxaMenuLoad** is:

(MxaMenuLoad *menu_file_name* *menu_group_name*)

menu_file_name is the name of the MNS file and *menu_group_name* is the group name defined in that menu. If you supply **nil** for the *menu_group_name*, **MxaMenuLoad** assumes that the group name is the same as the menu file name. Thus in the previous exercise, you could have entered:

(MXAMENULOAD "MAXAC" nil)

MxaMenuLoad relies on a "dummy" ***POP*x* menu section containing one menu item in order to detect whether the menu has been loaded already. The dummy section never gets displayed, but it must have a menu item with a name tag whose name corresponds to the menu group name in this way:

MENU_*menu_group_name* [Anything];

The menu item label and macro are unimportant. All that matters is that the name tag for the second item below the ***POP*x* section label be MENU_*menu_group_name* (the first item is the menu title and can be anything). The dummy menu we've defined in MAXAC.MNS is:

```
***POP99
ID_DUMMY        [Menu Identifier]
MENU_MAXAC      [Menu Id];
```

Again, the important thing is that the name tag for the first menu item is MENU_MAXAC, since MAXAC is the group name for this menu.

The syntax for **MxaInsertMenu** is:

(MxaInsertMenu *group_name.menu_section_name*)

MxaInsertMenu counts the number of existing pull-down menus (using **MxaCountMenus**, another function from MA_UTILS.LSP) and inserts the specified menu before the last one. The assumption is that the last pull-down menu is the Help menu and it should remain the right-most menu.

Unfortunately, if you are also loading poorly behaved applications, they may place their pull-down menus after the Help menu and mess up the pull-down menu order. If you can't fix their menu loading code, try to load your menus first, then let them load theirs.

In the next exercise, you'll incorporate **MxaMenuLoad** and **MxaInsertMenu** into ACAD.LSP and MAXAC.MNL.

Exercise

Attaching and Displaying a Partial Menu Automatically

Continue from the previous exercise, with all of the files from \MAXAC\BOOK\CH14 in your \MAXAC\DEVELOP directory.

Launch your text editor and open \MAXAC\DEVELOP\ACAD.LSP. Add the following **s::startup** function near the end of the file, just before (princ "\nACAD.LSP loaded.\n"):

Figure 14.4
The MaxAC customization toolbar from MAXAC.MNS.

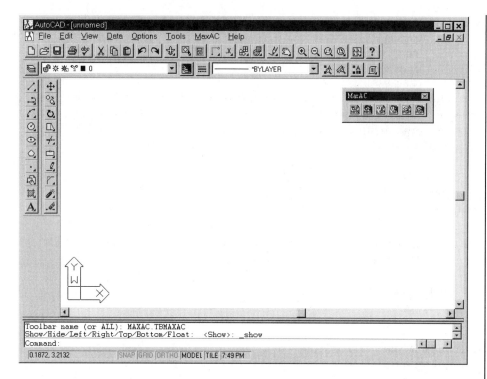

```
(defun s::startup () [Enter]
   (MxaMenuLoad "MAXAC" nil) [Enter]
   (princ) [Enter]
) [Enter]
```

Save ACAD.LSP and open \MAXAC\DEVELOP\MAXAC.MNL. Add the following line to the top of the file, before (princ "\nMaxAc.mnl loaded.\n"):

(MxaInsertMenu "MaxAC.pop1") [Enter]

Save MAXAC.MNL. Exit and relaunch AutoCAD to test the changes. Your pull-down menus should look like Figure 14.3.

Command: *Choose* **M**axAC, MaxAC **T**oolbar   Displays the MaxAC customization toolbar (see Figure 14.4), which is defined in MAXAC.MNS

Remain in AutoCAD for the next exercise.

Creating an Integrated Professional Menu System

Warning: Partial menu loading is new to R13, and it suffers from several significant bugs in R13c4. If you launch AutoCAD in a minimized state, AutoCAD loads the base menu *after* the partial menu, thus obliterating any partial menus that you might have loaded in ACAD.LSP. As a result, your partial menus won't be loaded when you unminimize AutoCAD. You can restore the partial menus by opening or starting a new drawing, which causes AutoCAD to re-execute the partial menu loading code in ACAD.LSP. Obviously this isn't a very elegant workaround. A better one is to force AutoCAD to delay partial menu loading until the user restores the AutoCAD program window, as described in an earlier warning. Put the expression `(while (not (menucmd "P1.1=?")))` before any partial menu loading code in the `s::startup` function. For example:

```
(while (not (menucmd "P1.1=?")))
(MxaMenuLoad "PARTMNS1" nil)
```

Another bug occurs if you attach more than one partial menu at a time, in which case only the MNL file associated with the last partially loaded menu loads when you launch AutoCAD. MNL files associated with any other partially loaded menus don't load until you reinitialize the drawing editor (by starting a new drawing or opening an existing one). If you anticipate using more than one partial menu at a time, load the MNL files explicitly in your ACAD.LSP file. To avoid loading the MNL file twice, you can use the technique described earlier in the chapter for ACAD.MNL: after using MENULOAD to load the partial menu file, check whether a function or variable that's defined in the MNL file is null. If it is null, you know that the MNL file didn't get loaded. For example:

```
(defun s::startup ()
    ;;delay until AutoCAD Window is unminimized:
    (while (not (menucmd "P1.1=?")))
    ;;make sure ACAD.MNL gets loaded:
    (or ai_tiledvp_chk (load "ACAD.MNL"))
    ;;load first partial menu:
    (MxaMenuLoad "PARTMNS1" nil)
    ;;check whetherMNL was loaded -
    ;;NOTE: must have(setq Mnl1Loaded T) in PARTMNS1.MNL:
    (or Mnl1Loaded
        (progn
            (princ "\nLoading PARTMNS1.MNL explicitly.")
            (load "PARTMNS1.MNL")))
    ;;load second [and final] partial menu:
    (MxaMenuLoad "PARTMNS2" nil)
    ;;AutoCAD loads last MNL automatically,
    ;;so don't need to check
    (princ)
)
```

Automatic Program Loading

Many applications rely on a large number of AutoLISP, ADS, and/or ARX programs. If the application loads all of these programs when it initializes, the user will have to endure a long delay at every initialization. The long delay is even more unnecessary if some of the programs don't get used in the current drawing session. A more efficient technique is to load each program automatically the first time the user executes it. Several short delays as each program loads usually are much less noticeable and disruptive than a single long delay.

Most techniques for "on demand" automatic loading of AutoLISP, ADS, and ARX programs check whether a function or global variable that's defined in the program file is null, and if so, load the file. One of the easiest ways to accomplish this feat is to use R13's **autoload**, **autoxload**, and **autoarxload** functions.

Using `autoload`, `autoxload`, and `autoarxload`

Autodesk provides three useful functions for loading AutoLISP, ADS, and ARX programs automatically the first time they're executed: **autoload**, **autoxload**, and **autoarxload**. These functions are defined in \R13\COM\ACADR13.LSP, which AutoCAD R13 loads automatically every time you change drawings (see Chapter 4).

The syntax for all three functions is the same; only the function name varies:

```
(autoload file_name '(command_name_1 [command_name_2 ...]))
```

- **autoload** is the function name that you use for AutoLISP program files. You would substitute **autoxload** or **autoarxload** for an ADS or ARX program file, respectively.

- *file_name* is the name of the LSP, EXE (EXP in DOS), or ARX program file.

- *command_name_1 [command_name_2 ...]* are the names of one or more commands defined in the program file. In the case of an AutoLISP program file, these command names will be (defun c:*xxx* () ...) style command definitions. If the file contains only one defined command, then you use only the *command_name_1* parameter. Notice that the *command_name_x* arguments must be contained in a list, even if there's only one of them, and the list must be preceded by an apostrophe to prevent evaluation.

ACADR13.LSP contains numerous examples of using **autoload**, **autoxload**, and **autoarxload**, including:

```
(autoload "ddmodify" '("ddmodify"))
```

Creating an Integrated Professional Menu System

AutoCAD's DDMODIFY is an AutoLISP program (DDMODIFY.LSP) and associated DCL file (DDMODIFY.DCL). This **autoload** expression is the mechanism whereby DDMODIFY gets loaded (① in Figure 14.5). The first time you enter the DDMODIFY command (that is, the command_name_1 parameter) in a drawing session, AutoCAD loads the associated AutoLISP file name (the file_name parameter). AutoCAD then executes the command name that you entered (the command_name_1 parameter again).

Here's a more complex example from ACADR13.LSP:

```
(autoload "3d" '("3d" "3d" "ai_box" "ai_pyramid" "ai_wedge"
            "ai_dome" "ai_mesh" "ai_sphere" "ai_cone"
            "ai_torus" "ai_dish")
)
```

3D.LSP defines 10 commands (3D, AI_BOX, AI_PYRAMID, and so on). This **autoload** expression makes sure that when the user enters any of the 10 command names, AutoCAD loads 3D.LSP and then executes the desired command.

The **autoload** function and its companion functions for ADS and ARX applications work by defining a temporary **c:command_name** function for each of the command names in the '(command_name_1 [command_name_2 ...]) list. The temporary

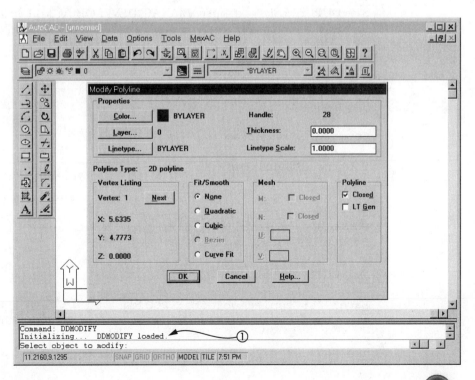

Figure 14.5
DDMODIFY gets loaded by autoload.

function instructs AutoCAD to search the library search path for the program file specified in the *file_name* parameter. If AutoCAD finds the file, the temporary function loads it and then executes the **c:***command_name* that you entered. If AutoCAD doesn't find the file, the temporary function issues an error message. After the program file is loaded, the temporary functions are replaced by their normal definitions, just as if you'd entered **(load** *file_name***)** at the command prompt.

In the following exercise, you'll add an **autoload** expression to ACAD.LSP for DUPLYR. You'll also create a shorter alias for DUPLYR.

Exercise

Creating an autoload Expression

Open \MAXAC\DEVELOP\DUPLYR.LSP in your text editor and add the following two lines at the end:

(defun c:DPL () (c:DUPLYR)) [Enter]
(princ) [Enter]

Save DUPLYR.LSP. Next, open \MAXAC\DEVELOP\ACAD.LSP and add the following line near the end, just before (princ "\nACAD.LSP loaded.\n"):

(autoload "DUPLYR" '("DUPLYR" "DPL")) [Enter]

Save ACAD.LSP and return to AutoCAD. Start a new drawing, which causes AutoCAD to load the modified ACAD.LSP.

Command: *Draw some lines*

Command: **DUPLYR** [Enter]

Initializing...   **autoload** loads DUPLYR.LSP

Duplicate objects on a different layer...
Select objects: *Select some of the lines*

Select objects: [Enter]

Layer for copied objects: **FOO** [Enter]   DUPLYR creates the layer FOO
                                              and moves the original objects to it

Command: **DPL** [Enter]

Creating an Integrated Professional Menu System

> Notice that there is no `Initializing...` prompt this time, because DUPLYR.LSP is already loaded.
>
> ```
> Duplicate objects on a different
> layer...
> Select objects:
> ``` *Complete the command as before*
>
> Remain in AutoCAD for the next exercise.

By adding **autoload** expressions to ACAD.LSP or MNL files in this way, you can make a large number of custom programs available without causing the overhead of loading all of them in every drawing session. In this example, we could have used just one command name:

```
(autoload "DUPLYR" '("DUPLYR"))
```

Sometimes, though, you want to provide a shorter alias for commonly used commands. **autoload** provides a simple way to run the command, whether the user enters the long name or the alias. In order for the alias to work, you must define a command that calls the long command name:

```
(defun c:DPL () (c:DUPLYR))
(princ)
```

This example defines a new command DPL whose only action is running the function **c:DUPLYR** (that is, our custom DUPLYR command). The **princ** ensures that DUPLYR.LSP loads quietly.

Warning: If you forget to define the alias in the program file and then try to use it in AutoCAD, **autoload** causes an infinite loop of repeated `Initializing...` prompts that you'll have to cancel out of.

When you've used **autoload** to make a command available, you can include the command name (or its alias, if any) in a menu, just as you would do with a native AutoCAD command:

```
[Duplicate Layer]^C^CDUPLYR
```

Automatic Loading from Menu Macros

If you want to make a command available only from a menu choice, you can load it automatically in this way:

```
[Duplicate Layer]^C^C^C^P(if (null c:DUPLYR) (load "DUPLYR"))^P +
DUPLYR
```

The `(if (null c:DUPLYR)` test determines whether DUPLYR.LSP has been loaded. If not (that is, if **c:DUPLYR** is **nil**), then `(load "DUPLYR")` loads the AutoLISP file. If **c:DUPLYR** is non-nil, that means that DUPLYR.LSP has been loaded already, and AutoLISP skips the rest of the expression, thus avoiding the overhead of reloading the file. In either case, the menu macro executes the DUPLYR command.

You can shorten the AutoLISP expression as follows:

```
[Duplicate Layer]^C^C^C^P(or c:DUPLYR (load "DUPLYR"))^P +
DUPLYR
```

Because AutoLISP stops evaluating an `or` expression when it reaches the first non-nil argument, it won't get to `(load "DUPLYR")` if **c:DUPLYR** is **nil**.

A small cosmetic problem with the previous two menu macros is that they display **nil** or **T** on the command line (the results of the **if** or **or** test). You can eliminate the problem by adding **princ** after the expression and wrapping all of the AutoLISP code in **progn**:

```
[Duplicate Layer]^C^C^C^P(progn (or c:DUPLYR (load "DUPLYR")) +
(princ))^P DUPLYR
```

A more serious problem is that entering DUPLYR at the command prompt doesn't work consistently. If you enter the command before choosing it from the menu (that is, before DUPLYR.LSP has been loaded by the menu macro), AutoCAD reports Unknown command. If you enter the command after having chosen it from the menu previously, the command works fine because DUPLYR.LSP was loaded already. **autoload** avoids this confusion.

Redefining Commands

AutoCAD allows you to redefine its native commands. Doing so is a two-step process:

1. Create a new definition with `(defun c:xxx`, where *xxx* is the original command's name.

2. Undefine the native command using AutoCAD's UNDEFINE command.

For example, the following code redefines the COPY command so that it always uses the Multiple option:

Creating an Integrated Professional Menu System

```
;;;Create new definition of COPY:
(defun c:COPY (/ SelSet)
    (prompt "\nUsing COPY Multiple...")
    (setq SelSet (ssget))
    (if SelSet (command "_.COPY" SelSet "" "_M"))
    (princ)
)

;;;Undefine native COPY:
(defun s::startup (/ OldCmd)
    (setq OldCmd (getvar "CMDECHO"))
    (setvar "CMDECHO" 0)
    (command "_.UNDEFINE" "COPY")
    (setvar "CMDECHO" OldCmd)
    (princ)
)
(princ)
```

You would add this code to ACAD.LSP or an MNL file for your menu. Notice that, in an ACAD.LSP file, the UNDEFINE part must go inside an **s::startup** function, since it uses **command**. Of course, you would need to coordinate this **s::startup** function with any existing **s::startup** function.

Warning: Professional applications should avoid redefining commands without compelling reasons to do so, especially when the redefined version behaves differently from the user's point of view. Thus this COPY example would be appropriate for personal customization, but not for a professional application. When a professional application redefines a command, the redefinition usually should be limited to performing some "hidden" action (for example, modifying the drawing database or updating an external file), without changing the original prompt sequence or results.

When you've redefined (or simply undefined) a command, entering the command name preceded by a period allows you to access the native, undefined command. This feature can be good or bad. On the one hand, it gives knowledgeable users quick access to the native command. As you've seen throughout the book, it also gives you a way to ensure that your menu macros and AutoLISP programs are using native commands. On the other hand, it enables clever users to circumvent your command redefinitions, which might cause problems for your application. AutoCAD doesn't provide any method of preventing this circumvention.

Redefining PLOT

The PLOT command is a frequent candidate for redefinition. Many users often perform some "preprocessing" steps before plotting: turn off construction layers, update a plot stamp with the current date, time, and drawing name, and so on. You can automate these steps by redefining the PLOT command. One complication is that calling the PLOT command using the AutoLISP **command** function causes AutoCAD to run the command-line version of PLOT (see Figure 14.6), which isn't what most users want to see when they're running PLOT interactively. (See Chapter 8 for fully automated and batch plotting techniques.) There are two ways around this problem.

 In R13c4 for Windows, you can start the Plot Dialog Box from AutoLISP using the following expression:

```
(menucmd "Gacad.ID_Print=|")
```

ID_Print is the name tag for the **F**ile, **P**rint pull-down menu choice in ACAD.MNS (whose menu group name is ACAD), and the vertical bar tells AutoCAD to execute the macro for that menu choice.

Here's an example of how to incorporate this expression into a custom MYPLOT command:

```
;;;Create custom version of PLOT command:
(defun c:MYPLOT ()
   preprocessing commands here...
   (if (or
            (= 0 (getvar "CMDDIA"))       ;CMDDIA=0
            (= 4 (getvar "CMDACTIVE"))    ;script is running
       )
       ;;CMDDIA=0 or script is running, so run command-line PLOT:
       (command "_.PLOT")
       ;;CMDDIA=1 and script not running, so use dialog:
       (menucmd "Gacad.ID_Print=|")
   )
   (princ)
)
```

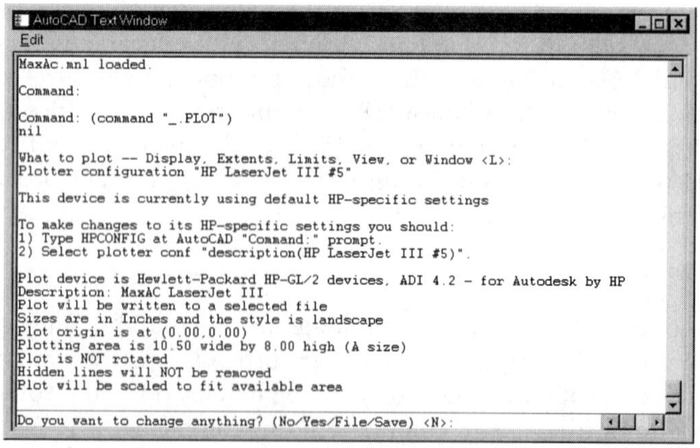

Figure 14.6
Plotting from AutoLISP runs the command-line version.

Notice that our **c:MYPLOTcommand** runs the command-line version of the native PLOT command if the CMDDIA system variable is zero or if a script is running (CMDACTIVE = 4). These features ensure that our custom command doesn't remove functionality that is part of the native PLOT command.

The (menucmd "Gacad.ID_Print=|") approach is easy to implement, but it requires R13 Windows, and it assumes that the **F**ile, **P**rint pull-down menu item hasn't been altered in the current menu file:

```
ID_Print            [&Print...      Ctrl+P]^C^C_plot
```

This approach also forces the user to enter a command name that's different from the standard AutoCAD PLOT. Unfortunately you can't use (menucmd "Gacad.ID_Print=|") inside a command redefinition of PLOT, because it causes an infinite loop (the redefined c:PLOT command calls the **P**rint menu item, which calls the redefined c:PLOT command, which calls the **P**rint menu item, and so on). You could circumvent this problem by adding a pull-down menu item that calls the native AutoCAD PLOT command:

```
ID_DotPlot    [Native Plot Command]^c^c^c .PLOT
```

Then you'd use the new menu item's name tag instead of ID_Print: (menucmd "Gacad.ID_DotPlot=|"). This workaround eliminates the infinite loop, but adds a menu item that you might not want to make accessible to users. You could make the menu choice inaccessible by prefixing its label with a tilde (~), but users might be confused by a custom menu item that's always grayed out.

The other approach to plotting from AutoLISP isn't limited to R13 Windows and doesn't generate the problems caused by (menucmd "Gacad.ID_Print=|"), but it requires an auxiliary ADS or ARX program. Such a program feeds keystrokes (in this case, the keystrokes **PLOT** followed by ENTER) to AutoCAD. Because AutoCAD sees the keystrokes as interactive input, rather than AutoLISP or script input, it displays the Plot Configuration dialog instead of the command-line version of PLOT. MAWIN.ARX, which is documented later in this chapter, includes a function that performs this trick: **AcadPlotDlg**. To use the function in an AutoLISP program, simply call it with no parameters: **(AcadPlotDlg)**. Note that it should be the last action in your AutoLISP program.

Another example of a so-called "keyboard stuffer" program is Mike Dickason's freeware program LISPPLOT, several versions of which are available in the libraries of the CompuServe ACAD forum. In order to use LISPPLOT and similar programs, you would load the program and call the appropriate function, as shown in the following example:

```
;;;Create new definition of PLOT:
(defun c:PLOT ()
   preprocessing commands here...
   (if (or
            (= 0 (getvar "CMDDIA"))       ;CMDDIA=0
```

```
                (= 4 (getvar "CMDACTIVE"))    ;script is running
         )
         ;;CMDDIA=0 or script is running, so run command-line PLOT:
         (command "_.PLOT")
         ;;CMDDIA=1 and script not running, so use dialog:
         (progn
            (or lispplot (xload "LISPPLOT"))   ;xload ADS program
            (lispplot)                         ;execute function
         )
      )
   )
   (princ)
)
```

Redefining File-Handling Commands

Another common need is to redefine AutoCAD's file-handling commands (SAVE, QUIT, and so on). Applications frequently want to perform some action before the user leaves the drawing. Unfortunately, there's no easy way to accomplish this task with AutoLISP. You would need to redefine seven commands (END, NEW, OPEN, QSAVE, QUIT, SAVE, and SAVEAS), each of which has command-line and dialog box versions and a variety of subtleties. Mimicking AutoCAD's responses to all of the permutations of actions that a user could take with these commands isn't feasible with AutoLISP.

AutoCAD's ADS and ARX programming interfaces provide better control through request code *notifications* that inform the application when the user is about to save a drawing, end, or quit. You need an in-depth knowledge of C and either ADS or ARX to take advantage of this feature.

Using DIESEL and AutoLISP in Menus

Although it's possible to write menu macros that consist of nothing more than AutoCAD commands, you'll find that many useful macros require some DIESEL and AutoLISP expressions. You've already developed quite a few such macros in earlier chapters. This section demonstrates some more sophisticated ways of incorporating DIESEL and AutoLISP in menu macros.

Controlling Scale Factors in Macros

Menu macros often insert or draw scale-dependent objects (text, title blocks, blocks representing drafting symbols, and so on). Your macros should be intelligent enough to handle drawings of different scales. AutoLISP or DIESEL expressions provide the means of calculating scale-dependent factors.

The standard technique is to set a system variable (often DIMSCALE) to the drawing's scale factor and then use AutoLISP or DIESEL arithmetic functions to calculate scales and sizes relative to the drawing scale factor. For example, in AutoLISP:

Creating an Integrated Professional Menu System

Figure 14.7
Automatic hatch pattern settings from the Concrete *macro*.

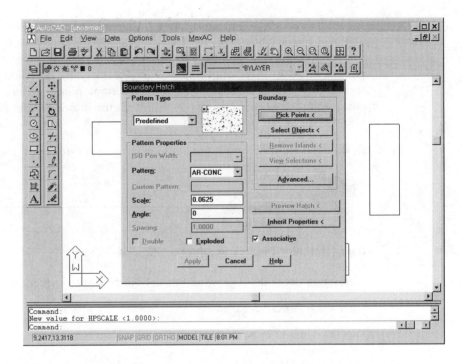

```
(* 0.5 (getvar "DIMSCALE"))
```

or in DIESEL:

```
$m=$(*,0.5,$(getvar,DIMSCALE))
```

Warning: Users who dimension in paper space might set DIMSCALE to 0.0, which will break macros that depend on DIMSCALE. Professional applications should store the drawing scale factor separately (in a hidden block containing application data, for example). For a personal menu, you could store the drawing scale factor in one of the USERR*x* system variables. Professional applications should avoid changing any of the USER system variables—they're called USER for a reason.

In the next exercise, you experiment with the following hatching submenu, which sets hatching scale factors based on the current DIMSCALE value:

```
MA_Hatch         [->&Hatch]
                 [Concrete...]^C^C^C^P_.HPNAME AR-CONC _.HPANG 0 +
_.HPSCALE (* 0.0625 (getvar "DIMSCALE")) ._BHATCH
                 [Earth...]^C^C^C^P_.HPNAME EARTH _.HPANG 45 +
_.HPSCALE (* 0.5 (getvar "DIMSCALE")) ._BHATCH
                 [Sand...]^C^C^C^P_.HPNAME AR-SAND _.HPANG 0 +
_.HPSCALE (* 0.0625 (getvar "DIMSCALE")) ._BHATCH
                 [<-Steel...]^C^C^C^P_.HPNAME ANSI32 HPANG 0 +
_.HPSCALE (* 0.375 (getvar "DIMSCALE")) ._BHATCH
```

Maximizing AutoCAD® R13

AutoCAD's concrete and sand hatch patterns (AR-CONC and AR-SAND) look good at about 1/16 (or 0.0625) of the drawing scale factor. The EARTH pattern needs to be a much larger percentage of the drawing scale factor, and we've chosen 0.5. The ANSI32 pattern with a scale factor of 3/8 (0.375) works well for hatching steel. You might choose to use different multipliers, but the important point is that these macros ensure a consistent plotted hatch size, no matter what the drawing scale is. Notice that each macro also sets the hatch pattern angle.

In the next exercise, you'll test these macros with different values of DIMSCALE.

Exercise

Using Hatching Macros at Different Scales

Start a new drawing, or erase all of the objects in your current drawing.

Command: **RECTANG** [Enter]

First corner: *Use the RECTANG command to draw a series of rectangles, similar to those shown in Figure 14.8*

Figure 14.8
Results of the hatching macros.

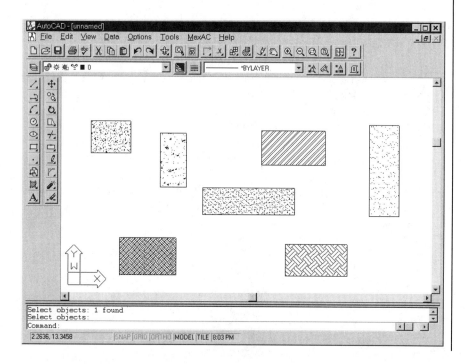

Command: **DIMSCALE** [Enter]	
New value for DIMSCALE <1.0000>: [Enter]	Verifies that DIMSCALE is 1.0
Command: *Choose* **M**AXAC, **H**atch, Concrete	Sets HPNAME to AR-CONC, HPANG to 0, and HPSCALE to 0.625 before displaying the Boundary Hatch dialog box (see Figure 14.7)
Choose the Select **O**bjects *button*	Temporarily hides the Boundary Hatch dialog
Select Objects: *Select one of the rectangles*	
Select Objects: [Enter]	Ends object selection
Choose the Apply *button*	Creates the hatching and closes the Boundary Hatch dialog

Next you'll repeat the macro, but with DIMSCALE set to 2.0. Of course, in real drawings you'd be more likely to have two different drawings with different DIMSCALE settings, rather than one drawing in which you changed the setting.

Command: **DIMSCALE** [Enter]	
New value for DIMSCALE <1.0000>: 2 [Enter]	Changes DIMSCALE to 2.0
Command: *Choose* **M**AXAC, **H**atch, Concrete	Now HPSCALE is 0.125
Hatch one of the rectangles	The pattern is half as dense

Experiment with the other three macros and with other values of DIMSCALE. The results should look something like Figure 14.8.

Remain in AutoCAD for the next exercise.

Note: These hatching macros put hatch objects on the current layer. In a production menu, you probably would want to set a hatching layer current at the beginning of each macro, so that the hatch objects go on an appropriate layer.

Controlling Cursor Menus with DIESEL

In Chapter 10, you saw examples of how DIESEL can be used in conjunction with AutoLISP to create dynamic popup cursor menus. Another interesting application of DIESEL is its use in pointing device button macros, to dynamically swap static cursor menus defined in a menu file when a pointing device button is pressed.

Most modern Windows applications use the right mouse button to activate a *context-sensitive* menu. It is called context-sensitive because the contents of the menu that appears when you press the button change depending on what you are currently doing, and/or what is currently selected. AutoCAD does not make use of context-sensitive cursor menus (with one exception: when you right-click on an OLE [Object Linking and Embedding] object that's embedded in a drawing in AutoCAD for Windows). In AutoCAD Release 12, Autodesk introduced a new feature called *automatic side-screen menu paging* (see Chapter 13). Controlled by the MENUCTL system variable, this feature allows you to associate side-screen menu pages with AutoCAD commands, by giving a menu page a name that is the same as the associated command. With the MENUCTL system variable set to 1, AutoCAD automatically swaps in the side-screen menu page that has the same name as the currently active command. MENUCTL is useful because it makes frequently used options for each command immediately available. But the side-screen menu has its disadvantages, mainly that it consumes a significant amount of precious display real estate, and doesn't work all that well in a windowed operating environment, where windows are resized as needed, which can conceal parts of the menu.

Creating a Context-Sensitive Cursor Menu

Consider the underlying mechanism AutoCAD uses to control automatic paging of the side-screen menu. AutoCAD searches for the name of a menu page which is the same as the name of the currently active command, and if it finds one with that name, it swaps it into the side-screen menu automatically. The CMDNAMES system variable holds the name of the currently active built-in command, making that information available to AutoLISP and DIESEL. Now consider that a DIESEL macro triggered by pressing a button on a pointing device can use CMDNAMES to determine what the name of the current command is. You've already seen how menu macros can instruct AutoCAD to swap menu pages for side screen, pull-down, cursor, tablet, and icon menus. Also consider how DIESEL can use the CMDACTIVE system variable to compose the name of a popup menu dynamically, swap it into the POP0 cursor menu and display it, every time the button that triggers the macro is pressed. Voilà! You have all of the ingredients needed to give the popup cursor menu the very same functionality that MENUCTL provides for side-screen menus. The following simple button menu macro is executed every time the button it is assigned to is pressed. (The ***AUX*x* and ***BUTTONS*x* and sections of the AutoCAD menu define the effects of the pointing device buttons, except for the pick button—see Chapter 13 for details.)

```
***AUX3
$P0=POP0 $M=$P0=$(getvar,cmdnames) $P0=*
```

Creating an Integrated Professional Menu System

Examine the macro closely. It is incredibly simple considering what it does. Its elegance is an example of the power that DIESEL can bring to menu macros. The macro in the preceding example has three separate parts that perform the following actions:

1. **$P0=POP0** swaps the default (***POP0) menu into the cursor menu (P0).

2. **$M=$P0=$(getvar,cmdnames)** instructs AutoCAD to swap the ***POPx menu whose name or alias is the same as the name of the currently active command, into the cursor menu. If AutoCAD does not find a menu having a name or alias that is the same as the currently active command, it simply ignores the request and does nothing, and the menu that's currently swapped into the cursor menu remains there (which is the one swapped in by the first part of the macro).

3. **$P0=*** Pops up the cursor menu.

Thus, by defining a POPx menu for any command you want to support, you can easily provide context-sensitive cursor menus for any command. The following listing shows two POPx menus, for the ARC and CHANGE/CHPROP commands, defined in 14CURSOR.MNS from this chapter's directory:

```
***POP5
**ARC
[ARC command]
[Center point]_C
[End point]_E
[Angle]_A
[Direction]_D
[Radius]_R
[Chord Length]_L

***POP9
**CHANGE
**CHPROP
[Change/Chprop demonstrates conditional gray out]
[Color]_C
[Layer]_La
[Linetype]_Lt
[Linetype Scale]_S
[$(if,$(eq,$(getvar,cmdnames),CHPROP),~)Elevation]_E
[Thickness]_T
[--]
[Match Color]_C;'Match C
[Match Layer]_La;'Match La
```

```
[Match Linetype]_Lt;'Match Lt
[Match Linetype Scale]_S;'Match S
[$(if,$(eq,$(getvar,cmdnames),CHPROP),~)Match Elev]_E;'Match E
[Match Thickness]_T;'Match T
```

In the preceding listing, the labels ***POP5 and ***POP9 are the names of the menus. The labels that start with two leading asterisks are *aliases*, which provide another means of referencing the menu whose name they follow in menu control language directives.

Hence, the three following directives all refer to the same menu (the second menu in the preceding listing).

```
$P0=CHANGE
$P0=CHPROP
$P0=POP9
```

A ***POP*x* menu can have any number of aliases, all of which can be used to refer to same menu. The aliases must follow the ***POP*x* label of the menu they reference in the menu source. Look at the preceding button macro again.

```
$P0=POP0  $M=$P0=$(getvar,cmdnames)  $P0=*
```

Assuming that the CHANGE command is currently active when you press the button this macro is assigned to, after the DIESEL expression is evaluated, the macro becomes:

```
$P0=POP0  $P0=CHANGE  $P0=*
```

Thus the button macro instructs AutoCAD to swap the ***POP9 menu shown in the preceding listing into the cursor menu, and display it (see Figure 14.9).

14CURSOR.MNS defines context-sensitive menus for a number of AutoCAD commands, along with button macros for activating them. When you attach the menu file, its associated MNL file replaces the base menu's ***AUX*x* and ***BUTTONS*x* sections with the corresponding sections from 14CURSOR.MNS. You can activate the context-sensitive cursor menus by holding down CTRL and pressing the first programmable pointing device button (Button1, the same button assigned to ENTER in the standard AutoCAD menu, or the right button on a 2- or 3-button mouse). If you use a 3-button mouse or a digitizer, you may want to activate the context-sensitive menus with a different button that doesn't require you to hold down the CTRL key to activate the menus. See Chapter 13 for more information about button customization options.

The commands that have context-sensitive menus defined for them in 14CURSOR.MNS are listed in Table 14.2.

Next, you'll test the context-sensitive cursor menus from 14CURSOR.MNS. This exercise assumes that you are using a 2- or 3-button mouse. Where you see the instruction *Click*

Creating an Integrated Professional Menu System

Figure 14.9
The CHPROP context-sensitive cursor menu.

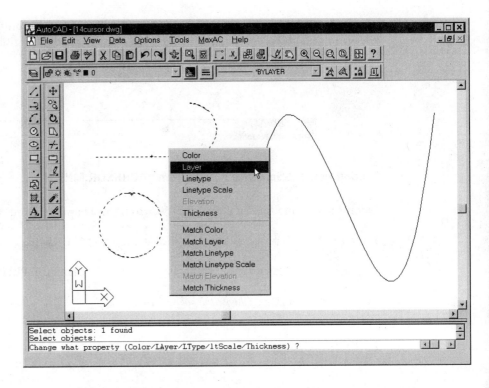

Table 14.2
Context-Sensitive Menus from 14CURSOR.MNS

Menu	Linked Commands
***POP3	GRIP_MOVE, GRIP_STRETCH, GRIP_ROTATE, GRIP_SCALE, GRIP_MIRROR
***POP4	PLINE
***POP5	ARC
***POP6	PEDIT
***POP7	INSERT, MINSERT
***POP8	TEXT, DTEXT
***POP9	CHANGE, CHPROP
***POP10	SPLINEDIT
***POP11	SPLINE
***POP12	DIM, DIM1, DIMLINEAR, DIMANGULAR, DIMORDINATE, DIMRADIUS, DIMALIGNED, DIMBASELINE, DIMCENTER, DIMCONTINUE, DIMDIAMETER

CTRL+Button1, the instruction refers to the first programmable button. This is the button that displays the context-sensitive menus. If you are using a digitizer or other pointing device, substitute the equivalent button.

Exercise

Testing Context-Sensitive Menus from 14CURSOR.MNS

Open the drawing file 14CURSOR.DWG from this chapter's directory.

From the **M**axAC pull-down menu, choose **M**enus, 14CURSOR.

Command: **PEDIT** [Enter]	Starts the PEDIT command
Select polyline: *Choose the polyline in the upper left part of the view*	
Click *CTRL+Button1*	PEDIT cursor menu appears (see Figure 14.10)
Close/Join/Width/Edit vertex/Fit/Spline/Decurve/Ltype gen/Undo/eXit <X>: *Choose* **W**idth	
Enter new width for all segments: **.5** [Enter]	Specifies polyline width of .5 units
Close/Join/Width/Edit vertex/Fit/Spline/Decurve/Ltype gen/Undo/eXit <X>: **EXIT** [Enter]	Exits the PLINE command
Command: **SPLINEDIT** [Enter]	Starts the SPLINEDIT command
Select spline: *Choose the spline on the right side of the view*	
Click *CTRL+Button1*	SPLINEDIT cursor menu appears
Fit Data/Close/Move Vertex/Refine/rEverse/Undo/eXit <X>: *Choose* **C**lose	Closes spline

Creating an Integrated Professional Menu System

Figure 14.10
The PEDIT context-sensitive cursor menu.

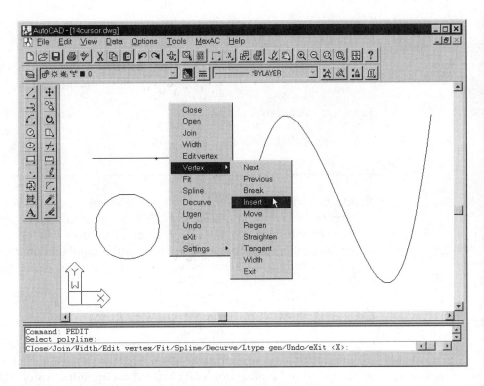

`Open/Move Vertex/Refine/rEverse/Undo/eXit <X>:` **EXIT** [Enter]	Exits SPLINEDIT command
`Command:` **ARC** [Enter]	Starts ARC command
`Center/<Start point>:` **8,4** [Enter]	Specifies 8,4 as first endpoint
Click CTRL+Button1	ARC cursor menu appears
`Center/End/<Second point>:` *Choose* **C**enter *point*	Specifies Center option
`Center:` **0,4** [Enter]	Specifies center point of 0,4
Click CTRL+Button1	ARC Cursor menu appears
`Angle/Length of chord/<End point>:` *Choose* **A**ngle	Specifies Angle option
`Included angle:` **90** [Enter]	Specifies angle of 90 degrees

Command: **LINE** [Enter]	Starts the LINE command
Hold down CTRL and right-click in the drawing view	Default OSNAP cursor menu appears
From point: *Press CANCEL*	Cancels LINE command

Remain in AutoCAD for the next exercise.

Notice that during each command, a different cursor menu appeared when you right-clicked while holding down the CTRL key, each containing options that are specific to the command that was active at the time. Also note that when you started the LINE command and pressed the same key and button combination, the standard object snap cursor menu appeared, because there is no context-sensitive menu defined for the LINE command.

Extending the Object Snap Menus with AutoLISP

Occasionally some clever use of AutoLISP results in menu enhancements that don't even require any changes to your menus. You might have noticed that AutoCAD's object snap cursor menu and toolbar don't do anything useful unless AutoCAD is requesting a point. For example, when you choose Endpoint from the cursor menu, AutoCAD enters **_endp** at the command prompt, which results in an Unknown command error message. The following AutoLISP code creates commands for each of the object snap abbreviations, so that entering these commands at the command prompt (whether from the keyboard or from menus) sets the corresponding running object snap mode.

```
(defun defosnapcmd (name bits)
   (eval
      (list 'defun (read (strcat "C:" name)) nil
         (list 'setvar
            "OSMODE"
            (if (eq 0 bits)
               0
               (list 'boole 6 '(getvar "OSMODE") bits)
            )
         )
         '(princ)
      )
   )
)

(setq ospairs)
```

Creating an Integrated Professional Menu System

```
'(  ("NON"    0)
    ("ENDP"   1)
    ("MID"    2)
    ("INT"    32)
    ("APPINT" 2048)
    ("CEN"    4)
    ("QUA"    16)
    ("PER"    128)
    ("TAN"    256)
    ("NOD"    8)
    ("INS"    64)
    ("NEA"    512)
  )
)

(foreach pair ospairs (apply 'defosnapcmd pair))
```

This is a relatively advanced example of AutoLISP, which should whet your appetite for this book's companion volume, *Maximizing AutoLISP*. But you can use it as is without worrying about the details of functions like **read** and **boole**.

Exercise

Testing TBOSNAP.LSP

Command: *Pop up the cursor menu and choose* Endpoint	
Command: _endp Unknown command "ENDP". Type ? for list of commands.	AutoCAD doesn't recognize the command
Command: *Choose* **T**ools, **T**oolbars, **O**bject Snap	Displays the Object Snap toolbar (see Figure 14.11)
Command: *Choose the* Snap to Endpoint *icon (see Figure 14.11)*	As with the cursor menu, AutoCAD doesn't recognize _endp

From the **M**axAC pull-down menu, choose **L**oad Chapter File, TBOSNAP.LSP.

Command: *Pop up the cursor menu and choose*

Figure 14.11
The Object Snap toolbar, enhanced by TBOSNAP.LSP.

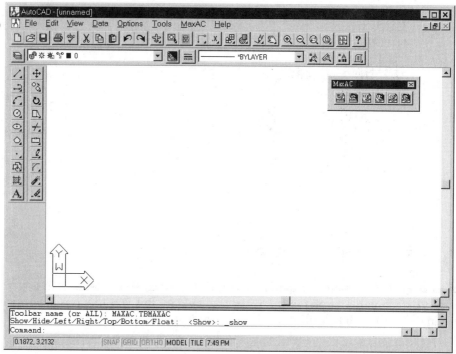

```
Endpoint

Command: _endp                    AutoCAD doesn't complain, but
                                  instead sets the running object
                                  snap to endpoint
```

If you pop up the cursor menu again or enter DDOSNAP, you'll see that the endpoint running object snap is now set.

Try picking other object snap modes from the cursor menu and Object Snap toolbar. You can type them as well. Notice that the running object snap modes are cumulative.

To finish, choose the None object snap mode from the toolbar or cursor menu in order to turn off all running object snaps.

Remain in AutoCAD for the next exercise.

Creating an Integrated Professional Menu System

If you want to activate the TBOSNAP.LSP feature in all of your drawing sessions, add the following lines to ACAD.LSP:

```
(if (eq "failed" (load "TBOSNAP" "failed"))
    (alert "   Can't find required file TBOSNAP.LSP   ")
)
```

Comparison of AutoLISP and DIESEL in Macros

In many cases, you will have a choice between using AutoLISP expressions or DIESEL expressions in menu macros. AutoLISP often gives you more flexibility, simply because it provides so many more functions. Also, you can store and retrieve values in AutoLISP variables, while DIESEL is limited to the USER system variables (integer, real, and string) for storage.

DIESEL retains one important advantage: AutoCAD can evaluate DIESEL expressions in menu macros even when another AutoLISP program is active. If you try to execute a menu macro containing AutoLISP expressions while another AutoLISP program is active, the macro won't work and you'll see the Can't reenter AutoLISP error message. Thus, when you have a choice, you should use DIESEL for macros that can be executed transparently (while another command is active) or are intended to supply responses to other programs.

Supporting Menus and AutoLISP Programs with an ARX Application

Custom menus and AutoLISP programs together give you a great deal of customization flexibility, but they can't do everything. For example, AutoLISP can't perform file management tasks or read and write binary files by itself. Also, Autodesk has been focusing more effort on improving its C-based programming interfaces, ADS and ARX, than on AutoLISP. Fortunately, AutoLISP programmers aren't completely out in the cold. ADS and ARX programmers can create libraries of useful AutoLISP functions. You can use these libraries to extend the power of AutoLISP without having to know anything about ADS or ARX.

One example, for R13 Windows, is MAWIN.ARX by Tony Tanzillo. MAWIN is a collection of functions that add Windows application programming interface (API) functionality to AutoLISP. With these functions, your AutoLISP programs and menu macros can perform tasks like launching applications, getting directory names, working with INI files, and creating dialog boxes on the fly.

Because MAWIN.ARX relies on the Windows API, it will not work with DOS AutoCAD.

Another useful AutoLISP function library is DOSLIB by Dale Fugier of Robert McNeel & Associates. It's an ADS program that works with Windows and DOS AutoCAD (R13 and R12). DOSLIB includes many functions for file and directory manipulation and INI file reading and writing. DOSLIB is freeware and is included on the MaxAC CD-ROM.

MAWIN.ARX Functions

This section documents the functions in MAWIN.ARX. Refer to MAWIN.TXT and MAWIN.LSP (in the MAXAC directory) for additional information.

ShellOpen

(ShellOpen *file_name***)**

Opens a file using the registered application that is associated with the file's type. Works with Windows 95 shortcuts (LNK files). Can also be used to invoke the Explorer on a specified folder by passing the name of the folder in **file_name**. *Returns* **T** *if the file was opened, or* **nil** *if the operation failed.*

Examples:

```
(ShellOpen "ACAD.LSP")
```

If ACAD.LSP exists in the current directory, this example opens it in the text editor associated with LSP files on your system and returns **T**. If ACAD.LSP does not exist in the current directory, this example returns **nil**.

```
(ShellOpen (findfile "ACAD.LSP"))
```

Same as the previous example, except that AutoCAD searches the library search path for ACAD.LSP.

```
(ShellOpen "C:\\WINDOWS\\DESKTOP\\TEXTPAD.LNK")
```

Launches TextPad, assuming that there's a shortcut to it called TEXTPAD on the Windows 95 desktop.

```
(ShellOpen "C:\\R13\\COM\\SUPPORT")
```

Opens an Explorer window containing the files in the C:\\R13\\COM\\SUPPORT directory (see Figure 14.12). When you double-click on a file, Windows opens the file

Creating an Integrated Professional Menu System

Figure 14.12
(ShellOpen "C:\\R13\\COM\\SUPPORT")

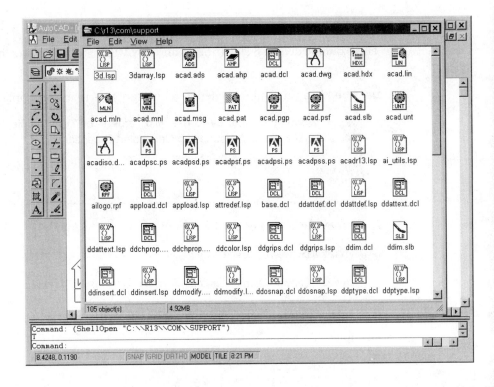

with its associated program. The Explorer window doesn't close automatically—you must close it.

ShellOpen is especially useful for creating macros that open text or program files with your preferred text editor. We used this function to create the macros in the MaxAC Customization toolbar.

DeleteFile

(**DeleteFile** *file_name* . . .)

Deletes one or more files. Returns the number of files deleted, or **nil** *if there are no such files.*

Examples:

```
(DeleteFile "$TEMP$.$$$")
```

If $TEMP$.$$$ exists in the current directory, this example deletes it and returns 1. If $TEMP$.$$$ does not exist in the current directory, this example returns **nil**.

```
(DeleteFile "$TEMP1$.$$$" "$TEMP2$.$$$")
```

This example deletes $TEMP1$.$$$ and $TEMP2$.$$$ (assuming they exist) and returns 2.

DeleteFile is useful for deleting temporary files that your application creates.

Warning: Do not try to use wild cards (* or ?) in the file specification. If you do, **DeleteFile** generates an Unhandled Exception.

GetWindowsDirectory

(GetWindowsDirectory)

Returns the path to the Windows directory.

Example:

```
(GetWindowsDirectory)
```

Returns "C:\\WINDOWS" on a system with Windows installed in the C:\WINDOWS directory.

GetTempDirectory

(GetTempDirectory)

Returns the path to the first of:
1. The path specified by the TMP environment variable.
2. The path specified by the TEMP environment variable, if TMP is not defined.
3. The current directory, if both TMP and TEMP are not defined.

Examples:

```
(GetTempDirectory)
```

Returns "C:\\WINDOWS\\TEMP" on a system with the TMP environment variable set to C:\WINDOWS\TEMP.

You can use **GetTempDirectory** to define a location for any temporary files that your application uses.

GetProfileString

(GetProfileString *ini_file_name section key default***)**

Returns the value of a configuration string variable in a Windows INI-type file (including ACAD.INI and ACADNT.CFG). **ini_file_name** *is the name of an INI-type file. If you don't supply a directory with the file name,* **GetProfileString** *defaults to the Windows directory.* section *is the name of a section in the INI name. (INI file sections appear in brackets: [General].)* **key** *is the name of the configuration setting. (An INI file key is the string before the equal sign:* PrototypeDwg=acad.dwg.*) default is the value to return if the configuration settings isn't present.*

Returns the value of the setting as a string, or **default** *if* **ini_file_name,** **section**, *or* **key** *doesn't exist.*

Examples:

```
(GetProfileString (strcat (getenv "ACADCFG") "\\ACADNT.CFG")
                  "AutoCAD" "DefaultLoginName" "")
```

Returns the default AutoCAD login name (for example, "MarkM").

```
(GetProfileString (strcat (getenv "ACADCFG") "\\ACAD.INI")
                  "General" "PrototypeDwg" "")
```

Returns "acad.dwg" in a standard AutoCAD configuration.

GetProfileInt

(GetProfileInt *ini_file_name section key default***)**

GetProfileInt *works the same as* **GetProfileString**, *except that it returns the configuration setting as an integer.* **default** *must be an integer.*

Example:

```
(GetProfileInt (strcat (getenv "ACADCFG") "\\ACAD.INI")
               "Drawing Window" "ScreenMenu" -1)
```

Returns 0 if the screen menu is turned off or 1 if it's turned on.

Tip: `GetProfileInt` and `GetProfileString` are useful for retrieving application settings from an application-specific INI file. You can create an INI file with your text editor and store settings for layers, sizes, and so on. Then you can use `GetProfileInt` and `GetProfileString` to retrieve the settings in menu macros and AutoLISP programs. In a professional application, you would also want to create a program that allowed users to view and modify the settings.

Beep

(Beep)

*Plays the Default sound defined in the Control Panel's Sound Properties applet — see Figure 14.13. Returns **nil**.*

Example:

(Beep)

Makes a short beep and returns **nil**.

Warning: If the Default sound event in the Control Panel's Sound Properties applet is set to (None), then **Beep** won't make any sound.

Sound

(Sound [*frequency duration*])

*Plays a sound on the system speaker (in Windows NT and Windows 3.1 only) at the specified frequency and duration. Returns **nil**.*

Warning: **Sound** works only in Windows NT and Windows 3.1. It does not work in Windows 95.

Examples:

(Sound 500 100)

Creating an Integrated Professional Menu System

Table 14.3
MessageBeep Sound Codes

Event	Code	AutoLISP Constant
Error	16	MT:ERROR
Question	32	MT:QUESTION
Exclamation	48	MT:EXCLAMATION
Asterisk	64	MT:ASTERISK

Plays a sound of 500 Hertz for 100 milliseconds and returns **nil**.

MessageBeep

(MessageBeep [code])

Plays registered event sounds (defined by the Control Panel's Sound Properties applet — see Figure 14.13), as listed in Table 14.3. You can use either the code number or the AutoLISP constant name. Any other values are not defined and can have unpredictable results. Returns **nil.**

Examples:

(MessageBeep 48)

Plays the event sound registered for Exclamation and returns **nil**.

 Note: If the sound event in the Control Panel's Sound Properties applet is set to (None), then **MessageBeep** will make any ordinary short beep.

MessageBox

(MessageBox *caption flags message_text***)**

Displays the standard Windows Message dialog box, with options for system icons (① in Figure 14.14), multi-line text messages (②), and pre-defined buttons (③— Yes, No, OK, Cancel, Abort, Retry, Ignore, etc.). **MessageBox** *also plays the same system event sounds as* **MessageBeep** *does.* **caption** *is the title or caption of the dialog.* **flags** is the sum of one or more of the Mx_xxx constants listed in Table 14.4. **message_text** is the text of the message, which should be no longer than 500 characters. You must include newlines (\n) to put explicit line breaks in the text.

The **flags** argument is the sum of four values (one from each category). Thus a **flags** value of zero displays a simple OK dialog with no icon. A **flags** value of 292 (= 4 + 256 + 32) displays a dialog with a question mark icon and Yes and No buttons, where No is the default button.

MessageBox returns an integer that indicates which button the user chose (the ESC key acts the same as choosing Cancel). The return codes are listed in Table 14.5.

Examples:

```
(MessageBox "MaxAC Sample" 64 "For your information...")
```

Displays a dialog box with the title `MaxAC Sample`, the Windows Information icon, the text `For your information...`, and an OK button. Returns 1 when the user chooses OK.

```
(MessageBox "MaxAC MessageBox Test"
   (+ MB:YESNO MT:QUESTION)
   (strcat
      "   I grow old ... I grow old ...\n"
      "I shall wear my trousers rolled.\n\n"
      "   Shall I part my hair behind? Do I dare to eat a peach?"))
```

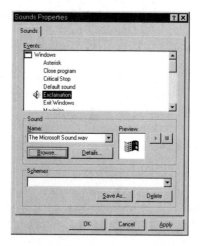

Figure 14.13
The Control Panel's Sound Properties applet.

Figure 14.14
MessageBox dialog components.

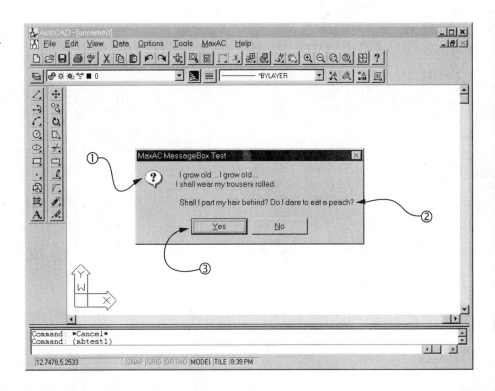

Displays the dialog box shown in Figure 14.14. Returns 6 if the user chooses Yes or 7 if the user chooses No.

```
(MessageBox "MaxAC Sample"
    (+ MB:OKCANCEL MD:BUTTON2 MT:EXCLAMATION)
    "Delete all DWG files?")
```

Displays the dialog box shown in Figure 14.15. It returns 1 if the user chooses OK or 2 if Cancel.

MAWIN.LSP contains several useful "wrapper" functions that demonstrate how to use **MessageBox**. (MAWIN.ARX does not require MAWIN.LSP.) See Chapter 11 for more information about **MessageBox**.

Figure 14.15
MessageBox sample.

Table 14.4
MessageBox Flag Codes

Code	AutoLISP Constant
Buttons to show (choose one):	
0	MB:OK
1	MB:OKCANCEL
2	MB:ABORTRETRYIGNORE
3	MB:YESNOCANCEL
4	MB:YESNO
5	MB:RETRYCANCEL
Default button (choose one):	
0	MD:BUTTON1
256	MD:BUTTON2
512	MD:BUTTON3
768	MD:BUTTON4
System icon to show (choose one):	
0	(no icon)
16	MT:HAND or MT:ERROR or MT:STOP
32	MT:QUESTION
48	MT:EXCLAMATION or MT:WARNING
64	MT:ASTERISK or MT:INFORMATION
Modal relationship between AutoCAD window and message dialog (choose one):	
0	MM:APPLMODAL
4096	MM:SYSTEMMODAL
8192	MM:TASKMODAL

Table 14.5
MessageBox Result Codes

Code	AutoLISP Constant
0	MR:NONE
1	MR:OK
2	MR:CANCEL
3	MR:ABORT
4	MR:RETRY
5	MR:IGNORE
6	MR:YES
7	MR:NO

Creating an Integrated Professional Menu System

> **AddToRecentDocs**
>
> **(AddToRecentDocs** *file_name*)
>
> Adds a link (shortcut) to **file_name** to the Windows 95 or NT 4.0 Start, Documents list (the list of recently opened documents). Returns the file name if successful or **nil** if not.

Example:

(AddToRecentDocs "PLAN.DWG")

If PLAN.DWG exists in the current directory, this example adds it to the Windows 95 or NT 4.0 Start, Documents list and returns "PLAN.DWG". If PLAN.DWG doesn't exist, this example returns **nil**.

Note: MAWIN.ARX automatically adds the name of the current drawing to the Start, Documents list each time you enter the editor with a named drawing. AutoCAD alone does not perform this action.

> **FileExists**
>
> **(FileExists** *file_name*)
>
> Returns **T** if **file_name** (a file or directory) exists.

Example:

(FileExists "PLAN.DWG")

If PLAN.DWG exists in the current directory, this example returns **T**, or otherwise, **nil**.

> **SysWindow**
>
> **(SysWindow** *code*)
>
> Instructs AutoCAD to minimize, maximize or restore itself. Table 14.6 lists the valid values for code. Returns **T**.

Table 14.6
SysWindow code *Values*

Code	AutoCAD Window Action
0	Minimize
1	Restore
2	Maximize

Example:

`(SysWindow 2)`

Maximizes the AutoCAD window and returns **T**.

AcadPlotDlg

(AcadPlotDlg)

Invokes the Plot Configuration dialog box. Returns **nil**.

Example:

`(AcadPlotDlg)`

Invokes the Plot Configuration dialog box and returns **nil**.

Warning: You should call this function as the last action in your AutoLISP program, just before it exits. Attempting to do anything in the AutoLISP program after calling this function may have unpredictable results.

Creating Help Files

Help files provide the text, formatting, and navigational structure for AutoCAD's on-line help systems. As you saw in Chapter 4, R13 supports two kinds of help files: Windows-specific HLP files (Figure 14.16) and platform-independent AHP files (Figure 14.17). AutoCAD creates a help index file (HDX) for each platform-independent AHP file.

You can use custom help files to document company standards or supply on-line help for your custom applications. Although you can modify the standard ACAD.AHP file, usually you'll want to create a separate application help file (HLP for Windows and/or AHP for all platforms).

Figure 14.16
Windows-specific help.

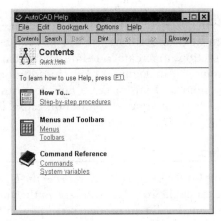

Windows-specific help files use the standard Windows help format and help engine, but aren't available in DOS or other operating systems. Platform-independent help files work the same in all AutoCAD platforms, but they are more primitive in terms of interface and features.

If you're developing a personal menu or menu system for a small number of users in your office, it might not be worthwhile to spend the time developing AutoCAD on-line help. Maintaining ordinary printed documentation is work enough, and you're close enough to the users that you can answer questions as they arise.

On the other hand, professional applications can benefit from custom on-line help in addition to printed documentation. Usually the custom help will be linked to Help buttons on application dialogs and custom application command names.

Figure 14.17
Platform-independent help.

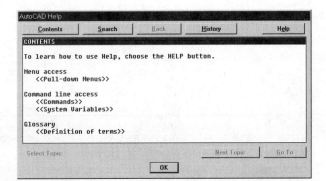

Other Means of Providing Application Help

Help files aren't the only custom help options. As you saw in Chapter 13, you can use the ***HELPSTRINGS menu section to create one-line status bar help strings for pull-down and toolbar menu items (① in Figure 14.18). In addition, you can define ToolTips for each toolbar menu item ②. Both of these help features are worth taking advantage of no matter what the level of your customization effort is. They're easy to implement and provide immediate feedback to users.

Developing Windows Help Files

AutoCAD doesn't include the tools for developing Windows-specific HLP files (sometimes called WinHelp). These tools are available from Microsoft, from the authors of compilers (for example, Borland), and from the developers of specialized help authoring tools. The usual procedure is to create help source files in Rich Text Format (RTF) using a word processor like Microsoft Word for Windows. Then you use a help compiler to create the HLP file that can be read by WINHELP.EXE (the Windows help engine).

Here are some sources of WinHelp authoring tools and information:

- Microsoft Windows Help Author Toolkit (on the Microsoft Developer Network CDs and available as WHAT6.EXE in the CompuServe MSL forum)

Figure 14.18
Help string and ToolTip.

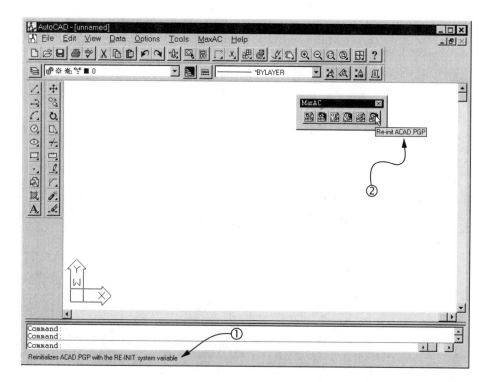

Creating an Integrated Professional Menu System

- CompuServe WINSDK forum, WINHELP library (#16)
- RoboHelp from BlueSky Software, La Jolla, CA
- Doc2Help from WexTech Systems, New York, NY
- ForeHelp from ForeFront, Boulder, CO

Understanding the Platform-Independent Help File Format

The R13 platform-independent help file format is quite different from the format used by R12 and earlier versions. The standard AutoCAD platform-independent help file, \R13\COM\SUPPORT\ACAD.AHP, is organized into sections. ACAD.AHP is an ASCII file, so you can examine or modify it with your text editor. Platform-independent help files contain, in addition to the text that you see in the help dialog, five kinds of *directives*. Each directive starts with a backslash and a special character. The directives are listed in Table 14.7.

Here is the help for the LINE command from ACAD.AHP. The first three lines are directives. Some lines are very long in the file but are printed on several lines here, so the [Enter] show where the returns are in the file:

```
\#line [Enter]
\$LINE Command [Enter]
\KLINE;line segments: straight;arcs: continuing;continuing a line or
arc;lines: continuing;line segments: deleting;lines: undo;deleting:
line segments;closing, polygon;polygons: closing;polygons: formed
from lines [Enter]
LINE [Enter]
Creates straight line segments [Enter]
The LINE command creates a line whose endpoints are specified using
```

Table 14.7
Platform-Independent Help File Directives

Directive	Purpose	Where displayed
\#	Topic ID (or context string)	Not displayed
\$	Title	History and Search lists
\K	Key word(s)	Search list (separate key words with semicolons)
\E	End of AHP file	Not displayed
\(space)	Comment	Not displayed

```
two-dimensional or three-dimensional coordinates. [Enter]
> From the Draw menu, choose Line [Enter]
> At the Command prompt, enter line [Enter]
    From point: Specify a point [Enter]
    To point: Specify a point [Enter]
    To point: Specify a point, enter <<u>>line_undoing_a_line>, or
enter <<c>>line_closing_a_polygon>, or press RETURN [Enter]
AutoCAD draws a line segment and continues to prompt for points,
enabling you to draw <<continuous
lines>>line_continuing_a_line_or_arc>. However, each line segment is
a separate object. Press RETURN to end the command. [Enter]
For example, the following command sequence draws a single line
segment. [Enter]
    Command: line [Enter]
    From point: Specify a point. [Enter]
    To point: Specify a point [Enter]
    To point: Press RETURN [Enter]
------------------------------------------\ [Enter]
See Also [Enter]
For more information on drawing lines, see "Drawing Line Objects" in
chapter 2, "Creating Objects," in the AutoCAD User's Guide. [Enter]
Commands: <<PLINE>>pline> creates two-dimensional polylines.
<<XLINE>>xline> creates an infinite line. <<RAY>>ray> creates a
semi-infinite line. [Enter]
```

Each section begins with the \# directive followed by a *topic ID* (also called a *context string* in Windows help terminology). In the preceding example, line is the topic ID. When you request help for the LINE command topic, AutoCAD looks up the \#line section in the help file and presents the information that follows the label, up to the next topic ID (see Figure 14.19). All topic IDs must be unique, and they cannot contain spaces.

After the topic ID directive comes the *title* directive, which is introduced by \$. The title appears in the Search and History lists, which you access with the buttons at the top of the AutoCAD Help dialog. Figure 14.20 shows the AutoCAD Help Search subdialog.

Next comes the *key words* directive, which is introduced by \K. You can list more than one key word by separating them with semicolons. Key words also appear in the Search list, and thus provide an alternative way of locating a topic.

The *end of file* directive (/E) occurs only once and signals the end of the AHP file to AutoCAD. Any line that begins with / and then a space is ignored by AutoCAD.

If a section exceeds one vertical screen page in length, AutoCAD's Help dialog allows you to scroll through the text with the vertical scroll bar. AutoCAD automatically wraps long text lines. It treats hard returns as paragraph separators (as do most word processors)

Creating an Integrated Professional Menu System

Figure 14.19
Platform-independent help for the LINE command.

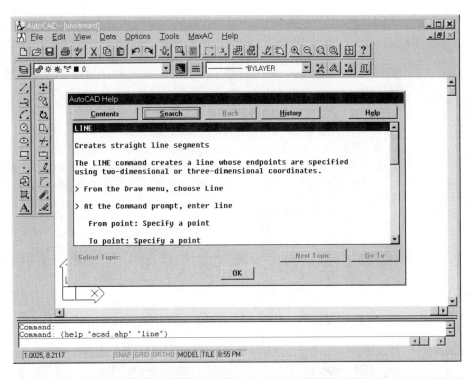

Figure 14.20
The AutoCAD Help Search sub-dialog displays title and key words directives.

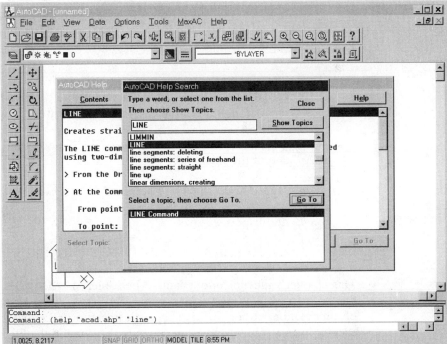

and adds a blank line when displaying the text. If you want to put a return in the AHP file, but don't want AutoCAD to add a blank line, end the line with a backslash (\).

The other special formatting enables hypertext links. The format is:

`<<jump word or phrase>>jump_to_topic_ID>`

You enclose the word or phrase that provides the link in double angle brackets, and follow it by the topic ID of the topic to jump to, and then a single angle bracket. The user sees the jump word or phrase in double angle brackets, but not the jump to topic ID. For example, the LINE example shown earlier includes this hypertext link:

`<<PLINE>>pline>`

A variation is to replace the final angle bracket with a square bracket:

`<<jump word or phrase>>jump_to_topic_ID]`

In this case, AutoCAD leaves the current topic in the main AutoCAD Help dialog and displays the topic in a More AutoCAD Help subdialog. When the user chooses Close, the original topic reappears.

Developing Platform-Independent Help Files

Application help files often come from documentation created in a word processor. The general procedure for converting a word processor file to an AutoCAD platform-independent help file is as follows:

1. Save the documentation as an ASCII text file with an AHP extension. Usually you do not want to save with line breaks—keep each paragraph as one long line in the text file and let AutoCAD wrap the lines.

2. Add comment lines at the top of the file and the \E directive at the end of the file.

3. Add topic ID, title, and key word directives for each section.

4. Add hypertext links.

5. Test the help file.

This chapter's files include a sample help file, MAWIN.AHP. This file documents the functions in MAWIN.ARX. We created MAWIN.AHP from the MAWIN.ARX function documentation earlier in this chapter, using the procedure just described. Here's an excerpt from MAWIN.AHP. Some lines are very long in the file but are printed on several lines here, so the [Enter] shows where the returns are in the file:

```
\ MAWIN.AHP: Maximizing AutoCAD sample help file [Enter]
\ by MM 14 June 1996 [Enter]
\ [Enter]
\#main_index [Enter]
\$Contents [Enter]
\Kcontents;TOC;table of contents [Enter]
CONTENTS [Enter]
This sample help file (MAWIN.AHP) documents the functions in
MAWIN.ARX. Refer to MAWIN.TXT and MAWIN.LSP for additional
information. [Enter]
Functions: \ [Enter]
 <<AddToRecentDocs>>AddToRecentDocs> \ [Enter]
 <<Beep>>Beep> \ [Enter]
 <<DeleteFile>>DeleteFile> \ [Enter]
 <<FileExists>>FileExists> \ [Enter]
 <<GetProfileString>>GetProfileString> \ [Enter]
 <<GetTempDirectory>>GetTempDirectory> \ [Enter]
 <<GetWindowsDirectory>>GetWindowsDirectory> \ [Enter]
 <<GetProfileInt>>GetProfileInt> \ [Enter]
 <<MessageBeep>>MessageBeep> \ [Enter]
 <<MessageBox>>MessageBox> \ [Enter]
 <<ShellOpen>>ShellOpen> \ [Enter]
 <<Sound>>Sound> [Enter]

...

\#DeleteFile [Enter]
\$DeleteFile function [Enter]
\KDeleteFile;delete a file;erase a file;temporary files [Enter]
DeleteFile function [Enter]
Syntax: [Enter]
  (DeleteFile file_name) or (DeleteFile file_name_list) [Enter]
Deletes one or more files. [Enter]
Returns the number of files deleted, or nil if there are no
such
files. [Enter]
Examples: [Enter]
  (DeleteFile "$TEMP$.$$$") [Enter]
If $TEMP$.$$$ exists in the current directory, this example de-
letes it and returns 1. If $TEMP$.$$$ does not exist in the cur-
rent directory, this example returns nil. [Enter]
```

```
        (DeleteFile '("$TEMP1$.$$$" "$TEMP2$.$$$")) [Enter]
This example deletes $TEMP1$.$$$ and $TEMP2$.$$$ (assuming they
exist) and returns 2. [Enter]
DeleteFile is useful for deleting temporary files that your
application creates. [Enter]
[Enter]
\E [Enter]
```

You can experiment with this help file by using the AutoLISP **help** function:

```
(help "MAWIN.AHP")
```

or:

```
(help "MAWIN.AHP" topic_ID)
```

Figure 14.21 shows the results of entering **(HELP "MAWIN.AHP" "DELETEFILE")**.

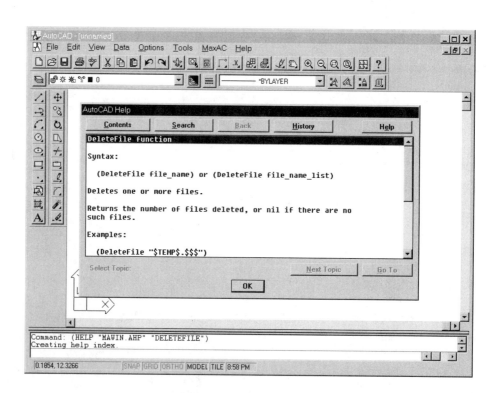

Figure 14.21
The custom MAWIN.AHP help file.

Finishing Up

Use the MENU command to change the menu back to \MAXAC\ACADMA.MNS, or to any other menu that you want to use for further customization work.

In this chapter, you modified ACAD.LSP and MAXAC.MNL. If you want to keep any of these changes or any of the other chapter support files, move them to the \MAXAC directory, then delete or move any remaining files out of \MAXAC\DEVELOP.

Conclusion

With the techniques shown in this chapter, you should be ready to develop sophisticated custom AutoCAD menus and other application components. Use ideas from all of the chapters in the book and files from the MaxAC CD-ROM to develop menus and other support files that fit your discipline and office standards. And see the appendixes for even more information.

As your AutoCAD customization experience grows, you'll find that a deeper understanding of AutoLISP and dialog box programming will be helpful. This book's companion volume, *Maximizing AutoLISP for AutoCAD R13*, will take you that next step when you're ready to go.

appendix

AutoLISP Function Quick Reference

Each AutoLISP function is listed alphabetically, with a brief description of the function's action, the results returned, and its syntax, showing number and data type of arguments.

KEY:

. . . means additional arguments may follow.

`identifer` is a required argument.

`[identifer]` is an optional argument.

(+ number...)

*+ (addition) returns the sum of 0 and all **number**s. An integer is returned if all **number**s are integers; a real is returned if any **number** is a real.*

(- number...)

*- (subtraction) returns the difference of 0 and the sum of all **number**s. An integer is returned if all **number**s are integers; a real is returned if any **number** is a real.*

(* number...)

** (multiply) returns the product of 1 and all **number**s. An integer is returned if all **number**s are integers; a real is returned if any **number** is a real.*

(/ number...)

*/ (division) returns the quotient of the first **number** divided by the product of the remaining **number**s and 1. An integer is returned if all **number**s are integers; a real is returned if any **number** is a real. If all **number**s are integers, the remainder is dropped.*

(= expr...)

= *(equality) returns* **T** *(true) if all* **expr(s)** *are equal (that is, evaluate to the same value), otherwise returns* **nil**. *Does not compare the contents of lists.*

(/= expr...)

/= *(not equal) returns* **T** *if all* **expr(s)** *are not equal, otherwise returns* **nil**. *Does not compare the contents of lists.*

(< expr...)

< *(less than) returns* **T** *if each* **expr** *is less than all subsequent* **expr** *arguments, otherwise returns* **nil**. *Only numbers and strings are valid.*

(<= expr...)

<= *(less than or equal to) returns* **T** *if each* **expr** *is less than or equal to all subsequent* **expr** *arguments, otherwise returns* **nil**. *Only numbers and strings are valid.*

(> expr...)

> *(greater than) returns* **T** *if each* **expr** *is greater than all subsequent* **expr** *arguments, otherwise returns* **nil**. *Only numbers and strings are valid.*

(>= expr...)

>= *(greater than or equal to) returns* **T** *if each* **expr** *is greater than or equal to all subsequent* **expr** *arguments, otherwise returns* **nil**. *Only numbers and strings are valid.*

(~ number)

~ *(tilde) returns the bitwise NOT of the integer* **number**.

(1+ number)

Returns **number** *incremented by 1.*

(1- number)

Returns **number** *decremented by 1.*

(abs number)

Returns the absolute value of **number**.

(acad_colordlg colorindex [flag])

Displays the standard AutoCAD color selection dialog, which allows the user to select a physical or logical color. The **colorindex** *is an integer from 0 to 256 (inclusive) that specifies* the default color. The value 0 specifies the logical color of BYBLOCK. The value 256 specifies the logical color of BYLAYER. If **flag** is supplied and **nil**, the Byblock and Bylayer buttons on the dialog box are disabled and **colorindex** may not be 0 or 256. If **flag** is omitted or non-nil, the Byblock and Bylayer buttons are enabled.

(acad_helpdlg *helpfile topic***)**

This function is obsolete, and is replaced by the built-in function **help**. See **help** for a description of this function.

(acad_strlsort *list***)**

Returns a copy of **list** (a list of strings) lexically sorted in ascending order.

(action_tile *key action***)**

Associates the DCL dialog box tile whose key attribute value is the string **key**, with the action expression **action**, a string containing an AutoLISP expression. Returns **T** on success, **nil** otherwise.

(add_list *item***)**

Adds to or replaces items in a DCL list box or popup list tile, depending on the mode argument in a previous call to **start_list**. The **item** is a string which is to be added to the list box, or is to replace an existing item in the list box. See the **start_list** and **end_list** functions.

(ads)

Returns a list of the names and paths of the currently loaded ADS applications. If no ADS applications are loaded, it returns **nil**.

(alert *string***)**

Displays **string** in an alert box. **string** can be multiple lines of text delimited by the newline (\n) character.

(alloc *nodes***)**

Specifies the number of **nodes** (an integer) to expand heap space by when requested by the **expand** function. Returns the previous value assigned or the default.

(and *expr...***)**

Returns **T** if all **expr**(s) evaluate to non-nil values, otherwise returns **nil**. Stops evaluating arguments at the first non-nil result.

(angle *point1 point2*)

> Returns an angle in radians, between the positive X-axis and a line from **point1** to **point2**, measured counterclockwise.

(angtof *string* [*mode*])

> Returns the floating-point value in radians of **string**. The angle display format of **string** is indicated by the optional **mode** argument: 0 = decimal degrees; 1 = degrees, minutes, seconds; 2 = grads; 3 = radians; 4 = surveyor's units. If **mode** is not specified, the current value of AUNITS is used. Returns **nil** if **string** is not a valid angle input expression.

(angtos *angle* [*mode* [*precision*]])

> Returns a string conversion of **angle** from radians to the units specified by **mode**. The conversion defaults to the current angular units and precision unless otherwise specified with the optional **mode** and **precision** arguments.

(append [*list*]...)

> Returns a list that is a concatenation of one or more **lists**.

(apply *function arglist*)

> Applies **function** to the arguments supplied in the list **arglist**. Returns the result of the call to **function**.

(arx)

> Returns a list of the names and paths of the currently loaded ARX applications. If no ARX applications are loaded, it returns **nil**.

(arxload *arxfile* [*onfailure*])

> Loads AutoCAD Runtime Extension (ARX) applications. **arxfile** (a string) is the file specification of the application file (an extension of ARX is assumed). **onfailure** is an expression that is returned unevaluated if the specified file could not be loaded.

(arxunload *arxfile* [*onfailure*])

> Unloads a curently loaded AutoCAD Runtime Extension (ARX) application. **arxfile** (a string) is the file specification of the application file. **onfailure** is an expression that is returned unevaluated if the specified file could not be unloaded.

(ascii *string*)

> Returns the integer ASCII code of the first character in **string**.

(assoc key list)

*Returns the first sublist in **list** (a list of lists), whose first element is equal to **key**. Returns **nil** if no sublist in **list** has a first element that is equal to **key**.*

(atan number1 [number2])

*Returns the arctangent of **number1**, from -pi to pi. If **number2** is provided, the arctangent of **number1/number2** is returned. If **number2** is 0, either -pi/2 or +pi/2 radians (-90 or +90 degrees) is returned, depending on the sign of **number1**.*

(atof string)

*Returns a real number converted from **string**. Returns 0.0 if **string** does not contain a real number.*

(atoi string)

*Returns an integer converted from **string**. Returns 0 if **string** does not contain an integer.*

(atom expr)

*Returns **T** if **expr** is an atom (**not** a list), otherwise returns **nil**. Note that the special value **nil** is both a list and an atom.*

(atoms-family formatmode [comparelist])

*Returns a list of non-nil AutoLISP symbols (atoms). It also can compare a list of atoms (with the optional **comparelist** argument) to determine if its atoms are on the atom list. If **formatmode** is 0, it returns the list as atoms (symbol names); if **formatmode** is 1, it returns the list as strings. **comparelist** is an optional list of strings.*

(autoarxload arxfile cmdlist)

*Defines stub commands that load a specified ARX application. **arxfile** is the file specification of the ARX file (the extension ARX is assumed). **cmdlist** is a list of strings that are the names of the commands to be defined, which automatically load the ARX program the first time one of the commands is invoked.*

(autoload lspfile cmdlist)

*Defines stub commands that demand-load an AutoLISP application when one of the commands is invoked. **lspfile** is the file specification of the AutoLISP file (the extension LSP is assumed). **cmdlist** is a list of strings that are the names of the commands to be defined, which automatically load the specified AutoLISP program the first time one of the commands is invoked.*

(autoxload *adsfile* **cmdlist)**

> *Defines stub commands that load a specified ADS application.* **adsfile** *is the file specification of the ADS file (the extension .EXE is assumed on all Windows platforms, and EXP on DOS platforms).* **cmdlist** *is a list of strings that are the names of the commands to be defined, which automatically load the ADS program the first time one of the commands is invoked.*

(boole *function integer...***)**

> *Applies one of 16 possible Boolean operations, determined by the integer* **function**, *to one or more* **integer** *arguments.*

(boundp *sym***)**

> *Returns* **T** *if the symbol* **sym** *is assigned to a non-nil value, otherwise returns* **nil**.

(car *list***)**

> *Returns the first element in* **list**. *Use* **car** *to extract the X coordinate of a point list.*

(cadr *list***)**

> *Returns the second element in* **list**. *Use* **cadr** *to extract the Y coordinate of a point list.*

(caddr *list***)**

> *Returns the third element in* **list**. *Use* **caddr** *to extract the Z coordinate of a point list.*

(cdr *list***)**

> *Returns a copy of* **list** *with the first element removed.*

(c????r *list***)**

> *Returns an element or list from* **list**, *specified by the combination of* **a** *and* **d** *characters in the* **c????r** *expression, up to four levels deep. For example,* **caadr**, **cddr**, **cadar**, *and so on.*

(chr *integer***)**

> *Returns a single character string containing the character specified by the ASCII code* **integer**.

(client_data_tile *key clientdata***)**

> *Associates the string* **clientdata** *with the DCL dialog box tile whose key attribute value is equal to the string* **key**. *Returns* **T** *on success, and* **nil** *otherwise.*

AutoLISP Function Quick Reference

(close *handle***)**

Closes the file specified by the **handle** *argument (the integer that was returned by* **open** *to open the file), and returns* **nil**. *The* **handle** *argument is an integer that must be a valid handle of a currently open file.*

(command *[expr...]***)**

Evaluates each **expr** *and sends the result to AutoCAD as input. Each* **expr** *argument must evaluate to a string, a number, or* **nil**. *If* **command** *is called with* **nil** *or with no arguments, the effect is the equivalent of pressing CANCEL (ESC or CTRL+C). Passing a null string* **(""")** *is the same as pressing ENTER. The symbol PAUSE (a string variable set to* **"\\"**), *causes AutoCAD to pause for interactive input by the user.*

(cond *clause...***)**

Evaluates the first expression in each **clause** *(a list consisting of one or more expressions) in the order supplied, until a non-nil result is returned or all clauses have been evaluated. If the result of the first expression in a* **clause** *is non-nil, all remaining expressions in the same* **clause** *are evaluated, any subsequent clauses are ignored, and the result of the last expression in the successful clause is returned.*

(cons *expr1 expr2***)**

If **expr2** *is a list,* **cons** *returns a new list with* **expr1** *as the new first element. If* **expr2** *is not a list,* **cons** *returns a dotted pair, in the form* **(expr1 . expr2)**.

(cos *angle***)**

Returns the cosine of **angle**, *an angle expressed in radians.*

(cvunit *number from to***)**

Returns a real in the **to** *system of measure, converted from the number in the* **from** *system of measure.* **From** *and* **to** *are strings corresponding to definitions in the ACAD.UNT file.*

(defun *name* **(***[argument]... [/ local...]***)** *expression...***)**

Creates a function with a given **name** *(a symbol). The* **argument** *list can supply variables to be passed to the function. Argument list variables following an optional slash are variables that are local to the function and do not affect other functions. The function evaluates the program expressions and returns the result of the last* **expression** *evaluated. Prefixing a C: to the function name creates a LISP command that acts like a standard AutoCAD command. Defining an* **S::STARTUP** *function in the ACAD.LSP file creates an automatic executing function.*

(dictadd *dictname key newobj***)**

> Adds a new entry to a dictionary. **dictname** is the entity name of the dictionary to which the object is to be added. The string **key** is the key name of the new object, which must be unique within the dictionary. **newobj** is the entity name of an existing non-graphical object that is to added to the dictionary. If an object already resides in the dictionary with the same key name as **key**, the operation fails. If **newobj** is added to the dictionary, **dictadd** returns **newobj**, and **nil** otherwise.

(dictnext *dictname* [*rewind*]**)**

> Successively returns each entry in a dictionary. **dictname** is the entity name of the dictionary to access. If **rewind** is supplied and non-nil, the first entry in the dictionary is returned. When called repeatedly without **rewind**, **dictnext** returns the dictionary entry that follows the one it returned last. When all dictionary entries have been returned, **dictnext** returns **nil**.

(dictremove *dictname key***)**

> Removes the dictionary whose key name is the string **key** from the dictionary specified by the entity name **dictname**.

(dictrename *dictname oldkey newkey***)**

> Renames the key name of the dictionary entry whose existing key name is the string **oldkey**, to the string **newkey**, in the dictionary specified by the entity name **dictname**.

(dictsearch *dictname key* [*setnext*]**)**

> Searches the dictionary specified by the entity name **dictname** for an entry whose key name is equal to the string **key**, and returns the entry if found, or **nil** if not found. If **setnext** is supplied and non-nil, the entry that follows the one returned is returned by the next call to **dictnext**.

(dimx_tile *key***)**

> Returns the X dimension, in screen pixels, of the DCL image tile whose key attribute value is the string **key**.

(dimy_tile *key***)**

> Returns the Y dimension, in screen pixels, of the DCL image tile whose key attribute value is the string **key**.

(distance *point point***)**

> Returns the 3D distance in drawing units between two 3D or 2D **point**s.

AutoLISP Function Quick Reference

(distof *string* [*mode*]**)**

Converts a string containing a valid distance expression to a real number. **string** is a string that contains a distance expression in the format accepted by AutoCAD on the command line, depending on the current units setting. **mode** is a positive, non-zero integer that specifies the units format accepted in **string**, which corresponds to the value of the LUNITS system variable. If **mode** is omitted, it defaults to the current value of the LUNITS system variable.

(done_dialog [*status*]**)**

Closes an active DCL dialog box. The optional integer **status** is returned by the call to **start_dialog** that activated the dialog box. If **status** is not present, a value of 1 is returned by **start_dialog** if the dialog was accepted, or 0 if the dialog was canceled. Returns the location of the upper left corner of the dialog box in left-hand screen pixel coordinates.

(end_image)

Ends output sequence to a DCL image tile that was started by a prior call to **start_image**.

(end_list)

Ends DCL list box or popup list transaction that was started by a prior call to **start_list**.

(entdel *ename***)**

Toggles the deleted state of the entity whose entity name is **ename.** Returns **ename**.

(entget *ename* [*applist*]**)**

Returns an entity data list describing the entity whose entity name is **ename** (an entity name). The optional **applist** argument is a list of registered applications for which extended entity data is to be returned along with normal entity data. If **ename** is the name of a deleted entity, **entget** returns **nil**.

(entlast)

Returns the last non-deleted entity name in the database.

(entmake *edata***)**

Creates a new entity described by the entity data list **edata**. Returns **edata** if successful, otherwise returns **nil.**

(entmakex edata)

Creates a new non-graphical database object described by the entity data list **edata**. Returns the entity name of the object if successful, otherwise returns **nil**.

(entmod edata)

Modifies an existing entity in the database in accordance with the entity data list **edata** and returns the new entity data list.

(entnext [ename])

Returns the entity name of the first non-deleted entity in the database. If the optional entity name **ename** is provided, the entity name of the first non-deleted entity immediately following the entity specified by **ename** is returned.

(entsel [prompt])

Displays a pickbox cursor and the prompt Select object: on the console, and waits for the user to select an object. If an object is selected, returns a list whose first element is the entity name of the selected object, which is followed by the selection point as a list of three real numbers. If the optional **prompt** string is provided, it is displayed in place of the default prompt. Returns **nil** if no object is selected.

(entupd ename)

Explicitly updates (regenerates) the entity whose entity name is **ename**. Used primarily to update a complex entity (POLYLINE or INSERT with attributes) after modifying one or more subentities.

(eq expr1 expr2)

Returns **T** if **expr1** is equal to **expr2**. Does not compare the contents of lists and should not be used with same. Otherwise, it returns **nil**.

(equal expr1 expr2 [accuracy])

Returns **T** if **expr1** is equal to **expr2**, and **nil** otherwise. Compares the contents of lists. The optional **accuracy** value determines the maximum difference between two numbers in order to qualify as equal.

(*error* string)

A user-definable error function. The **string** contains a message describing the error.

(eval expr)

Explicitly evaluates **expr**, an expression, and returns its result.

(exit)

Terminates the current application and generates an error condition that will cause AutoLISP to call the ***error*** function if defined.

(exp number)

Returns **e** (the base of natural logarithms, approximately 2.71828183), raised to the power of **number**, as a real.

(expand segments)

Attempts to expand the number of available nodes by **segments**. The number of nodes per segment is defined by calling the **alloc** function, and defaults to a value of 512. Returns the number of segments that were successfully acquired.

(expt base power)

Returns the number **base** raised to the number **power**. If **base** and **power** are integers, the result is an integer. Otherwise, the result is a real number.

(fill_image x1 y1 x2 y2 color)

Draws a rectangle in a DCL image tile in the specified **color** with the diagonal corners **x1, y1** and **x2, y2** respectively.

(findfile file)

If **file** is a filename without a drive or path prefix, **findfile** searches the AutoCAD library search path for **file**, and returns its full path if found, or **nil** if not found. If **file** includes a drive or path prefix, and the file exists in the specified location, its full path is returned, and **nil** otherwise.

(fix number)

Returns **number** converted to the next lowest integer.

(float number)

Converts **number** to a real number.

(foreach symbol list expression...)

Evaluates all **expression**(s) once for each element in **list**, each time assigning the successive element in **list** to the locally bound variable **symbol**. Returns the result of the last **expression**.

(gc)

Performs an explicit garbage collection causing all dirty nodes to be recycled and made available for use.

(gcd integer integer)

Returns the greatest common denominator of two **integer**s.

(get_attr key attr)

Returns the initial attribute value of the attribute **attr** (a string), for the DCL tile whose key attribute value is the case-sensitive string **key**. If no value exists for the specified tile and attribute, a null string is returned.

(get_tile key)

Returns the runtime (current) value of the DCL tile whose key attribute value is the case-sensitive string **key**. If the specified tile does not exist, **nil** is returned.

(getenv varname)

Returns the value of the environment variable specified by the string **varname**. Returns **nil** if the environment variable does not exist.

(getcfg key)

Searches for configuration variable in [APPDATA] section of ACAD.CFG, specified by the string **key**. Returns the variable's value if found, or **nil** if not found.

(getcname cmdname)

Translates between local and global command names for built-in commands. The string **cmdname** must be the local or global name of a built-in command. If **cmdname** is the local name, the global name of the same command is returned. If **cmdname** is the global name, the local name of the same command is returned.

getxxxx input functions

All the arguments to **get** functions are optional except for the **basepoint** argument to **getcorner**. All **get** functions accept a **prompt** argument that is displayed on the text console. When used in menu macros, all calls to **get** functions require a backslash to pause for input. When used after calling the **initget** function, all **get** functions except **getstring** can return option key word strings. Table A.1 shows the syntax of the **get** functions.

AutoLISP Function Quick Reference

(getfiled *title filename extension flags***)**

Displays a file dialog box with a title of **title** *with a default* **filename** *and* **extension***. The* **flags** *argument specifies bit-coded dialog box characteristics.*

(getvar *varname***)**

Returns the value of the AutoCAD system variable specified by the string **varname***.*

(graphscr)

Switches from the text screen to the graphics screen on single-screen DOS systems. Activates the main AutoCAD window on windowed systems. Returns **nil***.*

(grclear)

Clears the current viewport. A redraw refreshes the screen.

Table A.1
Get Functions

GET FUNCTION	EXPLANATION
(getangle [*basept*] [*prompt*]**)**	Acquires angle from two points or typed input based on current settings for ANGBASE and ANGDIR.
(getcorner *basept* [*prompt*]**)**	Acquires second corner of a rubber-banded rectangle.
(getdist [*basept*] [*prompt*]**)**	Acquires distance from two points or typed input.
(getint [*prompt*]**)**	Acquires an integer.
(getkword [*prompt*]**)**	Acquires one of a list of predefined key words.
(getorient [*basept*] [*prompt*]**)**	Like **getangle** but ignores non-East base angle setting.
(getpoint [*basept*] [*prompt*]**)**	Acquires a point with optional rubber-band vector.
(getreal [*basept*] [*prompt*]**)**	Acquires a real number.
(getstring [*flag*] [*prompt*]**)**	Acquires a string. If **flag** is supplied and non-nil, spaces are accepted, and input must be terminated by pressing ENTER.

(grdraw frompt topt color [mode])

Draws a vector between the two supplied points **frompt** and **topt** in the color specified by the integer **color**. A negative color argument complements any color it is drawn over (the XOR color). If drawn over itself, it erases itself. A non-zero optional **mode** argument highlights (usually dashing) the vector.

(grread [track] [allkeys [cursor]])

Reads the input to any device directly. If the optional **track** argument is present and non-nil, it returns the current pointing device (mouse or digitizer cursor) location without waiting for a point pick. The optional **allkeys** is bit-coded and indicates how various input keys are handled. The optional **cursor** argument is a binary-encoded integer that specifies the type of cursor displayed.

(grtext [box text [mode]])

Writes a string in the text portion of the graphics screen specified by the **box** number. An optional non-zero **mode** integer highlights the box of text, and zero de-highlights it. A box number of -1 writes to the status line, -2 writes to the coordinate status line, and 0, 1, 2, 3 and so on write to the side-screen menu labels, with 0 representing the top label.

(grvecs vlist [trans])

Batch-draws vectors on the graphics screen. **vlist** is a list of vector endpoints and color directives, in the form:

([color [frompoint topoint]...]...)

Trans is an optional transformation matrix that is applied to all vectors in **vlist**.

(handent handle)

Returns the entity name of the entity whose handle is the string **handle**.

(help [file [topic [command]]])

Invokes the help facility on all platforms. **File** is a string that specifies the name of the help file. If **file** is a null string, the default AutoCAD help file is used. **Topic** is a string that specifies the initial help topic to display. If **topic** is a null string, the contents are displayed. **Command** is a string that specifies the initial state of the help file, which must be one of the following strings:

AutoLISP Function Quick Reference

Command String	Meaning
HELP_CONTENTS	Displays help contents
HELP_HELPONHELP	Displays help for help
HELP_PARTIALKEY	Displays topic search dialog

(if *testexpr thenexpr* [*elseexpr*]**)**

If *testexpr* evaluates to a non-nil value, *thenexpr* is evaluated. If *test* is **nil**, *thenexpr* is not evaluated and the optional *elseexpr*, if supplied, is evaluated. The function returns the value of the evaluated expression or **nil** if no expression was evaluated.

(initget *bits string***)**

Establishes options for the **getxxx** functions. The **bits** set input filtering options, and the **string** sets key words. It returns **nil**.

1	Null input is not allowed.
2	Zero values are not allowed.
4	Negative values are not allowed.
8	Does not check limits.
16	Not used.
32	Dashed lines used for rubber-banding.
64	No Z value allowed.
128	Allows arbitrary keyboard input.

(inters *point1 point2 point3 point4* [*onseg*]**)**

Returns the point of intersection of a line between the first two **point**s and a line between the second two **point**s. If the optional **flag** is present and **nil**, the lines are infinitely projected to calculate the intersection. Otherwise, if flag is omitted or non-nil, the intersection must lie between the endpoints of both lines.

(itoa *integer***)**

Returns a string conversion of *integer*.

(lambda ([*argument*]... [**/** *local*...]**)** *expression*...**)**

Defines an anonymous function. See **defun** for format.

(last *list*)

Returns the last element from **list**.

(length *list*)

Returns the number of elements in **list**.

(list [*expr*]...)

Returns a list constructed from the supplied **expr**(s).

(listp *item*)

Returns **T** if **item** is a list. Otherwise, returns **nil**.

(load *filename* [*onfailure* [*verbose*]])

Loads the AutoLISP file specified by the **filename**. The optional **onfailure** argument is returned without evaluation if the operation fails. The optional **verbose** argument, if supplied, causes the file to be echoed to the console as it is loaded.

(load_dialog *filename*)

Loads the DCL file specified by filename, and returns a handle to the loaded file.

(log *number*)

Returns the natural log of the supplied **number** as a real.

(logand *integer*...)

Returns the bitwise logical AND of one or more integers.

(logior *integer*...)

Returns the bitwise logical OR of one or more integers.

(lsh *integer num*)

Returns **integer** with its bits shifted by **num** places. If **num** is positive, the bits in **integer** are shifted left by **num** places. If **num** is negative, the bits in **integer** are shifted right by **num** places.

(mapcar *function list*...)

Sequentially applies the specified **function** on each set of elements in one or more argument **list**s, and returns the results as a list.

AutoLISP Function Quick Reference

(max *number***...)**

    Returns the largest **number**.

(mem)

    Reports the usage of AutoLISP memory.

(member *expr list***)**

    If **expr** is found in **list**, returns the remainder of the **list** starting at the **expr**. Otherwise, it returns **nil**.

(menucmd *string***)**

    Issues the menu command specified by **string**. MENUCMD can load and display menu pages or issue menu label control codes such as for graying or checking of menu items. The **string** is exactly the same as the menu command you would put in a menu macro, except you omit the leading $ code, such as "S=NAME" to change the screen device page named NAME. Can also be used to evaluate DIESEL expressions from AutoLISP by prefixing the "M=" directive.

(min *number***...)**

    Returns the smallest **number**.

(minusp *number***)**

    Returns **T** if **number** is negative. Otherwise, returns **nil**.

(mode_tile *key mode***)**

    Sets the enabled, selected, and focus states of DCL tiles. **Key** is the value of the key attribute of the tile whose mode is to be changed, and **mode** is an integer from 0 to 4 that specifies the state to alter, as follows:

        0   Enable tile

        1   Disable tile

        2   Set focus to tile

        3   Select edit box contents

        4   Toggle image highlighting

(namedobjdict)

    Returns the entity name of the root dictionary object.

(nentsel *prompt***)**

> For simple entities, behaves exactly like **entsel**. For polylines, returns a list containing the vertex subentity name and the point coordinates used to select the polyline. For block inserts with attributes, if an attribute is selected, returns the attribute subentity name and pick point. For block inserts, if other than an attribute is selected, it returns a list containing the entity name of the selected entity within the block definition, a matrix of points used for coordinate system transformations, and the block definition or nested block definitions that the selected entity is within. The optional **prompt** string can provide specific instructions for entity selection.

(nentselp *prompt pt***)**

> Operates similarly to **nentsel** but accepts **pt**, an optional point argument which is substituted for the user selection point. See the AutoLISP Programmer's Reference Manual for more information.

(new_dialog *dlgname handle* [*action* [*position*]]**)**

> Creates a new instance of a dialog box. **dlgname**, a string, is the name of the dialog box. **handle**, an integer, is the handle of the loaded DCL file containing the definition of the dialog box. The optional **action** string argument is the default tile action expression for all tiles in the dialog box. **position** is a list of two integers which specifies the screen coordinates of the dialog box. If the function succeeds, it returns **T**, otherwise it returns **nil**.

(not *expr***)**

> Returns **T** if **expr** evaluates to **nil**, and **nil** otherwise.

(nth *index list***)**

> Returns the element of **list** specified by the integer **index**. A value of 0 returns the first element in **list**.

(null *item***)**

> Identical to **not**.

(numberp *item***)**

> Returns **T** if **item** is a number. Otherwise, returns **nil**.

(open *filename mode***)**

> Opens a file specified by the string **filename** for the use specified by the one-character string **mode**. The modes are **r** for read, **w** for write, and **a** for append. **Mode** must be a lowercase character. Returns an integer file handle if the file is opened, which is used in subsequent operations involving the open file. Returns **nil** if the file was not opened.

AutoLISP Function Quick Reference

(or *expr***...)**

> Returns **T** and ceases further evaluation at the first non-nil **expr**. Otherwise, if all expressions evaluate to **nil**, returns **nil**

(osnap *point mode***)**

> Returns a point value specified by the object snap **mode** string on the supplied **point** value. The **mode** argument is a string, like "END,INT".

pause

> The constant **pause** is used in the **command** function to wait for user input.

pi

> The constant **pi** is set to approximately 3.1415926.

(polar *point angle dist***)**

> Returns a point calculated at **angle** and **dist** from a supplied base **point**.

(prin1 *expr* **[***file***])**

> Prints the **expr** to the screen and returns the **expr**. Control characters in strings are not expanded. The optional **file** argument (an integer) is the handle of an open file. If **file** is supplied and is open for writing, output is redirected to file.

(princ *expr* **[***file***])**

> Prints the **expr** to the screen and returns the **expr**. Control characters in strings are expanded. The optional **file** argument (an integer) is the handle of an open file. If **file** is supplied and is open for writing, output is redirected to file.

(print *expr* **[***file***])**

> Prints the **expr** on a new line to the screen and returns the **expr**. Control characters in strings are not expanded. The optional **file** argument (an integer) is the handle of an open file. If **fileg** is supplied and is open for writing, output is redirected to file.

(progn *expr***...)**

> Evaluates any number of **expr**(s) and returns the value of the last. Grouping a series of expressions within a **progn** allows them to be used where only one expression is allowed.

(prompt *string***)**

> Displays a **string** statement in the screen's prompt area. Returns **nil**.

(quote expr) or 'expr

Returns **expr** without evaluation; the abbreviated ' performs the same function.

(read string)

Converts a string into an AutoLISP expression. Only the first expression in the string is returned.

(read-char file)

Returns the ASCII character code of a single character typed at the keyboard or reads a character from an open file if the optional **file** handle (an integer) is supplied.

(read-line file)

Returns a string typed at the keyboard or a line from an open file if the optional **file** handle (an integer) is included.

(redraw [ename [mode]])

Redraws the current viewport unless **ename** is provided, in which case the entity represented by **ename** is redrawn. The **mode** option redraws the entity in four possible ways:

    1 = standard redraw,

    2 = reverse redraw (blank),

    3 = highlight redraw, and

    4 = de-highlight.

Returns **nil**.

(regapp appid)

Registers an application name with AutoCAD. AutoCAD places the name in the APPID symbol table. If the registration succeeds, the function returns the name. Otherwise, **nil** is returned.

(rem number number...)

Returns the remainder of the first **number** divided by the product of the rest of the **numbers**.

(repeat number expr...)

Evaluates each **expr** the **number** of times specified. **Number** must be a non-negative integer.

AutoLISP Function Quick Reference

(reverse *list***)**

Returns a copy of list with its elements in reverse order.

(rtos *number* **[***mode* **[***precision***]])**

Returns a string conversion of the supplied **number** *in the format of the current UNITS setting. If the optional* **mode** *and* **precision** *arguments are supplied, they override the current units settings.* **Mode** *specifies the unit type as shown below.*

    1 *Scientific*   4 *Architectural*

    2 *Decimal*   5 *Fractional*

    3 *Engineering*

(set *expr1 expr2***)**

Evaluates **expr1** *and assigns* **expr2** *to the result.* **expr1** *must evaluate to a symbol. Returns* **expr2**.

(set_tile *key value***)**

Assigns **value** *(a string) to the DCL tile whose key attribute value is the case-sensitive string* **key**.

(setq *sym expr* **[***sym2 expr2***]...)**

Sets **sym** *(a symbol) to the result of the* **expr** *that immediately follows. Returns the result of the last* **expr**.

(setcfg *key value***)**

Assigns the string **value** *to the configuration variable string* **key** *in the [APPDATA] section of ACAD.CFG. Returns the variable's value if found, or* **nil** *if not found. The* **key** *must have the form:*

    "**AppData/appname/[subitem1/].../subitemN**"

(setfunhelp *C:FUNCTION* **[***helpfile* **[***topic* **[***command***]]])**

Registers the user-defined command function **C:FUNCTION** *(a string) with the specified* **helpfile**, **topic**, *and* **command**. *See the* **help** *function for detailed descriptions of the* **helpfile**, **topic**, *and* **command** *arguments. This function is not operational in DOS AutoCAD.*

(setvar *sysvar value***)**

Sets the AutoCAD system variable specified by **sysvar** *to* **valueg** *and returns* **value**.

(setview viewdata [viewport])

Changes the view in the current viewport, or the viewport whose viewport ID is the optional integer **viewport**. *The view parameters are defined in the list* **viewdata**, *which has the same format as the entity data list returned by* **tblsearch** *for a view table entry.*

(sin angle)

Returns the sine of a radian **angle**.

(slide_image x1 y1 x2 y2 slidename)

Displays a slide in a DCL image tile. The slide location is defined by corner points **x1**, **y1**, **x2**, *and* **y2**. *The slide file or slide library file entry displayed is specified in the string* **slidename**.

(snvalid string [flag])

Returns **T** *if* **string** *is a valid symbol table name (such as a block, layer, and so on), and* **nil** *otherwise. If* **flag** *is supplied and non-nil, the vertical bar (|) is permitted in* **string**.

(sqrt number)

Returns the square root of a **number** *as a real.*

(ssadd [ename [selset]])

Creates an empty selection set when no arguments are provided. If the optional entity name **ename** *is provided, a selection set is created with just that entity name. If an entity name is provided along with an existing selection set* **selset**, *it is added to the selection set. Returns the new selection set.*

(ssdel ename selset)

Deletes the entity specified by **ename** *(and entity name) from the selection set specified by* **selset** *(a selection set) and returns the selection set.*

(ssget [mode [point1 [point2]] [pointlist]] [filterlist])

Returns a selection set of entities. If **ssget** *fails, it returns* **nil**. *With no arguments (or a* **nil**), **ssget** *uses AutoCAD's standard object selection to get a user selection. The optional* **mode** *arguments are strings that specify options to automate selection without user input. Modes are P for Previous, L for Last, I for Implied pickfirst pending selection set, W for Window, C for Crossing, WP for WPoly, CP for CPoly, X (not A) for all entities, and F for Fence. The optional* **point1** *argument alone selects a single point. The W and C modes require* **point1** *and* **point2** *arguments to specify their corner points. The WP, CP, and F modes require* **pointlist**, *a list of points. Any of these modes can be followed by*

filterlist, *which is a filtering list of properties. Only entities matching the filtering list are selected.*

(ssgetfirst)

Returns a list containing two selection sets. The first selection set in the list contains the entities that currently have grips attached to them. The second selection set in the list contains the entities that currently have grips attached to them and are selected. If there are no gripped and/or selected entities, **nil** *appears in place of either or both selection sets.*

(sslength *selset*)

Returns the number of entities in the selection set specified by **selset***.*

(ssmemb *ename selset*)

Returns the entity name **ename** *if it is in the selection set* **selset***. Otherwise, returns* **nil***.*

(ssname *selset number*)

Returns the entity name from **selset***, a selection set, that corresponds to the position* **number** *argument. The first entity is number 0.* **Number** *may be an integer from 0 to 32767, or a real like 32999.0, if you need to go beyond the limits of integers.*

(ssnamex *selset [index]*)

Returns information about how entities were added to a selection set. The **selset** *argument is a selection set. The optional* **index** *argument is an integer that specifies the entity in the selection set from which information is to be retrieved.*

(sssetfirst *gripped [selected]*)

Attaches grips and/or selects objects to become the current selection. **Gripped** *is a selection set containing the entities that are to have grips attached to them.* **selected** *is the selection set containing the entities that are to be selected and have grips attached to them.*

(start_dialog)

Starts a DCL dialog box event processing loop. Returns the value that is passed as an argument in a call to **done_dialog** *to close the dialog. If one or more dialogs were terminated by a call to* **term_dialog,** *then* **start_dialog** *returns -1. If no value is passed to* **done_dialog,** *and* **term_dialog** *was not called to close the dialog,* **start_dialog** *returns 1 if the dialog box was accepted, or 0 if it was canceled.*

(start_image *key*)

Specifies the DCL image tile to which all subsequent **fill_image***,* **vector_image***,* and **slide_image** *operations are applied, up to the next call to* **end_image**. *Key is the value of the target image tile's key attribute.*

(start_list *key* **[***operation* **[***index***]])**

Specifies the DCL list box or popup list to which all subsequent calls to **add_list** *are applied.* **Key** *is the value of the target list box or popup list's key attribute.* **Operation** *is an optional integer that specifies the operation that is to be performed on the list box or popup list.* **Index** *is an optional integer that specifies an existing item in a list box or popup list to which a delete or change operation is to be applied. The first item in a list box or popup list is referenced by an* **index** *value of 0.*

The possible values for the **operation** *argument are enumerated here.*

Value	Description
1	*Change the text of an item in an existing list*
2	*Append new item to an existing list*
3	*Replace existing list with a new list (default)*

(startapp *appname* **[***params***])**

Launches a Windows application or a DOS application controlled by a PIF file. **appname** *is a string containing the executable file specification. If only a filename is supplied in* **appname**, *the file is searched for in all directories specified by the system's primary search path (the PATH environment variable in DOS and Windows). The optional* **params** *string argument is passed directly to the launched application as a command-line parameter.*

(strcase *string* **[***flag***])**

Returns **string** *converted to uppercase unless the optional* **flag** *evaluates to* **T**, *which converts to lowercase.*

(strcat *string...***)**

Concatenates one or more **string** *arguments and returns them as a single string.*

(strlen *string...***)**

Returns the total number of characters in one or more **string** *arguments.*

(subst *newitem olditem list***)**

Returns a copy of **list** *with all occurrences of* **olditem** *replaced by* **newitem**.

(substr string start [length])

*Returns the portion of a **string** from the **start** position number of the supplied **string** either to the end of the **string** or to the end of the number of characters specified by the optional **length** value. A **start** value of 1 is the first character in **string**.*

(tablet code [row1 row2 row3 direction])

*Returns the current tablet configuration if **code** is equal to zero. Do not specify the optional arguments if **code** = 0. If code is equal to 1 the tablet is configured with the **row1 row2 row3 direction** arguments. The optional arguments are required if **code** = 1.*

(tblnext table [rewind])

*Returns the data of table objects in the table specified by the string **table**. If the optional **rewind** argument is supplied and non-nil, the data of the first entry in the table is returned, otherwise the data of the table entry that follows the one most recently returned by **tblnext** is returned. After the last entry in a table has been returned, subsequent calls to **tblnext** return **nil**. The next table entry returned by **tblnext** can also be established using the **tblsearch** function.*

(tblobjname table symbol)

*Returns the object name of the table entry whose name is the string **symbol**, from the table specified by the string **table**, or **nil** if the entry and/or table do not exist.*

(tblsearch table symbol [setnext])

*Searches for an entry in the table specified by the string **table**, with a name that is the string **symbol**. If an entry with the specified name is found, the data for that table entry is returned. If the optional **setnext** argument is supplied and non-nil, the next call to the **tblnext** function will return the table entry that follows the one found by **tblsearch**.*

(term_dialog)

*If an application is terminated while any DCL files are open, AutoCAD automatically calls **term_dialog**. This function is used mainly for aborting nested dialog boxes. The **term_dialog** function always returns **nil**. If this function is called to terminate a dialog, the call to **start_dialog** that started the dialog returns a value of -1.*

(terpri)

*Prints a new line on the screen and returns **nil**.*

(textbox elist)

*Returns a list of two 3D points representing the lower left and upper right corners of a box that encloses a text entity described by **elist**. The first returned point indicates the offset* from the insertion point and the second returned point indicates the distance from the first returned point. Does not work on MTEXT.

(textpage)

> **DOS:** *Switches from the graphics screen to the text screen on single-screen.*
> **Windows**: *Activates the text window in Windows AutoCAD.*
> **All platforms:** *Clears the screen/window and locates the cursor at the top.*

(textscr)

> **DOS:** *Switches from the graphics screen to the text screen on single-screen.*
> **Windows**: *Activiates the text window in Windows AutoCAD.*

(trace function...)

> Activates tracing for the specified **functions**(s), causing every call to them to be displayed on the text console, along with their arguments and result.

(trans point from to [flag])

> Returns a coordinate transformed from one coordinate system to another. **point** is the coordinate to be translated. **from** is the source coordinate system which the point is expressed in. **to** is the target coordinate system which the point is to be transformed into. Both **from** and **to** can be an integer code specifying a predefined coordinate system, a normal unit vector specifying an arbitrary coordinate system or an entity name of an entity that defines a coordinate system. The range of integer values for **from** and **to** are:
>
> 0 for World Coordinate System (WCS),
>
> 1 for User Coordinate System (UCS),
>
> 2 for Display Coordinate System (DCS, screen),
>
> 3 for Paper Space (PSDCS)
>
> If **flag** is present and non-nil, the point value is treated as a 3D displacement instead of as a point.

(type expr)

> Returns a symbol indicating the data type of **expr**, such as REAL, INTEGER, STRING, LIST, and so on.

(unload_dialog handle)

> Unloads a currently loaded DCL file specified by the integer **handle**, which is the value returned by **load_dialog**.

(untrace function...)

Removes the trace function from the supplied **function**(s).

(vector_image x1 y1 x2 y2 color)

Draws a vector in a DCL image tile, from the tile pixel coordinate **x1,y1** to the point **x2,y2**, in the color specified by the integer **color**.

(ver)

Returns a string with the current AutoLISP version.

(vmon)

Turns on AutoLISP's virtual paging of functions. It makes function definitions eligible to be swapped to disk or in and out of RAM to allow for the loading of more programs. Generally not required on 32-bit versions of AutoCAD and is provided for backward-compatibility.

(vports)

Returns a list of the current TILEMODE or paper space viewport settings. The list contains sublists with the viewport numbers and display coordinate corner points for each port. The active viewport is first on the list.

(wcmatch string pattern)

Returns **T** if **string** matches wild-card **pattern**. Otherwise, it returns **nil**. **Pattern** is a string containing one or more wildcard patterns.

(while testexpr expr...)

Evaluates **testexpr** first and if its result is non-nil, then each **expr** is evaluated in the order supplied until all **expr**'s are evaluated, and the entire process is repeated until **testexpr** evaluates to **nil**. Returns the result of the last **expr** to be evaluated. Note that if **testexpr** evaluates to **nil** the first time, then the **expr**'s are never evaluated.

(write-char code [file])

Writes a character specified by the ASCII character **code** integer to the console or a file specified by the optional integer file handle **file**.

(write-line string [file])

Writes **string** to the text console, or a file specified by the optional integer file handle **file**.

(xdroom ename)

Returns the number of bytes available for extended data in the entity specified by the entity name **ename**.

(xdsize list)

Returns the number of bytes an extended entity data **list** will occupy.

(xload file [onfailure])

Loads an ADS program file whose name is **file**. If successful or if the application is already loaded, the name of the ADS file is returned. If **onfailure** is supplied, it is returned without evaluation if an error occurs; otherwise, an error message is issued.

(xunload file [onfailure])

Unloads an external ADS program whose file name is the string **file** from the current drawing session. If successful, the name of the external ADS file is returned. If not successful and the optional **onfailure** argument is supplied, it is returned unevaluated, otherwise an error message is issued.

(zerop number)

Returns **T** if **number** equals 0 (zero). Otherwise, returns **nil**.

appendix

Diesel Function Quick Reference

Each DIESEL function is listed alphabetically, with a brief description of the function's action, the results returned, and its syntax, showing number and data type of arguments.

KEY:

... *means additional arguments may follow.*

identifier *is a required argument.*

[*identifier*] *is an optional argument.*

`$(/,arg1,arg2,...,arg9)`

Returns **arg1** *divided by* **arg2** *divided by* **arg3** ...

`$(*,arg1,arg2,...,arg9)`

Returns **arg1** *times* **arg2** *times* **arg3** ...

`$(+,arg1,arg2,...,arg9)`

Returns **arg1** *plus* **arg2** *plus* **arg3** ...

`$(-,arg1,arg2,...,arg9)`

Returns **arg1** *minus* **arg2** *minus* **arg3** ...

`$(and,integer1,...,integer9)`

Returns the sum of all bit codes that are present in all **integer** arguments. Otherwise, it returns 0.

$(angtos,angle[,mode[,precision]])

Returns a string conversion of **angle** from radians to the units specified by **mode**. The conversion defaults to the current angular units and precision if the optional **mode** and **precision** are omitted. The mode values are:

0   Degrees

1   Degrees/minutes/seconds

2   Grads

3   Radians

4   Surveyor's units

$(edtime,date,format)

Formats a Julian **date** into any or all of the following: day of the week, month, year, hour, minute and seconds. **format** is any combination of the format codes shown in Table 10.1 in Chapter 10.

$(eq,string1,string2)

Returns 1 (true) if **string1** is identical to **string2**, otherwise it returns 0 (false).

$(=,num1,num2)

Returns 1 (true) if **num1** is equal to **num2**, otherwise it returns 0 (false).

$(eval,string)

Evaluates **string** and returns the results. **$eval** takes only one string argument, but it can be a series of DIESEL expressions that are not separated by commas. Any non-DIESEL text within **string** is treated literally, meaning that spaces and punctuation are returned unaltered. **$eval** returns all text and the results of DIESEL expressions as a single combined string.

$(fix,number)

Returns an integer value of the **number** and drops the remainder.

$(getenv,envar)

Returns the value of **envar**, an environment variable name. If **envar** does not exist, a null string is returned.

$(getvar,sysvar)

Returns the value of **sysvar**, a system variable name.

$(>,num1,num2)

> **$>** *(greater than) returns 1 (true) if* **num1** *is greater than* **num2**, *otherwise it returns 0 (false).*

$(>=,num1,num2)

> **$>=** *(greater than or equal to) returns 1 (true) if* **num1** *is greater than or equal to* **num2**, *otherwise it returns 0 (false).*

$(if,test,thenexpr[,elseexpr])

> *If* **test** *is true (not false or 0), the* **thenexpr** *is evaluated. If the test is false (0), the optional* **elseexpr** *is evaluated. The* **$if** *function returns the value of the evaluated expression.*

$(index,position,list)

> *Returns the item specified by* **position** *(integer) in* **list**, *a string containing a comma-delimited list. The first item in the list is item number 0, not 1.* **$index** *returns "" (a null string) if* **position** + 1 *exceeds the number of items in* **list**.

$(<,num1,num2)

> **$<** *(less than) returns 1 (true) if* **num1** *is less than* **num2**, *otherwise it returns 0 (false).*

$(<=,num1,num2)

> **$<=** *(less than or equal to) returns 1 (true) if* **num1** *is less than or equal to* **num2**, *otherwise it returns 0 (false).*

$(linelen)

> *Returns the number of characters that can be displayed on the status line.*

$(!=,num1,num2)

> **$!=** *(not equal) returns 1 (true) if* **num1** *is not equal to* **num2**, *otherwise it returns 0 (false).*

$(nthindex,arg1,arg2,...,arg8)

> *Returns the* **arg**ument *specified by its* **index** *integer position in a list of arguments. It counts 0,1,2... and returns an $ (NTH, ??) error if* **index** *exceeds the number of arguments.*

$(or,integer1,...,integer9)

Returns the sum of all bit codes that are present in any of the **integer** arguments. If all arguments are 0, it returns 0.

$(rtos,number[,mode[,precision]])

Returns a string conversion of the supplied **number** in the format and with the accuracy of the current linear units format and precision settings, unless the optional **mode** and **precision** arguments are supplied. The optional **mode** and **precision** arguments override the current linear units settings. The mode values are:

1 *Scientific*   4 *Architectural*

2 *Decimal*   5 *Fractional*

3 *Engineering*

$(strlen,string)

Returns the length of **string**.

$(substr,string,start[,length])

Returns the portion of **string**, beginning with **start** (an integer) character position and continuing for **length** (an integer) characters. If **length** is not specified, the rest of **string** is returned.

$(upper,string)

Returns **string**, changed to all uppercase.

$(xor,integer1,...,integer9)

With two arguments, returns the sum of all bit codes that are present in only one of the **integer** arguments. With multiple arguments, compares the first to the second, then compares the result to the third, and so on. If both of only two arguments, or the last two of multiple arguments, are 0 or have all bits in common, it returns 0.

appendix C

DCL Quick Reference

This quick reference provides a tabular listing of all predefined DCL attributes, reserved words, and tile syntax descriptions.

DCL Tile Attributes

Table C.1
DCL Tile Attribute Summary

Attribute name	Associated Tiles	Description
action	all active tiles	AutoLISP action expression
alignment	all tiles	horizontal or vertical position in a cluster
allow_accept	edit_box, image_button, list_box,	activates is_default button when this tile is selected
aspect_ratio	image, image_button	aspect ratio of an image
big_increment	slider	incremental distance to move
children_alignment	row, column, radio_row, radio_column, boxed_row, boxed_column, boxed_radio_row, boxed_radio_column	alignment of a cluster's children

Attribute name	Associated Tiles	Description
`children_fixed_height`	`row, column, radio_row, radio_column, boxed_row, boxed_column, boxed_radio_row, boxed_radio_column`	height of a cluster's children doesn't grow during layout
`children_fixed_width`	`row, column, radio_row, radio_column, boxed_row, boxed_column, boxed_radio_row, boxed_radio_column`	width of a cluster's children doesn't grow during layout
`color`	`image, image_button`	background color of an image
`edit_limit`	`edit_box`	maximum number of characters that can be entered in edit box
`edit_width`	`edit_box, popup_list`	visible width of tile
`fixed_height`	all tiles	inhibits vertical expansion
`fixed_width`	all tiles	inhibits horizontal expansion
`fixed_width_font`	`list_box, popup_list`	uses fixed-pitch font in list
`height`	all tiles	height of the tile
`initial_focus`	`dialog`	key of the tile with initial focus
`is_bold`	`text`	displayed as bold (platform must support bold text)
`is_cancel`	`button`	button activated when the CANCEL key is pressed

DCL Quick Reference

Attribute name	Associated Tiles	Description
`is_default`	`button`	button activated when accept key is pressed
`is_enabled`	all active tiles	tile is initially enabled
`is_tab_stop`	all active tiles	tile is a tab stop
`key`	all active tiles	used by application to reference tile
`label`	`boxed_row, boxed_column, boxed_ radio_row, boxed_radio_column, button, dialog, edit_box, list_box, popup_list, radio_button, text, toggle`	displayed label of the tile
`layout`	`slider`	Specifies orientation (horizontal or vertical) of slider
`list`	`list_box, popup_list`	initial values displayed in list
`max_value`	`slider`	maximum value of slider
`min_value`	`slider`	minimum value of slider
`mnemonic`	all active tiles	mnemonic character of tile (can also be specified by prefixing character with & in tile's label attribute)
`multiple_select`	`list_box`	allows multiple items to be selected
`password_char`	`edit_box`	masks characters entered in edit box with specified character
`small_increment`	`slider`	incremental distance to move
`tabs`	`list_box, popup_list`	tab stops for list display

Attribute name	Associated Tiles	Description
`tab_truncate`	`list_box, popup_list`	truncates text at tab if too long
`value`	`text`, all active tiles except buttons and image buttons	initial value of tile
`width`	all tiles	width of tile

DCL Color Reserved Words

Table C.2
Color reserved words

Identifier	Description
`dialog_line`	Current dialog box line color
`dialog_foreground`	Current dialog box foreground color (for text)
`dialog_background`	Current dialog box background color
`graphics_foreground`	White
`graphics_background`	Current background of the AutoCAD graphics screen
`black`	AutoCAD (ACI) color = 0 (black) (light on dark background)
`red`	AutoCAD (ACI) color = 1 (red)
`yellow`	AutoCAD (ACI) color = 2 (yellow)
`green`	AutoCAD (ACI) color = 3 (green)
`cyan`	AutoCAD (ACI) color = 4 (cyan)
`blue`	AutoCAD (ACI) color = 5 (blue)
`magenta`	AutoCAD (ACI) color = 6 (magenta)
`white`	AutoCAD (ACI) color = 7 (white) (dark on light background)

DCL Tile Reference

Key to DCL Tile Reference

`attribute = [value1|default_value2|value3];`

Attributes whose values are limited to a set of reserved words appear with those eligible reserved words separated by vertical bars and enclosed in brackets. A reserved word that is shown in boldface type is the default value for the attribute.

Alignment attributes have two sets of bracketed values nested within another pair of brackets, separated by a vertical bar. In this case, the set of values that apply depends on whether the tile is a child of a row or column tile.

`attribute = type (default);`

Attributes whose values are variable appear with the type of data required for the attribute in bold italics, and the default value in parenthesis. The range of types is as follows:

Type	Description
number	An integer or real number
real	A real number
integer	An integer
string	A string (must be enclosed in double quotes)
color	An integer from 0 to 255 (AutoCAD Color Index) or a reserved color word (see Table C.2)

`[tile]...`

Indicates tile may contain one or more children.

Boxed Column

```
: boxed_column {
   alignment = [[top|centered|bottom]|[left|centered|right]];
   children_alignment = [left|centered|right];
   children_fixed_height = [true|false];
   children_fixed_width = [true|false];;
   fixed_height = [true|false];
   fixed_width = [true|false];
   height = number;
```

```
    label = string (" ");
    width = number;
    [tile]...
}
```

Boxed Radio Column

```
: boxed_radio_column {
    alignment = [[top|centered|bottom] | [left|centered|right]];
    children_alignment = [left|centered|right];
    children_fixed_height = [true|false];
    children_fixed_width = [true|false];
    fixed_height = [true|false];
    fixed_width = [true|false];
    height = number;
    label = string (" ");
    width = number;
    [tile]...
}
```

Boxed Radio Row

```
: boxed_radio_row {
    alignment = [[top|centered|bottom] | [left|centered|right]];
    children_alignment = [top|centered|bottom];
    children_fixed_height = [true|false];
    children_fixed_width = [true|false];
    fixed_height = [true|false];
    fixed_width = [true|false];
    height = number;
    label = string (" ");
    width = number;
    [tile]...
}
```

Boxed Row

```
: boxed_row {
    alignment = [[top|centered|bottom] | [left|centered|right]];
    children_alignment = [top|centered|bottom];
    children_fixed_height = [true|false];
    children_fixed_width = [true|false];
```

```
    fixed_height = [true|false];
    fixed_width = [true|false];
    height = number;
    label = string (" ");
    width = number;
    [tile]...
}
```

Button

```
: button {
    action = string;
    alignment = [[top|centered|bottom] | [left|centered|right]];
    fixed_height = [true|false];
    fixed_width = [true|false];
    height = number;
    is_cancel = [true|false];
    is_default = [true|false];
    is_enabled = [true|false];
    is_tab_stop = [true|false];
    key = string;
    label = string;
    mnemonic = string;
    width = number;
}
```

Column

```
: column {
    alignment = [[top|centered|bottom] | [left|centered|right]];
    children_alignment = [left|centered|right];
    children_fixed_height = [true|false];
    children_fixed_width = [true|false];
    fixed_height = [true|false];
    fixed_width = [true|false];
    height = number;
    label = string;
    width = number;
    [tile]...
}
```

Concatenation

```
: concatenation {
    fixed_width = [true|false];
    fixed_height = [true|false];
    children_alignment = [left|centered|right];
    [text_part]...
}
```

Dialog

```
: dialog {
    initial_focus;
    label;
    value;
    [tile]...
}
```

Edit Box

```
: edit_box {
    action = string;
    alignment = [[top|centered|bottom]|[left|centered|right]];
    allow_accept = [true|false];
    edit_limit = integer;
    edit_width = integer;
    fixed_height = [true|false];
    fixed_width = [true|false];
    is_enabled = [true|false];
    is_tab_stop = [true|false];
    key = string;
    label = string;
    mnemonic = string;
    password_char = [true|false];
    value = string;
    width = number;
}
```

Error Tile

```
: errtile {
    alignment = [[top|centered|bottom]|[left|centered|right]];
```

```
      fixed_height = [true|false];
      fixed_width = [true|false];
      is_bold = [true|false];
      key = string ("error");
      label = string (" ");
      value = string;
      width = number (35);
}
```

Image

```
: image {
    action = string;
    alignment = [[top|centered|bottom]|[left|centered|right]];
    aspect_ratio = number;
    color = integer (7);
    fixed_height = [true|false];
    fixed_width = [true|false];
    height = number;
    is_enabled = [true|false];
    is_tab_stop = [true|false];
    key = string;
    label = string;
    mnemonic = string;
    value = string;
    width = number;
}
```

Image Button

```
: image_button {
    action = string;
    alignment = [[top|centered|bottom]|[left|centered|right]];
    allow_accept = [true|false];
    aspect_ratio = number;
    color = integer (7);
    fixed_height = [true|false];
    fixed_width = [true|false];
    height = number;
    is_enabled = [true|false];
    is_tab_stop = [true|false];
    key = string;
    label = string;
    mnemonic = string;
    width = number;
}
```

List Box

```
: list_box {
    action = string;
    alignment = [[top|centered|bottom] | left|centered|right]];
    allow_accept = [true|false];
    fixed_height = [true|false];
    fixed_width = [true|false];
    fixed_width_font = [true|false];
    height = number (10);
    is_enabled = [true|false];
    is_tab_stop = [true|false];
    key = string;
    label = string;
    list = string;
    mnemonic = string;
    multiple_select = [true|false];
    tabs = string;
    tab_truncate = [true|false];
    value = string;
    width = number (10);
}
```

Paragraph

```
: paragraph {
    fixed_height = [true|false];
}
```

Popup List

```
: popup_list {
    action = string;
    alignment = [[top|centered|bottom] | left|centered|right]];
    edit_width = integer;
    fixed_height = [true|false];
    fixed_width = [true|false];
    fixed_width_font = [true|false];
    height = number;
    is_enabled = [true|false];
    is_tab_stop = [true|false];
```

```
    key = string;
    label = string;
    list = string;
    mnemonic = string;
    tabs = string;
    tab_truncate = [true|false];
    value = string;
    width = number;
}
```

Radio Button

```
: radio_button {
    action = string;
    alignment = [[top|centered|bottom]|[left|centered|right]];
    fixed_height = [true|false];
    fixed_width = [true|false];
    height = number;
    is_enabled = [true|false];
    is_tab_stop = [true|false];
    key = string;
    label = string;
    mnemonic = string;
    value = string;
    width = number;
}
```

Radio Column

```
: radio_column {
    alignment = [[top|centered|bottom]|[left|centered|right]];
    children_alignment = [left|centered|right];
    children_fixed_height = [true|false];
    children_fixed_width = [true|false];
    fixed_height = [true|false];
    fixed_width = [true|false];
    height = number;
    label = string (" ");
    width = number;
    [tile]...
}
```

Radio Row

```
: radio_row {
    alignment = [[top|centered|bottom] | [left|centered|right]];
    children_alignment = [top|centered|bottom];
    children_fixed_height = [true|false];
    children_fixed_width = [true|false];
    fixed_height = [true|false];
    fixed_width = [true|false];
    height = number;
    label = string (" ");
    width = number;
    [tile]...
}
```

Row

```
: row {
    alignment = [[top|centered|bottom] | [left|centered|right]];
    children_alignment = [top|centered|bottom];
    children_fixed_height = [true|false];
    children_fixed_width = [true|false];
    fixed_height = [true|false];
    fixed_width = [true|false];
    height = number;
    label = string;
    width = number;
    [tile]...
}
```

Slider

```
: slider {
    action = string;
    alignment = [[top|centered|bottom] | [left|centered|right]];
    big_increment = integer (1000);
    fixed_height = [true|false];
    fixed_width = [true|false];
    height = number;
    key = string;
    label = string;
    layout = [horizontal|vertical];
    max_value = integer (10000);
```

```
    min_value = integer (0);
    mnemonic = string;
    multiple_select = [true|false];
    small_increment = integer (100);
    value = string;
    width = number;
}
```

Text

```
: text {
    alignment = [[top|centered|bottom] | [left|centered|right]];
    fixed_height = [true|false];
    fixed_width = [true|false];
    is_bold = [true|false];
    key = string;
    label = string;
    value = string;
    width = number;
}
```

Text_part

```
: text_part {
    label = string;
}
```

Toggle

```
: toggle {
    action = string;
    alignment = [[top|centered|bottom] | [left|centered|right]];
    fixed_height = [true|false];
    fixed_width = [true|false];
    height = number;
    is_enabled = [true|false];
    is_tab_stop = [true|false];
    key = string;
    label = string;
    mnemonic = string;
    value = string;
    width = number;
}
```

Spacer

```
: spacer {
    alignment = [[top|centered|bottom] | [left|centered|right]];
    fixed_height = [true|false];
    fixed_width = [true|false];
    height = number;
    width = number;
}
```

Spacer_0

```
spacer_0 : {
    height = number (0);
    width = number (0);
}
```

Spacer_1

```
spacer_1 : {
    height = number (1);
    width = number (1);
}
```

Predefined Buttons and Button Assemblies

```
ok_only;

ok_cancel;

ok_cancel_help;

ok_cancel_help_errtile;

ok_cancel_help_info;

: ok_button {
    label = string ("  OK  ");
    key = string ("accept");
    is_default = [true|false];
}
```

```
: cancel_button {
   label = string ("Cancel");
   key = string ("cancel");
   is_cancel = [true|false];
}

help_button : {
   label = string ("&Help...");
   key = string ("help");
}

info_button : {
   label = string ("&Info...");
   key = string ("info");
}
```

appendix D

Errors, Problems, and Solutions

This appendix identifies some of the errors and problems that you can encounter while customizing AutoCAD, as well as solutions and workarounds for when you do encounter them. No list of customization errors and problems can be exhaustive, but this appendix will help you avoid or deal with the most common ones in R13c4. You should also look at the *AutoLISP Error Codes and Error Messages* chapter of the *AutoCAD Customization Guide*, and at its other indexed error topics.

Each chapter in this book includes many highlighted warnings, notes, and tips to help you avoid customization problems. If you don't find your problem listed in this appendix, browse the warnings, notes, and tips in the appropriate chapter in the book.

Note: If you think your problem is specific to this book, see the end of the Introduction for steps to take. Also, see the README.TXT file on the MaxAC CD-ROM or in your MaxAC directory for corrections and last-minute information. And for even more current information, see Rusty's Group A web site on the Internet's World Wide Web at `http://www.group-a.com/~rusty`.

If you encounter other problems, experiment. Try to narrow down the problem by simplifying the custom component that's causing the problem (for example, a menu macro or AutoLISP program). Try alternative commands, other syntax, or different AutoCAD platforms. Check and double-check your typing—a surprisingly large number of errors are caused by misspelled command or variable names, which can be hard to spot when you've been staring at the screen for hours. Sometimes printing out the offending support file and looking at it away from the computer will help you spot errors.

CompuServe is an excellent source of help for customization problems. Post your questions in the Customization section (currently #6) of the ACAD forum and/or the AutoLISP section (currently #5) of the CADENCE forum. In addition, Steve Johnson's Bug Watch column in *CADalyst* magazine and his unofficial bug lists (13BUG.ZIP [ASCII text format] and 13BUGDOC.ZIP [Microsoft Word format] in Library 1 of the ACAD forum) are a good way to stay informed about bugs in AutoCAD. The current bug list documents are on the MaxAC CD-ROM, but you may find more more recent versions on CompuServe.

Customization Environment Problems

The following problems concern your AutoCAD customization or operating environment, as described in Chapter 2.

Support Files not Found, or Wrong Versions Found

Problem: AutoCAD reports that it can't find support files (for example, menus, blocks, or AutoLISP programs), or seems to load the wrong versions of support files (ACAD.PGP, ACAD.LSP, and so on).

Solution: Verify that you've configured the support path correctly and that support files are located in the proper locations in AutoCAD's library search path. (See Chapter 4 for details.) In order to see where AutoCAD is finding a support file, enter **(findfile file_name)** at the command prompt, where **file_name** is the name of the support file in quotation marks.

Can't Open Support Files or Drawings after a System Crash

Problem: After a system crash, AutoCAD reports that it can't open some support files (for example, ACAD.LOG) or drawing files because they're locked.

Solution: Unlock the files with the AutoCAD FILES command, or delete the lock files using Explorer, File Manager, or the DOS DEL command. If the locked files are on a network disk, make sure that they aren't in use by someone else before you unlock them. Assuming that you use AutoCAD to unlock files, read the confirmation message carefully. It will say something like: The file C:\MAXAC\ACAD.LOG was locked by MarkM at 15:19 on 6/1/1996. Do you still wish to unlock it? If the locked by name is your AutoCAD login name, then that's a good sign that the lock file is left over from your system crash. If it's someone else's login name, you should check with that person before unlocking the file.

If you prefer to delete lock files with an operating system utility such as Explorer, you'll need to be even more careful. Always delete lock files *before* you launch AutoCAD, and double-check that you're deleting your own lock files. Although AutoCAD's lock files use a binary format, if you open one of them in a text editor, you should be able to identify the login name. See Chapter 14 in the *AutoCAD User's Guide* for a list of lock file extensions.

If you never share drawings over a network, than you can turn off AutoCAD's file locking feature (use CONFIG, Operating Parameters, File locking) and avoid all of its attendant problems.

Menu Loading Problems

The following problems can occur when AutoCAD tries to load a menu file, either when you launch the program, open a drawing, or use the MENU command.

Windows Toolbar Customization Disappears

Problem: Toolbar changes that you made with the Toolbars dialog box (TBCONFIG command) or by editing the MNS file disappear after you load an MNU file with the MENU or MENULOAD command. This problem occurs because AutoCAD creates a new MNS file (overwriting any existing file) when you load an MNU file in Windows. That's why AutoCAD displays the warning: `Loading of a template menu file (MNU) overwrites and redefines the menu source file (MNS file), which results in the loss of any toolbar customization changes that have been made. Continue loading MNU file?`

Solution: Before you start customizing toolbars, move the MNU file to a place where AutoCAD won't find it (such as a floppy disk or a different directory on your hard disk). Then make all editing changes to the MNS file. Alternatively, remember to copy any customization changes, whether from the Toolbars dialog or from direct editing of the MNS file, into the MNU file.

Toolbar Icons Turn into Smiley Faces

Problem: After AutoCAD compiles your menu, the custom icons change to smiley faces.

Solution: This problem occurs when AutoCAD can't find the custom icons (BMP or DLL files—see Chapter 13) in the library search path. Move the BMP or DLL files to a directory on the library search path, or add the directory to the support directory path. Then delete the MNC file and recompile the menu.

Accelerator Key Definitions Don't Work

Problem: The accelerator key definitions in the ***ACCELERATORS section of your menu file don't work.

Solution: On the Preferences dialog's **S**ystem tab, change the Keystrokes setting to Menu **F**ile. The other setting (**A**utoCAD Classic) forces AutoCAD to ignore all accelerator key definitions. Also make sure that you're running R13 Windows. R13 DOS does not support customizable accelerator keys.

The Wrong Menu Loads

Problem: AutoCAD loads (or tries to load) a different menu when you change drawings.

Solution: On the Preferences dialog's **M**isc tab, turn off the **U**se menu in header setting. See Chapter 4 for details.

DOS Menus Look Wrong and Don't Work

Problem: In R13 DOS, the pull-down menus have the wrong labels (for example, labels that begin with ID_ and include ampersands) and/or don't work correctly when you pick each menu item.

Solution: Make sure that you haven't tried to use a Windows menu file in DOS. R13 DOS doesn't support name tags and other enhanced features that are included in most R13 Windows menus.

Windows Menus Are Missing or Incomplete

Problem: In R13 Windows, the pull-down menu labels don't include underlined accelerator keys, and Windows-only menus (toolbars, accelerators, and help strings) are missing completely.

Solution: Make sure that you haven't tried to use a DOS menu file in Windows. Although R13 Windows will load most DOS menus, DOS menus lack many of the new R13 menu customization features. If you're customizing an older DOS AutoCAD menu for which there isn't an R13 Windows version, then you'll need to add these new features. See Chapters 3 and 13 for details.

Menus Become Garbled in Windows 3.1

Problem: When you load a menu or after you've worked for a while in Windows 3.1 (or 3.11), menu items appear on the wrong pull-down menus, or pull-down menus get truncated. This problem can be caused by low system resources in Windows. (In Program Manager, use Help, About Program Manager to check the current amount of free system resources.)

Solution: The best solution is to upgrade your operating system to Windows 95 or NT. Stop-gap solutions involve making more base memory available. Close down other Windows applications. Exit and restart AutoCAD and then Windows. Remove drivers and other memory-resident programs from CONFIG.SYS and AUTOEXEC.BAT. Run MEMMAKER to optimize memory.

Menu Macro Problems

The following problems concern running menu macros.

A Menu Macro Crashes or Leaves a Command Active

Problem: A menu macro doesn't run properly. It crashes during execution, causes error messages on the command line, or leaves you in the middle of a command when it's finished.

Solution: Double-check the command sequence by typing it interactively at the command prompt. Count every press of the ENTER key and make sure that you've included one space or semicolon in the macro per ENTER. Check for extra trailing spaces at the end of a line and delete them.

A Menu Macro Stops Prematurely

Problem: A multi-line menu macro stops before it's finished.

Solution: Make sure that you've included the macro continuation character (+) at the end of each line, except for the last, of a multi-line macro.

A Menu Macro Stalls and Then Resumes

Problem: A menu macro stops in the middle, and then resumes later after you press ENTER.

Solution: The macro includes an extra pause character (\). Often this problem is the result of trying to use a backslash as a path separator. Use a forward slash instead. See Chapter 5 for more information.

Text and Font Problems

The following problems can occur when you work with text and font files.

AutoCAD Reports Missing Font Files

Problem: When you load a drawing, AutoCAD can't locate one or more font files.

Solution: There are many reasons that AutoCAD might not be able to find font files, especially if they're custom font files:

1. The files might be on your system, but not in the current library search path. See "Support Files not Found, or Wrong Versions Found" early in this appendix.

2. The styles might contain font file names with paths that are different from the paths on your system. Make sure that the font files are located in AutoCAD's library search path, and then strip the paths from the font file names. To strip the paths, edit the font file names in the Text Style dialog (DDSTYLE command) or run the STYLE command and use the Type It button in the Select Font File dialog.

3. If the missing font files are PostScript (PFB) or TrueType (TTF) fonts, you might not have installed these files with AutoCAD. A minimal instal-

lation omits the PostScript and TrueType fonts. Re-run the R13 installation program and install all of the fonts.

4. The font files might be custom ones that you don't have. Substitute a different font (for example, ROMANS.SHX) or ask the creator of the drawing to send the custom font files.

Warning: Custom fonts are often copyrighted, and you may need to purchase the right to use them legally. Substitute fonts may not align and fit in the drawing the same as the original font.

See Chapter 6 for more information about font file and custom problems.

Question Marks Appear in Text Strings

Problem: When you load a drawing, AutoCAD displays question marks where other text characters should be.

Solution: This problem usually is caused by mismatched font or big font files: the font files that the creator of the drawing used aren't the same as the ones on your system. As a result, some of the character codes refer to different characters (see Chapter 6 for details). Ask the creator of the drawing for copies of the drawing's font files, and put those files in a separate directory with the drawing (see the previous warning).

Script Problems

The following problems concern running scripts.

A Script Crashes or Leaves a Command Active

Problem: A script doesn't run properly. It crashes during execution, causes error messages on the command line, or leaves you in the middle of a command when it's finished.

Solution: As with menu macros, double-check the command sequence by typing it interactively at the command prompt, or by using LispPad to send it line-by-line to AutoCAD. Count every press of the ENTER key and make sure that you've included one space or carriage return in the script per ENTER. Check for extra trailing spaces at the end of a line or extra trailing carriage returns at the end of a file and delete them.

A Script That Refers to a Long File Name Crashes

Problem: A script that tries to use a Windows 95 or NT long file name containing spaces crashes in some circumstances. For example:

```
INSERT "LONG NAME BLOCK" 0,0 1 1 0
```

Errors, Problems, and Solutions

crashes without inserting LONG NAME BLOCK.DWG.

Solution: Substitute a short name. For example:

```
INSERT "SHRTNAME=LONG NAME BLOCK" 0,0 1 1 0
```

A Configuration Script Stalls

Problem: A script that runs the CONFIG command stops at a particular configuration question, even though the script contains a valid answer to that question.

Solution: Some drivers (for example, the HP-GL/2 plotter driver) ask questions that don't accept script input—see Chapter 8 for an example. You have to answer these questions interactively.

AutoLISP and DCL Problems

For standard AutoLISP error messages, see the *AutoCAD Customization Guide*, Chapter 17: "AutoLISP Error Codes and Error Messages".

Tip: Use LispPad to test AutoLISP code by sending it to AutoCAD a portion at a time, and to test DCL code by previewing it.

Incorrect Request for Command List Data

Problem: During the loading of ACAD.LSP, AutoCAD reports the error `incorrect request for command list data`. This error is caused by attempts to call AutoCAD-specific AutoLISP functions such as **command**, **setvar**, and so forth before the drawing editor has fully initialized.

Solution: Place calls to AutoCAD-specific functions inside the body of a function named **s::startup**, which, if defined, is automatically called by AutoCAD after the drawing editor is initialized. See Chapter 13 for details and examples.

TEXT Evaluation and the PAUSE symbol

Problem: AutoCAD commands that accept text with embedded spaces do not recognize the AutoLISP PAUSE constant when passed in the **command** function, and instead interpret the value of the symbol (`"\\"`) literally. Commands most frequently affected by this are INSERT (block name if long filenames are enabled and block insertion attributes), and the TEXT command (text value). In these cases, the value of the PAUSE symbol becomes the value of the text or attribute.

Solution: Set the TEXTEVAL system variable to 1 (ON). This system variable tells AutoCAD to evaluate all expressions passed from AutoLISP including those when text with embedded spaces are expected.

Globalization and Localization Unfriendly Warnings

Problem: AutoCAD displays a warning message indicating that a DCL dialog box is too large to fit on a Japanese VGA display.

Problem: AutoCAD rejects commands submitted to it via the AutoLISP **command** function if the commands are not prefixed with an underscore, and displays a dialog indicating that non-interactive commands must be preceded by an underscore.

Solution: Set the undocumented GLOBCHECK system variable to 0 (OFF), to suppress globalization-unfriendly warnings.

Mismatched Parentheses

Problem: `extra right paren` or `malformed list` errors appear when loading AutoLISP programs.

Mismatched parentheses are the most common source of AutoLISP errors. The error occurs when attempting to load an AutoLISP program file in which one or more open or close parentheses are missing. If there is an extra close parenthesis, loading is terminated as soon as the extra parenthesis is detected, causing an `extra right paren` error. If there are one or more missing close parentheses, AutoLISP will load the entire file and generate a `malformed list` error.

Solution: There are several ways to resolve mismatched parentheses. For `extra right paren` errors, try reloading the AutoLISP file a second time by calling the **load** function with an undocumented optional third argument of **T**, as shown in the following example.

```
(load filename nil T)
```

Using this undocumented form of the **load** function causes the contents of the file to be echoed to the text console as it is loaded. When an extra right parenthesis is encountered, loading stops, and the last expression displayed is where the extra parenthesis was detected.

Finding missing right parentheses (which causes the `malformed list` error) is more difficult, because AutoLISP can't detect the error until the entire file is loaded. For resolving missing right parentheses, the use of a text or code editor with parentheses matching (such as LispPad, which is included with this book) is the easiest solution.

Errors, Problems, and Solutions

Dot Notation Errors

Problem: `invalid dotted pair` and `misplaced dot` error messages. AutoLISP uses the period or "dot" for both literal real numbers, and for denoting a quoted *dotted pair*. Two errors that occur because of misplaced dots or invalid literal numbers are `invalid dotted pair` and `misplaced dot`. These two errors occur when you omit a leading digit from a literal real number, omit a leading and/or trailing space from a quoted dotted list, or have mismatched double quotes around strings that contain periods or commas, as the following examples demonstrate:

```
(setq x .223)        ; Error: invalid dotted pair (no leading digit)
(setq y '(1. 2))     ; creates list containing 1.0 and 2
(setq y '(1 .2))     ; Error: invalid dotted pair (no space after dot)
(setq y '(1 . 2))    ; creates dotted pair containing 1 and 2
(setq s this is a test.")  ; Error: invalid dotted pair and an invalid
                     ; (malformed) string - no opening double
                     ; quote
```

Solution: Load the AutoLISP file, and pass **T** as the undocumented optional third argument to the **load** function, as described for finding an extra right parenthesis.

Malformed Strings

Problem: `malformed string` and `exceeded maximum string length` errors. The error `malformed string` results from an odd number of double quotes on a single line, or failing to escape a double quote that is embedded in a literal string. An `exceeded maximum string length` error can also be caused by an odd number of double quotes or failing to escape an embedded quote.

In some cases, loading an AutoLISP file containing a missing double quote will *not* cause an error. Usually this happens when the missing double quote is near the end of the file, and causes AutoCAD to display a prompt consisting of a number followed by the greater-than sign, suggesting that there are mismatched parentheses. This is caused by the fact that all right parentheses that appear after a starting double quote are interpreted as part of the string.

Solution: In all cases, you must find the missing quote by using your editor's search function to find each pair of double quotes. You can also determine the general location of a missing double quote in a program file by loading it with the optional third argument to the **load** function, as recommended for mismatched parentheses.

Calling Redefined Commands

Problem: An `unknown command` error occurs when calling redefined AutoCAD commands via the AutoLISP **command** function.

Solution: Ensure that all calls to built-in AutoCAD commands made via the `command` function are prefixed by a period, which causes AutoCAD to always use the native versions of commands even if they have been redefined.

Problem: An unknown command error occurs when attempting to call a command defined by another AutoLISP program, via the `command` function.

Solution: The `command` function cannot be used to call AutoLISP and ADS-defined commands. They can be called as functions by prefixing their names with `C:`.

Displaying the Plot Dialog Box

Problem: AutoCAD provides no way for an AutoLISP program to activate the PLOT dialog box and wait for the user to complete interacting with it.

Solution: Use the `AcadPlotDlg` function from MAWIN.ARX, which is included with this book. Be sure that the call to this function is the very last action of the AutoLISP program that uses it, prior to exiting. See Chapter 14 for an example.

Debugging DCL callbacks

Problem: AutoLISP provides no easy way to view callback event data and callback variables when a DCL dialog box is active.

Solution: Use the AutoLISP `trace` function to trace the `lambda` function, which is called every time a callback event occurs. This will cause the values of all callback variables to be displayed on the text console. See *Maximizing AutoLISP for AutoCAD R13* or the *AutoCAD Customization Guide* for more information.

Debugging in DOS AutoCAD

Problem: DOS versions of AutoCAD provide no easy way to view more than a screenful of error trace-back information displayed when an AutoLISP error occurs.

Solution: While DOS AutoCAD is still a viable production CAD environment, it is not very friendly for developing AutoLISP programs. The best solution to this general problem is to do as much development and debugging as possible using AutoCAD for Windows. When you do need to debug in DOS AutoCAD, you can capture the error trace-back with AutoCAD's log file facility. Run the LOGFILEON command before the error and the LOGFILEOFF command after the error. DOS AutoCAD writes its log file output to the file ACAD.LOG in the current directory.

appendix E

System Variables

 This appendix contains a table of AutoCAD system variables. Variable names and features shown in bold are new in Release 13.

The first two columns in the table contain the system variable name and the default setting (assuming that you use the standard AutoCAD prototype drawing ACAD.DWG or no prototype drawing). The third column lists the commands that can change the system variable. The fourth column contains a brief description of the system variable, a description of acceptable values where applicable, and information about where the variable is stored and whether it's read-only.

AutoCAD saves most system variable values in the drawing file. Some variables are global, and are saved in the ACADNT.CFG or ACAD.INI file. We mark these variables (CFG) and (INI), respectively. Some variables aren't saved at all; AutoCAD reinitializes them to their default values each time you change drawings. We mark these variables (NS).

You can set most system variables by entering their names directly at the command prompt (or the dim prompt for dimension variables). A few variables, which are shown in italics, duplicate AutoCAD command names, and you must use the SETVAR command to set these. You can set any read-write system variable, italicized or not, with the AutoLISP **setvar** function. You can retrieve any system variable's value with the AutoLISP **getvar** function (see Chapter 9). Variables marked (RO) are read-only, which means that you cannot change them directly.

This table is intended as a concise guide to AutoCAD system variables, rather than a comprehensive treatise on every subtlety of every variable. Refer to Appendix A in the R13 *AutoCAD Command Reference* for additional information about some system variables.

Table E.1
AutoCAD System Variables

Variable Name	Default Setting	Command Names	Variable Description
ACADPREFIX	"C:\R13\COM\SUPPORT;..."	PREFERENCES	AutoCAD search path. (INI)(RO)
ACADVER	"13_c4"		AutoCAD version number. (NS)(RO)
AFLAGS	0	DDATTDEF, ATTDEF	Default values for attribute creation modes. The value is the sum of the following: 0 = No mode set 1 = Invisible 2 = Constant 4 = Verify 8 = Preset (NS)
ANGBASE	0.0	DDUNITS, UNITS	The direction of angle 0 in the current UCS
ANGDIR	0.0	DDUNITS, UNITS	The direction of angle measure: 0 = Counter-clockwise 1 = Clockwise
APERTURE	10	DDOSNAP, APERTURE	Half the OSNAP target size in pixels. (CFG)
AREA	0.0000	AREA, LIST	The last computed area in square drawing units. (NS)(RO)
ATTDIA	0		Attribute value entry method when inserting blocks: 0 = Attribute prompts 1 = DDATTE dialog box
ATTMODE	1	ATTDISP	Attribute display mode: 0 = Off 1 = Normal (use each attribute's setting) 2 = On
ATTREQ	1		Attribute value prompting when inserting blocks: 1 = Prompt for values 0 = Don't prompt; use defaults
AUDITCTL	0		Creation of an ADT log file containing AUDIT results: 0 = Don't create audit file 1 = Create audit (ADT) file (CFG)
AUNITS	0	DDUNITS, UNITS	Angular units: 0 = Decimal deg. 1 = Degrees/min/sec 2 = Grads 3 = Radians 4 = Surveyors units
AUPREC	0	DDUNITS, UNITS	The number of angular units decimal places

System Variables

Variable Name	Default Setting	Command Names	Variable Description
BACKZ	0.0000	DVIEW	The DVIEW back clipping plane offset in drawing units. *See* *VIEWMODE.* (RO)
BLIPMODE	1	BLIPMODE	Display of "blip" markers at each pick point: 0 = Don't show blips 1 = Show blips
CDATE	19960622.11252156 (for example)	TIME	Current date and time in YYYYMMDD.HHMMSSmsec format. (NS)(RO)
CECOLOR	"BYLAYER"	DDEMODES, COLOR	Current entity color
CELTSCALE	**1.0**	**DDLTYPE, DDEMODES**	**Current entity linetype scale**
CELTYPE	"BYLAYER"	**DDLTYPE, DDEMODES, LINETYPE**	Current entity linetype
CHAMFERA	0.0000	CHAMFER	Default first chamfer distance **(requires CHAMMODE = 0)**
CHAMFERB	0.0000	CHAMFER	Default second chamfer distance **(requires CHAMMODE = 0)**
CHAMFERC	**0.0000**	**CHAMFER**	**Default chamfer length (requires CHAMMODE = 1)**
CHAMFERD	**0.0**	**CHAMFER**	**Default chamfer angle (requires CHAMMODE = 1)**
CHAMMODE	**0**	**CHAMFER**	**Chamfer drawing mode** 0 = Use two distances 1 = Use length and angle
CIRCLERAD	0.0000	CIRCLE	Default radius for new circles. (NS)
CLAYER	0	DDLMODES, LAYER	Current layer
CMDACTIVE	0	Any command	Indicates whether an AutoCAD command is active: 0 = None 1 = Ordinary command 2 = Ordinary and transparent command 4 = Script 8 = Dialog box (NS)(RO)
CMDDIA	1		Controls whether the PLOT command and ASE commands issue dialog boxes or command-line prompts: 0 = Command-line prompts 1 = Dialog boxes (CFG)
CMDECHO	1		Controls AutoCAD command-prompt echoing for AutoLISP programs: 1 = Echo 0 = Don't echo (NS)

Variable Name	Default Setting	Command Names	Variable Description
CMDNAMES	" "	Any command	Names of any active commands and transparent commands (for example, "LINE'ZOOM"). (NS)(RO)
CMLJUST	0	MLINE	Default multiline justification: 0 = Top 1 = Middle 2 = Bottom
CMLSCALE	1.0	MLINE	Default multiline scale factor. 1.0 causes AutoCAD to use the width defined for the multiline scale. A negative number flips the order of the offset lines.
CMLSTYLE	"STANDARD"	MLINE, MLSTYLE	Default multiline style
COORDS	1	[^D] [F6]	Controls the updating of the coordinate display: 0 = Absolute upon picks 1 = Absolute continuously 2 = Absolute continuously plus relative during point prompts
CVPORT	2	VPORTS	The current viewport's number (1 is the overall paper space viewport)
DATE	2450259.50609445 (for example)	TIME	The current date and time in Julian format. (NS)(RO)
DBMOD	0	Most commands	Describes modifications to the current drawing database (sum of the following): 0 = None 1 = Objects 2 = Symbol table 4 = Database variable 8 = Window 16 = View (RO)(NS)
DCTCUST	" "	SPELL	Custom spelling dictionary file name, including path. (CFG)
DCTMAIN	"enu" (for American English)	SPELL	Main spelling dictionary file name (see *AutoCAD Command Reference* Appendix A for language keywords). (CFG)
DELOBJ	1		Controls whether objects that create other objects are deleted: 0 = Don't delete 1 = Delete
DIASTAT	1	DD*xxxx* commands	Exit code for the most recently used DD dialog box command: 0 = Cancel 1 = OK (NS)(RO)

System Variables

Variable Name	Default Setting	Command Names	Variable Description
DIMALT	Off (or 0)	DDIM	Controls whether dimension text includes a second measurement in alternative units: 0 (or Off) = Don't add alternate units 1 (or On) = Add alternate units
DIMALTD	2	DDIM	Number of decimal places for alternate dimension text (when DIMALT = On).
DIMALTF	25.4000	DDIM	Scale factor for alternate dimension text (when DIMALT = On).
DIMALTTD	**2**	**DDIM**	**Number of decimal places for alternate dimension tolerance text (when DIMALT = On).**
DIMALTTZ	**0**	**DDIM**	**Controls suppression of zero inches, zero feet, and decimal leading and trailing zeroes in alternate dimension tolerance text (when DIMALT = On).** See DIMZIN *for values.*
DIMALTU	**2**	**DDIM**	**Units format for alternate dimension text (when DIMALT = On).** See DIMUNIT *for values.*
DIMALTZ	**0**	**DDIM**	**Controls suppression of zero inches, zero feet, and decimal leading and trailing zeroes in alternate dimension text (when DIMALT = On).** See DIMZIN *for values.*
DIMAPOST	" "	DDIM	User-defined prefix and/or suffix for alternate dimension text (when DIMALT = On).
DIMASO	1		Controls whether each dimension is created as a single, associative object: 0 = Off 1 = On
DIMASZ	0.1800	DDIM	Controls the size of dimension and leader arrows, and of hook lines. Controls size of dimension blocks (when DIMBLK, DIMBLK1, or DIMBLK2 is set). Also affects the fit of dimension text inside dimension lines. DIMASZ has no effect when DIMTSZ is non-zero.
DIMBLK	" "	DDIM	User-defined block to insert, instead of the standard arrow or tick, at both ends of the dimension line.

Maximizing AutoCAD® R13

Variable Name	Default Setting	Command Names	Variable Description
DIMBLK1	" "	DDIM	User-defined block to insert, instead of the standard arrow or tick, at the first end of the dimension line. *See DIMSAH.*
DIMBLK2	" "	DDIM	User-defined block to insert, instead of the standard arrow or tick, at the second end of the dimension line. *See DIMSAH.*

DIMCEN	0.0900	DDIM	Controls center marks or center lines drawn by radial DIM commands:
			  0 = no center marks
			  positive value = draw center marks
			    (value = length)
			  negative value = draw center lines
			    (absolute value = length)

DIMCLRD	0	DDIM	Color for dimension lines, arrows, and leaders. A valid color number (1 - 255), or:
			  0 = BYBLOCK
			  256 = BYLAYER

DIMCLRE	0	DDIM	Color for dimension extension line. *See DIMCLRD for values.*
DIMCLRT	0	DDIM	Color for dimension text. *See DIMCLRD for values.*
DIMDEC	**4**	**DDIM**	**Number of decimal places for primary dimension text**
DIMDLE	0.0000	DDIM	Additional length of the dimension line beyond the extension lines (when DIMTSZ is nonzero).
DIMDLI	0.3800	DDIM	Offset distance between successive continuing or baseline dimensions
DIMEXE	0.1800	DDIM	Additional length of extension lines beyond the dimension line
DIMEXO	0.0625	DDIM	Gap from each dimension point to the beginning of its corresponding extension line

System Variables

Variable Name	Default Setting	Command Names	Variable Description
DIMFIT	3	DDIM	Controls how AutoCAD tries to fit dimension text and arrowheads for small dimensions, and whether the dimension line moves with dimension text: 0 = Put arrowheads or text outside if both won't fit 1 = Try to fit text inside 2 = Try to fit arrowheads inside 3 = Try to fit text, then arrowheads inside 4 = Try to fit arrowheads inside; move text above dimension line and draw a short leader to it if necessary 5 = Same as 4, but without the short leader In addition, codes 0 through 3 cause the dimension line to "follow" the text when you move the latter, while codes 4 and 5 let you move the text independently from the leader line.
DIMGAP	0.0900	DDIM	Gap between each end of dimension text and the adjacent end of dimension line. Also determines, along with DIMASZ and **DIMFIT**, when dimension text is forced outside of the extension lines. A negative DIMGAP value creates a "basic dimension" box around the dimension text.
DIMJUST	0	DDIM	Horizontal justification of dimension text: 0 = Centered 1 = Next to first extension line 2 = Next to second extension line 3 = Above and aligned with first extension line 4 = Above and aligned with second extension line
DIMLFAC	1.0000	DDIM	Scale factor to apply to dimension text values. Negative values affect dimensions in paper space only (AutoCAD uses the absolute value in paper space and 1.0 in model space).
DIMLIM	Off (or 0)	DDIM	Controls whether dimension text appears as two numbers representing upper and lower limits: 0 (or Off) = Show normal text 1 (or On) = Show limits text *See DIMTP and DIMTM.*
DIMPOST	" "	DDIM	User-defined prefix and/or suffix for dimension text
DIMRND	0.0000	DDIM	Rounding value for linear dimension text

Variable Name	Default Setting	Command Names	Variable Description
DIMSAH	Off (or 0)	DDIM	Enables the use of DIMBLK1 and DIMBLK2 rather than DIMBLK or a default terminator: 0 (or Off) = Use DIMBLK 1 (or On) = Use DIMBLK1 and DIMBLK2
DIMSCALE	1.0000	DDIM	Scale factor applied to all other dimension variables representing the sizes of objects making up the dimension symbol (everything except measured lengths, angles, coordinates, and tolerances). The special value 0.0 is for dimensioning with TILEMODE = 0. When you set DIMSCALE to 0.0 and dimension in a paper space viewport, AutoCAD computes a reasonable dimension scale factor based on the viewport's scale factor. In all other cases, AutoCAD treats DIMSCALE = 0.0 as though it were 1.0.
DIMSD1	**Off (or 0)**	**DDIM**	**Controls suppression of the first dimension line:** **0 (or Off) = Don't suppress** **1 (or On) = Suppress**
DIMSD2	**Off (or 0)**	**DDIM**	**Controls suppression of the second dimension line:** **0 (or Off) = Don't suppress** **1 (or On) = Suppress**
DIMSE1	Off (or 0)	DDIM	Controls suppression of the first extension line: 0 (or Off) = Don't suppress 1 (or On) = Suppress
DIMSE2	Off (or 0)	DDIM	Controls suppression of the second extension line: 0 (or Off) = Don't suppress 1 (or On) = Suppress
DIMSHO	On (or 1)		Determines whether the value of associative dimension text is updated during dragging: 0 (or Off) = Don't update 1 (or On) = Update continuously
DIMSOXD	Off (or 0)	DDIM	Controls suppression of dimension lines outside extension lines (when text is forced outside the extension lines): 0 (or Off) = Don't suppress 1 (or On) = Suppress
DIMSTYLE	**"STANDARD"**	DDIM	Current dimension style. (RO)

System Variables

Variable Name	Default Setting	Command Names	Variable Description
DIMTAD	0	DDIM	Controls vertical justification of dimension text: 0 = Centered inside dimension line 1 = Above dimension line **2 = On side of dimension line farthest from dimension points** **3 = Conforming to JIS representation**
DIMTDEC	**4**	**DDIM**	**Number of decimal places for primary dimension tolerance text**
DIMTFAC	1.0	DDIM	Scale factor for dimension tolerance text height **and for fraction height**
DIMTIH	On (or 1)	DDIM	Controls whether dimension text inside the extension lines is always horizontal or aligned with the dimension line: 0 (or Off) = Align with dimension line 1 (or On) = Always horizontal
DIMTIX	Off (or 0)	DDIM	Controls whether dimension text is forced inside extension lines, even if it doesn't fit there: 0 (or Off) = Don't force text inside 1 (or On) = Always force text inside
DIMTM	0.0000	DDIM	Negative (or lower) tolerance value used when DIMTOL or DIMLIM is on
DIMTOFL	Off (or 0)	DDIM	Controls whether AutoCAD draws a dimension line between extension lines when text is placed outside the extension lines: 0 (or Off) = Don't draw line inside 1 (or On) = Always draw line inside
DIMTOH	On (or 1)	DDIM	Controls whether dimension text outside the extension lines is always horizontal or aligned with the dimension line: 0 (or Off) = Align with dimension line 1 (or On) = Always horizontal
DIMTOL	Off (or 0)	DDIM	Controls whether dimension text includes tolerance values (DIMTP and DIMTM): 0 (or Off) = Don't add tolerance values 1 (or On) = Add tolerance values

Variable Name	Default Setting	Command Names	Variable Description
DIMTOLJ	1	DDIM	Vertical justification of tolerance text, relative to primary dimension text: 0 = Bottom 1 = Middle 2 = Top
DIMTP	0.0000	DDIM	Positive (or upper) tolerance value used when DIMTOL or DIMLIM is on
DIMTSZ	0.0000	DDIM	Size of tick marks to draw in place of arrowheads. When DIMTSZ is 0.0, AutoCAD draws arrowheads. When DIMTSZ is non-zero, AutoCAD draws tick marks instead. A non-zero DIMTSZ also affects the fit of dimension text inside dimension lines.
DIMTVP	0.0000	DDIM	Percentage of dimension text height to offset dimension text vertically. AutoCAD ignores DIMTVP unless DIMTAD is 0.
DIMTXSTY	"STANDARD"	DDIM	Text style for dimension text
DIMTXT	0.1800	DDIM	The dimension text height for text styles that don't have fixed heights
DIMTZIN	0	DDIM	Controls suppression of zero inches, zero feet, and decimal leading and trailing zeroes in dimension tolerance text (when DIMTOL = On). *See DIMZIN for values.*
DIMUNIT	2	DDIM	Units format for primary dimension text (when DIMALT = On): 1 = Scientific 2 = Decimal 3 = Engineering 4 = Architectural (stacked fractions) 5 = Fractional (stacked fractions) 6 = Architectural 7 = Fractional
DIMUPT	0 (Off)	DDIM	Controls whether AutoCAD allows 'User Positioned Text' during dimension creation: 0 = Dimension text always goes to its default position 1 = Dimension line pick point also controls location of text

System Variables

Variable Name	Default Setting	Command Names	Variable Description
DIMZIN	0	DDIM	Suppresses the display of zero inches or zero feet in architectural dimension text, and leading and/or trailing zeros in decimal dimension text: 0 = Suppress zero feet and zero inches 1 = Suppress neither 2 = Suppress zero inches 3 = Suppress zero feet 4 = Suppress leading zeros 8 = Suppress trailing zeros 12 = Suppress leading and trailing zeros DIMZIN also affects real-to-string conversions performed by the AutoLISP functions **rtos** and **angtos**.
DISPSILH	0		**Controls the display of silhouette curves on body objects in wireframe mode** 0 = Off 1 = On
DISTANCE	0.0000	DIST	The last distance computed by the DISTANCE command. (NS)(RO)
DONUTID	0.5000		The default inner diameter for new DONUT entities; may be 0. (NS)
DONUTOD	1.0000		The default outer diameter for new DONUT entities; must be nonzero. (NS)
DRAGMODE	2	DRAGMODE	Controls object dragging on screen during editing operations: 0 = Off 1 = On if requested 2 = Automatically on
DRAGP1	10		Regen-drag sampling rate. (CFG)
DRAGP2	25		Fast-drag sampling rate. (CFG)
DWGCODEPAGE	"ansi_1252" (in Windows 95)		**Code page used for the drawing. In R13, DWGCODEPAGE is given the same value as SYSCODEPAGE, and exists only for compatibility with earlier releases.** (RO)
DWGNAME	"UNNAMED"		The current drawing's name. DWGNAME might or might not include the drawing's path, depending on how the user opened the drawing (dialog box vs. command-line entry). (NS)(RO)
DWGPREFIX	"C:\R13\WIN"		The current drawing's drive and directory path. (NS)(RO)

Variable Name	Default Setting	Command Names	Variable Description
DWGTITLED	0	NEW, QSAVE, SAVE, SAVEAS	Indicates whether or not the current drawing has been named: 0 = No 1 = Yes (RO)(NS)
DWGWRITE	1	OPEN	Indicates whether the current drawing was opened in read-only mode: 0 = Read-only 1 = Read/write (NS)
EDGEMODE	**0**		**Controls whether EXTEND and TRIM consider boundary edges as extended to infinity** **0 = Don't consider edges as extended** **1 = Consider edges as extended**
ELEVATION	0.0000	ELEV	The current elevation in the current UCS for the current space
ERRNO	0		An error number generated by AutoLISP and ADS applications. (See Chapter 18 in the *AutoCAD Customization Guide*). Not listed by SETVAR.
EXPERT	0		Suppresses successive levels of Are you sure? warnings: 0 = None (display all messages) 1 = REGEN/LAYER 2 = BLOCK/WBLOCK/SAVE 3 = LINETYPE 4 = UCS/VPORT 5 = DIM (NS)
EXPLMODE	**1**		**Controls whether you can explode non-uniformly scaled blocks:** **0 = No** **1 = Yes**
EXTMAX	-1.0000E+20,-1.0000E+20		The X,Y coordinates of the drawing's upper right extents in the WCS. (RO)
EXTMIN	1.0000E+20,1.0000E+20		The X,Y coordinates of the drawing's lower left extents in the WCS. (RO)
FACETRES	**0.5**		**Influences the smoothness of hidden-line, shaded, and rendered views of objects. Valid values are 0.01 to 10.0.**
FFLIMIT	**0**		**Controls the number of PostScript and TrueType fonts that AutoCAD can keep in memory. Valid values are 0 to 100. 0 means don't impose any limit. (CFG)**

System Variables

Variable Name	Default Setting	Command Names	Variable Description
FILEDIA	1		Controls the display of the dialog box for file name requests: 0 = Only when a tilde (~) is entered 1 = Always display a file dialog (CFG)
FILLETRAD	0.0000	FILLET	The current fillet radius
FILLMODE	1	FILL	Controls the display of fills in traces, solids, and wide polylines: 0 = Off 1 = On
FONTALT	" "		**Alternate font to use when AutoCAD can't locate a font (and the font doesn't have a valid mapping in the FONTMAP file). (CFG)**
FONTMAP	" "		**Font mapping file name.** *See Chapter 6.*
FRONTZ	0.0000	DVIEW	The DVIEW front clipping plane's offset, in drawing units; *see* VIEWMODE. (RO)
GRIDMODE	0	DDRMODES, GRID	Controls grid display in the current viewport: 1 = On 0 = Off
GRIDUNIT	0.0000,0.0000	DDRMODES, GRID	X and Y grid increment for the current viewport
GRIPBLOCK	0	DDGRIPS	Controls the display of grips for entities in blocks: 1 = On 0 = Off (CFG)
GRIPCOLOR	5	DDGRIPS	Color of unselected (outlined) grips. Valid values are 1 to 255. (CFG)
GRIPHOT	1	DDGRIPS	Color of selected (filled) grips. Valid values are 1 to 255. (CFG)
GRIPS	1	DDSELECT	Controls the display of entity grips and grip editing 1 = On 0 = Off (CFG)
GRIPSIZE	5	DDGRIPS	Size of grip box in pixels. (CFG)
HANDLES	1	HANDLES	**Used to control whether AutoCAD created hexadecimal-numbered** *handles* **for each object. In R13, handles are always on, so the only valid value is 1.** (RO)

Maximizing AutoCAD® R13

Variable Name	Default Setting	Command Names	Variable Description
HIGHLIGHT	1		Determines whether selected objects are highlighted: 0 = Off 1 = On (NS)
HPANG	0	BHATCH, HATCH	Default angle for new hatches. (NS)
HPBOUND	**1**		**Controls whether the BOUNDARY command creates regions or closed polylines: 0 = Create regions 1 = Create polylines (NS)**
HPDOUBLE	0	BHATCH, HATCH	Controls whether user-defined hatches comprise single lines or double lines perpendicular to one another: 0 = Single lines 1 = Double lines (NS)
HPNAME	"ANSI31"	BHATCH, HATCH	Default name for new hatches. (NS)
HPSCALE	1.0000	BHATCH, HATCH	Default scale factor for new hatches; must be nonzero. (NS)
HPSPACE	1.0000	BHATCH, HATCH	Default spacing for user-defined hatches. must be nonzero. (NS)
INSBASE	0.0000,0.0000,0.0000	BASE	Insertion base point of current drawing in current space and current UCS
INSNAME	" "	DDINSERT, INSERT	Default block name for new insertions. (NS)
ISAVEBAK	**1**		**Controls whether AutoCAD creates a BAK file each time you save: 0 = Don't create BAK file 1 = Create BAK file (CFG)**
ISAVEPERCENT	**50**		**Controls AutoCAD's incremental save feature. ISAVEPERCENT is the percentage of wasted space that AutoCAD will tolerate before it performs a full save. Larger values allow more wasted space but usually reduce save times. ISAVEPERCENT = 0 turns off incremental saves and forces AutoCAD to perform a full save every time. (CFG)**
ISOLINES	**4**		**Number of isolines on the surfaces of solid objects. Valid values are 0 to 2047.**
LASTANGLE	0.0	ARC	End angle of the last arc in the current-space UCS. (NS)(RO)

System Variables

Variable Name	Default Setting	Command Names	Variable Description
LASTPOINT	0.0000,0.0000,0.0000		Coordinate of the last point entered, in the current space and UCS (recall with "@"). (NS)
LENSLENGTH	50.0000	DVIEW	Current viewport's perspective view lens length, in millimeters. (RO)
LIMCHECK	0	LIMITS	Controls limits checking for current space: 0 = Allow objects to be drawn outside limits 0 = Don't allow objects to be drawn outside limits
LIMMAX	12.0000,9.0000	LIMITS	Upper right X,Y limit of current space, relative to the WCS
LIMMIN	0.0000,0.0000	LIMITS	Lower left X,Y limit of current space, relative to the WCS
LOGINNAME	" "	CONFIG	Name entered by the user or configuration file during login to AutoCAD. (CFG)(RO)
LTSCALE	1.0000	LTSCALE	Global scale factor applied to linetypes
LUNITS	2	DDUNITS, UNITS	Linear units format: 1 = Scientific 2 = Decimal 3 = Engineering 4 = Architectural 5 = Fractional
LUPREC	4	DDUNITS, UNITS	Units precision. For architectural and fractional units, two to the power LUPREC defines the largest denominator that AutoCAD displays (for example, 2 to the power 4 = 16). For all other units, LUPREC defines the number of decimal places that AutoCAD displays.
MACROTRACE	0		Controls the DIESEL macro-debugging display. Not listed by SETVAR. 0 = Off 1 = On (NS)
MAXACTVP	16		Maximum number of viewports to regenerate. (NS)
MAXOBJMEM	0		**Amount of virtual memory the drawing can use before AutoCAD activates its object pager. MAXOBJMEM = 0 disables the object pager and lets AutoCAD use its normal pager. See Chapter 1. (NS)**

Variable Name	Default Setting	Command Names	Variable Description
MAXSORT	200		Maximum number of symbols and file names sorted in lists. AutoCAD doesn't take the time to sort lists (of layer names, for example) that contain more items than MAXSORT. (CFG)
MENUCTL	1		Command-line input-sensitive screen menu-page switching: 0 = Off 1 = On (CFG)
MENUECHO	0		Controls the display of menu macro actions on the command line; the value is the sum of the following: 0 = No suppression 1 = Suppresses menu input 2 = Suppresses command prompts 4 = Suppresses disable ^P toggling 8 = Displays DIESEL input/output strings (NS)
MENUNAME	ACAD	MENU	The current **base** menu name, **including the path**. (RO)
MIRRTEXT	1		Controls reflection of text by the MIRROR command: 0 = Retain text direction 1 = Reflect text
MODEMACRO	" "		A DIESEL language expression to control status-line display. (NS)
OFFSETDIST	-1.0000	OFFSET	The default distance for the OFFSET command; negative values enable the Through option
MTEXTED	"Internal"	PREFERENCES	Text editor to use with mtext objects. You can set MTEXTED to the names of two mtext editors: the default editor and one that appears when you choose the Full Editor button in the :LISPED dialog. You separate the names of the two editors with a pound sign (#). For example, setting MTEXTED to :LISPED#INTERNAL causes R13c4 for Windows to use the simple editing dialog by default and the Edit MText dialog when you choose the Full Editor button. (CFG)
OFFSETDIST	-1.0000	**OFFSET**	Default distance. Negative values cause AutoCAD to use Through mode.
ORTHOMODE	0	F8, CTRL+L	Sets the current Ortho mode state: 0 = Off 1 = On

System Variables

Variable Name	Default Setting	Command Names	Variable Description
OSMODE	0	DDOSNAP, OSNAP	The current object snap mode; the value is the sum of the following: 0 = NONe 1 = ENDPoint 2 = MIDpoint 4 = CENter 8 = NODe 16 = QUAdrant 32 = INTersection 64 = INSertion 128 = PERpendicular 256 = TANgent 512 = NEArest 1024 = QUIck 2048 = APPint
PDMODE	0		Controls the graphic display style of point entities
PDSIZE	0.0000		Controls the size of point graphic display: 0 = single pixel positive = absolute negative = relative to view
PELLIPSE	**0**		**Controls the type of object created by the ELLIPSE command:** **0 = True ellipse** **1 = Polyline approximation**
PERIMETER	0.0000	AREA, DBLIST, LIST	The last computed perimeter. (NS)(RO)
PFACEVMAX	4		Maximum number of vertices per face in a PFACE mesh. (NS)(RO)
PICKADD	1	DDSELECT	Controls whether selected entities are added to or replace the current selection set: 0 = Replace (SHIFT to add or remove) 1 = Added (SHIFT to remove only) (CFG)
PICKAUTO	0	DDSELECT	Controls the implied (AUTO) windowing feature for object selection: 0 = Implied windowing is off 1 = Implied windowing is on (CFG)
PICKBOX	3		Half the object-selection pick box size, in pixels. (CFG)
PICKDRAG	0	DDSELECT	Determines whether you must hold the pick button down in between picking the two corners of a selection box: 0 = Off (pick, release, pick) 1 = On (pick, hold, release) (CFG)

Variable Name	Default Setting	Command Names	Variable Description
PICKFIRST	1	DDSELECT	Controls whether noun/verb editing is allowed (object selection before command selection): 0 = Don't allow noun/verb editing 1 = Allow noun/verb editing (CFG)
PLATFORM	"Microsoft Windows Version 4.0 (x86)"		Indicates the hardware/operating system version of AutoCAD in a string such as "386 DOS Extender," "Sun 4/SPARCstation," and so on. (NS)(RO)
PLINEGEN	0		Determines control points for polyline generation of non-continuous linetypes: 0 = Vertices 1 = End points
PLINEWID	0.0000	PLINE	Default width for new polyline entities
PLOTID	" "	PLOT	Current plotter configuration description. (CFG)
PLOTROTMODE	**1**		**Controls the orientation of plots on the paper. When PLOTROTMODE = 1, the lower left corner of the plotting area always is at lower left corner of the paper. When PLOTROTMODE = 0, the rotation of plotting area relative to paper is as follows: for 0-degree plot rotation, rotation icon aligns with lower left corner of paper; for 90-degree rotation, icon aligns with upper left corner of paper; for 180-degree rotation, icon aligns with upper right corner of paper; for 270-degree rotation, icon aligns with lower right corner of paper.**
PLOTTER	0	PLOT	Current plotter configuration number. The first plotter configuration is #0. (CFG)
POLYSIDES	4	POLYGON	Default number of sides (3 to 1024) for new polygon entities. (NS)
POPUPS	1		Determines whether the Advanced User Interface (dialog boxes, menu bar, pull-down menus, icon menus) is supported: 1 = Yes 0 = No (NS)(RO)

System Variables

Variable Name	Default Setting	Command Names	Variable Description
PROJMODE	1		Controls the projection mode for EXTEND and TRIM. 0 = True 3D mode (no projection) 1 = Project to X,Y plane of current UCS 2 = Project to current view plane (CFG)
PSLTSCALE	1		Paper space scaling of model space linetypes: 0 = Off 1 = On
PSPROLOG	" "		Name of the PostScript post-processing section of ACAD.PSF to be appended to the PSOUT command's output. (CFG)
PSQUALITY	75		Controls rendering quality of PostScript images imported with the PSIN command: 0 = Disable image rendering positive = number of pixels per drawing unit negative = same as positive, but doesn't fill outlines (CFG)
QTEXTMODE	0	QTEXT	Controls quick text mode (rectangular boxes in place of text strings): 0 = Off 1 = On
RASTERPREVIEW	0		Controls generation of raster preview images in the drawing file: 0 = BMP preview only 1 = BMP and WMF previews 2 = WMF previews only 3 = No previews AutoCAD uses the BMP preview, if present, to display an image of the drawing in file dialog boxes. The WMF preview is of no use to AutoCAD.
REGENMODE	1	REGENAUTO	Controls whether certain regenerations happen automatically, without warning: 0 = Don't regenerate automatically 1 = Always regenerate automatically
RE-INIT	0	REINIT	Reinitializes various parts of AutoCAD. Not listed by SETVAR. The sum of: 0 = No reinitialization 1 = Digitizer port 2 = Plotter port 4 = Digitizer device 8 = Display device 16 = Reload ACAD.PGP (NS)

Variable Name	Default Setting	Command Names	Variable Description
RIASPECT	0.0		Pixel aspect ratio for raster images imported with GIFIN, PCXIN, and TIFFIN. (Not a true system variable, but a global variable used by the raster in commands.) (NS)
RIBACKG	0		Background color number for raster images imported with GIFIN, PCXIN, and TIFFIN. (Not a true system variable, but a global variable used by the raster in commands.) (NS)
RIEDGE	0		Controls edge detection for raster images imported with GIFIN, PCXIN, and TIFFIN. Valid values are 0 to 255, with 0 being no edge detection. (Not a true system variable, but a global variable used by the raster in commands.) (NS)
RIGAMUT	256		Determines the number of colors used to display raster images imported with GIFIN, PCXIN, and TIFFIN. (Not a true system variable, but a global variable used by the raster in commands.) (NS)

RIGREY	0		Controls whether raster images imported with GIFIN, PCXIN, and TIFFIN are shown in gray-scale or color:
			  0 = Color
			  >0 = Gray-scale
			(Not a true system variable, but a global variable used by the raster in commands.) (NS)

RITHRESH	0		Controls luminance (brightness) threshold when importing raster images with GIFIN, PCXIN, and TIFFIN. RITHRESH disables luminance threshold checking. Positive values cause AutoCAD to import only pixels whose luminance is greater than the TITHRESH value. (Not a true system variable, but a global variable used by the raster in commands.) (NS)
SAVEFILE	"AUTO.SV$"	CONFIG	Default file name for automatic file saves. (CFG) (RO)

System Variables

Variable Name	Default Setting	Command Names	Variable Description
SAVEIMAGES	0		Controls whether graphics metafile images for application-defined objects ("zombies") are saved with the drawing file: 0 = Allow application-defined object's definition to control whether metafile image is written 1 = Always save metafile images 2 = Never save metafile images A metafile image allows you to see a representation of the application-defined object when you load the drawing without the object's supporting application.
SAVENAME	" "	SAVEAS	The most recent drawing save name, including path. (NS)(RO)
SAVETIME	120	CONFIG, **PREFERENCES**	The default interval in minutes between automatic file saves. SAVETIME = 0 disables automatic saves. Note that AutoCAD performs automatic saves only when there is command-line activity. (CFG)
SCREENBOXES	0 (in Windows)		The number of available side-screen menu slots in the current graphics screen area. SCREENBOXES = 0 if the side-screen menu is disabled. (RO)(CFG)
SCREENMODE	3 (in Windows)		Indicates the active AutoCAD screen mode or window: 0 = Text 1 = Graphics 2 = Dual screen SCREENMODE is always 3 in Windows. (CFG)(RO)
SCREENSIZE	716.0000,383.0000 (Windows 95 800x600)		Size of current viewport in pixels. (RO)
SHADEDGE	3		Controls the display of edges and faces by the SHADE command: 0 = Faces shaded, edges unhighlighted 1 = Faces shaded, edges in background color 2 = Faces unfilled, edges in entity color 3 = Faces in entity color, edges in background
SHADEDIF	70		Specifies the ratio of diffuse-to-ambient light used by the SHADE command; expressed as a percentage of diffuse reflective light
SHPNAME	" "	SHAPE	Default shape name. (NS)
SKETCHINC	0.1000	SKETCH	Recording increment for SKETCH segments

Maximizing AutoCAD® R13

Variable Name	Default Setting	Command Names	Variable Description
SKPOLY	0		Controls the type of entities generated by SKETCH: 0 = Lines 1 = Polylines
SNAPANG	0.0	DDRMODES, SNAP	Angle of snap and grid rotation in the current viewport, for the current UCS.
SNAPBASE	0.0000,0.0000	DDRMODES, SNAP	X,Y base point for snap and grid in the current viewport, for the current UCS.
SNAPISOPAIR	0	DDRMODES, SNAP	Current isoplane for isometric drawing in the current viewport (when SNAPSTYL = 1): 0 = Left 1 = Top 2 = Right
SNAPMODE	0	DDRMODES, SNAP, F9, CTRL+B	Indicates the state of Snap for the current viewport: 0 = Off 1 = On
SNAPSTYL	0	DDRMODES, SNAP	Snap style for the current viewport: 0 = Standard 1 = Isometric
SNAPUNIT	1.0000,1.0000	DDRMODES, SNAP	Snap X,Y increment for the current viewport.
SORTENTS	96	DDSELECT	Bit code controlling the type(s) of operations for which AutoCAD should not use spatial database organization and instead sort entities in the order created; the value is the sum of the following: 0 = None (don't sort for anything) 1 = Object selection 2 = Object snaps 4 = Redraws 8 = Slide creation (MSLIDE) 16 = Regenerations 32 = PLOT 64 = PostScript output (PSOUT) The default value of 96 (= 32 + 64) sorts for plots and PSOUT. (CFG)
SPLFRAME	0		Controls the display of control polygons for spline-fit polylines, defining meshes of surface-fit polygon meshes, and invisible 3D face edges: 0 = Off 1 = On
SPLINESEGS	8		Number of line segments generated for each spline curve

System Variables

Variable Name	Default Setting	Command Names	Variable Description
SPLINETYPE	6		Controls the spline type generated by the PEDIT command's Spline option: 5 = Quadratic B-Spline 6 = Cubic B-Spline
SURFTAB1	6		Number of surface tabulations generated by RULESURF and TABSURF. Also the REVSURF and EDGESURF M-direction mesh density.
SURFTAB2	6		REVSURF and EDGESURF N-direction mesh density.
SURFTYPE	6		Controls surface type generated by the PEDIT Smooth option: 5 = Quadratic B-spline 6 = Cubic B-spline 8 = Bezier
SURFU	6		M-direction surface density of 3D polygon meshes
SURFV	6		N-direction surface density 3D polygon meshes
SYSCODEPAGE	"ansi_1252" (in Windows 95)		Code page used by AutoCAD on this platform. See Chapter 6 for more information on code pages. See Appendix A in the *AutoCAD Command Reference* for a list of some code pages. (RO)(NS)
TABMODE	0	TABLET, F4, CTRL+T	Controls tablet tracing mode: 0 = Off 1 = On (NS)
TARGET	0.0000,0.0000,0.0000	DVIEW	UCS coordinates of the current viewport's target point. (RO)
TDCREATE	2450259.46492766 (for example)	NEW, OPEN	Date and time of the current drawing's creation, in Julian format. (RO)
TDINDWG	0.00000000	NEW, OPEN	Total amount of editing time elapsed in the current drawing, in Julian days. (RO)
TDUPDATE	2450259.50306910 (for example)	QSAVE, SAVE, SAVEAS	Date and time when the file was last saved, in Julian format. (RO)
TDUSRTIMER	0.00000000	TIME	User-controlled elapsed time in Julian days. (RO)
TEMPPREFIX	" "	CONFIG	Directory for placement of AutoCAD's temporary files; defaults to the drawing directory. (NS)(RO)

Variable Name	Default Setting	Command Names	Variable Description
TEXTEVAL	0		Controls the checking of text input for AutoLISP expressions: 0 = Don't check for AutoLISP expressions 1 = Check for AutoLISP expressions when creating text (using the TEXT command) or attribute values (NS)
TEXTFILL	0		**Controls whether PostScript and TrueType text is displayed and plotted filled in:** **0 = Don't fill** **1 = Fill**
TEXTQLTY	50		**Controls the displayed and plotted resolution of PostScript and TrueType text. Valid values are 0 to 100, with higher numbers representing greater resolution.**
TEXTSIZE	0.2000	DTEXT, TEXT	Default height for text created with variable-height text styles
TEXTSTYLE	"STANDARD"	TEXT, STYLE	Current text style's name
THICKNESS	0.0000	DDEMODES	Current 3D extrusion thickness
TILEMODE	1	TILEMODE	Controls visibility of paper space: 0 = Show paper space and any viewports into model space 1 = Show model space only
TOOLTIPS	1		**Controls whether ToolTips appear for toolbar icons:** **0 = Don't display ToolTips** **1 = Display ToolTips** **TOOLTIPS is valid only on Windows platforms. (CFG)**
TRACEWID	0.0500	TRACE	Default width for traces
TREEDEPTH	3020	DDSELECT	A code number (four digits) representing the maximum number of divisions for spatial database index for model space (first two digits) and paper space (last two digits).
TREEMAX	10000000	TREEMAX	Maximum number of nodes for spatial database organization for the current memory configuration. (CFG)
TRIMMODE	1		**Controls whether CHAMFER and FILLET trims away edges:** **0 = Don't trim** **1 = Trim** **(NS)**

System Variables

Variable Name	Default Setting	Command Names	Variable Description
UCSFOLLOW	0		Controls automatic display of the plan view in the current viewport when switching to a new UCS: 0 = Off 1 = On
UCSICON	1	UCSICON	Controls the UCS icon's display; the value is the sum of the following: 0 = Off 1 = On 2 = At origin when possible
UCSNAME	" "	DDUCS, UCS	Name of the current UCS for the current space ('"' = unnamed). (RO)
UCSORG	0.0000,0.0000,0.0000	DDUCS, UCS	WCS origin of the current UCS for the current space. (RO)
UCSXDIR	1.0000,0.0000,0.0000	DDUCS, UCS	X direction of the current UCS. (RO)
UCSYDIR	0.0000,1.0000,0.0000	DDUCS, UCS	Y direction of the current UCS. (RO)
UNDOCTL	5	UNDO	Current undo state; the value is the sum of the following: 0 = UNDO disabled 1 = UNDO enabled 2 = Single command can be undone 4 = Auto mode enabled 8 = Undo group is currently active (NS)(RO)
UNDOMARKS	0	UNDO	Current number of marks in the UNDO command's history. (NS)(RO)
UNDOONDISK	**1**		**Controls whether undo data are written to disk or held in memory:** **0 = Use memory** **1 = use disk file**
UNITMODE	0		Controls the display of user input of fractions, feet and inches, and surveyor's angles: 0 = Per LUNITS 1 = As input
USERI1 - 5	0		Integer variables for user data. USERI1 through USERI5 are not listed by SETVAR.
USERR1 - 5	0.0000		Real number variables for user data. USERR1 through USERR5 are not listed by SETVAR.

Variable Name	Default Setting	Command Names	Variable Description
USERS1 - 5	" "		String variables (up to 460 characters long) for user data. USERS1 through USERS5 are not listed by SETVAR. (NS)
VIEWCTR	6.2433,4.5000	ZOOM, PAN, VIEW	X,Y center point coordinate of the current view in the current viewport, with respect to the UCS. (RO)
VIEWDIR	0.0000,0.0000,1.0000	DVIEW	Camera point offset from target in the WCS. (RO)
VIEWMODE	0	DVIEW, UCS	The current viewport's viewing mode; the value is the sum of the following: 0 = Disabled 1 = Perspective 2 = Front clipping on 4 = Back clipping on 8 = UCSFOLLOW on 16 = FRONTZ offset in use (RO)
VIEWSIZE	9.0000	ZOOM, VIEW	The current view's height, in drawing units. (RO)
VIEWTWIST	0	DVIEW	The current viewport's view-twist angle. (RO)
VISRETAIN	0	VISRETAIN	Controls retention of xref layer settings in the current drawing: 0 = Don't save xref layer settings (xref DWG's layer settings control) 1 = Save xref layer settings (parent DWG's layer settings control)
VSMAX	37.4600,27.0000,0.0000	ZOOM, PAN, VIEW	Upper right X,Y coordinate of the current viewport's virtual screen for the current UCS. (NS)(RO)
VSMIN	-24.9734,-18.0000,0.0000	ZOOM, PAN, VIEW	Lower left X,Y coordinate of the current viewport's virtual screen for the current UCS. (NS)(RO)
WORLDUCS	1	UCS	Indicates whether the current UCS is equivalent to the WCS: 0 = False 1 = True (NS)(RO)
WORLDVIEW	1	DVIEW,UCS	Controls the automatic changing of a UCS to the WCS during the DVIEW and VPOINT commands: 0 = Off 1 = On
XREFCTL	0		Controls the creation of an xref log (XLG) file each time AutoCAD resolves an xref: 0 = Don't create XLG file 1 = Create XLG file

appendix F

ASCII and Unicode Tables

This appendix contains an ASCII code table, showing the keystrokes, the hex, octal, and decimal codes, and the characters for all 256 ASCII characters. It also contains a table of the standard AutoCAD Unicode codes and characters. If you compare the hex codes of the ASCII and Unicode tables, you will see that the codes of the available characters from 0A or 000A to FF or 00FF are the same in both tables.

ASCII Table

Table F.1 shows all the ASCII (American Standard Code for Information Interchange) characters. The first 32 ASCII characters are control characters. The table shows their keystrokes; their hexadecimal, octal and decimal ASCII values; and their control sequence. The next 96 (32 through 127) characters are the standard ASCII printable characters. The table shows their keystrokes; their hexadecimal, octal, and decimal ASCII values; and their printable symbols. The last 128 (128 through 255) characters are extended ASCII codes. The printable symbols are shown for 128 through 168; above 168, the symbols are non-standard and vary from device to device.

Note: Many of the characters, particularly the control characters and extended ASCII codes, are entered with the SHIFT, CONTROL (CTRL) and ALT keys. The table shows these like SHIFT+TAB, CTRL+PAGEUP, and ALT+U keystroke. The keystroke column indicates the keystroke with SHIFT+x, where x is the lower case character, for all characters entered with SHIFT, even upper case letters or symbols that may be shown on your keyboard (so SHIFT+' indicates the keystroke for ~ and SHIFT+1 the keystroke for !). The *MINUS, *5, and *PLUS keystrokes indicate the MINUS, PLUS, and (with NUM LOCK off), the 5 keys on numeric keypad.

Tip: You can use the ALT key and numeric keypad to enter any ASCII character from the keyboard, even control characters. How the character is interpreted depends on the context. For example, a Carriage Return character is generally interpreted the same as the ENTER key, but some programs interpret it as only a Carriage Return and some interpret it as a Carriage Return + Line Feed combination. To enter an ASCII character from the keyboard, press ALT and type 0*xx* or 0*xxx* with the numeric keypad (where *xx* or *xxx* is the decimal code shown in the following table), then release the ALT key. For example, ALT+013 enters the Carriage Return character and ALT+0122 enters a lower case z. On some systems, the 0 (zero) may be omitted and you enter only ALT+*xx* or ALT+*xxx*.

The characters in the table are shown in the MS Windows Courier New TTF font, and the extended ASCII characters are those for that font, as printed on an HP LaserJet III printer. Characters which are undefined in that font are shown as a box, like ☐. Extended ASCII characters may appear different in different fonts or on different printers.

To access extended ASCII codes with DOS programs that use ANSI.SYS, you generally must use a zero, a semicolon, and a second code, such as 0;73, for the PG UP key. To determine the second code, you subtract 128 from the decimal value shown in the table, except for the following characters: ALT+9 (%9 decimal 128), ALT+10 (%10 decimal 129), ALT+MINUS (%- decimal 130), ALT+EQUAL (%= decimal 131), and CTRL+PAGEUP (^PgUp decimal 132). For these exceptions, you just use the decimal value, such as 0;129 for %10.

ASCII and Unicode Tables

Table F.1
ASCII CHARACTERS

Control Codes

Keystroke	Hex	Oct	Dec	Abbr	Control Sequence
CTRL+@	00	000	000	NUL	Null
CTRL+a	01	001	001	SOH	Start of Heading
CTRL+b	02	002	002	STX	Start of Text
CTRL+c	03	003	003	ETX	End of Text
CTRL+d	04	004	004	EOT	End of Transmission
CTRL+e	05	005	005	ENQ	Enquiry
CTRL+f	06	006	006	ACK	Acknowledge
CTRL+g	07	007	007	BEL	Bell
CTRL+h	08	010	008	BS	Backspace
CTRL+i	09	011	009	HT	Horizontal Tab
CTRL+j	0A	012	010	LF	Line Feed
CTRL+k	0B	013	011	VT	Vertical Tab
CTRL+l	0C	014	012	FF	Form Feed
CTRL+m	0D	015	013	CR	Carriage Return
CTRL+n	0E	016	014	SO	Shift Out
CTRL+o	0F	017	015	SI	Shift In
CTRL+p	10	020	016	DLE	Data Link Escape
CTRL+q	11	021	017	DC1	Device Control 1
CTRL+r	12	022	018	DC2	Device Control 2
CTRL+s	13	023	019	DC3	Device Control 3
CTRL+t	14	024	020	DC4	Device Control 4
CTRL+u	15	025	021	NAK	Negative Acknowledge
CTRL+v	16	026	022	SYN	Synchronous Idle
CTRL+w	17	027	023	ETB	End Transmission Block
CTRL+x	18	030	024	CAN	Cancel
CTRL+y	19	031	025	EM	End of Medium
CTRL+z	1A	032	026	SUB	Substitute
CTRL+[	1B	033	027	ESC	Escape
CTRL+\	1C	034	028	FS	File Separator
CTRL+]	1D	035	029	GS	Group Separator
CTRL+CTRL+	1E	036	030	RS	Record Separator
CTRL+_	1F	037	031	US	Unit Separator

Table F.1 (continued)
ASCII CHARACTERS

Printable Characters

Keystroke	Hex	Oct	Dec	Symbol	Keystroke	Hex	Oct	Dec	Symbol
SPACE	20	040	032	(space)	SHIFT+p	50	120	080	P
SHIFT+1	21	041	033	!	SHIFT+q	51	121	081	Q
SHIFT+QUOTE	22	042	034	"	SHIFT+r	52	122	082	R
SHIFT+3	23	043	035	#	SHIFT+s	53	123	083	S
SHIFT+4	24	044	036	$	SHIFT+t	54	124	084	T
SHIFT+5	25	045	037	%	SHIFT+u	55	125	085	U
SHIFT+7	26	046	038	&	SHIFT+v	56	126	086	V
' QUOTE	27	047	039	'	SHIFT+w	57	127	087	W
SHIFT+9	28	050	040	(	SHIFT+x	58	130	088	X
SHIFT+0	29	051	041	)	SHIFT+y	59	131	089	Y
SHIFT+8	2A	052	042	*	SHIFT+z	5A	132	090	Z
SHIFT+=	2B	053	043	+	[	5B	133	091	[
, COMMA	2C	054	044	,	\	5C	134	092	\
- MINUS	2D	055	045	-	]	5D	135	093	]
. PERIOD	2E	056	046	.	SHIFT+6	5E	136	094	^
/	2F	057	047	/	SHIFT+MINUS	5F	137	095	_
0	30	060	048	0	` BACKQUOTE	60	140	096	`
1	31	061	049	1	a	61	141	097	a
2	32	062	050	2	b	62	142	098	b
3	33	063	051	3	c	63	143	099	c
4	34	064	052	4	d	64	144	100	d
5	35	065	053	5	e	65	145	101	e
6	36	066	054	6	f	66	146	102	f
7	37	067	055	7	g	67	147	103	g
8	38	070	056	8	h	68	150	104	h
9	39	071	057	9	i	69	151	105	i
SHIFT+;	3A	072	058	:	j	6A	152	106	j
;	3B	073	059	;	k	6B	153	107	k
SHIFT+COMMA	3C	074	060	<	l	6C	154	108	l
=	3D	075	061	=	m	6D	155	109	m
SHIFT+.	3E	076	062	>	n	6E	156	110	n
SHIFT+/	3F	077	063	?	o	6F	157	111	o
SHIFT+2	40	100	064	@	p	70	160	112	p
SHIFT+a	41	101	065	A	q	71	161	113	q
SHIFT+b	42	102	066	B	r	72	162	114	r
SHIFT+c	43	103	067	C	s	73	163	115	s
SHIFT+d	44	104	068	D	t	74	164	116	t
SHIFT+e	45	105	069	E	u	75	165	117	u
SHIFT+f	46	106	070	F	v	76	166	118	v
SHIFT+g	47	107	071	G	w	77	167	119	w
SHIFT+h	48	110	072	H	x	78	170	120	x
SHIFT+i	49	111	073	I	y	79	171	121	y
SHIFT+j	4A	112	074	J	z	7A	172	122	z
SHIFT+k	4B	113	075	K	SHIFT+[	7B	173	123	{
SHIFT+l	4C	114	076	L	SHIFT+\	7C	174	124	|
SHIFT+m	4D	115	077	M	SHIFT+]	7D	175	125	}
SHIFT+n	4E	116	078	N	SHIFT+`	7E	176	126	~
SHIFT+o	4F	117	079	O	BACKSPACE	7F	177	127	□

Table F.1 (continued)
ASCII CHARACTERS

Extended Characters

Keystroke	Hex	Oct	Dec	Symbol	Keystroke	Hex	Oct	Dec	Symbol
ALT+9	80	200	128	□	ALT+`	A9	251	169	©
ALT+0	81	201	129	□	LEFT SHIFT	AA	252	170	ª
ALT+MINUS	82	202	130	‚	ALT+\	AB	253	171	«
ALT+=	83	203	131	ƒ	ALT+z	AC	254	172	¬
CTRL+PAGEUP	84	204	132	„	ALT+x	AD	255	173	-
F11	85	205	133	…	ALT+c	AE	256	174	®
F12	86	206	134	†	ALT+v	AF	257	175	¯
SHIFT+F11	87	207	135	‡	ALT+b	B0	260	176	°
SHIFT+F12	88	210	136	ˆ	ALT+n	B1	261	177	±
CTRL+F11	89	211	137	‰	ALT+m	B2	262	178	²
CTRL+F12	8A	212	138	Š	ALT+,	B3	263	179	³
ALT+F11	8B	213	139	‹	ALT+.	B4	264	180	´
ALT+F12	8C	214	140	Œ	ALT+/	B5	265	181	µ
	8D	215	141	□	RIGHT SHIFT	B6	266	182	¶
	8E	216	142	□	PRINTSCREEN	B7	267	183	·
SHIFT+TAB	8F	217	143	□	ALT	B8	270	184	¸
ALT+q	90	220	144	□	ALT+SPACE	B9	271	185	¹
ALT+w	91	221	145	'	CAPS LOCK	BA	272	186	º
ALT+e	92	222	146	'	F1	BB	273	187	»
ALT+r	93	223	147	"	F2	BC	274	188	¼
ALT+t	94	224	148	"	F3	BD	275	189	½
ALT+y	95	225	149	•	F4	BE	276	190	¾
ALT+u	96	226	150	–	F5	BF	277	191	¿
ALT+i	97	227	151	—	F6	C0	300	192	À
ALT+o	98	230	152	˜	F7	C1	301	193	Á
ALT+p	99	231	153	™	F8	C2	302	194	Â
ALT+[	9A	232	154	š	F9	C3	303	195	Ã
ALT+]	9B	233	155	›	F10	C4	304	196	Ä
ALT+ENTER	9C	234	156	œ	NUM LOCK	C5	305	197	Å
CTRL	9D	235	157	□	SCROLL LOCK	C6	306	198	Æ
ALT+a	9E	236	158	□	HOME	C7	307	199	Ç
ALT+s	9F	237	159	Ÿ	UP	C8	310	200	È
ALT+d	A0	240	160	(blank)	PAGE UP	C9	311	201	É
ALT+f	A1	241	161	¡	*MINUS	CA	312	202	Ê
ALT+g	A2	242	162	¢	LEFT	CB	313	203	Ë
ALT+h	A3	243	163	£	*5	CC	314	204	Ì
ALT+j	A4	244	164	¤	RIGHT	CD	315	205	Í
ALT+k	A5	245	165	¥	*PLUS	CE	316	206	Î
ALT+l	A6	246	166	¦	END	CF	317	207	Ï
ALT+;	A7	247	167	§	DOWN	D0	320	208	Ð
ALT+QUOTE	A8	250	168	¨	PAGE DOWN	D1	321	209	Ñ

Table F.1 (continued)
ASCII CHARACTERS

Extended Characters (continued)

Keystroke	Hex	Oct	Dec	Symbol	Keystroke	Hex	Oct	Dec	Symbol
INS	D2	322	210	Ò	ALT+F2	E9	351	233	é
DEL	D3	323	211	Ó	ALT+F3	EA	352	234	ê
SHIFT+F1	D4	324	212	Ô	ALT+F4	EB	353	235	ë
SHIFT+F2	D5	325	213	Õ	ALT+F5	EC	354	236	ì
SHIFT+F3	D6	326	214	Ö	ALT+F6	ED	355	237	í
SHIFT+F4	D7	327	215	×	ALT+F7	EE	356	238	î
SHIFT+F5	D8	330	216	Ø	ALT+F8	EF	357	239	ï
SHIFT+F6	D9	331	217	Ù	ALT+F9	F0	360	240	ð
SHIFT+F7	DA	332	218	Ú	ALT+F10	F1	361	241	ñ
SHIFT+F8	DB	333	219	Û	CTRL+PRTSCR	F2	362	242	ò
SHIFT+F9	DC	334	220	Ü	CTRL+LEFT	F3	363	243	ó
SHIFT+F10	DD	335	221	Ý	CTRL+RIGHT	F4	364	244	ô
CTRL+F1	DE	336	222	Þ	CTRL+END	F5	365	245	õ
CTRL+F2	DF	337	223	ß	CTRL+PAGEDN	F6	366	246	ö
CTRL+F3	E0	340	224	à	CTRL+HOME	F7	367	247	÷
CTRL+F4	E1	341	225	á	ALT+1	F8	370	248	ø
CTRL+F5	E2	342	226	â	ALT+2	F9	371	249	ù
CTRL+F6	E3	343	227	ã	ALT+3	FA	372	250	ú
CTRL+F7	E4	344	228	ä	ALT+4	FB	373	251	û
CTRL+F8	E5	345	229	å	ALT+5	FC	374	252	ü
CTRL+F9	E6	346	230	æ	ALT+6	FD	375	253	ý
CTRL+F10	E7	347	231	ç	ALT+7	FE	376	254	þ
ALT+F1	E8	350	232	è	ALT+8	FF	377	255	ÿ

ASCII and Unicode Tables

Unicode Table

The following is a table of the standard AutoCAD Unicode codes and characters. Unicode uses 16-bit numbers to represent characters, which enables it to keep track of 65,536 characters and thus to unify most languages and a large number of special symbols in a single encoding scheme. About 35,000 characters have been assigned numbers by the Unicode Consortium. The first 128 characters are the same as in ASCII. Compare its hex codes to the hex codes in the ASCII table, and you will see that the codes of the available characters from 000A through 00FF correlate. See the *Native AutoCAD SHX Fonts and Unicode* section of Chapter 6 for more on Unicode.

To enter a special character using Unicode in R13, type its hexadecimal code, prefaced by **\U+**. If you're editing in the Edit MText dialog box, you must also preface the code with the % sign; put **%\U+** before the four-character hexadecimal code. For example, to use the less-than-or-equal-to sign (≤), type **%\U+2264** in the Edit MText dialog, and **\U+2264** elsewhere.

In Table F2, the Line Feed character is indicated with [LF] and the space character with [ ]. Note that some codes and characters (typically the 00xE codes and characters, such as the 00BE code and ¾ character) are missing from the Unicode table. The missing characters do not print in the ROMANS font (instead printing a ? character), even though some, such as the ¾ character *do* show up in the Modify Text or Edit Text dialog boxes. The character set displayed by the Char. Set button in the Text Style dialog box (DDSTYLE command, in R13c4 or later) is a more reliable indicator than the Modify Text or Edit Text dialog boxes.

Table F.2
AutoCAD R13 Unicode Characters

000A	[LF]	0047	G	006F	o	00C9	É	00F5	õ	0170	Ű
0020	[ ]	0048	H	0070	p	00CA	Ê	00F6	ö	0171	ű
0021	!	0049	I	0071	q	00CB	Ë	00F7	÷	0179	Ź
0022	"	004A	J	0072	r	00CC	Ì	00F8	ø	017A	ź
0023	#	004B	K	0073	s	00CD	Í	00F9	ù	017B	Ż
0024	$	004C	L	0074	t	00CE	Î	00FA	ú	017C	ż
0025	%	004D	M	0075	u	00CF	Ï	00FB	û	017D	Ž
0026	&	004E	N	0076	v	00D1	Ñ	00FC	ü	017E	ž
0027	'	004F	O	0077	w	00D2	Ò	00FD	ý	0410	А
0028	(	0050	P	0078	x	00D3	Ó	00FF	ÿ	0411	Б
0029	)	0051	Q	0079	y	00D4	Ô	0104	Ą	0412	В
002A	*	0052	R	007A	z	00D5	Õ	0105	ą	0413	Г
002B	+	0053	S	007B	{	00D6	Ö	0106	Ć	0414	Д
002C	,	0054	T	007C	|	00D8	Ø	0107	ć	0415	Е
002D	-	0055	U	007D	}	00D9	Ù	010C	Č	0416	Ж
002E	.	0056	V	007E	~	00DA	Ú	010D	č	0417	З
002F	/	0057	W	00A1	¡	00DB	Û	010E	Ď	0418	И
0030	0	0058	X	00A2	¢	00DC	Ü	010F	ď	0419	Й
0031	1	0059	Y	00A3	£	00DD	Ý	0118	Ę	041A	К
0032	2	005A	Z	00A5	¥	00DF	ß	0119	ę	041B	Л
0033	3	005B	[	00A7	§	00E0	à	011A	Ě	041C	М
0034	4	005C	\	00AA	ª	00E1	á	011B	ě	041D	Н
0035	5	005D	]	00AB	«	00E2	â	0141	Ł	041E	О
0036	6	005E	^	00B0	°	00E3	ã	0142	ł	041F	П
0037	7	005F	_	00B1	±	00E4	ä	0143	Ń	0420	Р
0038	8	0060	`	00B5	µ	00E5	å	0144	ń	0421	С
0039	9	0061	a	00BA	º	00E6	æ	0147	Ň	0422	Т
003A	:	0062	b	00BB	»	00E7	ç	0148	ň	0423	У
003B	;	0063	c	00BC	¼	00E8	è	0150	Ő	0424	Ф
003C	<	0064	d	00BD	½	00E9	é	0151	ő	0425	Х
003D	=	0065	e	00BF	¿	00EA	ê	0158	Ř	0426	Ц
003E	>	0066	f	00C0	À	00EB	ë	0159	ř	0427	Ч
003F	?	0067	g	00C1	Á	00EC	ì	015A	Ś	0428	Ш
0040	@	0068	h	00C2	Â	00ED	í	015B	ś	0429	Щ
0041	A	0069	i	00C3	Ã	00EE	î	0160	Š	042A	Ъ
0042	B	006A	j	00C4	Ä	00EF	ï	0161	š	042B	Ы
0043	C	006B	k	00C5	Å	00F1	ñ	0164	Ť	042C	Ь
0044	D	006C	l	00C6	Æ	00F2	ò	0165	ť	042D	Э
0045	E	006D	m	00C7	Ç	00F3	ó	016E	Ů	042E	Ю
0046	F	006E	n	00C8	È	00F4	ô	016F	ů	042F	Я

0430	а										
0431	б										
0432	в										
0433	г										
0434	д										
0435	е										
0436	ж										
0437	з										
0438	и										
0439	й										
043A	к										
043B	л										
043C	м										
043D	н										
043E	о										
043F	п										
0440	р										
0441	с										
0442	т										
0443	у										
0444	ф										
0445	х										
0446	ц										
0447	ч										
0448	ш										
0449	щ										
044A	ъ										
044B	ы										
044C	ь										
044D	э										
044E	ю										
044F	я										
20A7	₧										
2126	Ω										
2205	∅										
221E	∞										
2264	≤										
2302	⌂										

appendix G

The Maximizing AutoCAD CD-ROM

This appendix describes the programs, utilities, and tools on the MaxAC CD-ROM disk that came packaged with this book. As mentioned in the introduction, the exercise support and code files on this disk can save you a lot of typing and debugging time. But that is only a part of the value of the contents of the CD-ROM. The MaxAC CD-ROM also includes a number of shareware, freeware, and demo programs, utilities, and tools that can enhance your customization and drafting efforts and make them quicker and easier. Some of these are unique new development tools, not previously available.

Note: For updated information and maybe even additional programs, utilities, and tools, see the README.TXT file on the MaxAC CD-ROM or in your MaxAC directory, or see Rusty's Group A web site on the Internet's World Wide Web at `http://www.group-a.com/~rusty`.

Except as noted in the related sections of this appendix, the programs and utilities on the MaxAC CD-ROM require AutoCAD Release 13, and many require R13c4.

The following list briefly describes the tools and utilities on the MaxAC CD-ROM:

Development Tools and Utilities:

Maximizing AutoCAD Windows Runtime Extension (MaWIN) adds useful functions to AutoLISP for accessing Windows API functionality.

LispPad for Maximizing AutoCAD simplifies editing, loading, and testing AutoLISP, menu, script, and DCL code.

TextPad is a powerful, full-featured, 32-bit Windows-based text editor.

Vital LISP Demo is a demo of a professional AutoLISP development environment.

Lisp2C compiles AutoLISP code into C language code ADS applications.

KELV13 adds protected symbol definitions for R13 to the Kelvinator, which helps protect AutoLISP code from theft and modification.

DOSLib provides DOS command-line functionality in AutoLISP.

HLP2AHP is an R12 Help to R13 Help conversion program.

IDSHX is an ADS-defined AutoLISP subroutine that identifies the type of a compiled shape or font file.

LispPlot enables you to invoke ACAD's plot dialog box from AutoLISP.

DCL Spy for Maximizing AutoCAD is a learning and debugging aid for understanding and debugging dialog box callback functions.

Maximizing AutoCAD Button Image Explorer for Windows 95 and NT 4.0 enables you to browse and extract the tool button images used in AutoCAD.

System and Project Management Tools:

TIDY TOOLS groups, catalogs, prints, views, lists, finds, edits, tests, and loads AutoLISP, EXP, ARX, Script, and Menu files, manages text and customization files, and imports text.

Mega-App Loader is an enhanced replacement for the APPLOAD.LSP and APPLOAD.DCL files in R13.

Album is a stand-alone application that displays directories of DWG files.

SlideManager extracts, manages, and prints slide files and libraries.

SuperPurge provides better drawing file cleanup and file size reduction.

ShowRefs lists all external files referenced by a drawing or group of drawings.

DWGList lists layers and blocks within drawings from the DOS prompt.

PlotStamp adds dynamic plot data stamping capability to AutoCAD R13 for Windows, dynamically specifying the value of an AutoCAD text object.

Release 13 Bug List documents known R13 bugs and documentation errors.

Application and Productivity Tools and Utilities:

BigDCL replaces a number of standard AutoCAD dialogs with reworked versions that make better use of larger format displays (800 x 600 minimum).

TbOsnap enables the Object Snap toolbar to toggle running object snap modes at the command prompt.

The Maximizing AutoCAD CD-ROM

BlockEdit automates the editing of a block defined in the current drawing using a second instance of AutoCAD.

DynaMode is a utility that uses ARX to define a custom, context-sensitive status line for any command.

ATTEXT duplicates the AutoCAD ATTEXT command from the DOS prompt.

DDEDTEXT provides an enhanced single-line dialog box for editing text, dimensions, and mtext objects.

DDCHTEXT provides a dialogue box for modifying the properties of various types of annotation (text-like) objects.

SectionMaster is collection of parametric section-generating routines that draw steel, concrete, and wood structural sections as 2D polylines.

Disclaimers, Copyrights, Restrictions, Licenses, and Payments

The publisher and authors of this book make no warranty, including but not limited to any implied warranties of merchantability or fitness for a particular purpose, regarding any of the software, files, and accompanying materials on the MaxAC CD-ROM.

Note: The software, files, and accompanying materials are provided on the MaxAC CD-ROM solely on an "as-is" basis, for your convenience, to be used at your discretion. Purchase of this book does not include support for software, files, and accompanying materials, except as specifically required for use in the exercises of the book, as described in the *Introduction*. The authors of this book do provide fee-based consulting. Additional support for many of the programs on the CD-ROM may be obtained from the developers of the software, generally upon shareware registration and payment, as described in the documentation and information files in the directory for the specific software on the MaxAC CD-ROM.

In no event shall publisher or authors of this book be liable to any special, collateral, incidental, or consequential damages in connection with or arising out of the use of the software, files, and accompanying materials. Additional disclaimers are contained in documentation and information files on the MaxAC CD-ROM.

All of the contents of the MaxAC CD-ROM are copyrighted and/or restricted in certain ways, as described in the following sections and in documentation and information files on the MaxAC CD-ROM.

MaxAC Exercise, Support, and Code Files

The Maximizing AutoCAD exercise, support, development, utility, and code files on the MaxAC CD-ROM (installed in the MAXAC\BOOK directory) are provided primarily for use in the book's exercises; however, you may find many of the support and code files useful in developing your own menus, programs, and applications. All of it is copyrighted, either as specified in documentation and information files on the MaxAC CD-ROM, or if not specified, then as part of the Maximizing AutoCAD R13 book and MaxAC CD-ROM itself. However, except for those programs that are more specifically restricted in their sections of this appendix or in documentation and information files on the MaxAC CD-ROM, you may freely make use of these files in your menus, programs, and applications, and may distribute those menus, programs, and applications to others, subject to the two following restrictions:

1. No material or code from the Maximizing AutoCAD exercise, support, development, utility, and code files on the MaxAC CD-ROM may be distributed in any products that are competitive to the books *Maximizing AutoCAD R13*, *Maximizing AutoLISP for AutoCAD R13*, or any other books, products, or services authored, created, or provided by Rusty Gesner, Mark Middlebrook, or Tony Tanzillo. In other words, you can't use our stuff to compete with us.

2. All menus, programs, applications, products, or services using material from the Maximizing AutoCAD exercise, support, development, utility, and code files on the MaxAC CD-ROM must prominently contain this notice: "Contains, with permission, copyrighted material from *Maximizing AutoCAD R13*, by Rusty Gesner, Mark Middlebrook, and Tony Tanzillo" and the specific code used must be commented as being "...copyrighted material from *Maximizing AutoCAD R13*, by Rusty Gesner, Mark Middlebrook, and Tony Tanzillo."

Shareware Programs and Utilities

Many of the programs and utilities included on the MaxAC CD-ROM are shareware and are copyrighted and restricted. Shareware is a "Try-Before-You-Buy" method of software distribution. It is not free software. The shareware authors would like you to try their programs before registration. Although these shareware products are functional and useful as-is, some of them have limited features and/or documentation. When you register and pay, you often get a more full-featured product, updates, and/or printed documentation.

Unless otherwise noted in the specific sections describing these shareware products in this appendix, our arrangements to provide these shareware programs and utilities on the MaxAC CD-ROM give you permission to make use of products without registering them in the context of working through the instructions, techniques, methods, and tutorials of the book *only*, and you *must* register and pay for them if you use them for any other purpose. Registration instructions are in the specific sections describing these shareware products, or provided by the product upon installation or use.

Freeware Programs and Utilities

Some of the programs and utilities included on the MaxAC CD-ROM are freeware. Freeware is free, but may be copyrighted. Generally, you may use and redistribute them (with all files intact as provided to you) freely and without registration or payment, but they may be copyrighted or restricted in specific ways. See the specific sections in this appendix describing these shareware products.

Demo and Limited Version Programs

Some of the programs and utilities included on the MaxAC CD-ROM are demo or limited versions of commercial programs. Generally, you may use them without registration or payment, but you are encouraged to purchase the full commercial versions if you find them useful. See the specific sections in this appendix describing these shareware products.

Installing the Programs and Utilities on the MaxAC CD-ROM

The entire contents of the MaxAC CD-ROM are accessible from both the Setup program and directly in the MAXAC directory. There are two steps to installing most of the programs and utilities from the MaxAC CD-ROM. First, you use the MaxAC Setup program or the DOS XCOPY command to copy the files to the program and utility's specific subdirectory within the MAXAC\TOOLS directory on your hard drive. Second, you follow the instructions in the specific program and utility's section of this appendix.

However, some of the programs on the MaxAC CD-ROM can be installed directly from their *d:*\MAXAC\TOOLS*component* directories on the CD-ROM (where *d:* is your CD-ROM drive letter, and *component* is the path of the component) and need not be first copied to C:\MAXAC\TOOLS*component* directories on your hard drive. Just follow the instructions in the specific program and utility's section of this appendix.

Note: The instructions in this appendix assume that you install the MaxAC CD-ROM in the default MAXAC directory on your hard drive. If you decide instead to install it in another directory, substitute the name you use for the MAXAC directory throughout the appendix.

Copying the Programs and Utilities in Windows

If you are using any version of Windows, use the following exercise to copy the needed files for the program or utility you want to install. If you are using DOS, skip to the next section.

Maximizing AutoCAD® R13

Exercise

Copying the MaxAC CD-ROM Programs and Utilities in Windows

Put the MaxAC CD-ROM in your CD-ROM drive.

If using Windows 95, from the Taskbar choose Start, Run and enter **d:SETUP** (where *d:* is your CD-ROM drive letter), then choose OK. Otherwise, in Program Manager, choose File, Run, and enter **d:SETUP**, then choose OK.

The Maximizing AutoCAD R13 setup window should appear. Follow its instructions and choose from its options until you get to the Setup Type dialog box.

From the Setup Type dialog box, choose Custom, *then* Next	Displays the Select Components dialog box
In the Select Components dialog box, select (check) the Selected Tools *component and clear the* Selected Chapters *and* Everything *components (to check or clear a component, you highlight it and press the space bar)*	Displays a list of tools in the list box on the right
In the list box of tools, select (check) the desired tools, then choose Next *and follow the rest of the instructions setup provides*	Copies (and extracts if needed) each selected tool to its own specific subdirectory within the MAXAC\TOOLS directory

Next, proceed to the specific program and utility's section of this appendix for further installation instructions.

Copying the Programs and Utilities in DOS

If you are using DOS, use the following instructions and exercise to copy the needed files for the program or utility you want to install. Otherwise, skip this section.

Warning: Many of the programs and utilities on the MaxAC CD-ROM are unusable in DOS. Check the specific program or utility's section of this appendix to determine its usability.

You use the DOS XCOPY command to copy the components you want to install. The syntax is as follows:

```
C:\ XCOPY d:\MAXAC\TOOLS\component\*.* C:\MAXAC\TOOLS\component /s /i
```
⏎

where **d:** is your CD-ROM drive letter, and **component** is the path of the component you want to copy.

For example, to copy the DOSLIB utilities, use:

```
C:\ XCOPY d:\MAXAC\TOOLS\DOSLIB\*.* C:\MAXAC\TOOLS\DOSLIB /s /i
```
⏎

Then, proceed to the specific program and utility's section of this appendix for further installation instructions.

Development Tools and Utilities

Maximizing AutoCAD Windows Runtime Extension Limited Edition

Maximizing AutoCAD Windows Runtime Extension (also called MaWIN in this book) and its MAWIN.ARX file add a number of useful functions to AutoLISP as ExRxSubrs (external ARX subroutines) for accessing Windows API functionality. Several of these functions have been used in exercises in this book.

The functions defined by the Maximizing AutoCAD Windows Runtime Extension Limited Edition include:

(ShellOpen *file*) Opens a file using the registered application that is associated with the file's type. Can also be used to invoke the Explorer on a specified folder by passing the name of the folder in *file*.

(DeleteFile *file* **...)** Deletes files (any number of files can be passed).

(GetWindowsDirectory) Returns the path to the Windows directory.

(GetTempDirectory) Returns the first directory found as follows:

   1. The path specified by the TMP environment variable.

   2. The path specified by the TEMP environment variable, if TMP is not defined.

   3. The current directory, if both TMP and TEMP are not defined.

(GetProfileString *filename section key default*) Returns the value of a configuration string variable in an INI-type file (including ACAD.CFG).

(GetProfileInt *filename section key default*) Returns the value of a configuration integer variable in an INI-type file (including ACAD.CFG).

(Beep) Issues a beep.

(Sound *[frequency [duration]]*) Plays a sound on the system speaker (NT and Win3.1 only) at the specified frequency and duration.

(MessageBeep *code*) Plays registered event sounds (defined by control panel sound applet).

(MessageBox *Caption flag MessageText*) provides AutoLISP access to the commonly used Windows Message dialog box, which can display a system icon, multi-line text messages and pre-defined buttons with labels such as yes, no, ok, cancel, abort, retry, ignore, and so on. Displaying this dialog will also play the same system sounds accessed by the **(MessageBeep)** function.

(AddToRecentDocs *filename*) On Windows 95 and NT 4.0, adds a link (shortcut) for *filename* to the recent document list (the Documents leaf of the Start menu). (Note: MAWIN.ARX automatically adds the name of the current drawing to the recent documents list each time you enter the editor with a named drawing. AutoCAD alone does not do this.)

(FileExists *filename*) Returns T if *filename* (a file or directory) exists.

(SysWindow *code*) Instructs the AutoCAD Window to minimize, maximize or restore itself.

(AcadPlotDlg) Invokes the PLOT command (dialog box).

(StartModal *appname [parameters]*) Executes the Windows application *appname,* passing it the optional *parameters*, and waits until the launched application exits before returning control to the calling AutoLISP program.

See the MAWIN.TXT file for more complete documentation on each of these functions.

The **MessageBox** function is a more robust replacement for the AutoLISP (alert) function's dialog, which is basically just a simple implementation of the same Windows message box. Maximizing AutoCAD Windows Runtime Extension Limited Edition includes the MAWIN.LSP file, which provides documentation and examples of using **MessageBox**, and defines the following predefined "wrapper" functions for commonly used **MessageBox** configurations, and also includes several functions to control the AutoCAD window:

(YesNoDialog *title default message*)

(WarningDlg *title default message*)

(ErrorDlg *title message*)

(RestoreAcad)

(MinimizeAcad)

(MaximizeAcad)

(Explore *folder*)

In addition, the full commercial version, the Maximizing AutoCAD Windows Runtime Extension Standard Edition, includes:

(FindFiles *pattern* [*attributeflags*]**)** Searches for files matching *pattern* and *attributeflags* and returns a list of all matching files, in the form:

((*filename filesize FileTime attributes*)...)

For example, to get a list of all subdirectories under D:\MAXAC you would use: (FindFiles "d:\\maxac*.*" fa:directory).

(CopyFile *source target*) Copies the file *source* to *target*.

(MoveFile *source target*) Moves (or renames) the file *source* to *target*.

(MakeDir *dirspec*) Creates the directory *dirspec*.

(RemDir *dirspec*) Removes the directory *dirspec*.

(SplitPath *filespec*) Takes the string *filespec*, a file specification that includes a drive, directory, filename, and extension, decomposes it into discrete component strings and returns them in a list of the form:

(*drive directory filename extension*)

(MakePath *drive directory filename extension*) Returns a string that contains a file specification constructed from the individual components supplied in the argument strings.

Installation and Use

For use with the book, use the MaxAC Setup program on the CD-ROM to install MaWIN. It will copy the files to the MAXAC directory, which is on the AutoCAD path in the book's configuration, and will add a line in an ACAD.RX file in the MAXAC directory to automatically load it when you start AutoCAD in the book's configuration. See the MAWIN.TXT and MAWIN.LSP files for information and examples on the use of the Maximizing AutoCAD Windows Runtime Extension Limited Edition.

For installation and loading in other configurations of AutoCAD, perform the following steps:

1. Copy all of the files in the TOOLS\MaWIN directory from the CD-ROM to a directory on the AutoCAD search path.

2. Start AutoCAD and, to determine the existence and location of the ACAD.RX file, enter (**findfile "acad.rx"**) at the AutoCAD command prompt. If the result is **nil**, then there is no ACAD.RX file on the library search path, and you must create one.

3. Using a text editor, edit the file ACAD.RX if it exists, or create a new file called ACAD.RX in a directory on the AutoCAD library search path. Add the full path to MAWIN.ARX to ACAD.RX:

d:\path\MAWIN.ARX

where *d:\path* is the actual path where MAWIN.ARX is located on your system.

4. Save ACAD.RX and exit your text editor.

5. Exit and restart AutoCAD with a new drawing and see the MAWIN.TXT and MAWIN.LSP files for information and examples on the use of the Maximizing AutoCAD Windows Runtime Extension Limited Edition.

If you want to use the predefined MessageBox "wrapper" functions or functions to control the AutoCAD window, you also must use the AutoLISP load function to load the MAWIN.LSP file.

Note: Maximizing AutoCAD Windows Runtime Extension Limited Edition is a special limited edition of Maximizing AutoCAD Windows Runtime Extension Standard Edition. Maximizing AutoCAD Windows Runtime Extension Limited Edition is a copyrighted Maximizing AutoCAD development utility. It is not shareware and may not be redistributed in part or in any form. If you like it and want to use it for development beyond the exercises in the book other than for your own personal use, you must register and pay for the full commercial version, MaWIN Standard Edition, which provides additional features and entitles you to develop and distribute applications using it. See MAWIN.TXT for more information, and see the ORDER.FRM file for ordering information.

Maximizing AutoCAD Windows Runtime Extension Limited Edition is provided courtesy of the authors of this book. See the *Introduction* for contact information.

LispPad for Maximizing AutoCAD

LispPad for Maximizing AutoCAD is used for many of the AutoLISP and DCL exercises in this book. It simplifies the process of editing, loading, and testing AutoLISP programs, allowing you to focus on the code you're writing rather than the distracting, tedious steps required to load and test it.

For example, with most normal editors that are not specifically designed to be used for AutoLISP code editing, you must edit code in the editor, save it to a file, switch to AutoCAD, and then call the AutoLISP **load** function to load the file. LispPad eliminates all of those steps by automatically loading the current selection or the entire contents of the editor buffer directly into AutoCAD with nothing more than a single keystroke or click of a button.

LispPad also simplifies the creation, editing, testing, and debugging of menu macros, script, and DCL files. The LispPad editor can automatically instruct AutoCAD to display a dialog box defined in a DCL file that is currently open in the editor. You can repeatedly edit the source for a dialog in the editor, then press a single key to view the resulting dialog box on your display directly, without the need to first save the DCL file to disk and load it into AutoCAD (which involves significantly more steps than loading AutoLISP programs). With LispPad's dialog preview facility, developing DCL code is far simpler than doing the same task with a standard text editor. Script files can also be executed directly from LispPad without having to first save them to disk, switch to AutoCAD, and issue the SCRIPT command.

LispPad for Maximizing AutoCAD does not include all of the features found in the full commercial versions of LispPad Standard Edition. It has a 16K file size limit, no on-line help, and does not include some advanced tools found in the Standard Edition versions. In addition to more advanced features, the full Standard Edition versions of LispPad offer larger file size limits: the 16-bit version has a 32K file size limit (Win3.1, Win95 and NT), and the 32-bit version has no file size limit (Win95 and NT only).

For the latest information on the features and prices of the Standard Edition versions of LispPad, see the ORDER.FRM file, or Tony's web site on the Internet at http://ourworld.compuserve.com/homepages/tonyt, or Rusty's Group A web site at http://www.group-a.com/~rusty.

LispPad Installation and Use

Before you can use LispPad, you must install it from the CD. There's no installation program or other setup utility to complicate things, and you can simply copy the files from the CD-ROM to your hard disk and place it in whatever location you want, then run the executable file (LISPPAD.EXE) or create a shortcut to it and execute the shortcut. In the following exercise, you set up LispPad on your Windows 95 desktop for easy access.

Maximizing AutoCAD® R13

 Exercise

Setting Up LispPad

Copy, or use the MaxAC Setup program to copy, all of the files from the TOOLS\LISPPAD directory on the CD-ROM to the MAXAC\TOOLS\LISPPAD directory (or to any other directory you prefer).

Open Explorer and navigate to the LispPad directory,
`C:\MAXAC\TOOLS\LISPPAD`

Drag `LISPPAD.EXE` *onto the desktop*   Creates LispPad's shortcut on the desktop

Double-click the shortcut icon   Runs LispPad

 Note: If your desktop is already too cluttered, you can create the LispPad shortcut in the AutoCAD R13 or MaxAC program group instead, which will also place it in the Start menu of that program group.

In Windows NT or Windows 3.1, make the AutoCAD R13 program group current in Program Manager and use File, New, Program Item to add a program item icon for LispPad.

See the *Using the Maximizing AutoCAD Code Editor* section of Chapter 9 for an annotated illustration of the LispPad interface and descriptions of the operations you'll need to know to open, edit, save, and load AutoLISP programs. See the exercises in Chapter 9 for examples of using LispPad for AutoLISP development.

See the *Using LispPad to Design and Test Dialog Boxes* exercise in the *DCL Learning and Development Tools* section of Chapter 11 for the use of LispPad in dialog box development.

See LispPad's documentation in the LISPPAD.TXT file for more detailed information on all of its features and how to use them.

 Note: LispPad for Maximizing AutoCAD is a special limited edition of LispPAD and does not include all features of the Standard Edition, which can be purchased from the author. LispPad for Maximizing AutoCAD is a copyrighted Maximizing AutoCAD development utility. See the LISPPAD.TXT file for copyright, license, disclaimers, and other information. In addition to using it in the exercises in the book, you may use LispPad for Maximizing AutoCAD for your own development work. However, we encourage you to register and pay for one of the commercial versions of LispPad

Standard Edition with their expanded file size limits, on-line help, and advanced feature sets. See the ORDER.FRM file for ordering information.

The special edition of LispPad for Maximizing AutoCAD is provided courtesy of the authors of this book. See the *Introduction* for contact information.

TextPad

TextPad 2.0 is designed to provide the power and functionality to satisfy the most demanding text-editing requirements. It is Windows-based, and comes in 16- and 32-bit editions. The 32-bit edition can edit files up to the limits of virtual memory, and it will work with Windows 95, Windows NT, and Windows 3.1 (with Win32s extensions).

TextPad has been implemented according to the Windows 95 user interface guidelines, so great attention has been paid to making it easy for both beginners and experienced users. In-context help is available for all commands, and in-context menus pop up with the right mouse button. The Windows multiple document interface allows multiple files to be edited simultaneously, with up to two views on each file. Text can be dragged and dropped between files. In addition to the usual cut-and-paste capabilities, you can correct the most common typing errors with commands to change case, and transpose words, characters and lines. Other commands let you indent blocks of text, split or join lines, and insert whole files. Any change can be undone or redone, right back to the first one made. Visible bookmarks can be put on lines, and edit commands can be applied to lines with bookmarks. Frequently used combinations of commands can be saved as keystroke macros, and the spelling checker has dictionaries for 10 languages. It also has a customizable tools menu, and integral file compare and search commands, with hypertext jumps from the matched text to the corresponding line in the source file (ideal for integrating compilers).

Note: TextPad is available in 16- and 32-bit editions. Only the 32-bit edition is on the MaxAC CD-ROM. The 32-bit edition requires Windows 95 or NT 3.51, or the Win32s 1.3 32-bit extensions in Windows 3.1. If you are using AutoCAD R13 in Windows 3.1, you already have the Win32s 32-bit extensions installed.

The MaxAC CD-ROM includes a TextPad's US English dictionary for checking spelling. If you need a dictionary in a different language, see the TextPad web site at http://www.textpad.com/.

TextPad Installation

Installation requires only two or three easy steps:

1. From the TOOLS\TEXTPAD directory on the CD-ROM, or the C:\MAXAC\TOOLS\TEXTPAD directory if you copied it to your hard drive, run SETUP.EXE and follow its instructions.

In Windows 95, TextPad installs itself into C:\PROGRAM FILES\TEXTPAD and adds itself to the Start, **P**rograms menu. In Windows 3.1 and NT, TextPad installs itself in C:\TEXTPAD and adds a new program group for itself called TextPad.

2. Copy the 100.SUP and 171.LEX files from the TOOLS\TEXTPAD directory on the CD-ROM, or the C:\MAXAC\TOOLS\TEXTPAD directory on your hard drive, to the program directory in which you installed TextPad.

3. If you ran SETUP from the MAXAC\TOOLS\TEXTPAD directory on your hard drive, you can now delete the MAXAC\TOOLS\TEXTPAD directory.

In the Start, Programs menu in Windows 95 or Program Manager in Windows 3.1 or NT, open the TextPad menu or program group and select TxtPad32 to start TextPad.

Tip: To make TextPad available directly from the Windows 95 desktop, create a shortcut for it. In Explorer, drag C:\PROGRAM FILES\TEXTPAD\TXTPAD32.EXE onto the desktop.

Note: TextPad is distributed as shareware, and a license must be purchased if you make use of it for other than use in the context of working through the instructions, techniques, methods, and tutorials of the book. See the README.TXT, LICENSE.TXT, and REGISTER.TXT files. Full details are in the on-line help, and updates can be obtained from the Internet at the TextPad web site at http://www.textpad.com/.

TextPad is provided courtesy of Keith MacDonald, Helios Software.

Vital LISP Demo

Vital LISP is an award-winning, modern, fully integrated, professional AutoLISP development environment that includes a LISP-aware syntax-oriented editor (with syntax highlighting and syntactic coloration), an AutoLISP source-code debugger, a compiler for AutoLISP, and a runtime support system for AutoLISP-based application development and deployment. Vital LISP dramatically improves efficiency and productivity of AutoLISP programming and protects the source code against idea theft and modification.

Vital LISP Demo Installation and Use

From the TOOLS\VLSPDEMO directory on the CD-ROM, or the C:\MAXAC\TOOLS\VLSPDEMO directory if you copied it to your hard drive, run SETUP.EXE and follow its instructions to install and run the demo.

Note: The Vital LISP Demo on the MaxAC CD-ROM is only a self-running demonstration of the features of Vital LISP. To obtain the actual Vital LISP program, in either the Lite, Standard, or Professional Edition, see the ORDER.TXT file in the MAXAC\TOOLS\VLSPDEMO directory or contact Basis Software, Inc.

Vital LISP is provided courtesy of Serguei Akimenko and Peter Petrov, of Basis Software, Inc., West Chester, Pennsylvania, USA. You can contact Basis Software on the Internet at http://www.access.digex.net/~basis or basis@access.digex.net, on CompuServe at 76101,1432, or by telephone at 1-800-324-5450 or 1-610-701-9790 in the USA, or ++41 61 606 9660 in Switzerland.

Lisp2C

Lisp2C compiles AutoLISP code into C language code ADS applications, which protects their contents and prevents modification. The advantages of Lisp2C include:

Protection of your investment. Lisp2C creates ADS applications, which are executable files, so no one can read your AutoLISP code and borrow your know-how. Yet, with Lisp2C, you don't need to learn C or throw away the AutoLISP code you've spent hundreds of hours writing.

Improved program execution speed. Without any recoding, you benefit up to 10+ time performance boost (measured on LISP-intensive code).

Ease of use. You don't have to understand C or ADS, and you can still take the benefits right away. Or, if you do understand C and ADS, you can deliver your application as ADS to the market today, and then gradually recode the critical spots for even greater speed. In addition to C source code (documented with LISP statements), Lisp2C provides you with the batch, make, and linker files necessary to build an executable file. Lisp2C compiles your LISP files as they are. You may want to make some changes to fully benefit from the new approach, but all this can be done in LISP. And finally, you don't need to worry about allocated memory in C. Lisp2C comes with its own garbage collector, and finds and releases all the memory for you.

Built-in debugger. Developing programs with an interpreter is generally easier than edit-compile-link-xload-fail loop. But AutoLISP has very limited capabilities to assist you in debugging. Now, there's a possibility to execute a program step-by-step, tracing into the functions, or stepping over them, monitoring and resetting variables on the fly, defining new functions, loading parts of the code, and more.

Applications can own Lisp variables. Has your application ever failed just because someone else has redefined your global variables? There is a chance now to keep all your variables just for yourself. Simply throw in the /L (local) switch and no other application will ever change your global variables again.

See the included manual (L2C-31.DOC in the MAXAC\TOOLS\LISP2C directory on your hard drive or the TOOLS\LISP2C directory on the CD-ROM) for complete information on Lisp2C.

Installation

From the TOOLS\LISP2C directory on the CD-ROM, or the C:\MAXAC\TOOLS\LISP2C directory if you copied it to your hard drive, run INSTALL.EXE and follow its instructions. After installation, if you first copied it to your hard drive, you can delete the temporary MAXAC\TOOLS\LISP2C directory.

See the Q-START.TXT file for quick-start instructions and see the full manual, L2C-31.DOC, for further instructions.

Note: The evaluation version on the MaxAC CD-ROM is the same as full version ($1,000 US), except: it has a limit of 100 lines per function (any number of functions), and displays "DEMO VERSION" when loaded or unloaded) with **xload** and **arxload**). Contact Basic d.o.o. for the full version.

LISP2C is provided courtesy of Jure Spiler, of Basic d.o.o (d.o.o. = Ltd), Ljubljana, Slovenia. Basic d.o.o.'s primary business is selling and developing applications for AutoCAD. They are known world wide as the developers of Lisp2C translator, a unique tool for converting Lisp programs into professional C-language ADS applications. Jure Spiler may be contacted via CompuServe at 70541,1765 or at Basic d.o.o.'s web site on the Internet at http://www.basic.si.

Kelv13

AutoLISP is an interpreted language, so applications are in source code form, exposing their developers to theft of their proprietary products. Encryption is possible, but not particularly secure. Since encryption cannot provide total security, Autodesk has made available the Kelvinator as a second level of defense. It translates an AutoLISP code into apparent gibberish that still executes identically to the original program, making theft and modification much more difficult.

Kelvination of a program does the following:

- Deletes comments.

- Translates variable names into gibberish.

- Removes indentation and line breaks.

You can obtain the Kelvinator (in the KELV.COM self-extracting archive) from the LISP/Menu/Source library of the ACAD forum on CompuServe or from the Autodesk web site on the Internet at http://www.autodesk.com.

The KELV.COM file doesn't include an up-to-date file of protected symbol definitions for R13, but the MaxAC CD-ROM does. After you obtain and extract KELV.COM, replace its KELV.DEF file with the MAXAC\TOOLS\KELV13.DEF file (rename it to KELV.DEF), and the Kelvinator will then work with R13. See the KELV.DOC file extracted from KELV.COM for instructions on using the Kelvinator.

KELV13.DEF is provided courtesy of Tom Stoekel, of MCG Software, who may be reached via CompuServe at 71521,450.

DOSLib

DOSLib, or DOS Library, is a library of AutoLISP-callable functions that provide DOS command-line functionality in AutoCAD. Developed as an AutoCAD Development System (ADS) application, DOSLib gives your AutoLISP functions and commands greater DOS capability than what is offered by the current suite of AutoLISP functions.

DOSLib extends the AutoLISP programming language by providing the following functionality:

- Drive-handling functions to change between drives and check disk space.

- Path-handling functions to manipulate path specifiers.

- Directory-handling functions to create, rename, remove, and change directories.

- File-handling functions to copy, delete, move, and rename files. Functions for getting directory listings, searching and finding multiple instances of files, and changing attributes are provided.

- Initialization file-handling functions to manipulate Windows-style initialization (INI) files.

- Process-handling functions to run internal DOS commands or other programs.

- Miscellaneous functions, like changing the system date and time.

DOSLib Installation

Use the DOS XCOPY command or the MaxAC Setup program on the MaxAC CD-ROM to copy the files to the MAXAC\TOOLS\DOSLIB directory.

Then, with your word processor, read and follow the instructions in the MAXAC\TOOLS\DOSLIB\DOSLIB.DOC file (a MS Word document).

Note: DOSLib is a trademark of Robert McNeel & Associates. DOSLib is freeware. See the DOSLIB.DOC file for trademark, copyright, and license information.

DOSLib is provided courtesy of Dale Fugier, Robert McNeel & Associates, Seattle, Washington, USA.

HLP2AHLP

HLP2AHP is an R12 Help to R13 Help conversion program. If you find R12 an easier format to write Help files in, or if you write applications that must support R12 and R13, you can use HLP2AHP and write only one set of help files, then translate them.

HLP2AHLP Installation

Use the DOS XCOPY command or the MaxAC Setup program on the MaxAC CD-ROM to copy the H2AORDER.TXT, HLP2AHP.LSP, and HLP2AHP.TXT files to the MAXAC\TOOLS\DLCLSP directory.

Then, read and follow the instructions in the MAXAC\TOOLS\DLCLSP\HLP2AHLP.TXT file.

Note: This version of HLP2AHP is offered for your free evaluation; however, it is a limited version and only translates the body of the help file. With this evaluation version, you need to manually compile a KEYWORD/HOTLINK list so you can construct a Table of Contents and HOTLINK your topics. If you register and pay for the full version of the software, you also get additional features, including Table of Contents generation and a KEYWORK/HOTLINK list to aid you in manually setting up HOTLINKS within your help file. See the H2AORDER.TXT file and the license information in the MAXAC\TOOLS\DLCLSP\HLP2AHLP.TXT file.

HLP2AHLP is provided courtesy of Derek L. Cromwell, Buchanan, Michigan, USA, who may be reached via CompuServe at 71173,1257.

IDSHX

IDSHX is an ADS subroutine that identifies the type of a compiled shape or font file. After you load IDSHX, you can call it from your AutoLISP programs. This enables your programs to determine the type of font file and respond appropriately.

IDSHX Installation and Use

The TOOLS\IDSHX directory on the CD-ROM or the MAXAC\TOOLS\IDSHX directory on your hard drive contains versions for R12 and R13, for DOS and Windows (with C source code included). See the !README.TXT file for the appropriate version and filename to load (with the APPLOAD command or **xload** function).

After you load IDSHX, you can call it from your AutoLISP programs. To use it, you call **(idshx "shxfile.shx")**, where *shxfile*.shx is the name of a an SHX file. It checks the SHX file and returns a code value:

0   unknown file

1   font file

2   bigfont file

3   unicode font file (R13 only)

4   shape file

nil   file not found or not accessible

For example, (IDSHX "/r13/com/fonts/complex.shx") returns 3, indicating a Unicode font file. You must specify the path; IDSHX doesn't search the support path.

 Note: IDSHX is freeware and may be freely distributed or modified, so long as it is identified as having been modified, and credit is given for the original.

IDSHX is provided courtesy of Owen Wengerd, of Manu-Soft Computer Services, which provides AutoCAD customization, support, consulting, training, CNC programming software, customization, manufacturing (CAM) software, in-house AutoCAD applications, and sheet-metal fabrication expertise. You can contact Manu-Soft by phone at (330) 695-5903, or visit the Internet web site http://ourworld.compuserve.com/homepages/owenw, where you will find several other useful AutoCAD utilities.

LispPlot

LispPlot is an R13 ADS application that enables you to invoke ACAD's plot dialog box from within a LISP routine, in AutoCAD R13 for Windows only. It has been tested with R13c4, and may or may not work with earlier versions of R13. For other versions, in the ACAD forum on CompuServe, see the files LSPPLT.ZIP for DOS R12 and R13 support and LSPPLW.ZIP for R12 support.

Normally, if you use AutoLISP to redefine ACAD's PLOT command so that you can perform some operations on the drawing, such as running a date stamp routine, and then invoke the PLOT command with the AutoLISP sequence (command "._plot"), ACAD disables the plot dialog boxes and then runs the PLOT command. With LISPPLOT, you can invoke ACAD's plot dialog box from within AutoLISP and run your datestamp routine (or whatever else you need to run), then invoke the PLOT command to have the plot dialog box appear for the user.

LISPPLOT can also be used as a keyboard stuffer to place any string you desire into the keyboard buffer.

LispPlot Installation and Use

If using AutoCAD for DOS, use the DOS XCOPY command to copy the files from MAXAC\TOOLS\LISPPLOT\DOS on the CD-ROM to the MAXAC\TOOLS\LISPPLOT directory. If using AutoCAD for Windows, use the MaxAC Setup program to copy the files from MAXAC\TOOLS\LISPPLOT\WIN to the MAXAC\TOOLS\LISPPLOT directory. Then see the LISPPLOT.TXT file for usage information.

Note: LispPlot is copyrighted freeware. For more information, see the LISPPLOT.TXT file, which gives you permission to use, copy, modify, and distribute this software and its documentation for any purpose and without fee.

LispPlot is provided courtesy of Mike Dickason, of MD Computer Consulting, who can be reached via CompuServe at 72711,3404. Check the ACAD forum on CompuServe for other utilities by Mike Dickason.

DCL Spy

DCL Spy for Maximizing AutoCAD is a learning and debugging aid that can help you understand and debug dialog box callback functions, using the following features:

Watch. A dynamic display showing the values of the special callback variables, which is updated each time a callback occurs.

Log. A log display showing all callback events, and the values of the callback variables for each event.

Attributes. An attribute page that shows each tile's design-time DCL attributes and their values.

With DCL Spy you can trace dialog callbacks as they occur, and easily see the values of callback variables. It aids in DCL development by dynamically displaying each tile's DCL attributes, synchronized with callback events so that when a callback occurs, the attributes of the tile that generated the callback are displayed in a list.

DCL Spy Installation and Use

From the TOOLS\DCLSPY directory on the CD-ROM, or the C:\MAXAC\TOOLS\DCLSPY directory if you copied it to your hard drive, copy or move all DCL Spy files into a directory on AutoCAD's support path.

The Maximizing AutoCAD CD-ROM

Then, start AutoCAD and enter **(load "DCLSPY")** at the command prompt to launch the DCL Spy dialog box.

To see an example of its use, try starting the DDUNITS command and when its Units Control dialog box appears, you should see callbacks being traced on the Watch page of the DCL Spy dialog box. If you switch to the Event Log page, you should see callbacks being logged, and if you switch to the Attributes page, you should see all tile attributes for the traced tiles.

DCLSpy requires the ADS Application DDELISP.EXE (from the R13\WIN\SAMPLE directory) to be moved to the support directory or a directory on the library path.

See the DCLSPY.TXT file included with DCL Spy, or choose Options, then About for more information.

Note: The DCLSpy is a copyrighted Maximizing AutoCAD development utility.

DCL Spy is provided courtesy of Tony Tanzillo, one of the authors of this book. See the *Introduction* for contact information.

Maximizing AutoCAD Button Image Explorer

The Maximizing AutoCAD Button Image Explorer for Windows 95 and NT 4.0 enables you to browse and extract the tool button images used in AutoCAD, so you can modify them to create custom buttons in your applications.

The Button Image Explorer lets you:

- Open ACADBTN.DLL and browse the button images
- Extract button images to BMP files
- Copy button images to the Clipboard

The Button Image Explorer Installation and Use

Copy, or use the MaxAC Setup program to copy, the ACADBTNS.EXE file from the TOOLS\BUTTONS directory on the CD-ROM to MAXAC\TOOLS\BUTTONS or another directory on your hard drive. Run the ACADBTNS.EXE program, or to create a shortcut for it, drag ACADBTNS.EXE onto the desktop, then double-click on it to run it. See the ACADBTNS.TXT file or choose View, then About for more information

The Button Image Explorer tries to locate and open ACADBTN.DLL when it is launched. Then it displays a list of ACADBTN.DLL's resources, and the image of the currently selected resource. If ACADBTN.DLL is not found, use File, then Open to locate and open it (its default location is \R13\WIN). The File menu contains three commands:

Open   Opens the DLL file.

Save as Bitmap Saves the selected image to a BMP file.

Copy Bitmap   Copies the selected image to the Clipboard.

Note: The Button Image Explorer is a copyrighted Maximizing AutoCAD development utility.

The Button Image Explorer is provided courtesy of the authors of this book. See the *Introduction* for contact information.

System and Project Management Tools

TIDY TOOLS

TIDY TOOLS is an ADS program designed to quickly group, catalog, print, view, list, find, edit, test, and load tool files, namely AutoLISP, EXP, ARX, Script, and Menu files. Plus, the same, fast features are available for managing text and customization files, and for text importation. Now, you can make very efficient use of a much larger inventory of tool and text files than by any other method. TIDY TOOLS works with AutoCAD R12 DOS and R13 DOS.

TIDY TOOLS Installation

In the TOOLS\TIDYTOOL directory on the CD-ROM, or the C:\MAXAC\TOOLS\TIDYTOOL directory if you copied it to your hard drive, read the READ.1ST and READ.ME files, then run the INSTALL.EXE program and follow its instructions.

Note: TIDY TOOLS is fully featured shareware. You may only evaluate Tidy Tools for a trial period *not* exceeding 30 days. After evaluation, you must either delete the full program or purchase a registered/licensed version of the program. To order, see the program file, TIDY.ORD. Price: Only $19.95 US for the first station, $14.95 US for each additional station.

The difference between the shareware and licensed versions is that the shareware version displays a randomly generated message that the program version is shareware after 24 files have been cataloged.

Tidy Tools is provided courtesy of Don Sigalet, TIDY TOOLS, Penticton, B. C. Canada, who may be reached via CompuServe at 73642,1375.

Mega-App Loader

Mega-App Loader for R13 consists of a pair of files that replace the APPLOAD.LSP and APPLOAD.DCL included in R13. Mega-App Loader includes Program Groups, for logical grouping and loading of programs. The Group Names appear in a Popup List, and the active Groups files are listed in the List Box below it. Group Names are saved in a revised version of the APPLOAD.DFS file, each followed by its entries, so the file is readable, even from a text editor.

Select All, Clear All, Invert Selection, Load All, and Unload All buttons have been added to further speed access to applications. Also, the APPLOAD.DFS file location will always be the same path as the APPLOAD.LSP file location, so previously located applications will not have to be re-located simply because AutoCAD was started from a different directory.

There is also an R12 version of this routine that does not recognize ARX files.

Mega-App Loader Installation

Use the DOS XCOPY command or the MaxAC Setup program on the MaxAC CD-ROM to copy all of the APPLOAD.* files and the APPORDER.TXT from MAXAC\TOOLS\DLCLSP file to the MAXAC\TOOLS\DLCLSP directory.

Then, read the instructions, license, and other information in the MAXAC\TOOLS\DLCLSP\APPLOAD.DOC file (a MS Word document). If you decide to use the Mega-App Loader, make backup copies of the APPLOAD.LSP and APPLOAD.DCL in your R13\COM\SUPPORT directory before you replace them with the APPLOAD.LSP and APPLOAD.DCL from the MAXAC\TOOLS\DLCLSP directory.

Note: This version of Mega-APP Loader is shareware, and limits you to three groups and five files per group. When you try to exceed these limits, an Alert Box is displayed. If you like Mega-APP Loader, you should register and pay for the full version, which removes these limitations. See the APPLOAD.DOC and APPORDER.TXT files.

Mega-App Loader is provided courtesy of Derek L. Cromwell, Buchanan, Michigan, USA, who may be reached via CompuServe at 71173,1257.

Dr. DWG Quickview

Dr. DWG QuickView, from CSWL Inc., is a compact, stand-alone viewer that allows you to view and annotate drawings saved in DWG, DXF, and Binary DXF file formats. Dr. DWG QuickView loads and displays AutoCAD R12 and R13, 2D and 3D drawings in a flash and has enhanced memory management. Dr. DWG QuickView is ideal for previewing and redlining drawings.

Dr. DWG QuickView supports common drawing manipulation functions including Zoom, Pan, Annotate, Redline, Toggle between PaperSpace and ModelSpace, and more. It supports viewport operations, block manipulation, scrolling, and measuring distances in different drawing units. It also prints the drawing with redline information.

These features make Dr. DWG QuickView not just a viewer in the absence of AutoCAD but also an ideal companion to AutoCAD for previewing files or seeing them as thumbnails. While you are designing a complicated drawing in AutoCAD, you can use Dr. DWG QuickView to view the entire drawing from a different angle or at a convenient zoom level, or view other drawings without loading another instance of AutoCAD.

Dr. DWG QuickView is a single small executable—easy to install and a useful tool for viewing or printing your DWG files anywhere. CSWL Inc. also offers Dr. DWG View Professional (OCX) and Dr. DWG Library. Dr. DWG View Professional (OCX) is an OLE custom control for viewing AutoCAD drawings. You can insert the OCX control as an object into any OLE container application and use it from within that application. Dr. DWG Library consists of a set of APIs for reading, viewing, and writing AutoCAD DWG files.

Dr. DWG QuickView Installation and Use

Separate versions of Dr. DWG QuickView, for 16- and 32-bit environments, are provided in the 16-BIT and 32-BIT subdirectories of the TOOLS\DRDWG directory on the CD-ROM, or the C:\MAXAC\TOOLS\DRDWG directory if you copied it to your hard drive. Run SETUP.EXE from either the 16-BIT or the 32-BIT subdirectory and follow its instructions.

Unless you specify another directory, Dr. DWG QuickView installs itself into C:\DRDWGVW. In Windows 95, Dr. DWG QuickView adds itself to the Start, **P**rograms menus, and in Windows 3.1 and NT, it adds a new program group for itself called Dr. DWG QuickView.

To run Dr. DWG QuickView after installation is complete: in Windows 95 or NT 4.x, choose Dr. DWG QuickView from the Start, **P**rograms menu or double-click on any DWG or DXF file in Explorer; or in Windows 3.1 or NT3.x, click on the Dr. DWG QuickView icon in the Dr. DWG QuickView program group. For more information, choose **H**elp in Dr. DWG QuickView or see the README.TXT file in the C:\DRDWGVW directory.

Dr. DWG QuickView is shareware. It will cease to function 60 days after installation. If you like it and want to continue to use it, you must register and pay for it. Registration is $49 and includes a manual and support. For more information and registration, choose **A**bout from the **H**elp menu or see the ORDER.TXT and README.TXT files in the directory in which you installed Dr. DWG QuickView (C:\DRDWGVW by default).

Album

Album Version 1.0 displays the DWG files of a selected directory as a stand-alone application. It runs in Windows, independently of AutoCAD, and displays the bitmap-thumbnail header image of all drawing files in the selected directory. You can drag and drop an image onto AutoCAD to open its drawing file, or double-click on an image to launch AutoCAD and open the drawing file.

Album is written to the Windows NT operating system and has been tested with Windows NT workstation 3.5 German. It also runs in Windows 95, but has not been tested extensively except with Windows NT. Source code (VC++) is available.

Album Installation and Use

Copy, or use the MaxAC Setup program to copy, the files from the TOOLS\ALBUM directory on the CD-ROM to the MAXAC\TOOLS\ALBUM directory.

Then, to use Album, run the ALB.EXE program executable, or to create a shortcut for it, drag ALB.EXE onto the desktop, then double-click on it to run it. Then choose Open from the File menu and select a drawing file to open. Album then displays images for all drawings in the selected directory, in addition to the selected file.

See the README.TXT and ALB.HLP files for more information. To display the ALB.HLP file, double-click on it, or in Album choose the Help item from the ? menu.

 Note: Album Version 1.0 is freeware. Permission to use, copy, modify, and distribute this software for any purpose and without fee is provided in the README.TXT file.

ALBUM is provided courtesy of H. C. Brueckner, of netnice, Adelsdorf, Germany, who may be reached via CompuServe at 100015,1534.

SlideManager

SlideManager is a DOS program that enables you to manage, manipulate, and print slide files and libraries. Its features include:

- Add slide file(s) to slide library
- Delete slide file from slide library
- Extract one or all slide files from slide library
- Remove slide file from slide library
- Update slide file in slide library
- Flip through slide library
- Display slide file and enable pans and zooms
- Create an HPGL plot file from slide file

- Output slide file to Epson printer
- View slide file from slide library
- Convert slide file to DXF format

SlideManager is an improvement over Autodesk's SLIDELIB program and enables you to add slides to the ACAD.SLB file for use by your custom applications.

SlideManager Installation and Use

Use the DOS XCOPY command or the MaxAC Setup program on the MaxAC CD-ROM to copy the files to the MAXAC\TOOLS\SLIDEMGR directory (or any directory you prefer). Start SlideManager by changing to that directory and entering **SLDMGR** at the DOS prompt. Starting SlideManager with no parameters will bring you into the SlideManager menu system. From there, use the F1 key to access the context-sensitive Help system.

You can get a brief explanation of SlideManager's command-line mode options by entering **SLDMGR ?** at the DOS prompt. You'll get even more help by choosing the "Line Mode" help topic from within the menu help system.

Note: SlideManager is shareware. See the READ.ME file. You may use SlideManager in the context of working through the instructions, techniques, methods, and tutorials of the book, but if you like it and use it for other purposes, after a 30-day trial period, you must make a registration payment of $25 to John Intorcio at the address listed in the READ.ME file.

SlideManager is provided courtesy of John Intorcio, of JMIcro, Woburn, Massachusetts, USA.

SuperPurge Lite

SuperPurge is "A Better PURGE Command" for AutoCAD Release 13 users, providing better drawing cleanup and file size reduction.

SuperPurge defines three new commands:

SPURGE. This runs SuperPurge. You get either a dialog box or the command-line version depending on the setting of CMDDIA.

-SPURGE. This forces use of the command-line version of SPURGE.

SPURGEALL. Purges *everything*—it could take a while on huge drawings.

SuperPurge Lite Installation and Use

See the !README.TXT and !MANUAL.TXT in the TOOLS\SPURGE directory on the

CD-ROM or the MAXAC\TOOLS\SPURGE directory on your hard drive for information on installing and using SuperPurge Lite.

 Note: SuperPurge Lite is shareware, so if you like it and continue to use it, you need to pay for it. See the !LICENSE.TXT file for more information.

SuperPurge Lite is provided courtesy of Owen Wengerd, of Manu-Soft Computer Services, which provides AutoCAD customization, support, consulting, training, CNC programming software, manufacturing (CAM) software, in-house AutoCAD applications, and applied sheet-metal fabrication expertise. You can contact Manu-Soft by phone at (330) 695-5903, or visit their Internet web site at http://ourworld.compuserve.com/homepages/owenw, where you will find several other useful AutoCAD utilities.

ShowRefs

Have you ever received a drawing that was missing a menu, font, or XREF drawing file that needed to accompany it, or sent such a drawing to someone? ShowRefs is a shareware program that shows all external files that are referenced by a drawing or group of drawings. Use it to check drawings before shipping or archiving them. The files it lists include menu and font files (MNU, MNX, SHX), and all XREFs, including nested XREFSs. SHOWREFS is command-line driven and fast. It can process single files, directories, or entire drives.

ShowRefs works directly with the DWG file from the DOS prompt. There is no need to convert the drawing(s) to a DXF format, or to load AutoCAD before running this program. ShowRefs in no way modifies or writes to any drawing file. ShowRefs version 2.0 reads drawings through R13c4.

This program is especially helpful in the preparation of script files. AutoCAD terminates the execution of a script file if any referenced menu (R11 or earlier), font, or bigfont isn't found in the search path. A command-line option (/A) is provided when running this program on a group of files to be included in a script routine, and this option will warn you of any potential problems due to missing files. This program can be used for the same purpose if you send your drawings to a plotting service and want to ensure that all necessary files are included on the disk to be sent to the plotting service.

ShowRefs Installation and Use

Use the DOS XCOPY command or the MaxAC Setup program on the MaxAC CD-ROM to copy the files to the MAXAC\TOOLS\SHOWREFS directory.

Then, to use ShowRefs, run the SHOWREFS.EXE program executable with the appropriate filenames and parameters. For the syntax and parameters, as well as more information on using ShowRefs, see the SHOWREFS.TXT file or execute SHOWREFS.EXE with no parameters.

Note: ShowRefs is shareware, so if you like it and continue to use it, you need to pay for it. See the SHOWREFS.FRM file, or GO SWREG #6429 on CompuServe to register.

ShowRefs is provided courtesy of Mike Dickason, of MD Computer Consulting, who can be reached via CompuServe at 72711,3404. Check the ACAD forum on CompuServe for other utilities by Mike Dickason.

DWGList

DWGList lists drawing layers and blocks within drawings from the DOS prompt, quickly processing entire directories and subdirectories. It accepts multiple command-line filespecs, as well as wildcards, and can be used to locate every drawing using a specified block, or containing a specified layer. DWGLIST is a DOS program that runs directly from the DOS prompt. There is no need to be running AutoCAD when running DWGLIST.

DWGList Installation and Use

Use the DOS XCOPY command or the MaxAC Setup program on the MaxAC CD-ROM to copy the files to the MAXAC\TOOLS\DWGLIST directory.

Then, to use DWGList, run the DWGLIST.EXE program executable with the appropriate filenames and parameters. For the syntax and parameters, as well as more information on using DWGList, see the DWGLIST.DOC file or execute DWGLIST.EXE with no parameters.

Note: DWGList is shareware, so if you like it and continue to use it, you need to pay for it. See the DWGLIST.FRM file for registration information.

DWGList is provided courtesy of Mike Dickason, of MD Computer Consulting, who can be reached via CompuServe at 72711,3404. Check the ACAD forum on CompuServe for other utilities by Mike Dickason.

PlotStamp

PlotStamp adds dynamic plot data stamping capability to AutoCAD R13 for Windows by allowing you to use DIESEL to dynamically specify what value an AutoCAD text object will have when the drawing that contains it is plotted. By using DIESEL to define the contents of the text, you can add the time and date when the drawing was plotted, along with other relevant information such as the drawing filename and user name. PlotStamp

The Maximizing AutoCAD CD-ROM

includes a custom dialog box (see Figure 10.11 in Chapter 10) for defining the stamp text, and you can select from a list of pre-defined DIESEL expressions for most common stamp data.

PlotStamp Installation and Use

Copy, or use the MaxAC Setup program to copy, the files in the TOOLS\PLOTSTMP directory on the CD-ROM to a directory on the AutoCAD search path. See the PLOTSTMP.TXT file for complete installation and usage instructions.

Warning: Because PlotStamp date and time data is updated at plot time, PlotStamp is not suitable for inserting static dates or times in text strings.

Note: PlotStamp for Maximizing AutoCAD is a limited demonstration version of PlotStamp, a copyrighted commercial utility. This limited demonstration version of PlotStamp is a "freeware" version that can be freely distributed, and may be upgraded to the full product. See the PLOTSTMP.TXT file. PlotStamp for Maximizing AutoCAD adds the prefix [PlotStamp Unregistered] to all stamp text objects. The standard edition of PlotStamp, which can be ordered directly from the author, does not impose this limitation. See the ORDER.FRM file for more information.

PlotStamp is provided courtesy of Tony Tanzillo, one of the authors of this book. See the *Introduction* for contact information.

R13 Bug List

The Release 13 Bug List, Seventh Edition, documents all known bugs and documentation errors in each revision of R13, from the initial shipping version through R13c4. Use the DOS XCOPY command or the MaxAC Setup program on the MaxAC CD-ROM to copy the files to the MAXAC\TOOLS\13BUG directory and read either the MS Word 13BUG07.DOC or 13BUG07.TXT file. The information in both files is the same, but the DOC file uses icon files from the MAXAC\TOOLS\13BUG directory to enhance the presentation.

Note: The Release 13 Bug List, Seventh Edition, is copyrighted. See the legal information in the 13BUG07.DOC or 13BUG07.TXT file.

The R13 bug list is provided courtesy of Steve Johnson, of cad nauseam, Leederville, WA, Australia, who can be reached via CompuServe at 100251,2544.

Application and Productivity Tools and Utilities

BigDCL

Most standard AutoCAD dialog boxes are designed to operate on VGA-class displays (640 x 480). Unfortunately, this constraint adversely affects the layout and clarity of many dialogs. BigDCL is a set of replacements for ACAD.DCL and BASE.DCL that replace a number of standard AutoCAD dialogs with reworked versions that make better use of larger format displays (800 x 600 minimum).

Included files:

ACAD.DCL   Replacement for standard ACAD.DCL

BASE.DCL   Replacement for standard BASE.DCL

BASEEX.DCL   Required for custom replacement dialogs

Warning: You may need to set the GLOBCHECK system variable to 0 in order to use these enhanced dialog boxes. If it is set to 1, it may display an alert box every time one of replacement dialogs appears, telling you the dialog is "globalization unfriendly" and/or that it will not fit on a Japanese VGA display.

Effects in affected dialogs include:

Layer (DDLMODES): More visible layers, more characters in layer and linetype names, and completely reworked control clusters.

Linetype (DDLTYPE): More visible linetypes and larger images.

Attribute Edit (DDATTE, INSERT, and so on): More visible attributes and wider fields.

Image (slide) menu dialog: Larger images.

BigDCL Installation and Use

Follow these steps to install BigDCL and see an example of its use:

1. *Back up* the existing ACAD.DCL and BASE.DCL files that are in your R13\COM\SUPPORT directory (note: if you are already using a customized version of either of these files, then you may not be able to use these replacements unless you are able to integrate them into the existing customized versions of those files).

2. From the TOOLS\BIGDCL directory on the CD-ROM, or the C:\MAXAC\TOOLS\BIGDCL directory if you copied it to your hard drive, copy the three included DCL files into the R13\COM\SUPPORT directory.

3. Start AutoCAD. You may need to set the GLOBCHECK system variable to 0.

4. Open a drawing that contains at least six existing layers and several block insertions with attributes.

5. Issue DDEMODES, DDLTYPE, and DDATTE.

6. Issue **(menucmd "I=ACAD.IMAGE_3DOBJECTS")** then **(menucmd "I=*")** to view the larger slide images.

You can now take advantage of the greater dialog box size provided by BigDCL.

Note: BigDCL is a copyrighted Maximizing AutoCAD utility.

BigDCL is provided courtesy of the authors of this book. See the *Introduction* for contact information.

TbOsnap

The TBOSNAP.LSP file enables you to use the Object Snap toolbar at the command prompt to toggle running object snap modes. With this file loaded, clicking a button on the ACAD menu's Object Snap toolbar or selecting an item from the popup cursor menu toggles the menu item or button's associated running object snap mode.

TbOsnap Installation and Use

Copy, or use the MaxAC Setup program on the MaxAC CD-ROM to copy, the TBOSNAP.LSP file from the TOOLS\TBOSNAP directory to a directory on AutoCAD's support path.

See the *Testing TBOSNAP.LSP* exercise in the *Extending the Object Snap Menus with AutoLISP* section of Chapter 14 for an example of using TbOsnap. To load it without needing to have Chapter 14's support files and menu present, enter **(load "tbosnap")** at the AutoCAD command prompt.

Note: TbOsnap is a copyrighted Maximizing AutoCAD utility.

TbOsnap is provided courtesy of Tony Tanzillo, one of the authors of this book. See the *Introduction* for contact information.

BlockEdit

 BlockEdit completely automates the editing of a block defined in the current drawing using a second instance of AutoCAD. The BLOCKEDIT command prompts for block selection, then opens the block for editing in another instance of AutoCAD, and waits until you've exited the block editing instance of AutoCAD. After you've edited the block and exited the block editing instance, the edited block is automatically updated in the initial editing session.

 Note: BlockEdit requires R13 c4 in Windows 95 or NT, and will not work with external references.

BlockEdit Installation and Use

Copy, or use the MaxAC Setup program on the MaxAC CD-ROM to copy, the BLOCKED.LSP file from the TOOLS\BLOCKED directory to a directory on AutoCAD's support path.

At the AutoCAD command prompt, enter **(load "blocked")**, then enter **BLOCKEDIT**.

At the `Block to edit/<select>:` prompt, you can enter the name of the block to be edited, or you can press ENTER and pick an insertion of it.

When you select the block, a second AutoCAD session containing the block's definition is started. Edit the block, then end that editing session, and the block definition in the first editing session will be updated automatically.

 Note: BlockEdit is a copyrighted Maximizing AutoCAD utility.

BlockEdit is provided courtesy of Tony Tanzillo, one of the authors of this book. See the *Introduction* for contact information.

DynaMode

The MaxAC CD-ROM includes a special limited version of DynaMode. DynaMode is a utility that uses the AutoCAD Runtime Extension (ARX), which enables you to define a custom, context-sensitive status line for any command. DynaMode adds functions to AutoLISP that allow you to easily define context-sensitive status lines for any built-in AutoCAD command, as well as commands that are defined in AutoLISP, ADS, or ARX. Once a context-sensitive status line is defined and associated with one or more commands, this status line appears only when its associated command is active. In the commercial version of DynaMode, a context-sensitive status line can be associated with one or several commands by the use of wildcards.

Additional information about DynaMode, along with examples and updates can also be found at the author's web site on the Internet at `http://ourworld.compuserve.com/homepages/tonyt`.

DynaMode Installation and Use

Use the DOS XCOPY command or the MaxAC Setup program on the MaxAC CD-ROM to copy the files to the MAXAC\TOOLS\DYNAMODE directory, or to a directory on your AutoCAD search path. Then, see the *Dynamic Context-Sensitive Status Lines* section of Chapter 10 for an exercise showing how to load and use DynaMode. See also the DYNAMODE.TXT file.

Note: DynaMode is a copyrighted Maximizing AutoCAD development utility. The version of DynaMode that's included on this book's CD-ROM enables you to define up to five context-sensitive status lines and has the wildcard command naming feature disabled. The commercial version of DynaMode, removes these restrictions and allows you to define an unlimited number of context-sensitive status lines and associate each with a different AutoCAD command or group of commands. See the ORDER.FRM file for ordering information.

DynaMode is provided courtesy of Tony Tanzillo, one of the authors of this book. See the *Introduction* for contact information.

ATTEXT

ATTEXT is a DOS-based shareware program that duplicates the AutoCAD ATTEXT command from the DOS prompt. There is no need for AutoCAD to be running, or to convert the DWG files to DXF format before processing. ATTEXT in no way modifies the DWG files. It simply reads the DWG file and creates an output file containing the specified information. In addition to the standard AutoCAD attribute extraction options, ATTEXT also allows the drawing name to be one of the extract fields. ATTEXT version 2.0 works with drawings through R13.

ATTEXT Installation and Use

Use the DOS XCOPY command or the MaxAC Setup program on the MaxAC CD-ROM to copy the files to the MAXAC\TOOLS\ATTEXT directory.

Then, to use ATTEXT, run the ATTEXT.EXE program executable with the appropriate filenames and parameters. For the syntax and parameters, as well as more information on using ATTEXT, see the ATTEXT.TXT file or execute ATTEXT.EXE with no parameters.

Note: ATTEXT is shareware, so if you like it and continue to use it, you need to pay for it. Online registration is available on CompuServe. GO SWREG #6431 to register, or see the ATTEXT.FRM file.

ATTEXT is provided courtesy of Mike Dickason, of MD Computer Consulting, who can be reached via CompuServe at 72711,3404. Check the ACAD forum on CompuServe for other utilities by Mike Dickason.

DDEDTEXT

DDEDTEXT is a program that provides a single-line dialog box for editing text, dimensions, and mtext objects. It makes editing all kinds of annotation objects easier in AutoCAD. DDEDTEXT may seem similar to the single-line edit dialog box you get as a default in R13c4 by setting the MTEXTED system variable to `:LISPED#INTERNAL`, but DDEDTEXT also provides several additional features:

- A button is provided to allow you to convert the text between all uppercase, each word mixed case, first word mixed case, and all lowercase.

- Buttons are provided to insert the symbols for diameter, degrees, and plus/minus into the text.

- Extra buttons can be enabled and modified to suit your requirements (registered users receive source code).

- When editing text, mtext, and dimensions, the original text value is displayed. This is particularly useful when editing dimensions, as it shows the dimension text and can prevent the need to zoom in to inspect dimension values.

- A Reset button allows you to revert to the original value and continue editing, rather than having to cancel the command and reselect the object.

- If an mtext entity is too long to edit in a single line, the full Mtext editor is automatically invoked.

- You can use the old %%c, %%d and %%p symbols, which will be automatically converted to the equivalent new Unicode control sequences.

- Registered users get a User button, which brings up a dialog box containing a list of user-definable text and symbols to insert into the text.

- Registered users can configure DDEDTEXT to automatically be invoked when creating mtext and dimensions.

DDEDTEXT Installation and Use

Use the DOS XCOPY command or the MaxAC Setup program on the MaxAC CD-ROM to copy the files to the MAXAC\TOOLS\DDEDTEXT directory, then see the DDEDTEXT.TXT file for installation and usage instructions.

Note: DDEDTEXT is shareware, so if you like it and continue to use it, you need to pay for it. See the DDEDTEXT.TXT file for license information. Online registration is available on CompuServe. GO SWREG #5138 to register, or see the DDEDTEXT.FRM file.

DDEDTEXT is provided courtesy of Steve Johnson, of cad nauseam, Leederville, WA, Australia, who can be reached via CompuServe at 100251,2544.

DDCHTEXT

DDCHTEXT is a program that provides a dialogue box for modifying the properties of various types of annotation (text-like) objects: TEXT, MTEXT, ATTDEF and ATTRI-BUTE. See the DDCHTEXT.TXT file for the advantages of using DDCHTEXT.TXT instead of AutoCAD's commands.

DDCHTEXT Installation and Use

Use the DOS XCOPY command or the MaxAC Setup program on the MaxAC CD-ROM to copy the files to the MAXAC\TOOLS\DDCHTEXT directory, then see the DDCHTEXT.TXT file for installation and usage instructions.

DDCHTEXT is shareware, so if you like it and continue to use it, you need to pay for it. See the DDCHTEXT.TXT file for license information. Online registration is available on CompuServe. GO SWREG #10333 to register, or see the DDCHTEXT.FRM file.

DDCHTEXT is provided courtesy of Steve Johnson, of cad nauseam, Leederville, WA, Australia, who can be reached via CompuServe at 10025,2544.

SectionMaster

SectionMaster is a collection of parametric section-generating routines that draw steel, concrete, and wood structural sections as 2D polylines. The entities created by SectionMaster are compatible with modeling tools such as AME or AutoCAD Designer. SectionMaster works with AutoCAD R13 for Windows 3.1, Windows NT 3.51, or Windows 95.

Installation and Use

Use the DOS XCOPY command or the MaxAC Setup program on the MaxAC CD-ROM to copy the files to the root directory of a diskette. Then, execute the INSTALL.EXE program from that diskette, and respond to its prompts, entering the path of a directory on AutoCAD's support path, and press ENTER at the remaining prompts.

To load SectionMaster in AutoCAD, enter **(load "sctmwin")** at the command prompt. Then, to display the SectionMaster dialog box, enter **SCTMWIN**. In the initial SectionMaster dialog box, select a material and product option, then choose OK to display a second dialog box in which you specify the section parameters and insertion point. For more information, choose Help in this second dialog box.

Note: Section Master is a shareware product, designed for use on a single machine. One set of files may be copied onto a single machine. Use on a network, whether placed on a file server or on a single machine connected to a network, does not constitute use on a single machine. Because Section Master is shareware, if you like it and continue to use it, you need to pay for it. The $25 US registration fee should be sent to:

> Elliptra Software Products
> 6281 S. Vivian St.
> Denver, CO 80127-4613

Please include your name, company name and address, phone and fax numbers, and email address. With registration, Elliptra will notify you of any product upgrades to Section Master that may occur in the future.

Section Master is provided courtesy of Bill Shanahan, of Elliptra Software, which specializes in structural utilities for AutoCAD. Visit Elliptra's web site to learn more about their products, including SteelBook and JoistPro. You can also download free utilities, obtain AutoLISP and ADS coding tips, and view classified ads. The Elliptra Software Internet address is http://www.csn.net/elliptra.

That's all for this book and the MaxAC CD-ROM, but for more goodies, updates, and information, look for the companion book, *Maximizing AutoLISP for AutoCAD R13,* and check the authors' web sites on the Internet.

Index

Note: All **exercises** are indexed under the main heading **Exercises**

A

ACAD.ADS, 120, 128–29
ACAD.AHP, 120, 132–33
ACAD.CFG, 119–22
ACAD.DCL, 896
ACAD.HDX, 120, 132–33
ACAD.HLP, 120, 132–33
ACAD.INI, 119–22
ACAD.LIN, 120, 129–30, 271
ACAD.LOG, 120, 133
ACAD.LSP, 120, 123
   conflict, avoiding, 69–70
   creating menus and, 716–19
   partial menus and, versus MNL files, 723–25
ACAD.MLN, 120, 129–30
ACAD.MNC, 120, 123–27
ACAD.MNL, 120, 723
   functions, 724
   partial menus and, 723
ACAD.MNR, 121, 123–27
ACAD.MNS, 121, 123–27
ACAD.MNU, 121, 123–27
ACAD.MNX, 121, 123–27
ACAD.PAT, 121, 129–30
ACAD.PGP, 76, 121, 122
   defining,
      aliases in, 77–82
      external commands in, 82–86
   reloading/testing, 81
ACAD.PSF, 121, 131
ACAD.RX, 121, 128–29
ACAD.SLB, adding hatching sides to, 307–8
ACAD.UNT, 121, 131
ACADAXPAGE, 25
ACADBTN.DLL, 120
acad_colordlg, color dialog boxes and, 535–36
acad_colordlg colorindex [flag], 776
acad_helpdlg helpfile topic, 777
ACADMA.LIN, 249
   linetypes and, 256–58
ACADMA.PAT, 271
ACADMAXMEM, 25
ACADMAXOBJMEM, 25
ACADMAXPAGE, 25
ACADNT.CFG, 120
ACADNT.CFG., 120
ACADPAGEDIR, 25
AcadPlotDlg, described, 874
ACADPREFIX, system variable, 834
ACADR13.LSP, 120, 123
acad_strlsort list, 777
ACADVER, system variable, 834
Accelerator keys
   defining, 91–98
      creating, 93–95
Accelerators, macro creation and, 98–100
action, DCL attribute, 807
Action expression variables, 593–601
action_tile, 592, 598
action_tile key action, 777
add_list, 575–76, 588–89
add_list item, 777
AddToRecentDocs, 763
   described, 874
ADI plotter, optimizing, 30–31
ads, 777
ADS program, 127–28
   loading, 136–38, 366
AFLAGS, system variable, 834
Album, 890–91
   described, 868

Index

installation/use of, 891
alert string, 777
Aliases
   deciding among, 98–100
   defining, in ACAD.PGP file, 77–82
   macro creation and, 98–100
alignment, action, DCL attribute, 807
ALL selection mode, 191–92
alloc nodes, 777
allow_accept, DCL attribute, 807
and expr, 777
and function, 413
ANGBASE, system variable, 834
ANGDIR, system variable, 834
Angle character
   font definition instruction codes, 223–25
   making, 221–26
   specification byte codes, 225
angle function, 400–401
Angle input, 612–13
angle point1 point2, 778
angtof, 612–13
angtof string [mode], 778
angtos angle [mode [precision]], 778
Annotations, AutoLISP coding and, 449–50
APERTURE, system variable, 834
append [list], 778
Application
   issues, third party, 140–42
   tools/utilities, 896–902
apply function arglist, 778
AREA, system variable, 834
Arguments
   AutoLISP, 381–84
   optional, 348–49
   repeating, 349–50
   required, 347–48
arx, 778
ARX programs, 120, 127–28
   loading, 136–38, 366
arxload arxfile [onfailure], 778
arxunload arxfile [onfailure], 778
ascii string, 778
ASCII table, 859–64
aspect_ratio, DCL attribute, 807
assoc key list, 779
ASTEROID, complex hatch patterns and, 277–89
ASTEROID. PAT, 249
atan number1 [number2], 779
atof string, 779
atoi, 603–4
atoi string, 779

atom expr, 779
atoms-family formatmode [comparelist], 779
13ATTACH.MNL, 628
ATTDIA, system variable, 834
ATTEXT
   described, 869
   installation/use of, 899–900
ATTFIND, 617–22
attfind_blklist, 618
ATTMODE, system variable, 834
ATTREQ, system variable, 834
Attribute
   boxes, finding with dialog boxes, 617–22
   edit, 896
attribute DCL tile reference, 811
AUDITCTL, system variable, 834
AUNITS, system variable, 834
AUPREC, system variable, 834
AUto selection mode, 190–91
autoarxload, 732–35
autoarxload arxfile cmdlist, 779
AutoCAD
   boot system, setting up, 10–11
   button image explorer, maximizing, 887–88
   commands,
      defining new, 378–79
      executions, 403–7
   configuration,
      in DOS, 58–61
      files, 48–52
         multiple, 51–52
   initialization sequence, 133–36
      drawing load initialization, 135–36
      load initialization, 134
   installing/configuring, 16–22
      streamlining, 17
   on a network, 14
   operating systems, maintaining multiple, 10
   organizing,
      benefits of, 40–41
      directories, 41–45
   pager settings, 24–25
   platform choices, 2–13
      customized vs. production, 9–10
      DOS versus Windows, 3–4
      Windows 3.1 versus NT and 95, 4–9
   prompt, using AutoLISP at, 355–63
   setup/organization of, 1–37
   support files,
      directing to, 54–56
      stock, 118–33
   text editor selection/using, 61–69

Index

Windows, maximizing runtime extension limited edition, 873–77
AutoLISP
  accessing tile attributes from, 585–88
  arguments, 381–84
    optional, 348–49
    repeating, 349–50
    required, 348–49
  bound variables, 381–84
  code,
    editor maximizing, 366–69
    formatting, 350
  coding practices/standards, 448–50
    comments/annotations, 449–50
    symbol naming conventions, 448–49
  command, 76
    defining new, 378–79
    definitions, creating, 86–91
    function and, 360–61, 403–7
  conditional logic, 407–27
  coordinates, working with, 359–60
  data types, 350–55
    entity names, 355
    file descriptors, 355
    integers, 352
    lists, 354–55
    reals, 353
    selection sets, 355
    strings, 352–53
    symbols, 351–52
  debugging, 445–48
  described, 335–36
  DIESEL and, 454–55
    calling expressions from, 491–94
  error handling, 440–48
  expressions and evaluation, 344–47
  file dialog boxes in, 528–32
  functions, 775–806
    user defined, 369–84
  geometric functions, 399–403
  limitations/alternatives of, 339–40
  list operations, 392–99
  loading, 136–38
    with MNL files, 365–66
  macro creation and, 98–100
  mainstream programming languages versus, 338–39
  math,
    functions, 385
    operations and, 384–87
  object snap menus, extending, 750–52
  partial menu loading with, 707–9
  prefix versus infix notation, 341
  problems, 829–32
  program,
    automatic loading of, 364–65
    files, 363–69
    logic, 407–27
  programming with, 337–40
  rules of scope, 381–84
  string operations, 387–90
  supporting, with ARX application, 753–64
  syntax/semantics of, 347–50
  system variables, 361–63
  user-defined functions, 369–84
    anatomy of, 371–78
    calling command functions transparently, 379–80
    defining new AutoLISP commands, 378–79
  user input,
    acquiring/validating, 428–40
    bit modes/keywords, 430–35
    formatting input prompts and, 437–39
    initget function, input control, 429–30
    selecting objects and, 435–37
    variables and expressions, 340–44
  using at AutoCAD prompt, 355–63
  variables, 342
    assignment of, 342–44
autoload, 732–35
autoload lspfile cmdlist, 779
AUTOSPOOL, plotting to a, 31–33
autoxload, 732–35
autoxload adsfile cmdlist, 780
Auxiliary menus, 652–59

B

BACKZ, system variable, 835
BASE.DCL, 896
BASEEX.DCL, 896
Batch plotting
  drawings, 325–28
    summary, 332–33
  sequence, 296
Beep, 758
  described, 874
BFSPECMA.SHP, 237
BFSPECMA.TXT, big fonts contained in, 238–39
BHATCH, 247
  release 13 and, 250–51
Big fonts, 207–8
  adding, 227–42
  character map, 237

Index

combining characters using subshapes, 229–40
escape characters, 228
extended, 241–42
file creation for, 228–29
BigDCL
described, 868
installation/use of, 896
big_increment, DCL attribute, 807
32 Bit file, optimizing, 34
Bitwise logical operations, 498–505
black, DCL color reserve word, 810
BLANK.SLD, 628
10BLANK.MNS, 453
BLIPMODE, 487–88
system variable, 835
BlockEdit
described, 869
installation/use of, 898
Blocks, using with path names, 170–72
blue, DCL color reserve word, 810
boole function integer, 780
Boot system, multiple, setting up, 10
Bound variables, AutoLISP, 381–84
Boundary hatch, release 13 and, 250–51
boundp sym, 780
BOX selection mode, 190–91
Boxed
column, DCL tile reference, 811
radio column, DCL tile reference, 812
radio row, DCL tile reference, 812
row, DCL tile reference, 812–13
Boxes, dialog. *See* Dialog boxes
Branching
conditional,
AutoLISP, 407–27
DIESEL, 466–68
BREAK command, 89–91
BrowseForFolder, 533–35
Bubble macros, 147
Bug list, 894
Button, DCL tile reference, 813
Button
DCL tile reference, 813
image explorer, maximizing, 887–88
menus, 652–59
predefined, 820–21
toolbars and, 672

C

c????r list, 780
caddr function, lists and, 394, 398
caddr list, 780
cadr function, lists and, 394–99
cadr list, 780
cancel_button, 821
Captions, menu, creating with DIESEL, 505–8
car function, lists and, 394–99
car list, 780
CDATE, system variable, 835
cdr function, lists and, 394–95, 399
cdr list, 780
CD-ROM, MaxAC, installing, 45–48
CECOLOR, system variable, 468–69, 835
CELTSCALE
linetype scale and, 260–61
release 13 and, 250, 261
system variable, 835
CELTYPE, system variable, 835
Center line character, 230–33
CHAMFERA, system variable, 835
CHAMFERB, system variable, 835
CHAMFERC, system variable, 835
CHAMFERD, system variable, 835
CHAMMODE, system variable, 835
Character definition, line-feed, 219–20
Characters
combining, using subshapes for, 229–40
custom fraction, 233–41
designing, center line, 230–33
Check marks, in menu labels, 480–86
CHECK.PAT, 249
children_alignment, DCL attribute, 807
children_fixed_height, DCL attribute, 808
children_fixed_width, DCL attribute, 808
CHPROP, linetype scale and, 261
chr integer, 780
CIRCLERAD, system variable, 835
CLAYER, system variable, 835
client_data_tile key clientdata, 780
close handle, 781
Clusters, dialog boxes and, 550–57
CMDACTIVE, system variable, 835
CMDDIA
scripts and, 309
system variable, 835
CMDECHO, system variable, 835
CMDNAMES, system variable, 836
CMLJUST, system variable, 836
CMLSCALE, system variable, 836
CMLSTYLE, system variable, 836
Code formatting, DCL, 545–46
color, DCL attribute, 808
Color dialog box, 535–36
[color [frompoint topoint]], 788

Index

Column, DCL tile reference, 813–14
Command definitions, AutoLISP, creating, 86–91
command [expr...], 781
Command functions
   angular units in, 407
   AutoLISP, 360–61, 403–7
   calling transparently, 379–80
   command-line echoing and, 649
Command-line echoing, controlling, 648–51
Commands, multiple, menu macros and, 183–86
Comments, AutoLISP coding and, 449–50
Complex linetypes. *See* Linetypes, complex
cond clause..., 781
COND function, conditional branching with, 418–22
Conditional branching
   AutoLISP, 407–27
      COND and, 418–22
      logical functions and, 413–14
      predicates/relational functions and, 209–13
      PROGN function and, 414–18
   DIESEL, 466–68
Configuration files
   ACAD.CFG, 119–22
   ACAD.INI, 119–22
   ACADNT.CFG, 119–22
   AutoCAD, 48–52
      multiple, 51–52
      copying to MAXAC directory, 51
Configuring AutoCAD, 21–22
cons expr1 expr2, 781
Context-sensitive, cursor menu, creating, 744–50
Control
   bits, initget, 430
   characters,
      setting modes with, 167–73
      string, 389
   structures, AutoLISP, 422–27
Coordinates, AutoLISP and, 359–60
COORDS, system variable, 836
CopyFile, described, 875
cos angle, 781
CPoly selection mode, 191–92
csnap function, 400
Cursor menu
   context-sensitive, creating, 744–50
   controlling, with DIESEL, 744–50
   labels, DIESEL and, 479–86
   structure of, 663–64

   special codes in, 664–69
   10CURSOR.MNS, 453
Custom fonts, 209–10
Customization environment problems, 824
Customizing, for portability, 142–43
CVPORT
   disabling menu items and, 495
   system variable, 836
cvunit number from to, 781
cyan, DCL color reserve word, 810

D

DASH3DOT, 252–56
Data
   storage and retrieval, with DIESEL, 470–72
   type,
      identifiers/output formats, 351
      AutoLISP, 350–55
DATE, system variable, 836
DBMOD, system variable, 836
DCL program, 120, 127–28
   color reserve words, 810
   problems, 829–32
   quick reference, 807–10
   tile reference, 811–21
      see also Dialog Control Language
DCL Spy
   described, 868
   dialog development and, 622–23
   installation/use of, 886–87
DCTCUST, system variable, 836
DCTMAIN, system variable, 836
DDATTE, 896
DDCHPROP
   color dialog boxes and, 535–36
   linetype scale and, 261
DDCHTEXT, 901
   described, 869
DDEDIT, 198
DDEDTEXT, 900–901
   described, 869
   installation/use of, 901
DDEMODES, color dialog boxes and, 535–36
DDLMODES, release 13 and, 250
DDLs, toolbars and, 675–76
DDLTYPE, 247
   linetype scale and, 261
   release 13 and, 250
DDSTYLE, 199, 201
Debugging, AutoLISP, 445–48
Default menu, setting the, 71–72

Index

Definition syntax, ACAD.PGP external commands, 84–86
defun function, 369–70, 371–72
defun name ([argument]... [/ local...]) expression...), 781
DeleteFile, 755
DELOBJ, system variable, 836
Developmental tools/utilities, 873–88
Dialog, DCL tile reference, 814
Dialog boxes
    accessing, tile data, 574–80
    calling/avoiding, in macros, 161–67
    color, 535–36
    common, 522–36
    control of, 570–74
        tile states, 580–85
    designing, 569–624
    directories/folders selection, 533–34
    event-driven programming and, 591
    events, 592–93
    file, 526–32
        using in AutoLISP, 528–32
    finding attribute boxes with, 617–22
    introduced, 518–19
    library search path, 532–33
    list boxes, 588–91
        handling, 613–17
    message, 523–26
    MTEXT edit, 536
    numerical input, validating, 601–5
    popup lists and, populating, 588–91
    primer on, 519–22
    process flow, 571
    tile,
        access/control, 574–90
        callbacks, handling edit box, 605–12
    validating distance/angle input, 612–13
    variables, action expression, 593–601
    versus image tile menus, 687–90
Dialog Control Language, 536–57
    basics, 537–38
    clusters, 550–57
    code formatting, 545–46
    learning/development tools, 543–46
    prototype tiles, 538–39
    subassemblies, 538
    syntax/semantics of, 539
    tile,
        attributes, 540
        parametrics, 546, 548–49
        references, 540–43
    user spacers, 549–50
    see also DCL
Dialog development, tools for, 622–23
dialog_background, DCL color reserve word, 810
dialog_foreground, DCL color reserve word, 810
dialog_line, 810
DIASTAT, system variable, 836
dictadd dictname key newobj, 782
dictnext dictname [rewind], 782
dictremove dictname key, 780
dictrename dictname oldkey newkey, 780
dictsearch dictname key [setnext], 780
DIESEL
    AutoLISP and, 454–55
        calling expressions from, 491–94
    bitwise logical operators, 498–505
    check marks, setting, 482–86
    creating custom status lines, 459–79
        color status, 468–69
        complex, 462–64
        complex data manipulation with, 469–70
        conditional branching in, 466–68
        data storage/retrieval, 470–72
        displaying complex, 464–66
        dynamic context-sensitive, 477, 479
        time/date format, 461–62
        for Windows, 472–77
    cursor menus, controlling with, 744–50
    custom applications, extending, 513–14
    debugging/error messages, 514–15
    described, 451–53
    dynamic menus and, 505–13
        AutoLISP and, 508–12
        creating captions, 505–8
        creating toolbar drop-down list with, 512–13
    expressions,
        evaluation, 488, 490–91
        testing, 455–58
    macros, debugging, 458–59
    mechanics of, 453–59
    menu,
        items, disabling, 454–505
        macros and, 487–91
    pull-down/cursor menu labels and, 479–86
Digitizers
    optimizing,
        for DOS, 35–36
        for Windows, 28–29
DIMALT, system variable, 837
DIMALTD, system variable, 837
DIMALTF, system variable, 837
DIMALTTD, system variable, 837

Index

DIMALTTZ, system variable, 837
DIMALTU, system variable, 837
DIMALTZ, system variable, 837
DIMAPOST, system variable, 837
DIMASO, system variable, 837
DIMASZ, system variable, 837
DIMBLK1, system variable, 838
DIMBLK2, system variable, 838
DIMBLK, system variable, 837
DIMCEN, system variable, 838
DIMCLRD, system variable, 838
DIMCLRE, system variable, 838
DIMCLRT, system variable, 838
DIMDEC, system variable, 838
DIMDLE, system variable, 838
DIMDLI, system variable, 838
DIMEXE, system variable, 838
DIMEXO, system variable, 838
DIMFIT, system variable, 839
DIMGAP, system variable, 839
DIMJUST, system variable, 839
DIMLFAC, system variable, 839
DIMLIM, system variable, 839
DIMPOST, system variable, 839
DIMRND, system variable, 839
DIMSAH, system variable, 840
DIMSCALE, system variable, 840
DIMSD1, system variable, 840
DIMSD2, system variable, 840
DIMSE1, system variable, 840
DIMSE2, system variable, 840
DIMSHO, system variable, 840
DIMSOXD, system variable, 840
DIMSTYLE, system variable, 840
DIMTAD, system variable, 841
DIMTDEC, system variable, 841
dimtext function, 372
DIMTFAC, 213
DIMTFAC, system variable, 841
DIMTIH, system variable, 841
DIMTIX, system variable, 841
DIMTM, system variable, 841
DIMTOFL, system variable, 841
DIMTOH, system variable, 841
DIMTOL, system variable, 841
DIMTOLJ, system variable, 842
DIMTP, system variable, 842
DIMTSZ, system variable, 842
DIMTVP, system variable, 842
DIMTXSTY, system variable, 842
DIMTXT, system variable, 842
DIMTZIN, system variable, 842
DIMUNIT, 213

DIMUNIT, system variable, 842
DIMUPT, system variable, 842
dimx_tile key, 780
dimy_tile key, 780
DIMZIN, system variable, 843
Directories
   organizing, 41–45
      AutoCAD R13, 43–45
   selecting, 533–35
Disk access, optimizing, 34
Display drivers
   optimizing,
      for DOS, 35
      for Windows, 28
DISPSILH, system variable, 843
DISTANCE, system variable, 843
distance function, 400–401
distance point point, 780
Distance, validating, 612–13
distof, 612–13
distof string [mode], 783
done_dialog, 609–10
done_dialog [status], 783
DONUTID, system variable, 843
DONUTOD, system variable, 843
DOS
   advantages of, 3
   AutoCAD in, 58–61
   basic configuration for, 22
   copying programs/utilities, 872–73
   installing, MaxAC CD-ROM for, 48
   MNU file structure, 633–34
   optimizing, 35–36
   running in Windows, 95, 12–13
   text editors, selecting/configuring for,
      67–68
DOSLib
   described, 868
   installation of, 883–84
Dr. DWG QuickView, 889–96
DRAGP1, system variable, 843
DRAGP2, system variable, 843
Drawing files, selecting with dialog box,
   532–34
Drawing load initialization, 135–36
Drawings, DIESEL macros and, 325–28
Dropdown menus, toolbars, creating with DIE-
   SEL and AutoLISP, 512–13
DTEXT, DIESEL macros and, 496
Duplicate layer macros, 148
DVIEW, disabling menu items and, 495–96
DWGCODEPAGE, system variable, 843
DWGList

Index

described, 868
installation/use of, 894
DWGNAME, system variable, 843
DWGPREFIX, system variable, 843
DWGTITLED, system variable, 844
DWGWRITE, system variable, 844
Dynamic menus, creating with DIESEL and AutoLISP, 508–12
DynaMode, 477, 898–99
described, 869
installation/use of, 899

E

Echoing, command line, controlling, 648–51
EDGEMODE, system variable, 844
Edit box, DCL tile reference, 814
EDIT command, 85
edit_action, 605–6
edit_limit, DCL attribute, 808
edit_width, DCL attribute, 808
ELEVATION, system variable, 844
Embedded
  shape, syntax for, 264
  text string, 265–68
    syntax for, 264
end_image, 783
end_list, 783, 588–89
entdel ename, 783
entget ename [applist], 783
Entity names, AutoLISP and, 355
entlast function, 346, 783
  lists and, 397
entmake edata, 783
entmakex edata, 784
entmod edata, 784
entnext [ename], 784
entsel function, 401, 435–36
entsel [prompt], 784
entupd ename, 784
eq expr1 expr2, 784
eq function, 410
equal expr1 expr1 [accuracy], 784
equal function, 410
ERRNO, system variable, 844
error, function, 442–45
error, string, 784
Error handling, AutoLISP, 440–48
Error tile, DCL tile reference, 814–15
errtile, 602
Escape characters, big font, 228
eval expr, 784
Evaluation, AutoLISP and, 344–47

Event-driven programming, dialog boxes and, 591
Events, dialog boxes, 592–93
EXE file extension, 120
Exercises
  ACAD.LSP file, making a dummy, 70
  ACAD.PGP,
    adding command aliases to, 79–80
    using external commands, 82–83
  ACADMA.LIN, adding linetypes to, 257
  adding,
    accelerator key definitions, 689
    image tile menu of flow symbols, 686
    partial menu, 705–7
    plotter configuration, 316–18
    side-screen menu, 697–98
    tablet menu, 701–3
  adding/modifying, tile callbacks, 599–601
  angle character, defining, 221–22
  assigning custom status lines, 460
  ASTEROID,
    adding diagonal lines to, 283
    adding dots/testing the pattern, 281–82
    adding first boulder lines to, 284–87
    planning the pattern, 278–79
    writing pattern header/first line family, 280–81
  AutoCAD fonts, looking at, 203–4
  AutoLISP,
    creating for
      BREAK, 89–90
      ZOOM Previous, 87–88
    drawing polyline rectangle with, 362
    program,
      file creation, 374
      program file loading, 375
    using,
      an expression with, 341–42
      math functions, 386–87
  big fonts, maximizing usage of, 239–40
  browse for folder dialog box, using, 534–35
  calculate input, using AutoLISP, 346–47
  character definition, creating 1/8 fraction, 236
  checkered plate pattern, 275–77
  compiling shape files, 216
  complex status line, displaying, 464
  controlling,
    Dialogs with AutoLISP, 578–80
    partial menu with AutoLISP, 708–9
    TEXTEVAL evaluation with, 358–59
    tile states with mode_tile, 582–83
  converting strings to integers, 603

Index

creating,
    accelerator key, 94–95
    autoload expression, 734–35
    button icons, 111–12
    custom date status display, 462
    slide library/viewing its slides, 305–6
    slides/slide libraries, 682–83
    toolbar with flyout, 678
custom MaxAC batch file, creating, 59–60
DASH3DOT linetype,
    creating, 253–54
    using, 254–55
designing, center line character, 230–31
dialog boxes, using modeless, 521–22
dialogs, modifying with prototype assemblies, 565–67
DIESEL, evaluation order, 489–90
edit boxes, handling, 606–8, 610–11
embedded shapes, creating linetypes with, 269–70
embedded text, adding offsets to, 267
graphic character inquiry, 217–19
installing, MaxAC CD-ROM, 46, 48
linetype,
    alignment/scale, 258–59
    with embedded text, 266, 267–68
LispPad, design/testing dialog boxes with, 544–45
list boxes,
    adding items to, 589–91
    handling multiple-selection, 615–16
ListPad, setting up, 878
lists, manipulating, 395–96
macros,
    dialog box use in, 162–66
    making commands/repeat indefinitely, 185
    making a leader, 168–69
    making pipe valve repeat, 183–84
    making a single-object leader <———, 176
    making a single-object leader———>, 173–74
    pipe valve with layer setup, 181–82
    planning a, 154
    spaces versus returns in text, 160
    splitting across multiple lines, 172
    using semicolons/object snaps in, 159–60
    writing with pauses for input, 155–56
MaxAC CD-ROM, copying to Windows, 872
menu file,
    creating new, 635–38

reloading/testing accelerator keys, 95–96
menus,
    creating simple, 151–52
    improving bubble, 177–79
    loading/starting a drawing, 149–50
MIDPT, using, 380
modifying,
    program item, 55–56
    program shortcut, 54–55
    support path, 57
new program,
    item, 53–54
    shortcut, 52–53
numbers, converting to distance strings, 391–92
partial menus,
    attaching with MxaMenuLoad and MxaInsertMenu, 726–28
    attaching/displaying automatically, 729–30
plotting script,
    creating/testing, 324
    developing/running a batch, 326–27
    planning, 319–22
point lists,
    accessing in system variables, 393
    working with, 359–60
PostScript fill pattern, adding and using, 290–92
programs, loading, listing, unloading, 137–38
pull-down/cursor codes, looking at, 667
rows and columns,
    modifying dialogs with, 557
    working with, 554–56
selections sets, building in advance, 186–87
setting,
    default menu, 71–72
    global system variables, 71–72
slide shows, creating scripted, 312–13
slides,
    making a sequence of, 300–301
    viewing, 301–2
spacer tiles, using, 549–50
testing,
    ACAD.PGP reloading, 80–81
    alert function, 524
    all standard menu sections, 641–42
    ATTFIND command, 621–22
    complex status line, 465
    COND function, 420

Index

context-sensitive menus from 14CUR-SOR.MNS, 748–50
dialog tile callbacks, 597–98
DIESEL evaluation order, 489–90
DIESEL functions, 456–57
DIESEL menu disabling, 497–98
DynaMode, 478–79
functions from 10DIESEL.LSP, 493–94
linear units cursor submenu, 485–86
MEAS function, 402–3
menu/command echoing, 651
MIDPOINT.LSP, 376–77
object snap enhanced cursor menu, 504–5
order of evaluation, 345
PopUpMenu function, 511–12
QHATCH, 475
RECT command, 439–40
RECTANG function, 397–98
TBOSNAP.LSP, 751–53
TUBESEC command, 416–17
variable scope, 384
text editors,
   associating registered file types with, 65–66, 67
   testing, 68–69
tile,
   attributes, accessing from AutoLISP, 586–87
   placement/size, using attributes to control, 547–48
   prototypes, using, 563
toolbars,
   adding new icons to, 112–14
   creating with existing icons, 106–8
   creating with new icons, 109–10
   showing/hiding, 103
tracing user-defined functions, 447–48
user-defined function, defining/calling a, 370–71
using,
   an error handler, 444–46
   free and bound variables, 381–83
   GET functions for input, 432–34
   hatching macros at different scales, 742–43
   name tags, 646–47
      repeat loop in AutoLISP, 423–24
variable,
   using, 343
   using at command line, 356–57
viewing, AutoLISP errors, 441
exit, 785

EXP file extension, 120
exp number, 785
expand segments, 785
EXPERT, system variable, 844
EXPLMODE, system variable, 844
EXPLODE, 272
expr functions, 349, 410–11, 776
Expressions, AutoLISP and, 344–47
expt base power, 785
Extended big fonts, 241–42
External commands
   ACAD.PGP and, 82–86
      definition syntax, 84–86
      definitions, 86
EXTMAX, system variable, 844
EXTMIN, system variable, 844

F

FACETRES, system variable, 844
Fence selection mode, 191–92
FFLIMIT, system variable, 844
File
   descriptors, AutoLISP and, 355
   dialog boxes, 526–32
      in AutoLISP programs, 528–32
   to edit command, 85
FILEDIA
   scripts and, 309
   system variable, 845
FileExists, 763
FileExits, described, 874
File-handling commands, redefining, 740
FILLETRAD, system variable, 845
fill_image x1 y1 x2 y2 color, 785
FILLMODE, system variable, 845
Fills, PostScript, 289–93
find file, library search path and, 532–33
findfile file, 785
FindFiles pattern, described, 875
fix number, 785
fixed_height, DCL attribute, 808
fixed_width, DCL attribute, 808
fixed_width_font, DCL attribute, 808
float number, 785
FLOW*.SLD, 628
FLOWSYMB.SLB, 628
Flyouts, toolbars and, 672–74
FMP, 131
   file extension, 120
Folders
   organizing, 41–45
      AutoCAD R13, 43–45

Index

selecting, 533–35
Font, problems, 827–28
FONTALT, 199, 210
    mapping fonts and, 210–11
    system variable, 845
FONTMAP, 199
    bugs in, 211
    mapping fonts and, 210–11
    system variable, 845
FONTMAP.PS, 121, 131
Fonts. *See* Text fonts/styles
foo function, 340
    variables and, 383
foreach symbol list expression..., 785
Formatting
    AutoCAD code, 350
    DCL code, 545–46
    input prompts, 437–39
Fractions, custom characters, 233–41
Freeware programs/utilities, 871
Functions
    user-defined,
        in AutoLISP, 369–84
        described, 371–78
    see also specific type of function

G

gc, 786
gcd integer integer, 786
Geometric functions, AutoLISP, 399–403
Get functions, Table A.1, 787
getangle [basept] [prompt], 787
get_attr, 585–88
get_attr key attr, 786
getcfg key, 786
getcname cmdname, 786
getcorner basept [prompt], 787
getdist [basept] [prompt], 787
getenv varname, 786
getfiled
    file dialog boxes and, 526–32
    library search path and, 532–33
getfiled title filename extension flags, 787
getint, 348
getint [prompt], 787
getkword [prompt], 787
getorient [basept] [prompt], 787
getpoint [basept] [prompt], 787
GetProfileInt, 757
    described, 874
GetProfileString, 757
    described, 873

getreal [basept] [prompt], 787
getstring [flag] [prompt], 787
GetTempDirectory, 756
get_tile, 574–75
get_tile key, 786
getvar, 361–62
getvar varname, 787
GetWindowsDirectory, 756
getxxxx input functions, 786–802
graphics_background, DCL color reserve
    word, 810
graphics_foreground, DCL color reserve
    word, 810
graphscr, 787
grclear, 787
grdraw frompt topt color [mode], 788
green, DCL color reserve word, 810
GRIDMODE, system variable, 845
GRIDUNIT, system variable, 845
GRIPBLOCK, system variable, 845
GRIPCOLOR, system variable, 845
GRIPHOT, system variable, 845
GRIPS
    menus and, 189
    system variable, 845
GRIPSIZE, system variable, 845
grread [track] [allkeys [cursor]], 788
grtext [box text [mode]], 788
grvecs vlist [trans], 788

H

handent handle, 788
HANDLES, system variable, 845
HATCH, 247, 271–72
    10HATCH.LSP, 476–77
Hatch patterns, 271–89
    creating,
        complex, 277–89
        simple, 274–77
    release 13 and, 249–51
    understanding, 273–74
    using, 271–73
HATCHEDIT, 272
    release 13 and, 251
height, DCL attribute, 808
hello_action, 597
help [file [topic [command]]], 788
Help files, 132–33
    creating, 764–72
Help strings, 690
help_button, 821
HIGHLIGHT, system variable, 846

Index

HLP2AHLP, installation of, 884
HLP2AHP, described, 868
HPANG, system variable, 846
HPDOUBLE, system variable, 846
HPNAME, system variable, 846
HPSCALE, system variable, 846
HPSOUND, system variable, 846

I

Icon menus. *See* Image menus
Icons, toolbars and, 675–76
IDSHX
    described, 868
    installation/use of, 884–85
if function, branching and, 408–9
if testexpr thenexpr [elseexpr], 789
Image, DCL tile reference, 815
Image button, DCL tile reference, 815
Image menus, 631
Image tile menus, 679–90
    making slides, 680–83
        designing, 683–86
    versus dialog boxes, 687–90
image_button, 586
Implied Windowing, menus and, 189
Indefinite repetitions, 185–86
Infix notation, 341
info_button, 821
Inheritance, prototypes and, 559–63
initget bits string, 789
initget input control, 429–30
    control bits, 430
    modes/keywords, 430–35
initial_focus, DCL attribute, 808
Initialization sequence, AutoCAD, 133–36
Input
    angle, 612–13
    functions, command line, 428
    numerical, validating, 601–5
    pausing scripts for, 314
    prompts, formatting, 437–39
    user,
        acquiring/validating, 428–40
        pausing for menu macros and, 154–57
INSBASE, system variable, 846
INSERT, 896
INSNAME, system variable, 846
Installing
    AutoCAD, 17–21
    MaxAC CD-ROM, 45–48
Integers, AutoLISP and, 352
inters function, 400–401

inters point1 point2 point3 point4 [onseg], 789
ISAVEBAK, 25, 26
ISAVEBAK, system variable, 846
ISAVEPERCENT, 25, 26
ISAVEPERCENT, system variable, 846
is_bold, DCL attribute, 808
is_cancel, DCL attribute, 808
is_default, DCL attribute, 809
is_enable, DCL attribute, 809
ISO line types, scaling, 262–63
ISOLINES, system variable, 846
is_tab_stop, DCL attribute, 809
Item labels, menus and, 643–45
itoa integer, 789

K

Kelv13, 882–83
    described, 868
key, DCL attribute, 809
Keyboard aliases. *See* Aliases
Keyboard macros
    ACAD.PGP, defining aliases in, 77–82
    external commands, defining, 82–86
    understanding, 76–77

L

label, DCL attribute, 809
lambda ([argument]... [/ local...]) expression...), 789
Languages, interpreted/compiled, 337–38
last list, 790
LASTANGLE, system variable, 846
LASTPOINT, 488, 490–91
    DIESEL expression evaluation and, 488, 490–91
    system variable, 847
Layer DDLMODES, 896
Layers, controlling with menu macros, 180–83
layout, DCL attribute, 809
Leader arrow <——macros, 148
Leader arrow——> macros, 148
Leader macros, 147
length list, 790
LENSLENGTH, system variable, 847
Library
    search path, 139–40
        dialog boxes and, 532–33
LIMCHECK, system variable, 847
LIMMAX, system variable, 847
LIMMIN, system variable, 847

Line-feed character definition, 219–20
LINETYPE, 247
    DDLTYPE, 896
    release 13 and, 249–51
Linetypes
    complex dash-dot, 263–71
        embedding shapes in, 268–71
        embedding text in, 265–68
        understanding, 264–65
    ISO scaling patterns, 262–63
    simple dash-dot, 251–63
        2D polyline generation, 263
        creating, 252–56
        endpoint alignment/scale settings, 258–60
        file format, 256–58
        scaling patterns, 260–63
        understanding, 251–52
Lisp2C, 881–82
    described, 867
: LISPED, 198
LispPad, 543–44
    dialog development and, 622
LispPlot, 885–86
    described, 868
    installation/use of, 886
List
    box, DCL tile reference, 816
    boxes,
        handling, 613–17
        populating, 588–91
list, DCL attribute, 809
List, operations, AutoLISP, 392–99
list [expr], 790
list function, list operations and, 392, 394–97
list_assign, 589
listp item, 790
ListPad
    described, 867
    editor, 367–69
    installation/use of, 877–79
    maximizing,
        AutoCAD and, 877–79
        AutoLISP code editor and, 366–69
Lists, AutoLISP and, 354–55
load, 363–64
load filename [onfailure [verbose]], 790
Load initialization, 134
load_dialog, 571–72
load_dialog filename, 790
Loading, AutoLISP/ADS/ARX programs, 136–38
log number, 790

logand integer, 790
Logical functions, 413–14
LOGINNAME, system variable, 847
logior integer, 790
Looping, AutoLISP and, 422–27
lsh integer num, 790
LSP program, 120, 127–28
LTSCALE
    linetype scale and, 260–61
    release 13 and, 249–50
    system variable, 847
LTYPE, 247
LTYPESHP.LIN, embedding shapes and, 268–71
LTYPESHP.SHX, embedding shapes and, 268–71
LUNITS, 483–85
    system variable, 847
LUPREC, system variable, 847

M

MacAC directory, coping configuration files to, 51
Macros
    ACAD.PGP,
        defining aliases in, 77–82
        defining external commands in, 82–86
    DIESEL,
        debugging, 458–59
        versus AutoLISP, 753
    menu, 147–98
        see also Menu macros
    menus, 648
    scale factors in, controlling, 740–43
    understanding keyboard, 76–77
MACROTRACE, system variable, 847
magenta, DCL color reserve word, 810
Main menu, configuring, 23–24
MakeDir, described, 875
MakePath, described, 875
mapcar function list..., 790
Mapping fonts, 210–13
    PostScript, 212–13
Math
    functions, AutoLISP, 385
    operations, AutoLISP and, 384–87
MaWin, installation/use of, 875–77
MAWIN.ARX function, 754–64
    dialog boxes and, 525–26
max number..., 791
MaxAC CD-ROM
    exercise, support, code files, 869–70

Index

installing, 45–48, 871–73
MAXAC.MNS, 453
MAXACTVP, system variable, 847
MAXOBJMEM, 25
MAXOBJMEM, system variable, 847
MAXSORT, system variable, 848
max_value, DCL attribute, 809
meas function, geometric functions and, 401, 403
Mega-App loader, 888–89
  described, 868
mem, 791
member expr list, 791
Memory reserve command, 85
Menu macros, 145–93
  automatic loading from, 735–36
  control characters, setting modes with, 167–73
  controlling layers with, 180–83
  defining, 147–48
  DIESEL, using, 487–91
  making long, 172–73
  multiple commands and, 183–86
    indefinite repetitions, 185–86
  object snaps in, 177–80
  polylines and, 173–77
  problems with, 826–27
  pulldown, 146
  repeating, 183–86
  scale factors in, controlling, 740–43
  selection sets used in, 186–92
    modes in, 190–92
    object settings/menus and, 188–90
  special characters used in, 157–67
    calling/avoiding dialog boxes in, 161–67
    semicolons, 159–61
    table of, 158
  user input, pausing for, 154–57
  writing, simple, 149–57
  see also Macros
Menu structure, 631–48
  combining menus, 703–10
  command-line echoing, controlling, 648–51
  DOS and, 625
    MNU files, 633–34
  help strings, 690
  interface use and, 710–11
  item,
    disabling/checking, 669
    labels, 643–45
    name tags, 645–48
  menus,
    button/Auxiliary, 652–59

cursor, 663–69
image tile, 679–90
macros and, 648
pull-down, 659–69
sections, 634–43
side-screen, 690–99
toolbar, 669–79
overview of, 148–49
R13 and, 625, 629–31
section details, 651–703
submenus, using, 642–43
tablet menus, 699–703
Windows MNS, 632–33
see also Menus
menucmd string, 791
MENUCTL, system variable, 848
MENUECHO, command, command-line echoing and, 648–51
MENUECHO, system variable, 848
MENULOAD, command, 704–7
MENUNAME, system variable, 848
Menus
  captions, creating with DIESEL, 505–8
  combining, 703–10
  controlling, aperture in, 180
  creating,
    ACAD.LSP and, 716–19
    s::startup function and, 719–20
  default, setting the, 71–72
  defining, 147–48
    accelerator keys in, 91–98
  DIESEL use in, 740–53
    AutoLISP use in, 740–53
  grips and, 189
  implied Windowing and, 189
  item labels of, 643–45
  items, DIESEL, disabling, 454–505
  loading problems, 825–26
  macros, 648
  MAWIN.ARX function and, 754–64
  MNL files and, 720–25
    "helper" functions, 721–23
    versus ACAD.LSP, 723–25
  noun/verb selection and, 189
  partial, ACAD.MNL and, 723
  press and drag and, 189
  reloading/testing file, 95–96
  section features of, 639
  sections of, 634–43
  submenus and, 642–43
  support files, 714
  supporting, with ARX application, 753–64
  title, slides and, 303–5

Index

use shift to add and, 189
see also Menu structure
Message dialog box, 523–26
    enhanced, 524–26
MessageBeep, 759
    described, 874
MessageBox, 759–61
    described, 874
    flag codes, 762
    result codes, 762
MIDPT function, 372–73, 375–79
min number..., 791
minusp number, 791
min_value, DCL attribute, 809
MIRRTEXT, system variable, 848
mnemonic, DCL attribute, 809
MNL files
    loading AutoLISP code with, 365–66
    menus and, 720–25
        "helper" functions, 721–23
        partial, 709–10
        versus ACAD.LSP, 723–25
MNS file structure, 632–33
MNU file structure, 633–34
10MODE1.LSP, 453
10MODE2.LSP, 453
MODEMACRO
    status line displays and, 460–66
    string length limit and, 471–72
    system variable, 848
Modes, setting with control characters, 167–73
mode_tile, 581–85
mode_tile key mode, 791
Modifying
    program items, 55–56
    shortcut properties, 54–55
    support path, 56–58
MOFSET command
    logic flow of, 427
    while function and, 425–27
MoveFile, described, 875
MSLIDE, 298–305
MTEXT, 198, 199
    edit dialog box and, 536
    font mapping and, 211–12
MTEXTED, system variable, 198, 848
Multiple commands, macros and, 183–86
multiple_select, DCL attribute, 809
MXAInsertMenu, 726–29
MxaMenuLoad, 726–29

N

namedobjdict, 791
nentsel prompt, 792
nentselp prompt pt, 792
Network
    AutoCAD on, 14
    optimization, 27–28
    storing files on, 13, 15
NEW, scripts and, 309
New program, item, creating, 52–54
new_dialog, 572–73
new_dialog dlgname handle [action [position]], 792
not expr, 792
not function, 413
Noun/verb selection, menus and, 189
nth index list, 792
null item, 792
number, 775, 776, 806
numberp item, 792
Numerical input, validating, 601–5

O

Object snap
    menus,
        extending with AutoLISP, 750–52
        macros, 177–80
Object sort method control, 179–80
OFFSETDIST, system variable, 848
ok_cancel, 820
ok_cancel_help, 820
ok_cancel_help_errtile, 820
ok_cancel_help_info, 820
ok_only,, 820
OPEN, scripts and, 309
open filename mode, 792
Operating systems, maintaining multiple, 10
Optimizing
    system performance, 22–37
        network, 37–38
        platform-independent, 23–27
        Windows, 28–36
Optional argument, AutoLISP, 348–49
or expr..., 793
or function, 413
ORTHOMODE, system variable, 848
OSMODE, system variable, 849
osnap point mode, 793

Index

P

Paragraph, DCL tile reference, 816
Partial menus
    ACAD.MNL and, 723
    combining menus with, 703–10
    loading techniques, 725–32
    MNL files and, 709–10
    program loading, automatic, 732–36
    redefining commands, 736–40
        file handling, 740
        PLOT, 737–40
password_char, DCL attribute, 809
Path names, using blocks with, 170–72
pause, 793
PCP files, 121, 132
    described, 295, 329–33
    scripts plotting and, 331–32
PDMODE, system variable, 849
PDSIZE, system variable, 849
PELLIPSE, system variable, 849
PERIMETER, system variable, 849
PFACEVMAX, system variable, 849
PFB
    file extension, 121
    font files, 130
PFM
    file extension, 121
    font files, 130
Phar lap swap files, optimizing, 36
pi, 793
PICKADD, system variable, 849
PICKAUTO, system variable, 849
PICKBOX, system variable, 849
PICKDRAG, system variable, 849
PICKFIRST, system variable, 850
PICKSTYLE, DIESEL and, 500–501
Pipe valve macros, 148
PLATFORM, system variable, 850
PLINEGEN, system variable, 850
PLINEWID, system variable, 850
PLOT
    redefining, 737–40
    scripts and, 309
Plot configuration parameters, 132
PLOTID, system variable, 850
PLOTROTMODE, system variable, 315, 850
PlotStamp, 894–95
    described, 868
    installation/use of, 894
PLOTTER, system variable, 315, 850
Plotters
    optimizing, 29

    drivers for, 30–31, 36
Plotting, with scripts, 314–25
polar function, 400–401
polar point angle dist, 793
Polylines
    2D linetype generation, 263
    menu macros and, 173–77
POLYSIDES, system variable, 850
Popup list
    DCL tile reference, 816–17
    populating, 588–91
POPUPS, system variable, 850
Port name, plotting to a, 31
Portability, customizing for, 142–43
PostScript
    fills, 289–93
    font mapping, 212–13
    fonts, 208–9
    support files, 131–33
Predicate functions, AutoLISP, 409–13
Prefix notation, 341
Press and drag, menus and, 189
prin1 expr [file], 793
princ expr [file], 793
print expr [file], 793
print function, string operations and, 388
Printers
    optimizing, 29
        drivers for, 30–31, 36
Problems
    AutoLISP, 829–32
    customization environmental, 824
    DCL, 829–32
    menu,
        loading, 825–26
        macros, 826–27
    script, 328–29
    text/font, 827–28
Productivity tools/utilities, 896–902
progn expr..., 793
progn function, 414–18
Program
    files, AutoLISP, 363–69
    items, modifying, 55–56
Programming, event-driven, dialog boxes and, 591
Project management tools, 888–89
PROJMODE, system variable, 851
prompt string, 793
Prototype tiles, Dialog Control Language and, 538–39
Prototypes, 558–67
    definition of, 558

Index

design techniques, structured, 563–64
  inheritance and, 559–63
PSFILL, 247
PSLTSCALE
  linetype scale and, 261
  system variable, 851
PSOUT, 289
pspace? function, 372
PSPROLOG, system variable, 851
PSQUALITY, system variable, 851
Pull-down menus
  displaying/swapping automatically, 662–63
  labels of, DIESEL and, 479–86
  structure of, 659–69
    special codes in, 664–69
    titles/item labels, 661–62

Q

QHATCH, 453
QTEXTMODE, system variable, 851
QUIT, scripts and, 309
quote expr, 794
quote function, 395

R

Radio
  button, DCL tile reference, 817
  column, DCL tile reference, 817
Radio row, DCL tile reference, 818
RASTERPREVIEW, system variable, 25, 26, 851
read string, 794
read-char file, 794
read-line file, 794
Reals, AutoLISP and, 353
rectang function, 397–98
red, DCL color reserve word, 810
redraw [ename [mode]], 794
regapp appid, 794
REGENMODE, system variable, 851
RE-INIT, system variable, 851
Relational functions
  AutoLISP, 409–13
    examples, 412
rem number number, 794
RemDir, described, 875
repeat function, 423–24
repeat number expr..., 794
Repeating arguments, AutoLISP, 349–50
Repetitions, indefinite, 185–86
Required arguments, AutoLISP, 347–48
Return code command, 85

reverse list, 795
RIASPECT, system variable, 852
RIBACKG, system variable, 852
RIEDGE, system variable, 852
RIGAMUT, system variable, 852
RIGREY, system variable, 852
RITHRESH, system variable, 852
ROMANS.SHP font, 214–17
Row, DCL tile reference, 818
rtos function, strings and, 391
rtos number [mode [precision]], 795
Rules of scope, AutoLISP, 381–84
Runtime extension limited edition, maximizing Windows with, 873–77

S

s::startup function, creating menus and, 719–20
sart, 347–48
SAVEAS, scripts and, 309
SAVEFILE, system variable, 852
SAVEIMAGES, system variable, 853
SAVENAME, system variable, 853
SAVETIME, system variable, 853
SCR file extension, 121, 129
SCREENBOXES, system variable, 853
SCREENMODE, system variable, 853
SCREENSIZE, system variable, 853
Script files, 328–29
  batch plotting drawings and, 325–29
  pausing for input, 314
Script problems, 828–29
Scripts
  creating, scripted slide shows and, 311–14
  described, 295
  planning/writing, 309–11
  plotting with, 314–25
    PCP files, 331–32
  understanding, 308
Search path, library, 139–40
Section details, menu structure, 651–703
SectionMaster, installation/use of, 901–2
seget function, 436
Selection sets
  AutoLISP and, 355
  used in menu macros, 186–92
    modes in, 190–92
  used in menu micros, object settings/menus and, 188–90
Semantics
  AutoLISP, 347–50
  DCL, 539

Index

Semicolons
    menu macros and, 159–61
        text strings and, 161
set expr1 expr2, 795
setcfg key value, 795
setfunhelp C:FUNCTION [helpfile [topic
        [command]]], 795
setq, variables and, 381
setq sym expr [sym2 expr2]..., 795
set_tile, 575–76, 588–89
set_tile key value, 795
setvar, 361–62
setvar sysvar value, 795
setview viewdata [viewpoint], 796
SHADEDGE, system variable, 853
SHADEDIF, system variable, 853
Shape files, discussed, 242–44
Shapes, embedded, syntax for, 264
Shareware programs/utilities, 870
ShellOpen, macros and, 755
Shortcut properties, modifying, 54–55
ShowRefs
    described, 868
    installation/use of, 893–94
SHP
    file extension, 121
    font files, 130
    shapes and, 242–44
SHPNAME, system variable, 853
SHX
    file extension, 121
    font, 205–7
        files, 130
Side-screen menus, 690–99
    context-sensitive paging of, 698–99
    overlaying, 694–98
Simple linetypes. *See* Linetypes, simple
sin angle, 796
SIngle selection mode, 190–91
SKETCHINC, system variable, 853
SKPOLY, system variable, 854
Slide shows, creating scripted, 311–14
slide_image_button, 586
slide_image x1 y1 x2 y2 slidename, 796
SlideManager, 891–92
    described, 868
    installation/use of, 892
Slider, DCL tile reference, 818–19
Slides, 297–308
    described, 295
    hatching, to ACAD.SLB, 307–8
    making/viewing, 298–305
    model space/paper space and, 302–3

title menus/dialog boxes and, 303–5
    using library files of, 305–7
small_increment, DCL attribute, 809
SNAPANG, system variable, 854
SNAPBASE, , system variable, 854
SNAPISOPAIR, system variable, 854
SNAPMODE, system variable, 470, 854
SNAPSTYL, system variable, 854
SNAPUNIT, system variable, 470, 854
snvalid string [flag], 796
SORTENTS, system variable, 854
Sorting, object method of controlling, 179–80
Sound, 758
    described, 874
Spacer, DCL tile reference, 820
Spacers, using, 549–50
Special characters
    fonts and, 207
    menu, 158
    menu macros and, 157–67
SPLFRAME, system variable, 854
SPLINESEGS, system variable, 854
SPLINETYPE, system variable, 855
SplitPath, described, 875
sqrt number, 796
ssadd [ename [selset]], 796
ssdel ename selset, 796
ssget [mode [point1 [point2]] [pointlist]]
        [filterlist], 796
ssgetfirst, 797
sslength selset, 797
ssmemb ename selset, 797
ssname selset number, 797
ssnamex selset [index], 797
sssetfirst gripped [selected], 797
Stacked text, 213–14
startapp appname [params], 798
start_dialog, 797
start_image key, 797
start_list, 575–76, 588–89
start_list key [operation [index]], 798
StartModal, described, 874
Status lines
    color status and, 468–69
    complex, displays of, 462–66
    creating custom,
        with DIESEL, 459–79
        simple, 460–61
    custom, for Windows, 472–77
    data manipulation and, 469–70
    dynamic context-sensitive, 477
    time/date format, 461–62
strcase string [flag], 798